Springer Collected Works in Mathematics

For further volumes:
http://www.springer.com/series/11104

Harald Cramér

Harald Cramér

Collected Works I

Editor

Anders Martin-Löf

Reprint of the 1994 Edition

Springer

Author
Harald Cramér
(Stockholm, 1893 – 1985)
Department of Mathematics
Stockholm University
Stockholm, Sweden

Editor
Anders Martin-Löf
Department of Mathematics
Stockholm University
Stockholm, Sweden

ISSN 2194-9875
ISBN 978-3-642-38358-8 (Softcover)
978-3-540-56671-7 (Hardcover)
DOI 10.1007/978-3-642-40986-8
Springer Heidelberg New York Dordrecht London

Library of Congress Control Number: 2012954381

Mathematics Subject Classification (1991): 01A55, 01A60, 11N05, 11M20, 6003, 60F05, 60F10, 60G10, 60G12, 60G15, 60G17, 60G25, 60H70, 62P05, 90A46

Printed on acid-free paper

Springer is part of Springer Science+Business Media (www.springer.com)

Preface

This is a collection of Harald Cramér's extensive works on pure mathematics, probability theory, mathematical statistics and insurance mathematics. Many of these are in publications which are not easy to find nowadays, and therefore the publication of this collection may well be particularly useful. Almost all his papers in languages other than Swedish are included, only a few duplicated survey articles and some obituaries are excluded. It gives me particular pleasure to make available his pioneering works on risk theory, no. 26 and 47, since these are not easily obtained. They were published in jubilee volumes of the Skandia Insurance Company in 1930 and 1955; despite their age, these eminent examples of his brilliant expository style remain highly readable, and make the reader feel unusually intelligent. It is also very pleasing to include the complete bibliography of Cramér's work compiled by Gunnar Blom and Bertil Matérn in 1984. Their work has greatly facilitated the editing of this collection. The generous support of the Filip Lundberg foundation has made its publication possible.

I think Cramér's works reflect beautifully the spirit of his ideal mathematician, which he himself eloquently described in a memorial speech to his friend Edward Phragmén in 1958: "He belonged to a generation of mathematicians for which it was self evident that mathematics constitutes one of the highest forms of human thought, perhaps even the highest. For these mathematicians numbers were a necessary form of human thought, and the science of numbers was a central humanistic discipline with a cultural value of its own, completely independent of its role as auxiliary science in technical or other areas. This does not however mean that they underestimated the importance of 'using theoretical knowledge to obtain practical know-how', as Phragmén once characterized the task of the actuarial mathematician."

Stockholm, 25 September 1993 *Anders Martin-Löf*

Table of contents volume I

Table of contents volume II

Harald Cramér 1893–1985

By Gunnar Blom

Ann. Stat. **15** (4), 1335–1350 (1987)

Harald Cramér died on October 5, 1985. He was a great scientist and a good man.

1. Harald Cramér—his life.

1.1. I had my last conversation with my teacher and friend, Harald Cramér in April 1985 when I visited him in his attic apartment in an old house in beautiful Djurgården, outside Stockholm. He had moved there after the death of his beloved wife. There were books in all corners. I admired the view across the water to the north and to the south. Ships passed on their way to and from the Stockholm harbour. However, Harald could not discern them, for his eyesight was failing. He was also nearly deaf. But his intellect was clear and his memory excellent. We spoke about things old and new. I showed him some research papers recently written by young Swedish research workers, and he was actively interested in everything. But there was melancholy in the air, and I think we both felt that it was the last time.

1.2. Harald Cramér was born on September 25, 1893, in Stockholm. In 1918 he married Marta Hanssow. She died in 1973. They had one daughter, Marie-Louise, who lives in Finland, and two sons, Tomas and Kim, who live in Stockholm.

1.3. Cramér began his studies at the University of Stockholm in 1912, and was particularly interested in chemistry and mathematics. During 1913–1914 he was a research assistant in biochemistry and published his first paper together with H. von Euler, later a Nobel laureate [1]. (The numbers in square brackets refer to the bibliography at the end of the paper.)

Cramér soon abandoned chemistry for mathematics and received a Ph.D. in 1917 for a thesis on Dirichlet series [6]. He was an assistant professor of mathematics from 1917–1929 at the University of Stockholm.

1.4. In the years before 1920, Cramér became interested in insurance, especially in problems connected with the risk of ruin of an insurance business. In 1919 he held a short appointment at Försäkringsinspektionen (Swedish Private Insurance Supervisory Service). I would guess it was there that he first become acquainted with the ruin problem. From 1920–1948 he was an actuary, first for the life insurance company Svenska livförsäkringsbolaget, and from 1929 on for the reinsurance company Sverige. From 1949–1961 he was a consulting actuary for Sverige.

Received October 1986; revised February 1987.

Insurance problems turned Cramér's attention to probability theory. His first paper on a probabilistic problem dates from 1923. In 1927 he was ready to publish an elementary textbook in Swedish, *Probability Theory and Some of Its Applications*.

1.5. In the middle of the 1920's, there were chairs in statistics at some Swedish universities, but none in mathematical statistics. Influential leaders of Swedish insurance companies recognized the need for such a chair. A professorship for Actuarial Mathematics and Mathematical Statistics was established at the University of Stockholm. In 1929, Harald Cramér became its first holder and head of the Institute of Actuarial Mathematics and Mathematical Statistics.

The institute attracted not only regular students, but also actuaries, biologists and other people interested in applications of probability theory. During 1934–1935 Cramér, possibly as the first scientist outside Russia, gave a series of lectures founded upon Kolmogorov's work. As H. Wold has remarked, this series became very important for the research of Cramér's students.

In 1934 W. Feller came to Stockholm and stayed for four years. It was of great value for Cramér personally and for his graduate students that Feller, with his mathematical roots in pre-1933 Germany and his skill as a researcher, was present at the institute during this period.

During this decade, Cramér was entrusted with the task of adjusting the bases of insurance premiums and reserves to the decreasing rates of interest and, also, with the broader task of constructing a new base ("the technical base of 1938"). In addition, he was appointed to a state commission for preparing a new insurance business act. These were time-consuming extra jobs for a professor at a newly founded institute.

1.6. Regarding the 1940's, I will be more detailed and more personal, which is natural, since in the autumn term of 1939 I had begun studying at the institute. One of my first memories is the move to new premises at Norrtullsgatan in northern Stockholm, built with funds obtained from insurance companies. The building was small and cosy; there was one room for the professor, a library, a small combined lecture and computing room and a few minor rooms—that was all. But we young students enjoyed the time we spent there and in the nearby café. We followed the vicissitudes of the war, performed long calculations using manual Odhner calculators and listened to Cramér's lectures on basic insurance mathematics. Touching the dull formulas with his magic wand, he turned them into poetry.

We progressed with our studies and were soon attending the seminars. Cramér was then busy with *Mathematical Methods* and provided us with subjects for our own seminars, taken from the third part of his book, in particular from the chapters on analysis of variance and regression analysis. It was a privilege to have this early contact with his book: The theory behind the standard statistical methods became clear and formed an essential part of our training.

During the war, Cramér showed great hospitality. Harald Bohr, the eminent mathematician, worked at the institute for some years, sheltered from the ill winds over Denmark, and Olav Reiersoel came from Norway.

After the war, Cramér spent a sabbatical year in the United States. For part of this time he was a visiting professor for the Bicentennial Year at Princeton University.

When Cramér returned to Sweden, guest speakers began to visit the institute, and we graduate students profited from their learning and experience. It was interesting to see people we had hitherto only read about. Indeed, they really existed! Still, most of all we liked Cramér's lectures given each Friday and Saturday morning, also attended by well-established actuaries and statisticians. After all these years, I can still feel the thrill in my mind when Cramér entered the lecture room and started speaking, distinctly, inspiringly, and always without a manuscript. It was sometimes said that Cramér was so clever that his listeners had a feeling of understanding his message even if, deep down, they did not grasp anything at all!

The seminars ended in May each year with a finale in Cramér's magnificent home in Djursholm, where Marta acted as a queen-like hostess. A lecture was given in Harald's library, we had a sumptuous meal and our hosts told us over coffee about their travels and encounters.

Cramér was a true gentleman. All his students liked him, but in a distant and respectful way. He gave the graduate students strong support and much encouragement. Niels Arley, the Danish scientist, writes in his *Memoirs* that Cramér was "one of the noblest and, at the same time, heartiest persons I have ever met."

Let me mention in passing that Cramér was fakultetsopponent (external examiner) when, in 1943, Arley defended his doctor's dissertation in public at the University of Copenhagen. In spite of the war, Cramér was allowed an entry permit to Denmark. His appearance on this occasion was, according to Arley, a display of Swedish academic elegance with doctor's tailcoat, doctor's hat and all that.

Let me return to the institute. To his students, Cramér was above all a theoretician, but he did not forget applications. Indeed, he was also an actuary! The written examinations revealed his dual interest in theory and real data. These exams were divided into one day for theoretical problems and one day for numerical and other applications. On the second day, one single large problem was given. For example: Given some raw mortality data, fit a Makeham curve by minimum chi-square. The second day was feared by some students including myself. If I cannot get started, what shall I do?

For some years in the 1940's, Cramér was occupied with a plan for adding to his institute a section for applied statistics. An organization was outlined by a committee chaired by Cramér, but the idea was never carried out. Instead, in 1948, we graduate students started an applied activity, known as the Statistical Research Group, which still exists. Cramér was the first Inspector of the Group and was helpful with both good advice and the acquisition of grants.

1.7. After this nostalgic retrospect of the 1940's, let me look, more briefly, at the 1950's.

Cramér was a good administrator and, as has already been said, an outstanding speaker. It was therefore not surprising that, in 1950, his colleagues elected him President of the University of Stockholm. After eight years Cramér became Chancellor of the Swedish universities. He was then 65 and at the same time retired from his professorship. I have a lively recollection of Harald's appointment as Chancellor, for we discussed it at the dinner given when I defended my Ph.D. thesis in May 1958. (By the way, this publication was the last of the 10 theses written during Cramér's time as a professor.) It was certainly an honour for Cramér to become Chancellor. It gave him many opportunities to express his views concerning the administration of the universities, but I think it was not an ideal task for an active scientist like him. He left the Chancellor's office after three years and returned to research.

In the 1950's Cramér again visited the United States. In 1953 he lectured at Berkeley for some time.

Harald's friends celebrated his 65th birthday with a Festschrift: *Probability and Statistics. The Harald Cramér Volume* (1959). (References with years in parentheses refer to the list of additional references at the end of the paper.)

1.8. In the 1960's Cramér was again very active in research. From 1962–1965 he worked for several periods of time at the Research Triangle Institute of North Carolina, with M. R. Leadbetter as his assistant and later as his coauthor.

1.9. In the 1970's Cramér continued to travel and to do research. In 1970 he gave the first S. S. Wilks lecture at Princeton, in 1976 he lectured (in French) at the European Meeting of Statisticians in Grenoble and in 1977 he lectured in Calcutta. He published 10 articles during this decade, at the age of 76 and over.

1.10. Even in the 1980's, after the age of 86, Harald spoke on several occasions in Sweden and elsewhere. In 1980 he was invited to give the Pfizer Colloquium Lecture at the University of Connecticut [112]; in the summer of 1983 he gave the main address at the opening ceremony of the Berkeley Conference in Honor of Jerzy Neyman and Jack Kiefer; in November of the same year he talked at the Royal Swedish Academy of Sciences in Stockholm about "Probability theory and statistical methodology—memories from sixty years." In June 1984, when almost 91, he delivered a lecture, memorable to all present, at the Conference on Stochastic Processes and Their Applications in Gothenburg: The subject was "Some remarks on the early development of the theory of stochastic processes."

1.11. Harald Cramér received many distinctions. He was awarded honorary university degrees at Princeton (1947), Copenhagen (1950), Stockholm (1964), Helsinki (1971), Edinburgh (1972), Calcutta (1977) and Paris (1977). The Royal Statistical Society elected him an Honorary Fellow and in 1972 awarded him The Guy Medal in Gold. He received a prize for actuarial mathematics from Accademia dei Lincei in Rome. He was an Honorary Member of the International Statistical Institute and of the American Academy of Arts and Sciences. He was a member of the Swedish, Danish, Finnish, Norwegian and Spanish Academies of Sciences.

1.12. Cramér was Chief Editor of *Skandinavisk Aktuarietidskrift* (now *Scandinavian Actuarial Journal*) from 1940–1963. He was the President of the Swedish Actuarial Society from 1935–1964 and in 1964 became its Honorary President.

2. Harald Cramér's scientific contributions—an overview. Cramér's scientific output is rich and varied. Some results belong to the history of mathematical statistics, while some bear his name and are used over and over again even today. All second and third year students of statistics know about the Cramér–Rao inequality; some also know about the Cramér–von Mises statistics. Graduate students hear about Cramér's contributions to the theory of the central limit theorem, some use the Cramér–Wold device and a few know the Cramér–Lévy theorem, stating that if a sum of two independent random variables is normally distributed, then so are the components. However, these are only a few examples of Cramér's contributions to different fields of probability and statistics.

As I have mentioned, Cramér worked as a chemist, pure mathematician, actuary, probabilist and statistician, and this is reflected in his publications spanning a period of 70 years.

Together with B. Matérn, I have compiled a bibliography of Cramér's works. Harald assisted us. On the occasion of his 90th birthday, we gave him a mimeographed version, which was later printed [Blom and Matérn (1984)]. It is divided into two parts, the first comprising 114 publications in foreign languages (from a Swedish point of view!) and the second containing 38 publications in Swedish.

The first part of the bibliography is reprinted at the end of this article. It contains four books, *Random Variables and Probability Distributions* [53], *Mathematical Methods of Statistics* [61], *The Elements of Probability Theory* [71] and *Stationary and Related Stochastic Processes* [96]; and 110 scientific papers and other publications, 19 of which may be classified as pure mathematics, 23 as actuarial mathematics including risk theory, 56 as probability and stochastic processes and 12 as other subjects.

The contributions to pure mathematics concern analytic functions and analytic number theory and will not be considered here, nor will his papers on actuarial mathematics, apart from some remarks on risk theory. Cramér's scientific actuarial work is commented upon in the obituaries by Bohman (1985) and Lundberg (1986).

3. Harald Cramér—the probabilist.

3.1. It is not altogether easy for me to assess the importance of Cramér's extensive work as a probabilist. Being myself a statistician, not a probabilist, I would hardly have undertaken to write this article were it not for the fact that (in [108]) Cramér had assembled personal recollections from his life as a probabilist. This is an admirable paper: Apart from what it tells us about Cramér's writings, it has a lasting value as a document in the history of probability.

With Cramér's recollections before me, the best thing I can do is, once more, to be subjective and convey to the reader some personal views formed during my years as a student at Cramér's institute and later as a university teacher.

When Cramér became interested in probability in the 1920's, this subject was not an accepted branch of mathematics. Isolated efforts to base probabilistic results on good mathematics had been made, but no general mathematical theory of probability existed.

Cramér felt the need for a radical change, as appeared in one of his Swedish papers written in 1926: "The probability concept should be introduced by a purely mathematical definition, from which its fundamental properties and the classical theorems are deduced by purely mathematical operations."

As we know, something like a revolution took place in the 1930's, brought about in the first place by Kolmogorov, but also by other scientists such as Khintchine, Lévy and Cramér. In this decade, probability theory became a respectable part of mathematics with probability defined as a set function, a stochastic process a well-defined concept used as a model for random changes in space or time and so forth.

In 1937 Cramér published *Random Variables and Probability Distributions*, a booklet read by a whole generation of probabilists. In a way, Cramér's dream of 1926 materialized in this book. Probability was given a clear foundation in the spirit of Kolmogorov and was studied by rigorous mathematical methods. Cramér's whole subsequent work was governed by this attitude, which proved immensely fruitful for the development of probability theory.

3.2. I will now describe some of Cramér's special achievements, beginning with the central limit theorem. Cramér is a great name in the history of this theorem. It was his work as an actuary that inspired him to study the behaviour of a large number of random variables when added and suitably standardized. In particular, he was interested in estimating the error of replacing the distribution function of a standardized sum by that of the standard normal distribution function.

At the beginning of the century, Charlier had expanded the error in an asymptotic series. Cramér showed that Charlier's proof was incorrect and replaced the series by a more satisfactory one. Generalizations to multidimensional random variables were given by Cramér and Wold in [50], where the well-known Cramér–Wold device for proving multidimensional normality is found. Cramér's first paper on the central limit theorem is [23] and the most important one is, I think, [30]. A pioneering paper [55] deals with a limit theorem for large deviations of standardized sums of random variables. A condensed version of many of Cramér's results can be found in *Random Variables*. Another good source of knowledge (on a less sophisticated level) is *Mathematical Methods*, Chapter 17, where I, like so many students, first learned about these things. I loved Cramér's way of proving the central limit theorem in this book by means of the continuity theorem.

Let me add in passing that I have always regarded the continuity theorem as the finest in probability theory. Lévy was first with the idea in *Calcul des*

Probabilités, but the final formulation is due to Lévy and Cramér. The theorem and its proof given in the first edition of *Random Variables* were not entirely correct and were corrected in the second edition. In *Mathematical Methods* a very clear proof is given that ought to be read by all serious students of probability, not only for its content but also as an example of Cramér's art of writing.

3.3. The second area I will dwell upon is the theory of stochastic processes. Two aspects are important.

First, Cramér wrote many papers and one book on stochastic processes. Second, by his enthusiasm and expert knowledge he stimulated others to work in the field. This influence was felt in the 1930's by his graduate students working with time series [Wold (1938)], and processes of importance for insurance [Segerdahl (1939) and Lundberg (1940)] and by those of my own generation working with inference for stochastic processes [Grenander (1950)] and with stationary processes on the line and in the plane [Hanner (1949) and Matérn (1960)]. Of course, this influence was felt also by many other of Cramér's readers and listeners.

I have a vivid memory of Cramér's authoritative lectures on stochastic processes given some time during the second half of the 1940's. It is a pity that these lectures did not result in a companion volume to *Mathematical Methods*.

In the sequel I will mention three of Cramér's specific contributions to the theory and development of stochastic processes.

3.4. First, Cramér's research in the field of risk theory greatly clarified this theory and stimulated its progress. The work of the father of risk theory, the Swede F. Lundberg, was difficult to read. Cramér took the trouble to penetrate it, and in *Skandinavisk Aktuarietidskrift* (1926) wrote a 22-page review (in Swedish) of one of Lundberg's papers. This article, written in a pungent style seldom used by the amiable Cramér, is worth quoting (the translation is mine): Lundberg "has an outstanding talent for putting forward even rather simple mathematical arguments in such a way that the readers get the very lively feeling of not understanding anything at all... The reader waits in vain for these small valuable hints in the form of comments to the proofs, these glimpses backwards and forwards that a benevolent author inserts in order to mitigate the troubles of the reader. Little help is to be expected from Lundberg in this respect; if such a remark is actually made, it is at times formulated with oracle-like abstruseness." Nevertheless, in the review Cramér extracted from Lundberg's work the essentials of what is now called the collective risk theory and gave it a modern formulation; see also the account of Lundberg's work which Cramér gave much later in [100].

Some of the most important publications on risk theory are [32], [69] and [72]. In the third paper, collective risk theory is treated as a special part of the general theory of stochastic processes. Among other things, Cramér studied the ruin functions for an insurance business, a task already begun by F. Lundberg. Cramér's estimate for ruin is often quoted. Later on, many other people, especially in Sweden, worked on these problems, and the technique has proved valuable even outside the insurance business; see, e.g., Feller (1971).

3.5. Second, Cramér was active in the field of stationary and related processes. The first wave of interest came in the 1940's, when Cramér wrote two now famous papers: In [58] he generalized the spectral representation of a covariance function of a univariate stationary process to vector processes; in [60] he derived a spectral representation of the process itself in terms of an orthogonal process. There was a second wave in the 1960's after Cramér's retirement. With M. R. Leadbetter he published *Stationary and Related Stochastic Processes* [96], an excellent book both for students and specialists, and another example of Cramér's interest in both teaching and research. The first part (Chapters 1–8) contains a very lucid account of the general theory of stochastic processes with special chapters on particular processes. In the second part (Chapters 9–15), the authors present a multitude of results concerning level crossings for stationary Gaussian processes. For example, they prove the fine result, due partly to Cramér, that under mild conditions the upcrossings by a stationary Gaussian process of a high level follow approximately a Poisson process. The second part of the book has given rise to much further research.

3.6. Third, Cramér and others developed the multiplicity theory of stochastic processes. For a comprehensive review of this field see *Random Processes, Multiplicity Theory and Canonical Decompositions* (1973). Cramér's work on this part of the theory of processes bears witness to his vitality after retirement. From 1960 onward he discussed multiplicity theory in about 10 papers, several of which are reprinted in the above-mentioned book; his last scientific paper [113], published when he was 89, treats this subject. The theory throws light on much work done in the past by many research workers, among them Wold, Hanner and Cramér himself, and completes, in a logical way, his activity as a probabilist.

4. Harald Cramér—the author of *Mathematical Methods*.

4.1. In 1945 Cramér published *Mathematical Methods of Statistics* [61]. This was a landmark in the history of mathematical statistics. Before this time, no unified statistical theory based on probability existed. Cramér saw the need and the possibility for building statistical science on the foundations of mathematical probability, and the written result of his efforts was *Mathematical Methods*. After the message of this book had been spread over the statistical world, our subject never was the same again. Let me comment upon some of the resulting changes.

4.2. First, education of statisticians has become different. Cramér convinced us that the ingredients corresponding to the three parts of the book, namely, mathematics, probability and statistics, are indispensable in the studies of a mathematical statistician. For several decades now, *Mathematical Methods* has been much used as a textbook by students in many countries. Some of these people later wrote books and papers themselves, in the same spirit, and this process will be repeated in years to come. In this way, the book has had, and will have, a tremendous influence on the training of mathematical statisticians.

Personally, I owe more to this book than I can briefly explain. About 40 years have passed since I first read *Mathematical Methods* at Cramér's institute, and I still use it now and then for contemplation and for reference. A single example of a section to which I have returned many times is 20.6, with its useful convergence theorem, sometimes called Cramér's theorem or the Cramér–Slutsky theorem.

4.3. Second, research in statistical theory has been profoundly affected by *Mathematical Methods*. Earlier, statistical ideas were often launched very loosely, even in scientific journals. With few exceptions, they are nowadays studied thoroughly, using rigorous mathematical methods, before they are published. It is my conviction that this new attitude toward research in statistics is largely due to Cramér's influence through this book.

A good example is the exposition in Chapter 27 of standard errors of sample moments and functions of such moments. Earlier, there was in this field a mess of calculations of doubtful validity, an Augean stable of formulas, which Cramér cleaned by letting in a stream of good quality mathematics.

4.4. Third, several new results in the third part of the book have led to further important research activities. One example concerns the asymptotic properties of maximum likelihood estimates in Chapter 33, where Cramér performed important research that was later continued by many people.

Another striking example is the introduction and proof in Chapter 32 of the Cramér–Rao inequality. Such a fundamental result is seldom given for the first time in a textbook. When reading this chapter in the 1940's, I was astonished that such a result was possible and I admired the proof immensely. Until then, statistical inference had been mainly a lot of approximations of unknown validity; here was a marvellous precise answer.

4.5. Finally, let me put forward the view that *Mathematical Methods* is unique among statistical textbooks for its formal qualities. Quite apart from its content, it is a joy to read such a clearly written book. In a way, I deplore that the development of statistics since 1945 has made it partly obsolete. Nevertheless, the book is still read by many people. A Russian mathematician has aptly said that *Mathematical Methods* is nowadays a necessary but not sufficient part of a statistician's education. In any case, this book will in the future be a rich source of knowledge regarding the history of mathematical statistics and will remain a masterpiece of its time.

Harald Cramér is dead but his memory will live on in the history of probability and statistics and in the hearts of his admirers and friends. He was a great scientist and a good man.

Acknowledgments. I wish to thank the following persons who have helped me with the preparation of this article: N. Arley, H. Bohman, L. Bondesson, K. L. Chung, J. Lanke, M. R. Leadbetter, O. Lundberg, B. Matérn, R. Sundberg and H. Wold. I am also indebted to the Editor and to the Managing Editor for many formal improvements of the original manuscript.

REFERENCES

BLOM, G. and MATÉRN, B. (1984). Bibliography. Publications by Harald Cramér. *Scand. Actuar. J.* **67** 1–10.

BOHMAN, H. (1985). Harald Cramér in memoriam. *Scand. Actuar. J.* **68** 3–4.

EPHREMIDES, A. and THOMAS, J. B., eds. (1973). *Random Processes, Multiplicity Theory and Canonical Decomposition*. Dowden, Hutchinson and Ross, Stroudsburg, Pa.

FELLER, W. (1971). *An Introduction to Probability Theory and Its Applications* **2**, 2nd ed. Wiley, New York.

GRENANDER, U. (1950). Stochastic processes and statistical inference. *Ark. Mat.* **1** 195–277.

GRENANDER, U., ed. (1959). *Probability and Statistics. The Harald Cramér Volume*. Almqvist and Wiksell, Stockholm.

HANNER, O. (1949). Deterministic and nondeterministic stationary random processes. *Ark. Mat.* **1** 161–177.

LUNDBERG, O. (1940). On random processes and their application to sickness and accident statistics. Thesis, Stockholm.

LUNDBERG, O. (1986). Obituary. Harald Cramér. *Astin Bull.* **16** 9–10.

MATÉRN, B. (1960). Spatial variation. *Medd. Statens Skogsforskningsinst.* **49** 1–144.

SEGERDAHL, C. O. (1939). On homogeneous random processes and collective risk theory. Thesis, Stockholm.

WOLD, H. (1938). A study in the analysis of stationary time series. Thesis, Stockholm.

VIDEOTAPE

A videotape containing a recording of a lecture by Harald Cramér on the history of probability is available from Continuing Education, American Statistical Association, 1429 Duke Street, Alexandria, Virginia 22314.

DEPARTMENT OF MATHEMATICAL STATISTICS
UNIVERSITY OF LUND
BOX 118
S-221 00 LUND
SWEDEN

1.

Sur une classe de séries de Dirichlet

Thesis, Stockholm University. Almquist & Wiksell, Uppsala, 51 pp. (1917)

INTRODUCTION.

C'est la démonstration élégante due à M. DE LA VALLÉE POUSSIN[1] du théorème fondamental dans la théorie de la fonction $\zeta(s)$ de RIEMANN qui a été le point de départ de ce travail.[2] Cette démonstration repose essentiellement sur le fait que la différence

$$\sum_{n=1}^{\infty} \frac{1}{n^s} - \int_1^{\infty} \frac{dx}{x^s}$$

représente une fonction entière de s, ce qu'on prouve par une suite d'intégrations partielles. En étudiant la théorie de la fonction $\zeta(s)$ et des séries de DIRICHLET, j'ai été amené à me poser la question suivante: cette méthode, peut-elle s'appliquer à des cas plus généraux?

Il me fallait donc comparer la série de DIRICHLET:

$$H(s) = \sum_{n=1}^{\infty} \frac{\varphi(n)}{n^s}$$

à l'intégrale

$$J(s) = \int_1^{\infty} \frac{\varphi(x)}{x^s} dx.$$

L'instrument analytique s'offrant tout naturellement pour

[1] Voir E. LANDAU: »Handbuch der Lehre von der Verteilung der Primzahlen», p. 271.

[2] Plusieurs des résultats de ce travail ont été présentés dans une conférence, faite en novembre 1916 à la Société Mathématique de Stockholm.

effectuer cette comparaison est la formule sommatoire d'EULER et de MACLAURIN. On a donné à cette formule, et en particulier au reste qui y figure, bien des formes différentes.[1] Je donnerai au § 1 la déduction de la formule sommatoire, en me rattachant à M. WIRTINGER,[2] qui en a fait dans un beau mémoire des applications intéressantes à plusieurs questions dans la théorie des fonctions.

En faisant ensuite l'application de la formule d'EULER aux séries de DIRICHLET, je donnerai au § 2 une classe de séries de la forme écrite ci-dessus, pour lesquelles la différence

$$\sum_{n=1}^{\infty} \frac{\varphi(n)}{n^s} - \int_1^{\infty} \frac{\varphi(x)}{x^s} dx$$

représente une fonction entière de s. Cette classe sera caractérisée par les deux conditions suivantes:

A. Pour $x \geqq 1$, $\varphi(x)$ sera continue ainsi que ses dérivées de tout ordre.

B. Il existe une constante réelle k, telle qu'on aura, pour x indéfiniment croissant:

$$\varphi^{(\mu)}(x) = O(x^{k-\mu})^{3}$$

$$\text{pour } \mu = 0, 1, 2 \ldots.$$

Ces conditions se trouvent vérifiées par des classes assez étendues de fonctions, comme on le verra dans la suite. Cependant, pour les séries qu'on étudie dans la théorie analytique des nombres, ces considérations ne semblent avoir aucune importance.

La fonction représentée par la série de DIRICHLET en vue étant maintenant partagée en deux éléments, je procé-

---

[1] Voir »Encyklopädie der mathematischen Wissenschaften», Bd. II. A. 12, p. 1324; et A. A. MARKOFF, »Differenzenrechnung», p. 112—140.

[2] »Einige Anwendungen der Euler-Maclaurin'schen Summenformel», Acta Math. 26,255.

[3] Je suppose connu le symbole O introduit par M. BACHMANN. Ces égalités ne sont pas toutes indépendantes. Voir HARDY and LITTLEWOOD: »Contributions to the arithmetic theory of series», Proc. Lond. Math. Soc. ser. 2, vol. 11, p. 411.

derai au § 3 à l'étude du premier élément, savoir de la fonction entière $G(s)$, définie par la relation

$$H(s) = G(s) + J(s).$$

Dans plusieurs des cas spéciaux traités plus tard, on verra que c'est cette fonction entière qui va déterminer la manière de se comporter de la fonction $H(s)$ à l'infini. Au contraire, les singularités à distance finie sont évidemment identiques à celles de $J(s)$.

Après avoir donné une application au cas où $\varphi(x)$, pour des valeurs très grandes de x, admet un certain développement, j'envisagerai au § 5 une série de Dirichlet sous la forme

$$H(s) = \sum_{n=1}^{\infty} \frac{\psi(\log n)}{n^s}.$$

En posant $\varphi(x) = \psi(\log x)$, voilà la condition B. prenant la forme

$$\psi^{(\mu)}(z) = O(e^{kz}), \quad \mu = 0, 1, 2 \ldots.$$

Je regarderai ici $\psi(z)$ comme une fonction de la variable complexe z, ce qui me permettra d'utiliser le calcul des résidus pour simplifier les conditions imposées à $\psi(z)$.

Comme une application immédiate, je traiterai dans le même paragraphe le cas où $\psi(z)$ est une fonction entière de z dont l'ordre ne surpasse pas l'unité.

Un des résultats ainsi obtenus est susceptible d'être généralisé, comme je le montrerai dans le § 6, où je traiterai une question analogue pour la série

$$\sum_{n=1}^{\infty} a_n e^{-\lambda_n s}$$

en supposant a_n comme fonction entière de λ_n.

Dans le paragraphe suivant, § 7, je vais me servir d'un théorème important dû à M. M. PHRAGMÉN et LINDELÖF.[1]

[1] »Sur une extension d'un principe classique de l'analyse etc.» Acta Math. 31, p. 381.

Or, le théorème donné par ces auteurs dans leur mémoire original étant d'une nature très générale et assez difficile à pénétrer, j'ai cru utile donner ici la démonstration explicite du théorème particulier dont j'aurai à me servir. La démonstration de ce théorème ainsi que son interprétation géométrique a été donnée par M. MARCEL RIESZ dans une conférence, faite à la Société Mathématique de Stockholm.

Appliqué aux séries de DIRICHLET, ce théorème nous fournit dans bien des cas un prolongement analytique de la fonction représentée par une telle série au delà de l'abscisse de convergence. Cela se fait p. ex. lorsque la fonction $\psi(z)$, mentionnée plus haut, est holomorphe dans un angle renfermant l'axe réel et positif et y vérifie uniformément une condition de la forme $\psi(z) = O(e^{k|z|})$. — Dans le même paragraphe, je donnerai encore deux théorèmes restant dans le même ordre d'idées, quoique ne se rattachant pas immédiatement au théorème de MM. PHRAGMÉN et LINDELÖF.

Enfin, dans les deux derniers paragraphes, je donnerai quelques applications des résultats précédents à des séries particulières et à une question de sommabilité.

En terminant cette introduction, c'est pour moi un agréable devoir d'exprimer ma vive reconnaissance à M. MARCEL RIESZ. Il a bien voulu m'indiquer la relation entre la formule d'EULER et mes résultats antérieurs, et je lui suis encore redevable de nombreuses observations, qui m'ont été très utiles. Mes plus sincères remercîments sont dûs encore à Mlle HILDA CRAMÉR, qui a bien voulu m'aider en revoyant mon manuscrit sous point de vue philologique.

§ 1. La formule sommatoire d'Euler.

Désignons par x une variable réelle et positive et posons

$$P_1(x) = [x] - x + \frac{1}{2}.$$

Pour toute valeur non-entière de x on a donc

$$P_1(x) = \sum_{n=1}^{\infty} \frac{\sin 2n\pi x}{n\pi}.$$

Introduisons maintenant une suite de fonctions $P_k(x)$, ($k = 2, 3 \ldots$), définies par les séries trigonométriques suivantes:

$$P_{2k}(x) = \sum_{n=1}^{\infty} \frac{\cos 2n\pi x}{2^{2k-1}(n\pi)^{2k}}$$

$$P_{2k+1}(x) = \sum_{n=1}^{\infty} \frac{\sin 2n\pi x}{2^{2k}(n\pi)^{2k+1}}.$$

Le développement de $P_k(x)$ est donc, pour $k > 1$, uniformément convergente pour toute valeur de x. On vérifie facilement que les $P_k(x)$ jouissent des propriétés suivantes:

$$(1) \qquad \begin{aligned} P_k(x) &= (-1)^k \frac{d}{dx} P_{k+1}(x) \\ P_k(x+1) &= P_k(x). \end{aligned}$$

Si x a une valeur entière, la première formule n'est pourtant valable que pour $k > 1$. Les $P_k(x)$ ainsi définies sont identiques aux fonctions de Bernoulli. Les valeurs que prennent ces fonctions, lorsque x est égal à un nombre entier m, sont données par les formules:

$$(2) \qquad \begin{aligned} P_{2k}(m) &= \frac{\zeta(2k)}{2^{2k-1}\pi^{2k}} = \frac{B_k}{(2k)!} \\ P_{2k+1}(m) &= 0, \end{aligned}$$

où $k = 1, 2 \ldots$, les B_k désignant les nombres bien connus de Bernoulli.

Soit maintenant $F(x)$ une fonction continue ainsi que ses dérivées de tout ordre pour $x \geqq 1$ et n un nombre entier positif. Par une intégration partielle, on obtient la formule

$$\int_n^{n+1} P_1(x)\,F'(x)\,dx = \int_n^{n+1}\left(n - x + \frac{1}{2}\right) F'(x)\,dx =$$

$$= -\frac{1}{2}F(n) - \frac{1}{2}F(n+1) + \int_n^{n+1} F(x)\,dx.$$

Puis, si l'on pose successivement $n = 1, 2 \ldots N-1$ et que l'on ajoute les résultats, il vient

$$\int_1^N P_1(x)\,F'(x)\,dx = -\sum_{n=1}^N F(n) + \frac{1}{2}[F(1) + F(N)] + \int_1^N F(x)\,dx,$$

d'où

$$(3)\quad \sum_1^N F(n) - \int_1^N F(x)\,dx = \frac{1}{2}[F(1) + F(N)] - \int_1^N P_1(x)\,F'(x)\,dx.$$

L'intégrale du second membre peut se transformer par une suite d'intégrations partielles, et nous aurons la formule sommatoire d'Euler dans la forme dont nous nous servirons, en tenant compte des relations (1) et (2):

$$(4)\quad \begin{aligned} \sum_{n=1}^N F(n) - \int_1^N F(x)\,dx &= \frac{1}{2}[F(1) + F(N)] + \\ &+ \sum_{\nu=1}^h \frac{(-1)^\nu B_\nu}{(2\nu)!}[F^{(2\nu-1)}(1) - F^{(2\nu-1)}(N)] + R_h, \end{aligned}$$

le reste R_h étant donné par l'une ou l'autre des expressions

$$(5)\quad \begin{aligned} R_h &= (-1)^h \int_1^N P_{2h}(x)\,F^{(2h)}(x)\,dx \\ &= (-1)^{h+1} \int_1^N P_{2h+1}(x)\,F^{(2h+1)}(x)\,dx. \end{aligned}$$

Quoique la déduction précédente de notre formule sommatoire en mette bien en évidence le caractère tout à fait élémentaire, elle paraît être un instrument analytique très fertile. Avant d'aborder notre sujet principal, indiquons ici, d'après une remarque de M. RIESZ, comment peut se déduire l'équation fonctionelle de la fonction $\zeta(s)$ même de la formule (3), qui ne constitue pourtant que la première étape de la formule d'EULER.

En effet, si l'on pose dans cette formule $F(x)=x^{-s}$, $s=\sigma+it$, et que l'on fait tendre N vers l'infini, en supposant d'abord $\sigma>1$, il vient

$$\zeta(s)=\frac{1}{2}+\frac{1}{s-1}+s\int_1^\infty\frac{P_1(x)}{x^{s+1}}dx$$

$$=\frac{1}{2}+\frac{1}{s-1}+\frac{s}{\pi}\int_1^\infty\sum_{n=1}^\infty\frac{\sin 2n\pi x}{n}\cdot\frac{dx}{x^{s+1}}$$

$$=\frac{1}{2}+\frac{1}{s-1}+\frac{s}{\pi}\sum_{m=1}^\infty\sum_{n=1}^\infty\frac{1}{n}\int_m^{m+1}\frac{\sin 2n\pi x}{x^{s+1}}dx.$$

Or, il est facile de prouver (par une intégration partielle) que le terme général de la série double est inférieur en valeur absolue à

$$\frac{|s+1|}{\pi n^2}\int_m^{m+1}\frac{dx}{x^{\sigma+2}},$$

il s'ensuit donc que cette série double converge absolument et uniformément dans toute partie finie du demiplan $\sigma>-1$.

La valeur de s étant comprise dans ce demiplan, on peut intervertir l'ordre des sommations et écrire

$$\zeta(s)=\frac{1}{2}+\frac{1}{s-1}+\frac{s}{\pi}\sum_{n=1}^\infty\frac{1}{n}\int_1^\infty\frac{\sin 2n\pi x}{x^{s+1}}dx. \tag{6}$$

Cependant, l'égalité

$$(7)\qquad \frac{1}{2}+\frac{1}{s-1}=s\int_0^1\frac{\frac{1}{2}-x}{x^{s+1}}\,dx$$

$$=s\int_0^1\frac{P_1(x)}{x^{s+1}}\,dx$$

$$=\frac{s}{\pi}\sum_{n=1}^{\infty}\frac{1}{n}\int_0^1\frac{\sin 2n\pi x}{x^{s+1}}\,dx$$

a lieu pour $\sigma<0$; en supposant $-1<\sigma<0$, on peut donc ajouter (6) et (7), d'où il suit:

$$\zeta(s)=\frac{s}{\pi}\sum_{n=1}^{\infty}\frac{1}{n}\int_0^{\infty}\frac{\sin 2n\pi x}{x^{s+1}}\,dx$$

$$=\frac{s}{\pi}\sum_{n=1}^{\infty}\frac{1}{n^{1-s}}\int_0^{\infty}\frac{\sin 2\pi y}{y^{s+1}}\,dy$$

$$=\frac{s}{\pi}\zeta(1-s)\int_0^{\infty}\frac{\sin 2\pi y}{y^{s+1}}\,dy.$$

L'intégrale peut s'évaluer par le calcul des résidus[1] et l'on obtient l'équation fonctionelle bien connue de la fonction $\zeta(s)$.

§ 2. Application aux séries de Dirichlet.

Passons à l'application de la formule d'Euler aux séries de Dirichlet. Nous allons les considérer d'abord dans la forme

[1] Voir N. Nielsen, »Handbuch der Gammafunktion», p. 153.

$$(8) \qquad H(s) = \sum_{n=1}^{\infty} \frac{\varphi(n)}{n^s},$$

plus tard, nous verrons que les calculs se simplifieront en regardant les coefficients non comme une fonction de leur indice n, mais comme fonction de $\log n$.

Posons encore

$$(9) \qquad J(s) = \int_1^{\infty} \frac{\varphi(x)}{x^s} dx,$$

et proposons-nous d'étudier la différence $H(s) - J(s)$, en faisant certaines hypothèses sur la fonction $\varphi(x)$. On peut remarquer qu'il n'importe point que nous avons pris l'unité pour limite inférieure dans $H(s)$ et $J(s)$, la formule d'Euler et tout le raisonnement qui va suivre ne subissant que des modifications évidentes, quand on change la limite inférieure. Quant aux hypothèses sur $\varphi(x)$, les voici:

A. Pour $x \geqq 1$, $\varphi(x)$ sera continue ainsi que ses dérivées de tout ordre.

B. Il existe une constante réelle k, telle qu'on aura, pour x indéfiniment croissant

$$\varphi^{(\mu)}(x) = O(x^{k-\mu})$$

$$\text{pour } \mu = 0, 1, 2 \ldots$$

Bien entendu, on ne suppose pas que ces égalités aient lieu uniformément pour toutes les valeurs de μ, ce qui restreindrait beaucoup la généralité.

En posant suivant l'usage un peu absurde $s = \sigma + it$, on voit que, pour $\sigma > k + 1$, (8) et (9) sont absolument convergentes. — Soit p un entier positif. Dans la formule sommatoire (4), nous posons

$$F(x) = \frac{\varphi(x)}{x^s}$$

et nous prenons suivant le cas $2h = p$ ou $2h + 1 = p$. En supposant d'abord $\sigma > k + 1$, on peut dans le premier membre faire tendre N vers l'infini. Pour voir comment se comportent alors les termes du second membre dépendant de N, remarquons que nous avons par la formule de LEIBNIZ:

$$F^{(r)}(x) = \sum_{\mu=0}^{r} (-1)^{r-\mu} \binom{r}{\mu} \frac{s(s+1)\ldots(s+r-\mu-1)}{x^{s+r-\mu}} \varphi^{(\mu)}(x).$$

En tenant compte de l'hypothèse B., on en conclut

$$\lim_{x \to \infty} F^{(r)}(x) = 0, \qquad (r = 0, 1, 2 \ldots)$$

$$\text{pour } \sigma > k + 1.$$

Il vient donc

$$G(s) = H(s) - J(s) = \frac{1}{2}\varphi(1) +$$

$$+ \sum_{\nu=1}^{h} \frac{(-1)^{\nu} B_{\nu}}{(2\nu)!} \sum_{\mu=0}^{2\nu-1} (-1)^{\mu+1} \binom{2\nu-1}{\mu} .$$

$$(10) \qquad . s(s+1)\ldots(s+2\nu-\mu-2)\varphi^{(\mu)}(1)$$

$$\pm \sum_{\mu=0}^{p} (-1)^{p-\mu} \binom{p}{\mu} s(s+1)\ldots(s+p-\mu-1) .$$

$$. \int_{1}^{\infty} \frac{P_p(x)\,\varphi^{(\mu)}(x)}{x^{s+p-\mu}} dx,$$

sans qu'il soit nécessaire de préciser le signe du reste. Cette égalité peut s'écrire sous une forme plus condensée en remarquant que la première somme est un polynome en s du degré $2h-1$, et que l'intégrale contenant $\varphi^{(\mu)}(x)$ est multipliée par un polynome en s du degré $p-\mu$. Ecrivons

$$G(s) = H(s) - J(s) =$$

$$(11) \qquad = u_{2h-1}(s) + \sum_{\mu=0}^{p} v_{p-\mu}(s) \int_{1}^{\infty} \frac{P_p(x)\,\varphi^{(\mu)}(x)}{x^{s+p-\mu}} dx,$$

les u_n et v_n désignant des polynomes dont le degré coïncide avec l'indice.

Cette égalité a donc certainement lieu lorsque $\sigma > k+1$, mais nous avons, en tenant compte de la condition B. et de l'expression de $P_p(x)$,

$$\frac{P_p(x)\,\varphi^{(\mu)}(x)}{x^{s+p-\mu}} = O\left(\frac{1}{x^{\sigma+p-k}}\right).$$

Donc, si nous prenons seulement $\sigma \geqq k+1-p+\varepsilon$, où $\varepsilon > 0$, toutes les intégrales du second membre de la relation (11) seront uniformément convergentes et bornées en valeur absolue. Il s'ensuit que la fonction $G(s)$ est holomorphe dans le demiplan $\sigma > k+1-p$. Par suite, p étant arbitraire, $G(s)$ est une fonction entière. Nous pouvons donc énoncer le théorème suivant:

Théorème I. *Si l'on pose* $H(s) = \sum_{n=1}^{\infty} \frac{\varphi(n)}{n^s}$, *la fonction* $\varphi(x)$ *satisfaisant aux conditions A. et B., on pourra écrire*

$$H(s) = G(s) + \int_1^{\infty} \frac{\varphi(x)}{x^s}\,dx,$$

$G(s)$ *désignant une fonction entière.*

Remarque. On pourrait évidemment remplacer l'égalité de la condition B. par l'égalité plus large

$$\varphi^{(\mu)}(x) = O(x^{k-\mu c}), \qquad 0 < c \leqq 1.$$

La différence $H(s) - J(s)$ est encore une fonction entière, et la démonstration se fait comme tout à l'heure, la formule (11) restant valable pour $\sigma > k+1-pc$.

Il pourrait avoir quelque intérêt de montrer ici par un exemple que $H(s) - J(s)$ n'est pas *toujours* une fonction entière, ce qu'on voit aisément en prenant p. ex. $\varphi(x) = \cos 2\pi x$. On aura alors

$$H(s) = \sum_{n=1}^{\infty} \frac{\cos 2n\pi}{n^s} = \sum_{n=1}^{\infty} \frac{1}{n^s} = \zeta(s)$$

$$J(s)=\int_1^\infty \frac{\cos 2\pi x}{x^s}\,dx;$$

$H(s)$ a donc un pôle au point $s=1$, tandis que l'intégrale $J(s)$ est uniformément convergente dans toute partie finie du demiplan $\sigma>0$, comme on le voit par une intégration partielle. $H(s)-J(s)$ aussi a donc un pôle au point $s=1$.

§ 3. Sur une transcendante entière.

Nous allons maintenant étudier les propriétés de la fonction entière $G(s)$. Comme on sait, dans la théorie des fonctions définies par une série de DIRICHLET, des études profondes ont été consacrées à l'ordre de grandeur des fonctions par rapport à une ordonnée indéfiniment croissante. Si sur la droite $\sigma=\sigma_0$ la valeur absolue d'une fonction $f(s)$, holomorphe sur cette droite,[1] ne croît pas plus vite que toute puissance de t, on dit que pour $\sigma=\sigma_0$ la fonction est d'ordre fini par rapport à l'ordonnée.

Les nombres α, pour lesquels l'égalité

$$f(\sigma_0+it)=O(|t|^\alpha)$$

est vérifiée, ont une limite inférieure μ, en général variable avec σ_0. En écrivant σ au lieu de σ_0, nous avons donc défini une fonction $\mu(\sigma)$ représentant dans une certaine mesure la croissance de $f(s)$. L'étude de cette fonction repose avant tout sur un théorème important dû à M. LINDELÖF.[2]

Dans ce travail, nous n'envisagerons que des fonctions pour lesquelles on a $\mu(\sigma)=0$ dès que σ surpasse une certaine valeur σ_1. Les fonctions définies par une série de DIRICHLET, ayant un demiplan de convergence absolue, font partie de cette classe; or, en étudiant d'une manière spéciale la fonction $\mu(\sigma)$ relative à de telles fonctions, on a fait usage principalement de la propriété à laquelle nous venons d'in-

[1] Sauf peut-être sur une portion finie de la droite.
[2] Pour tout ce qui concerne $\mu(\sigma)$ voir HARDY and RIESZ: »The general theory of Dirichlet's series», p. 11—18.

sister. Plusieurs des théorèmes démontrés à cet égard dans l'ouvrage cité de MM. HARDY and RIESZ s'étendent donc d'elles-mêmes à toutes les fonctions dont les $\mu(\sigma)$ correspondantes jouissent de cette propriété. Ainsi, on y trouve démontré que la fonction $\mu(\sigma)$ d'une telle fonction, tant qu'elle est finie et différente de zéro, est une fonction positive, décroissante, convexe et continue.[1]

Revenons cependant à la fonction entière $G(s)$. D'abord, nous introduirons la notation suivante: soit K_n un nombre positif tel que

$$(12) \qquad |\varphi^{(\mu)}(x)| \leqq K_n \, . \, x^{k-\mu}$$

pour $x \geqq 1$, $\mu = 0, 1 \ldots n$. Le nombre K_n est fini, puisque $\varphi(x)$ satisfait aux conditions A. et B.

Théorème II. *$\varphi(x)$ satisfaisant toujours aux conditions A. et B., la fonction entière $G(s)$ jouit des propriétés suivantes:*

1. *Elle ne peut pas être égale à une constante, si la série $\sum_{n=1}^{\infty} \frac{\varphi(n)}{n^s}$ ne se réduit pas à son premier terme.*[2]

2. *La fonction $\mu(\sigma)$ relative à $G(s)$ est toujours finie, et, tant qu'elle est différente de zéro, elle est une fonction positive, décroissante, convexe et continue. D'ailleurs, on a*

$$(13) \qquad \begin{aligned} &\mu(\sigma) = 0, \ \textit{pour} \ \sigma \geqq k+1 \\ &0 \leqq \mu(\sigma) \leqq k+1-\sigma, \ \textit{pour} \ \sigma \leqq k+1. \end{aligned}$$

3. *Dans le demiplan $\sigma \geqq k+1+\varepsilon > k+1$, le module de $G(s)$ a une limite supérieure finie. Encore, en posant $|s| = \varrho$, on a pour toute valeur suffisamment grande de ϱ:*

$$(14) \qquad |G(s)| < e^{\varrho \log \varrho} \, . \, K_{[\varrho+k]+3},$$

tandis que pour une infinité de valeurs de s aux modules indéfiniment croissants

---

[1] Une fonction continue est dite convexe si l'on a toujours $2f\left(\frac{x+y}{2}\right) \leqq f(x) + f(y)$. Voir HARDY and RIESZ, l. c., p. 17.

[2] On pourrait sans difficulté préciser encore cet énoncé, précision qui serait cependant dépourvue d'intérêt réel.

$$|G(s)| > e^{\varrho^{1-\varepsilon}}, \qquad (\varepsilon > 0),$$

en admettant que notre série ne se réduit pas à son premier terme.

1. Supposons que la série $\sum_{n=1}^{\infty} \frac{\varphi(n)}{n^s}$ ne se réduit pas à son premier terme, et que $\varphi(a)$ est le premier coefficient différent de zéro après $\varphi(1)$.[1]

Remarquons que, si $G(s)$ a une valeur constante, cette valeur est nécessairement égale à $\varphi(1)$. En effet, $H(s)$ tend uniformément vers $\varphi(1)$ lorsque σ tend vers l'infini positif, tandis que nous allons prouver dans un instant que $J(s)$ tend uniformément vers zéro. — Cependant nous allons prouver que $G(s)-\varphi(1)$ ne peut s'annuler identiquement.

En effet, on peut montrer aisément,[2] que la série de Dirichlet

$$H(s)-\varphi(1) = \sum_{n=a}^{\infty} \frac{\varphi(n)}{n^s}$$

peut s'écrire

$$H(s)-\varphi(1) = \frac{\varphi(a)}{a^s}(1+\psi(s)),$$

où $|\psi(s)| < \frac{1}{2}$ pour toute valeur de σ suffisamment grande.

Le module de $H(s)-\varphi(1)$ a donc une limite inférieure positive quand σ a une valeur fixe et suffisamment grande et que $|t|$ tend vers l'infini.

Ecrivons maintenant

$$J(s) = \int_1^{\infty} \frac{\varphi(x)}{x^s}\, dx = \frac{\varphi(1)}{s-1} + \frac{1}{s-1}\int_1^{\infty} \frac{\varphi'(x)}{x^{s-1}}\, dx,$$

---

[1] $\varphi(1)$ pouvant s'égaler à zéro.
[2] Hardy and Riesz, l. c., p. 17.

égalité ayant lieu pour toute valeur assez grande de σ et nous montrant que nous avons pour ces valeurs

$$|J(s)| < \frac{M}{|s-1|}.$$

$J(s)$ tend donc uniformément vers zéro avec $\frac{1}{\sigma}$, comme nous l'avons déjà prétendu, mais on voit aussi que $J(s)$ tend vers zéro lorsque, σ ayant une valeur fixe et très grande, $|t|$ tend vers l'infini.

Il s'ensuit donc que

$$G(s) - \varphi(1) = H(s) - \varphi(1) - J(s)$$

ne peut s'annuler identiquement, et le point 1. du théorème I est établi.

2. $H(s)$ et $J(s)$ étant bornées dans tout demiplan $\sigma \geqq k+1+\varepsilon > k+1$, il en est de même pour leur différence $G(s)$. Pour $\sigma \geqq k+1$, on a donc $\mu(\sigma) \leqq 0$. D'autre part, le raisonnement précédent nous montre que $\mu(\sigma) = 0$ pour $H(s)$ et $\mu(\sigma) \leqq -1$ pour $J(s)$, lorsque σ a une valeur très grande. Pour la différence $G(s) = H(s) - J(s)$ on aura alors $\mu(\sigma) = 0$, et l'on en conclut, comme dans la théorie des séries de Dirichlet,[1] que $\mu(\sigma) = 0$ pour $\sigma \geqq k+1$, et que $\mu(\sigma)$, tant qu'elle est finie et différente de zéro, est positive, décroissante, convexe et continue.

Pour évaluer $\mu(\sigma)$ lorsque σ est inférieur à $k+1$, retournons à la formule (11) du paragraphe précédent. Les intégrales qu'elle contient étant bornées en valeur absolue pour $\sigma \geqq k+1-p+\varepsilon$, elle fait voir que la croissance de $G(s)$ dans ce demiplan ne peut être supérieure à celle d'un polynome du degré p. Nous avons donc

$$0 \leqq \mu(\sigma) \leqq p$$

pour $\sigma > k+1-p$.

Or, en vertu de la convexité de $\mu(\sigma)$, il s'ensuit immédiatement

[1] Hardy and Riesz, l. c., p. 17—18.

H. Cramér. 17206 2

$$0 \leqq \mu(\sigma) \leqq k + 1 - \sigma,$$

et cette formule aura lieu pour tout σ inférieur à $k+1$, puisque p est arbitraire.

Nous avons donc démontré le point 2. du théorème II.

3. On a déjà montré que le module de $G(s)$ est borné dans tout demiplan $\sigma \geqq k + 1 + \varepsilon$. Or, $G(s)$ ne s'égalant pas à une constante, on voit immédiatement par un théorème dû à M. PHRAGMÉN[1] qu'on ne peut pas avoir toujours

$$|G(s)| < e^{\varrho^{1-\varepsilon}}$$

où $\varepsilon > 0$.

Pour étudier le module maximum de $G(s)$ sur le cercle $|s| = \varrho$, nous envisagerons la formule (10), en posant

$$p = [\varrho + k] + 3.$$

Si $k < 0$, on suppose ϱ suffisamment grand pour que p soit > 1. Le cercle $|s| = \varrho$ est situé tout entier à l'intérieur du demiplan $\sigma > k + 1 - p$, où cette formule est valable.

Sur ce cercle, on a donc:

$$G(s) = \frac{1}{2}\varphi(1) +$$

$$+ \sum_{\nu=1}^{h} \frac{(-1)^{\nu} B_{\nu}}{(2\nu)!} \sum_{\mu=0}^{2\nu-1} (-1)^{\mu+1} \binom{2\nu-1}{\mu} \cdot$$

$$\cdot s(s+1) \ldots (s+2\nu-\mu-2)\varphi^{(\mu)}(1) \pm$$

$$\pm \sum_{\mu=0}^{p} (-1)^{p-\mu} \binom{p}{\mu} s(s+1) \ldots (s+p-\mu-1) \int_{1}^{\infty} \frac{P_p(x)\,\varphi^{(\mu)}(x)}{x^{s+p-\mu}} dx$$

$$= A_1 + A_2 + A_3.$$

Nous allons faire une évaluation d'ailleurs assez superficielle de $G(s)$, en nous appuyant sur la formule (12) et sur les relations suivantes, faciles à vérifier:

[1] »Sur une extension d'un théorème classique de la théorie des fonctions», Acta Math. 28, p. 351.

$$\frac{B_\nu}{(2\nu!)} < \frac{1}{\pi^{2\nu}},$$

$$|P_p(x)| < \frac{1}{\pi^p}, \qquad (p > 1),$$

$$2h \leqq p, \quad \text{(voir p. 11)}$$

$$\varrho + p < 3\varrho$$

pour toute valeur suffisamment grande de ϱ, et

$$\sigma + p - k > 2$$

pour tout s sur le cercle $|s| = \varrho$.

Pour tout ϱ suffisamment grand, on a donc

$$|A_1| = \left|\frac{1}{2}\varphi(1)\right| < \frac{1}{3} K_p e^{\varrho \log \varrho};$$

$$|A_2| \leqq K_p \sum_{\nu=1}^{h} \frac{B_\nu}{(2\nu)!} \sum_{\mu=0}^{2\nu-1} \binom{2\nu-1}{\mu} |s(s+1)\ldots(s+2\nu-\mu-2)|$$

$$< K_p \sum_{\nu=1}^{h} \frac{1}{\pi^{2\nu}} \sum_{\mu=0}^{2\nu-1} \binom{2\nu-1}{\mu} (\varrho + 2h - 2)^{2\nu-\mu-1}$$

$$< K_p \sum_{\nu=1}^{h} \left(\frac{\varrho + 2h - 1}{\pi}\right)^{2\nu}$$

$$< K_p \sum_{\nu=1}^{h} \left(\frac{3\varrho}{\pi}\right)^{2\nu}$$

$$< K_p \left(\frac{3\varrho}{\pi}\right)^{2h+1}$$

$$\leqq K_p \left(\frac{3\varrho}{\pi}\right)^{\varrho+k+4}$$

$$< \frac{1}{3} K_p e^{\varrho \log \varrho};$$

$$|A_3| \leqq \sum_{\mu=0}^{p} \binom{p}{\mu} |s(s+1)\ldots(s+p-\mu-1)| \int_1^{\infty} \frac{|P_p(x)|\,|x^{\mu-k}|\,|\varphi^{(\mu)}(x)|}{x^{\sigma+p-k}}\, dx$$

$$< \frac{K_p}{\pi^p} \sum_{\mu=0}^{p} \binom{p}{\mu} (\varrho + p - 1)^{p-\mu} \int_1^{\infty} \frac{dx}{x^2}$$

$$= K_p \left(\frac{\varrho + p}{\pi}\right)^p$$

$$< K_p \left(\frac{3\varrho}{\pi}\right)^{\varrho+k+3}$$

$$< \frac{1}{3} K_p\, e^{\varrho \log \varrho}.$$

Enfin, dès que ϱ surpasse une certaine limite, nous aurons

$$|G(s)| < e^{\varrho \log \varrho} \,.\, K_{[\varrho+k]+3},$$

et la démonstration du théorème II est accomplie.

Le module maximum de $G(s)$ sur le cercle $|s| = \varrho$ dépend donc des constantes K_n qui mesurent pour ainsi dire la manière plus ou moins uniforme dont sont remplies les égalités

$$\varphi^{(\mu)}(x) = O(x^{k-\mu})$$

de la condition B. Dans des cas assez étendus, le facteur $K_{[\varrho+k]+3}$ n'est pas d'un ordre supérieur à l'ordre du premier facteur.[1]

Il y a lieu de remarquer ici qu'on ne peut espérer avoir dans le cas général une limite supérieure plus satisfaisante que celle qui vient d'être obtenue. En effet, prenant $\varphi(x) = 1$, on a évidemment pour tout n

$$K_n = 1$$

et

$$G(s) = \zeta(s) - \frac{1}{s-1}.$$

Alors, en donnant à s des valeurs négatives et impaires, on trouve pour des valeurs très grandes de ν:[2]

[1] Voir § 5.
[2] Voir LANDAU, »Handbuch etc.», p. 287.

$$|G(-(2\nu-1))| = \left|\frac{(-1)^\nu B_\nu + 1}{2\nu}\right| > e^{(1-\varepsilon)(2\nu-1)\log(2\nu-1)}$$

pour tout $\varepsilon > 0$.

Remarque. Pour voir comment se comporte la fonction $G(s)$ dans le cas plus général indiqué à la remarque du paragraphe précédent, nous introduirons les constantes K'_n telles que

$$|\varphi^{(\mu)}(x)| \leqq K'_n \,.\, x^{k-\mu c}$$

pour $x \geqq 1$, $\mu = 0, 1 \ldots n$. Tous les énoncés du théorème II restent valables, en remplaçant seulement (13) et (14) par

$$(13') \qquad o \leqq \mu(\sigma) \leqq \frac{k+1-\sigma}{c}, \qquad \text{pour } \sigma < k+1$$

et

$$(14') \qquad |G(s)| < e^{(1+\varepsilon)\varrho\log\varrho} \,.\, K'_{\left[\frac{\varrho+k+2}{c}\right]+1},$$

la démonstration se faisant comme tout à l'heure, sauf quelques modifications évidentes.

§ 4. Applications des résultats précédents.

Pour avoir une application des résultats précédents, nous traiterons le cas où $\varphi(x)$, pour $x \geqq R \geqq 1$, est donnée par l'égalité

$$\varphi(x) = x^\alpha (\log x)^\beta \sum_{\nu=0}^{\infty} \frac{c_\nu}{x^\nu},$$

α et β ayant des valeurs réelles ou complexes quelconques, le logarithme étant pris avec sa détermination réelle, et la série étant supposée absolument convergente pour $x \geqq R$. Dans l'intervalle $1 \leqq x \leqq R$, nous supposons $\varphi(x)$ donnée d'une manière quelconque, mais satisfaisant toujours à la condition A. du § 2.

Alors, en prenant $k > R(\alpha)$,[1] on voit que la condition

[1] $R(\alpha)$ désignant la partie réelle de α.

B. aussi est remplie, de manière que nous aurons

$$H(s) = \sum_1^\infty \frac{\varphi(n)}{n^s} = G(s) + \int_1^\infty \frac{\varphi(x)}{x^s} dx,$$

$G(s)$ désignant la fonction entière traitée au paragraphe précédent.

Pour étudier les propriétés de l'intégrale, posons

$$J(s) = \int_1^\infty \frac{\varphi(x)}{x^s} dx = \int_1^R + \int_R^\infty .$$

La première intégrale représente évidemment une fonction entière de s, qui sera désignée par $G_1(s)$. On voit sans difficulté que tous les énoncés du théorème II restent valables encore pour la fonction $G(s) + G_1(s)$.

Quant à l'intégrale

$$J_1(s) = \int_R^\infty \frac{\varphi(x)}{x^s} dx = \int_R^\infty \frac{(\log x)^\beta}{x^{s-\alpha}} \sum_{\nu=0}^\infty \frac{c_\nu}{x^\nu} dx,$$

elle a été traitée par PINCHERLE,[1] d'ailleurs d'une manière assez légère.

En supposant d'abord $\sigma > 1 + R(\alpha)$, on peut effectuer l'intégration terme par terme.[2] Il vient donc

$$J_1(s) = \sum_{\nu=0}^\infty c_\nu \int_R^\infty \frac{(\log x)^\beta}{x^{s+\nu-\alpha}} dx.$$

Remarquons avec PINCHERLE que la fonction $\psi(s) = \int_R^\infty \frac{(\log x)^\beta}{x^{s+\nu-\alpha}} dx$ satisfait à l'équation différentielle

[1] »Sur les fonctions déterminantes», Ann. de l'Ecole norm. sup. Ser. 3, T. 22, p. 20 et 49.
[2] Voir BROMWICH, »Infinite series», p. 452.

$$\psi' + \frac{\beta+1}{s+\nu-\alpha-1}\psi + \frac{(\log R)^{\beta+1}}{R^{\nu-\alpha-1}} \cdot \frac{R^{-s}}{s+\nu-\alpha-1} = 0.$$

Par suite, elle est de la forme

$$(15) \qquad g(s) + \frac{C.\log(s+\nu-\alpha-1)}{(s+\nu-\alpha-1)^{\beta+1}}$$

lorsque β est égal à un nombre entier négatif, et de la forme

$$(15') \qquad g(s) + \frac{C}{(s+\nu-\alpha-1)^{\beta+1}}$$

dans tous les autres cas, $g(s)$ désignant une fonction entière. Il n'y aurait d'ailleurs aucune difficulté à évaluer les constantes C; on trouverait $C = \frac{(-1)^{-\beta}}{(-\beta)!}$ dans le premier cas, et $C = \Gamma(\beta+1)$ dans le second cas, mais c'est là un point sans grande importance.

Ce dont il s'agit ici, c'est de rechercher si tous les points singuliers de la fonction $J_1(s)$ se trouvent parmi les points $s = \alpha + 1 - \nu$, ce qui n'est pas du tout une chose évidente.

Cependant, nous allons démontrer que $J_1(s)$ jouit des propriétés suivantes: si nous traçons une coupure indéfinie rectiligne du point $\alpha + 1$ parallèle à l'axe réel négatif, $J_1(s)$ devient holomorphe dans toute partie finie du plan. Dans le voisinage des points singuliers $s = \alpha + 1 - \nu$, $(\nu = 0, 1, 2 \ldots)$, $J_1(s)$ se comporte comme l'indiquent les expressions (15). Enfin, en excluant sur une ordonnée quelconque un petit segment des deux côtés de la coupure, on a toujours $J_1(s) = O(1)$.

En effet, nous avons

$$\begin{aligned} J_1(s) &= \sum_{\nu=0}^{\infty} c_\nu \int_R^\infty \frac{(\log x)^\beta}{x^{s+\nu-\alpha}} dx = \\ &= \sum_{\nu=0}^{\infty} c_\nu \int_{\log R}^\infty z^\beta e^{-(s+\nu-\alpha-1)z} dz \\ &= \sum_{\nu=0}^{N} + \sum_{\nu=N+1}^{\infty}, \end{aligned}$$

N étant un nombre entier arbitraire. La première somme d'un nombre fini de termes n'a évidemment que les points singuliers $\alpha+1$, $\alpha \ldots \alpha+1-N$ à distance finie, et représente bien une fonction holomorphe dans tout le plan où l'on a tracé la coupure qui vient d'être définie. On voit immédiatement que la fonction représentée par l'une quelconque des intégrales est $O(1)$ sur une ordonnée située dans le domaine de convergence de l'intégrale. Or, il résulte d'un théorème qui sera démontré plus loin (page 38) que cette propriété subsiste encore pour une ordonnée quelconque, en excluant toujours un petit segment des deux côtés de la coupure.

Donc, si nous pouvons démontrer que la série

$$\sum_{\nu=N+1}^{\infty} c_\nu \int_{\log R}^{\infty} z^\beta e^{-(s+\nu-\alpha-1)z} dz$$

soit uniformément convergente et bornée en valeur absolue pour $\sigma \geqq R(\alpha)+\frac{1}{2}-N$, toutes les propriétés annoncées de $J_1(s)$ seront établies.

En effet, pour $\nu \geqq N+1$ et $\sigma \geqq R(\alpha)+\frac{1}{2}-N$, on a

$$\left| \int_{\log R}^{\infty} z^\beta e^{-(s+\nu-\alpha-1)z} dz \right| < e^{-(\nu-N-1)\log R} \int_{\log R}^{\infty} z^{R(\beta)} e^{-\frac{1}{2}z} dz = \frac{K}{R^\nu}.$$

Par suite, la série $\sum \frac{|c_\nu|}{R^\nu}$ étant convergente, notre proposition est démontrée.

Enfin, la fonction $H(s)$ représentée par notre série de Dirichlet pourra dans ce cas s'écrire

$$H(s) = \overline{G}(s) + J_1(s),$$

$\overline{G}(s)$ désignant une fonction entière du même caractère que $G(s)$.

§ 5. Etude des séries $\sum \frac{\psi(\log n)}{n^s}$.

Envisageons maintenant une série de DIRICHLET sous la forme

$$H(s) = \sum_{n=1}^{\infty} \frac{\psi(\log n)}{n^s},$$

où la fonction $\psi(z)$, continue ainsi que toutes ses dérivées pour $z \geqq 0$, satisfait aux conditions

(16) $$\psi^{(\mu)}(z) = O(e^{kz}), \qquad (\mu = 0, 1, 2 \ldots).$$

Alors, il est facile de voir que la fonction $\varphi(x)$, définie par l'égalité

$$\varphi(x) = \psi(\log x),$$

satisfait aux conditions A. et B. du § 2, de manière qu'on aura

$$H(s) = G(s) + \int_1^{\infty} \frac{\psi(\log x)}{x^s} dx$$

$$= G(s) + \int_0^{\infty} \psi(z) e^{-(s-1)z} dz,$$

$G(s)$ ayant la signification usuelle.

Dès maintenant, $\psi(z)$ sera regardée comme une fonction de la variable complexe $z = x + iy$.[1] Il est alors naturel de supposer qu'une égalité de la forme $\psi(z) = O(e^{k|z|})$ ait lieu non seulement sur l'axe réel et positif, mais encore dans un domaine A qui le renferme. Pour la fonction $\psi(z) = e^{iz^2}$, p. ex., on a $\psi(z) = O(e^{2|z|})$ dans la bande limitée par les deux droites $y = \pm 1$.

On pourra alors recourir à l'intégrale de CAUCHY pour évaluer les dérivées $\psi^{(\mu)}(z)$. Le domaine A étant assujetti à

[1] Dans ce qui va suivre, la lettre x aura toujours cette signification

certaines conditions, on aura bien des relations de la forme (16). En effet, on peut énoncer le théorème suivant:

Théorème III. *Posons* $H(s)=\sum_{n=1}^{\infty}\frac{\psi(\log n)}{n^s}$. *Soit A un domaine renfermant l'axe réel et positif dans le plan des z, tel qu'à partir d'un certain point* $x_0=x_0(\varepsilon)$ *tout point x de cet axe peut être entouré d'un cercle du rayon* $e^{-\varepsilon x}$ *appartenant au domaine A, quelque petit que soit le nombre positif* ε. *Si l'on a*

$$\psi(z)=O(e^{k|z|})$$

uniformément dans A, $\psi(z)$ *étant holomorphe dans ce domaine, on aura*

$$H(s)=G(s)+\int_0^\infty \psi(z)e^{-(s-1)z}\,dz,$$

$G(s)$ *ayant la signification usuelle.*

En effet,[1] il ne s'agit que de montrer que nous avons pour des valeurs réelles et positives de z des relations de la forme (16). Supposons $x>x_0$, et posons

$$\psi^{(\mu)}(x)=\frac{\mu!}{2\pi i}\int_C\frac{\psi(z)\,dz}{(z-x)^{\mu+1}},$$

e contour d'intégration C étant le cercle du rayon $e^{-\varepsilon x}$ décrit autour du point x comme centre. Ce cercle appartenant par hypothèse au domaine A, nous aurons

$$|\psi^{(\mu)}(x)|<K\,.\,\mu!\,e^{\mu\varepsilon x}\,.\,e^{k(x+e^{-\varepsilon x})},$$

d'où

$$\psi^{(\mu)}(x)=O(e^{(k+\mu\varepsilon)x}).$$

Or, ε étant arbitraire, il faut seulement augmenter la constante k pour avoir des relations de la forme (16). Notre théorème se trouve donc démontré.

[1] Dans cette démonstration, je me suis servi d'une remarque de M. Riesz.

Remarque. 1. En utilisant les remarques faites aux paragraphes 2. et 3., on pourrait même remplacer le nombre arbitrairement petit ε du théorème précédent par un nombre fixe compris entre 0 et 1. La conclusion du théorème III sera encore valable.

2. Dans le théorème II, nous sommes parvenus à la relation

$$|G(s)| < e^{\varrho \log \varrho} . K_{[\varrho+k]+3},$$

valable pour tout ϱ suffisamment grand. Dans les conditions du théorème III, on pourra préciser ce résultat en évaluant les constantes K_n. Pour simplifier les calculs, supposons le domaine A tel que tout point de l'axe réel et positif peut être entouré d'un cercle du rayon 1 appartenant au domaine A. Soit dans ce domaine

$$|\psi(z)| < M e^{k|z|}.$$

Un calcul tout à fait analogue à celui fait pour démontrer le théorème III nous donne pour tout x réel et positif

$$|\psi^{(\mu)}(x)| < M e^{|k|} \mu ! \, e^{kx}.$$

Or, en posant $\varphi(u) = \psi(\log u)$, nous avons[1]

$$\varphi^{(n)}(u) = \frac{1}{u^n}(C_0 \psi^{(n)} - C_1 \psi^{(n-1)} + \cdots \pm C_{n-1}\psi'),$$

les C_ν étant les coefficients du développement

$$\lambda(\lambda+1)\ldots(\lambda+n-1) = C_0\lambda^n + C_1\lambda^{n-1} + \cdots C_{n-1}\lambda.$$

Par des considérations élémentaires, on trouve

$$C_\nu \leqq \binom{n-1}{\nu} \frac{n!}{(n-\nu)!},$$

d'où

$$|\varphi^{(n)}(u)| < M e^{|k|} u^{k-n} n! \sum_{\nu=0}^{n-1} \binom{n-1}{\nu}$$

$$= \frac{1}{2} M e^{|k|} u^{k-n} n! \, 2^n.$$

[1] Voir SCHLÖMILCH, »Übungsbuch z. Studium d. höh. Analysis«, 5 Aufl., Bd I, p. 60.

En tenant compte de la relation (12), on a donc

$$K_n < \frac{1}{2} M e^{|k|} n!\, 2^n,$$

et l'on en conclut aisément par la formule de STIRLING

$$K_{[\varrho+k]+3} < e^{\varrho \log \varrho},$$

formule valable pour toute valeur suffisamment grande de ϱ, de manière qu'on aura dans ce cas

$$|G(s)| < e^{2\varrho \log \varrho}.$$

Nous traiterons encore dans ce paragraphe le cas le plus simple qu'on pût imaginer pour appliquer le théorème précédent.

Théorème IV. *Soit* $\psi(z) = \sum_{\nu=0}^{\infty} \frac{c_\nu}{\nu!} z^\nu$ *une fonction entière, satisfaisant dans tout le plan à la relation*[1] $\psi(z) = O(e^{k|z|})$. *Avec les mêmes notations qu'au théorème précédent, nous aurons*

$$H(s) = G(s) + \sum_{\nu=0}^{\infty} \frac{c_\nu}{(s-1)^{\nu+1}},$$

la dernière série étant convergente pour $|s-1| > k$.

En effet, le domaine A embrassant ici tout le plan, on a par le théorème précédent

$$H(s) = G(s) + \int_0^\infty \psi(z) e^{-(s-1)z} dz = \tag{17}$$

$$= G(s) + \int_0^\infty e^{-(s-1)z} \sum_{\nu=0}^{\infty} \frac{c_\nu}{\nu!} z^\nu dz.$$

[1] On a donc nécessairement $k > 0$, si $\psi(z)$ n'est pas égale à une constante.

Or, on déduit d'un théorème connu dû à CAUCHY[1] que

$$c_\nu = O[(k+\varepsilon)^\nu] \tag{18}$$

pour tout $\varepsilon > 0$,

et il s'ensuit

$$|e^{-(s-1)z}| \cdot \sum_{\nu=0}^{\infty} \left| \frac{c_\nu}{\nu!} z^\nu \right| < K \, . \, e^{-(\sigma-k-1-\varepsilon)z}$$

pour des valeurs positives de z. En supposant d'abord $\sigma > k+1$, on peut dans la relation (17) effectuer l'intégration terme par terme,[2] d'où il suit

$$\begin{aligned} H(s) &= G(s) + \sum_{\nu=0}^{\infty} \frac{c_\nu}{\nu!} \int_0^\infty z^\nu e^{-(s-1)z} dz \\ &= G(s) + \sum_{\nu=0}^{\infty} \frac{c_\nu}{(s-1)^{\nu+1}} .^3 \end{aligned}$$

De la relation (18) on conclut immédiatement que la série $\sum \frac{c_\nu}{(s-1)^{\nu+1}}$ est uniformément convergente pour $|s-1| \geqq k' > k$. Le théorème IV est donc établi.

On peut tirer de ce théorème quelques conséquences intéressantes. Supposons d'abord la relation $\psi(z) = O(e^{k|z|})$ vérifiée pour toute valeur positive de la constante k. Alors, le cercle $|s-1| = k$, qui renferme toutes les singularités finies de $H(s)$, pourra être pris aussi petit qu'on le voudra. $H(s)$ n'admet donc à distance finie que le point singulier $s = 1$, qui est d'ailleurs un point singulier essentiel. On aura dans ce cas

$$H(s) = G(s) + G_1\left(\frac{1}{s-1}\right),$$

G_1 étant aussi une fonction entière.

[1] Pour des calculs analogues voir BOREL, »Leçons sur les fonctions entières», p. 62.
[2] Voir BROMWICH, l. c., p. 453.
[3] Voir NIELSEN, l. c., p. 151.

C'est ce qui arrivera quand on prend pour $\psi(z)$ une fonction entière dont l'ordre est inférieur à 1, ou une fonction d'ordre 1 mais du type minimum selon la terminologie de M. Pringsheim.[1] D'autre part, étant donné un cercle quelconque dans le plan des s, on peut construire une série de Dirichlet représentant une fonction holomorphe en tout point *extérieur* à ce cercle, mais pour laquelle le cercle sera une coupure essentielle. En effet, on peut toujours supposer le centre du cercle situé au point $s = 1$, ce qui peut se réaliser par une substitution linéaire. Soit donc k le rayon du cercle. Prenons une série de Taylor, $\sum_{\nu=0}^{\infty} c_\nu z^\nu$, ayant pour coupure essentielle son cercle de convergence, dont le rayon sera égal à $\frac{1}{k}$.

Posons $\psi(z) = \sum_{\nu=0}^{\infty} \frac{c_\nu}{\nu!} z^\nu$ et formons enfin la série de Dirichlet

$$H(s) = \sum_{n=1}^{\infty} \frac{\psi(\log n)}{n^s}.$$

Il s'ensuit immédiatement du théorème IV que cette série jouit des propriétés exigées.

§ 6. Un théorème sur les séries $\sum a_n e^{-\lambda_n s}$.

A distance finie, la fonction $H(s)$ traitée au théorème IV n'a aucun point singulier extérieur au cercle $|s-1| = k$.

Voilà un fait qui n'est qu'un cas spécial d'un théorème plus général, qui peut se déduire aisément sans recourir à la formule d'Euler.

Théorème V. *Soit* $h(s) = \sum_{n=0}^{\infty} e^{-\lambda_n s}$ *une série de* Dirichlet, *ayant une abscisse de convergence finie* l.

[1] »Elementare Theorie d. ganzen transcendenten Funktionen v. endl. Ordnung». Math. Ann. 58, p. 257.

La fonction entière $\psi(z)=\sum_{\nu=0}^{\infty}\frac{c_\nu}{\nu!}z^\nu$ *satisfaisant à la condition* $\psi(z)=O(e^{k|z|})$, *formons la série*

$$H(s)=\sum_{n=0}^{\infty}\psi(\lambda_n)e^{-\lambda_n s},$$

qui sera certainement convergente pour $\sigma>l+k$.

Supposons qu'on sait effectuer un prolongement analytique de $h(s)$, *de manière qu'on connaît un certain domaine connexe* D, *renfermant le demiplan* $\sigma>l$, *où la fonction* $h(s)$ *est holomorphe.*[1]

Appelons $D(k)$ *le domaine, formé de tous les points de* D *dont la distance à tout point du contour soit supérieure ou égale à* k. *Cela posé, le domaine* $D(k)$ *pourra se partager en plusieurs parties connexes, dont l'un,* $D'(k)$, *renfermera le demiplan* $\sigma>l+k$.

Enfin, la fonction $H(s)$ *sera holomorphe dans* $D'(k)$.

D'abord, nous avons comme auparavant

$$c_\nu=O[(k+\varepsilon)^\nu]$$

pour tout $\varepsilon>0$,

et nous obtiendrons

$$\begin{aligned}H(s)&=\sum_{n=0}^{\infty}\psi(\lambda_n)e^{-\lambda_n s}\\&=\sum_{n=0}^{\infty}\sum_{\nu=0}^{\infty}\frac{c_\nu}{\nu!}\lambda_n^\nu e^{-\lambda_n s}.\end{aligned}$$

Or, on conclut de l'évaluation précédente des c_ν:

$$\sum_{\nu=0}^{\infty}\left|\frac{c_\nu}{\nu!}\lambda_n^\nu e^{-\lambda_n s}\right|<K\,.\,e^{-\lambda_n(\sigma-k-\varepsilon)}.$$

En supposant d'abord $\sigma>l+k$, notre série double sera donc

[1] Nous dirons qu'une fonction est holomorphe dans un domaine, lorsqu'elle est holomorphe en tout point intérieur à ce domaine.

absolument convergente, et nous aurons en changeant l'ordre des sommations

$$H(s) = \sum_{\nu=0}^{\infty} \frac{c_\nu}{\nu!} \sum_{n=0}^{\infty} \lambda_n^\nu e^{-\lambda_n s}$$

$$= \sum_{\nu=0}^{\infty} \frac{(-1)^\nu c_\nu}{\nu!} h^{(\nu)}(s).$$

Prenons un nombre arbitraire $k_1 > k$. Nous allons montrer que la dernière série converge absolument et uniformément dans toute partie finie du domaine $D(k_1)$. En effet, posant $k' = \frac{1}{2}(k + k_1)$, tout point d'une telle partie pourra être entouré d'un cercle du rayon k' appartenant à un certain domaine fini intérieur à D, où la fonction $h(s)$ est bornée en valeur absolue. En un tel point, on aura, en intégrant le long de ce cercle:

$$h^{(\nu)}(s) = \frac{\nu!}{2\pi i} \int_C \frac{h(u)du}{(u-s)^{\nu+1}}$$

$$|h^{(\nu)}(s)| < M \frac{\nu!}{k'^\nu},$$

et enfin

$$\left| \frac{(-1)^\nu c_\nu}{\nu!} h^{(\nu)}(s) \right| < M_1 \left(\frac{k+\varepsilon}{k'} \right)^\nu.$$

Or, ε étant arbitraire, il s'ensuit que notre série converge absolument et uniformément à l'intérieur de toute partie finie de $D(k_1)$.

k_1 aussi étant arbitraire, on serait tenté de conclure que $H(s)$ soit holomorphe dans $D(k)$. Mais si $D(k)$ se partage en plusieurs parties connexes, cette conclusion serait évidemment incorrecte, et il n'est permis que d'affirmer que $H(s)$ est holomorphe dans la partie $D'(k)$, qui renferme le demiplan $\sigma > l + k$. Le théorème est donc établi.

Si la condition $\psi(z) = O(e^{k|z|})$ est remplie pour tout $k > 0$, nous voyons que $H(s)$ ne peut admettre d'autres singularités finies que celles de $h(s)$.[1]

[1] Cette proposition contient comme cas spécial un théorème sur les séries de TAYLOR. Voir HADAMARD: »La série de Taylor et son prolongement analytique», pp. 28 et 63.

Pour donner une idée de la portée du théorème qu'on vient de démontrer, proposons-nous de l'appliquer aux séries de TAYLOR, qui s'obtiennent en posant $\lambda_n = n$ et en faisant ensuite le changement de variables $e^{-s} = x$. Nous aurons

$$h(s) = \sum_{n=0}^{\infty} e^{-ns} = \frac{1}{1 - e^{-s}};$$

les seuls points singuliers de $h(s)$ sont donc les points $\pm 2n\pi i.(n = 0, 1 \ldots)$. Le domaine $D(k)$ s'obtient donc en

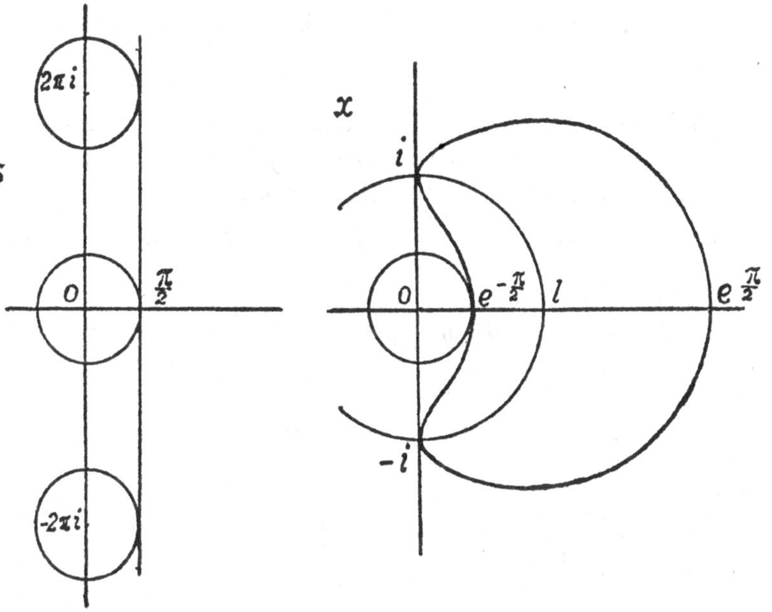

Fig. 1.

excluant du plan tous les cercles du rayon k ayant ces points pour centres. Si $k < \pi$, ce domaine est connexe, et $H(s) = = \Sigma \psi(n) e^{-ns}$ peut être prolongée analytiquement sur tout le plan en évitant les cercles qui viennent d'être assignés.

Au contraire, si $k \geqq \pi$, le domaine $D(k)$ se partagera en deux parties connexes dont nous avons désigné par $D'(k)$ celle qui est située à la droite de l'axe imaginaire dans le plan de s. La série $\Sigma \psi(n) e^{-ns}$ étant cependant dans le cas général divergente pour $\sigma < k$, on obtient pourtant un prolongement analytique de $H(s)$, quoique d'une nature beaucoup plus restreinte que dans le premier cas.

En effectuant ensuite le changement de variables $e^{-s} = x$, on voit sans difficulté comment se présente la chose pour les séries de TAYLOR dans leur forme ordinaire. Voir fig. 1, où l'on a pris $k = \frac{\pi}{2}\cdot$

Le résultat auquel on parvient ainsi est celui déjà donné par M. LINDELÖF dans ses »Leçons sur le calcul des résidus», pag. 135—6.[1]

§ 7. Un théorème de MM. Phragmén et Lindelöf et son application.

Théorème VI. *Soit une fonction $\psi(z)$ de la variable complexe $z = re^{i\varphi}$ jouissant des propriétés suivantes:*

1. $\psi(z) = O(e^{kz})$ *pour z réel et positif.*

2. $\psi(z) = O(e^{k_1 r})$ *pour* $\varphi = \varphi_1 = \frac{\pi}{\alpha}$, *k et* k_1 *étant deux constantes réelles et* $\alpha > 1$.

3. Dans l'angle $0 \leqq \varphi \leqq \varphi_1$, $\psi(z)$ *sera holomorphe, et on y aura uniformément*

$$e^{-r^{\alpha-\delta}}\psi(z) \to 0$$

lorsque $r \to \infty$.[2] $(1 < \alpha - \delta < \alpha)$.

Alors, on aura sur le vecteur $O\varphi$, *où* $0 \leqq \varphi \leqq \varphi_1$:

$$\psi(z) = O(e^{k(\varphi)\cdot r}),$$

où

$$k(\varphi) = \frac{k \sin(\varphi_1 - \varphi) + k_1 \sin\varphi}{\sin\varphi_1}.$$

On pourra encore avoir une interprétation géométrique de ce résultat. En effet, $k(\varphi)$ *peut être obtenue de la manière suivante: on porte la longueur* $\frac{1}{k}$ *sur l'axe réel et la longueur* $\frac{1}{k_1}$ *sur le vecteur* $O\varphi_1$, *puis on joint les points obtenus par une*

---

[1] Voir aussi S. WIGERT: »Sur une certaine classe de séries de puissances», Arkiv f. matematik etc., B. 12, n:o 7.

[2] On pourrait même remplacer cette condition par $e^{-\delta r^{\alpha}}\cdot\psi(z) \to 0$ pour tout $\delta > 0$.

droite. Le point d'intersection de cette droite avec le vecteur $O\varphi$ sera sur la distance $\frac{1}{k(\varphi)}$ de l'origine.

Envisageons d'abord le cas spécial où le module de $\psi(z)$ sur les vecteurs $\varphi = 0$ et $\varphi = \varphi_1$ a une limite supérieure finie M. Dans ces conditions on aura $|\psi(z)| \leqq M$ pour tout point situé dans l'angle considéré. Pour montrer cela, étudions la fonction

$$g(z) = e^{i\eta z^{a-\frac{\delta}{2}} e^{i\varphi_1 \frac{\delta}{4}}} \cdot \psi(z),$$

où η désigne un nombre positif arbitraire. On a pour $0 \leqq \varphi \leqq \varphi_1$:

$$|g(z)| = |\psi(z)| \cdot e^{-\eta r^{a-\frac{\delta}{2}} \sin\left[\frac{\delta}{4}\varphi_1 + \left(a - \frac{\delta}{2}\right)\varphi\right]}$$

$$\leqq |\psi(z)| e^{-\eta r^{a-\frac{\delta}{2}} \sin\frac{\delta\varphi_1}{4}}.$$

Cette inégalité nous fait voir que $|g(z)| \leqq M$ pour $\varphi = 0$ et $\varphi = \varphi_1$. De plus, on voit que le module de $g(z)$ tend uniformément vers zéro avec $\frac{1}{r}$, z restant dans l'angle considéré. Chaque point dans cet angle peut donc être entouré d'un contour sur lequel $|g(z)| \leqq M$. On a donc dans cet angle $|g(z)| \leqq M$, c. à d.

$$|\psi(z)| \leqq M e^{\eta r^{a-\frac{\delta}{2}}}.$$

Or, η étant arbitraire, on conclut de là que

$$|\psi(z)| \leqq M$$

lorsque z reste dans l'angle considéré.

Passons au cas général. Imaginons les constantes réelles p et τ déterminées de manière que

(19) $$|e^{pze^{i\tau}}| = e^{kr} \qquad \text{pour } \varphi = 0$$

et que

(19') $$\left|e^{pze^{i\tau}}\right| = e^{k_1 r}$$

pour $\varphi = \varphi_1$.

Alors, le module de la fonction $\psi(z)e^{-pze^{i\tau}}$ sera borné sur nos vecteurs $\varphi = 0$ et $\varphi = \varphi_1$, et nous pourrons en conclure que

(20) $$|\psi(z)| \leqq M \left|e^{pze^{i\tau}}\right|,$$

z restant dans l'angle considéré.

Cherchons donc à satisfaire aux relations (19). Nous en obtenons

$$R(pre^{i\tau}) = kr$$

$$R(pre^{i(\varphi_1+\tau)}) = k_1 r,$$

d'où

(21) $$p \cos \tau = k$$ $$p \cos (\varphi_1 + \tau) = k_1,$$

équations qui admettent toujours une solution. Il n'importe point d'avoir explicitement les valeurs de p et de τ, vu la formule (20) qui nous donne

$$\psi(z) = O(e^{pr\cos(\varphi+\tau)}).$$

D'autre part, en multipliant la première des équations (21 par $\dfrac{\sin(\varphi_1 - \varphi)}{\sin \varphi_1}$, la seconde par $\dfrac{\sin \varphi}{\sin \varphi_1}$ et en ajoutant les résultats, nous trouverons

$$p \cos (\varphi + \tau) = \frac{k \sin(\varphi_1 - \varphi) + k_1 \sin \varphi}{\sin \varphi_1}.$$

La première partie du théorème VI se trouve ainsi démontrée. Quant à l'interprétation géométrique que nous avons donnée, elle n'exige que des calculs purement élémentaires pour être vérifiée.

Il s'agit maintenant d'appliquer ce théorème aux séries de Dirichlet appartenant à la classe traitée dans ce travail. Pour abréger l'écriture, nous dirons dans ce qui va suivre qu'une fonction $\psi(z)$ satisfait aux conditions du § 5 lorsque pour z réel et positif les égalités

$$\psi^{(\mu)}(z) = O(e^{kz})$$

sont vérifiées.[1] En formant la série $H(s) = \sum_{n=1}^{\infty} \frac{\psi(\log n)}{n^s}$, absolument convergente pour $\sigma > k+1$, nous savons alors que

$$H(s) = G(s) + \int_0^\infty \psi(z) e^{-(s-1)z} dz,$$

$G(s)$ ayant la signification usuelle. Par le théorème III, on sait qu'il en est bien ainsi, lorsque la fonction $\psi(z)$ satisfait à une condition de la forme $\psi(z) = O(e^{k|z|})$ dans un domaine renfermant l'axe réel et positif et assujetti à certaines conditions, remplies p. ex. par un angle, une bande comprise entre deux parallèles, un domaine limité par des courbes telles que $y = \pm x^a$ etc.

Supposons maintenant que $\psi(z)$ satisfait à la fois aux conditions du § 5 et à celles du théorème VI, et cherchons d'obtenir un prolongement analytique de la fonction $H(s)$. Il est clair qu'il va s'agir de la discussion de la fonction

$$J(s) = \int_0^\infty \psi(z) e^{-(s-1)z} dz.$$

En effet, nous allons obtenir le prolongement cherché en changeant le chemin d'intégration. Posons

$$J_1(s) = \int_0^{\infty e^{i\varphi_1}} \psi(z) e^{-(s-1)z} dz,$$

l'intégrale étant prise le long du vecteur $\varphi = \varphi_1$. Par les hypothèses du théorème VI, on a sur ce vecteur

$$\psi(z) e^{-(s-1)z} = O[e^{k_1 r - R[(s-1)z]}]$$

[1] Presque tous les raisonnements qui vont suivre s'étendent d'elles-mêmes à la classe plus générale considérée aux pages 13 et 27.

ou, en posant $s-1=\varrho e^{i\theta}$,

$$\psi(z)\, e^{-(s-1)z} = O[e^{k_1 r - \varrho r \cos(\varphi_1+\theta)}].$$

Dans le domaine défini par l'inégalité

$$\varrho \cos(\varphi_1+\theta) \geqq k_1+\varepsilon > k_1,$$

l'intégrale dernière converge uniformément, étant en même temps bornée en valeur absolue; il s'ensuit donc que $J_1(s)$ est holomorphe dans le domaine $\varrho \cos(\varphi_1+\theta) > k_1$. Ce domaine peut être obtenu par la construction géométrique suivante. Par le point $s=k_1+1$, menons une parallèle à l'axe des imaginaires, et faisons ensuite tourner la figure un angle égal à $-\varphi_1$ autour du point $s=1$. Le demiplan $\sigma > k_1+1$ se transformera alors en un autre demiplan, qui constitue le domaine cherché.

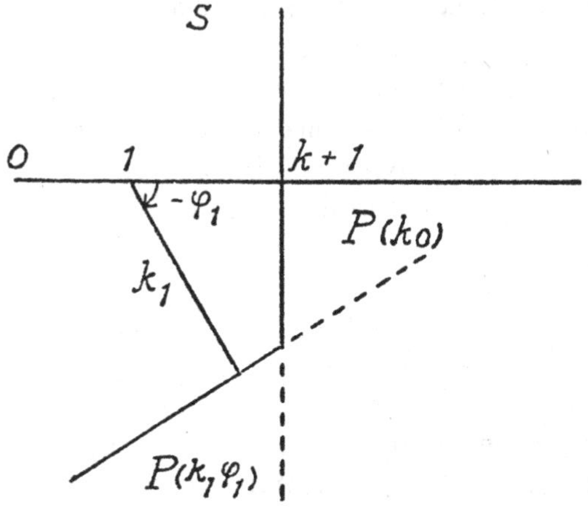

Fig. 2.

En désignant par $P(k_1, \varphi_1)$ le demiplan obtenu par cette construction, procédons maintenant à la démonstration du théorème suivant:

Théorème VII. *La fonction $\psi(z)$ satisfaisant à la fois aux conditions du § 5 et à celles du théorème VI, la fonction $H(s)$ est holomorphe dans le domaine obtenu par la réunion* *des deux demiplans $P(k, o)$ et $P(k_1, \varphi_1)$, dont le premier est identique au demiplan $\sigma > k + 1$ où converge la série $\sum \frac{\psi(\log n)}{n^s}$.*

De plus, la différence $H(s) - G(s)$ est bornée en valeur absolue dans tout domaine obtenu en remplaçant ci-dessus k et k_1 par $k + \varepsilon$ et $k_1 + \varepsilon$, où $\varepsilon > 0$.

En effet, il suffit de prouver que $J(s)$ et $J_1(s)$ sont identiques dans la partie commune des demiplans dont il s'agit. Ce point établi, le théorème se déduit immédiatement de la discussion précédente de $J_1(s)$.

Remarquons d'abord que dans le domaine considéré nous avons les inégalités

$$\sigma > k + 1$$

$$\varrho \cos(\varphi_1 + \theta) > k_1,$$

dont la première peut s'écrire aussi

$$\varrho \cos \theta > k.$$

Intégrons maintenant la fonction $\psi(z) e^{-(s-1)z}$ le long d'un secteur de cercle, formé par les vecteurs $\varphi = 0$ et $\varphi = \varphi_1$ et par un arc du cercle $|z| = R$.

Nous aurons

$$\int_0^R \psi(z) e^{-(s-1)z} dz - \int_0^{Re^{i\varphi_1}} \psi(z) e^{-(s-1)z} dz$$

$$= \int_R^{Re^{i\varphi_1}} \psi(z) e^{-(s-1)z} dz$$

$$= \int_0^{\varphi_1} \psi(Re^{i\varphi}) e^{-R\varrho e^{i(\varphi+\theta)}} i Re^{i\varphi} d\varphi.$$

Or, on a pour $0 \leqq \varphi \leqq \varphi_1$:

$$\psi(Re^{i\varphi}) e^{-R\varrho e^{i(\varphi+\theta)}} = O(e^{R[k(\varphi) - \varrho \cos(\varphi+\theta)]}),$$

et l'on conclut des inégalités précédentes que

$$k(\varphi) - \varrho \cos(\varphi + \theta) =$$

$$= \frac{k \sin(\varphi_1 - \varphi) + k_1 \sin \varphi - \varrho \cos(\varphi + \theta) \sin \varphi_1}{\sin \varphi_1}$$

$$= \frac{\sin(\varphi_1 - \varphi)(k - \varrho \cos \theta) + \sin \varphi (k_1 - \varrho \cos(\varphi_1 + \theta))}{\sin \varphi_1}$$

$$< 0$$

lorsque s reste dans le domaine considéré. L'intégrale le long de l'arc de cercle tend donc vers zéro lorsque R tend vers l'infini, et il vient à la limite:

$$J(s) = J_1(s)$$

dans ce domaine.

Le théorème VII, qui se trouve ainsi démontré, permet d'obtenir dans des cas étendus un prolongement de la fonction $H(s)$ au delà de l'abscisse de convergence. Dans des cas spéciaux, l'énoncé du théorème peut être varié de différentes manières. D'abord, il est clair en vertu de la symétrie que tous les résultats précédents restent valables quand on change le signe de l'angle φ_1, le demiplan $P(k_1 \varphi_1)$ se transformant en le demiplan symétrique par rapport à l'axe réel. En prenant une fonction $\psi(z)$, satisfaisant à la relation $\psi(z) = O(e^{k|z|})$ dans un angle renfermant l'axe réel, les conditions du § 5 sont remplies, et l'on obtient tout de suite une extension des deux côtés de l'axe réel du domaine d'existence de la fonction représentée par notre série de Dirichlet, Soit p. ex. $\psi(z) = O(e^{\varepsilon|z|})$ pour $x = R(z) \geqq 0$ et pour tout $\varepsilon > 0$. La fonction $H(s)$ sera alors holomorphe dans tout le plan où l'on aura tracé une coupure du point $s = 1$ le long de l'axe réel négatif.

D'autre part, il importe de remarquer que nous n'avons point supposé positives les constantes k et k_1. Soit donc $\psi(z)$ une fonction satisfaisant aux conditions du théorème VII, quel que soit le nombre réel k_1. Dans ces conditions, notre théorème montre immédiatement que $H(s)$ est une fonction entière. Un exemple nous est fourni par la série

$\sum \frac{e^{i(\log n)^2}}{n^s}$, comme on le verra plus loin. On pourrait imaginer sans difficulté encore d'autres cas où notre théorème conduirait à des conclusions intéressantes.

Evidemment, il ne convient pas toujours d'effectuer la déformation du chemin d'intégration comme nous l'avons fait ci-dessus. Au lieu de faire des hypothèses sur les propriétés de $\psi(z)$ dans un angle, envisageons p. ex. une bande comprise entre deux parallèles, dont l'une coïncide avec l'axe réel. Nous serons ainsi amenés au théorème suivant.

Théorème VIII. *$\psi(z)$ satisfaisant toujours aux conditions du § 5, supposons encore*

$$\psi(z) = O(e^{Mx})$$

uniformément dans la bande limitée par les droites $y = o$, $x = o$ et $y = p$ et située à la droite de l'axe des imaginaires, $\psi(z)$ étant holomorphe dans cette bande et sur son contour, et

$$\psi(z) = O(e^{k_1 x})$$

sur la droite $y = p$, k_1 étant inférieur à k. Cela posé, formons la série

$$H(s) = \sum_{n=1}^{\infty} \frac{\psi(\log n)}{n^s},$$

qui sera certainement convergente pour $\sigma > k + 1$. Alors, la fonction $H(s)$ sera encore régulière pour $\sigma > k_1 + 1$, et pour une valeur fixe de σ comprise entre $k_1 + 1$ et $k + 1$ on aura

$$H(s) = O(e^{[k(\sigma)+\varepsilon]\cdot|t|}),$$

où

$$k(\sigma) = \frac{k + 1 - \sigma}{k - k_1} |p|, \qquad \varepsilon > 0,$$

lorsque t croît indéfiniment, ayant le même signe que p. Au contraire, t et p ayant des signes opposés, on aura

$$H(s) = G(s) + O\left(\frac{1}{t}\right),$$

$G(s)$ ayant la signification usuelle.

Ce théorème diffère manifestement du précédent en ce qu'ici il nous faut avoir une évaluation *meilleure* de $\psi(z)$ sur la droite considérée, tandis que dans le théorème VII il n'importe quelle que soit la valeur de la constante k_1. D'autre part, le prolongement analytique effectué s'étend dans le théorème VII sur une région angulaire mais dans le théorème VIII sur une bande comprise entre deux parallèles à l'axe des imaginaires.

La démonstration du théorème VIII n'exige qu'un emploi du théorème de Cauchy. D'abord, on a évidemment

$$H(s) = G(s) + \int_0^{\infty} \psi(z) e^{-(s-1)z} dz$$

$$= G(s) + J(s).$$

En intégrant la fonction $\psi(z)e^{-(s-1)z}$ le long du rectangle aux sommets 0, ip, R, $R + ip$, et faisant ensuite tendre R vers linfini, on voit qu'il existe un domaine dans lequel on a

$$J(s) = \int_0^{ip} \psi(z) e^{-(s-1)z} dz + \int_{ip}^{ip+\infty} \psi(z) e^{-(s-1)z} dz. \tag{22}$$

En effet, sur le côté joignant les points R et $R + ip$, on a

$$\psi(z) e^{-(s-1)z} = O(e^{|tp| - R(\sigma - 1 - M)}).$$

Pour que l'intégrale le long de ce côté tende vers zéro, il suffit donc que σ soit assez grand. Les fonctions représentées par les deux membres de la relation (22) sont donc identiques dans tout leur domaine d'existence. Or, la première intégrale représente évidemment une fonction entière de s, et l'on a, σ ayant une valeur fixe quelconque,

$$(23) \qquad \left| \int_0^{ip} \psi(z)\, e^{-(s-1)z}\, dz \right| < K\, \frac{e^{pt}-1}{t}.$$

Quant à la seconde intégrale, on a sur le chemin d'intégration

$$\psi(z)\, e^{-(s-1)z} = O\left(e^{k_1 x - (\sigma-1)x + pt}\right);$$

l'intégrale est donc uniformément convergente dans toute partie finie du demiplan $\sigma > k_1 + 1$, ce qui prouve la première partie du théorème énoncé.

De plus, nous voyons que l'intégrale est $O(e^{pt})$ pour une valeur fixe de σ supérieure à $k_1 + 1$. En tenant compte du théorème II, on a donc pour $\sigma > k_1 + 1$, t ayant le même signe que p:

$$H(s) = O(e^{pt}):$$

De cette relation, on pourrait déduire une fonction $\nu(\sigma)$ de même que nous avons déduit plus haut (page 14) la fonction $\mu(\sigma)$ de la relation

$$f(s) = O(|t|^a).$$

La fonction $\nu(\sigma)$ étant supposée finie dans un certain intervalle, M. HARDY[1] a démontré quelle y est une fonction convexe et continue, ainsi que nous l'avons affirmé plus haut pour $\mu(\sigma)$.

Nous en concluons pour la fonction $H(s)$:

$$H(s) = O\left[e^{\left(\frac{k+1-\sigma}{k-k_1}|p|+\varepsilon\right)|t|}\right],$$

égalité ayant lieu lorsque t croît indéfiniment en étant du même signe que p, σ ayant une valeur fixe comprise entre $k_1 + 1$ et $k + 1$.

D'autre part, quand p et t sont de signes opposés, la relation (23) montre que nous avons

[1] »The application of Abel's method of summation to Dirichlet's series». Quarterly Journal, vol. 47.

$$H(s) = G(s) + O\left(\frac{1}{t}\right)$$

pour toute valeur fixe de σ supérieure à $k_1 + 1$. En ne considérant que ces valeurs de t, on obtiendrait donc la même fonction $\mu(\sigma)$ relative aux deux fonctions $H(s)$ et $G(s)$. Notre théorème se trouve démontré.

D'après les résultats de M. Hardy, une série de la classe considérée dans ce théorème sera sommable par la méthode d'Abel[1] ou sommable (A) tant que $\sigma > k_1 + 1$ et que

$$\frac{k + 1 - \sigma}{k - k_1} \cdot |p| < \frac{\pi}{2}.$$

Théorème IX. *Remplaçons dans l'énoncé du théorème précédent la droite $y = p$ par une courbe $y = h(x)$, et la bande considérée par la région comprise entre cette courbe et l'axe réel et positif, $h(x)$ jouissant des propriétés suivantes:*

1. $h(o) = 0$,

2. $h(x) > o$ pour $x > o$,

3. $h'(x) = O(1)$ lorsque $x \to \infty$. De plus, à partir d'une certaine valeur de x, la fonction $h(x)$, toujours continue, sera monotonément décroissante vers zéro. Dans cesconditions, la fonction $H(s)$ sera encore holomorphe dans le demiplan $\sigma > k_1 + 1$. Lorsque σ a une valeur fixe supérieure à $k_1 + 1$, on aura

$$H(s) = O(e^{\varepsilon t}), \quad (t > 0),$$

pour tout $\varepsilon > 0$, et

$$H(s) = G(s) + O(1), \quad (t < 0).$$

[1] M. Hardy (l. c.) dit qu'une série Σa_n est sommable (A) si la série $\Sigma a_n e^{-ny}$ est convergente pour tout $y > 0$ et que $\Phi(y) = \Sigma a_n e^{-ny} \to l$ quand $y \to 0$. Le domaine de sommabilité (A) d'une série de Dirichlet $\Sigma \frac{a_n}{n^s}$ est le demiplan où la fonction représentée par cette série est holomorphe et $O(e^{H \cdot |t|})$ avec $H < \frac{\pi}{2}$.

La série de Dirichlet $\sum \frac{\psi(\log n)}{n^s}$ *sera donc certainement sommable* (A) *pour* $\sigma > k_1 + 1$.

Nous pouvons nous dispenser de donner la démonstration de ce théorème, qui est absolument pareille à celle du théorème précédent.

Remarque. La méthode employée dans ce paragraphe s'applique sans modification au cas plus général traité aux remarques faites aux paragraphes 3, 4 et 5. Les hypothèses que nous avons faites ne sont pas toutes indispensables; ainsi, il est bien permis de changer les limites inférieures dans $H(s)$ et $J(s)$. Aussi serait-il facile de voir comment se modifieraient les théorèmes précédents en admettant des pôles ou des points singuliers essentiels à l'intérieur de nos contours d'intégration.

§ 8. Sur quelques séries particulières.

Nous allons maintenant appliquer les résultats précédents à quelques séries particulières. Prenons d'abord la série $\sum_1^\infty \frac{e^{i(\log n)^\alpha}}{n^s}$, l'exposant α ayant une valeur réelle quelconque. Si $\alpha < 0$, on doit évidemment supprimer le premier terme. Cette série appartenant bien à notre classe, considérons la fonction $\psi(z)$ correspondante:

$$\psi(z) = e^{iz^\alpha}.$$

Prenons d'abord $\alpha < 1$; sur chaque vecteur issu de l'origine, nous aurons $\psi(z) = O(e^{\varepsilon r})$ pour tout $\varepsilon > 0$. Si nous traçons une coupure du point $s = 1$ le long de l'axe réel négatif, notre fonction devient donc par le théorème VII holomorphe dans tout le plan. En excluant un petit segment des deux côtés de la coupure, on aura pour $\sigma < 1$:

$$0 \leqq \mu(\sigma) \leqq 1 - \sigma.$$

Pour $\alpha = 1$ nous obtenons la fonction $\zeta(s - i)$. Le cas $\alpha > 1$

a été traité par M. Hardy.[1] En posant $z = r e^{i\varphi}$, on aura

$$|\psi(z)| = e^{-r^{\alpha} \sin \alpha \varphi};$$

l'argument φ étant pris suffisamment petit, les conditions du théorème VII sont donc remplies quelle que soit la constante k_1, et la fonction représentée par notre série est une transcendante entière. L'étude de la croissance de cette fonction par rapport à t peut se faire, dans le cas $\alpha \geqq 2$, par le théorème VIII. Prenons d'abord $\alpha = 2$; sur la droite $y = p$, on a alors

$$|\psi(z)| = e^{-2px}.$$

Le théorème cité nous donne

$$H(s) = O\left(e^{\left(\frac{1-\sigma}{2}+\varepsilon\right)t}\right), \qquad t > 0$$

$$H(s) = G(s) + O\left(\frac{1}{t}\right), \qquad t < 0,$$

σ ayant une valeur fixe quelconque < 1. M. Hardy a donné pour $t > 0$ le résultat plus précis

$$H(s) \backsim \sqrt{\pi} \, . \, e^{\frac{1-\sigma}{2} t},$$

tandis que, pour $t < 0$, notre résultat est plus précis que celui de M. Hardy. En supposant $\alpha > 2$, on pourra appliquer soit le théorème VIII, soit le théorème IX. On aura dans ce cas

$$H(s) = O(e^{\varepsilon t}), \qquad t > 0$$

$$H(s) = G(s) + O\left(\frac{1}{t}\right), \qquad t < 0$$

pour tout $\varepsilon > 0$ et pour toute valeur fixe de $\sigma < 1$.

Dans le cas $1 < \alpha < 2$, on pourrait étudier la croissance par une méthode analogue en intégrant le long de la courbe $y = x^{2-\alpha}$, et l'on trouverait alors des types de croissance encore plus élevés.

---

[1] »Example to illustrate a point in the theory of Dirichlet's series.» The Tôhoku Math. Journal, Vol. 8, n:o 2; »The application of Abel's method of summation to Dirichlet's series.» Quarterly Journal, Vol. 47.

Prenant comme second exemple la série $\sum\limits_{1}^{\infty}\dfrac{e^{i\,n^{\alpha}}}{n^s}$, où $0 < \alpha < 1$, nous aurons une série appartenant à la classe plus générale mentionnée pp. 14 et 31. Nous avons ici $\psi(z) = e^{i\,e^{\alpha z}}$, $|\psi(z)| = e^{-e^{\alpha x}\sin\alpha y}$; en appliquant le théorème VIII, on voit donc tout de suite que la série représente une fonction entière, et que nous avons

$$H(s) = O(e^{\varepsilon t})^1 \qquad\qquad t > 0,$$

$$H(s) = G(s) + O\left(\frac{1}{t}\right), \qquad t < 0.$$

Au lieu de multiplier les exemples, ce qui n'aurait aucun sens, disons encore quelques mots sur la possibilité d'appliquer nos résultats au célèbre problème de RIEMANN concernant la distribution des zéros de la fonction $\zeta(s)$. Il s'agit ici de rechercher si la fonction représentée par la série $\dfrac{1}{\zeta(s)} = \sum\limits_{n=1}^{\infty}\dfrac{\mu(n)}{n^s}$ est holomorphe dans le demiplan $\sigma > \dfrac{1}{2}$. En appliquant notre théorème VII, nous sommes amenés à la conclusion négative que voici: Il n'existe pas de fonction analytique $\psi(z)$ holomorphe dans un angle renfermant l'axe réel et positif et jouissant des propriétés:

1. $\psi(\log n) = \mu(n)$

2. $\psi(z) = O(e^{k\,r})$

uniformément dans l'angle mentionné il y a un instant. En effet, l'existence d'une telle fonction $\psi(z)$ aurait pour conséquence la régularité de la fonction $\dfrac{1}{\zeta(s)}$ dans une région renfermant une partie infinie de la »bande critique».

Au contraire, les théorèmes VIII et IX fournissent des systèmes de conditions suffisantes pour la vérité de l'hypothèse de RIEMANN. On est amené à chercher à former une fonction analytique $\psi(z)$ satisfaisant pour des valeurs réelles et positives de z aux égalités $\psi^{(\mu)}(z) = O(e^{k\,z})$, et jouissant

---

[1] Ce résultat pourrait encore être précisé. Voir HARDY and RIESZ, l. c., p. 25.

de plus des propriétés suivantes:

$$1.\quad \psi(\log n) = \mu(n)$$

$$2.\quad \psi(z) = O\left(e^{-\frac{1}{2}|z|}\right)$$

sur une courbe telle que celle traitée au théorème VIII ou bien sur une parallèle à l'axe réel et positif.

$$3.\quad \psi(z) = O(e^{M|z|})$$

uniformément dans la région comprise entre la dite courbe et l'axe réel et positif.

L'existence d'une telle fonction suffirait, d'après nos résultats, pour démontrer l'hypothèse de RIEMANN.

Bien entendu, je tiens à assurer que je ne crois avoir fait par là aucun pas véritable vers la solution du problème. En effet, cherchant à former une telle fonction, on se retrouverait devant le problème original.

§ 9. Application à un problème de sommabilité.

Procédons maintenant à une application d'une toute autre nature. Soit $\sum_{\nu=0}^{\infty} c_\nu z^\nu$ une fonction entière. Posons

$$\psi(z) = \sum_{\nu=0}^{\infty} \frac{c_\nu}{\nu!} z^\nu,$$

et formons la série de Dirichlet

$$H(s) = \sum_{n=1}^{\infty} \frac{\psi(\log n)}{n^s}, \tag{24}$$

absolument convergente pour $\sigma > 1$. Par le théorème IV, nous savons que $H(s)$ n'a dans tout le plan que le point singulier $s = 1$ à distance finie, et que nous avons

$$H(s) = G(s) + \sum_{\nu=0}^{\infty} \frac{c_\nu}{(s-1)^{\nu+1}},$$

$G(s)$ ayant le signification usuelle, et la série infinie représentant une fonction entière de $\frac{1}{s-1}$.

Il s'ensuit que, pour $\sigma < 1$, la série (24) ne peut être sommable ni par la méthode typique de M. Riesz, ni par celles d'Abel ou de M. Borel,[1] les régions de sommabilité étant dans tous les cas limitées par des parallèles à l'axe des imaginaires.

Sur la droite $\sigma = 1$, il en est autrement. Soit k un entier positif. Nous allons ici chercher à décider, si, au point singulier essentiel $s = 1$, la série (24) peut être sommable $(\log n, k)$ selon la terminologie de M. Riesz.[2] Formons la moyenne typique de M. Riesz:

$$\omega^{-k} C^k(\omega) = \omega^{-k} \sum_{\log n < \omega} \frac{\psi(\log n)}{n} (\omega - \log n)^k.$$

Nous avons alors,[3] en posant

$$h(s) = H(s+1) = \sum_{n=1}^{\infty} \frac{\psi(\log n)}{n} \cdot n^{-s},$$

$$\frac{2\pi i}{k!} \omega^{-k} C^k(\omega) = \omega^{-k} \int_{1-i\infty}^{1+i\infty} \frac{e^{\omega s}}{s^{k+1}} h(s)\, ds. \tag{25}$$

Pour transformer cette expression, intégrons la fonction

$$\frac{e^{\omega s}}{s^{k+1}} h(s)$$

le long du rectangle aux sommets $1 \pm iT$, $-\frac{1}{2} \pm iT$. A l'intérieur de ce contour, la fonction à intégrer a un seul point singulier, savoir $s = 0$. Il faut donc calculer le résidu correspondant.

Le résidu de la fonction $\frac{e^{\omega s}}{s^{k+1}} G(s+1)$ au point $s = 0$ est

[1] Hardy, »The application to Dirichlet's series of Borel's exponential method of summation», Proc. Lond. Math. Soc., ser. 2, vol. 8.
[2] Hardy and Riesz, l. c., p. 21.
[3] Hardy and Riesz, l. c. p. 50.

évidemment un polynome du degré k en ω; désignons-le par $p(\omega)$ et observons que le coefficient de ω^k dans $p(\omega)$ est égal à $\frac{1}{k!}\cdot \mathring{G}(1)$. D'autre part, le résidu de

$$\frac{e^{\omega s}}{s^{k+1}}\sum_{\nu=0}^{\infty}\frac{c_\nu}{s^{\nu+1}}$$

à l'origine est égal à

$$\sum_{\nu=0}^{\infty}\frac{c_\nu}{(\nu+k+1)!}\,\omega^{\nu+k+1}.$$

Nous avons donc

$$\int_{(\mathring{R})} = 2\pi i\, p(\omega) + 2\pi i\sum_{\nu=0}^{\infty}\frac{c_\nu}{(\nu+k+1)!}\,\omega^{\nu+k+1}.$$

Faisons ensuite tendre T vers l'infini; les intégrales prises le long des côtés horizontales du rectangle tendront vers zéro, comme on le voit aisément. En effet, en vertu du théorème II, la fonction à intégrer est sur ces côtés $O\left(T^{-\frac{3}{2}}\right)$.

A la limite, il vient donc

$$\int_{1-i\infty}^{1+i\infty}\frac{e^{\omega s}}{s^{k+1}}h(s)\,ds = \int_{-\frac{1}{2}-i\infty}^{-\frac{1}{2}+i\infty}\frac{e^{\omega s}}{s^{k+1}}h(s)\,ds +$$

$$+ 2\pi i\, p(\omega) + 2\pi i\sum_{\nu=0}^{\infty}\frac{c^\nu}{(\nu+k+1)!}\,\omega^{\nu+k+1}.$$

Or, nous avons évidemment

$$\left|\int_{-\frac{1}{2}-i\infty}^{-\frac{1}{2}+i\infty}\frac{e^{\omega s}}{s^{k+1}}h(s)\,ds\right| < e^{-\frac{1}{2}\omega}\left|\int_{-\frac{1}{2}-i\infty}^{-\frac{1}{2}+i\infty}\left|\frac{h(s)}{s^{k+1}}\right|ds\right|,$$

d'où, en introduisant dans (25) et faisant ensuite tendre ω vers l'infini:

$$\omega^{-k} C^k(\omega) \backsim G(1) + k! \sum_{\nu=0}^{\infty} \frac{c_\nu}{(\nu + k + 1)!} \omega^{\nu+1}.$$

Pour que la série $\sum_{n=1}^{\infty} \frac{\psi(\log n)}{n}$ soit sommable $(\log n, k)$, il faut et il suffit donc que la fonction entière

$$\sum_{\nu=0}^{\infty} \frac{c_\nu}{(\nu + k + 1)!} z^{\nu+1}$$

tende vers une limite finie lorsque z s'éloigne indéfiniment suivant l'axe réel et positif.

On peut remarquer que cette fonction n'est autre chose que

$$z^{-k} \left(\int_0^z dz \right)^{k+1} \psi(z),$$

ce qui rend notre résultat très intuitif. Probablement, on parviendrait au même résultat en appliquant la formule d'Euler à des séries sommables, comme l'a fait M. Hardy.[1]

On aura un exemple en prenant $k = 1$ et

$$c_\nu = (-1)^\nu \frac{(\nu + 2)!}{(2\nu + 4)!}.$$

La fonction entière considérée est alors égale à $\dfrac{\cos \sqrt{z} - 1 + \frac{1}{2} z}{z}$, qui tend vers $\frac{1}{2}$ lorsque $z \to \infty$.

On pourrait employer la même méthode pour étudier la sommabilité en un point quelconque de la droite $\sigma = 1$, mais les expressions des résidus sont alors très compliquées.

[1] Theorems connected with Maclaurin's test for the convergence of series», Proc. Lond. Math. Soc., ser. 2, vol. 9.

2.

Etudes sur la sommation des séries de Fourier

Ark. Mat. Astr. Fys. **13** (20), 1–21 (1918)

Communiqué le 13 Fevrier 1918 par IVAR BENDIXSON et H. VON KOCH.

Dans cette note, je me propose de développer les résolutions de quelques problèmes relatifs à la sommation des séries de FOURIER par des moyennes arithmétiques d'ordre non entier. Ces problèmes furent posés par M. MARCEL RIESZ au cours de ses leçons sur les séries trigonométriques, et c'est lui qui m'a encouragé à m'en occuper. Cependant les résultats, surtout en ce qui concerne la partie dernière de la note, ont été un peu étonnants pour lui comme pour moi.

Je commencerai par rappeler quelques propriétés de la méthode de sommation par des moyennes arithmétiques d'ordre arbitraire. Ensuite, j'étudierai les quantités que j'appelle par extension des »constantes de M. LEBESGUE», et dans la partie dernière de la note je traiterai l'analogue du phénomène de GIBBS pour la méthode de sommation adoptée. Dans cet ordre d'idées, je prouverai le théorème suivant, qui me semble assez inattendu: Il existe un nombre positif $k < 1$, tel que les moyennes arithmétiques d'ordre inférieur à k montrent le phénomène de GIBBS, tandis qu'il n'en est pas ainsi pour les moyennes d'ordre supérieur à k. Il ne m'a pas encore réussi à déterminer la valeur de k.

La méthode de sommation.

1. Pour définir la méthode de sommation qui sera adoptée ici, nous introduirons la notation suivante[1]

$$(1-x)^{-k-1} = \sum_{n=0}^{\infty} C_n^{(k)} x^n,$$

où $k > -1$. On a donc

$$C_n^{(k)} = \frac{(k+1)(k+2)\cdots(k+n)}{n!},$$

de manière que

$$C_n^{(k)} = \frac{n^k}{\Gamma(k+1)} + O(n^{k-1}) \qquad \text{lorsque } n \to \infty.$$

Soit $\sum_0^\infty u_n$ une série quelconque; posons d'une manière tout à fait formelle

$$\sum_{n=0}^{\infty} S_n^{(k)} x^n = (1-x)^{-k-1} \sum_{n=0}^{\infty} u_n x^n = (1-x)^{-k} \sum_{n=0}^{\infty} s_n x^n,$$

où $s_n = u_0 + u_1 + \cdots u_n$. On a visiblement

$$S_n^{(k)} = \sum_{\nu=0}^{n} C_\nu^{(k)} u_{n-\nu} = \sum_{\nu=0}^{n} C_\nu^{(k-1)} s_{n-\nu}.$$

Le quotient

$$M_n^{(k)} = \frac{S_n^{(k)}}{C_n^{(k)}}$$

s'appelle la moyenne arithmétique n-ième d'ordre k relatif à la série donnée, et lorsque cette moyenne tend vers une limite finie L quand $n \to \infty$, la série Σu_n est dite sommable par des moyennes arithmétiques d'ordre k avec la somme L.

---

[1] Voir p. ex. S. Chapman: »On non-integral orders of summability of series and integrals». Lond. M. S. Proc., Ser. 2, Vol. 9, p. 369.

De l'identité

$$S_n^{(k+h)} = \sum_{\nu=0}^{n} C_\nu^{(h-1)} S_{n-\nu}^{(k)},$$

on conclut sans difficulté, en utilisant un théorème connu dû à CESÀRO,[1] que Σu_n est aussi sommable d'ordre $k+h$ avec la même somme L, h désignant une quantité positive arbitraire.

En modifiant légèrement le raisonnement, on peut même démontrer que, les limites d'oscillation[2] des moyennes d'ordre k étant supposées finies et égales à m et M, les mêmes limites des moyennes d'ordre $k+h$ sont aussi finies et sont comprises dans l'intervalle (m, M), avec les extrémités duquel elles peuvent aussi coïncider.

Je finirai par indiquer une autre généralisation de ce résultat, qui va nous être utile dans la suite. Supposons que les termes u_n dépendent d'une variable x, ce que nous indiquerons en écrivant $u_n(x)$, $M_n^{(k)}(x) \ldots$ au lieu de u_n, $M_n^{(k)} \ldots$. Dans la moyenne n-ième d'ordre k, faisons tendre x vers une certaine valeur μ, qui sera un point de continuité pour tout $u_n(x)$, au même temps que n tend vers l'infini. Supposons encore

$$\limsup_{\substack{x \to \mu \\ n \to \infty}} M_n^{(k)}(x) = l,$$

l désignant une quantité finie. Par un raisonnement absolument pareil à celui indiqué tout à l'heure, on parvient à la relation

$$\limsup_{\substack{x \to \mu \\ n \to \infty}} M_n^{(k+h)}(x) \leqq l.$$

[1] Voir CHAPMAN, l. c. p. 377.
[2] pour $n \to \infty$.

Les constantes de M. Lebesgue.

2. D'après une remarque bien connue de M. LEBESGUE,[1] l'existence d'une fonction continue dont la série de FOURIER ne converge pas partout tient à ce fait que

$$\frac{2}{\pi}\int_0^{\frac{\pi}{2}}\frac{|\sin(2n+1)t|}{\sin t}dt \to \infty$$

lorsque $n \to \infty$. Nous désignerons ici par $\varrho_n^{(0)}$ la valeur de cette quantité, qui constitue la valeur maximum des sommes $(n+1)$-ièmes s_n des séries de FOURIER des fonctions $f(x)$ telles qu'on ait constamment $|f(x)| \leq 1$. — Sous le nom de »constantes de M. LEBESGUE» les quantités $\varrho_n^{(0)}$ ont été étudiées par MM. FEJÉR[2] et GRONWALL.[3] Parmi d'autres résultats, ces auteurs ont montré que l'on a

$$\varrho_n^{(0)} \backsim \frac{4}{\pi^2}\log n.$$

Quant aux moyennes arithmétiques de M. FEJÉR, il en est autrement. On a

$$M_n^{(1)}(x) = \frac{s_0 + s_1 + \cdots s_n}{n+1} = \frac{1}{(n+1)\pi}\int_0^{\frac{\pi}{2}}[f(x+2t) + $$
$$+ f(x-2t)]\frac{\sin^2(n+1)t}{\sin^2 t}dt.$$

La valeur maximum $\varrho_n^{(1)}$ de cette moyenne pour toutes les fonctions non supérieures à 1 en valeur absolue est égale à

$$\frac{2}{(n+1)\pi}\int_0^{\frac{\pi}{2}}\frac{\sin^2(n+1)t}{\sin^2 t}dt = 1.$$

[1] »Leçons sur les séries trigonométriques», p. 86.
[2] Journal f. Mathematik, 138, p. 22.
[3] Math. Annalen, 72, p. 244.

D'ailleurs, cette limite est atteinte quand $f(x)$ est partout égal à 1.

On voit la différence profonde entre ces deux procédés de sommation: le procédé ordinaire et la méthode de M. Fejér, ou bien la sommation par des moyennes arithmétiques d'ordre 0 et d'ordre 1. Il est donc bien naturel de se demander comment se présentent ces choses pour les méthodes d'ordre non entier.[1]

3. Je me propose donc de résoudre le problème suivant: parmi l'ensemble des fonctions sommables $f(x)$ de période 2π telles que $|f(x)| \leqq 1$, trouver celle qui donne à la moyenne n-ième d'ordre k sa valeur maximum $\varrho_n^{(k)}$; ensuite étudier les quantités $\varrho_n^{(k)}$ pour des valeurs indéfiniment croissantes de n. Le cas $k > 1$ étant le plus souvent dépourvu d'intérêt, le nombre k sera toujours compris dans l'intervalle $(0, 1)$.

Sur ces constantes $\varrho_n^{(k)}$, qui peuvent s'appeler encore par extension des constantes de M. Lebesgue, je démontrerai d'abord le théorème suivant.

La limite $\varrho^{(k)} = \lim_{n \to \infty} \varrho_n^{(k)}$ *a une valeur finie et déterminée pour* $0 < k \leqq 1$. *Considérée comme fonction de* k, *cette valeur limite varie toujours dans le même sens de* $+\infty$ *jusqu'à* 1, *lorsque* k *croît de* 0 *à* 1.

D'ailleurs, on a

$$\varrho^{(k)} = \frac{2\Gamma(k+1)}{\pi} \int_0^\infty \frac{dz}{z^{k+1}} |I_k(\sin z)|\, dz,$$

$I_k(\sin z)$ *désignant l'intégrale* k*-ième de* $\sin z$, *c. à d.*

$$I_k(\sin z) = \frac{1}{\Gamma(k)} \int_0^z (z-u)^{k-1} \sin u\, du.$$

[1] On sait que les moyennes convergent presque partout vers la valeur de la fonction, du moment qu'on suppose positif l'ordre de sommation. G. H. Hardy: »On the summability of Fourier's series». Lond. M. S. Proc., Ser. 2, Vol. 12, 1913.

On a ici

$$s_n(x) = \frac{1}{\pi}\int_0^{\frac{\pi}{2}} [f(x+2t) + f(x-2t)] \frac{\sin(2n+1)t}{\sin t} dt,$$

d'où

$$M_n^{(k)}(x) = \frac{1}{\pi C_n^{(k)}} \int_0^{\frac{\pi}{2}} [f(x+2t) + f(x-2t)]\, S_n^{(k)}(t)\, dt,$$

si l'on pose

$$S_n^{(k)}(t) = \frac{1}{\sin t} \sum_{\nu=0}^{n} C_\nu^{(k-1)} \sin(2\overline{n-\nu}+1)\,t.$$

Il s'ensuit que l'on a pour la valeur maximum cherchée

$$\varrho_n^{(k)} = \frac{2}{\pi C_n^{(k)}} \int_0^{\frac{\pi}{2}} |S_n^{(k)}(t)|\, dt,$$

cette limite étant atteinte p. ex. pour $x = 0$ par la fonction $f(x)$ telle que

$$f(2t) = f(-2t) = \operatorname{sgn}(S_n^{(k)}(t)).^{1}$$

Pour prouver l'existence de $\lim_{n\to\infty} \varrho_n^{(k)}$, nous divisons l'intervalle d'intégration en deux, à savoir $(0, \delta)$ et $\left(\delta, \frac{\pi}{2}\right)$, en désignant par δ un nombre positif arbitrairement petit. La suite $C_\nu^{(k-1)}$, $\nu = 0, 1 \ldots$, étant constamment décroissante, on a dans $\left(\delta, \frac{\pi}{2}\right)$

[1] On pose sgn $\alpha = 1$, $= 0$, $= -1$, suivant que $\alpha \gtreqless 0$.

$$|S_n^{(k)}(t)| < \frac{C_0^{(k-1)}}{\sin\delta}\,\mathrm{Max}\left|\sum_{\nu=0}^{\nu\leq n}\sin(2\overline{n-\nu}+1)t\right|$$

$$< \frac{1}{\sin^2\delta},$$

d'où

$$\varrho_n^{(k)} = \frac{2}{\pi C_n^{(k)}}\int_0^{\delta}|S_n^{(k)}(t)|\,dt + o(1).$$

Posons maintenant dans l'expression de $S_n^{(k)}(t)$

$$C_\nu^{(k-1)} = \frac{\nu^{k-1}}{\Gamma(k)} + O(\nu^{k-2}),$$

ce qui equivaut, en réalité, à un passage aux moyennes arithmétiques de M. Riesz.[1] Il s'ensuit

$$S_n^{(k)}(t) = \frac{1}{\Gamma(k)\sin t}\sum_{\nu=0}^{n-1}(n-\nu)^{k-1}\sin(2\nu+1)t +$$

$$+\frac{\sin(2n+1)t}{\sin t} + O\left(\sum_{\nu=0}^{n-1}(n-\nu)^{k-2}\frac{|\sin(2\nu+1)t|}{\sin t}\right).$$

k étant inférieur à 1, la somme $\sum_0^{n-1}(n-\nu)^{k-2}$ est $O(1)$, et l'on s'assure facilement, en tenant compte de la relation

$$\int_0^{\delta}\frac{|\sin\mu t|}{\sin t}\,dt = O(\log\mu)$$

qui est une conséquence immédiate des travaux déjà cités de MM. Fejér et Gronwall, que la contribution des deux

[1] Voir p. ex. Hardy and Riesz: »The general theory of Dirichlet's series», p. 22. Presque tous les résultats établis dans cette note restent valables pour les moyennes de M. Riesz.

derniers termes dans $\varrho_n^{(k)}$ tend vers zéro avec $\frac{1}{n}$. On a donc

$$\varrho_n^{(k)} = \frac{2}{\pi\Gamma(k)C_n^{(k)}}\int_0^\delta\left|\sum_{\nu=0}^{n-1}(n-\nu)^{k-1}\sin(2\nu+1)t\right|\frac{dt}{\sin t} + o(1).$$

L'erreur commise en remplaçant dans cette intégrale $\sin(2\nu+1)t$ par $\sin 2\nu t$ est inférieure en valeur absolue à

$$\frac{2}{\pi\Gamma(k)C_n^{(k)}}\int_0^\delta\frac{2\sin\frac{t}{2}}{\sin t}\left|\sum_{\nu=0}^{n-1}(n-\nu)^{k-1}\cos\left(2\nu+\frac{1}{2}\right)t\right|dt <$$

$$< \frac{2\sum\limits_1^n \nu^{k-1}}{\pi\Gamma(k)C_n^{(k)}\cos\frac{\delta}{2}}\cdot\delta < M\delta,$$

M désignant une quantité indépendante de n. On pourra donc représenter cette erreur par $O(\delta)$, le signe O étant valable uniformément pour toute valeur de n supérieure à une certaine limite.

Dans l'expression de $\varrho_n^{(k)}$, qui s'écrit maintenant

$$\varrho_n^{(k)} = \frac{2}{\pi\Gamma(k)C_n^{(k)}}\int_0^\delta\left|\sum_{\nu=0}^{n-1}(n-\nu)^{k-1}\sin 2\nu t\right|\frac{dt}{\sin t} + o(1) + O(\delta),$$

nous remplacerons la somme sous le signe intégrale par l'intégrale analogue:

$$\int_0^n (n-v)^{k-1}\sin 2vt\,dv,$$

et nous chercherons une borne supérieure pour l'erreur commise. On a

$$\left|\int_0^n (n-v)^{k-1}\sin 2vt\,dv - \sum_{\nu=0}^{n-1}(n-\nu)^{k-1}\sin 2\nu t\right| \leqq$$

$$\leqq \sum_{\nu=0}^{n-1}\int_\nu^{\nu+1} |(n-v)^{k-1}\sin 2vt - (n-\nu)^{k-1}\sin 2\nu t|\,dv \leqq$$

$$\leqq \sum_{\nu=0}^{n-1}\int_\nu^{\nu+1}(n-v)^{k-1}|\sin 2vt - \sin 2\nu t|\,dv +$$

$$+ \sum_{\nu=0}^{n-1}|\sin 2\nu t|\int_\nu^{\nu+1}((n-v)^{k-1}-(n-\nu)^{k-1})\,dv <$$

$$< \frac{n^k}{k}\cdot 2\sin t + (1-k)\sum_{\nu=0}^{n-2}(n-\nu-1)^{k-2}|\sin 2\nu t| +$$

$$+ \frac{1-k}{k}\cdot|\sin 2(n-1)t|,$$

$\sin t$ étant positif, parce que $0<t<\delta$. Après l'introduction dans notre expression de $\varrho_n^{(k)}$, le premier terme donne évidemment $O(\delta)$, cette notation ayant la même signification que tout à l'heure. D'autre part, la relation déjà utilisée

$$\int_0^\delta \frac{|\sin \mu t|}{\sin t}\,dt = O(\log \mu)$$

montre que la contribution des deux derniers termes dans $\varrho_n^{(k)}$ tend vers zéro avec $\frac{1}{n}$, ce qui nous permet d'écrire

$$\varrho_n^{(k)} = \frac{2}{\pi\Gamma(k)\,C_n^{(k)}}\int_0^\delta \frac{dt}{\sin t}\left|\int_0^n (n-v)^{k-1}\sin 2vt\,dv\right| + o(1) + O(\delta).$$

En remplaçant dans cette intégrale $\sin t$ par t, on voit sans peine que l'erreur commise est $O(\delta)$. En introduisant la notation

$$I_k(\sin z) = \frac{1}{\Gamma(k)} \int_0^z (z-u)^{k-1} \sin u \, du,$$

on obtient par un changement de variables

$$\varrho_n^{(k)} = \frac{2n^k}{\pi C_n^{(k)}} \int_0^{2n\delta} \frac{dz}{z^{k+1}} | I_k(\sin z) | + o(1) + O(\delta).$$

Enfin, en faisant tendre n vers l'infini, le premier terme a une limite déterminée indépendante de δ, ce qu'on va montrer dans un instant, tandis que le deuxième terme tend vers zéro. Or, δ étant arbitrairement petit, il en résulte

$$\varrho^{(k)} = \lim_{n \to \infty} \varrho_n^{(k)} = \frac{2\Gamma(k+1)}{\pi} \int_0^\infty \frac{dz}{z^{k+1}} | I_k(\sin z) |.$$ [1]

La convergence de l'intégrale, qui a besoin d'être mise en évidence, résulte immédiatement des formules

$$I_k(\sin z) = z^{k+1} \left(\frac{1}{\Gamma(k+2)} - \frac{z^2}{\Gamma(k+4)} + \cdots \right) =$$

$$= \sin \left(z - \frac{k\pi}{2} \right) + \frac{1}{\Gamma(k)} \int_0^\infty (z+t)^{k-1} \sin t \, dt,$$

dont la seconde, qui m'a été communiquée par M. Riesz s'obtient sans peine en écrivant

$$\Gamma(k) I_k(\sin z) = \int_{-\infty}^{z} - \int_{-\infty}^{0}.$$

[1] Comme je m'en suis aperçu pendant la rédaction de ce travail, cette expression de $\rho^{(k)}$ aurait pu se déduire plus facilement d'une formule intéressante due à M. W. H. Young (Quarterly Journal 43, p. 177). Cependant, la formule de M. Young montre seulement que cette expression est une borne supérieure et ne met nullement en évidence ni la propriété de $\rho^{(k)}$ d'être le vrai maximum des moyennes d'ordre k, ni les autres propriétés démontrées dans le texte. M. Young lui-même ne dit rien sur l'application de sa formule au sujet qui nous occupe ici.

4. Si dans l'expression de $\varrho^{(k)}$ on fait tendre k vers zéro, la fonction sous le signe intégrale tend vers $z^{-1}|\sin z|$, et par suite $\varrho^{(k)} \to \infty$. D'autre part, en faisant tendre k vers 1, on obtient

$$\varrho^{(k)} \to \frac{2}{\pi} \int_0^\infty \frac{1-\cos z}{z^2}\, dz = \frac{2}{\pi} \int_0^\infty \frac{\sin z}{z}\, dz = 1.$$

Il reste encore de prouver la monotonité de $\varrho^{(k)}$, considéré comme fonction de k; nous reviendrons tout à l'heure à cette question. Cependant, on peut montrer tout de suite que l'on a toujours $\varrho^{(k)} > 1$ pour $0 < k < 1$. Par un lemme de M. W. H. YOUNG,[1] concernant l'inversion de l'ordre des integrations, on a

$$\frac{2\Gamma(k+1)}{\pi} \int_0^\infty \frac{dz}{z^{k+1}} I_k(\sin z) = \frac{2k}{\pi} \int_0^\infty \frac{dz}{z} \int_0^1 (1-u)^{k-1} \sin uz\, du =$$

$$= \frac{2k}{\pi} \int_0^1 (1-u)^{k-1}\, du \int_0^\infty \frac{\sin uz}{z}\, dz = 1.$$

En comparant cette formule avec l'expression de $I_k(\sin z)$ à la fin du numéro précédent, qui met en évidence l'oscillation de I_k, on conclut

$$\varrho^{(k)} = \frac{2\Gamma(k+1)}{\pi} \int_0^\infty \frac{dz}{z^{k+1}} |I_k(\sin z)| >$$

$$> \frac{2\Gamma(k+1)}{\pi} \int_0^\infty \frac{dz}{z^{k+1}} I_k(\sin z) = 1.$$

Notre démonstration au numéro précédent met en evidence qu'il n'est pas indispensable de supposer $|f(x)| \leqq 1$ dans tout l'intervalle considéré.

Au contraire, en supposant seulement les limites d'oscillation de f au point x comprises entre 1 et -1, les moyennes

[1] l. c., p. 165.

d'ordre k sont au même point pour toutes les valeurs assez grandes de l'indice comprises entre $\varrho^{(k)} + \varepsilon$ et $-\varrho^{(k)} - \varepsilon$. (Il est évident comment se modifie cet énoncé quand on suppose les limites d'oscillation de f égales à m et M.)

Comme nous venons de le voir, n, k et x étant donnés, on peut toujours assigner une fonction de l'ensemble considéré, dont la moyenne n-ième d'ordre k prend au point x la valeur $\varrho_n^{(k)}$. On n'a pas le droit d'en conclure immédiatement l'existence d'une fonction f du même ensemble telle que pour des indices $n_1, n_2 \ldots$ convenablement choisis $M_{n_i}^{(k)}(x)$ tend vers $\varrho^{(k)}$. Cependant, on va maintenant établir ce point par une construction explicite.

Dans ce but, remarquons d'abord que nous avons démontré plus haut que

$$I(n, \delta) = \frac{2}{\pi C_n^{(k)}} \int_{\delta}^{\frac{\pi}{2}} |S_n^{(k)}(t)| \, dt$$

tend vers zéro avec $\frac{1}{n}$, δ ayant une valeur positive arbitrairement petite. Soit n_1 un entier quelconque. Déterminons l'entier $n_2 > n_1$ tel que $I\left(n_2, \frac{1}{n_1^2}\right) < \frac{1}{n_1}$, et, d'une manière générale, n_{i-1} étant donné, déterminons $n_i > n_{i-1}$ tel que

$$I\left(n_i, \frac{1}{n_{i-1}^2}\right) < \frac{1}{n_{i-1}}.$$

Définissons une fonction f par les conditions suivantes

$$f(2t) = \operatorname{sgn}\left(S_{n_1}^{(k)}(t)\right) \quad \text{pour} \quad \frac{1}{n_1^2} < t < \frac{\pi}{2}$$

. . . .

$$f(2t) = \operatorname{sgn}\left(S_{n_i}^{(k)}(t)\right) \quad \text{pour} \quad \frac{1}{n_i^2} < t \leq \frac{1}{n_{i-1}^2}$$

. . . .

$$f(0) = f(\pi) = 0, \quad f(t) = f(-t) = f(t + 2\pi).$$

La fonction f ainsi définie fait bien partie de notre ensemble, et l'on a

$$M_{n_i}^{(k)}(0) = \frac{1}{\pi C_{n_i}^{(k)}} \int_0^{\frac{\pi}{2}} (f(2t) + f(-2t)) S_{n_i}^{(k)}(t)\, dt =$$

$$= \frac{2}{\pi C_{n_i}^{(k)}} \int_0^{\frac{\pi}{2}} |S_{n_i}^{(k)}(t)|\, dt + \frac{2}{\pi C_{n_i}^{(k)}} \int_0^{\frac{1}{n_i^2}} (f(2t) - sgn\, S_{n_i}^{(k)}(t))\, S_{n_i}^{(k)}(t)\, dt$$

$$+ \frac{2}{\pi C_{n_i}^{(k)}} \int_{\frac{1}{n_{i-1}^2}}^{\frac{\pi}{2}} (f(2t) - sgn\, S_{n_i}^{(k)}(t))\, S_{n_i}^{(k)}(t)\, dt.$$

Dans l'intervalle $\left(0, \frac{1}{n_i^2}\right)$, on obtient par un calcul peu difficile $|S_n^{(k)}(t)| < 3n C_n^{(k)}$; d'après les propriétés des nombres n_i les deux derniers termes sont donc inférieurs en valeurs absolues à $\frac{4}{n_i}$ et $\frac{2}{n_{i-1}}$ respectivement. Pour $i \to \infty$, on a donc

$$\lim M_{n_i}^{(k)}(0) = \lim \varrho_{n_i}^{(k)} = \varrho^{(k)}.$$

Puisque $\varrho^{(k)} > 1$, ces moyennes sortent donc effectivement de l'intervalle d'oscillation de f au point 0.

Maintenant, on peut montrer sans difficulté que $\varrho^{(k)}$ est une fonction non croissante de k dans l'intervalle (0, 1). En effet, formons une fonction f, dont une infinité de moyennes d'ordre k_1 tendent vers $\varrho^{(k_1)}$. Les moyennes d'ordre $k_2 < k_1$ d'indices assez grands sont toutes inférieures à $\varrho^{(k_2)} + \varepsilon$, et l'on en conclut, d'après un théorème général cité au n:o 1, que $\varrho^{(k_1)} \leqq \varrho^{(k_2)}$.

Le phénomène de Gibbs.[1]

5. Soit $f(x)$ une fonction, analytique dans l'intervalle $(-\pi, \pi)$ sauf au point $x=0$, qui sera un point de discontinuité de première espèce. Le phénomène de GIBBS consiste dans une certaine singularité des sommes s_n de la série de FOURIER de f autour du point de discontinuité. En donnant à x une valeur fixe tandis qu'on fait tendre n vers l'infini, on aura $\lim s_n(x) = f(x)$ pour $x \neq 0$ et $= \frac{1}{2}(f(+0) + f(-0))$ pour $x = 0$. On serait donc tenté de croire que la proposition suivante soit exacte: »En joignant à la courbe $y = f(x)$ (pour $x \neq 0$) le segment de droite aux extrémités $(0, f(\pm 0))$ on obtient la courbe limite des courbes $y = s_n(x)$ pour $n \to \infty$.»

Or, il n'en est pas ainsi: pour avoir cette courbe limite il faut encore, dans la courbe qui vient d'être définie, que le segment de droite se dilate dans la proportion $\frac{2}{\pi}\int\limits_0^\pi \frac{\sin t}{t}\,dt : 1$.[2] Ce fait d'apparence singulière s'explique aisément en étudiant de plus près les courbes $y = s_n(x)$, p. ex. en les comparant à des intégrales définies, comme on va le faire ici. — Au contraire, les moyennes de M. FEJÉR étant toutes comprises entre les bornes de la fonction, on est assuré qu'elle ne montrent pas le même phénomène.

Voilà encore une fois le contraste entre ces deux procédés de sommation qui se présente et qui nous fait reprendre l'étude des moyennes généralisées de ce nouveau point de vue. Il est évident que tout se ramène à l'étude d'un cas particulier, parce que la série obtenue en retranchant membre à membre les séries de FOURIER de deux fonctions de l'espèce considérée, ayant la même discontinuité au point $x = 0$, converge uniformément dans le voisinage de ce point.

---

[1] Voir FEJÉR: Math. Ann. 64, p. 273, GRONWALL: ibid. 72, p. 228, et surtout WEBER: ibid. 73, p. 286.

[2] Approximativement 1,18 : 1.

Envisageons donc la fonction particulière représentée par la série $\sum_{1}^{\infty}\frac{\sin nx}{n}$, c. à d.

$$f(x)=\begin{cases}\frac{\pi-x}{2} & \text{pour} \quad 0<x<2\pi\\ 0 & \text{»} \quad x=0\\ \frac{-\pi-x}{2} & \text{»} \quad -2\pi<x<0.\end{cases}$$

En posant

$$\varphi_k(x)=\int_0^x\left(1-\frac{t}{x}\right)^k\frac{\sin t}{t}\,dt=\Gamma(k+1)\int_0^x\frac{dt}{t^{k+1}}I_k(\sin t)$$

pour $0\leqq k\leqq 1$, commençons par établir le lemme suivant.

A tout nombre positif ε, *aussi petit qu'on le suppose, on peut faire correspondre deux autres nombres positifs* η *et* N *tels que pour* $|x|\leqq\eta$, $n>N$ *on a*

$$|M_n^{(k)}(x)-\varphi_k(nx)|<\varepsilon.$$

D'abord, on a ici

$$M_n^{(k)}(x)=\frac{1}{C_n^{(k)}}\sum_{\nu=1}^{n}C_{n-\nu}^{(k)}\frac{\sin\nu x}{\nu}.$$

En passant aux moyennes de M. Riesz, on a par un calcul analogue à celui du n:o 3

$$M_n^{(k)}(x)-\sum_{\nu=1}^{n}\left(1-\frac{\nu}{n}\right)^k\frac{\sin\nu x}{\nu}\to 0,$$

lorsque $n\to\infty$, et cela uniformément pour toutes les valeurs réelles de x. Il nous reste donc de prouver que

$$\sum_{\nu=1}^{n}\left(1-\frac{\nu}{n}\right)^k\frac{\sin\nu x}{\nu}-\varphi_k(nx)\to 0$$

uniformément pour toute valeur de n, lorsque $x\to 0$.

En prenant les dérivées par rapport à x, étudions d'abord la différence

$$\Delta(n, x) = \sum_{\nu=1}^{n} \left(1 - \frac{\nu}{n}\right)^k \cos \nu x - n \varphi'_k(nx),$$

qui s'écrit par un changement de variable

$$\Delta(n, x) = n^{-k} \sum_{\nu=1}^{n} (n-\nu)^k \cos \nu x - n^{-k} \int_0^n (n-t)^k \cos tx\, dt.$$

Par la formule sommatoire d'Euler, on a

$$\Delta(n, x) = -\frac{1}{2} + kn^{-k} \int_0^n (n-t)^{k-1} \cos tx\, P(t)\, dt +$$

$$+ xn^{-k} \int_0^n (n-t)^k \sin tx\, P(t)\, dt,$$

$P(t)$ désignant la fonction représentée par $\sum_{\nu=1}^{\infty} \frac{\sin 2\nu\pi t}{\nu\pi}$.

Or, on a

$$kn^{-k} \left| \int_0^n (n-t)^{k-1} \cos tx\, P(t)\, dt \right| < 1,$$

et par la seconde formule de la moyenne

$$xn^{-k} \int_0^n (n-t)^k \sin tx\, P(t)\, dt = x \int_0^{\theta n} \sin tx\, P(t)\, dt =$$

$$= \frac{x}{\pi} \sum_{\nu=1}^{\infty} \frac{1}{\nu} \int_0^{\theta n} \sin 2\nu\pi t \,.\, \sin tx\, dt,$$

l'intégration terme par terme se justifiant par des propriétés connues de la série $\sum \frac{\sin \nu t}{\nu}$. Pour $|x| < \pi$ on en conclut immédiatement

$$x n^{-k} \left| \int_0^n (n-t)^k \sin tx \, P(t) \, dt \right| < \frac{2\pi}{9},$$

d'où

$$|\varDelta(n, x)| < \frac{3}{2} + \frac{2\pi}{9} < 3.$$

En multipliant par dx et en intégrant de 0 à x, on obtient donc, $|x|$ étant $< \pi$,

$$\left| \sum_{\nu=1}^{n} \left(1 - \frac{\nu}{n}\right)^k \frac{\sin \nu x}{\nu} - \varphi_k(nx) \right| < 3x$$

ce qui prouve notre lemme.

6. Dans la discussion suivante, il suffit de considérer des valeurs positives de x. Prenons d'abord $k = 0$ et regardons la courbe correspondante

$$y = \varphi_0(x) = \int_0^x \frac{\sin t}{t} \, dt.$$

Pour $x > 0$, cette courbe est située au-dessus de l'axe des x; elle coupe une infinité de fois la droite $y = \frac{\pi}{2}$ et vient s'y confondre à l'infini. Les abscisses $x = \pi$, 3π etc. correspondent à des maxima, dont les valeurs forment une suite décroissante, où le premier terme est $\int_0^\pi \frac{\sin t}{t} \, dt = 1{,}18 \frac{\pi}{2}$. D'autre part, les abscisses $x = 2\pi$, 4π etc. correspondent à des minima, dont les valeur croissent toujours vers $\frac{\pi}{2}$.

En passant maintenant à la courbe $y = \varphi_0(nx)$ on a la même courbe resserrée dans la proportion $1:n$ vers l'axe des y.

Pour de grandes valeurs de n, les premiers maxima de la courbe sont donc situés à une distance très petite de cet axe, et la courbe limite des courbes $y = \varphi_0(nx)$ se compose des segments de droite $y = \frac{\pi}{2}$, $x > 0$ et $x = 0$, $0 < y < \int_0^\pi \frac{\sin t}{t}\, dt$. Or, d'après le lemme précédent, cette courbe représente $M_n^{(0)}(x)$ dans l'intervalle $(0, \eta)$ pour $n > N$ à moins de ε près; voilà donc l'origine du phénomène de GIBBS pour la sommation ordinaire.

Prenons maintenant $k = 1$. On a

$$\varphi_1(x) = \int_0^x \left(1 - \frac{t}{x}\right) \frac{\sin t}{t}\, dt = \varphi_0(x) - \frac{1 - \cos x}{x};$$

$$\varphi'_1(x) = \frac{1 - \cos x}{x^2}.$$

On voit que φ'_1 est toujours $\geqq 0$; $\varphi_1(x)$ ne va donc jamais en décroissant, et la courbe correspondante n'a aucun point commun avec son asymptote $y = \frac{\pi}{2}$. Les valeurs $x = 2\pi$, 4π etc. correspondent à des points d'inflexion de la courbe, où elle est tangente à la courbe $y = \varphi_0(x)$. La courbe $y = \varphi_1(nx)$, obtenue par le procédé expliqué il y a un instant, n'a donc jamais des points au-dessus de $y = \frac{\pi}{2}$, et il s'ensuit que les moyennes d'ordre premier ne montrent aucun phénomène de GIBBS.

7. Pour des valeurs de k entre 0 et 1, notre problème se trouve donc ramené au suivant: décider si la fonction $y = \varphi_k(x)$ prend ou non des valeurs supérieures à $\frac{\pi}{2}$. En effet, les parties de la courbe $y = \varphi_k(nx)$, situées au-dessus de $y = \frac{\pi}{2}$ viennent s'approcher indéfiniment de l'axe des y,

lorsque n croît sans cesse; la courbe limite aura donc également une partie au-dessus de cette droite. En désignant par η_k la borne supérieure de $\varphi_k(x)$ pour $x > 0$, il est évident que la courbe limite se compose des segments de droite $y = \frac{\pi}{2}$, $x > 0$ et $x = 0$, $0 < y < \eta_k$.

Tout se ramène donc à l'étude de η_k, considéré comme fonction de k. Evidemment, on a toujours $\eta_k \geqq \frac{\pi}{2}$, puisque $\lim_{x \to \infty} \varphi_k(x) = \frac{\pi}{2}$. Il s'ensuit immédiatement de ce qui a été dit au n:o 1 que η_k ne croît jamais avec k dans l'intervalle $(0, 1)$, car on a

$$\eta_k = \limsup_{\substack{n \to \infty \\ x \to 0}} M_n^{(k)}(x).$$

D'autre part, η_k est aussi une fonction continue de k, car la courbe $y = \varphi_0(x)$ doit se déformer d'une manière continue pour coïncider avec $y = \varphi_1(x)$. Tant que $\eta_k > \frac{\pi}{2}$, les moyennes d'ordre k montrent le phénomène de GIBBS. Il en est bien ainsi pour de petites valeurs positives de k; en est-il de même dans tout l'intervalle $(0, 1)$?

Il n'en est pas ainsi, comme on va le montrer maintenant. On a

$$\varphi_k(x) - \frac{\pi}{2} = \int_0^x \left(1 - \frac{t}{x}\right)^k \frac{\sin t}{t}\, dt - \int_0^\infty \frac{\sin t}{t}\, dt =$$

$$= \int_0^1 \frac{(1-t)^k - 1}{t} \sin tx\, dt - \int_x^\infty \frac{\sin t}{t}\, dt =$$

$$= \int_0^1 \frac{(1-t)^{k+1} - 1}{t} \sin tx\, dt + \int_0^1 (1-t)^k \sin tx\, dt -$$

$$- \int_x^\infty \frac{\sin t}{t}\, dt.$$

Par des intégrations partielles, on obtient

$$\int_0^1 \frac{(1-t)^{k+1}-1}{t} \sin tx\, dt = \frac{\cos x - k - 1}{x} + \frac{2\Theta_1}{x^2};$$

$$\int_0^1 (1-t)^k \sin tx\, dt = \frac{1}{x} - \frac{\Gamma(k+1)\cos\left(x - \frac{k\pi}{2}\right)}{x^{k+1}} + \frac{4\Theta_2}{x^2};$$

$$\int_x^\infty \frac{\sin t}{t} dt = \frac{\cos x}{x} + \frac{2\Theta_3}{x^2},$$

d'où

$$\varphi_k(x) = \frac{\pi}{2} - \frac{k}{x} - \frac{\Gamma(k+1)\cos\left(x - \frac{k\pi}{2}\right)}{x^{k+1}} + \frac{8\Theta}{x^2},$$

la valeur absolue des Θ étant inférieure à 1.

Donc, pour tout $k > 0$, c'est le terme $-\frac{k}{x}$ qui va déterminer le signe de la différence $\varphi_k(x) - \frac{\pi}{2}$ pour des valeurs assez grandes de x, et la courbe $y = \varphi_k(x)$ est située entièrement au-dessous de $y = \frac{\pi}{2}$ à partir d'un certain point. C'est donc seulement la courbe $y = \varphi_0(x)$ qui a une infinité de points communs avec son asymptote.

D'autre part, la courbe se déformant d'une manière continue lorsque k varie, on peut évidemment prendre la différence $1 - k$ suffisamment petite pour que la courbe $y = \varphi_k(x)$ soit située *toute entière* au-dessous de $y = \frac{\pi}{2}$. Il existe donc des valeurs de k inférieures à 1, pour lesquelles on a $\eta_k = \frac{\pi}{2}$, et où par suite il n'y a aucun phénomène de Gibbs.

Le résultat de notre étude du phénomène de GIBBS peut donc s'exprimer ainsi: *en passant de la sommation ordinaire au procédé par des moyennes arithmétiques, le phénomène de Gibbs devient toujours plus faible et finit par s'évanouir pour une certaine valeur inférieure à l'unité de l'ordre de sommation.*

Tryckt den 16 oktober 1918.

3.

Un théorème sur les séries de Dirichlet et son application

Ark. Mat. Astr. Fys. **13** (22), 1–14 (1918)

Communiqué le 10 Avril 1918 par IVAR BENDIXSON et H. VON KOCH.

Introduction.

Dans ma Thèse de Doctorat,[1] j'ai démontré un théorème sur les séries de DIRICHLET, dont je me propose de présenter ici une généralisation et quelques conséquences, qui ne me semblent pas tout à fait dépourvues d'intérêt.

En supposant qu'on sache prolonger analytiquement la fonction représentée par la série $\Sigma e^{-\lambda_n s}$ au dehors du demi-plan de convergence de la série, le théorème cité donne des renseignements sur la nature de la fonction représentée par la série $\Sigma\varphi(\lambda_n)e^{-\lambda_n s}$, où φ désigne une fonction entière, assujettie à une certaine condition relative à son ordre de grandeur. La démonstration du théorème, reposant sur l'emploi d'une série double absolument convergente, est extrêmement simple, et la généralisation obtenue en échangeant $\Sigma e^{-\lambda_n s}$ contre $\Sigma a_n e^{-\lambda_n s}$ ne présente aucune difficulté si la dernière série a un domaine de convergence absolue.

Dans ce qui va suivre, je vais cependant me dispenser de cette hypothèse restrictive, ce qui exigera des considérations beaucoup plus délicates que dans le cas précédent.

[1] »Sur une classe de séries de Dirichlet», Upsala 1917, p. 30.

Après avoir démontré au n:o 1. mon théorème fondamental, je passerai aux applications, dont les plus importantes seront celles qui concernent les fonctions entières. Je n'en citerai ici que les deux propositions suivantes que je crois les plus importantes, et dont la première est une forme plus précise d'un théorème énoncé par M. WIGERT.[1]

1. Une fonction entière dont le module finit par rester inférieur à $e^{k|z|}$, où $k < \pi$, ne peut s'annuler pour tout nombre entier positif sans s'annuler identiquement.

2. Si le module d'une fonction entière $\varphi(z)$ finit par rester inférieur à $e^{k|z|}$, il y a sur chaque vecteur issu de l'origine une infinité de points aux modules indéfiniment croissants où $|\varphi(z)| > e^{-(k+\varepsilon)|z|}$, et cela a lieu pour tout $\varepsilon > 0$.

Démonstration du théorème.

1. Je désignerai toujours par $\lambda_1, \lambda_2 \ldots$ une suite de nombres positifs tels que $\lambda_{n+1} > \lambda_n$, $\lambda_n \to \infty$, et par $\varphi(z) = \sum_0^\infty \frac{c_\nu}{\nu!} z^\nu$ une fonction entière telle que

$$(1) \qquad |\varphi(z)| \leqq e^{k|z|}$$

pour toute valeur assez grande de $|z|$. Par un procédé bien connu, on obtient alors

$$(2) \qquad |c_\nu| < M \nu^{\frac{1}{2}} k^\nu,$$

M étant une constante. Pour abréger, je poserai $\varphi(\lambda_n) = \varphi_n$.

Théorème. *Soit la série de Dirichlet*

$$h(s) = h(\sigma + it) = \sum_1^\infty a_n e^{-\lambda_n s}$$

[1] »Sur un théorème concernant les fonctions entières», Arkiv för matematik etc, B. 11, N:o 21, 1916.

convergente pour $\sigma > l \geqq 0$. *Supposons qu'on sait prolonger analytiquement la fonction* $h(s)$ *au dehors du demiplan* $\sigma > l$, *de manière qu'on connait un certain domaine connexe* D, *renfermant ce demiplan, où* $h(s)$ *est holomorphe.*[1]

Appelons $D(k)$ *le domaine formé de tous les points de* D *dont la distance à tout point du contour est supérieure ou égale à* k. *En général, ce domaine se partagera en plusieurs parties connexes, dont l'un* $D'(k)$, *renfermera le demiplan* $\sigma > l + k$.

Cela posé, la série de Dirichlet

$$H(s) = \sum_{1}^{\infty} a_n \varphi_n e^{-\lambda_n s}$$

sera convergente pour $\sigma > l + k$, *et la fonction* $H(s)$ *sera holomorphe dans* $D'(k)$.

Démontrons d'abord la convergence de la série pour $\sigma > l + k$. On a, en posant $A_\nu = a_1 + a_2 + \cdots + a_\nu$:

$$\sum_{1}^{n} a_\nu \varphi_\nu = \sum_{1}^{n-1} A_\nu (\varphi_\nu - \varphi_{\nu+1}) + A_n \varphi_n.$$

Mais $A_\nu = O(e^{\lambda_\nu (l+\varepsilon)})$, d'où

$$\sum_{1}^{n} a_\nu \varphi_\nu = O\left(e^{\lambda_n (l+\varepsilon)} \sum_{1}^{n-1} |\varphi_{\nu+1} - \varphi_\nu|\right) + O(e^{\lambda_n (l+k+\varepsilon)}).$$

Cependant $\sum_{1}^{n-1} |\varphi_{\nu+1} - \varphi_\nu|$ est inférieur à la variation totale de $\varphi(z)$ dans l'intervalle $(0, \lambda_n)$, c. à d. inférieur à

$$\int_0^{\lambda_n} |\varphi'(z)| dz < \sum_{0}^{\infty} \frac{|c_\nu|}{\nu!} \lambda_n^\nu = O(e^{\lambda_n (k+\varepsilon)}).$$

[1] Nous dirons qu'une fonction est holomorphe dans un domaine lorsqu'elle est holomorphe en tout point intérieur à ce domaine.

Donc

$$\sum_{1}^{n} a_\nu \varphi_\nu = O\left(e^{\lambda_n (l+k+2\varepsilon)}\right),$$

et il s'ensuit que $\Sigma a_n \varphi_n e^{-\lambda_n s}$ converge pour $\sigma > l + k$.

Passons à la dernière partie de notre énoncé, qui est la plus essentielle. En supposant d'abord $\sigma > l + k$, nous aurons par le raisonnement précédent

$$H(s) = \sum_{n=1}^{\infty} a_n \varphi_n e^{-\lambda_n s}$$

$$= \sum_{n=1}^{\infty} \sum_{\nu=0}^{\infty} \frac{c_\nu}{\nu!} a_n \lambda_n^\nu e^{-\lambda_n s}.$$

Je dis qu'on peut intervertir l'ordre des sommations et écrire

$$H(s) = \sum_{\nu=0}^{\infty} \sum_{n=1}^{\infty} \frac{c_\nu}{\nu!} a_n \lambda_n^\nu e^{-\lambda_n s} \tag{3}$$

$$= \sum_{\nu=0}^{\infty} \frac{(-1)^\nu c_\nu}{\nu!} h^{(\nu)}(s).$$

En effet, considérons d'abord la série double

$$\sum_{n=1}^{\infty} \sum_{\nu=0}^{\infty} \frac{c_\nu}{\nu!} a_n x^n \lambda_n^\nu e^{-\lambda_n s}$$

où $0 < x < 1$. Elle est absolument convergente, car on a

$$\sum_{n=1}^{\infty} \sum_{\nu=0}^{\infty} \left| \frac{c_\nu}{\nu!} a_n x^n \lambda_n^\nu e^{-\lambda_n s} \right| < \sum_{n=1}^{\infty} K x^n |a_n| e^{-\lambda_n (\sigma - k - \varepsilon)},$$

K désignant une constante et ε une quantité positive arbitraire. Le coefficient de x^n étant la valeur absolue du terme général d'une série convergente (en prenant ε assez petit), on conclut que la dernière série est convergente.

On a donc

$$\sum_{n=1}^{\infty} a_n \varphi_n e^{-\lambda_n s} x^n = \sum_{\nu=0}^{\infty} \frac{c_\nu}{\nu!} \sum_{n=1}^{\infty} a_n \lambda_n^\nu e^{-\lambda_n s} x^n.$$

En faisant tendre dans cette relation x vers 1, le premier membre tend vers $H(s)$, d'après le théorème d'ABEL.

Dans le second membre, on a pour $\nu = 1, 2 \ldots$

$$\lim_{x=1} \sum_{n=1}^{\infty} a_n \lambda_n^\nu e^{-\lambda_n s} x^n = (-1)^\nu h^{(\nu)}(s);$$

pour parvenir à la relation (3), il suffit donc de prouver que

$$\sum_{\nu=N}^{\infty} \frac{c_\nu}{\nu!} \sum_{n=1}^{\infty} a_n \lambda_n^\nu e^{-\lambda_n s} x^n$$

tend vers zéro avec $\frac{1}{N}$, uniformément pour $0 \leqq x \leqq 1$.

En effet, on a par le lemme d'ABEL

$$\left| \sum_{n=1}^{\infty} a_n \lambda_n^\nu e^{-\lambda_n s} x^n \right| \leqq \text{Max} \left| \sum_{n=1}^{m} a_n \lambda_n^\nu e^{-\lambda_n s} \right|.$$

Or, nous avons supposé $\sigma > l + k$; on pourra donc entourer le point s d'un cercle du rayon $k_1 > k$ situé tout entier à l'intérieur du demiplan $\sigma > l$. Sur le contour de ce cercle, la série $\Sigma a_n e^{-\lambda_n s}$ est uniformément convergente; toutes les expressions $\sum_{1}^{m} a_n e^{-\lambda_n s}$ sont donc, sur ce contour, inférieures en valeurs absolues à une certaine constante B.

En intégrant le long de ce cercle, on a donc

$$\left| \sum_{n=1}^{m} a_n \lambda_n^\nu e^{-\lambda_n s} \right| = \left| \frac{\nu!}{2\pi i} \int_C \frac{\sum_{1}^{m} a_n e^{-\lambda_n u} du}{(u-s)^{\nu+1}} \right| < \frac{B\nu!}{k_1^\nu}.$$

Donc, pour $0 \leq x \leq 1$, on a d'après (2)

$$\left|\sum_{\nu=N}^{\infty} \frac{c_\nu}{\nu!} \sum_{n=1}^{\infty} a_n \lambda_n^\nu e^{-\lambda_n s} x^n\right| < \sum_{\nu=N}^{\infty} \frac{M \nu^{\frac{1}{2}} k^\nu}{\nu!} \cdot \frac{B \nu!}{k_1^\nu},$$

et cette expression tend bien vers zéro avec $\frac{1}{N}$, puisque $k_1 > k$.

Le développement

$$H(s) = \sum_{\nu=0}^{\infty} \frac{(-1)^\nu c_\nu}{\nu!} h^{(\nu)}(s)$$

est donc certainement valable pour $\sigma > l + k$; cependant les deux membres de cette relation représentent des fonctions analytiques qui coïncideront ainsi dans tout leur domaine d'existence.

Pour prouver notre théorème, il suffit donc de prouver que la série au second membre de la dernière relation converge uniformément dans toute région finie intérieure au domaine désigné par $D(k)$. En effet, à une telle région P on peut toujours faire correspondre un nombre $k_1 > k$ tel que tout point de P peut-être entouré d'un cercle du rayon k_1 situé tout entier à l'intérieur d'une certaine partie finie de D, où la fonction $h(s)$ est holomorphe et bornée en valeur absolue.

En intégrant le long de ce cercle, on a pour un point arbitraire de P:

$$h^{(\nu)}(s) = \frac{\nu!}{2\pi i} \int_C \frac{h(u)\,du}{(u-s)^{\nu+1}},$$

d'où

$$\left|\frac{(-1)^\nu c_\nu}{\nu!} h^{(\nu)}(s)\right| < K \,.\, \nu^{\frac{1}{2}} \left(\frac{k}{k_1}\right)^\nu.$$

Notre série converge donc uniformément dans P. P étant une partie finie quelconque intérieure au domaine $D(k)$, on serait tenté de conclure qu'elle représente une fonction holomorphe dans $D(k)$. Or, ce domaine pouvant se partager en plusieurs parties connexes, comme on l'a déjà signalé, cette conclusion serait évidemment incorrecte, car les fonctions représentées par la série dans ces diverses parties peuvent être absolument indépendantes l'une de l'autre.

Cependant, dans la partie $D'(k)$ qui renferme le demi-plan $\sigma > l + k$, la fonction représentée par la série coïncide nécessairement avec le prolongement analytique de $H(s)$, les deux fonctions étant identiques pour $\sigma > l + k$.

Notre théorème est donc établi.

Applications.

2. En premier lieu, supposons la relation $|\varphi(z)| < e^{k|z|}$ satisfaite pour toute valeur positive de k, ce qui sera le cas p. ex. lorsqu'on prend pour $\varphi(z)$ une fonction d'ordre inférieur à 1. *Dans ce cas,* $H(s) = \Sigma a_n \varphi_n e^{-\lambda_n s}$ *ne peut admettre d'autres points singuliers à distance finie que ceux de* $h(s) = \Sigma a_n e^{-\lambda_n s}$. En effet, le domaine D s'obtient en excluant du plan les points singuliers de $h(s)$; la fonction $H(s)$ est donc holomorphe dans le domaine obtenu en excluant tous les cercles décrits autours de ces points avec un rayon commun k, qui peut être pris arbitrairement petit.

Proposons-nous d'établir maintenant une relation entre l'infini de $h(s)$ quand s s'approche du contour de D, et celui de $H(s)$ au voisinage du contour de $D'(k)$. — Supposons, seulement pour fixer les idées et pour éviter des complications inutiles, que $s = s_0$ est un pôle d'ordre α pour $h(s)$, situé sur la droite de convergence de la série, et qu'à l'intérieur du cercle $|s - s_0| = R$ il n'y a aucune autre singularité.

Alors, si l'on prend $k < \frac{1}{2} R$, le domaine $k < |s - s_0| < R - k$ fait partie de $D'(k)$. Cela posé, on peut affirmer que l'on a, le point s tendant vers le contour $|s - s_0| = k$, tout en restant à son extérieur:

$$|H(s)| < \frac{M}{(|s - s_0| - k)^{\alpha + \frac{3}{2}}}.$$

En effet, en reprenant le raisonnement final du numéro précédent, on voit sans peine que dans ce cas

$$\left| \frac{(-1)^\nu c_\nu}{\nu!} h^{(\nu)}(s) \right| < \frac{K}{(|s - s_0| - k)^a} \, \nu^{\frac{1}{2}} \left(\frac{k}{k + \frac{1}{2}(|s - s_0| - k)} \right)^\nu.$$

Or, lorsque $x \to 1$, on a $\displaystyle\sum_1^\infty \nu^{\frac{1}{2}} x^\nu \sim \frac{\Gamma\left(\frac{3}{2}\right)}{(1-x)^{\frac{3}{2}}}$ et l'on en conclut, d'après (3):

(4)
$$|H(s)| < \frac{M}{(|s - s_0| - k)^{a + \frac{3}{2}}}.$$

L'ordre d'un pôle situé sur $|s - s_0| = k$ ne peut donc dépasser $\alpha + 1$.

Cependant dans le cas déjà indiqué où k peutêtre pris arbitrairement petit, on ne peut même pas en conclure qu'un pôle de $h(s)$ soit aussi un pôle de $H(s)$, car rien ne prouve que (4) soit remplie uniformément dans le voisinage de $k = 0$. Au contraire, on verra dans un instant qu'un tel point est, en général, un point singulier essentiel de $H(s)$.

3. Si, dans le voisinage du point singulier s_0 situé sur la droite de convergence, la fonction $h(s) = \Sigma a_n e^{-\lambda_n s}$ peut être développée en série de LAURENT

(5)
$$h(s) = \sum_{n=0}^\infty \frac{p_n}{n!} (s - s_0)^n + \sum_{n=1}^\infty \frac{(n-1)! \, q_n}{(s - s_0)^n},$$

convergente pour $0 < |s - s_0| < R$, et si l'on prend $k < \frac{1}{2} R$, la couronne circulaire $k < |s - s_0| < R - k$ appartient au domaine $D'(k)$. La série

$$\sum_{\nu=0}^\infty \frac{(-1)^\nu c_\nu}{\nu!} h^{(\nu)}(s) = \sum_{\nu=0}^\infty \frac{(-1)^\nu c_\nu}{\nu!} \left(\sum_{n=0}^\infty \frac{p_{n+\nu}}{n!} (s - s_0)^n + \right.$$

$$\left. + (-1)^\nu \sum_{n=1}^\infty \frac{(n + \nu - 1)! \, q_n}{(s - s_0)^{n+\nu}} \right)$$

y est donc convergente et représente $H(s)$. Dans l'expression (5) de $h(s)$, la première série représente évidemment une fonction holomorphe pour $|s-s_0|<R$; on en conclut donc comme au n:o 1 que la première des deux séries doubles qui viennent d'être écrites, converge absolument pour $|s-s_0|< <R-k$. En supposant $k<|s-s_0|<R-k$, on peut donc transformer la dernière formule en écrivant

$$(6) \qquad H(s)=\sum_{n=0}^{\infty}\frac{(s-s_0)^n}{n!}\sum_{\nu=0}^{\infty}\frac{(-1)^\nu c_\nu p_{n+\nu}}{\nu!}+$$

$$+\sum_{\mu=1}^{\infty}\frac{(\mu-1)!\sum\limits_{\nu=0}^{\mu-1}\frac{c_\nu q_{\mu-\nu}}{\nu!}}{(s-s_0)^\mu},$$

de manière que les coefficients des puissances négatives de $s-s_0$ s'expriment par des sommes finies.

Cette formule, qui me fut indiqué d'abord par M. GERT BONNIER pour le cas spécial considérée dans ma Thèse, donne lieu à une remarque intéressante. En effet, les coefficients des puissances négatives sont, au facteur $(\mu-1)!$ près, les mêmes qu'on obtiendrait en formant le produit des deux séries

$$\sum_0^{\infty}\frac{c_\nu}{\nu!}z^\nu \quad \text{et} \quad \sum_{\mu=1}^{\infty}q_\mu z^\mu,$$

ils ne peuvent donc tous s'annuler s'il y a du moins un des c_ν et un des q_μ différent de zéro. *Donc, si l'on a pris* $k<\frac{1}{2}R$, *la fonction* $H(s)$ *présentera certainement au moins un point singulier à l'intérieur ou sur le contour de* $|s-s_0|=k$, *à moins que* $\varphi(z)$ *ne se réduise à zéro.* Cette remarque va nous être utile dans la suite.

Pour faire une application immédiate de la formule (6), prenons $h(s)=\sum\limits_1^{\infty}\frac{1}{n^s}=\zeta(s)$. En désignant par F et F_1 des fonctions entières, on aura par (6):

$$h(s) = \frac{1}{s-1} + F(s)$$

$$H(s) = \sum_1^\infty \frac{\varphi(\log n)}{n^s} =$$

$$= \sum_{\mu=0}^\infty \frac{c_\mu}{(s-1)^{\mu+1}} + F_1(s),$$

la dernière série étant convergente pour $|s-1| > k$. Dans ce domaine, $H(s)$ n'admet donc d'autre singularité que le point à l'infini, théorème que j'ai déduit d'une manière différente dans ma Thèse. — Toute série ordonnée suivant les puissances négatives de $s-1$ et convergente pour $|s-1| > k$, représente donc une fonction pouvant s'exprimer comme la somme d'une série de Dirichlet ordinaire convergente pour $\sigma > k+1$ et d'une fonction entière.

En choisissant les c_ν d'une manière convenable, on pourrait p. ex. former une série de Dirichlet représentant une fonction admettant $|s-1| = k$ comme coupure essentielle.

4. *Applications à la théorie des fonctions entières.*

Les conséquences qu'on vient de tirer de la formule (6) permettent de démontrer sans peine le théorème suivant: *La suite $\lambda_1, \lambda_2 \ldots$ étant donné, supposons que l'on puisse trouver une série $\Sigma a_n e^{-\lambda_n s}$ telle que la fonction qu'elle représente admette sur la droite de convergence un point singulier s_0, au voisinage duquel elle peut être développée en série de Laurent, convergente pour $0 < |s-s_0| < R$. Cela posé, une fonction entière $\varphi(z)$ telle que $|\varphi(z)| < e^{k|z|}$ où $k < \frac{1}{2}R$, ne peut s'annuler pour $z = \lambda_n$, $n = 1, 2 \ldots$, sans être identiquement nulle.*

En effet, formons avec cette fonction $\varphi(z)$ la fonction

$$H(s) = \Sigma a_n \varphi_n e^{-\lambda_n s}$$

qui se réduit évidemment à zéro. Dans l'hypothèse que $\varphi(z)$ ne soit pas identiquement nulle, on vient de prouver que $H(s)$ à nécessairement une singularité pour $|s-s_0| \leqq k$, voilà donc une contradiction.

En prenant p. ex. $h(s) = \Sigma e^{-ns} = (e^s - 1)^{-1}$ on aura $s_0 = 0$, $R = 2\pi$, d'où la proposition suivante, qui est une forme plus précise d'un théorème énoncé par M. WIGERT:[1] *Une fonction entière* $\varphi(z)$ *dont le module finit par rester inférieur à* $e^{k|z|}$, *où* $k < \pi$, *ne peut s'annuler pour tout nombre entier positif sans s'annuler identiquement.* Chez M. WIGERT, on avait la condition plus restreinte $k < \frac{1}{2}\pi$. On peut remarquer que nous avons trouvé ici la vraie valeur de la constante, ce qu'on trouve en considérant la fonction $\sin \pi z$.

Notre théorème est susceptible d'être généralisé beaucoup en considérant aussi des singularités algébriques et logaritmiques etc., les domaines désignés par D, $D(k)$ et $D'(k)$ dans l'énoncé du théorème fondamental pouvant bien se recouvrir eux-mêmes.

Le même théorème nous permet de démontrer aussi un théorème sur les séries de DIRICHLET. *Si la série*

$$\frac{1}{\lambda_1} + \frac{1}{\lambda_2} + \cdots$$

est convergente, la fonction

$$h(s) = \Sigma a_n e^{-\lambda_n s}$$

n'admet sur la droite de convergence ni des pôles, ni des points singuliers essentiels.

Supposons en effet qu'il n'en soit pas ainsi. Par des théorèmes connus sur le genre des fonctions entières, on peut former une telle fonction admettant tous les λ_n comme zéros, telle que son module reste inférieur à $e^{k|z|}$ pour tout $k > 0$, ce qui serait impossible d'après notre théorème.

En utilisant une remarque faite tout à l'heure, la dernière proposition pourrait aussi être notablement étendue. Il semble que les séries considérées ici ne puissent admettre

[1] »Sur un théorème concernant les fonctions entières», Arkiv för matematik etc., B. 11, N:o 21, 1916.

que des points singuliers d'une espèce très compliquée. Par analogie avec des théorèmes connus sur les séries de puissances, on serait peut-être tenté de croire que la droite de convergence d'une telle série constitue toujours une coupure essentielle, mais il n'en est pas ainsi. On le voit p. ex. en considérant la série $\Sigma(e^{-n^2 s} - e^{-(n^2+e^{-n^3})s})$, qui représente une fonction entière.

Prenons maintenant $h(s) = \Sigma e^{-ns} = (e^s - 1)^{-1}$ et supposons $k < \pi$. La fonction $H(s) = \Sigma \varphi(n) e^{-ns}$ ayant nécessairement un point singulier pour $|s| < k$, on ne peut pas avoir pour tout n

$$|\varphi(n)| < e^{-(k+\varepsilon)n}$$

avec $\varepsilon > 0$, car alors la série $H(s)$ serait convergente pour $\sigma > -(k+\varepsilon)$. Un simple changement de variables suffit pour démontrer:

Si le module de la fonction entière $\varphi(z)$ finit par rester inférieur à $e^{k|z|}$, où $k > 0$, il y a sur chaque vecteur arg. $z =$ const. une infinité de points aux modules indéfiniment croissants, pour lesquels

$$|\varphi(z)| > e^{-(k+\varepsilon)|z|},$$

et cela a lieu pour tout $\varepsilon > 0$.

Pour finir, remarquons qu'en se servant d'un théorème de M. Fekete,[1] on est immédiatement amené au résultat suivant: *Si $\Sigma a_n e^{-\lambda_n s}$ représente une fonction entière sans être partout convergente, et si $F(z)$ est une fonction entière telle que les points d'affixes $a_n F(\lambda_n)$ sont tous situés à l'intérieur d'un angle d'ouverture $< \pi$ excluant l'origine, $F(z)$ ne peut satisfaire pour toute valeur de z à une relation de la forme*

$$|F(z)| < e^{K|z|}.$$

5. Dans ma Thèse déjà cité, j'ai montré qu'un théorème de M. Lindelöf[2] sur les séries de puissances peut se déduire d'une manière simple à l'aide du théorème fondamental dont on s'est occupé ici.

[1] »Sur les séries de Dirichlet», Comptes rendus, T. 150.
[2] »Le calcul des résidus», p. 136.

Par la forme généralisée obtenue dans cette note, on peut encore généraliser ces résultats, en démontrant p. ex. la proposition suivante. *Supposons connue l'étoile principale de la fonction* $f(x) = f(re^{iv}) = \Sigma a_n x^n$. *Si* $x_0 = r_0 e^{iv_0}$ *est un point quelconque n'appartenant pas à cette étoile, on exclura du plan tout point situé à l'intérieur de la courbe*

$$r = r_0 e^{\pm\sqrt{k^2 - (v - v_0)^2}}.$$

Dans l'étoile restante, la fonction $F(x) = \Sigma a_n \varphi(n) x^n$ *est holomorphe.*

La démonstration de cette proposition repose entièrement sur l'étude des domaines D et $D(k)$; elle est tout à fait analogue à ma démonstration précédente.

Resumé.

Föreliggande arbete utgör en utvidgning af en sats om Dirichlet'ska serier, gifven i förf:s doktorsafhandling våren 1917. Satsens innebörd är i korthet följande:

Om en funktion $h(s)$, framställd genom en Dirichlet's serie $\Sigma a_n e^{-\lambda_n s}$, antages känd utöfver det område, där serien konvergerar, så är det möjligt att utföra en liknande fortsättning af funktionen $H(s) = \Sigma a_n \varphi(\lambda_n) e^{-\lambda_n s}$ utöfver det område, där man kan vara säker om den senare seriens konvergens.

φ betecknar här en hel funktion, underkastad ett visst inskränkande villkor med afseende på sin storleksordning.

De områden, där funktionerna $h(s)$ och $H(s)$ på detta sätt blifva kända, stå i ett enkelt geometriskt sammanhang med hvarandra.

Af den i n:o 1 bevisade hufvudsatsen göras sedan några tillämpningar, bl. a. på teorien för hela funktioner.

Följande båda satser, hvilka ej torde vara alldeles utan intresse, må här anföras:

1. Om absoluta beloppet för en hel funktion $\varphi(z)$ växer mindre hastigt än $e^{k|z|}$, där $k < \pi$, så kan $\varphi(z)$ ej ha nollställen i *alla hela positiva tal* utan att bli identiskt noll.

(Detta är en skarpare form af en sats af S. Wigert, publicerad i Arkiv för matematik etc., B. 11.)

2. Om absoluta beloppet för en hel funktion $\varphi(z)$ växer mindre hastigt än $e^{k|z|}$, så finns det på hvarje vektor oändligt många punkter (med hur stora absoluta belopp som helst), i hvilka

$$|\varphi(z)| > e^{-(k+\varepsilon)|z|}$$

och detta gäller för hvarje $\varepsilon > 0$.

Tryckt den 16 oktober 1918.

Uppsala 1918. Almqvist & Wiksells Boktryckeri-A.-B.

4.

Über die Herleitung der Riemannschen Primzahlformel

Ark. Mat. Astr. Fys. **13** (24), 1–7 (1918)

Mitgeteilt am 24. April 1918 durch IVAR BENDIXSON und H. VON KOCH.

---

1. Seit dem ersten, im Jahre 1895 durch v. MANGOLDT gegebenen, strengen Beweis der berühmten RIEMANN'schen Primzahlformel ist es üblich,[1] bei der Herleitung dieser Formel von dem Studium der Funktion

$$F(x, r) = \sum_{p^m < x} \frac{\log p}{p^{mr}} \qquad (x \neq p^m)$$

(die Summe ist über alle Primzahlpotenzen $p^m < x$ zu erstrecken) auszugehen, um dann durch Integration nach dem Parameter r die RIEMANN'sche Formel für $f(x)$ abzuleiten.

Ich werde nun im folgenden zeigen, dass bei dieser Art der Herleitung die Einführung des Parameters r fast überflüssig ist. Die RIEMANN'sche Formel — zwar ohne Wertbestimmung der darin auftretenden additiven Konstanten — wird sich nämlich direkt aus der v. MANGOLDT'schen Formel für $F(x, 0)$ ableiten lassen.

Unterwegs wird sich auch das folgende neue Resultat ergeben: *Die Reihe* $\sum \frac{x^\varrho}{\varrho}$, *wo die Summe in unten näher zu*

---

[1] Vgl. z. B. LANDAU, Handbuch usw., S. 333.

beschreibenden Weise über die komplexen Nullstellen ϱ *der Zetafunktion zu nehmen ist, zeigt in der Umgebung einer beliebigen Primzahlpotenz* $x = p^m$ *die Gibbs'sche Erscheinung in genau entsprechender Weise wie die Reihe* $\sum_1^\infty \frac{\sin nt}{n}$ *in der Umgebung von* $t = 0$. — *Ähnliches gilt für die Reihe* $\Sigma Li(x^\varrho)$.

2. Wenn im folgenden das Zeichen $\sum_\varrho$ vorkommt, soll immer darunter verstanden werden, dass die Summe über alle komplexen Wurzeln ϱ der Gleichung $\zeta(s) = 0$ so zu nehmen ist, dass immer zwei konjugierte ϱ zusammengefasst, und die Paare nach absolut wachsenden Ordinaten geordnet werden. Jedes ϱ wird mit seiner Multiplizität berücksichtigt. Unter einer Teilsumme einer solchen Reihe verstehen wir, wenn $\varrho = \beta + i\gamma$ gesetzt wird, eine Summe der Form $\sum_{|\gamma| < T}$, wo $T > 0$ ist. Es gilt dann der

Hilfssatz I. *In der Umgebung der Primzahlpotenz* $x = p^m$, *der ein Punkt ungleichmässiger Konvergenz für die Reihe*

$$\sum_\varrho \frac{x^\varrho}{\varrho}$$

ist, schwanken die Teilsummen dieser Reihe zwischen den Unbestimmtheitsgrenzen (für unendlich wachsendes T*):*

$$S \pm \frac{\log p}{\pi} \int_0^\pi \frac{\sin t}{t} dt = S \pm 0{,}59 \log p, \qquad \text{(annäherungsweise)}$$

wo S *die Summe der Reihe für* $x = p^m$ *ist. Da der Sprung der Reihe bei dem Übergang von* p^m *zu* $p^m \pm 0$ *nur* $\frac{1}{2} \log p$ *beträgt, zeigt sie also die Erscheinung von Gibbs.*

Aus dem Texte der Seiten 364—368 von LANDAU's grossem Primzahl-Handbuch geht nämlich ohne Schwierigkeit hervor, dass

$$\sum_{|\gamma|<T} \frac{x^\varrho}{\varrho} = -\frac{\log p}{2\pi i} \int_{a-Ti}^{a+Ti} \frac{\left(\frac{x}{p^m}\right)^s}{s} ds + \varphi(T, x)$$

ist, wo $a > 1$, $T > 0$ und $\varphi(T, x)$ eine stetige Funktion ist, die für $T \to \infty$ gleichmässig im Intervalle $p^m - \frac{1}{2} < x < p^m + \frac{1}{2}$ gegen eine stetige Funktion von x konvergiert. — In den folgenden Formeln soll $\varphi(T, x)$ immer diese Bedeutung beibehalten, ohne notwendig dieselbe Funktion zu bezeichnen. — Es gilt

$$\int_{a-Ti}^{a+Ti} \frac{\left(\frac{x}{p^m}\right)^s}{s} ds = i \left(\frac{x}{p^m}\right)^a \int_0^T \left(\frac{e^{it \log \frac{x}{p^m}}}{a + it} + \frac{e^{-it \log \frac{x}{p^m}}}{a - it} \right) dt =$$

$$= 2ai \left(\frac{x}{p^m}\right)^a \int_0^T \frac{\cos\left(t \log \frac{x}{p^m}\right)}{a^2 + t^2} dt +$$

$$+ 2i \left(\frac{x}{p^m}\right)^a \int_0^T \frac{t \sin\left(t \log \frac{x}{p^m}\right)}{a^2 + t^2} dt,$$

woraus man, da

$$\frac{t}{a^2 + t^2} = \frac{1}{t} - \frac{a^2}{t(a^2 + t^2)}$$

ist, leicht folgert, dass

$$\sum_{|\gamma|<T} \frac{x^\varrho}{\varrho} = -\frac{\log p}{\pi} \left(\frac{x}{p^m}\right)^a \int_0^{T \log \frac{x}{p^m}} \frac{\sin t}{t} dt + \varphi(T, x)$$

gilt.

Da für $x = p^m$ der Ausdruck $\left(\frac{x}{p^m}\right)^a - 1$ verschwindet, kann man auch schreiben

$$\sum_{|\gamma| < T} \frac{x^\varrho}{\varrho} = -\frac{\log p}{\pi} \int_0^{T \log \frac{x}{p^m}} \frac{\sin t}{t} dt + \varphi(T, x).$$

In einigen neueren Untersuchungen[1] über die GIBBS'sche Erscheinung bei den FOURIER'schen Reihen wird aber gezeigt, dass diese im wesentlichen eben auf dem Verhalten des Integrals $\int \frac{\sin t}{t} dt$ beruht. Wie nahe z. B. oberhalb p^m man x auch nimmt, so lässt sich doch immer ein T so angeben, dass $T \log \frac{x}{p^m} = \pi$ wird; da $\int_0^z \frac{\sin t}{t} dt$ seinen grössten Wert für $z = \pi$ annimmt, werden offenbar die Unbestimmtheitsgrenzen der Teilsummen, wenn $T \to \infty$ und $x \to p^m$, gleich

$$\sum_\varrho \frac{p^{m\varrho}}{\varrho} \pm \frac{\log p}{\pi} \int_0^\pi \frac{\sin t}{t} dt,$$

was zu beweisen war.

Hilfssatz II.[2] *Die Reihe* $\sum_\varrho Li(x^\varrho)$ *konvergiert für jeden positiven, von 1 verschiedenen, Wert von* x.

Es gilt, wenn $\varrho = \beta + i\gamma$,

$$Li(x^\varrho) = \int_{-\infty + i\gamma \log x}^{\varrho \log x} \frac{e^z}{z} dz \pm \pi i$$

je nachdem $\gamma \log x \gtrless 0$ ist, wobei man geradlinig zu inte-

---

[1] Vgl. WEBER, Math. Ann. B. 73, S. 286; CRAMÉR, Arkiv för matematik, B. 13, N:o 20, 1918.
[2] Dieser Satz ist selbstverständlich nicht neu.

grieren hat. Bei unserer Art der Summation aber verschwinden die $\pm \pi i$, so dass

$$\sum_{|\gamma|<T} Li(x^\varrho) = \sum_{|\gamma|<T} \int_{-\infty}^{\beta \log x} \frac{e^{u+i\gamma \log x}}{u + i\gamma \log x} du.$$

Durch Anwendung der Identität

$$\frac{1}{u+i\gamma \log x} = \frac{1}{\varrho \log x} - \frac{u-\beta \log x}{\varrho \log x (u+i\gamma \log x)}$$

bekommt man aber

$$\sum_{|\gamma|<T} Li(x^\varrho) = \frac{1}{\log x} \sum_{|\gamma|<T} \frac{x^\varrho}{\varrho} - \frac{1}{\log x} \sum_{|\gamma|<T} \frac{x^{i\gamma}}{\varrho} \int_{-\infty}^{\beta \log x} \frac{u-\beta \log x}{u+i\gamma \log x} e^u du$$

und wenn man hier T ins Unendliche wachsen lässt, so ist die erste Reihe rechts für jedes positive x konvergent, die zweite für jedes von 1 verschiedene $x > 0$ sogar absolut konvergent. — Die letzte Formel zeigt auch, dass die Reihe $\Sigma Li(x^\varrho)$ bezüglich der GIBBS'schen Erscheinung ein entsprechendes Verhalten zeigt wie $\sum \frac{x^\varrho}{\varrho}$.

3. Wir setzen wie gewöhnlich

$$\psi(x) = \sum_{p^m<x} \log p$$

$$f(x) = \sum_{p^m<x} \frac{1}{m},$$

wenn x nicht einer Primzahlpotenz gleich ist; wenn aber $x = p^m$ ist, soll immer

$$\psi(x) = \frac{1}{2}[\psi(x+0) + \psi(x-0)]$$

$$f(x) = \frac{1}{2}[f(x+0) + f(x-0)]$$

sein.

Zwischen diesen Funktionen besteht die durch direkte Anwendung der Definitionen unmittelbar zu beweisende Identität

$$f(x) = \frac{\psi(x)}{\log x} + \int_2^x \frac{\psi(u)}{u(\log u)^2} du.$$

Unter Anwendung dieser Identität werden wir jetzt aus der v. MANGOLDT'schen Formel für $\psi(x)$:

$$\psi(x) = x - \sum_\varrho \frac{x^\varrho}{\varrho} - \frac{1}{2} \log\left(1 - \frac{1}{x^2}\right) - \log 2\pi$$

eine Formel für $f(x)$ ableiten.

Indem man diesen Ausdruck in unsere Identität einsetzt, findet man nach einigen leichten Umformungen durch partielle Integration:

$$f(x) = \int_2^x \frac{du}{\log u} - \left[\frac{1}{\log x} \sum_\varrho \frac{x^\varrho}{\varrho} + \int_2^x \frac{du}{u(\log u)^2} \sum_\varrho \frac{u^\varrho}{\varrho}\right] +$$
$$+ \int_x^\infty \frac{du}{u(u^2-1)\log u} + K,$$

wo K (sowie im folgenden K_1, K_2 ...) eine Konstante ist.

Die Reihe $\sum_\varrho \frac{u^\varrho}{\varrho}$ ist bekanntlich[1] im Integrationsintervalle $(2, x)$ mit Ausnahme der Umgebungen der Primzahlpotenzen gleichmässig konvergent; da aber nach Hilfssatz I in der Umgebung jedes solchen Punktes die Teilsummen der Reihe gleichmässig beschränkt sind, so ist die gliedweise Integration von 2 bis x erlaubt.[2] Man hat also

$$f(x) = \int_2^x \frac{du}{\log u} - \sum_\varrho \int_2^x \frac{u^{\varrho-1}}{\log u} du + \int_x^\infty \frac{du}{u(u^2-1)\log u} + K_1.$$

[1] LANDAU, l. c. S. 364.
[2] Vgl. HOBSON, Theory of functions of a real variable, S. 540.

Nun aber ist

$$\int_2^x \frac{du}{\log u} = Li(x) - Li(2),$$

und durch eine leichte Anwendung des CAUCHY'schen Satzes findet man

$$\int_2^x \frac{u^{\varrho-1}}{\log u} du = \int_{\varrho \log 2}^{\varrho \log x} \frac{e^z}{z} dz = Li(x^\varrho) - Li(2^\varrho).$$

Also, da nach Hilfssatz II die Reihe $\sum_\varrho Li(x^\varrho)$ konvergiert:

$$f(x) = Li(x) - \sum_\varrho Li(x^\varrho) + \int_x^\infty \frac{du}{u(u^2-1)\log u} + K_2,$$

d. h. die RIEMANN'sche Primzahlformel, bis auf die Wertbestimmung $K_2 = -\log 2$.

Tryckt den 25 september 1918.

Uppsala 1918. Almqvist & Wiksells Boktryckeri-A.-B.

5.

Über die Nullstellen der Zetafunktion

Math. Z. **2** (3/4), 237–241 (1918)

I. Die von Riemann angegebene, erst 46 Jahre später durch von Mangoldt streng begründete Formel

$$N(T) = \frac{T}{2\pi}\left(\log\frac{T}{2\pi} - 1\right) + R(T), \tag{1}$$

wo in üblicher Bezeichnungsweise $R(T) = O(\log T)$ ist, ist bekanntlich in sehr eleganter Weise von Backlund[1]) wieder abgeleitet worden. — $N(T)$ bezeichnet hier die Anzahl der Wurzeln von $\zeta(s) = 0$, deren Ordinaten der Beziehung $0 < t \leqq T$ genügen.

Unter Voraussetzung der Richtigkeit der Riemannschen Hypothese, daß alle diese Wurzeln den reellen Teil $\frac{1}{2}$ haben, ist es nun Bohr[2]) vor einigen Jahren gelungen zu beweisen, daß sogar $R(T) = o(\log T)$ ist; ein Resultat, dem für unsere Vorstellung von der asymptotischen Verteilung der Nullstellen wegen der daraus entspringenden Formel

$$\lim_{T\to\infty} \frac{N(T+1) - N(T)}{\log T} = \frac{1}{2\pi}$$

eine gewisse Bedeutung zukommt. — Durch eine geeignete Abänderung der Backlundschen Beweismethode werde ich nun im folgenden dieses Resultat und sogar etwas mehr in einfacher Weise ableiten, indem ich nämlich beweisen werde: *Die Abschätzung $R(T) = o(\log T)$ bleibt richtig, wenn nur die Wahrheit der sogenannten Lindelöfschen Hypothese vorausgesetzt wird.* — Es besagt diese Hypothese, daß $\mu(\sigma) = 0$ für $\sigma > \frac{1}{2}$

---

[1]) „Über die Nullstellen der Zetafunktion", Dissertation, Helsingfors 1916 und „Sur les zéros de la fonction $\zeta(s)$ de Riemann", Comptes rendus, Bd. 158 (1914), S. 1979.

[2]) Bohr, Landau, Littlewood, „Sur la fonction $\zeta(s)$ dans le voisinage de la droite $\sigma = \frac{1}{2}$", Bulletin de l'Academie Royale de Belgique, Bd. 15 (1913), S. 1144 bis 1175.

ist, oder mit anderen Worten, daß gleichmäßig für $\sigma > \frac{1}{2}$, $t > 1$, $\zeta(s) = O(t^{\varepsilon})$ ist. Dieses Resultat enthält in der Tat dasjenige von Bohr, da nach dem bekannten Satze von Littlewood[3]) jene Vermutung eine Folge der Riemannschen ist. Der Beweis von Bohr macht in erschöpfender Weise von dem Nichtverschwinden von $\zeta(s)$ für $\sigma > \frac{1}{2}$ Gebrauch.

Es ist schon lange bekannt, daß in (1), T wurzelfrei vorausgesetzt,

$$R(T) = \frac{1}{\pi} \arg \zeta(\tfrac{1}{2} + iT) + O(1),$$

wenn man z. B. vom Punkte $s = 2$ mit dem Argumente 0 von $\zeta(s)$ ausgeht und zunächst geradlinig nach $s = 2 + iT$ und von hier aus geradlinig weiter nach $\frac{1}{2} + iT$ fortschreitet. Der Kern der Backlundschen Methode besteht nun in der Anwendung der leicht zu erkennenden Tatsache

$$\left|\frac{1}{\pi} \arg \zeta(\tfrac{1}{2} + iT)\right| < l + \tfrac{1}{2},$$

wo l die Anzahl der innerhalb des Kreises $|s - 2| = \frac{3}{2}$ gelegenen Nullstellen von

$$f(s) = \tfrac{1}{2}[\zeta(s + iT) + \zeta(s - iT)]$$

ist. Es sei nun ε eine beliebig kleine positive Größe, und es werde die Lindelöfsche Vermutung als wahr angenommen. Alsdann gibt es bekanntlich ein t_0, so daß für $t > t_0$

$$|\zeta(s)| < |t|^{\varepsilon^2} \quad \text{für } \sigma \geqq \tfrac{1}{2},$$

$$|\zeta(s)| < |t|^{\frac{1}{2} - \sigma + \varepsilon^2} \quad \text{für } 0 < \sigma \leqq \tfrac{1}{2}$$

gilt. Ich wende nun den Jensenschen Satz auf die Funktion $f(s)$ und auf den Kreis $|s - 2| = r = \frac{3}{2}(1 + \varepsilon)$ an. Hierbei wird $\varepsilon < \frac{1}{3}$ und $T > t_0 + 2$ vorausgesetzt. Indem ich nur die l Nullstellen innerhalb $|s - 2| = \frac{3}{2}$ beibehalte, erhalte ich

$$l \cdot \log(1 + \varepsilon) < \frac{1}{2\pi} \int_0^{2\pi} \log|f(2 + re^{i\varphi})|\, d\varphi - \log|f(2)|.$$

Da aber $|f(2)| = R[\zeta(2 + iT)] > \frac{1}{4}$ ist, so ergibt sich hieraus für ein hinreichend kleines ε

$$(2) \qquad l < \frac{1}{2\pi \log(1 + \varepsilon)} \int_0^{2\pi} \log|f(2 + re^{i\varphi})|\, d\varphi + \frac{2}{\varepsilon}.$$

Um nun das Integral abzuschätzen, bemerken wir, daß der aus der Halbebene $\sigma > \frac{1}{2}$ herausragende Bogen des Kreises $|s - 2| = \frac{3}{2}(1 + \varepsilon)$

[3]) Comptes rendus, Bd. 154 (1912), S. 263–266.

mit ε unendlich klein wird, und zwar von der Ordnung $\sqrt{\varepsilon}$. Für die auf diesen Bogen bezüglichen Werte von φ ist aber

$$\log|f(2+re^{i\varphi})| < \log(T+2)^{\frac{1}{2}-\sigma+\varepsilon^2} \leqq \varepsilon(\tfrac{3}{2}+\varepsilon)\log(T+2).$$

Daher ist

$$l < \frac{1}{\varepsilon}[M\sqrt{\varepsilon}\cdot\varepsilon(\tfrac{3}{2}+\varepsilon)+2\pi\varepsilon^2]\log(T+2)+\frac{2}{\varepsilon},$$

was für alle hinreichend großen Werte von T kleiner als $M_1\sqrt{\varepsilon}\cdot\log T$ ist, unter M und M_1 absolute Konstanten verstanden. Der Beweis des Satzes ist hiermit für jedes wurzelfreie T geliefert; für stetig wachsendes T kann er nunmehr unmittelbar gefolgert werden.

II. Es wird sich jetzt ergeben, daß unter Voraussetzung der Riemannschen Hypothese eine noch bessere Abschätzung von $R(T)$ erhalten werden kann.

Wenn diese Hypothese wahr ist, so ist nämlich[4])

$$R(T) = O\left[\log T\left(\frac{\log_3 T}{\log_2 T}\right)^{\frac{1}{2}}\right].$$

In der oben erwähnten Abhandlung gibt Littlewood ohne vollständige Angabe seines Beweises, der aber nicht schwer zu ergänzen ist, die Abschätzung:

$$\log\zeta(\sigma+it) = O\left[\left(\frac{\log t\log_2 t}{\log_3 t}\right)^{2(1-\sigma)}\log_3 t\right]$$

gleichmäßig für

$$\tfrac{1}{2}+\frac{\delta}{\log_2 t}\leqq\sigma\leqq 1,\quad t>0.$$

Unter Berücksichtigung bekannter Sätze schließt man hieraus: es ist für alle $\sigma \geqq \frac{1}{2}+2\frac{\log_3 t}{\log_2 t}$, $t>0$ gleichmäßig

$$\log\zeta(\sigma+it) = O\left[\left(\frac{\log t\log_2 t}{\log_3 t}\right)^{1-4\frac{\log_3 t}{\log_2 t}}\cdot\log_3 t\right] = \tag{3}$$

$$= O\left[\log t\cdot\log_2 t\cdot e^{-\frac{4\log_3 t}{\log_2 t}[\log_2 t+\log_3 t-\log_4 t]}\right] = O\left[\frac{\log t}{(\log_2 t)^3}\right].$$

Speziell auf der Kurve $\sigma=\frac{1}{2}+2\frac{\log_3 t}{\log_2 t}$ ist also für $t>t_0$

$$|\zeta(s)| < e^{K\frac{\log t}{(\log_2 t)^2}},\tag{4}$$

also nach der Funktionalgleichung auf der Kurve $\sigma=\frac{1}{2}-2\frac{\log_3 t}{\log_2 t}$, $t>t_0$

$$|\zeta(s)| < K_1\cdot t^{\frac{1}{2}-\sigma}e^{K\frac{\log t}{(\log_2 t)^2}} < e^{K_2\frac{\log t\cdot\log_3 t}{\log_2 t}}.\tag{5}$$

---

[4]) Ich setze $\log_2 t = \log\log t$, $\log_3 t = \log\log_2 t$.

Betrachten wir nun die für große t_0 im Gebiete

$$(6) \qquad \tfrac{1}{2} - 2\,\frac{\log_3 t}{\log_2 t} \leqq \sigma \leqq \tfrac{1}{2} + 2\,\frac{\log_3 t}{\log_2 t}, \quad t > t_0$$

reguläre Funktion

$$\varphi(s) = \frac{\zeta(s)}{e^{K_2 \frac{\log(-is)\cdot\log_3(-is)}{\log_2(-is)}}},$$

wo bei der Definition der vieldeutigen Funktionen jedem Logarithmus sein Hauptwert zu erteilen ist. Man sieht nun ohne Schwierigkeit, daß für $t \to \infty$

$$\frac{\log(-is)\cdot\log_3(-is)}{\log_2(-is)} = \frac{\log t\cdot\log_3 t}{\log_2 t} + O\left(\frac{1}{t}\right),$$

gilt, und zwar in bezug auf σ in jedem endlichen Intervall gleichmäßig.

Auf dem Rande des Gebietes (6) ist also nach (4) und (5) die Funktion $\varphi(s)$ beschränkt, und da sie im Innern des Gebietes jedenfalls nicht schneller als eine endliche Potenz von t wächst, so ist sie nach einem bekannten Phragmén-Lindelöfschen Satze auch im Innern beschränkt[5]).

Es ist also im Gebiete (6)

$$(7) \qquad \log|\zeta(s)| < K_3\,\frac{\log t\cdot\log_3 t}{\log_2 t}.$$

Wir setzen nun in (2)

$$\varepsilon = \frac{\log_3 T}{\log_2 T}.$$

Da $f(s) = \frac{1}{2}[\zeta(s+iT) + \zeta(s-iT)]$ ist und da über den Kreis $|s-2| = \frac{3}{2}(1+\varepsilon)$ integriert wird, so kommen bei der Integration für große Werte von T nur solche Werte des Argumentes in $\zeta(s+iT)$ in Betracht, die im Gebiete (6) oder rechts davon liegen. Für $\zeta(s-iT)$ gilt natürlich Entsprechendes in der unteren Halbebene.

Zuerst wird derjenige Teil des Integrals in (2) abgeschätzt, in dem entweder bei $\zeta(s+iT)$ oder bei $\zeta(s-iT)$ ein Argumentwert auftritt, der im Gebiete (6) oder in dem hierzu symmetrischen Gebiete der unteren Halbebene liegt. Der diesbezügliche Teil des Kreisbogens wird wieder schließlich unendlich klein, und zwar von der Ordnung $\left(\frac{\log_3 T}{\log_2 T}\right)^{\frac{1}{2}}$; für $f(s)$ gilt aber hier nach (7)

$$\log|f| = O\left(\frac{\log T\cdot\log_3 T}{\log_2 T}\right).$$

Für den ganzen übrigbleibenden Teil des Kreises gilt nach (3)

$$\log|f| = O\left[\frac{\log T}{(\log_2 T)^3}\right],$$

---

[5]) Vgl. Landau, Handbuch, II, S. 849.

so daß schließlich

$$l < \frac{1}{2\pi \log(1+\varepsilon)} \int_0^{2\pi} \log |f(2+re^{i\varphi})|\, d\varphi + \frac{2}{\varepsilon} =$$

$$= O\left\{\frac{\log_2 T}{\log_3 T}\left[\left(\frac{\log_3 T}{\log_2 T}\right)^{\frac{1}{2}} \cdot \frac{\log T \cdot \log_3 T}{\log_2 T} + \frac{\log T}{(\log_2 T)^3} + 1\right]\right\} = O\left[\log T \left(\frac{\log_3 T}{\log_2 T}\right)^{\frac{1}{2}}\right]$$

gilt. Hieraus kann nunmehr ähnlich wie oben die zu beweisende Abschätzung für $R(T)$ gefolgert werden.

Stockholm, den 23. Februar 1918.

(Eingegangen am 21. März 1918.)

6.

Studien über die Nullstellen der Riemannschen Zetafunktion

Math. Z. 4 (1/2), 104–130 (1919)

Einleitung.

Die über alle nicht reellen Nullstellen $\varrho = \beta + \gamma i$ der Riemannschen Zetafunktion erstreckte Summe

$$\sum_{\varrho} \frac{x^{\varrho}}{\varrho}, \qquad (x > 0) \tag{1}$$

wo die ϱ nach wachsendem $|\gamma|$ geordnet sind, spielt bekanntlich in der analytischen Primzahltheorie eine hervorragende Rolle. Diese Reihe stellt nämlich eine Funktion von x dar, welche in den Punkten $x = 1$, p^m, $\frac{1}{p^m}$ unstetig, sonst aber für jedes positive x stetig ist. — Unter p^m verstehen wir hier, wie überall in dieser Arbeit, eine beliebige ganzzahlige Potenz der beliebigen Primzahl p.

Nachdem diese Tatsachen — sowie die Konvergenz von (1) für alle $x > 0$ — zuerst von Herrn von Mangoldt[1]) bewiesen waren, wurde die Untersuchung durch Herrn Landau[2]) auf die allgemeinere[3]) Reihe

$$\sum_{\gamma > 0} \frac{x^{\varrho}}{\varrho^k} \qquad (0 < k \leqq 1) \tag{2}$$

ausgedehnt. Hierbei bedeutet ϱ^k den Wert der für $s > 0$ positiven analytischen Funktion s^k bei Fortsetzung in der von 0 bis $-\infty$ aufgeschnittenen Ebene, und die Reihe ist wieder nach wachsendem $|\gamma|$ geordnet; auf die Reihenfolge der Glieder mit gleichem $|\gamma|$ kommt es bei Herrn Landau, wie auch in dieser Arbeit, nicht an.

---

[1]) H. v. Mangoldt, Journ. f. Mathematik, Bd. 114 (1895), S. 255–305.

[2]) E. Landau, Math. Annalen, Bd. 71 (1912), S. 548–564.

[3]) Der Ausdruck (1) stimmt nämlich, wenn immer zwei konjugierte ϱ zusammengefaßt werden, mit der doppelten Summe der reellen Teile von (2) für $k = 1$ überein.

Über die Reihe (2) hat nun Herr Landau folgendes bewiesen:

a) *Konvergenz für* $x \neq 1,\ p^m,\ \frac{1}{p^m}$;

b) *gleichmäßige Konvergenz in jedem von diesen Punkten freien Intervall*;

c) *Divergenz*[4]) *für* $x = 1,\ p^m,\ \frac{1}{p^m}$.

Diese Tatsachen deuten auf einen arithmetischen Zusammenhang zwischen den Nullstellen ϱ und den Primzahlen p hin; Herr Landau erklärt aber, er habe keine Ahnung, worin derselbe besteht.

Es liegt nun sehr nahe, und mag vielleicht auch nicht ganz uninteressant erscheinen, die Reihe (2) auch für komplexe Werte der Veränderlichen x zu untersuchen. Dadurch wird man zwar nicht das Wesen des obengenannten Zusammenhanges erkennen, wohl aber wird man diesen Zusammenhang von einem neuen Gesichtspunkte aus erblicken. Es ist zweckmäßig, zuerst eine neue Veränderliche z durch die Gleichung

$$x = e^z$$

einzuführen und also die Reihe

$$\sum_{\gamma > 0} \varrho^{-k} e^{\varrho z}$$

zu studieren.

Wird für einen Augenblick angenommen, daß die sogenannte Riemannsche Vermutung wahr ist, d. h. daß alle ϱ den reellen Teil $\frac{1}{2}$ haben, so ergibt sich für $k = 1$ nach Multiplikation mit $e^{-\frac{1}{2}z}$ die Reihe

$$\sum_{\gamma > 0} \frac{e^{\gamma i z}}{\varrho} = \sum_{\gamma > 0} \frac{\cos \gamma z}{\varrho} + i \sum_{\gamma > 0} \frac{\sin \gamma z}{\varrho}. \tag{3}$$

Aus den Untersuchungen des Herrn Landau geht nun hervor, daß die erste Reihe rechts in (3) in den kritischen Punkten $z = \pm \log p^m$ divergiert, die zweite dagegen überall konvergiert, in der Umgebung der kritischen Punkte aber nicht gleichmäßig. Dies erinnert an das Verhalten der Reihe

$$\sum_{n=1}^{\infty} \frac{e^{niz}}{n} = \sum_{n=1}^{\infty} \frac{\cos nz}{n} + i \sum_{n=1}^{\infty} \frac{\sin nz}{n}$$

in den Punkten $z = \pm 2\nu\pi$, wo genau dieselben Konvergenzeigenschaften auftreten. Wenn aber in dieser Reihe die Veränderliche auch komplexe Werte annehmen darf, so wird durch sie eine analytische Funktion von z definiert, die in den obigen Punkten logarithmische Singularitäten hat.

[4]) Und selbst mehr als dies, nämlich Nichtbeschränktheit und im Falle $0 < k < 1$ sogar Nichtbeschränktheit der Summe der reellen Teile.

Man könnte also — jetzt wieder ohne die Riemannsche Vermutung — glauben, daß die Reihe

$$\sum_{\gamma>0} \frac{e^{\varrho z}}{\varrho},$$

welche für $\Im(z) > 0$ unbedingt konvergiert, eine in den Punkten $z = \pm \log p^m$ logarithmisch singuläre Funktion von z darstellt. Die Ableitung dieser Funktion

(4) $$V(z) = \sum_{\gamma>0} e^{\varrho z}$$

würde dann in jenen Punkten Pole erster Ordnung besitzen.

Dies trifft nun in der Tat zu. In dem ersten Kapitel dieser Arbeit wird nämlich die Funktion $V(z)$ für beliebige komplexe Werte von z dargestellt, und es werden über sie u. a. die folgenden Sätze abgeleitet:

1. *Wenn längs der negativen imaginären Achse ein Schnitt gemacht wird, so ist die Funktion $V(z)$ im Innern der ganzen aufgeschnittenen z-Ebene meromorph und hat nur die singulären Stellen $z = \pm \log p^m$, welche Pole erster Ordnung sind.*

2. *Die Funktion*

$$2\pi i V(z) - \frac{\log z}{1-e^{-z}} - \frac{C + \log 2\pi - \frac{1}{2}\pi i}{z}$$

ist in der ganzen Ebene eindeutig und — wenn in der Konvergenzhalbebene von (4), $\Im(z) > 0$, $0 < \arg z < \pi$ gesetzt wird — im Ursprung regulär.

Das zweite Kapitel bringt einige Anwendungen der obigen Untersuchung. Zunächst werden im ersten Abschnitt die schon oben angeführten Sätze von Landau über die Konvergenzverhältnisse der Reihe $\sum_{\gamma>0} \frac{x^\varrho}{\varrho^k}$ wieder abgeleitet. Wollte man die Riemannsche Vermutung als wahr annehmen, so würde es ein allgemeiner Konvergenzsatz von M. Riesz[5]) gestatten, unmittelbar von unseren obigen Sätzen über $V(z)$ auf die Landauschen Sätze zu schließen. Will man jedoch den Beweis ohne einschränkende Annahmen führen, so wird es notwendig, den Satz von M. Riesz etwas zu modifizieren. Wenn man den veränderten Satz nur für den uns interessierenden speziellen Fall der Reihe $\sum_{\gamma>0} \varrho^{-k} e^{\varrho z}$ ausspricht, so gestaltet sich sein Beweis sehr kurz und wird deshalb hier vollständig ausgeführt. Man wird erkennen, daß der Kern des Rieszschen Beweises von den unbedeutenden Änderungen vollständig unberührt geblieben ist.

[5]) M. Riesz, Acta Mathematica, Bd. 40 (1916), S. 349–361.

Wenn ich nicht irre, so lassen alle bisher bekannten, exakten (d. h. nicht asymptotischen) Formeln, welche die Nullstellen ϱ enthalten, auch die Primzahlen explizit auftreten. Es wird daher vielleicht nicht ohne Interesse sein, wenn im zweiten Abschnitt (unter Voraussetzung der Richtigkeit der Riemannschen Vermutung) eine Formel abgeleitet wird, bei welcher dies nicht der Fall ist, nämlich die Formel

$$\lim_{y \to 0} \sum_{\gamma > 0} e^{-\gamma y} \cos \gamma x = \frac{e^{\frac{3}{2}x} - e^{-\frac{3}{2}x} \mp e^{\mp\frac{1}{2}x}}{2(e^x - e^{-x})},$$

wo die oberen oder unteren Zeichen gelten, je nachdem x positiv oder negativ ist. Diese Formel ist indessen nur gültig, wenn x von 0 und von den $\pm \log p^m$ verschieden ist.

Schließlich wird im dritten Abschnitt eine Untersuchung über das Restglied $R(T)$ der Formel

$$N(T) = \frac{T}{2\pi}\left(\log \frac{T}{2\pi} - 1\right) + R(T)$$

angestellt. Es bedeutet hier $N(T)$ die Anzahl der ϱ, welche der Bedingung $0 < \gamma \leqq T$ genügen. Dann gilt bekanntlich[6])

$$R(T) = O(\log T)$$

und sogar im Falle der Richtigkeit der Riemannschen Vermutung[7])

$$R(T) = O\left(\log T \sqrt{\frac{\log\log\log T}{\log\log T}}\right);$$

dagegen gibt es unter derselben Voraussetzung eine positive Konstante c, so daß sowohl

$$\limsup_{T \to \infty} \frac{R(T)}{(\log T)^c} = +\infty,$$

als auch

$$\liminf_{T \to \infty} \frac{R(T)}{(\log T)^c} = -\infty$$

gilt. (Vgl. Bohr und Landau, Math. Annalen, Bd. 74 (1913), S. 3—30.)

Es wird hier die Funktion von s

$$\int_0^\infty R(v) e^{-vs} dv$$

untersucht, und es wird *vorläufig* die Riemannsche Vermutung als wahr angenommen. Alsdann hat die obige Funktion in den Punkten $s = \pm i \log p^m$ einfache Pole und ist im Ursprung wie der Ausdruck

[6]) E. Landau, Handbuch der Lehre von der Verteilung der Primzahlen, Leipzig und Berlin 1909, Bd. I, S. 368.

[7]) H. Cramér, Mathematische Zeitschrift, Bd. 2 (1918), S. 237–241.

$$\frac{7}{8s} + p(s)\log s$$

singulär, unter $p(s)$ eine in der Umgebung von $s=0$ konvergente Potenzreihe verstanden. Hieraus läßt sich unter Benutzung einer Verallgemeinerung des obengenannten Satzes von M. Riesz der Mittelwertsatz

$$\lim_{T\to\infty}\frac{1}{T}\int_0^T R(v)\,dv = \tfrac{7}{8}$$

und sogar

$$\int_0^T R(v)\,dv = \tfrac{7}{8}T + O(T^{\varepsilon})$$

für jedes $\varepsilon > 0$ ableiten.

Sodann wird aber gezeigt, daß dieses Resultat von der Frage nach der Richtigkeit oder Unrichtigkeit der Riemannschen Vermutung in der Tat gänzlich unabhängig ist, so daß die beiden letzten Gleichungen unbedingt bestehen. Daß die dort auftretende Konstante gerade den Wert $\frac{7}{8}$ hat, wird vielleicht besser verständlich nach einem Vergleich mit der für „wurzelfreies" T geltenden Formel[8])

$$N(T) = \frac{T}{2\pi}\left(\log\frac{T}{2\pi} - 1\right) + \tfrac{7}{8} + O\left(\frac{1}{T}\right) + \frac{1}{\pi}\Delta_{abc}\arg\zeta(s),$$

wo a reell und >1, $b = a + iT$ und $c = \frac{1}{2} + iT$ ist. Unter $\Delta_{abc}\arg\zeta(s)$ ist der Zuwachs von $\arg\zeta(s)$ gemeint, wenn s den gebrochenen Linienzug abc beschreibt.

Auch in diesem Falle wird die Befreiung von den einschränkenden Voraussetzungen durch eine Änderung der M. Rieszschen Sätze ermöglicht. Ich verdanke Herrn M. Riesz mehrere liebenswürdige Mitteilungen über die nötigen Änderungen, den Geltungsbereich und die Anwendbarkeit dieser wichtigen Sätze, die zum Teil von ihm noch nicht veröffentlicht sind. Erst durch diese Mitteilungen ist es mir möglich geworden, meine Sätze in der angedeuteten Weise zu verallgemeinern.

Erstes Kapitel.

Definition und Darstellung der Funktion $V(z)$.

Es sei $z = x + yi$ eine komplexe Veränderliche, und es bezeichne $\varrho = \beta + \gamma i$ eine beliebige komplexe Nullstelle der Riemannschen Zetafunktion[9]). Wir nehmen nun vorläufig $y > 0$ an; die Reihe $\Sigma e^{\varrho z}$, wo ϱ

[8]) R. J. Backlund, Acta Mathematica, Bd. 41 (1918), S. 351.

[9]) Es ist dann bekanntlich $0 < \beta < 1$, $|\gamma| > 12$.

nur diejenigen Nullstellen zu durchlaufen hat, deren Ordinaten γ positiv sind (mehrfache mehrfach gezählt), wird alsdann unbedingt konvergieren. Wir setzen

$$V(z) = \sum_{\gamma > 0} e^{\varrho z}$$

und wollen die Funktion $V(z)$ für beliebige Werte von z darstellen und untersuchen.

Um zuerst $V(z)$ durch ein bestimmtes Integral darzustellen, wenden wir den Cauchyschen Integralsatz auf die Funktion

$$e^{sz} \frac{\zeta'(s)}{\zeta(s)}$$

und das Rechteck mit den Ecken $0,\ 1,\ 1 + T_\nu i,\ T_\nu i$ in der Ebene der Veränderlichen $s = \sigma + it$ an; jedoch werden die Ecken 0 und 1 durch kleine nach oben gerichtete Kreisbögen vom Radius $\varepsilon < \frac{1}{2}$ umgangen. Hierbei bezeichnet T_ν eine beliebige der von Landau[10]) eingeführten Zahlen, welche für jedes ganzzahlige ν den Bedingungen

$$\nu < T_\nu < \nu + 1,$$

(5)
$$\frac{\zeta'(s)}{\zeta(s)} = O(\log^2 |s|)$$

gleichmäßig für $t = T_\nu$ und alle ν, genügen.

Der Integrand ist auf dem Integrationswege überall regulär, hat aber im Innern für jedes ϱ, das die Bedingung $0 < \gamma < T_\nu$ erfüllt, einen Pol erster Ordnung. Wird der Integrationsweg mit R bezeichnet, so ist also

$$2\pi i \sum_{0 < \gamma < T_\nu} e^{\varrho z} = \int_R e^{sz} \frac{\zeta'(s)}{\zeta(s)} ds.$$

Läßt man nun ν (durch ganze Zahlen) unendlich werden, so ist leicht ersichtlich, daß das Integral, erstreckt über die obere horizontale Rechteckseite, gegen den Grenzwert 0 strebt. Der Integrationsweg hat nämlich die feste Länge 1, nach (5) gilt auf ihm

$$e^{sz} \frac{\zeta'(s)}{\zeta(s)} = O(e^{-yT_\nu} \log^2 T_\nu).$$

Da, wie schon oben bemerkt, die Reihe $\sum_{\gamma > 0} e^{\varrho z}$ für $y > 0$ konvergiert, so ist

(6)
$$2\pi i\, V(z) = \int_{ABCDEF} e^{sz} \frac{\zeta'(s)}{\zeta(s)} ds,$$

---

[10]) Handbuch, I, S. 340.

wo die großen Buchstaben die folgenden Punkte bezeichnen (ε ist der gemeinsame Radius der kleinen Kreisbögen um 0 und 1):

$$A = +i\infty, \quad B = +i\varepsilon, \quad C = \varepsilon,$$
$$D = 1 - \varepsilon, \quad E = 1 + i\varepsilon, \quad F = 1 + i\infty.$$

Jetzt wollen wir (6) durch teilweise Integration so umformen, daß anstatt $\frac{\zeta'(s)}{\zeta(s)}$ die Funktion $\log \zeta(s)$ auftritt; da aber diese Funktion unendlich vieldeutig ist, so wollen wir uns zuerst für einen bestimmten Zweig entscheiden. Wir schneiden die s-Ebene von $s = 1$ bis $s = -\infty$ geradlinig auf und entfernen außerdem das durch die Ungleichungen

$$0 < \sigma < 1, \qquad |t| > 1$$

bestimmte Gebiet. In dem übrigbleibenden Teil der aufgeschnittenen Ebene ist jeder Zweig von $\log \zeta(s)$ eindeutig. Wir wählen denjenigen, der für $\sigma > 1$ durch die Reihe

$$\sum_{p,m} \frac{1}{m p^{ms}}$$

dargestellt wird. Für das Folgende ist wichtig zu bemerken, daß diese Reihe auch noch für $\sigma = 1$, $t \neq 0$ konvergiert, und zwar in jedem endlichen, den Punkt 1 ausschließenden Intervall der Geraden $\sigma = 1$ gleichmäßig[11]). — Für $0 < s < 1$ hat der gewählte Zweig von $\log \zeta(s)$, wie man leicht findet, den imaginären Teil $-\pi i$ auf dem oberen, $+\pi i$ auf dem unteren Ufer des Schnittes, eine Bemerkung, die uns auch später nützlich sein wird.

Wenn wir jetzt (6) teilweise integrieren, so finden wir aus bekannten Sätzen über die Zetafunktion, daß der integrierte Teil an den Grenzen verschwindet. Es ist nämlich[12])

$$\log \zeta(1 + it) = O(\log^2 t),$$

und nach der Funktionalgleichung (s. unten)

$$\log \zeta(it) = O(t \log t),$$

so daß wir

$$2\pi i\, V(z) = -z \int_{ABCDEF} e^{sz} \log \zeta(s)\, ds$$

---

[11]) Konvergenz für festes $t \neq 0$: Landau, *Handbuch*, Bd. I, S. 234—238. Die Aussage über gleichmäßige Konvergenz wird durch geeignete Abänderung des dortigen Gedankenganges bewiesen, folgt aber auch unmittelbar aus einem später anzuführenden Satz von M. Riesz.

[12]) Landau, *Handbuch*, Bd. I, S. 327.

erhalten. Diese Formel schreiben wir in der Form

(7) $$2\pi i V(z) = z\int\limits_{CBA} - z\int\limits_{CD} - z\int\limits_{DEF}$$

und transformieren das erste Integral unter Anwendung der Funktionalgleichung der Zetafunktion:

$$\zeta(s) = \frac{(2\pi)^s}{\pi} \sin\frac{s\pi}{2} \Gamma(1-s)\zeta(1-s).$$

Es ergibt sich dann

(8) $$\begin{cases} z\int\limits_{CBA} e^{sz} \log\zeta(s)\,ds = z\int\limits_{CBA} e^{sz}\Big(s\log 2\pi - \log\pi + \\ \qquad + \log\sin\frac{s\pi}{2} + \log\Gamma(1-s) + \log\zeta(1-s)\Big)\,ds. \end{cases}$$

Es entsteht nun die Frage: wie sollen wir in der letzten Formel die vieldeutigen Funktionen definieren, damit das Funktionszeichen $\log\zeta()$ immer den vorhin gewählten Zweig dieser Funktion bezeichne? Betrachten wir zuerst die linke Seite von (8); nach einer früheren Bemerkung hat man im Punkte C (d. h. $s=\varepsilon$) $\log\zeta(s)$ den imaginären Teil $-\pi i$ zu erteilen, um ihn dann stetig variieren zu lassen. Wenn nun auf der rechten Seite von (8) die vier ersten Glieder in dem Klammerausdruck für $s=\varepsilon$ reell sind, so muß man in diesem Punkte auch $\log\zeta(1-s)$ den imaginären Teil $-\pi i$ erteilen. Die Ebene ist nun von $s=1$ bis $s=-\infty$ geradlinig aufgeschnitten, und da der Punkt $s=\varepsilon$ unseres Integrationsweges an dem oberen Ufer des Schnittes liegt, so liegt $1-\varepsilon$ an dem unteren Ufer. Unser Zweig der Funktion $\log\zeta(s)$ hat aber hier der imaginäre Teil $+\pi i$; wenn also $\log\sin\frac{s\pi}{2}$ und $\log\Gamma(1-s)$ für $s=\varepsilon$ reell sein sollen, so muß man (8) in folgender Weise berichtigen:

(8a) $$\begin{cases} z\int\limits_{CBA} e^{sz} \log\zeta(s)\,ds = z\int\limits_{CBA} e^{sz}\Big(s\log 2\pi - \log\pi + \\ \qquad + \log\sin\frac{s\pi}{2} + \log\Gamma(1-s) + \log\zeta(1-s) - 2\pi i\Big)\,ds. \end{cases}$$

Wir führen nun diesen Ausdruck in (7) ein und lassen ε gegen Null abnehmen. Die Integrale über die kleinen Kreisbögen CB und DE werden dann ebenfalls gegen Null abnehmen, da jedes Glied des Integranden für $s=0$ sowie für $s=1$ entweder regulär ist oder höchstens eine logarithmische Singularität hat. Wir können also ruhig $\varepsilon=0$ setzen, wenn wir uns nur bezüglich der vieldeutigen Funktionen folgendes merken:

1. Das Funktionszeichen $\log\zeta()$, das nunmehr nur mit einem Argumentwert von der Form $1+it$ auftreten wird, bezeichnet die durch die Reihe

$$\sum_{m,p} \frac{1}{p^{m(1+it)}}$$

dargestellte Funktion;

2. $\log \Gamma(1-s)$ soll für $s=0$ reell sein;

3. $\log \sin \frac{s\pi}{2}$ war für $s=\varepsilon$ reell; wenn nun s den Kreisbogen CB beschreibt, so wird man im Punkte B (d. h. $s=i\varepsilon$) mit dem imaginären Teil $\frac{1}{2}\pi i$ eintreffen; da $\sin \frac{s\pi}{2}$ für rein imaginäres s selbst rein imaginär ist, so müssen wir $\log \sin \frac{s\pi}{2}$ auf der ganzen Strecke AB den festen imaginären Teil $\frac{1}{2}\pi i$ erteilen.

Es ergibt sich nun endlich aus (7)

$$(9) \begin{cases} 2\pi i V(z) = -z\int\limits_1^{1+i\infty} e^{sz} \log \zeta(s)\, ds + z\int\limits_0^{i\infty} e^{sz} \log \zeta(1-s)\, ds - \\ \quad - z(\log \pi + 2\pi i)\int\limits_0^{i\infty} e^{sz}\, ds + z \log 2\pi \int\limits_0^{i\infty} s e^{sz}\, ds - \\ \quad - z\int\limits_0^{1} e^{sz} \log \zeta(s)\, ds + z\int\limits_0^{i\infty} e^{sz} \log \sin \frac{s\pi}{2}\, ds + z\int\limits_0^{i\infty} e^{sz} \log \Gamma(1-s)\, ds. \end{cases}$$

Das erste Glied dieser Formel läßt sich auch in der Form

$$-z\int\limits_1^{1+i\infty} e^{sz} \log \zeta(s)\, ds = -z \lim_{\varepsilon \to 0} \int\limits_{1+i\varepsilon}^{1+i\infty} e^{sz} \sum_{p,m} \frac{1}{m p^{ms}}\, ds$$

schreiben. Die unendliche Reihe läßt sich, einer früheren Bemerkung zufolge, unter Benutzung eines bekannten Dinischen Satzes[13]) gliedweise integrieren, wodurch man erhält

$$-z\int\limits_1^{1+i\infty} e^{sz} \log \zeta(s)\, ds = z e^{z} \lim_{\varepsilon \to 0} \sum_{p,m} \frac{1}{m p^{m(1+i\varepsilon)}(z-\log p^m)} =$$

$$= z e^{z} \sum_{p,m} \frac{1}{m p^m (z-\log p^m)}.$$

Die unbedingte Konvergenz der letzten Reihe sowie die Zulässigkeit des Grenzüberganges $\varepsilon \to 0$ ist hier durch die Konvergenz der Reihe

---

[13]) Wenn $\Sigma f_n(x)$ für jedes $b>a$ gleichmäßig im Intervalle $a<x<b$, $\Sigma \int\limits_a^x f_n(t)\, dt$ aber gleichmäßig für alle $x>a$ konvergiert, so gilt $\Sigma \int\limits_a^\infty f_n(t)\, dt = \int\limits_a^\infty \Sigma f_n(t)\, dt$.

$$\sum_{p,m} \frac{1}{mp^m \log p^m}$$

gesichert. Das zweite Glied von (9) läßt sich in entsprechender Weise umformen, und man erhält nach einiger weiterer Rechnung

$$(10)\quad \left\{\begin{aligned} 2\pi i V(z) &= ze^z \sum_{p,m} \frac{1}{mp^m(z-\log p^m)} - z\sum_{p,m} \frac{1}{mp^m(z+\log p^m)} + \\ &+ \frac{\log 2\pi}{z} + \log\pi + \pi i(1+e^z) - z\int_0^1 e^{sz}\log|\zeta(s)|\,ds + \\ &+ z\int_0^{i\infty} e^{sz}\log\sin\frac{s\pi}{2}\,ds + z\int_0^{i\infty} e^{sz}\log\Gamma(1-s)\,ds. \end{aligned}\right.$$

Die beiden ersten Glieder des Ausdruckes rechts stellen in der ganzen Ebene meromorphe Funktionen von z dar, welche keine anderen Singularitäten als die Pole der einzelnen Glieder der Reihen, $z = \pm \log p^m$, besitzen. Die drei folgenden Glieder sind triviale Funktionen, und das Integral von 0 bis 1 repräsentiert eine ganze Funktion von z. Wir wollen jetzt die zwei letzten Glieder etwas näher untersuchen. Zunächst ist

$$(11)\quad \left\{\begin{aligned} z\int_0^{i\infty} e^{sz}\log\sin\frac{s\pi}{2}\,ds &= z\int_0^{i\infty} e^{sz}\log\left[\tfrac{1}{2}i\left(e^{-\frac{1}{2}s\pi i} - e^{\frac{1}{2}s\pi i}\right)\right]ds = \\ &= z(\tfrac{1}{2}\pi i - \log 2)\int_0^{i\infty} e^{sz}\,ds + \tfrac{1}{2}\pi i z\int_0^{i\infty} se^{sz}\,ds + z\int_0^{i\infty} e^{sz}\log\left(e^{-\pi i s}-1\right)ds = \\ &= \log 2 - \tfrac{1}{2}\pi i + \frac{\frac{1}{2}\pi i}{z} + iz\int_0^{\infty}\log t\, e^{itz}\,dt + iz\int_0^{\infty}\log\frac{e^{\pi t}-1}{t}\,e^{itz}\,dt\,^{14}) = \\ &= C + \log 2 - \pi i + \frac{\frac{1}{2}\pi i}{z} + \log z - \log\pi - \int_0^{\infty}\left(\frac{1}{e^t-1} - \frac{1}{t} + 1\right)e^{-\frac{z}{\pi i}t}\,dt = \\ &= C + \log 2 + \tfrac{1}{2}\pi i\left(\frac{1}{z} - 1\right) + \Psi\left(\frac{z}{\pi i}\right), \end{aligned}\right.$$

wo C die Eulersche Konstante bezeichnet und $\Psi(\alpha) = \dfrac{\Gamma'(\alpha)}{\Gamma(\alpha)}$ gesetzt ist.

Betrachten wir jetzt das letzte Glied von (10). In $z = x + iy$ haben wir schon $y > 0$ angenommen; wir wollen nun für einen Augenblick auch $x > 0$ annehmen. In dem Gebiete $\frac{\pi}{2} \leqq \arg s \leqq \pi$ gilt die Abschätzung

$$|e^{sz}\log\Gamma(1-s)| = e^{\sigma x - ty}\,|\log\Gamma(1-s)| = O\left(e^{-k|s|}\,|s|\,\log|s|\right),$$

[14]) Das letzte Integral wird durch teilweise Integration umgeformt.

unter k eine positive Konstante verstanden, so daß eine leichte Anwendung des Cauchyschen Satzes das Resultat

$$z\int_0^{i\infty} e^{sz}\log\Gamma(1-s) = z\int_0^{-\infty} e^{sz}\log\Gamma(1-s)\,ds =$$

$$= -ze^z\int_1^{\infty} e^{-sz}\log\Gamma(s)\,ds = -e^z\int_1^{\infty}\Psi(s)\,e^{-sz}ds.$$

ergibt. In diesem Ausdruck führen wir nun die folgende Integraldarstellung der Funktion $\Psi(s)$ ein:

$$\Psi(s) = -C + \int_0^{\infty}\frac{e^{-t}-e^{-ts}}{1-e^{-t}}\,dt.$$

Die Ordnung der beiden Integrationen wird dann vertauschbar, da alles auch nach Einführung der absoluten Beträge konvergent bleibt, und man erhält

$$(12)\quad \begin{cases} z\int_0^{i\infty} e^{sz}\log\Gamma(1-s) = \frac{C}{z} - e^z\int_0^{\infty}\frac{dt}{1-e^{-t}}\int_1^{\infty}\left(e^{-t-sz}-e^{-(t+z)s}\right)ds = \\ = \frac{C}{z} - \frac{1}{z}\int_0^{\infty}\frac{t}{e^t-1}\cdot\frac{dt}{t+z}. \end{cases}$$

Nach Einführung der Ausdrücke (11) und (12) und nach Multiplikation mit $\frac{1}{2\pi i}$ nimmt (10) nun endlich die Gestalt

$$(13)\quad \begin{cases} V(z) = \frac{z}{2\pi i}e^z\sum_{p,m}\frac{1}{mp^m(z-\log p^m)} - \frac{z}{2\pi i}\sum_{p,m}\frac{1}{mp^m(z+\log p^m)} + \\ + \left(\frac{1}{4}+\frac{C+\log 2\pi}{2\pi i}\right)\left(1+\frac{1}{z}\right) + \frac{1}{2\pi i}\Psi\left(\frac{z}{\pi i}\right) + \\ + \frac{1}{2}e^z - \frac{z}{2\pi i}\int_0^1 e^{sz}\log|\zeta(s)|\,ds - \frac{1}{2\pi i z}\int_0^{\infty}\frac{t}{e^t-1}\cdot\frac{dt}{t+z}. \end{cases}$$

Das letzte Integral konvergiert, wie unmittelbar ersichtlich, in jedem abgeschlossenen Teil der z-Ebene, das weder die Null noch eine negative reelle Zahl enthält, gleichmäßig. Jeder innere Punkt der Strecke $(0, -\infty)$ ist aber für die durch das Integral dargestellte Funktion regulär, wie man leicht durch Veränderung des Integrationsweges sieht. In der Tat findet man nach einer einfachen Abschätzung des Integranden durch Anwendung des Cauchyschen Satzes

$$\int_0^{\infty}\frac{t}{e^t-1}\cdot\frac{dt}{t+z} = \int_0^{\infty e^{i\alpha}}\frac{t}{e^t-1}\cdot\frac{dt}{t+z},$$

wo rechts über den Vektor mit dem Argumente α in der t-Ebene integriert wird. Hierbei wird $-\frac{\pi}{2} < \alpha < \frac{\pi}{2}$ und $-\frac{\pi}{2} < \arg z < \frac{\pi}{2}$ vorausgesetzt. Das letzte Integral hat aber den Vektor

$$\arg z = \pi + \alpha$$

als singuläre Linie und konvergiert in jedem außerhalb dieser Linie gelegenen Gebiet gleichmäßig. Da α beliebig nahe an $\frac{\pi}{2}$ genommen werden kann, sind wir nunmehr imstande, über die Funktion $V(z)$ den folgenden Satz auszusprechen:

1. *Wenn längs der negativen imaginären Achse ein Schnitt gemacht wird, so ist die Funktion $V(z)$ im Innern der ganzen aufgeschnittenen z-Ebene meromorph und hat nur die singulären Stellen $z = \pm \log p^m$, welche Pole erster Ordnung sind. Im Punkte $z = \log p^m$ ist das Residuum gleich $\frac{\log p}{2\pi i}$; für $z = -\log p^m$ aber ist es $\frac{\log p}{2\pi i p^m}$.*

Es möge hier bemerkt werden, daß die Funktion $e^{-\frac{1}{2}z} V(z)$, deren Reihenentwicklung unter Voraussetzung der Richtigkeit der Riemannschen Hypothese die einfache Gestalt $\sum\limits_{\gamma > 0} e^{i\gamma z}$ hat, auch eine mehr symmetrische Form der Residuen aufweist. Für $z = \pm \log p^m$ sind ihre Residuen nämlich $\frac{\log p}{2\pi i \sqrt{p^m}}$.

Um die Fortsetzung von $V(z)$ über den Schnitt hinaus auszuführen und die übrigen Zweige zu untersuchen, haben wir nach (13) nur das letzte Glied dieser Formel zu betrachten, da alle die übrigen Glieder rechts überall eindeutige Funktionen darstellen. Man überzeugt sich nun durch eine einfache Überlegung, z. B. durch Anwendung eines bekannten Satzes von Hermite[15]), daß die Funktion

$$\frac{1}{z}\int\limits_0^\infty \frac{t}{e^t - 1} \cdot \frac{dt}{t + z} - \frac{\log z}{e^{-z} - 1}$$

überall eindeutig ist. Wir wollen jetzt eine Rechnung anstellen, die uns Auskunft über das Verhalten dieser Funktion im Nullpunkt geben wird. Wir setzen

$$(14) \qquad \int\limits_0^\infty \frac{t}{e^t - 1} \cdot \frac{dt}{t + z} = \int\limits_0^1 + \int\limits_1^\infty,$$

wo das letzte Integral eine in der von -1 bis $-\infty$ aufgeschnittenen

[15]) Hermite, Journal f. Mathematik, Bd. 91 (1881), S. 55. Vgl. auch Goursat, Cours d'Analyse, 2. Auflage, Paris 1911, Bd. 2, S. 252.

8*

Ebene eindeutige und reguläre Funktion von z darstellt, die für $z=0$ den Wert

$$\int_1^\infty \frac{dt}{e^t-1} \tag{15}$$

annimmt. Für das erste Integral rechts in (14) gilt, wenn für $|t|<2\pi$

$$\frac{t}{e^t-1}=\sum_{\nu=0}^\infty c_\nu t^\nu$$

gesetzt wird,

$$\int_0^1 \frac{t}{e^t-1}\cdot\frac{dt}{t+z}=\sum_{\nu=0}^\infty c_\nu \int_0^1 \frac{t^\nu\,dt}{t+z}.$$

Schreiben wir hier $(t+z-z)^\nu$ für t^ν und entwickeln nach der Binomialformel, so ergibt sich nach zweckmäßiger Zusammenfassung der verschiedenen Glieder

$$(16)\quad\left\{\begin{aligned}\int_0^1 \frac{t}{e^t-1}\cdot\frac{dt}{t+z}&=\sum_{\nu=2}^\infty c_\nu \sum_{\mu=1}^{\nu-1}\binom{\nu}{\mu}(-z)^\mu\int_0^1 (t+z)^{\nu-\mu-1}\,dt+\\&+\sum_{\nu=1}^\infty c_\nu\int_0^1 (t+z)^{\nu-1}\,dt+\sum_{\nu=0}^\infty c_\nu(-z)^\nu\int_0^1\frac{dt}{t+z}.\end{aligned}\right.$$

Betrachten wir das Verhalten der drei Glieder rechts für $z=0$. Die Doppelreihe stellt — wie das folgende Glied — eine für $|z|<1$ reguläre Funktion dar und verschwindet für $z=0$. Das zweite Glied nimmt dort den Wert

$$\int_0^1\sum_{\nu=1}^\infty c_\nu t^{\nu-1}\,dt=\int_0^1\left(\frac{1}{e^t-1}-\frac{c_0}{t}\right)dt=\int_0^1\left(\frac{1}{e^t-1}-\frac{1}{t}\right)dt \tag{17}$$

an. Das dritte Glied endlich ist gleich

$$\log\frac{z+1}{z}\sum_{\nu=0}^\infty c_\nu(-z)^\nu=\frac{z\,[\log z-\log(z+1)]}{e^{-z}-1}=\frac{z\log z}{e^{-z}-1}+z\,\mathfrak{P}(z), \tag{18}$$

unter $\log z$ den für $z>0$ reellen Zweig und unter $\mathfrak{P}(z)$ eine für $|z|<1$ konvergente Potenzreihe verstanden. Setzt man

$$\beta=\int_0^1\left(\frac{1}{e^t-1}-\frac{1}{t}\right)dt+\int_1^\infty\frac{dt}{e^t-1},$$

so erhält man nach (14) bis (18)

$$\frac{1}{z}\int_0^\infty\frac{t}{e^t-1}\cdot\frac{dt}{t+z}=\frac{\log z}{e^{-z}-1}+\frac{\beta}{z}+\mathfrak{P}_1(z), \tag{19}$$

wo $\mathfrak{P}_1(z)$ eine zu $\mathfrak{P}(z)$ analoge Bedeutung hat. Wir bemerken beiläufig, daß die Koeffizienten von $\mathfrak{P}_1(z)$ sämtlich reell sind. Es wird sich aber jetzt ergeben, daß hier

$$\beta = 0 \tag{20}$$

ist. In der Tat gilt ja

$$C = \int_0^\infty \left(\frac{1}{e^t - 1} - \frac{1}{t e^t}\right) dt,$$

woraus

$$\beta - C = \int_0^1 \frac{dt}{t}(e^{-t} - 1) + \int_1^\infty \frac{dt}{t} e^{-t} = \int_0^1 \log t \, e^{-t} dt + \int_1^\infty \log t \, e^{-t} dt$$
$$= \Gamma'(1) = -C,$$

d. h. $\beta = 0$ folgt.

Der Ausdruck

$$\frac{1}{z}\int_0^\infty \frac{t}{e^t - 1} \cdot \frac{dt}{t + z} - \frac{\log z}{e^{-z} - 1}$$

stellt eine überall eindeutige Funktion von z dar. Sein erstes Glied bleibt für $-\pi < \arg z < \pi$ regulär; es ist leicht ersichtlich, daß dieses erste Glied in den Punkten $z = \pm 2\nu\pi i$ einfache Pole bekommt, sobald man $\arg z$ unbeschränkt variieren läßt. Das Residuum der durch

$$\frac{1}{z}\int_0^\infty \frac{t}{e^t - 1} \cdot \frac{dt}{t + z}$$

dargestellten Funktion für $z = \pm 2\nu\pi i$ ist offenbar gleich

$$-i\left(\arg z \mp \tfrac{1}{2}\pi\right).$$

Aus diesen Betrachtungen und den Formeln (13), (19) und (20) ergibt sich, daß unsere frühere Aussage über die Singularitäten der Funktion $V(z)$ in der folgenden Weise ergänzt werden kann.

2. *Die Funktion*

$$V(z) - \frac{1}{2\pi i}\left(\frac{\log z}{1 - e^{-z}} + \frac{C + \log 2\pi - \frac{1}{2}\pi i}{z}\right) \tag{21}$$

ist in der ganzen Ebene eindeutig und — wenn in der Konvergenzhalbebene der Reihe (4) $0 < \arg z < \pi$ *gesetzt wird — im Ursprung regulär.*

3. *Die Punkte* $z = \pm 2\nu\pi i$ $(\nu = 1, 2, \ldots)$ *sind für* $V(z)$ *einfache Pole mit dem Residuum* $\frac{1}{2\pi}(\arg z - \frac{1}{2}\pi)$. *Die Punkte* $z = -\pi i$, $-3\pi i, \ldots$ *sind gleichfalls einfache Pole mit dem Residuum* $-\frac{1}{2}$.

Da wir später den Wert der Funktion (21) für $z = 0$ — oder vielmehr ihres reellen Teiles — brauchen, wollen wir diesen Wert jetzt berechnen. Indem wir die Glieder erster und höherer Ordnung außer Betracht lassen, erhalten wir

$$\frac{1}{2\pi i}\Psi\left(\frac{z}{\pi i}\right) = -\frac{1}{2z} - \frac{C}{2\pi i} + \cdots, \qquad \tfrac{1}{2}e^z = \tfrac{1}{2} + \cdots,$$

$$-\frac{1}{2\pi i z}\int_0^\infty \frac{t}{e^t - 1}\cdot\frac{dt}{t+z} = \frac{\log z}{2\pi i(1-e^{-z})} + \frac{\alpha_0}{2\pi i} + \cdots,$$

wo nach einer früheren Bemerkung α_0 reell ist. Unter Benutzung von (13) ergibt sich also

$$V(z) = \frac{1}{2\pi i}\left(\frac{\log z}{1-e^{-z}} + \frac{C + \log 2\pi - \frac{1}{2}\pi i}{z}\right) + \tfrac{3}{4} + i\eta + z\,\mathfrak{P}_2(z), \tag{22}$$

wo η reell ist und die Potenzreihe $\mathfrak{P}_2(z)$ für $|z| < \log 2$ konvergent ist.

Zweites Kapitel.

Anwendungen.

I. Der Landausche Satz über $\sum\limits_{\gamma>0}\frac{x^\varrho}{\varrho^k}$.

Die Reihe $\sum\limits_{\gamma>0}\frac{x^\varrho}{\varrho^k}$ $(0 < k \leqq 1,\ x > 0)$ *ist*

a) *für* $x \neq 1,\ p^m,\ \frac{1}{p^m}$ *konvergent;*

b) *in jedem von diesen Punkten freien Intervalle gleichmäßig konvergent;*

c) *für* $x = 1,\ p^m,\ \frac{1}{p^m}$ *divergent.*

Die Substitution $x = e^z$ führt die Punkte $x = p^m,\ 1,\ \frac{1}{p^m}$ in $z = \log p^m,\ 0,\ -\log p^m$ über. Es gilt nun

$$V_k(z) = \sum_{\gamma>0}\frac{e^{\varrho z}}{\varrho^k} = \frac{e^{-\frac{1}{2}k\pi i}}{\Gamma(k)}\sum_{\gamma>0}\int_0^\infty t^{k-1}e^{\varrho(z+it)}\,dt \tag{23}$$

$$= \frac{e^{-\frac{1}{2}k\pi i}}{\Gamma(k)}\int_0^\infty t^{k-1}V(z+it)\,dt$$

und

$$V(z) = \frac{e^{-\frac{1}{2}(1-k)\pi i}}{\Gamma(1-k)}\cdot\frac{d}{dz}\int_0^\infty t^{-k}V_k(z+it)\,dt,$$

was in analoger Weise bewiesen wird. Aus diesen Gleichungen läßt sich ohne weiteres ablesen, daß die Funktionen $V(z)$ und $V_k(z)$ für reelle

Werte von z gleichzeitig regulär oder singulär sind; die oben aufgezählten Punkte sind also eben die reellen Singularitäten von $V_k(z)$.

Wenn nun die Riemannsche Vermutung bewiesen wäre, so würde man nach Multiplikation von $V_k(z)$ mit $e^{-\frac{1}{2}z}$ durch die Substitution $z = is$ eine Dirichletsche Reihe $\sum\limits_{\gamma>0} \varrho^{-k} e^{-\gamma s}$ erhalten. Auf diese Reihe könnte man dann einen allgemeinen Konvergenzsatz von M. Riesz[16]) anwenden und dadurch den Satz von Landau beweisen. Wir wollen aber den Beweis ohne einschränkende Annahmen führen und werden darum nötig haben, in den Voraussetzungen und dem Beweis des Rieszschen Satzes einige geringfügige Änderungen vorzunehmen. Wenn nämlich der reelle Teil von ϱ nicht durchweg gleich $\frac{1}{2}$ ist, so kommt man auf eine Dirichletsche Reihe mit komplexen Exponenten, und der allgemeine Satz von M. Riesz läßt sich nicht ohne weiteres für diesen Fall verallgemeinern. In unserem speziellen Falle aber, wo die Koeffizienten der Reihe von der Größenordnung ϱ^{-k} sind, gestaltet sich der Beweis ziemlich einfach.

Es bezeichne $\varrho_1, \varrho_2, \ldots$ die Nullstellen in der oberen Halbebene, nach wachsenden Ordinaten geordnet (mehrfache mehrfach gezählt, die Reihenfolge der ϱ mit gleicher Ordinate bleibt beliebig). Setzen wir $\varrho_n = \beta_n + i\gamma_n$, so ist bekanntlich die Anzahl der ϱ_n, welche der Bedingung $T < \gamma_n \leqq T + 1$ genügen, gleich $N(T+1) - N(T) = O(\log T)$. Um nun die Teile a und b des Landauschen Satzes zu beweisen, bemerken wir, daß eine abgeschlossene Strecke der reellen z-Achse, welche die Punkte $z = 0$, $\pm \log p^m$ ausschließt, immer in einer größeren Strecke enthalten ist, welche dieselbe Eigenschaft hat. Es seien z_1 und $z_2 > z_1$ die Endpunkte der größeren Strecke; dann läßt sich offenbar die positive Zahl c so wählen, daß die Funktion $V_k(z)$ auch noch im Rechteck z_1, z_2, $z_2 - ci$, $z_1 - ci$ regulär ist. Wir setzen der Kürze halber $z_1' = z_1 - ci$, $z_2' = z_2 - ci$, $z_1'' = z_1 + ci$, $z_2'' = z_2 + ci$. Wir betrachten jetzt die Funktionen

$$g_n(z) = e^{-\gamma_n iz}(z - z_1)(z - z_2)\left[V_k(z) - \sum_{\nu=1}^{n} \varrho_\nu^{-k} e^{\varrho_\nu z}\right]$$

und werden zeigen, daß bei hinreichend großem n auf der ganzen Begrenzung des Rechtecks z_1', z_1'', z_2'', z_2'

$$|g_n(z)| < \varepsilon$$

wird, wie klein auch die positive Zahl ε gewählt sein mag.

Für positives y konvergiert die Reihe $\sum\limits_1^\infty \varrho_n^{-k} e^{\varrho_n z}$, und es gilt

---

[16]) M. Riesz, Acta Mathematica, Bd. 40 (1916), S. 349—361.

$$V_k(z) - \sum_1^n \varrho_\nu^{-k} e^{\varrho_\nu z} = \sum_{n+1}^\infty \varrho_\nu^{-k} e^{\varrho_\nu z}.$$

In dem Intervalle $\gamma_n + \mu < \gamma \leqq \gamma_n + \mu + 1$ gibt es nun $O[\log(\gamma_n + \mu)]$ Nullstellen, so daß man erhält

$$(24) \quad V_k(z) - \sum_1^n \varrho_\nu^{-k} e^{\varrho_\nu z} = O\left[\sum_{\mu=0}^\infty \frac{\log(\gamma_n + \mu)}{(\gamma_n + \mu)^k} e^{-(\gamma_n + \mu) y}\right] = O\left(\gamma_n^{-k} \log \gamma_n \frac{e^{-\gamma_n y}}{y}\right],$$

gleichmäßig für alle $y > 0$. Ist nun z ein beliebiger Punkt der Strecke z_1, z_1'', so ist $|z - z_1| = y$; demnach ist nach (24) gleichmäßig auf dieser Strecke

$$(25) \qquad g_n(z) = O\left[e^{\gamma_n y} \cdot y \gamma_n^{-k} \log \gamma_n \frac{e^{-\gamma_n y}}{y}\right] = O(\gamma_n^{-k} \log \gamma_n).$$

Dieselbe Ungleichung gilt aus Symmetriegründen auch für die Strecke z_2, z_2''. Es sei jetzt z ein Punkt der auf der Geraden $y = c$ gelegenen Strecke z_1'', z_2''; dann ist wegen (24)

$$(26) \qquad g_n(z) = O\left(e^{\gamma_n c} \gamma_n^{-k} \log \gamma_n \frac{e^{-\gamma_n c}}{c}\right) = O(\gamma_n^{-k} \log \gamma_n).$$

Um $g_n(z)$ auch für $y < 0$ abzuschätzen, bezeichnen wir mit M das Maximum von $|V_k(z)|$ in unserem Rechtecke. Dann gilt gleichmäßig für $-c \leqq y < 0$, $z_1 \leqq x \leqq z_2$

$$V_k(z) - \sum_1^n \varrho_\nu^{-k} e^{\varrho_\nu z} = O\left(M + \sum_{\mu=1}^{\gamma_n} \frac{\log \mu}{\mu^k} e^{-\mu y}\right) = o\left(\frac{e^{-\gamma_n y}}{y}\right).$$

Hieraus kann nunmehr ähnlich wie oben für die in der Halbebene $y < 0$ gelegenen Seiten des Rechteckes die Abschätzung

$$(27) \qquad g_n(z) = o(1)$$

gefolgert werden.

Bei hinreichend großem n ergeben nun (25), (26) und (27) die Ungleichung

$$|g_n(z)| < \varepsilon.$$

für die ganze Begrenzung des Rechtecks. Da die Funktionen $g_n(z)$ im Rechtecke regulär sind, so gilt die letzte Ungleichung auch im Innern, also ist insbesondere auf der Strecke z_1, z_2 der reellen Achse

$$\left| V_k(z) - \sum_1^n \varrho_\nu^{-k} e^{\varrho_\nu z} \right| < \frac{\varepsilon}{|z - z_1| \cdot |z - z_2|},$$

d. h. die Reihe $\sum_{\gamma > 0} \varrho^{-k} e^{\varrho z}$ konvergiert gleichmäßig auf jeder in $z_1 < z < z_2$ gänzlich enthaltenen abgeschlossenen Strecke.

In den ausgeschlossenen Punkten $z = 0$, $\pm \log p^m$ aber kann unsere Reihe nicht konvergent sein. Für den Punkt $z = 0$ folgt dies unmittelbar, da

$$\sum \left(\frac{1}{\varrho_n^k} - \frac{1}{(i\gamma_n)^k}\right)$$

unbedingt konvergiert, während $\sum \gamma_n^{-k}$ divergiert. Nehmen wir aber an, die Reihe $\sum\limits_{\gamma>0} \varrho^{-k} e^{\varrho z}$ sei in einem beliebigen Punkte $z = x_0$ der reellen Achse konvergent, so gilt für positives y, wenn wir

$$S_{n,m} = \sum_n^m \varrho_\nu^{-k} e^{\varrho_\nu x_0}$$

setzen,

$$V_k(x_0 + yi) = \sum_1^\infty \varrho_\nu^{-k} e^{\varrho_\nu (x_0 + yi)} = \sum_1^{n-1} \varrho_\nu^{-k} e^{\varrho_\nu (x_0 + yi)} + \sum_n^\infty S_{n,\nu} (e^{\varrho_\nu yi} - e^{\varrho_{\nu+1} yi}).$$

Wenn nun ε eine beliebig kleine positive Größe bezeichnet, können wir n so groß wählen, daß für alle $m \geqq n$

$$|S_{n,m}| < \varepsilon$$

wird. Es gilt dann, wenn y gegen Null abnimmt, unter K_1, K_2, K_3 absolute Konstanten verstanden,

$$V_k(x_0 + yi) < K_1 + \varepsilon \sum_n^\infty \left| e^{\varrho_\nu yi} - e^{\varrho_{\nu+1} yi} \right|$$

$$< K_1 + K_2 y \varepsilon \sum_1^\infty e^{-\gamma_\nu y} < K_1 + K_3 y \varepsilon \sum_1^\infty \log n \, e^{-ny} = o\left(\log \frac{1}{y}\right).$$

Bezeichnet aber x_0 einen der reellen Pole von $V(z)$, so findet man leicht unter Benutzung von (23), daß $V_k(z)$ dort wie $(z - x_0)^{k-1}$ oder wie $\log (z - x_0)$, je nachdem $k < 1$ oder $k = 1$ ist, unendlich wird. In diesen Punkten muß also unsere Reihe divergieren.

II. Ein Grenzwertsatz.

In diesem Abschnitt wird die Riemannsche Vermutung als wahr angenommen. — Betrachten wir nun unter dieser Annahme die Gleichung (13) des vorigen Kapitels. Nach Multiplikation mit $e^{-\frac{1}{2}z}$ nimmt die linke Seite die Gestalt $\sum\limits_{\gamma>0} e^{i\gamma z} = \sum\limits_{\gamma>0} e^{i\gamma(x+yi)}$ an, wenn wir $y > 0$ voraussetzen.

Nehmen wir nun auch $x > 0$ und von den Werten $\log p^m$ verschieden an, vergleichen wir die reellen Teile der beiden Seiten von (13), und lassen wir dann y gegen Null abnehmen. Dann ergibt sich

$$\lim_{y\to 0}\sum_{\gamma>0} e^{-\gamma y}\cos\gamma x = e^{-\frac{1}{2}x}\left[\tfrac{1}{4}\left(1+\frac{1}{x}\right)+\frac{1}{2\pi}\Im\left(\Psi\left(\frac{x}{\pi i}\right)\right)+\tfrac{1}{2}e^{x}\right].$$

Es gilt nun

$$\Psi\left(\frac{x}{\pi i}\right) = -C - \frac{\pi i}{x} + \pi i \sum_{n=1}^{\infty}\left(\frac{1}{n\pi i} - \frac{1}{x+n\pi i}\right)$$

$$= -C - \frac{\pi i}{x} + x \sum_{n=1}^{\infty}\frac{x-n\pi i}{n(x^2+n^2\pi^2)},$$

also

$$\frac{1}{2\pi}\Im\left(\Psi\left(\frac{x}{\pi i}\right)\right) = -\tfrac{1}{2}\left(\frac{1}{x} + x\sum_{n=1}^{\infty}\frac{1}{x^2+n^2\pi^2}\right)$$

$$= -\tfrac{1}{2}\left(\tfrac{1}{2}+\frac{1}{2x}+\frac{1}{e^{2x}-1}\right)$$

und

$$\lim_{y\to 0}\sum_{\gamma>0} e^{-\gamma y}\cos\gamma x = \tfrac{1}{2}e^{-\frac{1}{2}x}\left(e^{x}-\frac{1}{e^{2x}-1}\right) = \frac{e^{\frac{3}{2}x}-e^{-\frac{3}{2}x}-e^{-\frac{1}{2}x}}{2(e^{x}-e^{-x})}$$

Diese Formel gilt für positives x. Ist aber $x<0$, so ist der Grenzwert des letzten Gliedes von (13) für $y\to 0$ nicht mehr rein imaginär, und es ergibt sich, wenn man die für negatives x in ähnlicher Weise gefundene Formel mit der vorigen zusammenfaßt,

$$(28)\qquad \lim_{y\to 0}\sum_{\gamma>0} e^{-\gamma y}\cos\gamma x = \frac{e^{\frac{3}{2}x}-e^{-\frac{3}{2}x}\mp e^{\mp\frac{1}{2}x}}{2(e^{x}-e^{-x})},$$

wo das obere oder das untere Zeichen gilt, je nachdem x positiv oder negativ ist. Man kann diese Formel auch aus der vorigen ableiten, wenn man nur bemerkt, daß die linke Seite eine gerade Funktion von x darstellt.

Die Gleichung (28) erscheint dadurch interessant, daß in ihr die Primzahlen gar nicht explizit auftreten; sie bleibt indessen, wie schon oben betont, nur richtig, wenn x von 0 und den $\pm\log p^m$ verschieden ist. Wenn $x=\pm\log p^m$ ist, so strebt die linke Seite von (28) gegen den Grenzwert $\mp\infty$.

III. Mittelwertsatz für das Restglied $R(T)$.

Es werde *vorläufig* die Riemannsche Vermutung als wahr angenommen. Dann lautet die Gleichung (22) für positives y

$$(29)\qquad e^{\frac{1}{2}z}\sum_{\gamma>0} e^{i\gamma z} = \frac{1}{2\pi i}\left(\frac{\log z}{1-e^{-z}}+\frac{C+\log 2\pi-\frac{1}{2}\pi i}{z}\right)+\tfrac{3}{4}+i\eta+z\,p(z),$$

wo η reell ist und die Potenzreihe $p(z)$ — wie im folgenden $p_1, p_2, \ldots$ — für $|z|<\log 2$ konvergent bleibt. Der Logarithmus hat für rein imaginäres z den imaginären Teil $\frac{1}{2}\pi i$. Wird hier

$$z = si = s e^{\frac{1}{2}\pi i}$$

gesetzt, so ergibt sich

$$(30) \quad \begin{cases} \sum\limits_{\gamma>0} e^{-\gamma s} = \dfrac{e^{-\frac{1}{2}is}}{2\pi i}\left(\dfrac{\log s + \frac{1}{2}\pi i}{1-e^{-is}} + \dfrac{C+\log 2\pi - \frac{1}{2}\pi i}{is}\right) + \frac{3}{4} + i\eta + s p_1(s) \\ = -\dfrac{1}{2\pi}\left(\dfrac{\log s + \frac{1}{2}\pi i}{2\sin\frac{s}{2}} + \dfrac{C+\log 2\pi - \frac{1}{2}\pi i}{s}\right) + \frac{7}{8} + i\eta_1 + s p_2(s) \\ = -\dfrac{C+\log 2\pi s}{2\pi s} + \dfrac{1}{2\pi}\log s\left(\dfrac{1}{s} - \dfrac{1}{2\sin\frac{s}{2}}\right) + \frac{7}{8} + i\eta_1 + s p_3(s), \end{cases}$$

wo auch η_1 reell ist. Läßt man nun s durch positive Werte gegen Null abnehmen, so ist sofort ersichtlich, daß $\eta_1 = 0$ sein muß, und wir erhalten endlich

$$(31) \qquad \sum_{\gamma>0} e^{-\gamma s} = -\frac{C+\log 2\pi s}{2\pi s} + \frac{7}{8} + s p_4(s)\log s + s p_3(s).$$

Lassen wir nun für einen Augenblick $\gamma_1 < \gamma_2 < \gamma_3 < \ldots$ die Ordinaten der Nullstellen ϱ, *von ihrer Vielfachheit abgesehen*, bezeichnen. Die obige Reihe nimmt dann folgende Form an:

$$(32) \quad \begin{cases} \sum\limits_{n=1}^{\infty}\left[N(\gamma_n) - N(\gamma_{n-1})\right] e^{-\gamma_n s} = \sum\limits_{n=1}^{\infty} N(\gamma_n)\left(e^{-\gamma_n s} - e^{-\gamma_{n+1} s}\right) = s\int\limits_0^{\infty} N(v) e^{-v} \\ = s\int\limits_0^{\infty} \dfrac{v}{2\pi}\left(\log\dfrac{v}{2\pi} - 1\right) e^{-vs}\, dv + s\int\limits_0^{\infty} R(v) e^{-vs}\, dv \\ = -\dfrac{C+\log 2\pi s}{2\pi s} + s\int\limits_0^{\infty} R(v) e^{-vs}\, dv, \end{cases}$$

und es ergibt sich[17])

$$(33) \qquad \int\limits_0^{\infty} R(v) e^{-vs}\, dv = \frac{7}{8s} + p_4(s)\log s + p_3(s),$$

oder, wenn wir

$$(34) \qquad R(v) = \frac{7}{8} + \bar{R}(v)$$

setzen,

$$(35) \qquad f(s) = \int\limits_0^{\infty} \bar{R}(v) e^{-vs}\, dv = p_4(s)\log s + p_3(s).$$

---

[17]) Die von den Primzahlen abhängigen Singularitäten $s = \pm i \log p^m$ der Funktion $V(is)$ sind also auch noch für die Funktion $\int\limits_0^{\infty} R(v) e^{-vs}\, dv$ Pole erster Ordnung.

Der im ersten Abschnitte dieses Kapitels angeführte Satz von M. Riesz gilt, wie sein Entdecker besonders hervorhebt, nicht nur für Dirichletsche Reihen, sondern unter geeigneter Abänderung der Voraussetzungen auch für Integrale von der Form

$$\int_0^\infty a(v)e^{-vs}dv.$$

M. Riesz hat in seinen Arbeiten über diesen Gegenstand auch bedeutende Verallgemeinerungen seines Satzes in Aussicht gestellt[18]. Auf meine diesbezügliche Anfrage hat er mir freundlichst mitgeteilt, daß er u. a. den folgenden Satz beweisen kann.

Es sei $\lim\limits_{v\to\infty} a(v)=0$ *und es möge die Funktion*

$$\varphi(s)=\varphi(\sigma+it)=\int_0^\infty a(v)e^{-vs}\,dv$$

im Gebiete

$$0<|s|\leqq\delta,\quad \sigma\geqq 0$$

regulär sein und daselbst die Ungleichung

$$|\varphi(s)-A|<|s|^\eta, \tag{36}$$

---

[18]) Vgl. M. Riesz, loc. cit.[5]), S. 350, sowie Journal f. Mathematik, Bd. 140 (1911), S. 89—99 und Comptes rendus, Bd. 149 (1909), S. 909—912. Über den noch nicht veröffentlichten Beweis des folgenden Satzes mögen hier einige Andeutungen Platz finden. Es gilt für jedes s im Innern des angedeuteten Gebietes $\varphi(s)=\frac{1}{2\pi i}\left[\int_{C_1}\frac{\varphi(z)\,dz}{z-s}+\int_{C_2}\frac{\varphi(z)\,dz}{z-s}\right]=J_1+J_2$, wo C_1 die Strecke $|t|<\delta$ der imaginären Achse, C_2 der übrigbleibende Teil der Begrenzung des Gebietes bezeichnen. Unter Benutzung der Gleichung $\frac{1}{z-s}=-\int_0^\infty e^{(z-s)v}\,dv$ findet man $a(v)=a_1(v)+a_2(v)$, $J_1=\int_0^\infty a_1(v)\,e^{-vs}\,dv$, $\lim\limits_{v\to\infty} a_1(v)=0$ sowie analoge Beziehungen für J_2 und $a_2(v)$. Es gilt übrigens $a_1(v)=-\frac{1}{2\pi i}\int_{C_1} e^{vz}\,\varphi(z)\,dz$. Die Konvergenz von $\int_0^\infty a_1(v)\,dv$ läßt sich nun genau so beweisen, wie die Konvergenz einer Fourierschen Reihe in einem Punkte, wo die zugehörende Funktion eine Lipschitzsche Bedingung befriedigt, d. h. durch Anwendung eines bekannten Riemannschen Satzes (Gesammelte Mathematische Werke, zweite Auflage, Leipzig 1892, S. 254), aus dem auch die Behauptung $\lim a_1(v)=0$ folgt. Für das Glied J_2 aber, das im Ursprung regulär ist, gilt der im Abschnitt I angeführte Satz aus Acta Math. 40 (loc. cit. [5]), S. 350 und 360), womit alles bewiesen ist.

wo $\eta > 0$ ist und A eine Konstante bezeichnet, erfüllen. Dann konvergiert das Integral

$$\int_0^\infty a(v)\,dv.$$

Wir wollen nun zeigen, daß die Voraussetznngen dieses Satzes bei der Funktion

$$f_k(s) = \int_0^\infty v^{-k}\bar{R}(v)\,e^{-vs}\,dv, \qquad 0 < k \leqq 1$$

erfüllt sind. Zunächst ist für jedes $k > 0$

$$\lim_{v\to\infty} v^{-k}\bar{R}(v) = 0.$$

Es gilt nun weiter

$$f_k(s) = \frac{1}{\Gamma(k)}\int_0^\infty \bar{R}(v)\,e^{-vs}\,dv\int_0^\infty z^{k-1}e^{-vz}\,dz = \frac{1}{\Gamma(k)}\int_0^\infty z^{k-1}f(s+z)\,dz;$$

für jedes $\delta < \log 2$ ist also $\varphi(s)$ im Gebiete $0 < |s| \leqq \delta$, $\sigma \geqq 0$ regulär, Schreiben wir nun unter Benutzung von (35)

$$\Gamma(k)f_k(s) = \int_0^\delta z^{k-1}p_4(s+z)\log(s+z)\,dz + \int_0^\delta z^{k-1}p_3(s+z)\,dz + \int_\delta^\infty z^{k-1}f(s+z$$

so haben wir nur noch zu zeigen, daß für hinreichend kleines δ jedes der drei Glieder rechts eine Ungleichung von der Form (36) befriedigt. Für die beiden letzten Glieder gilt dies in der Tat, da sie, wie leicht ersichtlich, für $s = 0$ regulär sind. Wir wollen jetzt beweisen, daß dies auch für das erste Glied gilt, wenn man $0 < \eta < k$ und

$$A = \int_0^\delta z^{k-1}p_4(z)\log z\,dz$$

annimmt. Schreiben wir

$$p_4(z) = c + zp_5(z),$$

so wird

$$\int_0^\delta z^{k-1}p_4(s+z)\log(s+z)\,dz - A = c\int_0^\delta z^{k-1}\log\frac{s+z}{z}\,dz +$$

$$+\int_0^\delta z^{k-1}[(s+z)p_5(s+z)\log(s+z) - zp_5(z)\log z]\,dz = \varDelta_1(s) + \varDelta_2(s).$$

Im Gebiete $0 < |s| \leqq \delta$, $\sigma \geqq 0$ gilt nun, wenn δ hinreichend klein und $0 < \eta < k$ ist,

$$|\Delta_1(s)| < |c| \int_0^\delta z^{k-1}\left(\log\frac{z+|s|}{z} + \operatorname{arctg}\frac{|s|}{z}\right) dz$$

$$< |c|\cdot|s|^k \int_0^\infty z^{k-1}\left(\log\frac{z+1}{z} + \operatorname{arctg}\frac{1}{z}\right) dz < \tfrac{1}{2}|s|^\eta$$

und

$$|\Delta_2(s)| < \int_0^\delta z^{k-1}\,dz \left|\int_z^{z+s} \frac{d}{dx}(x p_5(x)\log x)\,dx\right| < M\cdot|s| \int_0^\delta z^{k-1}\log\frac{1}{z}\,dz < \tfrac{1}{2}|s|^\eta .$$

Hieraus ergibt sich nun

$$\left|\int_0^\delta z^{k-1} p_4(s+z)\log(s+z)\,dz - A\right| < |s|^\eta;$$

die Funktion $f_k(s)$ befriedigt also eine Ungleichung von der Form (36), und es konvergiert nach dem Satze von M. Riesz das Integral

$$\int_0^\infty v^{-k}\bar{R}(v)\,dv$$

für jeden positiven Wert von k.

Es gilt also für $0 < k < 1$

$$\int_1^T v^{-k}\bar{R}(v)\,dv = \alpha_k(T), \tag{37}$$

wo $\lim\limits_{T\to\infty} \alpha_k(T)$ existiert, und also insbesondere $\alpha_k(T)$ für $T \geqq 1$ beschränkt ist. Es sei etwa

$$|\alpha_k(T)| < M_k.$$

Wir setzen

$$\int_1^T \bar{R}(v)\,dv = T^k g(T) \tag{38}$$

und erhalten nach einer teilweisen Integration aus (37)

$$g(T) = \alpha_k(T) - k\int_1^T g(v)\frac{dv}{v}.$$

Diese Integralgleichung wird nach dem gewöhnlichen iterierenden Verfahren gelöst, und es ergibt sich

$$g(T) = \alpha_k(T) - k\int_1^T \alpha_k(v)\frac{dv}{v} + k^2\int_1^T \frac{dv}{v}\int_1^v \alpha_k(v_1)\frac{dv_1}{v_1} - \dots, \tag{36}$$

also

$$|g(T)| < M_k\left(1 + \frac{k}{1}\log T + \frac{k^2}{2!}\log^2 T + \ldots\right) = M_k T^k,$$

oder nach (38)

$$\left|\int_1^T \bar{R}(v)\,dv\right| < M_k T^{2k}.$$

Da nun dies für $0 < k < 1$ gilt, so ist für beliebig kleines $\varepsilon > 0$ wegen (34)

$$\int_0^T R(v)\,dv = \tfrac{7}{8}T + O(T^\varepsilon). \tag{39}$$

Bis jetzt ist beim Beweise die Riemannsche Vermutung als wahr angenommen worden; wir wollen nun sehen, ob für das letzte Resultat diese Annahme auch wesentlich ist.

Setzen wir

$$\varrho = \tfrac{1}{2} - \alpha + \gamma i,$$

so daß immer

$$-\tfrac{1}{2} < \alpha < \tfrac{1}{2}$$

gilt, so geht die linke Seite der Gleichung (29) über in

$$e^{\frac{1}{2}z} \sum_{\gamma > 0} e^{(-\alpha + i\gamma)z},$$

und nach der Substitution $z = is$ erhalten wir an Stelle der linken Seite von (30)

$$\sum_{\gamma > 0} e^{-(\gamma + \alpha i)s}.$$

Da die Werte $\varrho = \beta + \gamma i$ zur Geraden $\beta = \frac{1}{2}$ symmetrisch liegen, so ist dieser Ausdruck für reelle $s > 0$ selbst reell und positiv; es wird also wie bei (30) auch hier $\eta_1 = 0$ sein müssen, und es ergibt sich

$$\sum_{\gamma > 0} e^{-(\gamma + \alpha i)s} = -\frac{C + \log 2\pi s}{2\pi s} + \tfrac{7}{8} + s p_4(s)\log s + s p_3(s). \tag{31a}$$

Es sei jetzt die Funktion $n(v)$ der komplexen Veränderlichen $v = u + wi$ in der folgenden Weise definiert:

$$n(v) = n(u + wi) = \text{Anzahl der } \varrho,$$

welche den Bedingungen

$$\gamma < u \quad \text{oder} \quad \begin{cases} \gamma = u \\ \alpha \leq w \end{cases} \quad \text{genügen.}$$

Für jedes reelle v, das von den γ verschieden ist, stimmt $n(v)$ mit der gewöhnlichen Funktion $N(v)$ überein.

Setzen wir

$$v_\nu = \gamma_\nu + \alpha_\nu i, \qquad \nu = 1, 2, \ldots,$$

wo die v_ν erst nach wachsenden γ_ν, die v_ν mit gleichem γ_ν aber nach wachsendem α_ν geordnet sind, *und wo jedes* $\gamma_\nu + \alpha_\nu i$ *von seiner Vielfachheit abgesehen nur einmal gezählt wird,* so läßt sich die linke Seite von (31a) auch schreiben

$$(32\text{a}) \qquad \sum_{\nu=1}^{\infty} [n(v_\nu) - n(v_{\nu-1})] e^{-v_\nu s} = \sum_{\nu=1}^{\infty} n(v_\nu)(e^{-v_\nu s} - e^{-v_{\nu+1} s}) =$$

$$= s \sum_{\nu=1}^{\infty} \int_{v_\nu}^{v_{\nu+1}} n(v) e^{-vs} dv = s \int_W n(v) e^{-vs} dv,$$

wo das Integral rechts über den Weg W zu erstrecken ist, das aus den geradlinigen Strecken $(0, \gamma_1 - \frac{1}{2}i)$, $(\gamma_1 - \frac{1}{2}i, \gamma_1 + \frac{1}{2}i), \ldots, (\gamma_{\nu-1} + \frac{1}{2}i, \gamma_\nu - \frac{1}{2}i)$, $(\gamma_\nu - \frac{1}{2}i, \gamma_\nu + \frac{1}{2}i), \ldots$ zusammengesetzt ist. Durch eine leichte Anwendung des Cauchyschen Satzes ergibt sich nun

$$s \int_W \frac{v}{2\pi}\left(\log \frac{v}{2\pi} - 1\right) e^{-vs} dv = -\frac{C + \log 2\pi s}{2\pi s},$$

und wir erhalten wegen (31a) und (32a)

$$(35\text{a}) \qquad \int_W \left[n(v) - \frac{v}{2\pi}\left(\log \frac{v}{2\pi} - 1\right) - \frac{7}{8}\right] e^{-vs} dv = p_4(s) \log s + p_3(s).$$

Genau wie oben läßt sich nun hieraus ableiten, daß die Funktion von s

$$\int_W v^{-k} \left[n(v) - \frac{v}{2\pi}\left(\log \frac{v}{2\pi} - 1\right) - \frac{7}{8}\right] e^{-vs} dv, \qquad (0 < k \leqq 1)$$

eine Lipschitzsche Ungleichung im Sinne von (36) befriedigt. Es liegt nun hier ein Integral der Form

$$\int a(v) e^{-vs} dv$$

mit $\lim_{v \to \infty} a(v) = 0$ vor, das aber nicht längs der reellen Achse, sondern über den Weg W genommen ist. Der oben angeführte Satz von M. Riesz läßt sich also nicht unmittelbar anwenden, kann aber ohne Schwierigkeit so abgeändert werden, daß er auch diesen Fall umfaßt.

In der Tat haben wir schon im ersten Abschnitt dieses Kapitels den Rieszschen Konvergenzsatz aus Acta Math. 40 dahin abgeändert, daß er auch eine gewisse Dirichletsche Reihe mit komplexen Exponenten, deren imaginäre Teile aber beschränkt sind, nämlich die Reihe $\sum_{\gamma > 0} \varrho^{-k} e^{-(\gamma + \alpha i)s}$, umfaßt. Die jetzt erforderliche Abänderung ist nun jener ganz analog, indem es hier verlangt wird, den für Integrale $\int a(v)e^{-vs}dv$ längs der reellen Achse gültigen Satz dahin zu verallgemeinern, daß er auch für gewisse Integrale über den Weg W, dessen Abstand von der reellen Achse beschränkt ist, gilt. Es läßt sich dies auch mit genau denselben Mitteln erreichen. Das Glied J_1 der Fußnote[18]) läßt sich nämlich sowohl als Integral über den Weg W, als auch in der Form $\int_0^\infty a_1(v)e^{-vs}dv$ darstellen und kann auch hier wie eine Fouriersche Reihe behandelt werden. Das Glied J_2 aber, das im Ursprung regulär ist, nimmt hier die Form $\int_W a_2(v)e^{-vs}$ an, und seine Konvergenz für $s=0$ können wir Wort für Wort wie in dem Abschnitt I beweisen, wenn wir nur überall die Summen durch Integrale ersetzen. Es ergibt sich also die Konvergenz von

$$\int_W v^{-k}\left[n(v)-\frac{v}{2\pi}\left(\log\frac{v}{2\pi}-1\right)-\frac{7}{8}\right]dv.$$

Auf dem Wege W gilt nun

$$n(v)=n(u+wi)=N(u)+O(\log u)=\frac{u}{2\pi}\left(\log\frac{u}{2\pi}-1\right)+O(\log u),$$

$$\frac{v}{2\pi}\left(\log\frac{v}{2\pi}-1\right)=\frac{u}{2\pi}\left(\log\frac{u}{2\pi}-1\right)+\frac{wi}{2\pi}\log\frac{u}{2\pi}+O\left(\frac{1}{u}\right),$$

$$v^{-k}=u^{-k}+O(u^{-k-1}).$$

Also konvergiert auch

$$\int_W u^{-k}\left[n(v)-\frac{u}{2\pi}\left(\log\frac{u}{2\pi}-1\right)+\frac{wi}{2\pi}\log\frac{u}{2\pi}-\frac{7}{8}\right]dv. \tag{40}$$

Wir behaupten nun, daß der reelle Teil von diesem Integral sich nur um eine unbedingt konvergente Reihe von dem Integral

$$\int_0^\infty u^{-k}\bar{R}(u)\,du$$

unterscheidet, so daß auch dieses Integral konvergiert.

Der Weg W setzt sich nämlich aus schrägen und senkrechten Strecken zusammen, und das Integral (40), über eine der senkrechten Strecken genommen, hat den reellen Teil Null. Es sei nun $(\gamma_\nu+\frac{1}{2}i, \gamma_{\nu+1}-\frac{1}{2}i)$ eine beliebige der schrägen Strecken; dann gilt

$$\Re\left[\int_{\gamma_\nu+\frac{1}{2}i}^{\gamma_{\nu+1}-\frac{1}{2}i} u^{-k}wi\log\frac{u}{2\pi}dv\right]=\gamma_\nu^{-k}\log\frac{\gamma_\nu}{2\pi}\Re\left[i\int_{\gamma_\nu+\frac{1}{2}i}^{\gamma_{\nu+1}-\frac{1}{2}i} w\,dv\right]+O(\gamma_\nu^{-k-1}\log\gamma_\nu)=$$

$$=\gamma_\nu^{-k}\log\frac{\gamma_\nu}{2\pi}\int_{-\frac{1}{2}}^{\frac{1}{2}} w\,dw+O(\gamma_\nu^{-k-1}\log\gamma_\nu)=O(\gamma_\nu^{-k-1}\log\gamma_\nu),$$

was unbedingt konvergent ist, wenn über alle ν summiert wird, und

$$\Re\left\{\int\limits_{\gamma_\nu+\frac{1}{2}i}^{\gamma_{\nu+1}-\frac{1}{2}i} u^{-k}\left[n(v)-\frac{u}{2\pi}\left(\log\frac{u}{2\pi}-1\right)-\frac{7}{8}\right]dv\right\}=\int\limits_{\gamma_\nu}^{\gamma_{\nu+1}} u^{-k}\bar{R}(u)\,du.$$

Die Summe aller solcher den schrägen Strecken zugehörenden Integrale, d. h. eben

$$\int\limits_0^\infty u^{-k}\bar{R}(u)\,du,$$

muß also konvergent sein. Hieraus kann man nun genau wie oben die Gleichung (39) wieder ableiten, so daß also der Mittelwertsatz

$$\int\limits_0^T R(v)\,dv=\tfrac{7}{8}T+O(T^\varepsilon)$$

für jedes $\varepsilon>0$ auch in diesem Fall richtig ist. Dieser Satz ist also von der Richtigkeit oder Unrichtigkeit der Riemannschen Vermutung unabhängig.

Stocksund bei Stockholm, den 8. September 1918.

(Eingegangen am 28. September 1918.)

Nachtrag.

Während der Drucklegung dieser Abhandlung richtete Herr Professor E. Lindelöf freundlichst meine Aufmerksamkeit auf die Arbeit „Über die Nullstellen der Zetafunktion" von Hj. Mellin (Annales Academiæ Scientiarium Fennicæ, Serie A, Bd. 10, (1917), Nr. 11). Es kommen dort einige zur Gleichung (28) der vorliegenden Arbeit analogen Formeln vor, und zwar wird — in etwas veränderter Bezeichnung — die Beziehung

$$\lim_{s\to 0}\sum_{\gamma>0}\varrho^{-s}\cos\gamma x=\frac{e^{\frac{3}{2}x}-e^{-\frac{3}{2}x}\mp e^{\mp\frac{1}{2}x}}{2\left(e^{x}-e^{-x}\right)}$$

hergeleitet. Diese Gleichung läßt wie meine Gleichung (28) die Primzahlen nicht explizit auftreten, fällt aber nicht mit ihr zusammen.

7.

Nombres premiers et équations indéterminées

Ark. Mat. Astr. Fys. **14** (13), 1–11 (1919)

Communiqué le 22 Janvier 1919 par L. E. PHRAGMÉN et H. VON KOCH.

---

1. Dans sa conférence sur »Les problèmes futurs des Mathématiques», M. HILBERT[1] a posé le problème suivant. — Soit donnée l'équation indéterminée

$$ax + by + c = 0, \tag{1}$$

a, b et c désignant des entiers premiers entre eux deux à deux. Peut-on toujours satisfaire à une telle équation par deux nombres premiers x et y? Peut-elle admettre plus d'une solution de ce genre?

Parmi les cas spéciaux de ce problème extrêmement difficile, les deux suivants ont été étudiés par bien des savants:

1°. GOLDBACH a affirmé sans démonstration que tout nombre entier et pair $2n$ est égal à la somme de deux nombres premiers, c. à d. que l'équation $x + y = 2n$ admet toujours une solution du genre indiqué ci-dessus. Cette proposition n'a pas été démontrée jusqu'ici.[2]

2°. L'équation $x - y = 2$, admet-elle une infinité de solutions en nombres premiers?

---

[1] Compte rendu du deuxième congrès international des mathématiciens tenu à Paris, p. 58. Voir aussi Göttinger Nachrichten, 1900, p. 253.

[2] Voir p. ex. STÄCKEL, »Über Goldbachs empirisches Theorem», Göttinger Nachrichten, 1896, p. 292.

En parlant de ces problèmes, M. LANDAU, dans sa conférence à Cambridge,[1] dit qu'il les croit inattaquables à l'état actuel de la science.

Tout en restant très éloigné d'en pouvoir donner la résolution, je veux présenter dans cette note quelques réflexions sur le problème de M. HILBERT et même sur des problèmes un peu plus généraux. Mes considérations vont d'ailleurs laisser parfaitement incontesté le jugement de M. LANDAU; elles iraient plutôt le confirmer, s'il y en avait besoin.

2. Je vais me servir d'une interprétation géométrique pour énoncer mon résultat principal. Supposons marqués dans le plan des coordonnées rectangulaires x et y tous les points P dont les deux coordonnées sont des nombres premiers positifs ou négatifs. L'équation (1), où les coefficients a, b, c désigneront désormais des nombres réels quelconques, représente une droite D dans ce plan, et c'est la position de cette droite relative aux points P qui nous intéresse. Si elle contient un certain nombre de ces points, chacun d'eux donne une solution en nombres premiers de notre équation (1); encore, s'il y a des points P à une distance très petite de D, il y a des systèmes de deux nombres premiers x, y satisfaisant *à peu près* à cette équation. Dans cet ordre d'idées, je vais démontrer le théorème suivant:

I. *Soit A un point quelconque de la droite D. Quelque petit que soit le nombre positif ε, on peut toujours trouver une droite D' passant par A et faisant avec D un angle inférieur à ε, tel qu'il existe une infinité de points P dont les distances à D' sont inférieures à ε.*

On voit sans difficulté comment s'exprimerait cela sans avoir recours au langage géométrique. En particulier, on peut toujours trouver un nombre a_1 tel que $|a - a_1| < \varepsilon$ et tel que l'inégalité

$$|a_1 x + by + c| < \varepsilon$$

admet une infinité de solutions en nombres premiers. L'application aux équations particulières $x - y = 2$ et $x + y = 2n$ est immédiate.

[1] »Gelöste und ungelöste Probleme aus der Theorie der Primzahlverteilung und der Riemann'schen Zetafunktion». Jahresberichte d. deutschen Math.-Ver. Bd. 21 (1912).

Par des raisons de symétrie, nous pouvons nous borner à considérer une demidroite D issue du point A, dont l'angle φ avec la direction positive de l'axe des x remplit la condition $0 \leq \varphi \leq \frac{\pi}{2}$. Il n'est pas difficile de voir qu'on peut même supprimer les cas $\varphi = 0$ et $\varphi = \frac{\pi}{2}$. Les coordonnées du point A étant (α, β), l'équation de la droite D peut donc se mettre sous la forme

$$y = \eta(x - \alpha) + \beta,$$

où $\eta > 0$. Si la distance d'un des points P à cette droite est inférieure à ε, on aura, les coordonnées p et q de ce point étant des nombres premiers positifs,

$$\eta(p - \alpha) + \beta - \varepsilon\sqrt{1 + \eta^2} < q < \eta(p - \alpha) + \beta + \varepsilon\sqrt{1 + \eta^2}.$$

Posons maintenant, x étant positif,

$$\vartheta(x) = \sum_{p < x} \log p$$

(la somme s'étendant à tous les nombres premiers positifs inférieurs à x), lorsque x n'est pas un nombre premier, et

$$\vartheta(p) = \frac{1}{2}[\vartheta(p + 0) + \vartheta(p - 0)].$$

La différence

$$\vartheta[\eta(p - \alpha) + \beta + \varepsilon] - \vartheta[\eta(p - \alpha) + \beta - \varepsilon]$$

ne peut donc être différente de zéro que lorsqu'il y a un point $P(p, q)$, dont la distance à la droite D soit inférieure à ε; dans ce cas elle est au plus égale à $\log q$. Formons enfin la fonction des deux variables positives ξ, η

$$h(\xi, \eta) = \sum_{p < \xi} [\vartheta(\eta(p - \alpha) + \beta + \varepsilon) - \vartheta(\eta(p - \alpha) + \beta - \varepsilon)];$$

et proposons-nous de démontrer que dans tout intervalle (η_1, η_2) avec $0 < \eta_1 < \eta_2$ il existe un nombre η_0 tel que

$$\lim_{\xi \to \infty} h(\xi, \eta_0) = \infty .$$

Ce point établi, le théorème I sera évidemment prouvé.

3. Envisageons d'abord la fonction analogue

$$g(\xi, \eta) = \sum_{p < \xi} [\psi(\eta(p - \alpha) + \beta + \varepsilon) - \psi(\eta(p - \alpha) + \beta - \varepsilon)],$$

$\psi(x)$ désignant la fonction connue définie par l'égalité

$$\psi(x) = \sum_{p^m < x} \log p,$$

lorsque x n'est pas égal à la puissance p^m d'un nombre premier p; dans ce cas dernier, on pose

$$\psi(p^m) = \frac{1}{2} [\psi(p^m + 0) + \psi(p^m - 0)].$$

Par une formule de M. VON MANGOLDT,[1] on aura alors pour $x > 1$

$$\psi(x) = x - \sum_{\varrho} \frac{x^\varrho}{\varrho} + \int_x^\infty \frac{dt}{t(t^2 - 1)} - \frac{1}{2} \log 2\pi ,$$

d'où il suit que

[1] Voir LANDAU, Handbuch der Lehre von der Verteilung der Primzahlen, Bd. I. p. 365. — Sans nuire à la généralité, on pourra supposer que l'on a pour $\eta_1 \leqq \eta \leqq \eta_2$ et pour tout nombre premier p considéré

$$\eta(p - \alpha) + \beta - \varepsilon > 1 .$$

S'il n'en est pas ainsi, il suffit de supprimer dans $g(\xi, \eta)$ un nombre suffisamment grand de termes au début.

$$g(\xi,\eta)=2\varepsilon\pi(\xi)-\sum_{p<\xi}\sum_{\varrho}\frac{(\eta(p-\alpha)+\beta+\varepsilon)^{\varrho}-(\eta(p-\alpha)+\beta-\varepsilon)^{\varrho}}{\varrho}$$

$$-\sum_{p<\xi}\int\limits_{\eta(p-\alpha)+\beta-\varepsilon}^{\eta(p-\alpha)+\beta+\varepsilon}\frac{dt}{t(t^2-1)}$$

Dans ces relations, $\pi(\xi)$ désigne combien il y a de nombres premiers inférieurs à ξ, tandis que le signe $\sum\limits_{\varrho}$ indique que la somme doit être étendue à tous les zéros complexes ϱ de la fonction $\zeta(s)$ de RIEMANN, rangés par ordre des ordonnées absolument croissantes. Dans un travail antérieur,[1] j'ai démontré que la série $\sum\limits_{\varrho}\frac{x^{\varrho}}{\varrho}$ peut être intégrée terme à terme entre des limites finies et supérieures à 1. On a donc

$$\int\limits_{\eta_1}^{\eta_2}g(\xi,\eta)d\eta=2\varepsilon\pi(\xi)(\eta_2-\eta_1)-$$

$$-\sum_{p<\xi}\sum_{\varrho}\int\limits_{\eta_1}^{\eta_2}d\eta\int\limits_{\beta-\varepsilon}^{\beta+\varepsilon}[\eta(p-\alpha)+z]^{\varrho-1}dz-\int\limits_{\eta_1}^{\eta_2}d\eta\sum_{p<\xi}\int\limits_{\eta(p-\alpha)+\beta-\varepsilon}^{\eta(p-\alpha)+\beta+\varepsilon}\frac{dt}{t(t^2-1)}=$$

$$=2\varepsilon\pi(\xi)(\eta_2-\eta_1)-R_1(\xi)-R_2(\xi).$$

Lorsque dans cette relation on fait tendre ξ vers l'infini, les deux derniers termes du second membre deviennent infiniment petits par rapport au premier. En effet, nous avons

$$R_1(\xi)=\sum_{p<\xi}\frac{1}{p-\alpha}\sum_{\varrho}\frac{1}{\varrho}\int\limits_{\beta-\varepsilon}^{\beta+\varepsilon}[(\eta_2(p-\alpha)+z)^{\varrho}-(\eta_1(p-\alpha)+z)^{\varrho}]dz=$$

$$=\sum_{p<\xi}\frac{1}{p-\alpha}\int\limits_{\beta-\varepsilon}^{\beta+\varepsilon}\sum_{\varrho}\frac{(\eta_2(p-\alpha)+z)^{\varrho}-(\eta_1(p-\alpha)+z)^{\varrho}}{\varrho}dz.$$

[1] H. CRAMÉR, »Über die Herleitung der Riemann'schen Primzahlformel», Arkiv för matematik etc., B. 13, n:o 24.

Or, c'est bien connu[1] que la quantité $\frac{1}{x}\sum_{\varrho}\frac{x^{\varrho}}{\varrho}$ tend vers zéro avec $\frac{1}{x}$; le terme général de la somme $\sum_{p<\xi}$ tend donc vers zéro, et il s'ensuit

$$\lim_{\xi\to\infty}\frac{R_1(\xi)}{\pi(\xi)}=0.$$

Encore, un calcul facile montre que l'on a

$$R_2(\xi)<K\sum_{p<\xi}\frac{1}{p^3}<K_1$$

la constante K_1 étant indépendante de ξ; il vient donc de même

$$\lim_{\xi\to\infty}\frac{R_2(\xi)}{\pi(\xi)}=0,$$

et l'on en conclut

$$\int_{\eta_1}^{\eta_2} g(\xi,\eta)d\eta \backsim 2\varepsilon\pi(\xi)(\eta_2-\eta_1).$$

Dans le but de montrer que cette relation reste vraie encore quand on y remplace $g(\xi,\eta)$ par $h(\xi,\eta)$, cherchons une limite supérieure de la différence de ces deux fonctions, lorsque ξ tend vers l'infini, η restant compris dans l'intervalle (η_1,η_2). En désignant par $\psi_1(t)$ la fonction $\psi(t)-\vartheta(t)$, nous aurons

$$g(\xi,\eta)-h(\xi,\eta)=\sum_{p<\xi}[\psi_1(\eta(p-\alpha)+\beta+\varepsilon)-\psi_1(\eta(p-\alpha)+\beta-\varepsilon)].$$

Le terme général de la somme du second membre n'est

[1] LANDAU, Handbuch, Bd. I, p. 195 et 365.

différente de zéro que lorsqu'il y a un nombre premier q et un nombre entier $m \geqq 2$ tels que

$$\eta(p-\alpha)+\beta-\varepsilon \leqq q^m \leqq \eta(p-\alpha)+\beta+\varepsilon,$$

et alors il est au plus égal à $\log q$. Cependant, si ε est choisi suffisamment petit (ce qu'on peut évidemment supposer sans nuire à la généralité) un nombre donné q^m ne peut satisfaire qu'à une seule inégalité de cette forme, η ayant une valeur fixe dans (η_1, η_2). On a donc, pour $\eta_1 \leqq \eta \leqq \eta_2$

$$g(\xi, \eta) - h(\xi, \eta) < \sum_{\substack{q^m < \eta_2(\xi-\alpha)+\beta \\ m \geqq 2}} \log q < K\sqrt{\xi}\log \xi,$$

K étant indépendant de ξ. Il s'ensuit que l'on a aussi pour $h(\xi, \eta)$

$$\int_{\eta_1}^{\eta_2} h(\xi, \eta)\,d\eta \sim 2\varepsilon\pi(\xi)(\eta_2 - \eta_1).$$

Cette relation nous apprend en particulier que l'intégrale

$$\int_{\eta_1}^{\eta_2} h(\xi, \eta)\,d\eta$$

tend vers l'infini avec ξ, η_1 et $\eta_2 > \eta_1$ étant deux nombres positifs quelconques. Nous allons montrer que ce résultat suffit pour établir l'existence d'un nombre η_0 compris entre η_1 et η_2 tel que $h(\xi, \eta_0)$ tend vers l'infini avec ξ.

En effet, il est évident qu'on ne peut pas avoir toujours $h(\xi, \eta) \leqq 1$ pour $\eta_1 < \eta < \eta_2$; il y a donc un nombre η' compris entre η_1 et η_2 et un nombre positif ξ' tel que

$$h(\xi', \eta') > 1.$$

Or, la définition de la fonction $h(\xi, \eta)$ met en évidence qu'elle jouit de la propriété suivante: ξ' et η' étant positifs et quelconques, on pourra trouver sur l'axe des η un intervalle I aboutissant au point η' tel qu'on aura

$$h(\xi, \eta) \geqq h(\xi', \eta')$$

pour $\xi \geqq \xi'$, tant que η appartient à I (en y comprenant les extrémités).

A l'intérieur de (η_1, η_2), on peut donc trouver un intervalle $(\eta_1^{(1)}, \eta_2^{(1)})$, tel qu'on a constamment

$$h(\xi, \eta) > 1$$

pour toute valeur suffisamment grande de ξ et pour $\eta_1^{(1)} \leqq \eta \leqq \eta_2^{(1)}$. Cependant l'intégrale

$$\int_{\eta_1^{(1)}}^{\eta_2^{(1)}} h(\xi, \eta) d\eta$$

doit encore tendre vers l'infini avec ξ; on voit donc par le même raisonnement qu'il est possible de trouver un intervalle $(\eta_1^{(2)}, \eta_2^{(2)})$ appartenant tout entier à $(\eta_1^{(1)}, \eta_2^{(1)})$, tel que

$$h(\xi, \eta) > 2$$

pour toute valeur suffisamment grande de ξ et pour $\eta_1^{(2)} \leqq \eta \leqq \eta_2^{(2)}$. Cela se continue indéfiniment, et on est amené à considérer la suite des intervalles $I_n = (\eta_1^{(n)}, \eta_2^{(n)})$, dont chacun est situé tout entier à l'intérieur du précédent et qui satisfont à la condition

$$h(\xi, \eta) > n$$

pour toute valeur suffisamment grande de ξ et pour $\eta_1^{(n)} \leqq \eta \leqq \eta_2^{(n)}$. Or, on sait qu'il y a certainement un point η_0 appartenant à tous les I_n, et dans ce point on doit avoir

$$h(\xi, \eta_0) > n$$

pour toute valeur suffisamment grande de ξ et pour tout entier positif n; il s'ensuit donc

$$\lim_{\xi \to \infty} h(\xi, \eta_0) = \infty,$$

ce qui démontre notre théorème I.

4. La méthode qui vient d'être appliquée conduit à la résolution de bien d'autres questions du même genre. Ayant appris qu'une bande comprise entre deux parallèles dont la distance peut être prise arbitrairement petite renferme dans certains cas une infinité des points P, il est naturel de se demander comment vont se distribuer ces points dans une telle bande. Voici un théorème qui répond dans une certaine mesure à cette question.[1]

II. *Soit* $F(x) > 0$ *une fonction non croissante de la variable positive* x, *tendant vers zéro lorsque* x *tend vers l'infini. Désignons par* $\sum\limits_{p}$ *une somme étendue à tous les nombres premiers positifs, par* $\sum\limits_{p,\eta}$ *une somme s'étendant seulement aux abscisses des points* P *aux coordonnées positives dont la distance à la droite* $y - \eta x = 0$ *soit inférieure au nombre fixe et positif* ε. $\left(\textit{Si } \varepsilon < \frac{1}{\sqrt{2}}, \textit{ on a donc } \sum\limits_{p} = \sum\limits_{p,1}.\right)$ *Cela posé, deux cas peuvent se présenter:*

A. Si la série $\sum\limits_{p} F(p)$ *est convergente,* $\sum\limits_{p,\eta} F(p) \log p$ *converge »presque partout» dans le domaine* $\eta > 0$.

B. Au contraire, $\sum\limits_{p} F(p)$ *étant divergente,* $\sum\limits_{p,\eta} F(p) \log p$ *diverge dans un ensemble de valeurs de* η *partout dense dans le même domaine.*

Pour établir ce théorème, on considère la fonction[2]

$$\bar{h}(\xi, \eta) = \sum_{p < \xi} F(p)[\vartheta(\eta p + \varepsilon) - \vartheta(\eta p - \varepsilon)]$$

[1] Pour simplifier, nous ne considérons ici que les droites passant par l'origine.

[2] Le fait que nous avons remplacé ici $\varepsilon\sqrt{1+\eta^2}$ par ε n'a aucune importance.

et la fonction analogue $\overline{g}(\xi, \eta)$. En faisant les mêmes calculs que tout à l'heure, on voit sans difficulté que dans le premier cas l'intégrale

$$\int_{\eta_1}^{\eta_2} \overline{g}(\xi, \eta)$$

reste finie lorsque ξ tend vers l'infini, tandis qu'il n'en est pas ainsi dans le second cas. Dans le premier cas, on peut donc voir immédiatement que l'intégrale obtenue en remplaçant $\overline{g}$ par $\overline{h}$ reste finie pour $\xi \to \infty$.

Or, $\overline{h}(\xi, \eta)$ étant une fonction non décroissante et non négative de ξ pour chaque valeur constante de η, on doit avoir[1]

$$\lim_{\xi\to\infty} \int_{\eta_1}^{\eta_2} \overline{h}(\xi, \eta)d\eta = \int_{\eta_1}^{\eta_2} \lim_{\xi\to\infty} \overline{h}(\xi, \eta)d\eta,$$

la dernière integrale étant comprise dans le sens qu'à donné à ce signe M. LEBESGUE. La fonction $\lim\limits_{\xi\to\infty} \overline{h}(\xi, \eta)$ est donc sommable dans tout intervalle (η_1, η_2); elle doit donc avoir presque partout une valeur finie.

Au contraire, dans le second cas, le passage de $\overline{g}$ à $\overline{h}$ dans l'intégrale exige un peu d'attention; elle réussit cependant grâce à la monotonie de $F(x)$. La fin de la démonstration est cette fois tout à fait analogue à celle du théorème I. — Les résultats obtenus sur la fonction $\overline{h}(\xi, \eta)$ sont évidemment équivalents au théorème II.

Enfin, par un changement léger, la méthode suivie conduit à ce dernier résultat:

III. *Etant donné un nombre positif incommensurable η inférieur à 1, on sait*[2] *qu'il est possible de trouver une infinité de fractions irréductibles $\frac{m}{n}$ telles que $\left|\eta - \frac{m}{n}\right| < \frac{1}{n^2}$. Si l'on exige encore que m et n soient des nombres premiers, cela n'est plus possible que pour un ensemble de mesure nulle de valeurs de η.*

[1] Voir C. DE LA VALLÉE POUSSIN, »Sur l'intégrale de Lebesgue», Trans. of the American Math. Soc., Vol. 16 (1915), p. 447.

[2] Voir MINKOWSKI, »Diophantische Approximationen», p. 2.

Pour la démonstration, on formera la fonction

$$\bar{\bar{h}}(\xi, \eta) = \sum_{p < \xi} \frac{1}{\log p}\left[\vartheta\left(\eta p + \frac{1}{p}\right) - \vartheta\left(\eta p - \frac{1}{p}\right)\right],$$

et l'on procédera comme au cas A du théorème précédent.

Les résultats qui viennent d'être exposés sont très incomplets et très insuffisants. Cependant elles concernent des questions d'un grand intérêt; j'espère pouvoir y revenir un jour avec plus de détails et plus de résultats.

Tryckt den 11 april 1919.

Uppsala 1919. Almqvist & Wiksells Boktryckeri-A.-B.

8.

Bemerkung zu der vorstehenden Arbeit des Herrn E. Landau

Math. Z. **6** (1/2), 155–157 (1920)

In der vorstehenden Arbeit[1]) verschärft Herr E. Landau eine früher von mir[2]) gegebene Abschätzung für das Restglied $R(T)$ der Formel

$$N(T) = \frac{T}{2\pi}\left(\log\frac{T}{2\pi} - 1\right) + R(T).$$

Es bezeichnet hier $N(T)$ wie gewöhnlich die Anzahl der Wurzeln $\beta + \gamma i$ der Gleichung $\zeta(s) = 0$, die der Beziehung $0 < \gamma \leqq T$ genügen, und es wird bei Herrn Landau, wie auch in dieser Arbeit, die Riemannsche Vermutung als wahr angenommen, so daß immer $\beta = \frac{1}{2}$ ist.

Es erscheint mir nun recht eigentümlich, daß diese Resultate, welche wesentlich aus einem Littlewoodschen Satze über die Größenordnung der Funktion $\log\zeta(s)$ im kritischen Streifen (vgl. Gleichung (2) unten) gefolgert werden, umgekehrt eine Verschärfung jenes Satzes zulassen. Dasselbe gilt für meine frühere Abschätzung von $R(T)$, wie ich schon vor mehr als einem Jahre anläßlich meiner diesbezüglichen Arbeit gefunden habe. Durch die Anwendung der Gleichung (6) der Landauschen Arbeit wird es aber möglich, der angedeuteten Verschärfung eine besonders einfache Form zu geben, wie ich hier ausführen will.

Es lautet jene Gleichung

$$N\left(t + 2\frac{\log\log\log t}{\log\log t}\right) - N(t) = O\left(\log t \frac{\log\log\log t}{\log\log t}\right).$$

Wie Herr Landau nehme ich hier t wurzelfrei und $> e^e$ an und teile das Intervall $(t-1, t+1)$ von t ausgehend nach beiden Seiten in

[1]) E. Landau, Über die Nullstellen der Zetafunktion, Mathematische Zeitschrift, 6 (1920), S. 151–154.

[2]) H. Cramér, Über die Nullstellen der Zetafunktion, Mathematische Zeitschrift, 2 (1918), S. 237–241.

Teilintervalle der Länge $\delta = \delta(t) = \frac{\log\log\log t}{\log\log t}$ ein. In jedem Teilintervall wird dann die Anzahl der γ

$$O(\delta \log t).$$

Wird nun $\frac{1}{2} + \frac{1}{\log\log t} \leqq \sigma \leqq 2$ angenommen, so gilt gleichmäßig in σ

$$\frac{\zeta'}{\zeta}(s) = \frac{\zeta'}{\zeta}(\sigma + it) = \sum \frac{1}{s - \frac{1}{2} - i\gamma} + O(\log t),$$

wo die Summe über die der Bedingung $|t - \gamma| < 1$ genügenden γ zu erstrecken ist[3]). Ich fasse nun wie Herr Landau die den einzelnen Teilintervallen zugehörenden Glieder zusammen, und es ergibt sich weiter:

$$\begin{aligned}\frac{\zeta'}{\zeta}(s) &= O\left(\sum_{|t-\gamma|<1} \frac{1}{\sqrt{\left(\frac{1}{\log\log t}\right)^2 + (t-\gamma)^2}}\right) + O(\log t)\\
&= O\left[\sum_{|t-\gamma|<1} \operatorname{Min}\left(\log\log t, \frac{1}{|t-\gamma|}\right)\right] + O(\log t)\\
&= O\left[\delta \log t\left(\log\log t + \frac{1}{\delta} + \frac{1}{2\delta} + \dots + \frac{1}{\left[\frac{1}{\delta}\right]\delta}\right) + \log t\right]\\
&= O\left(\log t \cdot \log\log\log t + \log t \cdot \log \frac{1}{\delta}\right)\\
&= O(\log t \cdot \log\log\log t).\end{aligned}$$

Wegen

$$\log \zeta\left(\frac{1}{2} + \frac{1}{\log\log t} + it\right) = \int_2^{\frac{1}{2} + \frac{1}{\log\log t}} \frac{\zeta'}{\zeta}(\sigma + it)\, d\sigma + O(1)$$

gilt aber für $\sigma = \frac{1}{2} + \frac{1}{\log\log t}$ auch

$$\log \zeta(s) = O(\log t \cdot \log\log\log t).$$

Andererseits gilt für $\sigma = 1$ nach dem Littlewoodschen Satze (vgl. Gleichung (2) unten)

$$\log \zeta(s) = O(\log\log\log t).$$

Die Hilfsfunktion

$$\frac{\log \zeta(s)}{[\log(-is)]^{2(1-s)} \log\log\log(-is)}$$

ist also für hinreichend große t_0 auf dem Rande des Gebietes $\frac{1}{2} + \frac{1}{\log\log t} \leqq \sigma \leqq 1$, $t > t_0$, beschränkt. Da sie weiter im Innern dieses Gebietes regulär bleibt und jedenfalls nicht schneller als eine endliche

[3]) Vgl. die vorstehende Arbeit des Herrn Landau.

Potenz von t wächst, ist sie nach dem bekannten Phragmén-Lindelöfschen Satze auch dort beschränkt. Hieraus folgert man aber durch eine einfache Rechnung

$$\log\zeta(s) = O\left[(\log t)^{2(1-\sigma)}\log\log\log t\right] \tag{1}$$

gleichmäßig für $\frac{1}{2} + \frac{1}{\log\log t} \leqq \sigma \leqq 1, \quad t > 0.$

Der mehrfach erwähnte Littlewoodsche Satz besagt nun

$$\log\zeta(s) = O\left[\left(\frac{\log t \log\log t}{\log\log\log t}\right)^{2(1-\sigma)}\log\log\log t\right] \tag{2}$$

gleichmäßig in demselben Gebiete, so daß in der Tat (1) für jedes $\sigma < 1$ ein schärferes Resultat ergibt.

Mörby, Stocksund, den 20. Juli 1919.

(Eingegangen am 24. Juli 1919.)

9.

Some theorems concerning prime numbers

Ark. Mat. Astr. Fys. **15** (5), 1–33 (1920)

Communicated November 26th 1919 by I. BENDIXSON and H. VON KOCH.

Contents.

Introduction.

The present paper contains some applications of the theory of the RIEMANN ζ-function to various questions concerning primes. It is well known that a very important part is played in the Theory of Prime Numbers by certain infinite series, whose terms involve the complex roots $\varrho = \beta + \gamma i$ of the function $\zeta(s)$. Thus the function $\psi(x)$ defined as[1]

$$\psi(x) = \sum_{p^m \leqq x} \log p$$

---

[1] The notation p^m should always be so understood that m is an arbitrary integer and p an arbitrary prime.

is given by the expression[1]

$$\psi(x) = x - \sum_{\varrho} \frac{x^{\varrho}}{\varrho} - \frac{1}{2}\log\left(1 - \frac{1}{x^2}\right) - \log 2\pi$$

for any $x \neq p^m$. There is an analogous, but more complicated expression for the function

$$f(x) = \sum_{p^m \leqq x} \frac{1}{m} = \pi(x) + \frac{1}{2}\pi\left(x^{\frac{1}{2}}\right) + \frac{1}{3}\pi\left(x^{\frac{1}{3}}\right) + \cdots,$$

where $\pi(x)$ denotes the number of primes $\leqq x$.

The infinite series occurring in these expressions are not absolutely convergent, and this makes it a very difficult task to obtain any definitive results from their consideration. It might be expected that the introduction of complex variables should make them more easy to deal with, and this point of view has in fact been adopted by Messrs HARDY and LITTLEWOOD[2] in their proof of an important theorem concerning $\psi(x)$.

In two recently published papers[3], I have treated the function

$$V(z) = V(x + yi) = \sum_{\gamma \geqq 0} e^{\varrho z},$$

(the sum contains only the roots ϱ situated in the upper half-plane), which is readily seen to be closely related to the series occurring in the expression for $\psi(x)$. This function has poles of the first order in the points $z = \pm \log p^m$, and the function

$$V(z) - \frac{\log z}{2\pi i(1 - e^{-z})}$$

---

[1] See E. LANDAU, »Handbuch der Lehre von der Verteilung der Primzahlen», Leipzig und Berlin 1909, p. 365.

[2] HARDY and LITTLEWOOD, »Contributions to the theory of the Riemann zeta-function and the theory of the distribution of primes». Acta Math., Vol. 41, p. 184.

[3] CRAMÉR, »Sur les zéros de la fonction $\zeta(s)$», Comptes rendus, T. 168, p. 539, and Studien über die Nullstellen der Riemann'schen Zetafunktion, Mathem. Zeitschrift, B. 4, p, 104.

is meromorphic all over the plane. From the expression of $V(z)$ given in the quoted papers, it is easy to deduce the following formula which is of a fundamental importance for the first three chapters of this paper:

$$\sum_{p,\,m} \frac{1}{m p^m (z - \log p^m)} = -\frac{\pi i}{z} + \frac{2\pi i}{z} \sum_{\gamma > 0} e^{(\varrho - 1) z} +$$

$$+ \int_0^1 e^{-tz} \log |\zeta(1-t)|\, dt + O(e^{-x}),$$

uniformly for $0 < y < 1$, as $x \to +\infty$. I shall now pass on to a brief account of the applications made by the aid of this formula.

If p_n denotes the n:th prime, it follows from the 'Prime Number Theorem'

$$\pi(x) \backsim \frac{x}{\log x}$$

that the average order of the difference $p_{n+1} - p_n$ is equal to $\log p_n$. But the problems concerning the maximum and the minimum order of this difference are extremely difficult and only very little investigated. Though it seems highly probable that $p_{n+1} - p_n = 2$ for an infinity of values of n, this has not yet been proved, and, as far as I know, it has not even been proved that

$$p_{n+1} - p_n < \frac{1}{2} \log p_n$$

or

$$p_{n+1} - p_n > 2 \log p_n$$

for an infinity of values of n. The only result which is rigorously proved is that

$$p_{n+1} - p_n = O\left(p_n e^{-\alpha \sqrt{\log p_n}}\right),$$

where α is a positive constant, and that, if the RIEMANN hypothesis is true,

$$p_{n+1} - p_n = O(\sqrt{p_n} \log^2 p_n).$$

This follows from known results concerning the function $\pi(x)$.[1] In the first chapter of this paper, it is shown that the last relation may be replaced by

$$p_{n+1} - p_n = O(\sqrt{p_n} \log p_n);$$

thus there is a constant c such that the interval $[n^2, (n + c \log n)^2]$ always contains at least one prime.

In his well-known Cambridge lecture[2], Prof. LANDAU mentions some unsolved problems belonging to this range of ideas. One of them is the following: '*is there always at least one prime between* n^2 *and* $(n+1)^2$?' — It is readily seen that this would be proved, if it were possible to show that we always have

$$p_{n+1} - p_n < 2\sqrt{p_n}.$$

I am far away from being able to prove such a deep result, but I can show that, *if* there are integers n such that the interval $[n^2, (n+1)^2]$ contains no prime, and if we denote by $q(x)$ the number of such integers $n < x$, then we have, assuming the thruth of the RIEMANN hypothesis,

$$q(x) = O\left(x^{\frac{2}{3}+\varepsilon}\right).$$

for any $\varepsilon > 0$. The proof, which is given in the second chapter, is based upon a lemma obtained by an arithmetical method first employed by WEYL.[3]

The third chapter is devoted to a study of the average order of the function

$$\left|\frac{\psi(x) - x}{\sqrt{x}}\right|,$$

---

[1] LANDAU, Handbuch etc., p. 329 and 388.

[2] »Gelöste und ungelöste Probleme aus der Theorie der Primzahlvertelung und der Riemannschen Zetafunktionen«; Jahresber. d. deutschen Math. Ver., B. 21 (1912).

[3] WEYL, »Über Gleichverteilung von Zahlen modulo Eins«, Math. Ann. B. 77, p. 344. Cf. the additional note at the end of the paper.

assuming the truth of the RIEMANN hypothesis. It has been proved by VON KOCH[1], that this function is of the form

$$O(\log^2 x),$$

and by HARDY and LITTLEWOOD[2] that it is *not* of the form

$$o(\log\log\log x).$$

The principal result obtained here is expressed by the relation

$$\frac{1}{x}\int_1^x \left|\frac{\psi(t)-t}{\sqrt{t}}\right| dt = O(1)$$

and by the following analogous relation, where the meanvalue is formed for the interval $(x, x+h)$, h being an arbitrary increasing function of x such that $h(x) = O(x)$:

$$\frac{1}{h}\int_x^{x+h} \left|\frac{\psi(t)-t}{\sqrt{t}}\right| dt = o\left(\sqrt{\frac{x}{h}}\right).$$

The latter result is trivial, if the increase of h is not greater than that of $\frac{x}{(\log x)^4}$. — From the expression given above for $\psi(x)$ it follows that, if the RIEMANN hypothesis is true,

$$\psi(x) = x - 2\sqrt{x}\sum_{\gamma>0}\frac{\sin(\gamma\log x)}{\gamma} + O(\sqrt{x}),$$

and the mean-value theorems concerning $\psi(x)$ may be deduced from this formula. This requires, however, a somewhat delicate discussion of a certain double series which is neither absolutely nor uniformly convergent, and thus I have perferred to deduce them as corollaries of the following theorem which may be found to present also some interest in itself: We have

---

[1] VON KOCH, »Sur la distribution des nombres premiers», Acta Mathematica, Vol. 24, p. 159.

[2] HARDY and LITTLEWOOD, l. c., p. 184.

$$\psi(x) = x - \sum_{\varrho} \frac{x^{\varrho}}{\varrho} e^{-\frac{|\gamma|}{x^3}} + O(\log^2 x)$$

and, if the RIEMANN hypothesis is true,

$$\psi(x) = x - 2\sqrt{x} \sum_{\gamma > 0} \frac{\sin(\gamma \log x)}{\gamma} e^{-\frac{\gamma}{x^3}} + O(\sqrt{x}).$$

The introduction of convergence factors into the series occurring in the expression for $\psi(x)$ has been proposed by several writers[1], but I believe that the two last relations will be found to be particularly simple.

Finally, the last chapter of the paper brings some generalizations of theorems proved in my above quoted paper in the Mathematische Zeitschrift. The general DEDEKIND zeta function $\zeta(s)$ corresponding to an arbitrary algebraic field is considered, and the function

$$V(z) = \sum_{\gamma > 0} e^{\varrho z}$$

is formed. Here, ϱ denotes any complex root of the equation $\zeta(s) = 0$. A formula representing $V(z)$ for all values of z is deduced, and by the use of this formula, generalizations of different theorems are given.

CHAPTER I.

On the maximum order of the difference between two consecutive primes.

In a paper quoted in the Introduction, I have deduced the following formula[2], where $z = x + iy$ and $y > 0$,

[1] See VON KOCH, »Contribution à la théorie des nombres premiers», Acta Mathematica, Vol. 33, p. 293; LANDAU, »Über einige Summen, die von den Nullstellen der Riemann'schen Zetafunktion abhängen», Acta Mathematica, Vol. 35, p. 271; HAMMERSTEIN, »Zwei Beiträge zur Zahlentheorie», Inauguraldissertation, Göttingen 1919.

[2] See formula (13), page 114 of my paper published in the 'Mathematische Zeitschrift'. The outlines of a proof of the analogous formula corresponding to a more general case are given in the last chapter of the present paper.

$$\sum_{\gamma>0} e^{\varrho z} = \frac{z}{2\pi i} e^{z} \sum_{p,m} \frac{1}{m p^m (z - \log p^m)} -$$

$$- \frac{z}{2\pi i} \sum_{p,m} \frac{1}{m p^m (z - \log p^m)} + \left(\frac{1}{4} + \frac{C + \log 2\pi}{2\pi i}\right)\left(1 + \frac{1}{z}\right) +$$

$$+ \frac{1}{2\pi i} \frac{\Gamma'}{\Gamma}\left(\frac{z}{\pi i}\right) + \frac{1}{2} e^{z} - \frac{z}{2\pi i} \int_0^1 e^{sz} \log |\zeta(s)| \, ds -$$

$$- \frac{1}{2\pi i z} \int_0^\infty \frac{t}{e^t - 1} \cdot \frac{dt}{t + z}.$$

Hence it may be deduced by some easy calculation that we have

$$(1) \qquad \sum_{p,m} \frac{1}{m p^m (z - \log p^m)} = - \frac{\pi i}{z} + \frac{2\pi i}{z} \sum_{\gamma \geq 0} e^{(\varrho - 1) z} +$$

$$+ \int_0^1 e^{-tz} \log |\zeta(1 - t)| \, dt + O(e^{-x}),$$

uniformly for $0 < y < 1$, as $x \to +\infty$. Taking the imaginary part on both sides of this equation, we obtain

$$\sum_{p,m} \frac{y}{m p^m [(x - \log p^m)^2 + y^2]} = \frac{\pi x}{x^2 + y^2} - 2\pi \Re \sum_{\gamma>0} \frac{e^{(\varrho - 1) z}}{z} +$$

$$+ \int_0^1 e^{-tx} \sin ty \log |\zeta(1 - t)| \, dt + O(e^{-x}).$$

It is, however, evident that the first and third terms of the right hand side are of the form

$$\frac{\pi}{x} + O\left(\frac{y^2}{x^3}\right)$$

and

$$O\int_0^1 yt \log\frac{1}{t}\, e^{-tx} dt = O\left(\frac{y\log x}{x^2}\right)$$

respectively; hence

$$(2)\qquad \sum_{p,m}\frac{y}{mp^m[(x-\log p^m)^2+y^2]}=\frac{\pi}{x}-2\pi\Re\sum_{\gamma>0}\frac{e^{(\varrho-1)z}}{z}+$$

$$+O\left(\frac{y\log x}{x^2}+e^{-x}\right),$$

where y may denote any positive function of x which tends to zeroas x tends to infinity.

The further treatment of this relation will be quite different according as the Riemann hypothesis is assumed or not. Our object will only be to prove the following theorem, corresponding to the former case.

Theorem I. *Suppose the* Riemann *hypothesis is true. Then it is possible to find a positive constant* c *such that*

$$\pi(x+c\sqrt{x}\log x)-\pi(x)>\sqrt{x}$$

for $x\geqq 2$. *Thus if* p_n *denotes the n:th prime, we have*

$$p_{n+1}-p_n=O(\sqrt{p_n}\log p_n).$$

The Riemann hypothesis being assumed, it follows from (2) that

$$\left|\sum_{p,m}\frac{y}{mp^m[(x-\log p^m)^2+y^2]}-\frac{\pi}{x}\right|<$$

$$<2\pi\frac{e^{-\frac{1}{2}x}}{\sqrt{x^2+y^2}}\sum_{\gamma>0}e^{-\gamma y}+O\left(\frac{y\log x}{x^2}+e^{-x}\right).$$

If in this relation we put

(3) $$y = x e^{-\frac{1}{2}x}$$

and use the relation[1]

$$\sum_{\gamma > 0} e^{-\gamma y} \backsim \frac{1}{2\pi y} \log \frac{1}{y},$$

then we obtain

(4) $$\sum_{p,m} \frac{y}{m p^m [(x - \log p^m)^2 + y^2]} = \frac{\pi + \Theta(x)}{x},$$

where $|\Theta(x)| < 1$ for all sufficiently great values of x. Hence in particular

$$\sum_{|\log p^m - x| \leq y} \frac{y}{m p^m [(x - \log p^m)^2 + y^2]} < \frac{5}{x},$$

$$\frac{y}{e^x . 2 y^2} \sum_{|\log p^m - x| \leq y} \frac{1}{m} < \frac{6}{x},$$

(5) $$\sum_{|\log p^m - x| \leq y} \frac{1}{m} < 12 \frac{y}{x} e^x = 12 e^{\frac{1}{2}x}.$$

On the other hand, it follows from (4) that

$$\sum_{p,m} \frac{y}{m p^m [(x - \log p^m)^2 + y^2]} > \frac{2}{x}$$

for all sufficiently great values of x. To estimate the contribution to the left hand side of the terms corresponding to $|\log p^m - x| > 1$, we group together the terms where $\log p^m$ belongs to one of the intervals $(0, 1)$, $(1, 2)$, ... $([x]-1, [x])$, $([x]+1, [x]+2)$, $([x]+2, [x]+3)$, ... Observing that

[1] See p. 123 of my paper quoted above. Compare also the last chapter of the present paper.

$$\sum_{\nu \leqq \log p^m \leqq \nu+1} \frac{1}{m} < f(e^{\nu+1}) = O\left(\frac{e^\nu}{\nu}\right),$$

we obtain

$$\sum_{|\log p^m - x| > 1} \frac{y}{m p^m[(x - \log p^m)^2 + y^2]} =$$

$$(6) \qquad = O\left(\frac{y e^1}{1(x-1)^2 e^1} + \frac{y e^2}{2(x-2)^2 e^2} + \cdots\right.$$

$$+ \frac{y e^{[x]-1}}{([x]-1)(x-[x]+1)^2 e^{[x]-1}} + \frac{y e^{[x]}}{[x] \,.\, 1^2 e^{[x]}} +$$

$$\left. + \frac{y e^{[x]+1}}{([x]+1) \,.\, 1^2 \,.\, e^{[x]+1}} + \frac{y e^{[x]+2}}{([x]+2)([x]+2-x)^2 e^{[x]+2}} + \cdots\right) =$$

$$= O(y).$$

Thus

$$\sum_{|\log p^m - x| \leqq 1} \frac{y}{m p^m[(x - \log p^m)^2 + y^2]} > \frac{1{,}5}{x}$$

for all sufficiently great values of x.

Now, we divide the interval $(x-1, x+1)$ into sub-intervals of the length $y = x e^{-\frac{1}{2}x}$, starting from the point x, (the two extreme intervals being possibly $< y$) and then all the terms where $\log p^m$ belongs to the same sub-interval are grouped together. Then (5) shows that the expression $\Sigma \frac{1}{m}$ corresponding to anyone of the sub-intervals will be less than $24 \frac{y}{x} e^x$. Let us exclude the first ν sub-intervals in both directions from the point x and try to estimate the sum of all the other terms. This will give us

$$\sum_{\nu y \leqq |x - \log p^m| \leqq 1} \frac{y}{m p^m[(x - \log p^m)^2 + y^2]} <$$

$$< 24 \frac{y}{x} e^x \cdot \frac{y}{e^{x-1}} \left(\frac{1}{\nu^2 y^2} + \frac{1}{(\nu+1)^2 y^2} + \cdots\right)$$

$$= \frac{24\, e}{x}\left(\frac{1}{\nu^2} + \frac{1}{(\nu+1)^2} + \cdots\right).$$

Thus it is possible so to choose ν that

$$\sum_{|\log p^m - x| < \nu y} \frac{y}{m p^m[(x - \log p^m)^2 + y^2]} > \frac{1,2}{x},$$

whence

$$\sum_{|\log p^m - x| < \nu y} \frac{1}{m} > 1,2 \frac{y}{x} e^{x - \nu y} > e^{\frac{1}{2}x}$$

and

$$f(e^{x+\nu y}) - f(e^{x-\nu y}) > e^{\frac{1}{2}x}.$$

By writing x in the place of e^x, it is readily seen that the last relation implies

$$f(x + 2\nu\sqrt{x}\log x) - f(x - 2\nu\sqrt{x}\log x) > \sqrt{x}.$$

The same inequality will hold for $\pi(x)$ because of the relation

$$f(x) - \pi(x) = O\left(\frac{\sqrt{x}}{\log x}\right).$$

Finally, a new change of the independent variable shows that

$$\pi(x + 5\nu\sqrt{x}\log x) - \pi(x) > \sqrt{x};$$

thus theorem I is proved.

If the RIEMANN hypothesis is not assumed, it is still possible to obtain from (2) some definite result, and the upper limit which is obtained in this case for the difference $p_{n+1} - p_n$ is a little better than what may be deduced from the inequalities for $\pi(x)$ given in LANDAU's Handbuch. But as this result does not seem to be of much real interest, it is omitted.

CHAPTER II.

On the existence of primes between n^2 and $(n+1)^2$.

Theorem II.[1] *If the* Riemann *hypothesis is true, and if $q(x)$ denotes the number of positive integers $n < x$, such that there is no prime between the limits n^2 and $(n+1)^2$, then we have*

$$q(x) = O\left((\log x)^3 x^{\frac{2}{3}}\right).$$

For the proof of this theorem, we shall make use of some lemmas.

Lemma I. *Whether we assume the* Riemann *hypothesis or not, there are two constants x_0 and v_0 such that the relation*

$$\int_{x-2}^{x+2} \left| \sum_{0<\gamma\leqq v} e^{\gamma i t} \right| dt < x v^{\frac{2}{3}} \log v$$

holds for $x > x_0$, $v > v_0$.

Proof. We put

(7) $$\sum_{0<\gamma\leqq v} e^{\gamma i t} = \sum_{\nu=1}^{n} e^{\gamma_\nu i t} = \varphi_n(t),$$

where $\gamma_1 \leqq \gamma_2 \leqq \cdots \leqq \gamma_n \leqq v$, and

(8) $$n = N(v) = \frac{v}{2\pi}\left(\log \frac{v}{2\pi} - 1\right) + O(\log v).$$

Then, we put

(9) $$\mu_n = \left[\left(\frac{n}{\log n}\right)^{\frac{1}{3}}\right]$$

and determine a succession of positive integers $\lambda_1^{(n)}, \lambda_2^{(n)} \dots \lambda_n^{(n)}$ such that

$$\left|\gamma_\nu - \frac{\lambda_\nu^{(n)}}{\mu_n}\right| \leqq \frac{1}{2\mu_n}.$$

[1] Cf. the additional note at the end of the paper.

Writing

$$\psi_n(t) = \sum_{\nu=1}^{n} e^{\frac{\lambda_\nu^{(n)} it}{\mu_n}},$$

we thus obtain

$$|\varphi_n(t) - \psi_n(t)| \leqq \sum_{\nu=1}^{n} \left| e^{\gamma_\nu it} - e^{\frac{\lambda_\nu^{(n)} it}{\mu_n}} \right|$$

$$\leqq \sum_{\nu=1}^{n} \left| \gamma_\nu - \frac{\lambda_\nu^{(n)}}{\mu_n} \right| \cdot t$$

$$\leqq \frac{n}{2\mu_n} \cdot t.$$

Hence it follows that

$$(10) \quad \int_{x-2}^{x+2} |\varphi_n(t)|\, dt \leqq \int_{x-2}^{x+2} |\varphi_n(t) - \psi_n(t)|\, dt + \int_{x-2}^{x+2} |\psi_n(t)|\, dt$$

$$\leqq \frac{2n}{\mu_n}(x+2) + \int_{x-2}^{x+2} |\psi_n(t)|\, dt.$$

Further, by the use of SCHWARZ's inequality, we have

$$(11) \quad \int_{x-2}^{x+2} |\psi_n(t)|\, dt \leqq 2 \sqrt{\int_{x-2}^{x+2} |\psi_n(t)|^2\, dt}.$$

From a certain n onwards, μ_n is greater than 1, and thus

$$\int_{x-2}^{x+2} |\psi_n(t)|^2\, dt < \int_{x-\mu_n\pi}^{x+\mu_n\pi} |\psi_n(t)|^2\, dt = \int_{0}^{2\mu_n\pi} |\psi_n(t)|^2\, dt =$$

$$= 2\mu_n\pi \int_0^1 \left| \sum_{\nu=1}^{n} e^{2\pi i \lambda_\nu^{(n)} t} \right|^2 dt = 2\mu_n\pi \int_0^1 \sum_{h,k=1}^{n} e^{2\pi it(\lambda_h^{(n)} - \lambda_k^{(n)})}\, dt.$$

Suppose now, that the integers $\lambda_1^{(n)} \dots \lambda_n^{(n)}$ divide up into groups of $h_1, h_2 \dots . h_p$ members, two numbers of the same groups being always, two numbers of different groups being never equal. Then the last relation gives

$$\int_{x-2}^{x+2} |\psi_n(t)|^2 dt < 2\mu_n \pi (h_1^2 + h_2^2 + \cdots h_p^2)$$

$$< 2 n \mu_n \pi \operatorname{Max}(h_1, h_2, \dots h_p).$$

But as the number of γ:*s* in the interval $(v, v+1)$ is of the form $O(\log v)$, we have for sufficiently great values of n

$$\int_{x-2}^{x+2} |\psi_n(t)|^2 dt < K n \mu_n \log n,$$

where K is an absolute constant. Hence by (9), (10) and (11)

$$\int_{x-2}^{x+2} |\varphi_n(t)| dt < \frac{2n}{\mu_n}(x+2) + 2\sqrt{K n \mu_n \log n}$$

$$< 3x(n^2 \log n)^{\frac{1}{3}}$$

for all sufficiently great values of n and x. Finally, the truth of the lemma follows by the use of (7) and (8).

Lemma II. *Let us assume the* RIEMANN *hypothesis. If we put* $y = y(t) = \lambda e^{-\frac{1}{2}t}$, *where* λ *is a positive constant, then the set* S_x *of points* t *inside* $(x-2, x+2)$, *such that at least one of the relations*

$$f(e^{t+y}) - f(e^t) < \frac{10y}{t} e^t \tag{12}$$

and

$$\frac{2\pi}{t} e^{-\frac{1}{2}t} \left| \sum_{\gamma>0} \cos \gamma t e^{-\gamma y} \right| < \frac{1}{t} \tag{13}$$

is not satisfied, is of measure M_x *such that*

$$M_x = O\left(x^2 e^{-\frac{1}{6}x}\right),$$

where the constant implied by the O may depend on λ. Further, we are going to show that the truth of (13) *implies the truth of* (12), *so that* S_x *may be defined as the set of points inside* $(x-2,\ x+2)$ *such that* (13) *is not satisfied.*[1]

Proof. We have

$$\left|\sum_{\gamma>0} \cos \gamma t\, e^{-\gamma y}\right| \leqq \left|\sum_{\gamma>0} e^{\gamma i t} . e^{-\gamma y}\right|$$

$$< y \int_0^\infty \left|\sum_{0<\gamma\leqq v} e^{\gamma i t}\right| e^{-vy} dv.$$

Hence by Lemma I (if we put $y_1 = y(x-2)$ and $y_2 = y(x+2)$)

$$\int_{x-2}^{x+2} \left|\sum_{\gamma>0} \cos \gamma t e^{-\gamma y}\right| dt = O \int_0^\infty x y_1 v^{\frac{2}{3}} \log v e^{-v y_2} dv$$

$$= O\left(x y_1 y_2^{-\frac{5}{3}} \log \frac{1}{y_2}\right)$$

$$= O\left(x^2 e^{\frac{1}{3}x}\right).$$

Thus, the measure of the set of points t inside $(x-2,\ x+2)$, such that (13) is not satisfied, is of the form

$$O\left(x^2 e^{\frac{1}{3}x} . e^{-\frac{1}{2}x}\right) = O\left(x^2 e^{-\frac{1}{6}x}\right).$$

To prove Lemma II, it is now only necessary to show that the truth of (13) implies the truth of (12).

Since the Riemann hypothesis is supposed to be true, the relation (2) shows that

[1] All the sets of points we shall have to deal with consist of a finite number of intervals; hence they are always measurable.

$$(14)\quad \sum_{p,m} \frac{y}{mp^m[(t-\log p^m)^2+y^2]} =$$

$$= \frac{\pi}{t} - \frac{2\pi}{|t+iy|} e^{-\frac{1}{2}t} \sum_{\gamma>0} \cos\left(\gamma t - \frac{1}{2}y - \arg(t+iy)\right) e^{-\gamma y} + $$
$$+ O\left(e^{-\frac{1}{2}t}\right)$$

$$= \frac{\pi}{t} - \frac{2\pi}{t} e^{-\frac{1}{2}t} \sum_{\gamma>0} \cos \gamma t e^{-\gamma y} + O\left(t e^{-\frac{1}{2}t}\right).$$

Hence it follows that, if (13) is true,

$$\sum_{p,m} \frac{y}{mp^m[(t-\log p^m)^2+y^2]} < \frac{4{,}5}{t},$$

and thus

$$\sum_{t<\log p^m \leqq t+y} \frac{1}{m} < \frac{9y}{t} e^{t+y} < \frac{10y}{t} e^t$$

as soon as t is sufficiently great. The last relation is identical with (12)

Lemma III. *If we assume the* RIEMANN *hypothesis, and if s_x denotes the set of points t inside $(x-1, x+1)$, such that*

$$\pi\left(e^{t+\frac{1}{2}e^{-\frac{1}{2}t}}\right) - \pi\left(e^{t-\frac{1}{2}e^{-\frac{1}{2}t}}\right) \leqq c\frac{e^{\frac{1}{2}t}}{t},$$

then the positive constant c may be so chosen that the measure m_x of s_x satisfies

$$m_x = O\left(x^3 e^{-\frac{1}{6}x}\right).$$

Proof. We denote by y the same function as in Lemma II, and we write as before $y_1 = y(x-2)$, $y_2 = y(x+2)$.

Let ξ_0 be a point inside $(x-2, x+2)$, and let us mark all the points $\xi_\nu = \xi_0 + \nu y_2$, $(\nu = 0, \pm 1, \pm 2, \ldots)$ which belong to this interval. It is plain that ξ_0 may be so chosen that at most $\frac{M_x}{y_2}$ of the points ξ_ν belong to the set S_x of Lemma II. We suppose this to have been done, and we exclude from the interval $(x-2, x+2)$:

Firstly: The set S_x.

Secondly: The interval $(\xi_\nu - \mu x y_2, \xi_\nu + \mu x y_2)$ for every ξ_ν belonging to S_x. (μ denotes a positive constant, which will be fixed later.)

The measure of the set thus excluded is less than

$$2\,\mu x y_2 \cdot \frac{M_x}{y_2} + M_x = O\left(x^3 e^{-\frac{1}{6}x}\right).$$

If we can show that, in the part of the complementary set which belongs to $(x-1, x+1)$, the relation

$$\pi\left(e^{t+\frac{1}{2}e^{-\frac{1}{2}t}}\right) - \pi\left(e^{t-\frac{1}{2}e^{-\frac{1}{2}t}}\right) > \frac{1}{2}\lambda\frac{e^{\frac{1}{2}t}}{t}$$

is satisfied as soon as suitable values are given to the constants λ and μ, then our lemma will be proved.

For such a value of t, (13) is satisfied, since t does not belong to S_x. Thus, we have by (6) and (14)

$$\sum_{|\log p^m - t| \leqq 1} \frac{y}{m p^m[(t-\log p^m)^2 + y^2]} > \frac{2}{t}$$

for all sufficiently great values of x. Hence

(15) $$\sum_{|\log p^m - t| < \frac{1}{2\lambda}y} \frac{y}{m p^m[(t-\log p^m)^2 + y^2]} > \frac{2}{t} -$$

$$- \sum_{\frac{1}{2\lambda}y \leqq |\log p^m - t| \leqq 1} \frac{y}{m p^m[(t-\log p^m)^2 + y^2]}.$$

Let us now consider the division of the interval $(t-1, t+1)$ into sub-intervals which is effected by the points ξ_ν. Since we have excluded from $(x-2, x+2)$ the interval $(\xi_\nu - \mu x y_2, \xi_\nu + \mu x y_2)$ for every ξ_ν belonging to S_x, there are at least $[\mu x]$ of the first ξ_ν:s on each side of the point t which do not belong to S_x, and for anyone of them we have by Lemma II

$$\sum_{\xi_\nu < \log p^m \leqq \xi_\nu + y_2} \frac{1}{m} < \sum_{\xi_\nu < \log p^m \leqq \xi_\nu + y(\xi_\nu)} \frac{1}{m} < \frac{10\, y(\xi_\nu)}{\xi_\nu} e^{\xi_\nu}.$$

This gives us, if we group together the terms belonging to the same sub-interval

$$(16) \qquad \sum_{\frac{1}{2\lambda} y \leqq |\log p^m - t| \leqq [\mu x] \cdot y_2} \frac{y}{m p^m [(t - \log p^m)^2 + y^2]} <$$

$$< \frac{10\, y_1}{x-2} \cdot \frac{e^{x+2}}{e^{x-2}} \left(\frac{2y}{\left(y\left[\frac{1}{2\lambda}\right]\right)^2} + \frac{2y}{\left(y\left[\frac{1}{2\lambda}\right] + y_2\right)^2} + \right.$$

$$\left. + \frac{2y}{\left(y\left[\frac{1}{2\lambda}\right] + 2y_2\right)^2} + \cdots \right) < \frac{20\, e^{10}}{x-2} \left(\frac{1}{\left(e^2\left[\frac{1}{2\lambda}\right]\right)^2} + \right.$$

$$\left. + \frac{1}{\left(e^2\left[\frac{1}{2\lambda}\right] + 1\right)^2} + \frac{1}{\left(e^2\left[\frac{1}{2\lambda}\right] + 2\right)^2} + \cdots \right) <$$

$$> \frac{30\, e^{10}}{e^2\left[\frac{1}{2\lambda}\right] - 1} \cdot \frac{1}{t}.$$

On the other hand we have always, whether ξ_ν belongs to S_x or not,

$$\sum_{\xi_\nu < \log p^m \leqq \xi_\nu + y_2} \frac{1}{m} < \sum_{\xi_\nu < \log n \leqq \xi_\nu + \lambda e^{-\frac{1}{2}\xi_\nu}} 1 < 2\lambda e^{\frac{1}{2}\xi_\nu}$$

as soon as x is sufficiently great; hence we obtain in the same way

$$(17)\qquad \sum_{[\mu x]\cdot y_2 < |\log p^m - t| \leqq 1} \frac{y}{m p^m[(t-\log p^m)^2 + y^2]} <$$

$$< 2\lambda y \frac{e^{\frac{1}{2}(x+2)}}{e^{x-2}} \left(\frac{1}{([\mu x] y_2)^2} + \frac{1}{([\mu x]+1)^2 y_2^2} + \frac{1}{([\mu x]+2)^2 y_2^2} + \cdots\right)$$

$$< \frac{2\lambda e^6}{\lambda([\mu x]-1)}$$

$$< \frac{3e^6}{\mu t}\,.$$

Now, it is seen from (15), (16) and (17) that, if λ and μ are appropriately chosen, (λ sufficiently small, μ sufficiently great) then

$$\sum_{|\log p^m - t| < \frac{1}{2\lambda} y} \frac{y}{m p^m[(t-\log p^m)^2 + y^2]} > \frac{1{,}5}{t},$$

and thus, since we have $y = \lambda e^{-\frac{1}{2}t}$,

$$(18)\quad f\left(e^{t+\frac{1}{2}e^{-\frac{1}{2}t}}\right) - f\left(e^{t-\frac{1}{2}e^{-\frac{1}{2}t}}\right) \geqq \sum_{|\log p^m - t| < \frac{1}{2\lambda} y} \frac{1}{m} > \frac{\lambda}{t} e^{\frac{1}{2}t}$$

for any t belonging to the complementary set and for all sufficiently great values of x.

But we have

$$f(e^x) = \pi(e^x) + \frac{1}{2}\mathrm{Li}\left(e^{\frac{1}{2}x}\right) + O\left(e^{\frac{1}{3}x}\right);$$

thus it follows from (18) that

$$\pi\left(e^{t+\frac{1}{2}e^{-\frac{1}{2}t}}\right) - \pi\left(e^{t-\frac{1}{2}e^{-\frac{1}{2}t}}\right) > \frac{\lambda}{2t} e^{\frac{1}{2}t},$$

and this proves Lemma III.

Proof of theorem II. Let us write t in the place of e^t and x in the place of e^x in Lemma III. Then it follows that, if a set of points t inside the interval $\left(\frac{x}{e}, ex\right)$ is defined by the condition

$$(19) \qquad \pi\left(t+\frac{2}{3}\sqrt{t}\right)-\pi\left(t-\frac{2}{3}\sqrt{t}\right)\leq\frac{c}{\log t}\sqrt{t},$$

then the positive constant c may be so chosen that the measure of this set is of the form

$$O\left((\log x)^3 x^{\frac{5}{6}}\right).$$

Suppose now that n is a positive integer, such that there is no prime between n^2 and $(n+1)^2$. Then it is easily seen that the left-hand side of (19) is equal to zero (and thus (19) is satisfied) in an interval contained in $[n^2, (n+1)^2]$, whose length is greater than $\frac{1}{4}n$. Thus if we denote by $n_1, n_2, \ldots$ all the integers of this kind, and by $q(x)$ the number of the n_ν:s which are less than x, we obtain

$$\sum_{\sqrt{\frac{x}{e}}\leq n_\nu<\sqrt{x}} n_\nu = O\left((\log x)^3 x^{\frac{5}{6}}\right);$$

hence

$$q(\sqrt{x})-q\left(\sqrt{\frac{x}{e}}\right)=O\left((\log x)^3 x^{\frac{5}{6}}.x^{-\frac{1}{2}}\right)$$

or

$$q(x)-q\left(\frac{x}{\sqrt{e}}\right)=O\left((\log x)^3 x^{\frac{2}{3}}\right),$$

$$q\left(\frac{x}{\sqrt{e}}\right)-q\left(\frac{x}{\sqrt{e^2}}\right)=O\left[(\log x)^3\left(\frac{x}{\sqrt{e}}\right)^{\frac{2}{3}}\right],$$

$$q\left(\frac{x}{\sqrt{e^2}}\right)-q\left(\frac{x}{\sqrt{e^3}}\right)=O\left[(\log x)^3\left(\frac{x}{\sqrt{e^2}}\right)^{\frac{2}{3}}\right],$$

. . . .

and finally by addition

$$q(x) = O\left((\log x)^3 x^{\frac{2}{3}}\right).$$

CHAPTER III.

On the average order of $\left|\frac{\psi(x)-x}{\sqrt{x}}\right|$.

For the proof of the principal result of this chapter, we shall need the following theorem, which may be found to present also some interest in itself.

Theorem III. *We have*

$$\psi(x) = x - \sum_{\varrho} \frac{x^\varrho}{\varrho} e^{-\frac{|\gamma|}{x^3}} + O(\log x)^2$$

and, if the Riemann *hypothesis is true,*

$$\psi(x) = x - 2\sqrt{x} \sum_{\gamma>0} \frac{\sin(\gamma \log x)}{\gamma} e^{-\frac{\gamma}{x^3}} + O(\sqrt{x}).$$

The fundamental relation (1) shows that we have uniformly for $0 < y < 1$

$$(20)\quad F(z) = z e^z \sum_{p,m} \frac{1}{m p^m (z - \log p^m)} = -\pi i e^z + 2\pi i \sum_{\gamma>0} e^{\varrho z} +$$

$$+ z \int_0^1 e^{tz} \log|\zeta(t)|\,dt + O(x).$$

where $z = x + iy$ and $x \to +\infty$. Now let us denote by ξ a positive quantity which differs from all the $\log p^m$ and let us put

$$\eta = e^{-3\xi}.$$

If we apply CAUCHY's Theorem to the integral

$$\int F(z)dz,$$

taking the contour of integration to be the rectangle

$$\left(\frac{1}{2} \pm i\eta,\ \xi \pm i\eta\right),$$

then we obtain, since $F(z)$ is real for real values of z,

$$2\pi i\psi(e^{\xi}) = -2i\Im\int_{\frac{1}{2}+i\eta}^{\xi+i\eta} F(z)dz + \int_{\xi-i\eta}^{\xi+i\eta} F(z)dz + O(e^{-3\xi}).$$

Hence, using the two different expressions of $F(z)$ given by (20), we obtain

$$(21)\quad \psi(e^{\xi}) = e^{\xi} - 2\Re\sum_{\gamma>0}\frac{e^{\varrho(\xi+i\eta)}}{\varrho} +$$

$$+\frac{1}{2\pi i}\int_{\xi-i\eta}^{\xi+i\eta} ze^{z}\sum_{p,m}\frac{1}{mp^{m}(z-\log p^{m})}dz + O(\xi^{2}).$$

But we have

$$\frac{1}{2\pi i}\int_{\xi-i\eta}^{\xi+i\eta} ze^{z}\sum_{\log p^{m}\leqq\xi-e^{-\xi}}\frac{1}{mp^{m}(z-\log p^{m})}dz =$$

$$= O\left(\xi\eta\frac{e^{\xi}}{e^{-\xi}}\sum_{\log p^{m}<\xi}\frac{1}{m}\right) = O(1).$$

and

$$\frac{1}{2\pi i}\int_{\xi-i\eta}^{\xi+i\eta} ze^{z}\sum_{\log p^{m}\geqq\xi+e^{-\xi}}\frac{1}{mp^{m}(z-\log p^{m})}dz =$$

$$= \frac{1}{2\pi i}\int_{\xi-i\eta}^{\xi+i\eta} z e^z \sum_{\log p^m \geqq \xi + e^{-\xi}} \left(\frac{z}{m p^m \log p^m (z - \log p_m)} - \right.$$

$$\left. - \frac{1}{m p^m \log p^m}\right) dz$$

$$= O\left(\xi^2 \eta \frac{e^\xi}{e^{-\xi}}\right) = O(1);$$

thus in the sum occurring under the sign of integration in (21) we may neglect all the terms which do not satisfy

$$|\xi - \log p^m| < e^{-\xi}.$$

The number of the remaining terms will certainly be less than the number of integers between $e^{\xi + e^{-\xi}}$ and $e^{\xi - e^{-\xi}}$, and thus necessarily less than 3 for all sufficiently great values of ξ. For anyone of them, we now exchange the straight line of integration for a demi-circle bent to the side opposite the pole of the integrand. We thus obtain

$$\frac{1}{2\pi i}\int_{\xi-i\eta}^{\xi+i\eta} z e^z \sum_{|\xi - \log p^m| < e^{-\xi}} \frac{1}{m p^m (z - \log p^m)} dz = O\left(\xi \frac{\eta e^\xi}{\eta e^\xi}\right)$$

$$= O(\xi),$$

and hence

$$\psi(e^\xi) = e^\xi - 2\Re \sum_{\gamma \geqq 0} \frac{e^{\varrho(\xi + i\eta)}}{\varrho} + O(\xi^2).$$

Substituting x for e^ξ and observing that

$$e^{\varrho i \eta} = [1 + O(\eta)] e^{-\gamma\eta},$$

we obtain the first part of Theorem III.

To prove the second part, it is enough to observe that, if the RIEMANN hypothesis is true,

$$\sum_\varrho \frac{x^\varrho}{\varrho} e^{-\frac{|\gamma|}{x^3}} = \sqrt{x} \sum_{\gamma > 0} \left(\frac{x^{i\gamma}}{\frac{1}{2} + i\gamma} + \frac{x^{-i\gamma}}{\frac{1}{2} - i\gamma}\right) e^{-\frac{\gamma}{x^3}} =$$

$$= 2\sqrt{x} \sum_{\gamma > 0} \frac{\sin(\gamma \log x)}{\gamma} e^{-\frac{\gamma}{x^3}} + O(\sqrt{x}).$$

Theorem IV. *If the* RIEMANN *hypothesis is true, then*

$$\frac{1}{x}\int_1^x \left|\frac{\psi(t)-t}{\sqrt{t}}\right| dt = O(1)$$

and

$$\frac{1}{h}\int_x^{x+h} \left|\frac{\psi(t)-t}{\sqrt{t}}\right| dt = o\left(\sqrt{\frac{x}{h}}\right),$$

where $h = h(x)$ denotes any positive function of x, such that $h(x) = O(x)$. The latter result is trivial, if the increase of $h(x)$ is not greater than that of $\frac{x}{(\log x)^4}$.

It follows from the preceding theorem that, if we put

$$\frac{\psi(t)-t}{\sqrt{t}} = -2\sum_{\gamma>0} \frac{\sin(\gamma \log t)}{\gamma} e^{-\frac{\gamma}{t^3}} + \lambda(t), \tag{22}$$

then there exists a positive constant K such that

$$|\lambda(t)| < K \tag{23}$$

for $t \geqq 1$. From (22) we obtain

$$\begin{aligned}\left(\frac{\psi(t)-1}{\sqrt{t}} - \lambda(t)\right)^2 &= 4\left(\sum_{\gamma>0} \frac{\sin(\gamma\log t)}{\gamma} e^{-\frac{\gamma}{t^3}}\right)^2 \\ &\leqq 4\left|\sum_{\gamma>0} \frac{t^{i\gamma}}{\gamma} e^{-\frac{\gamma}{t^3}}\right|^2 \\ &= 4\sum_{\gamma>0}\sum_{\gamma'>0} \frac{t^{i(\gamma-\gamma')}}{\gamma\gamma'} e^{-\frac{\gamma+\gamma'}{t^3}}.\end{aligned}$$

Hence, since the double series is absolutely and uniformly convergent for $1 \leqq t \leqq x$,

$$\int_1^x \left(\frac{\psi(t)-t}{\sqrt{t}} - \lambda(t)\right)^2 dt \leqq 4\sum_{\gamma>0}\sum_{\gamma'>0} \frac{1}{\gamma\gamma'}\int_1^x t^{i(\gamma-\gamma')} e^{-\frac{\gamma+\gamma'}{t^3}} dt. \tag{24}$$

Now, if we associate each term of the double series with the conjugated complex term, it follows from the second mean-value theorem that we may write

$$\sum_{\gamma>0}\sum_{\gamma'\geq 0}\frac{1}{\gamma\gamma'}\int_1^x t^{i(\gamma-\gamma')}e^{-\frac{\gamma+\gamma'}{t^3}}\,dt=\sum_{\gamma>0}\sum_{\gamma'>0}\frac{1}{\gamma\gamma'}e^{-\frac{\gamma+\gamma'}{x^3}}\int_{\xi(\gamma,\gamma')}^x t^{i(\gamma-\gamma')}dt=$$

$$=\sum_{\gamma>0}\sum_{\gamma'>0}\frac{x^{1+i(\gamma-\gamma')}-\xi^{1+i(\gamma-\gamma')}}{\gamma\gamma'[1+i(\gamma-\gamma')]}e^{-\frac{\gamma+\gamma'}{x^3}},$$

where $1<\xi(\gamma,\gamma')<x$ and $\xi(\gamma,\gamma')=\xi(\gamma',\gamma)$.

To show that the left hand side of (24) is of the form $O(x)$, it will thus be sufficient that we can prove that the double series

$$\sum_{\gamma>0}\sum_{\gamma'>0}\frac{1}{\gamma\gamma'[1+i(\gamma-\gamma')]}$$

is absolutely convergent. This is readily seen to be the case, if we observe that, since the number of $\gamma's$ between n and $n+1$ is of the form $O(\log n)$, we have

$$\sum_{\gamma'>0}\frac{1}{\gamma\gamma'[1+i(\gamma-\gamma')]}=$$

$$=O\left(\frac{\log 1}{\gamma.1[1+(\gamma-2)]}+\frac{\log 2}{\gamma.2[1+(\gamma-3)]}+\cdots\right.$$

$$+\frac{\log([\gamma]-1)}{\gamma([\gamma]-1)\{1+(\gamma-[\gamma])\}}+\frac{\log[\gamma]}{\gamma[\gamma]\{1+(1+[\gamma]-\gamma)\}}+$$

$$\left.+\frac{\log([\gamma]+1)}{\gamma([\gamma]+1)\{1+(2+[\gamma]-\gamma)\}}+\cdots\right)=O\left(\frac{\log\gamma}{\gamma}\right)^2,$$

and the series $\sum_{\gamma>0}\left(\frac{\log\gamma}{\gamma}\right)^2$ is certainly convergent. Thus we have

$$\int_1^x\left(\frac{\psi(t)-t}{\sqrt{t}}-\lambda(t)\right)^2dt=O(x),$$

further by the use of SCHWARZ's inequality

$$\int_1^x \left| \frac{\psi(t)-t}{\sqrt{t}} - \lambda(t) \right| dt = O(\sqrt{x}\,.\,\sqrt{x}) = O(x),$$

and finally by (23)

$$\int_1^x \left| \frac{\psi(t)-t}{\sqrt{t}} \right| dt = O(x),$$

which proves the first part of Theorem IV.

Quite in the same way we obtain, supposing $h = h(x) > 0$ and $h(x) = O(x)$,

$$\int_x^{x+h} \left(\frac{\psi(t)-t}{\sqrt{t}} - \lambda(t) \right)^2 dt = O(x);$$

hence by SCHWARZ's inequality

$$\int_x^{x+h} \left| \frac{\psi(t)-t}{\sqrt{t}} - \lambda(t) \right| dt = O(\sqrt{h}\,.\,\sqrt{x}),$$

and by (23)

$$\int_x^{x+h} \left| \frac{\psi(t)-t}{\sqrt{t}} \right| dt = O(\sqrt{hx}).$$

Thus Theorem IV is proved.

If we try to prove the same theorem by the aid of the ordinary formula for $\psi(x)'$, without the convergence factors introduced by Theorem III, we shall have to deal with the relation

$$(25) \qquad \left[\frac{\psi(t)-t}{\sqrt{t}} + O\left(\frac{1}{\sqrt{t}}\right) \right]^2 = \sum_{\varrho} \sum_{\varrho'} \frac{t^{\varrho+\varrho'-1}}{\varrho\varrho'},$$

which is obtained from

$$\frac{\psi(t)-t}{\sqrt{t}} = -\sum_{\varrho} \frac{t^{\varrho-\frac{1}{2}}}{\varrho} + O\left(\frac{1}{\sqrt{t}}\right).$$

The double series occurring in (25) is neither absolutely nor uniformly convergent, and thus the question arises whether it may be integrated term-by-term. This is in fact so, since the term-by-term integration of the series $\sum_{\varrho} \frac{t^{\varrho}}{\varrho}$ is known to be permitted[1] and the double series obtained by the integration has just been proved to be absolutely convergent. We thus obtain

$$\frac{1}{x}\int_1^x \left(\frac{\psi(t)-t}{\sqrt{t}}\right)^2 dt = \sum_{\varrho}\sum_{\varrho'} \frac{x^{\varrho+\varrho'-1}}{\varrho\,\varrho'\,(\varrho+\varrho')} + O(1),$$

and hence Theorem IV may be immediately deduced. The last relation suggests that we should have (by taking out the terms where $\varrho = \varrho'$)

$$\lim_{x\to\infty} \frac{1}{x}\int_1^x \left(\frac{\psi(t)-t}{\sqrt{t}}\right)^2 dt = \sum_{\varrho} \frac{1}{|\varrho|^2},$$

but I have not been able to find out whether this is true or not.

CHAPTER IV.

On the zeros of the Dedekind zeta function and the theory of prime ideals.

In this chapter, we shall denote by $\zeta(s)$ the DEDEKIND zeta function corresponding to an arbitrary algebraic field.

[1] See CRAMÉR, »Über die Herleitung der Riemann'schen Primzahlformel», Arkiv för Matematik, etc., B. 13, N:o 24.

By Np we shall denote the norm of an arbitrary prime ideal belonging to this field, and for the sake of brevity we shall write Np^{ms} for $(Np)^{ms}$.

After HECKE's discovery[1] of the general functional equation for $\zeta(s)$, a complete account of the theory of this function has been given by LANDAU[2]. It appears that all the theorems are quite analogous to the corresponding theorems in the theory of the RIEMANN zeta function. Thus if we decide to examine the function

$$V(z) = \sum_{\gamma > 0} e^{\varrho z}, \tag{26}$$

where $\varrho = \beta + \gamma i$ is a complex root of the equation $\zeta(s) = 0$, the investigation will be quite similar to the corresponding one given in my papers quoted in the Introduction. We shall here give merely the outlines of the transformations.

If we put $z = x + iy$, where $y > 0$, the series (26) is absolutely convergent and we have

$$2\pi i V(z) = -z \int e^{sz} \log \zeta(s) ds, \tag{27}$$

the contour of integration being the rectangle

$$(+ i\infty, 0, 1, 1 + i\infty),$$

it being understood that the points 0, 1 and the real zeros of $\zeta(s)$ between 0 and 1 (if there are any) are avoided by means of small arcs turned upwards. The expression $\log \zeta(s)$ denotes here and in the sequel the branch represented for $\sigma > 1$ by the series

$$\sum_{p, m} \frac{1}{m N p^{ms}}.$$

Now it has been discovered by HECKE that $\zeta(s)$ satisfies the functional equation

---

[1] HECKE, »Über die Zetafunktion beliebiger algebraischer Zahlkörper», Göttinger Nachrichten, Math.-Phys. Klasse, 1917.

[2] LANDAU, Einführung in die elementare und analytische Theorie der algebraischen Zahlen und der Ideale. Leipzig und Berlin (Teubner), 1918.

$$\zeta(s)=\left(\frac{(2\pi)^s}{\pi}\right)^n |\Delta|^{\frac{1}{2}-s}\left(\sin\frac{s\pi}{2}\right)^{r_1+r_2}\left(\cos\frac{s\pi}{2}\right)^{r_2}(\Gamma(1-s))^n\zeta(1-s),$$

where n denotes the degree of the field, Δ the »Grundzahl», r_1 the number of the real roots and $2r_2$ the number of the imaginary roots of the irreducible equation of the n'th degree satisfied by the algebraic number which generates the field. Thus we have

$$r_1+2r_2=n. \tag{28}$$

By means of the functional equation, (27) is transformed into

$$\begin{aligned}2\pi i V(z) = &-z\int_1^{1+i\infty} e^{sz}\log\zeta(s)ds + z\int_0^{i\infty} e^{sz}\log\zeta(1-s)ds - \\ &-z\left(\log\frac{\pi^n}{\sqrt{|\Delta|}}+(2-h)\pi i\right)\int_0^{i\infty} e^{sz}ds + z\log\frac{(2\pi)^n}{|\Delta|}\int_0^{i\infty} s e^{sz}ds - \\ &-z\int_0^1 e^{sz}\log\zeta(s)ds + z(r_1+r_2)\int_0^{i\infty} e^{sz}\log\sin\frac{s\pi}{2}ds + \\ &+zr_2\int_0^{i\infty} e^{sz}\log\cos\frac{s\pi}{2}ds + nz\int_0^{i\infty} e^{sz}\log\Gamma(1-s)ds,\end{aligned}$$

which corresponds to formula (9) of my paper in the Mathematische Zeitschrift. The multiform function $\log\sin\frac{s\pi}{2}$ is to be taken with the imaginary part $\frac{1}{2}\pi i$, and the integrands in the two last terms are to be given the value 0 for $s=0$. Here, h denotes the number of the real zeros s_ν of $\zeta(s)$, such that $0<s_\nu<1$. Among all the terms occurring in this expression, there is only one, namely the one involving $\log\cos\frac{s\pi}{2}$, having no corresponding term in the formula (9) of my earlier paper. This term is easily calculated, and we obtain

$$z\int_0^{i\infty} e^{sz}\log\cos\frac{s\pi}{2}\,ds = -\frac{\frac{1}{2}\pi i}{z} + \pi i\sum_{n=1}^{\infty}\frac{(-1)^{n-1}}{z+n\pi i}.$$

Thus, the further transformations give

(29) $2\pi i V(z) =$

$$= ze^z\sum_{p,m}\frac{1}{mNp^m(z-\log Np^m)} - z\sum_{p,m}\frac{1}{mNp^m(z+\log Np^m)} -$$

$$-z\int_0^1 e^{sz}\log\zeta(s)ds + \alpha + \frac{\beta}{z} + (r_1+r_2)\Psi\left(\frac{z}{\pi i}\right) +$$

$$+\pi i r_2\sum_{n=1}^{\infty}\frac{(-1)^{n-1}}{z+n\pi i} - \frac{n}{z}\int_0^{\infty}\frac{t}{e^t-1}\cdot\frac{dt}{t+z},$$

where

$$\alpha = n\log\pi - \frac{1}{2}\log|\Delta| + (2-h)\pi i + (r_1+r_2)\left(C+\log 2-\frac{1}{2}\pi i\right),$$

$$\beta = n(C+\log 2\pi) - \log|\Delta| + \frac{1}{2}r_1\pi i.$$

This corresponds to formula (13) of my earlier paper, which is quoted at the beginning of the first chapter of the present one. Hence we deduce the relation

$$(30)\quad ze^z\sum_{p,m}\frac{1}{mNp^m(z-\log Np^m)} = z\int_0^1 e^{sz}\log\zeta(s)ds + $$
$$+ 2\pi i\sum_{\gamma>0}e^{\varrho z} + O(x)$$

(uniformly for $0<y<1$, as $x\to+\infty$), corresponding to (1) and (20) of the present paper. Here, the imaginary part of $\log\zeta(s)$ is to be given the value $-\pi i$ as s tends to 1, and the real zeros of $\zeta(s)$ are to be avoided by means of small arcs turned upwards; thus (30) may be written in the form

$$(31)\quad z e^z \sum_{p,m} \frac{1}{m N p^m (z - \log N p^m)} = -\pi i e^z + 2\pi i \sum_{\gamma \geqq 0}{}' e^{\varrho z} + + z \int_0^1 e^{sz} \log |\zeta(s)|\, ds + O(x)$$

precisely analogous to (20). Here, $\varrho = \beta + \gamma i$ denotes any zero of $\zeta(s)$ such that $0 < \beta < 1$, $\gamma \geqq 0$, and the notation $\sum\limits_{\gamma=0}'$ means that the real ϱ's — if there are any — should be counted with the coefficient $\frac{1}{2}$. From (31) it is possible to deduce theorems concerning prime ideals analogous to the theorems concerning prime numbers given in the first three chapters of this paper.

From (29) it is deduced by some calculation that we have

$$2\pi i V(z) = \frac{n \log z}{1 - e^{-z}} + \frac{n(C + \log 2\pi) - \log |\varDelta| - \frac{1}{2} n \pi i}{z} + + \left(2 - h - \frac{r_1 + r_2}{2}\right) \pi i + b_0 + z p(z),$$

where b_0 is real and the power series $p(z)$ is convergent for sufficiently small values of $|z|$. Let us suppose for a moment that the 'generalized RIEMANN hypothesis' is true, *i. e.* that all the *complex* zeros of $\zeta(s)$ have their real part equal to $\frac{1}{2}$. Then it is found, quite in the same way as in my previous paper, that

$$(32)\quad \sum_{\gamma > 0} e^{-\gamma s} = -n \frac{C + \log 2\pi s}{2\pi s} + \frac{\log |\varDelta|}{2\pi s} + 1 - \frac{1}{8} r_1 - \frac{1}{2} h + + s p_1(s) \log s + s p_2(s),$$

the power series $p_1(s)$ and $p_2(s)$ being convergent for sufficiently small values of $|s|$. Now we introduce the function $N(T)$ equal to the number of γ's in the interval $0 < \gamma \leqq T$. Then we have (LANDAU, l. c., p. 86)

$$N(T)=\frac{nT}{2\pi}\left(\log\frac{T}{2\pi}-1\right)+\frac{\log|\Delta|}{2\pi}T+R(T)$$

and

$$R(T)=O(\log T).$$

Hence

$$\sum_{\gamma>0} e^{-\gamma s}=s\int_0^\infty N(v)e^{-vs}dv=$$

$$=-n\frac{C+\log 2\pi s}{2\pi s}+\frac{\log|\Delta|}{2\pi s}+s\int_0^\infty R(v)\,e^{-vs}\,dv$$

and by (32)

$$\int_0^\infty R(v)\,e^{-vs}dv=\frac{1-\frac{1}{8}r_1-\frac{1}{2}h}{s}+p_1(s)\log s+p_2(s).$$

Finally, it is proved in the same way as in the case of the RIEMANN $\zeta(s)$ that we have (h denotes the number of the real zeros s_ν of $\zeta(s)$, such that $0<s_\nu<1$)

$$\lim_{T\to\infty}\frac{1}{T}\int_0^T R(v)dv=1-\frac{1}{8}r_1-\frac{1}{2}h,$$

and that this relation is independent of the 'generalized RIEMANN hypothesis'.

Additional note. (August 23^d 1920) — While the present paper was being printed, I have found that some of the results given here are capable of being considerably improved. In particular, I have found the following theorem, which is a generalized and improved form of Theorem II of the present paper:

Let us denote by $h(x)$ the number of primes $p_n<x$, which satisfy the inequality

$$p_{n+1}-p_n>p_n^k,$$

where p_n *denotes the n:th prime and* $0 < k \leqq \frac{1}{2}$. *Then we have, if the Riemann hypothesis is true,*

$$h(x) = O\left(x^{1-\frac{3}{2}k+\varepsilon}\right)$$

for any $\varepsilon > 0$.

By trivial arguments, it is possible to obtain the relation

$$h(x) = O\left(x^{1-k}\right)$$

but no more. The proof of this theorem will be published in the Proceedings of the Cambridge Philosophical Society.

Tryckt den 8 september 1920.

10.

On the distribution of primes

Proc. Cambridge Phil. Soc. **20**, 272–280 (1921)

(Communicated by Prof. G. H. Hardy.) [*Received* 10 August 1920 : *read* 25 October.]

Throughout the whole of this paper, I shall assume the truth of the Riemann hypothesis concerning the roots of the Zeta function, viz. $\zeta(\sigma + it) \neq 0$ *for* $\sigma > \frac{1}{2}$.

This being so, it is known that

$$\pi(x) = Li(x) + O(x^{\frac{1}{2}} \log x) \quad \text{..................(1)}$$

and

$$p_n \sim n \log n \quad \text{..........................(2),}$$

where $\pi(x)$ denotes the number of primes less than or equal to x, and p_n denotes the nth prime. The last relation is independent of the Riemann hypothesis. But very little is known as to the behaviour of the difference

$$\Delta_n = p_{n+1} - p_n$$

between two successive primes, for large values of n. It follows from the "Prime Number Theorem" (1) or (2) that

$$\frac{\Delta_1 + \Delta_2 + \ldots \Delta_n}{n} \sim \log n \sim \log p_n$$

and from (1) that

$$\Delta_n = O(p_n^{\frac{1}{2}} \log^2 p_n).$$

I have recently shown* that the last relation may be replaced by

$$\Delta_n = O(p_n^{\frac{1}{2}} \log p_n).$$

So far as I know, this is all that is actually known about Δ_n. It is very probable that $\Delta_n = 2$ for an infinity of values of n; but this has not yet been proved, and it has not even been proved that $\Delta_n < \frac{1}{2} \log p_n$ or $\Delta_n > 2 \log p_n$ for an infinity of values of n. The object of the present paper is to prove the following theorem, which gives an upper limit for the frequency of certain large values of Δ_n.

Theorem. *Let* $h(x)$ *be the number of primes* $p_n \leqq x$ *satisfying the inequality*

$$p_{n+1} - p_n > p_n^k,$$

where $0 < k \leqq \frac{1}{2}$. *Then*

$$h(x) = O(x^{1 - \frac{3}{2}k + \epsilon})$$

for every positive ϵ.

* "Some theorems concerning prime numbers," *Arkiv för Matematik, Astronomi och Fysik*, Band 15 (1920), No. 5, pp. 1—32.

It is interesting to remark that we may obtain by a very trivial argument (*viz.* that the sum of all the $h(x)\Delta_n$'s which are greater than p_n^k must be less than $x+x^k$) the evaluation

$$h(x)=O(x^{1-k}),$$

but it seems impossible to improve this even by direct deduction from the Prime Number Theorem.

The proof of the theorem given here depends on the theory of the function $\Sigma e^{\rho z}$, which I have studied in some recent papers*. We denote here and in the sequel by $\rho=\frac{1}{2}+i\gamma$ an arbitrary zero of the function $\zeta(s)$, situated in the *upper half* of the plane of the complex variable $s=\sigma+it$.

In order to prove the theorem, we shall require a set of lemmas. It seems convenient to remark that all the sets of points we shall have to deal with in the proof consist of a finite number of finite intervals (and perhaps a finite number of isolated points). Hence their measure may be taken to be the measure in the elementary sense.

Lemma 1. *If*

$$\phi(v,t)=\sum_{\gamma\leqq v} e^{\gamma i t},$$

we have

$$\int_{x-2}^{x+2}|\phi(v,t)|\,dt=O\left(v^{\frac{1}{2}}(\log v)^{\frac{3}{2}}\right)$$

uniformly for $x>2$.

Proof. We have

$$|\phi(v,t)|^2=\sum_{\gamma\leqq v}\sum_{\gamma'\leqq v} e^{(\gamma-\gamma')it},$$

and thus

$$\int_{x-2}^{x+2}|\phi(v,t)|^2dt=\sum_{\gamma\leqq v}\sum_{\gamma'\leqq v}\int_{x-2}^{x+2} e^{(\gamma-\gamma')it}\,dt$$

$$<\sum_{\gamma\leqq v}\sum_{\gamma'\leqq v}\operatorname{Min}\left(\frac{2}{|\gamma-\gamma'|},4\right).$$

The number of numbers γ' in the interval $(\gamma+\nu,\gamma+\nu+1)$ is $O(\log(\gamma+\nu))$†. It follows that our sum is

$$O\left(\sum_{\gamma\leqq v}\log\gamma\right)+O\left[\sum_{\leqq v-1}\left(\frac{\log(\gamma+1)}{1}+\frac{\log(\gamma+2)}{2}+\dots+\frac{\log(\gamma+[v-\gamma])}{[v-\gamma]}\right)\right]$$

$$=O(v(\log v)^3),$$

* *l.c.* (footnote [1]); *Comptes rendus*, t. 168, p. 539; and *Mathematische Zeitschrift*, Band 4, pp. 104—

† Landau, *Handbuch*, p. 337.

uniformly in x. Hence the truth of the lemma follows by the use of Schwarz's inequality.

Lemma 2. *Let us put*

$$y = \frac{1}{t}\, e^{-\alpha t}$$

and consider the interval $x - 2 \leqq t \leqq x + 2$, *where* $x > 2$ *and* $\frac{1}{2} \leqq \alpha < 1$. *Then the set* S_x *of points* t *belonging to this interval such that*

$$\left| \sum_\gamma \cos \gamma t\, e^{-\gamma y} \right| \geqq \tfrac{1}{8} e^{\frac{1}{2}t}$$

is of measure M_x *such that*

$$M_x = O\left(x^2 e^{-\frac{1-\alpha}{2}x}\right).$$

Proof. We have

$$\left|\sum_\gamma \cos \gamma t\, e^{-\gamma y}\right| \leqq \left|\sum_\gamma \left(e^{i\gamma t} \,.\, e^{-\gamma y}\right)\right| = \left| y \int_0^\infty \phi(v, t)\, e^{-vy}\, dv \right|$$
$$\leqq y \int_0^\infty |\phi(v, t)|\, e^{-vy} dv.$$

It follows by Lemma 1 that, if we denote by y_1 and y_2 the values of y at the points $x - 2$ and $x + 2$ respectively,

$$\int_{x-2}^{x+2} \left|\sum_\gamma \cos \gamma t\, e^{-\gamma y}\right| dt = O\left(\int_0^\infty y_1 v^{\frac{1}{2}} (\log v)^{\frac{3}{2}} e^{-vy_2} dv\right)$$
$$= O\left(y_1 y_2^{-\frac{3}{2}} \left(\log \frac{1}{y_2}\right)^{\frac{3}{2}}\right)$$
$$= O\left(x^2 e^{\frac{1}{2}\alpha x}\right).$$

Thus the measure of the set of points t belonging to the interval of integration, such that

$$\left|\sum_\gamma \cos \gamma t\, e^{-\gamma y}\right| \geqq \tfrac{1}{8} e^{\frac{1}{2}t},$$

must be of the form

$$M_x = O\left(x^2 e^{\frac{1}{2}\alpha x} \,.\, e^{-\frac{1}{2}x}\right)$$
$$= O\left(x^2 e^{-\frac{1-\alpha}{2}}\right).$$

Lemma 3. *In the set* $\bar{S}_x$, *complementary to the set* S_x *of Lemma 2, we have*

$$\sum_1^\infty \frac{\eta(n)}{n\left((t - \log n)^2 + y^2\right)} = \frac{\pi + \theta}{ty},$$

where $|\theta| < 1$ *for all sufficiently large values of* x. *Here* $\eta(n)$ *denotes the arithmetical function defined for integral values of* $n \geqq 1$ *by*

$$\eta(n) = \frac{1}{m} \quad (n = p^m,\ p \text{ prime}),$$
$$\eta(n) = 0 \quad (\text{otherwise}).$$

Proof. In a recently published paper*, I have proved the formula

$$y\sum_{1}^{\infty}\frac{\eta(n)}{n((t-\log n)^2+y^2)}=\frac{\pi}{t}-2\pi\,\mathfrak{R}\sum_{\gamma}\frac{e^{(\rho-1)(t+iy)}}{t+iy}+o\left(\frac{1}{t}\right),$$

where y may denote any positive decreasing function of t which tends to zero as t tends to infinity. On the Riemann hypothesis, and assuming y to be the function of Lemma 2, this formula reduces to

$$y\sum_{1}^{\infty}\frac{\eta(n)}{n((t-\log n)^2+y^2)}$$
$$=\frac{\pi}{t}-\frac{2\pi}{t}e^{-\frac{1}{2}t}\left(1+O\left(\frac{y}{t}\right)\right)\sum_{\gamma}\cos\left(\gamma t-\tfrac{1}{2}y-\arg(t+iy)\right)e^{-\gamma y}$$
$$=\frac{\pi}{t}-\frac{2\pi}{t}e^{-\frac{1}{2}t}\left(1+O\left(\frac{y}{t}\right)\right)\left(\sum_{\gamma}\cos\gamma t\,e^{-\gamma y}+O\left(y\sum_{\gamma}e^{-\gamma y}\right)\right)+o\left(\frac{1}{t}\right).$$

But we have

$$\sum_{\gamma}e^{-\gamma y}=O\left(\sum_{1}^{\infty}\log n\,e^{-ny}\right)=O\left(\frac{1}{y}\log\frac{1}{y}\right).$$

Hence

$$y\sum_{1}^{\infty}\frac{\eta(n)}{n((t-\log n)^2+y^2)}=\frac{\pi}{t}\left(1-2e^{-\frac{1}{2}t}\sum_{\gamma}\cos\gamma t\,e^{-\gamma y}\right)+o\left(\frac{1}{t}\right).$$

By the definition of the set S_x, we have

$$\left|2e^{-\frac{1}{2}t}\sum_{\gamma}\cos\gamma t e^{-\gamma y}\right|<\tfrac{1}{4}$$

for all values of t belonging to the complementary set $\overline{S}_x$. Since $\overline{S}_x$ is contained in the interval $(x-2, x+2)$, we conclude that

$$\sum_{1}^{\infty}\frac{\eta(n)}{n((t-\log n)^2+y^2)}=\frac{\pi+\theta}{ty},$$

where $|\theta|<1$ for all sufficiently large values of x.

Lemma 4. *We denote by $f(v)$ the function*

$$f(v)=\sum_{n\leqq v}\eta(n)$$

introduced by Riemann, and we consider the interval $(x-2, x+2)$ of the two preceding lemmas. The set of points t belonging to this interval such that

$$\sum_{t<\log n\leqq t+y}\eta(n)=f(e^{t+y})-f(e^t)\geqq\frac{9y}{t}e^t$$

is, for all sufficiently large values of x, a subset of the set S_x of Lemma 2, so that its measure is of the form

$$O\left(x^2e^{-\frac{1-a}{2}x}\right).$$

* *l.c.*, footnote *, p. 272.

18—5

Proof. To prove this lemma, it is only necessary to show that

$$\sum_{t<\log n \leqq t+y} \eta(n) < \frac{9y}{t} e^t$$

if t belongs to the complementary set $\bar{S}_x$. This follows immediately from Lemma 3, for we have, since $\eta(n) \geqq 0$,

$$\sum_{t<\log n \leqq t+y} \frac{\eta(n)}{n((t-\log n)^2+y^2)} < \frac{\pi+1}{ty}$$

if x is sufficiently great and t belongs to $\bar{S}_x$. Hence

$$\sum_{t<\log n \leqq t+y} \eta(n) < \frac{\pi+1}{ty} \cdot 2y^2 \cdot e^{t+y} < \frac{9y}{t} e^t$$

for all sufficiently large values of x.

Lemma 5. *Let S be a set of points of measure M, situated in a finite interval ab. Then it is possible to divide ab into sub-intervals of length δ (the two extreme intervals being possibly less than δ) in such a way that not more than M/δ of the points of division belong to S.*

Proof. Consider any such division of ab, and denote by $\alpha\beta$ an arbitrary sub-interval of length δ. Let x be a point in $\alpha\beta$, and denote by $\phi(x)$ the number of points of S which are "congruent" to x according to the adopted division of ab. Then it is clear that

$$\int_\alpha^\beta \phi(x)\,dx = M.$$

Thus there must be at least one point x_0 in $\alpha\beta$, such that $\phi(x_0) \leqq M/\delta$. Starting the division of ab from this point, we see that the conditions stated in the lemma are fulfilled.

Lemma 6. *Let us denote bg σ_x the set of points t, belonging to the interval $(x-1, x+1)$, such that*

$$\pi(e^{t+cy}) - \pi(e^{t-cy}) \leqq \frac{y}{t} e^t,$$

where c is a positive constant. Here $\pi(v)$ denotes as usual the number of primes less than or equal to v. Then it is possible to give such a value to c that the measure μ_x of σ_x satisfies the relation

$$\mu_x = O\left(x^4 e^{-\frac{1-\alpha}{2}x}\right).$$

Proof. We shall prove this lemma by first excluding from the interval $(x-2, x+2)$ a certain set, the measure of which satisfies the relation just stated for μ_x, and then proving that we may choose c so that

$$\pi(e^{t+cy}) - \pi(e^{t-cy}) > \frac{y}{t} e^t \quad \text{.................(3)}$$

in the part of the remaining set which belongs to $(x-1, x+1)$.

By Lemma 5, we are able to divide $(x-2, x+2)$ into sub-intervals of the length

$$y_2 = \frac{1}{x+2} e^{-a(x+2)}$$

in such a way that not more than

$$\frac{M_x}{y_2}$$

of the points of division belong to the set S_x of Lemma 2.

Supposing this to have been done, we exclude from the interval $(x-2, x+2)$ the set Σ_x defined in the following way: we take first the whole set S_x and then, denoting by t_0 any of the just mentioned points of division belonging to S_x, the interval $(t_0 - 2x^2 y_2, t_0 + 2x^2 y_2)$. The measure of Σ_x is thus less than

$$M_x + 2x^2 y_2 . \frac{M_x}{y_2} = O\left(x^4 e^{-\frac{1-a}{2}x}\right).$$

In order to prove the lemma, we now have to show that c may be so chosen that (3) is valid for any t in $(x-1, x+1)$ not belonging to Σ_x. It is to be noticed that the definition of Σ_x in no way involves c.

Since t does not belong to Σ_x, it does not belong to S_x. Thus we have by Lemma 3

$$\sum_1^\infty \frac{\eta(n)}{n((t-\log n)^2 + y^2)} > \frac{\pi - 1}{ty} \quad \text{..............(4)}$$

for all sufficiently large values of x. We put

$$\sum_1^\infty = \sum_{\log n \leqq t-1} + \sum_{t-1 < \log n \leqq t - x^2 y_2} + \sum_{t - x^2 y_2 < \log n \leqq t - cy_2} + \sum_{t - cy_2 < \log n \leqq t + cy_2}$$
$$+ \sum_{t + cy_2 < \log n \leqq t + x^2 y_2} + \sum_{t + x^2 y_2 < \log n \leqq t+1} + \sum_{t+1 < \log n}$$
$$= A_1 + A_2 + \dots A_7 \quad \text{....................(5).}$$

Then we have

$$A_1 < \sum_{\log n \leqq t-1} \frac{\eta(n)}{n} < \sum_{n \leqq e^t} \frac{1}{n} = o\left(\frac{1}{ty}\right) \text{...........(6),}$$

and

$$A_7 = \sum_{\log n > t+1} \frac{\eta(n)}{n((t-\log n)^2+y^2)} < \sum_{\log n > t+1} \frac{\eta(n)}{n(\log n - t)^2}$$

$$= \sum_{\log n > t+1} \frac{\eta(n)}{n}\left(\frac{1}{(\log n)^2+t^2} + \frac{2t\log n}{((\log n)^2+t^2)(\log n - t)^2}\right)$$

$$< \sum_{\log n > t+1} \frac{\eta(n)}{n}\left(\frac{1}{(\log n)^2} + \frac{2t}{\log n}\right) = o\left(\frac{1}{ty}\right) \quad \ldots\ldots\ldots\ldots(7),$$

since the series $\sum \frac{\eta(n)}{n \log n}$ is convergent.

In order to obtain similar evaluations for A_2 and A_3, we consider the division of $(x-2, x+2)$ into sub-intervals of the length y_2 which we have used for the definition of Σ_x. Since t does not belong to Σ_x, it follows from this definition that none of the points of division situated in the interval $(t - x^2y_2, t + x^2y_2)$ belong to the set S_x. Hence, denoting by t_0 any such point of division, and by y_0 the corresponding value of y, we obtain, by Lemma 4,

$$\sum_{t_0 < \log n \leqq t_0+y_2} \eta(n) \leqq \sum_{t_0 < \log n \leqq t_0+y_0} \eta(n) < \frac{9y_0}{t_0} e^{t_0}.$$

Thus, if we consider first A_3, and group together the terms belonging to the same sub-interval (considered as interval of variation of $\log n$), we obtain

$$A_3 < K\frac{y}{t}e^t \cdot e^{-t}\left(\frac{1}{c^2y^2} + \frac{1}{(c+1)^2y^2} + \frac{1}{(c+2)^2y^2} + \ldots\right) < \frac{2K}{cty} \ldots(8),$$

supposing $c > 1$ and denoting by K a constant independent of c and t.

Grouping together the terms of A_2 in a similar way, we have

$$A_2 < e^{-(t-1)} \sum_{t-1 < \log n \leqq t - x^2y_2} \frac{1}{(t-\log n)^2}$$

$$= O\left[e^{-t}\cdot ye^t\left(\frac{1}{x^4y^2} + \frac{1}{(x^2+1)^2y^2} + \frac{1}{(x^2+2)^2y^2} + \ldots\right)\right]$$

$$= O\left(\frac{1}{x^2y}\right) = o\left(\frac{1}{ty}\right) \quad \ldots\ldots\ldots\ldots\ldots\ldots\ldots\ldots\ldots\ldots\ldots\ldots(9),$$

since the number of terms in each group is of the form $O(ye^t)$.

Of the remaining terms, A_6 may obviously be treated in the same way as A_2, and A_5 in the same way as A_3. Thus we see, from (4)—(9), that it is possible to determine two absolute constants c_0 and x_0 such that

$$A_4 = \sum_{t-cy_2 < \log n \leqq t+cy_2} \frac{\eta(n)}{n((t-\log n)^2+y^2)} > \frac{\pi-2}{ty}$$

for all values of t in $(x-1, x+1)$ not belonging to Σ_x, so long as $c > c_0$ and $x > x_0$. Hence, *a fortiori*,

$$\sum_{t-cy<\log n\leqq t+cy} \frac{\eta(n)}{n((t-\log n)^2+y^2)} > \frac{\pi-2}{ty}$$

and

$$\frac{e^{-(t-cy)}}{y^2} \sum_{t-cy<\log n\leqq t+cy} \eta(n) > \frac{\pi-2}{ty},$$

$$\sum_{t-cy<\log n\leqq t+cy} \eta(n) > (\pi-2)\frac{y}{t}e^{t-cy} \quad \ldots\ldots\ldots(10).$$

But we have

$$\sum_{t-cy<\log n\leqq t+cy} \eta(n) = \pi(e^{t+cy}) - \pi(e^{t-cy})$$

$$+ \sum_{\nu=2}^{\frac{t+cy}{\log 2}} \frac{1}{\nu}\left[\pi\left(e^{\frac{1}{\nu}(t+cy)}\right) - \pi\left(e^{\frac{1}{\nu}(t-cy)}\right)\right],$$

and

$$\sum_{\nu=2}^{\frac{t+cy}{\log 2}} \frac{1}{\nu}\left[\pi\left(e^{\frac{1}{\nu}(t+cy)}\right) - \pi\left(e^{\frac{1}{\nu}(t-cy)}\right)\right]$$

$$= O\left[\sum_2^t \frac{1}{\nu}\left(e^{\frac{1}{\nu}(t+cy)} - e^{\frac{1}{\nu}(t-cy)} + 1\right)\right] = O\left[\sum_2^t \frac{1}{\nu}\left(\frac{y}{\nu}e^{\frac{1}{2}t} + 1\right)\right]$$

$$= O(\log t) = o\left(\frac{y}{t}e^t\right).$$

Hence, if c is fixed and greater than c_0, we obtain from (10),

$$\pi(e^{t+cy}) - \pi(e^{t-cy}) > \frac{y}{t}e^t,$$

for all sufficiently large values of x, and for all values of t in $(x-1, x+1)$ not belonging to Σ_x.

Proof of the theorem.

Consider the set of points e^t, where t passes through all the points t of the set σ_x of Lemma 6. This set belongs to the interval (e^{x-1}, e^{x+1}), and its measure is less than $e^{x+1}\mu_x$. Hence if, in Lemma 6, we write x in the place of e^x, t in the place of e^t and k in the place of $1-\alpha$, this lemma will take the following form:

Let us denote by σ_x' the set of points t, belonging to the interval $\left(\frac{x}{e}, ex\right)$, such that

$$\pi\left(t+\frac{c}{\log t}t^k\right) - \pi\left(t-\frac{c}{\log t}t^k\right) \leqq \frac{t^k}{\log^2 t},$$

where c is a positive constant and $0 < k \leqq \frac{1}{2}$. *Then it is possible to give such a value to c that the measure* μ_x' *of* σ_x' *satisfies the relation*

$$\mu_x' = O\left(x^{1-\frac{1}{2}k} \log^4 x\right).$$

Suppose now that c has been properly fixed. Then it is clear that, if x is sufficiently large, and if p_n denotes a prime belonging to the interval $(\frac{1}{2}x, x)$ and satisfying the inequality

$$p_{n+1} - p_n > p_n^k,$$

then the interval (p_n, p_{n+1}) will contribute more than $\frac{1}{2}p_n^k$, and *a fortiori* more than $\frac{1}{2}(\frac{1}{2}x)^k$, to σ_x'. Hence we obtain

$$\tfrac{1}{2}(\tfrac{1}{2}x)^k \left(h(x) - h(\tfrac{1}{2}x)\right) = O\left(x^{1-\frac{1}{2}k+\epsilon}\right),$$

$$h(x) - h(\tfrac{1}{2}x) = O\left(x^{1-\frac{3}{2}k+\epsilon}\right).$$

If we replace in the last relation x first by $\frac{1}{2}x$, then by $\frac{1}{4}x$, and so on, and add together all the relations obtained in this way, we get

$$h(x) = O\left(x^{1-\frac{3}{2}k+\epsilon}\right).$$

Hence our theorem is proved.

CAMBRIDGE, 20 *July*, 1920.

11.

with E. Landau

Über die Zetafunktion auf der Mittellinie des kritischen Streifens

Ark. Mat. Astr. Fys. **15** (28), 1–4 (1921)

Mitgeteilt am 23. Februar 1921 durch IVAR BENDIXSON und H. VON KOCH.

Hilfssatz: *Für* $\sigma = \frac{1}{2}$ *und alle hinreichend grossen* t *ist*

$$|\zeta'(s)| \geqq |\zeta(s)|.$$

Vorbemerkung: Falls alle Wurzeln von $\zeta(s)$ auf $\sigma = \frac{1}{2}$ einfach sind, hat also $\zeta'(s)$ dort nicht unendlich viele Wurzeln.

Beweis: $\pi^{-\frac{s}{2}} \Gamma\left(\frac{s}{2}\right) \zeta(s)$ ist für $\sigma = \frac{1}{2}$ reell; für $\sigma = \frac{1}{2}$, $\zeta(s) \neq 0$ ist also $-\frac{1}{2} \log \pi + \frac{1}{2} \frac{\Gamma'}{\Gamma}\left(\frac{s}{2}\right) + \frac{\zeta'}{\zeta}(s)$ rein imaginär, somit

$$\left|\frac{\zeta'}{\zeta}(s)\right| \geqq -\Re \frac{\zeta'}{\zeta}(s) = -\frac{1}{2} \log \pi + \frac{1}{2} \Re \frac{\Gamma'}{\Gamma}\left(\frac{s}{2}\right) =$$

$$= \frac{1}{2} \log t + O(1) \to \infty.$$

Satz: *Ist die* MERTENS*sche Vermutung*

$$(1)\qquad M(x)=\sum_{n\leq x}\mu(n)=O(\sqrt{x})$$

wahr, so ist bei jedem $\delta>0$ *für* $t>b(\delta)$

$$(2)\qquad \frac{1}{\left|\zeta'\left(\frac{1}{2}+ti\right)\right|}<e^{\delta\log^2 t}.$$

Beweis: (1) sei wahr, also für $x>0$, wenn a_1 (desgl. a_2, a_3, a_4 nachher) absolut konstant ist,

$$|M(x)|<a_1\sqrt{x}.$$

1) Es bezeichne $\frac{1}{2}+\gamma i$, $\gamma>0$, eine Wurzel von $\zeta(s)$. Für $\varepsilon>0$ ist

$$\left|\frac{\varepsilon}{\zeta\left(\frac{1}{2}+\varepsilon+\gamma i\right)}\right|=\left|\varepsilon\sum_{n=1}^{\infty}\frac{\mu(n)}{n^{\frac{1}{2}+\varepsilon+\gamma i}}\right|=$$

$$=\left|\varepsilon\sum_{n=1}^{\infty}M(n)\left(\frac{1}{n^{\frac{1}{2}+\varepsilon+\gamma i}}-\frac{1}{(n+1)^{\frac{1}{2}+\varepsilon+\gamma i}}\right)\right|=$$

$$=\left|\varepsilon\left(\frac{1}{2}+\varepsilon+\gamma i\right)\int_1^{\infty}\frac{M(u)du}{u^{\frac{3}{2}+\varepsilon+\gamma i}}\right|<$$

$$<\varepsilon\left|\frac{1}{2}+\varepsilon+\gamma i\right|a_1\int_1^{\infty}\frac{du}{u^{1+\varepsilon}}=\left|\frac{1}{2}+\varepsilon+\gamma i\right|a_1;$$

$\varepsilon\to 0$ lehrt $\zeta'\left(\frac{1}{2}+\gamma i\right)\neq 0$ und sogar

$$(3)\qquad \frac{1}{\left|\zeta'\left(\frac{1}{2}+\gamma i\right)\right|}\leq\left|\frac{1}{2}+\gamma i\right|a_1<a_2\gamma.$$

2) Es sei $t_0 > 2$ und das Intervall $t_0 - \frac{1}{t_0^3} \leqq t \leqq t_0 + \frac{1}{t_0^3}$ nicht wurzelfrei. Wegen

$$\left|\zeta''\left(\frac{1}{2} + ti\right)\right| < a_3 t \text{ für } t > 1$$

(die bekannten schärferen Abschätzungen sind hier unerheblich) ist, wenn (3) auf ein γ jenes Intervalls angewendet wird,

$$\left|\zeta'\left(\frac{1}{2} + t_0 i\right)\right| = \left|\zeta'\left(\frac{1}{2} + \gamma i\right) + \int\limits_{\frac{1}{2}+\gamma i}^{\frac{1}{2}+t_0 i} \zeta''(s)ds\right| >$$

(4) $$> \frac{1}{a_2\left(t_0 + \frac{1}{t_0^3}\right)} - \frac{1}{t_0^3} a_3 \left(t_0 + \frac{1}{t_0^3}\right) > \frac{1}{2a_2 t_0},$$

falls überdies $t_0 > a_4$ ist.

3) $\delta > 0$ sei gegeben. Es sei $t_0 > 2$ und das Intervall $t_0 - \frac{1}{t_0^3} \leqq t \leqq t_0 + \frac{1}{t_0^3}$ wurzelfrei. Bekanntlich ist für wurzelfrei wachsendes t und $\frac{1}{2} \leqq \sigma \leqq 2$ gleichmässig

$$\frac{\zeta'}{\zeta}(s) = \sum_{|t-\gamma|<1} \frac{1}{s-\varrho} + O(\log t) = \sum_{|t-\gamma|<\frac{\delta}{10}} \frac{1}{s-\varrho} + O(\log t),$$

also für unsere t_0

$$\log\left|\zeta\left(\frac{1}{2} + t_0 i\right)\right| = \Re \log \zeta\left(\frac{1}{2} + t_0 i\right) = \Re \log \zeta(2 + t_0 i) +$$

$$+ \Re \int\limits_{2+t_0 i}^{\frac{1}{2}+t_0 i} \frac{\zeta'}{\zeta}(s)ds = \sum_{|t_0-\gamma|<\frac{\delta}{10}} \log|t_0 - \gamma| + O(\log t_0).$$

Nach einem Lemma von Herrn BOHR[1] ist die Gliederzahl der letzten Summe für $t_0 > b_1(\delta)$ kleiner als $\frac{3}{10}\delta \log t_0$; jedes Glied ist $\geqq -3 \log t_0$; daher ist unter Benutzung des Hilfssatzes für unsere $t_0 > b_2(\delta)$

$$-\log\left|\zeta'\left(\frac{1}{2}+t_0 i\right)\right| \leqq -\log\left|\zeta\left(\frac{1}{2}+t_0 i\right)\right| <$$

$$(5) \qquad < \frac{9}{10}\delta \log^2 t_0 + O(\log t_0) < \delta \log^2 t_0.$$

Mit (4) und (5) ist (2) bewiesen.

---

[1] BOHR, LANDAU und LITTLEWOOD, *Sur la fonction $\zeta(s)$ dans le voisinage de la droite $\sigma = \frac{1}{2}$* [Bulletins de l'Académie royale des Sciences, Lettres et Beaux-Arts de Belgique, Classe des sciences, 1913, S. 1144—1175], S. 1145.

Tryckt den 28 april 1921.

Uppsala 1921. Almqvist & Wiksells Boktryckeri-A.-B.

12.

Über das Teilerproblem von Piltz

Ark. Mat. Astr. Fys. **16** (21), 1–40 (1922)

Mitgeteilt am 26. Oktober 1921 durch E. L. PHRAGMÉN und I. BENDIXSON.

Einleitung.

Seit der Aufstellung der berühmten RIEMANN'schen Primzahlformel sind verschiedene analoge explizite Formeln für zahlentheoretische Funktionen gefunden worden. Sie sind alle von ähnlicher Bauart wie die RIEMANN'sche: zuerst kommt eine endliche Summe elementarer Funktionen, die für das Verhalten des ganzen Ausdrucks im Unendlichen ausschlaggebend ist, dann kommt ein Restglied, welches die Form einer unendlichen Reihe hat, deren Konvergenz und übrigen Eigenschaften nur durch sehr verwickelte Betrachtungen auf. geklärt werden können. Was man bis jetzt über die asymptotischen Eigenschaften dieser Funktionen kennt, ist daher grösstenteils unabhängig von den expliziten Formeln erreicht worden, die deshalb vielfach mit einem gewissen Misstrauen betrachtet worden sind; man vergleiche z. B. die Ausführungen von HADAMARD in der Encyclopédie des Sciences Mathématiques (Tome I, Vol. 3, p. 318).

In den letzten Jahren sind jedoch einige Resultate auf diesem Wege erhalten worden. Herr LITTLEWOOD[1] bewies mit Hilfe einer von Herrn v. MANGOLDT herrührenden, der

[1] HARDY und LITTLEWOOD, Contributions to the theory of the Riemann zeta-function and the theory of the distribution of primes, Acta Mathematica, *41* (1917), S. 119—196.

RIEMANN'schen verwandten, Formel einen früher gänzlich unerwarteten Satz über die Verteilung der Primzahlen, welcher sodann von Herrn LANDAU[1] für die Primideale eines algebraischen Zahlkörpers verallgemeinert wurde. Einige neue Eigenschaften der Primzahlen wurden auch von mir[2] mit einer hierhergehörenden Methode gefunden.

Ausserhalb der Primzahltheorie in engerem Sinne liegen die betreffenden Arbeiten des Herrn HARDY[3], die den Ausgangspunkt des vorliegenden Aufsatzes bilden. Herr HARDY bezeichnet mit $d(n)$ die Teilerzahl der ganzen Zahl n, mit $r(n)$ die Anzahl ihrer Darstellungen als Summe von zwei Quadraten. Er setzt dann für $x > 0$

$$D(x) = \sum_{n \leqq x} d(n) = x \log x + (2C - 1)x + \Delta(x),$$

$$R(x) = \sum_{n \leqq x} r(n) = \pi x + P(x),$$

wo C die EULER'sche Konstante bedeutet, und zeigt wie man $\Delta(x)$ und $P(x)$ durch unendliche Reihen mit BESSEL'schen Funktionen ausdrücken kann. Die Formel für $\Delta(x)$ war schon früher von VORONOÏ[4] gefunden worden; seine Beweismethoden waren jedoch viel komplizierter als die HARDY-schen. Da wir in dieser Arbeit nur von Teilerproblemen sprechen werden, wollen wir hier die wichtigsten Eigenschaften von $\Delta(x)$ kurz erwähnen. Man hat

$$\Delta(x) = O(x^{\frac{1}{3}} \log x)$$

aber

$$\Delta(x) \neq o(x^{\frac{1}{4}}).$$

---

[1] Über Ideale und Primideale in Idealklassen, Math. Zeitschr., *2* (1918), S. 52—154.

[2] Studien über die Nullstellen der Riemannschen Zetafunktion, Math. Zeitschr., *4* (1919), S. 104—130; Some theorems concerning prime numbers, Arkiv för Matematik, B. *15*, Nr. 5 (1920), 33 S.; On the distribution of primes, Proceedings of the Cambridge Phil. Soc., *20*: 2, (1921) S. 272—280.

[3] On the expression of a number as the sum of two squares, Quarterly Journal, *46* (1915), S. 263—283; On Dirichlet's divisor problem, Proc. Lond. Math. Soc. (2) *15* (1916), S. 1—25, On the average order of the arithmetical funktions $P(x)$ and $\Delta(x)$, Ebenda, S. 192—213.

[4] Sur une fonction transcendante et ses applications à la sommation de quelques séries, Annales de l' Ecole Normale, (3) *21*, S. 207—268 und S. 459—534.

Die erste Beziehung ist von VORONOÏ[1], die zweite von HARDY gefunden; HARDY skizziert auch wie man aus der expliziten Formel eine noch schärfere untere Grenze der Grössenordnung von $\Delta(x)$ ableiten kann. Dann beweist er auch den Mittelwertsatz:

$$\frac{1}{x}\int_1^x |\Delta(t)|\,dt = O\left(x^{\frac{1}{4}+\varepsilon}\right) \tag{1}$$

für jedes $\varepsilon > 0$. In dieser Formel kann übrigens, wie ich in einer noch ungedruckten Arbeit gezeigt habe, das ε weggenommen werden.

Die unendliche Reihe für $\Delta(x)$ enthält, wie gesagt, in ihren Gliedern BESSEL'sche Funktionen. Wendet man auf diese Reihe die bekannte asymptotische Entwicklung jener Funktionen an, so kann sie in der folgenden Form geschrieben werden

$$\Delta(x) = \frac{x^{\frac{1}{4}}}{\pi\sqrt{2}} \sum_1^\infty \frac{d(n)}{n^{\frac{3}{4}}} \cos\left(4\pi\sqrt{nx} - \tfrac{1}{4}\pi\right) + O(1), \tag{2}$$

wo $x > 1$ und nicht ganzzahlig ist. Die hier auftretende Reihe, die den »kritischen Teil» von $\Delta(x)$ darstellt, ist für alle $x > 1$ konvergent, und zwar gleichmässig in jedem Intervall, das keine ganze Zahl enthält.

Das jetzt behandelte sog. DIRICHLET'sche Teilerproblem ist von PILTZ[2] verallgemeinert worden. Es sei für ganzes $k \geqq 2$ und für $s > 1$

$$(\zeta(s))^k = \left(\sum_1^\infty \frac{1}{n^s}\right)^k = \sum_1^\infty \frac{d_k(n)}{n^s}$$

---

[1] Sur un problème du calcul des fonctions asymptotiques, Journal für Math., *126* (1903), S. 241—282. Die fast ebenso scharfe Abschätzung $\Delta(x) = O\left(x^{\frac{1}{3}+\varepsilon}\right)$ war schon 1886 von PFEIFFER gegeben, der von ihm gegebene Beweis war aber nicht ganz einwandfrei. Vgl. hierüber LANDAU, Die Bedeutung der Pfeifferschen Methode für die analytische Zahlentheorie, Wiener Sitzungsber., B. 121 (1912), Abt. II a, S. 2195—2332, wo ein korrekter Beweis nach der PFEIFFER'schen Methode gegeben wird.

[2] Über das Gesetz, nach welchem die mittlere Darstellbarkeit der natürlichen Zahlen als Produkte einer gegebenen Anzahl Faktoren mit der Grösse der Zahlen wächst, Inauguraldissertation, Berlin 1881, 31 S.

gesetzt, so dass $d_k(n)$ die Anzahl der Darstellungen von n als Produkt von k Faktoren bedeutet. Dann gilt

$$D_k(x) = \sum_{n \leqq x} d_k(n) = x\, p_{k-1}(\log x) + \varDelta_k(x),$$

wo p_{k-1} ein gewisses Polynom $(k-1)$ten Grades bedeutet, dessen Koeffizienten von x unabhängig sind, und wo nach LANDAU[1]

$$\varDelta_k(x) = O\left(x^{\frac{k-1}{k+1}+\varepsilon}\right)$$

für jedes $\varepsilon > 0$ gilt. — Nach einem LANDAU'schen Hilfssatz, den wir widerholt anwenden werden, ist $d_k(n) = O(n^\varepsilon)$.

Zweck dieser Arbeit ist nun, die explizite Formel (2) für $\varDelta(x) = \varDelta_2(x)$ auf beliebiges k zu verallgemeinern und daraus einige Folgerungen über $\varDelta_k(x)$ zu ziehen. Die verallgemeinerte Formel wird nach der HARDY'schen Methode abgeleitet; doch werden natürlich hier die analytischen Schwierigkeiten etwas grösser, und es tritt eine neue Erscheinung auf: die erhaltenen Reihen sind nicht konvergent, sie sind aber mit den sog. typischen Mitteln des Herrn M. RIESZ summierbar. Ich werde nämlich in § 3 den folgenden Satz beweisen.

Satz A. *Für nicht ganzzahliges $x > 1$ gilt*

$$\varDelta_k(x) = \sum_{\mu=0}^{\left[\frac{k-1}{2}\right]} x^{\frac{k-1}{2k}-\frac{\mu}{k}} (\alpha_\mu C_\mu + \beta_\mu S_\mu) + O(1),$$

wo α_μ und β_μ Konstanten sind, und

$$C_\mu = \sum_{n=1}^{\infty} \frac{d_k(n)}{n^{\frac{k+1}{2k}+\frac{\mu}{k}}} \cos 2k\pi (nx)^{\frac{1}{k}},$$

---

[1] Über die Anzahl der Gitterpunkte in gewissen Bereichen, Göttinger Nachrichten, 1912, S. 687—771. Der Hilfssatz über $d_k(n)$ findet sich auf S. 718. PILTZ hatte für $\varDelta_k(x)$ eine unschärfere Abschätzung gegeben. LANDAU bemerkt (S. 691) dass, wenn die RIEMANN'sche Vermutung über die Zetafunktion wahr ist, sogar $\varDelta_k(x) = O\left(x^{\frac{1}{2}+\varepsilon}\right)$ gilt. LANDAU hat übrigens später gezeigt, dass in Satz B. der Faktor x^ε sogar durch $(\log x)^{k-1}$ ersetzt werden kann: Über DIRICHLETS Teilerproblem, Münchener Berichte 1915, S. 317—328.

$$S_\mu = \sum_{n=1}^{\infty} \frac{d_k(n)}{n^{\frac{k+1}{2k}+\frac{\mu}{k}}} \sin 2k\pi (nx)^{\frac{1}{k}}$$

gesetzt ist. Die Reihen C_μ und S_μ sind summierbar $(n^{\frac{1}{k}}, \varkappa)$ [1], *wenn $\varkappa$ eine ganze Zahl grösser als $\frac{k-3}{2} - \mu$ ist. Die Summabilität ist eine gleichmässige in jedem x-Intervall, das keine ganze Zahl enthält. Wenn aber $\varkappa \leqq \frac{k-3}{2} - \mu$ ist, so ist wenigstens eine der Reihen C_μ und S_μ nicht summierbar $(n^{\frac{1}{k}}, \varkappa)$.*

Der Fall $k = 2$ ist somit der einzige, wo nur konvergente Reihen auftreten, da schon im Falle $k = 3$ entweder C_0 oder S_0 nach dem Obigen divergent ist.

Für den Beweis dieses Satzes sind verschiedene Hilfsbetrachtungen nötig, die in den beiden ersten Paragraphen gegeben werden. Zuerst wird in § 1 eine DIRICHLET'sche Reihe behandelt, die als erzeugende Funktion der Koeffizienten $d_k(n)$ dienen wird. In § 2 spreche ich von einigen wichtigen Sätzen des Herrn M. RIESZ[2], die bei dem HARDY'schen Beweis der Formel für $\Delta_2(x)$ angewandt werden. RIESZ hat in seinen Arbeiten bedeutende Verallgemeinerungen seiner Sätze in Aussicht gestellt, und es ist eben eine von diesen Verallgemeinerungen, die ich hier gebrauchen werde. Der Beweis, den ich mit freundlicher Zustimmung des Herrn RIESZ in § 2 für diesen verallgemeinerten Satz gebe, verläuft dem Seinigen ganz parallel.

In § 4 werden schliesslich als Anwendung der vorhergehenden Untersuchung einige Sätze über $D_k(x)$ bewiesen, nämlich zuerst den obengenannten

Satz B. (LANDAU) *Für jedes $\varepsilon > 0$ gilt*

$$\Delta_k(x) = O\left(x^{\frac{k-1}{k+1}+\varepsilon}\right).$$

Dann werde ich den HARDY'schen Mittelwertsatz (1) verallgemeinern, indem ich als letzte Anwendung beweisen werde:

[1] Die Erklärung dieser Bezeichnungsweise wird unten in § 2 gegeben.

[2] Ein Konvergenzsatz für DIRICHLET'sche Reihen, Acta Mathematica *40*, 1916, S. 349—361, und Sätze über Potenzreihen, Arkiv för Matematik, *11*, Nr. 12 (1916). Diese zwei Arbeiten sind für die vorliegenden Anwendungen besonders wichtig; man findet dort die übrigen Arbeiten des Verfassers über denselben Gegenstand angeführt.

Satz C. *Für $k > 2$ und jedes $\varepsilon > 0$ gilt*

$$\frac{1}{x}\int_1^x |\Delta_k(t)|\,dt = O\left(x^{\frac{k-2}{k}+\varepsilon}\right).$$

Wegen $\frac{k-2}{k} < \frac{k-1}{k+1}$ ist dieser Satz nicht eine Folgerung von Satz B.

Die vorliegende Arbeit wurde im Mai 1921 fertiggestellt. Bald danach erhielt ich briefliche Mitteilungen über zwei — noch nicht gedruckten — denselben Gegenstand berührenden Arbeiten. Erstens hat Herr A. WALFISZ in einer Göttinger Dissertation die Reihendarstellungen der Funktion $D_k(x)$ und verschiedener anderen Funktionen untersucht. Zweitens haben die Herren HARDY und LITTLEWOOD in einer an die London Mathematical Society eingereichten Abhandlung sogar

$$\Delta_k(x) = O\left(x^{\frac{k-2}{k}+\varepsilon}\right)$$

für alle $k \geq 4$ bewiesen. Hieraus folgt für alle diese Werte von k unmittelbar mein Satz C, der aber für $k = 3$ ein neues Ergebniss liefert. — Hierdurch wird zwar der Wert meiner Arbeit wesentlich reduciert; ich glaube aber, dass besonders die Methoden der Herren HARDY und LITTLEWOOD von den Meinigen so durchaus verschieden sind, dass die Veröffentlichung dieser Abhandlung sich doch rechtfertigen lässt.

§ 1.

Über die Funktion $\Sigma d_k(n) e^{-n^{\frac{1}{k}}s}$.

In diesem ersten Paragraphen werden wir die Funktion $f_k(s) = f_k(\sigma + it)$ untersuchen, die für $\sigma > 0$ durch die Reihe

$$f_k(s) = \sum_{n=1}^{\infty} d_k(n)\, e^{-n^{\frac{1}{k}}s}$$

dargestellt wird. Für diese Funktion werden wir zuerst den folgenden Ausdruck aufstellen, der den Ausgangspunkt unserer späteren Entwicklungen bildet:

$$(3)\quad f_k(s) = s^{-k} P_{k-1}(\log s) + \alpha + \frac{1}{(2\pi i)^k} \sum_{\mu=0}^{k} \binom{k}{\mu} (-1)^\mu M_k\left(-\frac{s e^{\frac{\mu \pi i}{k}}}{2\pi i}\right),$$

wo α eine Konstante, P_{k-1} (wie unten P'_{k-1}, $P''_{k-1} \cdots$) ein Polynom $(k-1)$sten Grades und

$$(4)\qquad M_k(s) = \sum_{n=1}^{\infty} \frac{d_k(n)}{n} L_k\left(\frac{s}{n^{\frac{1}{k}}}\right),$$

$$(5)\qquad L_k(s) = (k-2)! \int \cdots \int \left[\left(1 - (x_1 x_2 \cdots x_{k-1} y)^{\frac{1}{k}} s\right)^{-k} - 1\right] dx_1 dx_2 \ldots dx_{k-1}$$

ist. Wir setzen hier, wie überall im Folgenden

$$(6)\qquad y = 1 - x_1 - x_2 - \ldots - x_{k-1}$$

und erstrecken das $(k-1)$-fache Integral über das Gebiet $x_1 > 0$, $x_2 > 0$, $\ldots x_{k-1} > 0$, $y > 0$, das wir durch G bezeichnen. Das Integral konvergiert offenbar gleichmässig in jedem endlichen abgeschlossenen Gebiet in der s-Ebene, das keinen Teil der reellen Achse von $s = k$ bis $s = +\infty$ enthält, und (3) stellt also $f_k(s)$ wenigstens in der Halbebene $\sigma > 0$ dar.

Um diese Formeln abzuleiten werden wir zunächst das $(k-1)$-fache Integral der Funktion $f_k(s)$ betrachten. Wir setzen

$$F_k(s) = \sum_{n=1}^{\infty} \frac{d_k(n)}{n^{\frac{k-1}{k}}} e^{-n^{\frac{1}{k}} s},$$

so dass

$$(7)\qquad f_k(s) = (-1)^{k-1} \frac{d^{k-1}}{ds^{k-1}} F_k(s)$$

gilt, und erhalten aus der bekannten Formel

$$e^{-s} = \frac{1}{2\pi i}\int_{a-i\infty}^{a+i\infty} s^{-z}\Gamma(z)\,dz, \qquad (\sigma > 0,\ a > 0)$$

für $F_k(s)$ die Gleichung

$$F_k(s) = \frac{1}{2\pi i}\int_{2-i\infty}^{2+i\infty} s^{-z}\Gamma(z)\left[\zeta\left(\frac{k-1}{k} + \frac{1}{k}z\right)\right]^k dz.$$

Die Singularitäten der Funktion hinter dem Integralzeichen sind hier teils der von dem ζ-Faktor herrührende k-fache Pol in $z = 1$, teils die einfachen Pole $z = 0, -1, -2, \ldots$ der Gammafunktion. Durch eine leichte Anwendung des CAUCHY'schen Satzes erhalten wir daher für $\sigma > 0$

$$(8) \qquad F_k(s) = \frac{P'_{k-1}(\log s)}{s} + \sum_{\nu=0}^{p} \frac{(-s)^\nu}{\nu!}\left[\zeta\left(\frac{k-1}{k} - \frac{\nu}{k}\right)\right]^k + J_p,$$

wo p eine ganze positive Zahl und

$$J_p = \frac{1}{2\pi i}\int_{-p-\frac{1}{2}-i\infty}^{-p-\frac{1}{2}+i\infty} s^{-z}\Gamma(z)\left[\zeta\left(\frac{k-1}{k} + \frac{1}{k}z\right)\right]^k dz$$

ist.

Wir nehmen nun vorläufig s reell und $0 < s < 2k\pi$ an. Lassen wir dann p (durch ganze Zahlen) unendlich werden, so strebt J_p gegen Null. Es gilt nämlich, wenn wir $z = -p - \frac{1}{2} + iu$ setzen,

$$|s^{-z}| = s^{p+\frac{1}{2}},$$

$$|\Gamma(z)| = \left|\Gamma\left(-p - \frac{1}{2} + iu\right)\right| = \frac{\pi}{\cos h\pi u \left|\Gamma\left(p + \frac{3}{2} - iu\right)\right|},$$

$$\left|\zeta\left(\frac{k-1}{k} + \frac{1}{k}z\right)\right|^k = 2^k(2\pi)^{-p-\frac{3}{2}}\left|\sin\frac{\pi}{2}\left(\frac{k-p-\frac{3}{2}+iu}{k}\right)\right|^k.$$

$$\cdot\left|\Gamma\left(\frac{1}{k}+\frac{p+\frac{1}{2}-iu}{k}\right)\right|^k\left|\zeta\left(\frac{1}{k}+\frac{p+\frac{1}{2}-iu}{k}\right)\right|^k <$$

$$< K(2\pi)^{-p} e^{\frac{1}{2}\pi|u|}\left|\Gamma\left(\frac{2p+3-2iu}{2k}\right)\right|^k,$$

wo $p>k$ angenommen ist, und K (wie unten $K_1, K_2, \ldots$) eine von p und u unabhängige Konstante bedeutet. Hieraus folgt nun

$$\left|s^{-z}\Gamma(z)\zeta\left(\frac{k-1}{k}+\frac{1}{k}z\right)\right| < K_1\left(\frac{s}{2\pi}\right)^p e^{-\frac{1}{2}\pi|u|}\left|\frac{\left[\Gamma\left(\frac{2p+3-2iu}{2k}\right)\right]^k}{\Gamma\left(\frac{2p+3-2iu}{2}\right)}\right|$$

$$= K_1\left(\frac{s}{2\pi}\right)^p e^{-\frac{1}{2}\pi|u|}\left|\int\cdots\right.$$

$$\left.\cdots\int(x_1x_2\ldots x_{k-1}y)^{\frac{2p+3-2iu}{2k}-1}dx_1dx_2\ldots dx_{k-1}\right|$$

$$< K_1\left(\frac{s}{2\pi}\right)^p e^{-\frac{1}{2}\pi|u|}\int\cdots$$

$$\cdots\int(x_1x_2\ldots x_{k-1}y)^{\frac{2p+3}{2k}-1}dx_1dx_2\ldots dx_{k-1}$$

wo rechts das Integral über das oben erklärte Gebiet G erstreckt ist. Im Gebiete G ist nun überall $x_1 x_2 \ldots x_{k-1} y \leqq k^{-k}$; der letzte Ausdruck ist also kleiner als

$$K_2\left(\frac{s}{2k\pi}\right)^p e^{-\frac{1}{2}\pi|u|},$$

und man schliesst hieraus

$$|J_p| < K_2\left(\frac{s}{2k\pi}\right)^p\int_{-\infty}^{\infty}e^{-\frac{1}{2}\pi|u|}du < K_3\left(\frac{s}{2k\pi}\right)^p,$$

so dass tatsächlich $J_p \to 0$ und nach (8)

$$F_k(s) = \frac{P'_{k-1}(\log s)}{s} + \sum_{\nu=0}^{\infty}\frac{(-s)^\nu}{\nu!}\left[\zeta\left(\frac{k-\nu-1}{k}\right)\right]^k.$$

Diese Gleichung ist freilich nur für reelle s zwischen 0 und $2k\pi$ bewiesen, da aber beiderseitig analytische Funktionen von s stehen, die für $0 < s < 2k\pi$ regulär sind, bleibt sie natürlich im ganzen Existenzbereich der Funktionen gültig. Wir wollen jetzt die letzte Potenzreihe umformen und erhalten aus

$$\zeta\left(\frac{k-\nu-1}{k}\right) = 2(2\pi)^{-\frac{\nu+1}{k}} \cos\frac{\nu+1}{2k}\pi\, \Gamma\left(\frac{\nu+1}{k}\right)\zeta\left(\frac{\nu+1}{k}\right)$$

und

$$\left(\cos\frac{\nu+1}{2k}\pi\right)^k = 2^{-k}\left(e^{\frac{\nu+1}{2k}\pi i} + e^{-\frac{\nu+1}{2k}\pi i}\right)^k = 2^{-k}\sum_{\mu=0}^{k}\binom{k}{\mu} e^{\frac{(2\mu-k)(\nu+1)}{2k}\pi i}$$

unmittelbar

$$F_k(s) = \frac{P'_{k-1}(\log s)}{s} + P''_{k-1}(s) +$$

$$\frac{1}{2\pi i}\sum_{\mu=0}^{k} e^{\frac{\mu\pi i}{k}}\binom{k}{\mu}\sum_{\nu=k}^{\infty}\frac{\left[\Gamma\left(\frac{\nu+1}{k}\right)\right]^k}{\Gamma(\nu+1)}\left[\zeta\left(\frac{\nu+1}{k}\right)\right]^k\left(-\frac{se^{\frac{\mu\pi i}{k}}}{2\pi i}\right)^{\nu}.$$

Setzen wir also

$$\overline{M}_k(s) = \sum_{\nu=k}^{\infty}\frac{\left[\Gamma\left(\frac{\nu+1}{k}\right)\right]^k}{\Gamma(\nu+1)}\left[\zeta\left(\frac{\nu+1}{k}\right)\right]^k s^{\nu} \tag{9}$$

$$= \sum_{n=1}^{\infty}\frac{d_k(n)}{n^{\frac{1}{k}}}\sum_{\nu=k}^{\infty}\frac{\left[\Gamma\left(\frac{\nu+1}{k}\right)\right]^k}{\Gamma(\nu+1)}\left(\frac{s}{n^{\frac{1}{k}}}\right)^{\nu}$$

$$= \sum_{n=1}^{\infty}\frac{d_k(n)}{n^{\frac{1}{k}}}\overline{L}_k\left(\frac{s}{n^{\frac{1}{k}}}\right),$$

so haben wir

$$F_k(s) = \frac{P'_{k-1}(\log s)}{s} + P''_{k-1}(s) + \tag{10}$$

$$+\frac{1}{2\pi i}\sum_{\mu=0}^{k} e^{\frac{\mu\pi i}{k}}\binom{k}{\mu}\overline{M}_k\left(-\frac{se^{\frac{\mu\pi i}{k}}}{2\pi i}\right)$$

und

$$(11)\quad \overline{L}_k(s) = \sum_{\nu=k}^{\infty} \frac{\left[\Gamma\left(\frac{\nu+1}{k}\right)\right]^k}{\Gamma(\nu+1)} s^\nu$$

$$= \sum_{\nu=k}^{\infty} \int \cdots \int (x_1 x_2 \ldots x_{k-1} y)^{\frac{\nu+1}{k}-1} s^\nu \, dx_1 \, dx_2 \ldots dx_{k-1}$$

$$= P_{k-1}'''(s) + \int \cdots \int \frac{(x_1 x_2 \ldots x_{k-1} y)^{\frac{1}{k}-1}}{1-(x_1 x_2 \ldots x_{k-1} y)^{\frac{1}{k}} s} \, dx_1 \, dx_2 \ldots dx_{k-1},$$

die Integrale wieder über das Gebiet G genommen.

Jetzt wollen wir unter Benutzung von (7) zur Funktion $f_k(s)$ übergehen. Wir erhalten aus (9) bis (11)

$$f_k(s) = s^{-k} P_{k-1}(\log s) + \alpha + \frac{1}{(2\pi i)^k} \sum_{\mu=0}^{k} \binom{k}{\mu} (-1)^\mu M_k\left(-\frac{s e^{\frac{\mu \pi i}{k}}}{2\pi i}\right),$$

$$M_k(s) = \frac{d^{k-1}}{d s^{k-1}} \overline{M}^k(s) = \sum_{n=1}^{\infty} \frac{d_k(n)}{n} L_k\left(\frac{s}{n^{\frac{1}{k}}}\right),$$

$$L_k(s) = \frac{d^{k-1}}{d s^{k-1}} \overline{L}_k(s) = K_4 + (k-2)! \int \cdots$$

$$\cdots \int \frac{dx_1 \, dx_2 \ldots dx_{k-1}}{(1-(x_1 x_2 \ldots x_{k-1} y)^{\frac{1}{k}} s)^k}.$$

Die in dem letzten Ausdruck auftretende Konstante, die nur von k abhängt, bestimmt sich daraus, dass nach der Potenzreihe (11) $L_k(0) = 0$ ist; wir haben demnach

$$L_k(s) = (k-2)! \int \cdots$$

$$\cdots \int \left[(1-(x_1 x_2 \ldots x_{k-1} y)^{\frac{1}{k}} s)^{-k} - 1\right] dx_1 dx_2 \ldots dx_{k-1}$$

und die Formeln (3), (4) und (5) sind hierdurch vollständig bewiesen.

§ 2.

Summabilitätseigenschaften der Reihe $\sum \frac{d_k(n)}{n^a} e^{-n^{\frac{1}{k}} s}$.

Wir beweisen zuerst einen allgemeinen Satz über DIRICHLET'sche Reihen, der von Herrn M. RIESZ gefunden wurde (vgl. hierüber die Einleitung), und ziehen denn daraus einige spezielle Folgerungen, welche für die Darstellung der Funktion $D_k(x)$ wichtig sind.

Wenn $a_1, a_2, \ldots$ irgend eine gegebene Zahlenfolge und $\lambda_1, \lambda_2, \ldots$ eine monoton wachsende Folge positiver Zahlen mit $\lim_{n \to \infty} \lambda_n = \infty$ ist, so setzen wir für $\varkappa \geqq 0$, $\omega \geqq 0$

$$A_\varkappa(\omega) = \sum_{\lambda_n < \omega} a_n (\omega - \lambda_n)^\varkappa .$$

Gilt nun für unendlich wachsendes ω

$$\omega^{-\varkappa} A_\varkappa(\omega) \to A ,$$

so heisst die Reihe Σa_n *summierbar* $(\lambda, \varkappa)$ *mit der Summe* A. Sie ist dann auch summierbar $(\lambda, \varkappa')$, wo $\varkappa' > \varkappa$, mit derselben Summe.

Wenn die beiden Reihen $F(x) = \Sigma f_n(x)$ und $\Phi(x) = f_n'(x)$ in einem Intervall $a \leqq x \leqq b$ *gleichmässig* summierbar $(\lambda, \varkappa)$ sind, so gilt dort

$$\Phi(x) = F'(x).$$

Der oben erwähnte Satz von M. RIESZ lautet nun folgendermaassen.

Hilfssatz 1. *Es sei*

$$f(s) = a_1 e^{-\lambda_1 s} + \cdots + a_n e^{-\lambda_n s} + \cdots$$

eine DIRICHLET'*sche Reihe, deren Koeffizienten die Bedingung*

(12) $$\lim_{\omega \to \infty} \omega^{-\varkappa} e^{-c\omega} A_\varkappa(\omega) = 0$$

erfüllen, wo $\varkappa$ *eine nichtnegative ganze Zahl und* $c > 0$ *ist. Wenn die infolge der Bedingung für* $\sigma > c$ *reguläre*[1] *Funktion* $f(s)$ *auch in gewissen Punkten der Geraden* $\sigma = c$ *regulär ist, so ist die Reihe in diesen Punkten summierbar* $(\lambda, \varkappa)$. *Die Summabilität ist eine gleichmässige auf jeder abgeschlossenen Strecke der Geraden, welche ausschliesslich aus Regularitätsstellen besteht. Die Bedingung* (12) *ist auch notwendig, damit die Gerade* $\sigma = c$ *wenigstens einen Punkt enthalte, wo die Reihe summierbar* $(\lambda, \varkappa)$ *ist.*

Es bezeichne λ_p das grösste λ_n unterhalb ω. Dann haben wir

$$\sum_{\lambda_n < \omega} a_n e^{-\lambda_n s} (\omega - \lambda_n)^\varkappa = \sum_1^{p-1} (a_1 + \cdots + a_n) [e^{-\lambda_n s} (\omega - \lambda_n)^\varkappa -$$

$$- e^{-\lambda_{n+1} s} (\omega - \lambda_{n+1})^\varkappa] + (a_1 + \cdots + a_p) e^{-\lambda_p s} (\omega - \lambda_p)^\varkappa =$$

$$= - \int_0^\omega A_0(v) \frac{d}{dv} (e^{-sv} (\omega - v)^\varkappa) \, dv$$

und nach $\varkappa$-maliger partieller Integration

$$(13) \quad \sum_{\lambda_n < \omega} a_n e^{-\lambda_n s} (\omega - \lambda_n)^\varkappa =$$

$$= \frac{(-1)^{\varkappa+1}}{\varkappa!} \int_0^\omega A_\varkappa(v) \frac{d^{\varkappa+1}}{dv^{\varkappa+1}} (e^{-sv} (\omega - v)^\varkappa) \, dv + A_\varkappa(\omega) e^{-s\omega}.$$

Durch Entwicklung der Ableitung erhalten wir

$$(14) \quad \frac{d^{\varkappa+1}}{dv^{\varkappa+1}} (e^{-sv} (\omega - v)^\varkappa) = (-s)^{\varkappa+1} \omega^\varkappa e^{-sv} + e^{-sv} \sum H_{ijk} s^i \omega^j v^k,$$

wo H_{ijk} konstant und

$$(15) \qquad i = j + k + 1, \qquad j \leq \varkappa - 1$$

ist. Einführung in (13) ergibt nun, nach Division durch $\omega^\varkappa$,

---

[1] Nach Gleichung (17) durch den Summabilitätswert der Reihe definierte.

$$(16)\qquad \omega^{-\varkappa}\sum_{\lambda_n<\omega} a_n e^{-\lambda_n s}(\omega-\lambda_n)^{\varkappa} = \frac{s^{\varkappa+1}}{\varkappa!}\int_0^{\omega} A_{\varkappa}(v)\,e^{-sv}\,dv +$$

$$+\frac{(-1)^{\varkappa+1}}{\varkappa!}\sum H_{ijk}\,s^i\,\omega^{j-\varkappa}\int_0^{\omega} v^k A_{\varkappa}(v)\,e^{-sv}\,dv + \omega^{-\varkappa}A_{\varkappa}(\omega)\,e^{-s\omega}.$$

Es sei nun $\sigma > c$; wenn dann ω ins Unendliche wächst, so folgt aus (12), dass das erste Glied rechts in (16) einen Grenzwert hat, während das letzte Glied gegen Null strebt. Beides gilt sogar gleichmässig in jedem endlichen abgeschlossenen Teil der Halbebene $\sigma > c$. Es kann nun bewiesen werden[1], dass jedes einzelne Glied der rechts in (16) auftretenden dreifachen Summe in gleicher Weise gegen Null strebt; wir haben also für $\sigma > c$

$$(17)\qquad \lim_{\omega\to\infty}\omega^{-\varkappa}\sum_{\lambda_n<\omega} a_n e^{-\lambda_n s}(\omega-\lambda_n)^{\varkappa} = \frac{s^{\varkappa+1}}{\varkappa!}\int_0^{\infty} A_{\varkappa}(v)\,e^{-sv}\,dv,$$

d. h. die Reihe $\sum a_n e^{-\lambda_n s}$ stellt durch ihren Summabilitätswert eine für $\sigma > c$ reguläre Funktion $f(s)$ dar.

Es möge nun die Funktion $f(s)$ auch auf der Geraden $\sigma = c$ Regularitätsstellen besitzen, und es sei $s_0 = c + t_0 i$ eine solche Stelle. Wir bestimmen zwei Punkte $s_1 = c + t_1 i$ und $s_2 = c + t_2 i$ derart, dass $t_1 < t_0 < t_2$ sei, und die Funktion $f(s)$ auf der ganzen Strecke s_1, s_2 (mit Einschluss der Endpunkte) regulär sei. Weiter können wir die positive Grösse η so klein wählen, dass $f(s)$ auch noch im Rechteck $s_1 \pm \eta$, $s_2 \pm \eta$ dieselbe Eigenschaft besitze. Dabei können wir auch $c - \eta > 0$ annehmen.

Wir betrachten jetzt die Funktion

$$(18)\qquad g_{\omega}(s) = e^{\omega(s-c)}(s-s_1)^{2\varkappa+1}(s-s_2)^{2\varkappa+1}\Bigg(f(s) +$$

$$+(-1)^{\varkappa-k+1}\sum H_{ijk}\,s^i\,\omega^{j-\varkappa}\frac{d^k}{ds^k}\left(\frac{f(s)}{s^{\varkappa+1}}\right) - \omega^{-\varkappa}\sum_{\lambda_n<\omega} a_n e^{-\lambda_n s}(\omega-\lambda_n)^{\varkappa}\Bigg)$$

[1] Der betreffende Beweis findet sich vollständig ausgeführt auf S. 40—41 des Werkes: »The general theory of Dirichlet's series» von G. H. Hardy und M. Riesz (Cambridge 1915), weshalb wir ihn hier fortlassen können.

und werden zeigen, dass bei hinreichend grossen ω auf der ganzen Begrenzung des eben genannten Rechtecks

$$|g_n(s)| < \varepsilon$$

wird, wie klein auch die positive Zahl ε sein mag.

Es sei zuerst $\sigma > c$; dann gilt nach (14), (16) und (17)

$$(19) \quad g_\omega(s) = e^{\omega(s-c)}(s-s_1)^{2\varkappa+1}(s-s_2)^{2\varkappa+1}\left(\frac{s^{\varkappa+1}}{\varkappa!}\int_\omega^\infty A_\varkappa(v)e^{-sv}dv + \right.$$

$$\left. + \frac{(-1)^{\varkappa+1}}{\varkappa!}\sum H_{ijk}s^i\omega^{j-\varkappa}\int_\omega^\infty v^k A_\varkappa(v)e^{-sv}dv - \omega^{-\varkappa}A_\varkappa(\omega)e^{-s\omega}\right) =$$

$$= e^{\omega(s-c)}(s-s_1)^{2\varkappa+1}(s-s_2)^{2\varkappa+1}\left(\frac{(-1)^{\varkappa+1}}{\varkappa!}\omega^{-\varkappa}\int_\omega^\infty A_\varkappa(v)\frac{d^{\varkappa+1}}{dv^{\varkappa+1}}\cdot\right.$$

$$\left. \cdot(e^{-sv}(\omega-v)^\varkappa)dv - \omega^{-\varkappa}A_\varkappa(\omega)e^{-s\omega}\right).$$

Wegen (12) können wir für ein beliebig kleines positives δ eine Zahl ω_0 finden, so dass für $\omega > \omega_0$

$$(20) \qquad |A_\varkappa(\omega)| < \delta\,\omega^\varkappa e^{c\omega}$$

gilt. Das in dem letzten Ausdruck auftretende Integral ist aber gleich einer Summe von Gliedern von der Form

$$c_\nu s^{\varkappa+1-\nu}\int_\omega^\infty(\omega-v)^\nu A_\varkappa(v)e^{-sv}dv,$$

wo c_ν konstant und $0 \leqq \nu \leqq \varkappa$ ist, und wir haben, wenn s auf der Begrenzung des Rechtecks liegt, so dass $\sigma - c$ nach oben beschränkt bleibt,

$$\left|\int_\omega^\infty(\omega-v)^\nu A_\varkappa(v)e^{-sv}dv\right| < \delta\int_\omega^\infty(v-\omega)^\nu v^\varkappa e^{-v(\sigma-c)}dv =$$

$$= \delta e^{-\omega(\sigma-c)}\int_0^\infty v^\nu(\omega+v)^\varkappa e^{-v(\sigma-c)}dv <$$

$$< \delta\, 2^{\varkappa}\, e^{-\omega(\sigma-c)} \int_0^{\infty} v^{\nu}\,(\omega^{\varkappa} + v^{\varkappa})\, e^{-v(\sigma-c)}\, dv$$

$$< \frac{K\,\delta\,\omega^{\varkappa}}{(\sigma - c)^{2\varkappa+1}}\, e^{-\omega(\sigma-c)},$$

wo K nur von $\varkappa$ und von der oberen Grenze für $\sigma - c$ abhängt. Durch Einsetzung in (19) erhalten wir dann

$$|g_{\omega}(s)| < M\,\delta\,\frac{|(s - s_1)^{2\varkappa+1}\,(s - s_2)^{2\varkappa+1}|}{(\sigma - c)^{2\varkappa+1}},$$

wo M nur von $\varkappa$, η und c abhängt.

Wenn nun s auf einer der Geraden $t = t_1$ bez. $t = t_2$ liegt, so gilt $|s - s_1| = \sigma - c$ bez. $|s - s_2| = \sigma - c$, und man folgert hieraus ohne Schwierigkeit dass bei hinreichend kleinem δ

$$(21) \qquad |g_{\omega}(s)| < \varepsilon$$

auf der ganzen rechten Hälfte der Begrenzung des Rechtecks ausfällt.

Es gehöre nun s der linken Hälfte der Begrenzung an, so dass $\sigma < c$ gilt. Da $f(s)$ dort regulär ist und also $|f(s)|$ eine endliche obere Grenze G hat, folgt aus (15) für alle hinreichend grosse ω

$$\left| f(s) + (-1)^{\varkappa-k+1} \sum H_{ijk}\, s^i\, \omega^{j-k}\, \frac{d^k}{ds^k}\left(\frac{f(s)}{s^{\varkappa+1}}\right)\right| < 2G$$

und weiter nach (13), 15), (18) und (20)

$$(22) \quad |g_{\omega}(s)| < e^{-\omega(c-\sigma)}\, |(s - s_1)^{2\varkappa+1}\,(s - s_2)^{2\varkappa+1}|\,.$$

$$\cdot\left(2G + \frac{1}{\varkappa!\,\omega^{\varkappa}} \int_0^{\omega} \left| A_{\varkappa}(v)\, \frac{d^{\varkappa+1}}{dv^{\varkappa+1}}\left(e^{-sv}\,(\omega - v)^{\varkappa}\right)\right| dv + \delta\, e^{\omega(c-\sigma)}\right).$$

Nach Entwicklung der $(\varkappa + 1)$-fachen Ableitung haben wir hier analog wie im Falle $\sigma > c$ ein Integral von der Form

$$\delta \int_0^\omega v^\varkappa (\omega - v)^\nu e^{v(c-\sigma)} dv$$

abzuschätzen, wo wieder $0 \leqq \nu \leqq \varkappa$ gilt. (Die Abschätzung (20) gilt freilich nur für $\omega > \omega_0$; das Integral von 0 bis ω_0, dividert durch $\omega^\varkappa$, ist aber beschränkt und bewirkt also nur eine Vergrösserung der Konstante $2G$.) Es ergibt sich nun

$$\delta \int_0^\omega v^\varkappa (\omega - v)^\nu e^{v(c-\sigma)} dv < \delta \omega^\varkappa \int_0^\omega (\omega - v)^\nu e^{v(c-\sigma)} dv =$$

$$= \delta \omega^\varkappa e^{\omega(c-\sigma)} \int_0^\omega v^\nu e^{-v(c-\sigma)} dv < \frac{\delta \omega^\varkappa \nu!}{(c-\sigma)^{\nu+1}} e^{\omega(c-\sigma)}.$$

Weiter findet man unmittelbar durch Differentiation, dass der Ausdruck

$$e^{-\omega(c-\sigma)} \left| (s - s_1)^{2\varkappa+1} (s - s_2)^{2\varkappa+1} \right|$$

auf der linken Hälfte der Begrenzung unseres Rechtecks mit wachsendem ω gleichmässig gegen Null strebt. Für alle hinreichend grossen ω folgt also aus (22)

$$|g_\omega(s)| < L\delta \left(1 + \frac{|(s-s_1)^{2\varkappa+1}(s-s_2)^{2\varkappa+1}|}{(c-\sigma)^{\varkappa+1}}\right),$$

wo nun L von G, $\varkappa$, η und c abhängt, und schliesslich, da auf der Geraden $t = t_1$ bez. $t = t_2$ jetzt $|s - s_1| = c - \sigma$ bez. $|s - s_2| = c - \sigma$ gilt,

$$|g_\omega(s)| < \varepsilon$$

bei hinreichend kleinem δ auf der ganzen linken Hälfte der Begrenzung. Dieselbe Ungleichung gilt also nach (21) auf der ganzen Begrenzung, und folglich auch im Innern des Rechtecks. Insbesondere haben wir auf der zwischen s_1 und s_2 gelegenen Strecke der Geraden $\sigma = c$

$$\Bigg| f(s) - \omega^{-k} \sum_{\lambda_n < \omega} a_n e^{-\lambda_n s} (\omega - \lambda_n)^{\varkappa} + (-1)^{\varkappa - k + 1} \sum H_{ijk} s^i \omega^{j - \varkappa} \frac{d^k}{ds^k} \cdot$$

$$\cdot \left(\frac{f(s)}{s^{\varkappa+1}} \right) \Bigg| < \frac{\varepsilon}{|(s - s_1)^{2\varkappa+1} (s - s_2)^{2\varkappa+1}|}$$

und hieraus nach (15) für hinreichend grosses ω

$$(23) \quad \Bigg| f(s) - \omega^{-\varkappa} \sum_{\lambda_n < \omega} a_n e^{-\lambda_n s} (\omega - \lambda_n)^{\varkappa} \Bigg| < \frac{2\varepsilon}{|(s - s_1)^{2\varkappa+1} (s - s_2)^{2\varkappa+1}|}.$$

Die Reihe ist also im Punkte s_0 summierbar $(\lambda, \varkappa)$, und die Summabilität ist eine gleichmässige auf jeder im Innern des Intervalles s_1, s_2 enthaltene Strecke. Jede Strecke der Geraden $\sigma = c$, die ausschliesslich aus Regularitätsstellen besteht, ist in einer grösseren Strecke enthalten, welche dieselbe Eigenschaften besitzt. Nennen wir die Endpunkte dieser grösseren Strecke s_1 und s_2, so folgt aus (23) die gleichmässige Konvergenz der Reihe auf der gegebenen Strecke.

Der Beweis für den zweiten Teil unsres Satzes, dass die Bedingung (12) auch in gewissem Sinne notwendig ist, schliesst sich dem entsprechenden Beweise von RIESZ in genau derselben Weise an wie der obige Beweis an dem Konvergenzbeweis von RIESZ. Wir können uns also damitbegnügen, auf den Seiten 356—357 seiner oben zitierten Arbeit in den Acta Mathematica zu verweisen.

Durch die Anwendung dieses Satzes auf den uns interessierenden Reihen ergibt sich nun das folgende Resultat.

Hilfssatz 2. *Es sei $o < \alpha < 1$, $\varkappa$ eine nichtnegative ganze Zahl, x reell und nicht von der Form $\pm h^{\frac{1}{k}}$, wo h eine ganze positive Zahl oder Null ist. Damit die Reihe*

$$\sum_{1}^{\infty} \frac{d_k(n)}{n^{\alpha}} e^{2k\pi i x n^{\frac{1}{k}}}$$

summierbar $(n^{\frac{1}{k}}, \varkappa)$ sei, ist notwendig und hinreichend, dass

$$\varkappa > k - k\alpha - 1$$

gilt. Wenn dies der Fall ist, so ist die Summabilität eine gleichmässige in jedem x-Intervall, das von den $\pm h^{\frac{1}{k}}$ frei ist.

Um diesen Hilfssatz beweisen zu können, müssen wir zuerst die Lagen der auf der imaginären Achse gelegenen Singularitäten von $\sum \frac{d_k(n)}{n^a} e^{-n^{\frac{1}{k}}s}$ kennen. Es ist hinreichend, diese Frage nur für einen besonderen Wert von α zu untersuchen, und wir wählen dann die Funktion $F_k(s)$ des vorigen Paragraphen. Aus Gleichung (10) geht nun hervor, dass $F_k(s)$ ausser $s = 0$ nur dieselben Singularitäten wie die Funktionen $\overline{M}_k\left(-\frac{se^{\frac{\mu\pi i}{k}}}{2\pi i}\right)$, $\mu = 0, 1, \ldots k$, hat. Ferner ergibt sich aus (9) und (11), dass, wenigsten für den durch die DIRICHLET-sche Reihe dargestellten Zweig der Funktion, nur in den Fällen $\mu = 0$ und $\mu = k$ Singularitäten auf der imaginären Achse auftreten können, die alsdann von den Funktionen $\overline{L}_k\left(\pm \frac{s}{2\pi i\, n^{\frac{1}{k}}}\right)$ herrühren. Es gilt also, die Singularitäten von $\overline{L}_k(s)$ zu untersuchen.

Nach (11) erhält man aus $\overline{L}_k(s)$ durch Addition eines gewissen Polynoms $(k-1)$ ten Grades die Funktion

$$\lambda(s) = \sum_0^\infty \frac{\left[\Gamma\left(\frac{\nu+1}{k}\right)\right]^k}{\Gamma(\nu+1)} s^\nu,$$

und man verifiziert leicht, dass $\lambda(s)$ der Differentialgleichung

$$k^k \lambda^{(k)}(s) = \left(\frac{d}{ds}s\right)^k \lambda(s)$$

genügt, welche auch in der Form

$$\lambda^{(k)}(s) + \frac{a_1 s^{k-1}}{s^k - k^k}\lambda^{(k-1)}(s) + \frac{a_2 s^{k-2}}{s^k - k^k}\lambda^{(k-2)}(s) + \cdots \frac{a_k}{s^k - k^k}\lambda(s) = 0$$

mit konstanten a_μ gesetzt werden kann. Im Endlichen können also nur die Wurzeln der Gleichung $s^k - k^k = 0$ für $\lambda(s)$

singulär sein, d. h. für $\bar{L}_k\left(\pm \frac{s}{2\pi i n^{\frac{1}{k}}}\right)$ und damit für $F_k(s)$ gibt es auf der imaginären Achse nur die Singularitäten $s = \pm 2 k \pi i n^{\frac{1}{k}}$, $n = 0, 1, 2, \cdots$.

Der vorige Hilfssatz kann nun nicht ganz direkt auf unsere Reihe angewandt werden, da dort $c > 0$ vorausgesetzt wurde; wir betrachten demnach die Reihe

$$\sum_1^\infty \frac{d_k(n)}{n^a} e^{n^{\frac{1}{k}} c} \cdot e^{-n^{\frac{1}{k}} s^1},$$

wo $s^1 = s + c$ gesetzt ist. Nach dem vorigen Hilfssatz brauchen wir nur zu beweisen, dess für jedes $\varkappa \geqq 0$ und für jedes $\varepsilon > 0$

$$C_\varkappa(\omega) = \sum_{n^{\frac{1}{k}} < \omega} \frac{d_k(n)}{n^a} e^{n^{\frac{1}{k}} c} \left(\omega - n^{\frac{1}{k}}\right)^\varkappa = O\left(\omega^{k - ka - 1 + \varepsilon} e^{c\omega}\right)$$

gilt, während dagegen

$$C_\varkappa(\omega) = o\left(\omega^{k - ka - 1} e^{c\omega}\right)$$

falsch ist.

Wir haben nun

$$C_\varkappa(\omega) = \varkappa \int_1^\omega C_0(v)(\omega - v)^{\varkappa - 1} dv \tag{24}$$

und

$$C_0(v) = \sum_{n < v^k} \frac{d_k(n)}{n^a} e^{n^{\frac{1}{k}} c},$$

also

$$\sum_{n < v^k} \frac{1}{n^a} e^{n^{\frac{1}{k}} c} < C_0(v) = O\left(\sum_{n < v^k} \frac{1}{n^{a - \varepsilon}} e^{n^{\frac{1}{k}} c}\right). \tag{25}$$

Die positive Zahl c konnte nun ganz beliebig gewählt werden, nehmen wir an, es sei

$$c > ka\,,$$

so ist die Funktion $x^{-a}\, e^{x^{\frac{1}{k}} c}$ für $x > 1$ beständig wachsend, und es wird

$$\int_1^{v^k} e^{c x^{\frac{1}{k}}} \frac{dx}{x^a} - v^{-ka}\, e^{cv} < \sum_{n < v^k} n^{-a}\, e^{n^{\frac{1}{k}} c} < \int_1^{v^k} e^{c x^{\frac{1}{k}}} \frac{dx}{x^a} + v^{-ka}\, e^{cv}\,.$$

Da aber

$$\int_1^{v^k} e^{c x^{\frac{1}{k}}} \frac{dx}{x^a} = k \int_1^{v} u^{k-ka-1}\, e^{cu}\, du = \frac{k}{c}\, v^{k-ka-1}\, e^{cv} \left[1 + O\left(\frac{1}{v}\right)\right],$$

so ergibt sich hieraus

$$\sum_{n < v^k} \frac{1}{n^a}\, e^{n^{\frac{1}{k}} c} = \frac{k}{c}\, v^{k-ka-1}\, e^{cv} \left[1 + O\left(\frac{1}{v}\right)\right]$$

und weiter nach (25)

$$\frac{k}{c}\, v^{k-ka-1}\, e^{cv} \left[1 + O\left(\frac{1}{v}\right)\right] < C_0(v) = O\left(v^{k-ka-1+k\varepsilon}\, e^{cv}\right).$$

Durch Einsetzung in (24) erhalten wir erstens für hinreichend grosses ω

$$C_\varkappa(\omega) > \frac{k\varkappa}{2c} \int_{\omega-2}^{\omega-1} v^{k-ka-1} (\omega - v)^{\varkappa-1}\, e^{cv}\, dv$$

$$> \frac{k\varkappa}{2^{k-ka}\, c}\, e^{-2c} \cdot \omega^{k-ka-1}\, e^{c\omega}\,,$$

zweitens

$$C_\varkappa(\omega) = O\left(\int\limits_1^\omega v^{k-ka-1+k\varepsilon}(\omega - v)^{\varkappa-1} e^{cv} dv\right)$$

$$= O\left(\int\limits_1^{\omega-\omega^\varepsilon} + \int\limits_{\omega-\omega^\varepsilon}^{\omega}\right) = O\,(\omega^Q e^{c(\omega-\omega^\varepsilon)} +$$

$$+ \omega^{k-ka-1+(k+\varkappa)\varepsilon} e^{c\omega})$$

$$= O\,(\omega^{k-ka-1+(k+\varkappa)\varepsilon} e^{c\omega}).$$

Hierdurch ist unser Hilfssatz bewiesen, und daraus folgt nun die Richtigkeit der in der Einleitung gemachten Behauptungen über die Reihen C_μ und S_μ.

§ 3.

Darstellung der Funktion $D_k(x)$.

Wir betrachten zuerst die Funktion

$$B(x) = \frac{1}{k!} \sum_{n \leqq x^k} (x - n^{\frac{1}{k}})^k d_k(n)$$

und nehmen an, dass x positiv, aber nicht von der Form $n^{\frac{1}{k}}$ ist. Dann gilt die Gleichung

$$D_k(x^k) = \frac{d^k}{dx^k} B(x). \tag{26}$$

Nach einer bekannten Formel aus der Theorie der DIRICHLET'schen Reihen haben wir

$$B(x) = \frac{1}{2\pi i} \int\limits_{1-i\infty}^{1+i\infty} \frac{e^{xs}}{s^{k+1}} f_k(s)\, ds$$

und weiter unter Benutzung der Darstellungsformeln (3), (4) und (5) des ersten Paragraphen

$$(27)\quad B(x)=\frac{1}{2\pi i}\int\limits_{1-i\infty}^{1+i\infty}\frac{e^{xs}}{s^{2k+1}}P_{k-1}(\log s)\,ds+\frac{\alpha}{2\pi i}\int\limits_{1-i\infty}^{1+i\infty}\frac{e^{xs}}{s^{k+1}}ds+$$

$$+\frac{1}{(2\pi i)^{k+1}}\sum_{\mu=0}^{k}\binom{k}{\mu}(-1)^{\mu}\int\limits_{1-i\infty}^{1+i\infty}\frac{e^{xs}}{s^{k+1}}M_k\left(-\frac{se^{\frac{\mu\pi i}{k}}}{2\pi i}\right)ds$$

$$=x^{2k}P_{k-1}^{''''}(\log x)+\frac{\alpha}{k!}x^k+\frac{(k-2)!}{(2\pi i)^{k+1}}\sum_{\mu=0}^{k}\binom{k}{\mu}(-1)^{\mu}C_{\mu}(x),$$

wo

$$(28)\quad C_{\mu}(x)=\int\limits_{1-i\infty}^{1+i\infty}\frac{e^{xs}}{s^{k+1}}ds\sum_{n=1}^{\infty}\frac{d_k(n)}{n}\int_G\!\!\cdots\!\int\left[\left(1+(x_1x_2\cdots x_{k-1}y)^{\frac{1}{k}}\cdot\frac{se^{\frac{\mu\pi i}{k}}}{2\pi i n^{\frac{1}{k}}}\right)^{-k}-1\right]dx_1\cdots dx_{k-1}$$

ist.

Es muss jetzt bewiesen werden, dass die in $C_{\mu}(x)$ auftretende unendliche Reihe gliedweise integriert werden kann. Da die Reihe offenbar auf jeder endlichen Strecke der Geraden $\sigma=1$ gleichmässig konvergiert, genügt es zu zeigen, dass das Integral

$$(29)\quad \int_4^{\infty}\frac{dt}{t^{k+1}}\sum_{n=1}^{\infty}\frac{d_k(n)}{n}\int_G\!\!\cdots\!\int\left|\left(1+(x_1x_2\ldots x_{k-1}y)^{\frac{1}{k}}\frac{(1+it)e^{\frac{\mu\pi i}{k}}}{2\pi i n^{\frac{1}{k}}}\right)^{-k}-1\right|dx_1dx_2\ldots dx_{k-1}$$

und das analoge Integral von -4 bis $-\infty$ beide konvergieren. Wir nehmen also $|t|>4$ an und werden jetzt den Fall $\mu=0$ betrachten; aus dem Beweis geht hervor, dass die anderen Fälle ganz analog, und für $\mu=1,2,\ldots k-1$ sogar noch viel einfacher, behandelt werden können.

Es sei zunächst $n>\left(\frac{|t|}{k\pi}\right)^k$. Wie man leicht findet, gilt für jedes komplexe α, dessen reeller Teil grösser als $-\frac{1}{2}$ ist,

$$|(1+\alpha)^{-k}-1| < k\,2^{k+1}|\alpha|$$

und hieraus folgt, da in G

$$(x_1 x_2 \ldots x_{k-1} y)^{\frac{1}{k}} \frac{|t|}{2\pi n^{\frac{1}{k}}} < \frac{|t|}{2k\pi n^{\frac{1}{k}}} < \frac{1}{2}$$

ist,

$$(30) \qquad \int_G \cdots \int \left| \left(1+(x_1 x_2 \cdots x_{k-1} y)^{\frac{1}{k}} \frac{1+it}{2\pi i n^{\frac{1}{k}}}\right)^{-k} - 1 \right| \cdot dx_1 dx_2 \cdots dx_{k-1} < K_5 \frac{|t|}{n^{\frac{1}{k}}},$$

wo von nun an die Konstanten K höchstens von k abhängen.

Für $n \leqq \left(\frac{|t|}{k\pi}\right)^k$ spalten wir das Integral in zwei Teilen, je nachdem

$$(31) \qquad \left|1+(x_1 x_2 \ldots x_{k-1} y)^{\frac{1}{k}} \frac{t}{2\pi n^{\frac{1}{k}}}\right| \begin{cases} \geqq |t|^{-\frac{1}{2}} & \text{(Bereich } G_1) \\ < |t|^{-\frac{1}{2}} & \text{(Bereich } G_2); \end{cases}$$

dabei braucht G_2 natürlich nicht zu existieren. Zunächst ist

$$(32) \qquad \int_{G_1} \cdots \int \left| \left(1+(x_1 x_2 \ldots x_{k-1} y)^{\frac{1}{k}} \frac{1+it}{2\pi i n^{\frac{1}{k}}}\right)^{-k} - 1 \right| \cdot dx_1 dx_2 \ldots dx_{k-1} < K_6 |t|^{\frac{1}{2}k},$$

weiter haben wir

$$\int_{G_2} \cdots \int \left| \left(1+(x_1 x_2 \ldots x_{k-1} y)^{\frac{1}{k}} \frac{1+it}{2\pi i n^{\frac{1}{k}}}\right)^{-k} - 1 \right| dx_1 dx_2 \ldots dx_{k-1} <$$

$$(33) \qquad < \int_{G_2} \cdots \int \left(\frac{(2\pi)^k n}{x_1 x_2 \ldots x_{k-1} y} + 1 \right) dx_1 dx_2 \ldots dx_{k-1} <$$

$$< K_7 n \int_{G_2} \cdots \int \frac{dx_1 dx_2 \ldots dx_{k-1}}{x_1 x_2 \ldots x_{k-1} y}.$$

Aus (31) folgt nun, da $|t| > 4$ ist, dass in G_2

$$x_1 x_2 \ldots x_{k-1} y > \left(\frac{\pi}{|t|}\right)^k n \tag{34}$$

gilt, und hieraus folgt weiter

$$n \int_{G_2} \cdots \int \frac{dx_1\, dx_2 \ldots dx_{k-1}}{x_1 x_2 \ldots x_{k-1} y} < \left(\frac{|t|}{\pi}\right)^k \int_{G_2} \cdots \int dx_1\, dx_2 \ldots dx_{k-1}. \tag{35}$$

Das letzte Integral kann in folgender Weise abgeschätzt werden. Es sei $x_1, x_2, \ldots x_{k-1}$ irgend ein Punkt in G. Wir denken uns $x_2, x_3, \ldots x_{k-1}$ fest gewählt und wollen die Werte von x_1 bestimmen, für welche der Punkt in G_2 fällt, führen also zuerst die Integration nach x_1 aus. Damit der Punkt in G_2 fällt, muss nach (31)

$$-1 - |t|^{-\frac{1}{2}} < (x_1 x_2 \ldots x_{k-1} y)^{\frac{1}{k}} \frac{t}{2\pi n^{\frac{1}{k}}} < -1 + |t|^{-\frac{1}{2}}$$

sein; G_2 existiert also jedenfalls nur wenn $t < 0$ und

$$(x_1 x_2 \ldots x_{k-1} y)^{\frac{1}{k}} = \frac{2\pi n^{\frac{1}{k}}}{|t|} \left(1 + \theta |t|^{-\frac{1}{2}}\right)$$

ausfällt, wo $|\theta| < 1$ ist. Setzen wir $1 - x_2 - x_3 - \cdots - x_{k-1} = \alpha$, so muss notwendig $\alpha > 0$ sein und es muss $x_1 y = x_1(\alpha - x_1)$ zwischen den Grenzen

$$\frac{(2\pi)^k n}{|t|^k x_2 x_3 \ldots x_{k-1}} \left(1 \pm |t|^{-\frac{1}{2}}\right)^k$$

liegen. Aus (34) folgt aber

$$x_2 x_3 \ldots x_{k-1} > \left(\frac{\pi}{|t|}\right)^k n;$$

der Unterschied zwischen den beiden Grenzen für $x_1(\alpha - x_1)$ ist also kleiner als

$$2^k\left[(1+|t|^{-\frac{1}{2}})^k-(1-|t|^{-\frac{1}{2}})^k\right]<k\,2^{2k}\,|t|^{-\frac{1}{2}}.$$

Wenn $x_1(\alpha-x_1)$ zwischen diesen Grenzen liegen soll, so muss x_1, das jedenfalls dem Intervall $0<x_1<\alpha$ angehört, zwischen gewissen Grenzen liegen, und es ist klar, dass die zugelassenen Werte von x_1 im Allgemeinen zwei Intervalle bilden, welche zum Punkte $x_1=\frac{1}{2}\alpha$ symmetrisch liegen, sowie dass die Summe der Längen dieser Intervalle möglichst gross wird, wenn die obere Grenze für $x_1(\alpha-x_1)$ mit dem Maximalwert $\frac{1}{4}\alpha^2$ zusammenfällt. Die Summe der beiden Intervalle ist also nicht grösser als

$$2\sqrt{k\,2^{2k}\,|t|^{-\frac{1}{2}}},$$

so dass wir erhalten

$$(36)\quad \int\limits_{G_2}\!\!\cdots\!\int dx_1\,dx_2\ldots\,dx_{k-1}<K_8\,|t|^{-\frac{1}{4}}\int\limits_0^1\!\!\cdots\!\int\limits_0^1 dx_2\ldots dx_{k-1}=$$
$$=K_8\,|t|^{-\frac{1}{4}}.$$

Aus (30), (32), (33), (35) und (36) erhalten wir endlich für (29) und für das analoge Integral zwischen negativen Grenzen die Majoranten

$$K_9\int\limits_{\pm 4}^{\pm\infty}\frac{dt}{|t|^{k+1}}\left(\sum_{n\leq\left(\frac{|t|}{k\pi}\right)^k}\frac{d_k(n)}{n}\cdot|t|^{k-\frac{1}{4}}+\sum_{n>\left(\frac{|t|}{k\pi}\right)^k}\frac{d_k(n)}{n^{1+\frac{1}{k}}}\cdot|t|\right),$$

wo die beiden oberen oder die beiden unteren Zeichen zu nehmen sind. Da $d_k(n)=O(n^{\varepsilon})$ und $k>1$ ist, sind beide Integrale konvergent.

Wenn $\mu>0$ ist, geht der Beweis genau analog und wird im Falle $\mu=1,2,\ldots k-1$ sogar noch einfacher, da alsdann bei der Abschätzung des Integrals über G_2 in (33) ein Faktor $|t|^k$ im Nenner auftritt.

Aus dem jetzt geführten Beweise folgt, dass wir in (28) gliedweise integrieren und auch die Reihenfolge der Integrationen vertauschen können; wir haben also

$$(37)\quad C_\mu(x) = \sum_1^\infty \frac{d_k(n)}{n} \int_G \cdots \int dx_1\, dx_2 \ldots dx_{k-1} \int_{1-i\infty}^{1+i\infty} \frac{e^{xs}}{s^{k+1}} \cdot$$

$$\cdot \left[\left(1 + (x_1 x_2 \ldots x_{k-1} y)^{\frac{1}{k}} \frac{s e^{\frac{\mu\pi i}{k}}}{2\pi i n^{\frac{1}{k}}} \right)^{-k} - 1 \right] ds.$$

Betrachten wir zunächst das Integral von $1 - i\infty$ bis $1 + i\infty$. Links von dem Integrationsweg liegen die beiden Pole $s = 0$ und

$$s = \lambda = -2\pi i \left(\frac{n}{e^{\mu\pi i}\, x_1 x_2 \ldots x_{k-1} y} \right)^{\frac{1}{k}},$$

und wir erhalten durch eine einfache Anwendung des CAUCHY'schen Satzes

$$\int_{1-i\infty}^{1+i\infty} \frac{e^{xs}}{s^{k+1}} \left[\left(1 - \frac{s}{\lambda} \right)^{-k} - 1 \right] ds = (-1)^\mu \frac{x_1 x_2 \ldots x_{k-1} y}{n}.$$

$$.\left(P_{k-1}^{(5)}(\lambda x) + e^{\lambda x} P_{k-1}^{(6)}(\lambda x) \right),$$

also durch Einsetzung in (27) und (37)

$$B(x) = x^{2k} P_{k-1}^{(4)}(\log x) + \frac{\alpha}{k!} x^k + P_{k-1}^{(7)}(x) + \sum_{\mu=0}^{k} \binom{k}{\mu} E(x)$$

und nach (26)

$$D_k(x^k) = x^k p_{k-1}(\log x^k) + \alpha + \sum_{\mu=0}^{k} \binom{k}{\mu} \frac{d^k}{dx^k} E_\mu(x),$$

wo

$$E_\mu(x) = \sum_1^\infty \frac{d_k(n)}{n^2} \int_G \cdots \int x_1\, x_2 \ldots x_{k-1} y\, P_{k-1}^{(8)}(\lambda x)\, e^{\lambda x}\, dx_1\, dx_2 \ldots dx_{k-1}$$

ist. Der Exponent von e im allgemeinen Glied der Reihe $E_\mu(x)$ hat den reellen Teil

$$-2\pi x\left(\frac{n}{x_1 x_2 \ldots x_{k-1} y}\right)^{\frac{1}{k}} \sin\frac{\mu\pi}{k};$$

wenn daher μ von 0 und k verschieden ist, so ist die Reihe unbedingt konvergent und von der Form

$$E_\mu(x) = O\left(x^{k-1} e^{-2k\pi x \sin\frac{\pi}{k}}\right) = O(e^{-4\pi x}), \ (\mu = 1, 2, \ldots k-1).$$

Dasselbe gilt nach k-maliger Differentiation, und wir können daher schreiben

$$(38) \quad B(x) = x^{2k} P_{k-1}^{(4)}(\log x) + \frac{\alpha}{k!} x^k + P_{k-1}^{(7)}(x) + \\ + \sum_{\nu=0}^{k-1} c_\nu x^\nu G_\nu(x) + O(e^{-4\pi x})$$

und

$$(39) \quad D_k(x^k) = x^k p_{k-1}(\log x^k) + \\ + \sum_{p=0}^{k-1} x^p \sum_{\nu=p}^{k-1} c_{p\nu} G_\nu^{(k+p-\nu)}(x) + \alpha + O(e^{-4\pi x}),$$

wo die c_ν und die $c_{p\nu}$ konstant sind, und

$$(40) \quad G_\nu(x) = \sum_1^\infty \frac{d_k(n)}{n^{2-\frac{\nu}{k}}} \int \cdots \int_G (x_1 x_2 \ldots x_{k-1} y)^{1-\frac{\nu}{k}}$$

$$\left(e^{2\pi i x\left(\frac{n}{x_1 x_2 \ldots x_{k-1} y}\right)^{\frac{1}{k}}} + (-1)^\nu e^{-2\pi i x\left(\frac{n}{x_1 x_2 \ldots x_{k-1} y}\right)^{\frac{1}{k}}}\right) dx_1 dx_2 \ldots dx_{k-1}$$

gesetzt ist.

Wir müssen jetzt das hier auftretende Integral

$$J(\nu, n, x) = \int \cdots \int_G (x_1 x_2 \ldots x_{k-1} y)^{1-\frac{\nu}{k}} \cdot \\ \cdot e^{2\pi i x\left(\frac{n}{x_1 x_2 \ldots x_{k-1} y}\right)^{\frac{1}{k}}} dx_1 dx_2 \ldots dx_{k-1}$$

etwas näher untersuchen. Durch die Substitution

$$\begin{aligned} x_1 + x_2 + \dots x_{k-1} &= \xi_1 \\ x_2 + \dots x_{k-1} &= \xi_1 \xi_2 \\ &\cdots\cdots \\ x_{k-2} + x_{k-1} &= \xi_1 \xi_2 \dots \xi_{k-2} \\ x_{k-1} &= \xi_1 \xi_2 \dots \xi_{k-1} \end{aligned}$$

geht dieses Integral über in

$$J(\nu, n, x) = \int_0^1 \dots \int_0^1 U e^{2\pi i x n^{\frac{1}{k}}} V \, d\xi_1 d\xi_2 \dots d\xi_{k-1}, \tag{41}$$

wo

$$U = \prod_{\mu=1}^{k-1} \xi_\mu^{(k-\mu)\left(2-\frac{\nu}{k}\right)-1} (1-\xi_\mu)^{1-\frac{\nu}{k}}$$

und

$$V = \prod_{\mu=1}^{k-1} \left(\frac{1}{\xi_\mu^{k-\mu}(1-\xi_\mu)}\right)^{\frac{1}{k}}$$

gesetzt ist.

Wenn ξ_μ von 0 bis 1 wächst, so nimmt der μ.te Faktor des Produktes V zuerst monoton ab, erreicht dann im Punkte

$$\xi_\mu = \eta_\mu = \frac{k-\mu}{k-\mu+1},$$

wo die Ableitung eine einfache Nullstelle besitzt, sein Minimum

$$\zeta_\mu = \left(\frac{(k-\mu+1)^{k-\mu+1}}{(k-\mu)^{k-\mu}}\right)^{\frac{1}{k}} \tag{42}$$

und wächst schliesslich wieder monoton im Intervalle $\eta_\mu \leqq \xi_\mu \leqq 1$. Es gibt also eine Entwicklung

$$\left(\frac{1}{\xi_\mu^{k-\mu}(1-\xi_\mu)}\right)^{\frac{1}{k}} = \zeta_\mu + b_2(\xi_\mu - \eta_\mu)^2 + b_3(\xi_\mu - \eta_\mu)^3 + \ldots$$

mit $b_2 \neq 0$, und wenn wir

$$(43) \qquad \left(\frac{1}{\xi_\mu^{k-\mu}(1-\xi_\mu)}\right)^{\frac{1}{k}} = \zeta_\mu + t_\mu$$

setzen, so gibt es zu jedem positiven t_μ genau zwei entsprechende ξ_μ zwischen 0 und 1, eins im Intervalle $0 < \xi_\mu < \eta_\mu$ und eins in $\eta_\mu < \xi_\mu < 1$. Es kann auch $\xi_\mu - \eta_\mu$ in einer Reihe

$$(44) \qquad \xi_\mu - \eta_\mu = g_1 t_\mu^{\frac{1}{2}} + g_2 t_\mu + \ldots$$

mit $g_1 \neq 0$ entwickelt werden, und den beiden möglichen Werten von $t_\mu^{\frac{1}{2}}$ entsprechen hier die beiden Werte von ξ_μ. Die Reihe ist jedenfalls für hinreichend kleine t_μ konvergent.

Wir setzen nun in (41) für $\mu = 1, 2 \ldots k-1$

$$\int_0^1 = \int_0^{\eta_\mu} + \int_{\eta_\mu}^0$$

und führen in jedes Integral die Substitution (43) ein. Da

$$\frac{k-\mu+1}{k}(\xi_\mu - \eta_\mu)\,\xi_\mu^{k-\mu-1}\left(\xi_\mu^{k-\mu}(1-\xi_\mu)\right)^{-\frac{1}{k}-1} d\xi_\mu = dt_\mu$$

gilt, so ergibt sich

$$J(\nu, n, x) = \frac{k^{k-1}}{k!}\int_0^\infty \cdots \int_0^\infty \prod_{\mu=1}^{k-1}\left[\left(\frac{1}{\xi'_\mu - \eta_\mu} - \frac{1}{\xi''_\mu - \eta_\mu}\right)\frac{1}{(\zeta_\mu + t_\mu)^{2k-\nu+1}}\right] \cdot$$

$$. \, e^{2\pi i x n^{\frac{1}{k}} \prod_1^{k-1}(\zeta_\mu + t_\mu)} \, dt_1 \, dt_2 \ldots dt_{k-1},$$

wo ξ'_μ die grössere, ξ''_μ die kleinere der beiden zwischen 0 und 1 gelegenen Wurzeln von (43) oder (44) bedeutet. Für $0 \leqq r \leqq k$ haben wir also

$$(45) \quad \frac{d^r}{dx^r} J(\nu, n, x) = \frac{k^{k-1}(2\pi i)^r}{k!} n^{\frac{r}{k}} \int_0^\infty \cdots$$

$$\cdots \int_0^\infty \prod_1^{k-1} \left[\left(\frac{1}{\xi'_\mu - \eta_\mu} - \frac{1}{\xi''_\mu - \eta_\mu} \right) \frac{1}{(\zeta_\mu + t_\mu)^{2k-r-\nu+1}} \right] .$$

$$. e^{2\pi i x n^{\frac{1}{k}} \prod\limits_1^{k-1} (\zeta_\mu + t_\mu)} dt_1 \, dt_2 \ldots dt_{k-1} .$$

Da $\nu \leqq k-1$ ist, bleibt das Integral immer absolut konvergent. Es sei jetzt l eine von n und x unabhängige ganze Zahl derart, dass

$$l \geqq \nu + r - k + 1 ,$$

und es sei

$$P = 2\pi x n^{\frac{1}{k}} \prod_{\mu=2}^{k-1} (\zeta_\mu + t_\mu)$$

gesetzt; wir wollen in (45) zuerst nach t_1 integrieren und haben dann das Integral

$$\int_0^\infty \left(\frac{1}{\xi'_1 - \eta_1} - \frac{1}{\xi''_1 - \eta_1} \right) \frac{1}{(\zeta_1 + t_1)^{2k-r-\nu+1}} e^{P i t_1} dt_1 ,$$

zu berechnen. In der Umgebung von $t_1 = 0$ gilt eine Entwicklung

$$\varphi(t_1) = \left(\frac{1}{\xi'_1 - \eta_1} - \frac{1}{\xi''_1 - \eta_1} \right) \frac{1}{(\zeta_1 + t_1)^{2k-r-\nu+1}} =$$

$$= t_1^{-\frac{1}{2}} (h_0 + h_1 t_1 + h_2 t_1^2 + \cdots)$$

und wir haben

$$\int_0^\infty \varphi(t_1)\, e^{Pit_1} d t_1 = \frac{h_0}{P^{\frac{1}{2}}} \int_0^\infty e^{it} \frac{dt}{t^{\frac{1}{2}}} + \int_0^\infty (\varphi(t_1) - h_0 t_1^{-\frac{1}{2}})\, e^{Pit_1} dt_1 .$$

Nach einer partiellen Integration erhält man durch l-malige Widerholung derselben Umformungen

$$\int_0^\infty \varphi(t_1)\, e^{Pit_1} dt_1 = \frac{h_0'}{P^{\frac{1}{2}}} + \frac{h_1'}{P^{\frac{3}{2}}} + \cdots \frac{h_{l-1}'}{P^{l-\frac{1}{2}}} + \frac{h_l'}{P^l} \int_0^\infty \varphi_l(t_1) e^{Pit_1} dt_1 ,$$

wo $\varphi_l(t_1)$ in der Umgebung von $t_1 = 0$ eine Entwicklung von derselben Art wie $\varphi(t_1)$ gestattet, und das letzte Integral absolut konvergiert.

Hierdurch ergibt sich die linke Seite von (45) gleich einer Summe von l $(k-2)$-fachen Integralen und einem $(k-1)$-fachen, welches letztere, absolut genommen, kleiner als

$$(46) \qquad K_{10}\, n^{\frac{r}{k}} \left(x n^{\frac{1}{k}}\right)^{-l}$$

ist. Die $(k-2)$-fachen Integrale werden nun ebenso behandelt, indem man z. B. nach t_2 integriert. Durch fortgesetzte Anwendung dieses Verfahrens erhält man, da nach (42)

$$\prod_{\mu=1}^{k-1} \zeta_\mu = k$$

gilt,

$$(47) \qquad \frac{d^r}{dx^r} J(\nu, n, x) = n^{\frac{r}{k}} e^{2k\pi i x n^{\frac{1}{k}}} \sum_{\varkappa} \gamma_\varkappa \left(x n^{\frac{1}{k}}\right)^{-\frac{k-1}{2}-\varkappa} + \Omega$$

$$= \sum_{\varkappa} \gamma_\varkappa\, x^{-\frac{k-1}{2}-\varkappa} \frac{e^{2k\pi i x n^{\frac{1}{k}}}}{n^{\frac{k-1}{2k}+\frac{\varkappa-r}{k}}} + \Omega ,$$

wo die Konstanten $\gamma_\varkappa$ noch von r, ν und k, nicht aber von n und x abhängen. Die Summe wird über alle ganzen $\varkappa$ genommen, die den Beziehungen

$$(48) \qquad \begin{cases} \varkappa \geqq 0 \\ \frac{k-1}{2} + \varkappa < l \end{cases}$$

genügen, und für das Zusatzglied Ω gibt es eine obere Grenze von der Form (46).

Nach (39) und (40) gilt nun

$$(49) \qquad D_k(x^k) = x^k p_{k-1}(\log x^k) + \sum_{p=0}^{k-1}\sum_{\nu=p}^{k-1} c_{p\nu}\, x^p\, G_\nu^{(k+p-\nu)}(x) + \alpha + O(e^{-4\pi x})$$

und

$$(50) \qquad G_\nu(x) = \sum_{n=1}^{\infty} \frac{d_k(n)}{n^{2-\frac{\nu}{k}}} \left(J(\nu, n, x) + (-1)^\nu \bar{J}(\nu, n, x)\right),$$

wo $\bar{J}$ die zu J konjugiert komplexe Grösse bedeutet. Versucht man jetzt $G_\nu^{(r)}(x)$ durch die gliedweise Differentiation darzustellen, indem man (47) benutzt, so erhält man zunächst durch Einführung des Zusatzgliedes Ω immer absolut und gleichmässig konvergente Reihen, die übrigens von der Form $O(x^{-l})$ sind. Die übrigen Glieder von (47) geben Reihen von der Form

$$\sum_{n=1}^{\infty} \frac{d_k(n)}{n^{2+\frac{k-1}{2k}+\frac{\varkappa-\nu-r}{k}}}\, e^{\pm 2k\pi i x n^{\frac{1}{k}}},$$

und nach den getroffenen Festsetzungen ist hier der Exponent von n im Nenner immer positiv. Nach dem vorhergehenden Paragraphen sind also diese Reihen immer mit RIESZ'schen Mitteln vom Typus $n^{\frac{1}{k}}$ und von hinreichend hoher Ordnung summierbar, sogar gleichmässig in jedem abgeschlossenen Intervall der positiven x-Achse das keine k-te Wurzel aus einer ganzen Zahl enthält. Man kann also, wieder nach dem vorhergehenden Paragraphen, in (50) r Mal gliedweise differentiieren und dabei (47) anwenden, wenn man nur auftretende divergente Reihen in der angegebenen Weise summiert. Wir setzen dann $r = k + p - \nu$, $l = p + 1$, und erhalten aus (49), indem wir noch x an der Stelle von x^k schreiben,

$$D_k(x) = x\,p_{k-1}(\log x) + \sum_{p=0}^{k-1}\sum_{\nu=p}^{k-1}\sum_{\varkappa} x^{\frac{k-1}{2k} - \frac{k-p+\varkappa-1}{k}} (a_{p\nu\varkappa} C_{k-p+\varkappa-1} + b_{p\nu\varkappa} S_{k-p+\varkappa-1}) + \alpha + O(x^{-\frac{1}{k}}), \tag{51}$$

wo wieder

$$C_\mu = \sum_{n=1}^{\infty} \frac{d_k(n)}{n^{\frac{k+1}{2k}+\frac{\mu}{k}}} \cos 2k\pi (nx)^{\frac{1}{k}},$$

$$S_\mu = \sum_{n=1}^{\infty} \frac{d_k(n)}{n^{\frac{k+1}{2k}+\frac{\mu}{k}}} \sin 2k\pi (nx)^{\frac{1}{k}}$$

ist, und die Konstanten ausser von p, ν, $\varkappa$ nur noch von k abhängen. Aus (48), wo jetzt $l = p + 1$ ist, folgt dass der Ausdruck $k - p + \varkappa - 1$ genau alle ganzzahlige Werte μ durchläuft, die der Beziehung

$$0 \leqq \mu < \frac{k+1}{2}$$

genügen. Aus (51) folgt somit Satz A. (Im Falle eines geraden k würde zwar in Satz au Ach noch das Glied $\mu = \frac{1}{2}k$ mitzunehmen sein; man sieht aber unmittelbar dass dieses Glied absolut konvergente Reihen C_μ und S_μ enthält und von der Form $O(x^{-\frac{1}{2k}})$ ist, es kann deshalb fortgelassen werden.) Es geht aus dem Beweis hervor, dass man auch noch für das Restglied in Satz A eine asymptotische Entwicklung erhalten könnte.

§ 4.

Sätze über $D_k(x)$.

Nach (38), (50) und (47) haben wir, wenn in der letztgenannten Formel $r = 0$ gesetzt wird,

$$B(x) = \frac{1}{k!} \sum_{n \leqq x^k} (x - n^{\frac{1}{k}})^k d_k(n) = x^{2k} P_{k-1}^{(4)}(\log x) + \frac{\alpha}{k!} x^k + P_{k-1}^{(7)}(x) + \sum_{\nu=0}^{k-1} c_\nu x^\nu G_\nu(x) + O(e^{-4\pi x}), \tag{52}$$

$$G_\nu(x) = \sum_{n=1}^{\infty} \frac{d_k(n)}{n^{2-\frac{\nu}{k}}} (J(\nu, n, x) + (-1)^\nu \bar{J}(\nu, n, x)), \tag{53}$$

$$J(\nu, n, x) = \sum_{\varkappa=0}^{l} \gamma_\varkappa \, x^{-\frac{k-1}{2}-\varkappa} \frac{e^{2k\pi i x n^{\frac{1}{k}}}}{n^{\frac{k-1}{2k}+\frac{\varkappa}{k}}} + O\left((x n^{\frac{1}{k}})^{-l-\frac{k-1}{2}}\right), \tag{54}$$

wo die ganze positive Zahl l beliebig gross gewählt werden kann. Diese Gleichungen gelten aus Stetigkeitsgründen für alle $x > 1$.

Beweis von Satz B.

Wir setzen

$$\begin{aligned} \delta_k F(x) &= \sum_{\nu=0}^{k} \binom{k}{\nu} (-1)^{k-\nu} F(x+\nu h) \\ &= \int_0^h d x_1 \int_{x_1}^{x_1+h} d x_2 \ldots \int_{x_{k-1}}^{x_{k-1}+h} F^{(k)}(x+x_k)\, d x_k \end{aligned} \tag{55}$$

wenn $F(x)$ irgend eine Funktion bezeichnet, die, ausser in gewissen isolierten Punkten ohne Häufungsstelle im Endlichen eine stetige k-te Ableitung besitzt, und wir nehmen

$$h = x^{-\frac{k-1}{k+1}}.$$

Man findet nun, wenn $\varepsilon > 0$ beliebig klein ist,

$$\begin{aligned} h^{-k} \delta_k B(x) &= D_k(x^k) + O\left(h^{-k} \sum_{x^k < n \leqq (x+kh)^k} h^k d_k(n)\right) \\ &= D_k(x^k) + O(h\, x^{k-1+\varepsilon}), \end{aligned}$$

$$h^{-k} \delta_k \, x^{2k} P_{k-1}^{(4)}(\log x) = x^k p_{k-1}(\log x^k) + O(h\, x^{k-1+\varepsilon}),$$

und hieraus

$$\Delta_k(x^k) = D_k(x^k) - x^k p_{k-1}(\log x^k) =$$

$$= \sum_{\nu=0}^{k-1} c_\nu h^{-k} \delta_k x^\nu G_\nu(x) + O\left(x^{k\frac{k-1}{k+1}+\varepsilon}\right).$$

Nun folgt aus (53) und (54)

$$\left| h^{-k} \delta_k x^\nu G_\nu(x) \right| \leqq 2 \sum_{n=1}^{\infty} \frac{d_k(n)}{n^{2-\frac{\nu}{k}}} \left| h^{-k} \delta_k x^\nu J(\nu, n, x) \right|$$

und

$$\left| h^{-k} \delta_k x^\nu J(\nu, n, x) \right| \leqq n^{-\frac{k-1}{2k}} \sum_{\varkappa=0}^{l} |\gamma_\varkappa| \cdot \left| h^{-k} \delta_k x^{\nu-\varkappa-\frac{k-1}{2}} e^{2k\pi i x n^{\frac{1}{k}}} \right| +$$

$$+ O\left[h^{-k} \left(x n^{\frac{1}{k}} \right)^{\frac{k-1}{2}-l} \right]$$

$$= O\left(x^{\frac{k-1}{2}} n^{-\frac{k-1}{2k}} \operatorname{Min}(h^{-k}, n) \right),$$

wenn nur die ganze Zahl l hinreichend gross gewählt ist. Durch Einsetzung ergibt sich

$$h^{-k} \delta_k x^\nu G_\nu(x) = O\left[x^{\frac{k-1}{2}} \left(\sum_{1}^{h^{-k}} n^{\frac{k-1}{2k}+\varepsilon-1} + h^{-k} \sum_{h^{-k}}^{\infty} n^{\frac{k-1}{2k}+\varepsilon-2} \right) \right]$$

$$= O\left(x^{\frac{k-1}{2}} h^{-\left(\frac{k-1}{2}+k\varepsilon\right)} \right) = O\left(x^{k\frac{k-1}{k+1}+k\varepsilon} \right),$$

also

$$\Delta_k(x^k) = O\left(x^{k\frac{k-1}{k+1}+k\varepsilon} \right)$$

oder

$$\Delta_k(x) = O\left(x^{\frac{k-1}{k+1}+\varepsilon} \right).$$

Beweis von Satz C. Jetzt nehmen wir

$$h = \frac{1}{x}, \quad \frac{1}{2} x \leqq t \leqq x$$

und erhalten wie bei dem vorhergehenden Beweis

$$\Delta^k(t^k) = \sum_{\nu=0}^{k-1} c_\nu h^{-k} \delta_k t^\nu G_\nu(t) + O(x^{k-2+\varepsilon})$$

gleichmässig für alle t im Intervalle $\frac{1}{2}x, x$. Wir wollen zuerst zeigen, dass

$$(56) \qquad \int_{\frac{1}{2}x}^{x} (\Delta_k(t^k))^2 dt = O(x^{2k-3+\varepsilon})$$

gilt. Auf Grund der für beliebige komplexe $a_1, a_2, \ldots a_n$ geltenden Identität

$$|a_1 + a_2 + \ldots a_n|^2 \leqq n(|a_1|^2 + |a_2|^2 + \ldots |a_n|^2)$$

genügt es, die Gleichung

$$h^{-2k} \int_{\frac{1}{2}x}^{x} |\delta_k t^\nu G_\nu(t)|^2 dt = O(x^{2k-3+\varepsilon})$$

für $\nu = 0, 1, \ldots k-1$, oder, nach (53) und (54),

$$(57) \qquad h^{-2k} \int_{\frac{1}{2}x}^{x} \left| \delta_k t^{\nu - \varkappa - \frac{k-1}{2}} \sum_{n=1}^{\infty} \frac{d_k(n)}{n^{2+\frac{k-1}{2k}-\frac{\nu-\varkappa}{k}}} e^{2k\pi i t n^{\frac{1}{k}}} \right|^2 dt =$$

$$= O(x^{2k-3+\varepsilon})$$

für $\nu = 0, 1, \ldots k-1$ und $\varkappa = 0, 1, \ldots l$ zu beweisen, wenn nur die von x unabhängige Zahl l hinreichend gross gewählt ist. Die »schlimmsten» Werte von ν und $\varkappa$ sind natürlich $\nu = k-1$, $\varkappa = 0$, und wir werden deshalb nur das Integral

$$(58) \qquad h^{-2k} \int_{\frac{1}{2}x}^{x} \left| \delta_k t^{\frac{k-1}{2}} \sum_{1}^{\infty} \frac{d_k(n)}{n^{1+\frac{k+1}{2k}}} e^{2k\pi i t n^{\frac{1}{k}}} \right|^2 dt =$$

$$= h^{-2k} \sum_{m=1}^{\infty} \sum_{n=1}^{\infty} \frac{d_k(m)\, d_k(n)}{(mn)^{1+\frac{k+1}{2k}}} \int_{\frac{1}{2}x}^{x} \delta_k t^{\frac{k-1}{2}} e^{2k\pi i t m^{\frac{1}{k}}} \cdot \delta_k t^{\frac{k-1}{2}} e^{-2k\pi i t n^{\frac{1}{k}}} dt$$

behandeln. Je nachdem die erste oder zweite Form der Darstellung einer k-ten Differenz nach (55) angewendet wird, erhält man durch eine partielle Integration, wenn $m \neq n$ vorausgesetzt wird,

$$\int_{\frac{1}{2}x}^{x} \delta_k t^{\frac{k-1}{2}} e^{2k\pi i t m^{\frac{1}{k}}} . \delta_k t^{\frac{k-1}{2}} e^{-2k\pi i t n^{\frac{1}{k}}} dt = O\left(\frac{x^{k-1}}{\left|m^{\frac{1}{k}} - n^{\frac{1}{k}}\right|}\right)$$

oder

$$\int_{\frac{1}{2}x}^{x} \delta_k t^{\frac{k-1}{2}} e^{2k\pi i t m^{\frac{1}{k}}} . \delta_k t^{\frac{k-1}{2}} e^{-2k\pi i t n^{\frac{1}{k}}} dt = O\left(\frac{x^{k-1} m n h^{2k}}{\left|m^{\frac{1}{k}} - n^{\frac{1}{k}}\right|}\right),$$

und für die Glieder mit $m = n$

$$\int_{\frac{1}{2}x}^{x} \delta_k t^{\frac{k-1}{2}} e^{2k\pi i t m^{\frac{1}{k}}} . \delta_k t^{\frac{k-1}{2}} e^{-2k\pi i t m^{\frac{1}{k}}} dt = O(x^k m n h^{2k})$$

Die Summe der Glieder mit $m = n$ in (58) ist also, da $k \geqq 3$ gilt, von der Form

$$(59) \qquad \sum_{m=1}^{\infty} \frac{(d_k(m))^2}{m^{\frac{k+1}{k}}} . O(x^k) = O(x^k) = O(x^{2k-3}),$$

und wir brauchen nur noch

$$(60) \qquad x^{k-1} \sum_{m>n} \frac{d_k(m)\, d_k(n)}{(m\,n)^{1+\frac{k+1}{2k}}} . \frac{\text{Min}\,(m n, h^{-2k})}{m^{\frac{1}{k}} - n^{\frac{1}{k}}} = O(x^{2k-3+\varepsilon})$$

zu beweisen.

Da $d_k(n) = O(n^{\varepsilon})$ ist, werden wir die beiden Summen

$$S_1 = \sum_{\substack{m>n \\ mn \leqq h^{-2k}}} \frac{1}{(m\,n)^{\frac{k+1}{2k}-\varepsilon}\left(m^{\frac{1}{k}} - n^{\frac{1}{k}}\right)}$$

und

$$S_2 = \sum_{\substack{m>n \\ mn > h^{-2k}}} \frac{1}{(m\,n)^{1+\frac{k+1}{2k}-\varepsilon}\left(m^{\frac{1}{k}} - n^{\frac{1}{k}}\right)}$$

abzuschätzen haben. Es gilt nun

$$S_1 < k \sum_{\substack{m>n \\ mn \leqq h^{-2k}}} \frac{m^{\frac{k-1}{k}}}{(mn)^{\frac{k+1}{2k}-\varepsilon}(m-n)} < k \sum_{m \leqq 2h^{-k}} m^{\frac{k-3}{2k}+\varepsilon} \sum_{n=1}^{m-1} \frac{1}{n^{\frac{k+1}{2k}-\varepsilon}(m-n)} +$$

$$+ k \sum_{2h^{-k}<m \leqq h^{-2k}} m^{\frac{k-3}{2k}+\varepsilon} \sum_{n \leqq \frac{h^{-2k}}{m}} \frac{1}{n^{\frac{k+1}{2k}-\varepsilon}(m-n)} =$$

$$= \sum_{n \leqq 2h^{-k}} m^{\frac{k-3}{2k}+\varepsilon} \cdot O\left(m^{-\frac{k+1}{2k}+\varepsilon}\right) + \sum_{2h^{-k}<m \leqq h^{-2k}} m^{\frac{k-3}{2k}+\varepsilon} O\left(\frac{1}{m} \sum_{n \leqq \frac{h^{-2k}}{m}} n^{-\frac{k+1}{2k}+\varepsilon}\right) =$$

$$= O\left(h^{-k+2-\varepsilon}\right) + O\left(h^{-k+1-\varepsilon} \sum_{m>2h^{-k}} m^{-1-\frac{1}{k}+\varepsilon}\right) =$$

$$= O\left(h^{-k+2-\varepsilon}\right) = O\left(x^{k-2+\varepsilon}\right)$$

und

$$S_2 < k \sum_{m>h^{-k}} m^{-\frac{k+3}{2k}+\varepsilon} \sum_{\frac{h^{-2k}}{m} \leqq n \leqq m-1} \frac{1}{n^{1+\frac{k+1}{2k}-\varepsilon}(m-n)}$$

$$= O\left(\sum_{m>h^{-k}} m^{-\frac{k+3}{2k}-1-\frac{k+1}{2k}+\varepsilon}\right) + O\left\{\sum_{m>h^{-k}} m^{-1-\frac{k+3}{2k}+\varepsilon}\right.\cdot$$

$$\left. \cdot \operatorname{Min}\left[\left(\frac{h^{-2k}}{m}\right)^{-\frac{k+1}{2k}+\varepsilon}, 1\right]\right\} = O\left(h^{k+2+\varepsilon}\right) + O\left(h^{k+1-\varepsilon} \sum_{h^{-k}<m \leqq h^{-2k}} m^{-1-\frac{1}{k}+\varepsilon}\right) +$$

$$+ O\left(\sum_{m>h^{-2k}} m^{-1-\frac{k+3}{2k}+\varepsilon}\right) = O\left(h^{k+2-\varepsilon}\right).$$

Hieraus folgt nun

$$x^{k-1}\left(S_1 + h^{-2k} S_2\right) = O\left(x^{2k-3+\varepsilon}\right),$$

d. h. (60), somit (57) und schliesslich auch (56). Aus (56) schliessen wir weiter

$$\int_{\frac{1}{2}x^k}^{x^k} t^{\frac{1}{k}-1} (\Delta_k(t))^2 \, dt = O(x^{2k-3+\varepsilon}),$$

$$\int_{\frac{1}{2}x}^{x} (\Delta_k(t))^2 \, dt = O\left(x^{\frac{3k-4}{k}+\varepsilon}\right).$$

Für jedes ganze $k \geqq 3$ und jedes $\varepsilon > 0$ gibt es also ein $K = K(k, \varepsilon)$ derart, dass

$$\int_{\frac{1}{2}x}^{x} (\Delta_k(t))^2 \, dt < K x^{\frac{3k-4}{k}+\varepsilon}$$

für alle $x > 1$ gilt. Es sei nun $2^n < x \leqq 2^{n+1}$. Wird x der Reihe nach durch $\frac{1}{2}x, \frac{1}{4}x, \ldots \frac{1}{2^n}x$ ersetzt, so folgt

$$\int_{1}^{x} (\Delta_k(t))^2 \, dt < 2 K x^{\frac{3k-4}{k}+\varepsilon},$$

also nach der SCHWARZ'schen Ungleichung

$$\int_{1}^{x} |(\Delta_k(t)| \, dt = O\left(\sqrt{x \cdot x^{\frac{3k-4}{k}+\varepsilon}}\right)$$

oder

$$\frac{1}{x}\int_{1}^{x} |\Delta_k(t)| \, dt = O\left(x^{\frac{k-2}{k}+\varepsilon}\right),$$

womit Satz *C* bewiesen ist.

Tryckt den 3 mars 1922.

Uppsala 1922. Almqvist & Wiksells Boktryckeri-A.-B.

13.

Sur un problème de M. Phragmén

Ark. Mat. Astr. Fys. **16** (27), 1–5 (1922)

Communiqué le 7 Decembre 1921 par E. L. PHRAGMÉN et IVAR FREDHOLM.

1. En désignant par $\pi(x)$ le nombre des nombres premiers non supérieurs à x, posons avec RIEMANN

$$f(x) = \pi(x) + \frac{1}{2}\pi(x^{\frac{1}{2}}) + \frac{1}{3}\pi(x^{\frac{1}{3}}) + \cdots.$$

On connait l'hypothèse célèbre, énoncée par RIEMANN, sur les zéros complexes de la fonction $\zeta(s)$: tous ces zéros ont la partie réelle $\frac{1}{2}$. D'après M. VON KOCH[1], la relation

$$f(x) = Li(x) + O(\sqrt{x}\log x)$$

est une conséquence de cette hypothèse; on en déduit immédiatement la convergence de l'intégrale

$$\int_2^{\infty} \frac{|f(x) - Li(x)|}{x^{\frac{3}{2}}\log^3 x}\,dx.$$

Dans un travail de l'année 1901, M. PHRAGMÉN[2] prouve (sans admettre l'hypothèse de RIEMANN) que cette intégrale

[1] »Sur la distribution des nombres premiers», Acta Mathematica *24* (1901), p. 159.

[2] »Sur une loi de symétrie relative à certaines formules asymptotiques», Ofversigt af Kungl. Vetenskapsakademiens Förhandlingar, *58* (1901), p. 189.

ne peut converger si l'on omet le facteur $\log^3 x$; cependant il remarque (en admettant cette fois l'hypothèse) qu'il est assez vraisemblable qu'on peut remplacer $\log^3 x$ par $\log x$ sans détruire la convergence de l'intégrale.

En effet, M. PHRAGMÉN montre que la relation

$$(1)\qquad \pi(x) < Li(x) + K\frac{\sqrt{x}}{\log x} \quad (K \text{ étant constante})$$

entrainerait la convergence de l'intégrale

$$\int_2^\infty \frac{|f(x) - Li(x)|}{x^{\frac{3}{2}}\log x}\,dx.$$

En 1901, on regardait cette relation (1), et même l'inégalité plus restreinte

$$(2)\qquad \pi(x) < Li(x)$$

comme très probablement vraie. Or, il n'en est pas ainsi, car M. LITTLEWOOD[1] a prouvé l'existence des valeurs arbitrairement grandes de x ne satisfaisant ni à (1), ni à (2); il est donc impossible d'y fonder une démonstration.

Mais le théorème que M. PHRAGMÉN veut établir est néanmoins vrai. En effet, je vais démontrer, *en admettant toujours l'hypothèse de* RIEMANN, que l'intégrale

$$(3)\qquad \int_2^\infty \frac{|f(x) - Li(x)|}{x^{\frac{3}{2}}\log^\alpha x}\,dx$$

converge pour tout $\alpha > 0$. J'obtiendrai cela comme une conséquence du théorème suivant: on a, B désignant une certaine constante,

$$(4)\qquad \lim_{x\to\infty} \frac{1}{\log x}\int_2^x \left(\frac{f(t) - Li(t)}{Li(t)}\right)^2 dt = B.$$

[1] HARDY and LITTLEWOOD: »Contributions to the theory of the Riemann zeta-function and the theory of the distribution of primes», Acta Mathematica, *41* (1917), p. 184.

D'ailleurs on a

$$B = \sum_{\varrho} \left| \frac{n_\varrho}{\varrho} \right|^2,$$

la somme étant étendue à tous les zéros complexes ϱ de la fonction $\zeta(s)$, et n_ϱ désignant l'ordre de multiplicité de ϱ.

2. Dans un travail actuellement en cours d'impression[1], j'ai démontré un théorème analogue pour la fonction

$$\psi(x) = \sum_{p^m \leqq x} \log p,$$

à savoir

(5) $$\lim_{x \to \infty} \frac{1}{\log x} \int_2^x \left(\frac{\psi(t) - t}{t} \right)^2 dt = B.$$

Nous verrons qu'on peut en déduire la relation (4) en se servant de l'identité

$$f(x) = \frac{\psi(x)}{\log x} + \int_2^x \frac{\psi(t)}{t \log^2 t}$$

qu'on vérifie immédiatement sur la definition des fonctions $f(x)$ et $\psi(x)$. Il s'ensuit d'abord

$$f(x) - Li(x) = \frac{\psi(x) - x}{\log x} + \int_2^x \frac{\psi(t) - t}{t \log^2 t} + K$$

et, en posant

$$g(x) = \int_2^x \frac{\psi(t) - t}{t} dt,$$

on obtient

$$f(x) - Li(x) = \frac{\psi(x) - x}{\log x} + \frac{g(x)}{\log^2 x} + 2 \int_2^x \frac{g(t)}{t \log^3 t} dt + K.$$

---

[1] »Ein Mittelwertsatz in der Primzahltheorie», Mathematische Zeitschrift, *12* (1922), p. 147.

Or on a, l'hypothèse de RIEMANN étant admise,

$$g(x) = \sum_{\varrho} \frac{x^{\varrho}}{\varrho^2} + O(\log x) = O(\sqrt{x}),$$

et ensuite

$$f(x) - Li(x) = \frac{\psi(x) - x}{\log x} + O\left(\frac{\sqrt{x}}{\log^2 x}\right).$$

C'est par cette formule qu'on peut passer de (5) à (4). On aura maintenant

$$\frac{f(x) - Li(x)}{Li(x)} = \frac{\psi(x) - x}{x} \cdot \frac{x}{\log x \, Li(x)} + O\left(\frac{1}{\sqrt{x}\log x}\right)$$

$$= \frac{\psi(x) - x}{x}\left[1 + O\left(\frac{1}{\log x}\right)\right] + O\left(\frac{1}{\sqrt{x}\log x}\right),$$

$$\left(\frac{f(x) - Li(x)}{Li(x)}\right)^2 = \left(\frac{\psi(x) - x}{x}\right)^2\left[1 + O\left(\frac{1}{\log x}\right)\right] +$$

$$+ \frac{\psi(x) - x}{x} \cdot O\left(\frac{1}{\sqrt{x}\log x}\right) + O\left(\frac{1}{x\log^2 x}\right)$$

et

$$\frac{1}{\log x}\int_2^x \left(\frac{f(t) - Li(t)}{Li(t)}\right)^2 dt = \frac{1}{\log x}\int_2^x \left(\frac{\psi(t) - t}{t}\right)^2 \cdot \left[1 + O\left(\frac{1}{\log t}\right)\right] dt +$$

$$+ \frac{1}{\log x}\int_2^x \frac{\psi(t) - t}{t} \cdot O\left(\frac{1}{\sqrt{t}\log t}\right) dt + O\left(\frac{1}{\log x}\right).$$

Par l'emploi de l'inégalité de M. SCHWARZ, on conclut

$$\int_2^x \frac{\psi(t) - t}{t} \cdot O\left(\frac{1}{\sqrt{t}\log t}\right) dt = O\left(\sqrt{\int_2^x \left(\frac{\psi(t) - t}{t}\right)^2 dt \cdot \int_2^x \frac{dt}{t\log^2 t}}\right)$$

$$= O(\sqrt{\log x}),$$

on obtient donc[1] enfin par (5)

$$(6)\quad \frac{1}{\log x}\int_2^x \left(\frac{f(t)-Li(t)}{Li(t)}\right)^2 dt = \frac{1}{\log x}\int_2^x \left(\frac{\psi(t)-t}{t}\right)^2 \cdot$$

$$\cdot\left[1+O\left(\frac{1}{\log t}\right)\right]dt + O\left(\frac{1}{\sqrt{\log x}}\right) = B + o(1).$$

Du résultat dernier, on peut évidemment déduire la convergence de l'intégrale

$$\int_2^\infty \left(\frac{f(t)-Li(t)}{Li(t)}\right)^2 \frac{dt}{(\log t)^{1+a}}$$

pour tout $a>0$. Il vient donc, d'après l'inégalité de M. Schwarz,

$$\left(\int_2^x \frac{|f(t)-Li(t)|}{t^{\frac{3}{2}}\log^a t}\,dt\right)^2 \leq \int_2^x \left(\frac{f(t)-Li(t)}{Li(t)}\right)^2 \frac{dt}{(\lg t)^{1+a}}\cdot$$

$$\cdot\int_2^x \frac{(Li(t))^2}{t^3(\log t)^{a-1}}\,dt < K,$$

ce qui montre que l'intégrale (3) est bornée pour $x>2$. Par suite, la fonction à intégrer étant essentiellement positive, cette intégrale est convergente.

[1] Si l'hypothèse de Riemann n'est pas vraie, le premier membre de la relation (6) ne peut rester borné quand x tend vers l'infini.

Tryckt den 18 februari 1922.

Uppsala 1922. Almqvist & Wiksells Boktryckeri-A.-B.

14.

Ein Mittelwertsatz in der Primzahltheorie

Math. Z. **12**, 147–153 (1922)

Die Riemannsche Vermutung wird in den drei ersten Nummern von dieser Arbeit als wahr angenommen.

1. Es sei $x > 2$, und es bezeichne $\psi(x)$ die bekannte Tschebyscheffsche Primzahlfunktion

$$\psi(x) = \sum_{p^m \leqq x} \log p,$$

wo p die Primzahlen, m die ganzen Zahlen durchläuft. Dann gilt nach v. Koch[1])

$$\psi(x) = x + O(\sqrt{x} \log^2 x);$$

nach Littlewood[2]) gilt aber *nicht*

$$\psi(x) = x + o(\sqrt{x} \log\log\log x).$$

In einer neulich erschienenen Arbeit[3]) habe ich bemerkt, daß die Funktion

$$\frac{\psi(x) - x}{\sqrt{x}},$$

die also jedenfalls nicht beschränkt ist, doch einen beschränkten Quadratmittelwert hat, d. h. daß

$$\frac{1}{x} \int_2^x \left(\frac{\psi(t) - t}{\sqrt{t}} \right)^2 dt = O(1)$$

---

[1]) Sur la distribution des nombres premiers, Acta Mathematica **24** (1901), S. 159–182.

[2]) Hardy and Littlewood, Contributions to the theory of the Riemann zeta-function and the theory of the distribution of primes, Acta Mathematica **41** (1917), S. 119–196.

[3]) Some theorems concerning prime numbers, Arkiv för Matematik **15**, Nr. 5 (1920), 33 S.

10*

gilt. Ich habe auch daselbst die Frage aufgeworfen, ob nicht dieser Ausdruck sogar einen Limes für $x \to \infty$ hat. Daß aber diese Frage zu verneinen ist, geht aus dem zweiten Teile des folgenden jetzt zu beweisenden Satzes hervor, von dem der erste Teil gewissermaßen ein Analogon des bekannten Parsevalschen Satzes aus der Theorie der Fourierschen Reihen bildet.

Satz. *Wenn die Riemannsche Vermutung wahr ist, so gilt für* $x \to \infty$

$$\frac{1}{\log x}\int_2^x \left(\frac{\psi(t)-t}{t}\right)^2 dt \to \sum_\varrho \left|\frac{n_\varrho}{\varrho}\right|^2, \tag{1}$$

wo ϱ *die komplexen Nullstellen der Funktion* $\zeta(s)$ *durchläuft und* n_ϱ *den Grad der Vielfachheit von* ϱ *angibt. — Dagegen schwankt für jedes* $\alpha < 1$ *der Ausdruck*

$$\frac{1}{x^{2(1-\alpha)}}\int_2^x \left(\frac{\psi(t)-t}{t^\alpha}\right)^2 dt \tag{2}$$

zwischen zwei verschiedenen endlichen Unbestimmtheitsgrenzen.

Wenn wir in dieser Arbeit die Bezeichnung $\sum\limits_\varrho$ anwenden, so denken wir uns immer die $\varrho = \frac{1}{2} + i\gamma$ nach absolut wachsenden Ordinaten geordnet und, etwas abweichend vom gewöhnlichen Gebrauche, jedes ϱ von seiner Vielfachheit abgesehen nur genau einmal gezählt.

2. Für den Beweis werden wir die v. Mangoldtsche Formel[4])

$$\psi(t) = t - \sum_\varrho n_\varrho \frac{t^\varrho}{\varrho} - \frac{1}{2}\log\left(1 - \frac{1}{t^2}\right) - \log 2\pi$$

benutzen, die für alle $t > 2$ gilt, welche nicht von der Form $t = p^m$ sind. Die hier auftretende Reihe ist in jedem endlichen abgeschlossenen Intervall solcher t-Werte gleichmäßig konvergent, und ihre „Teilsummen" $\sum\limits_{|\gamma|<A} n_\varrho \frac{t^\varrho}{\varrho}$ sind in jedem Intervall $2 \leqq t \leqq x$ gleichmäßig beschränkt.

Schreiben wir also

$$\begin{aligned}\left(\frac{\psi(t)-t}{t}\right)^2 &= \frac{1}{t^2}\left(-\sum_\varrho n_\varrho \frac{t^\varrho}{\varrho} + O(1)\right)^2 \\ &= \sum_\varrho n_\varrho \sum_{\varrho'} n_{\varrho'} \frac{t^{\varrho+\varrho'-2}}{\varrho\varrho'} + O\left(\frac{1}{t^2}\right) + O\left(\left|\sum_\varrho n_\varrho \frac{t^{\varrho-2}}{\varrho}\right|\right) \\ &= \sum_\varrho n_\varrho \sum_{\varrho'} n_{\varrho'} \frac{t^{\varrho+\varrho'-2}}{\varrho\varrho'} + O(t^{-\frac{3}{2}}\log^2 t),\end{aligned}$$

[4]) Vgl. z. B. E. Landau, Handbuch der Lehre von der Verteilung der Primzahlen, Leipzig und Berlin 1909, 1, S. 365.

so kann hier von 2 bis x gliedweise integriert werden, und wir erhalten

$$\int\limits_2^x \left(\frac{\psi(t)-t}{t}\right)^2 dt = \sum_{\varrho} \frac{n_\varrho}{\varrho} \sum_{\varrho'} \frac{n_{\varrho'}}{\varrho'} \int\limits_2^x t^{\varrho+\varrho'-2}\, dt + O(1),$$

wo, bei der angegebenen Summationsordnung, die Doppelreihe konvergent ist. Es gilt nun

$$\varrho(1-\varrho) = |\varrho|^2$$

und nach bekannten Sätzen über die Verteilung der ϱ ist die Reihe

$$\sum_{\varrho} \left|\frac{n_\varrho}{\varrho}\right|^2$$

konvergent. Hieraus folgt weiter

(3) $$\int\limits_2^x \left(\frac{\psi(t)-t}{t}\right)^2 dt = \log x \sum_{\varrho} \left|\frac{n_\varrho}{\varrho}\right|^2$$
$$+ \sum_{\varrho} \frac{n_\varrho}{\varrho} \sum_{\varrho' \neq 1-\varrho} \frac{n_{\varrho'}}{\varrho'} \cdot \frac{x^{\varrho+\varrho'-1} - 2^{\varrho+\varrho'-1}}{\varrho+\varrho'-1} + O(1).$$

Wir wollen beweisen, daß das erste Glied rechts für das asymptotische Verhalten des ganzen Ausdrucks ausschlaggebend ist, d. h. daß die Doppelreihe von der Form $o(\log x)$ ist.

Es sei nun η eine beliebig kleine, aber von x unabhängige, positive Zahl, kleiner als das kleinste positive γ. Dann werden wir zuerst zeigen, daß die Summe, die man erhält, wenn man in der Doppelreihe nur die Glieder mit

$$|\varrho+\varrho'-1| \geqq \eta$$

nimmt, absolut konvergent und von der Form $O(1)$ ist. Aus Symmetriegründen genügt es zu zeigen, daß die beiden Reihen

$$\sum_{\gamma>0} \frac{n_\varrho}{|\varrho|} \sum_{\gamma'>0} \frac{n_{\varrho'}}{|\varrho'|(\gamma+\gamma')}$$

und

(4) $$\sum_{\gamma>0} \frac{n_\varrho}{|\varrho|} \sum_{0<\gamma'\leqq\gamma-\eta} \frac{n_{\varrho'}}{|\varrho'|(\gamma-\gamma')}$$

konvergieren, und man findet sogar, daß es hinreichend ist, nur die letzte Reihe zu betrachten. Nach der bekannten Formel

$$N(T) = \frac{T}{2\pi}\left(\log\frac{T}{2\pi} - 1\right) + O(\log T),$$

wo $N(T)$ die Anzahl der $\varrho = \frac{1}{2} + i\gamma$ angibt, die der Beziehung $0 < \gamma \leqq T$ genügen[5]), findet man nun für hinreichend große γ

[5]) Vgl. E. Landau, Handbuch der Lehre von der Verteilung der Primzahlen, Leipzig und Berlin 1909, 1, S. 368.

$$\sum_{0<\gamma'\leqq\gamma-\eta}\frac{n_{\varrho'}}{|\varrho'|(\gamma-\gamma')}=\sum_{0<\gamma'\leqq\gamma^{\frac{2}{3}}}+\sum_{\gamma^{\frac{2}{3}}<\gamma'\leqq\gamma-\gamma^{\frac{2}{3}}}+\sum_{\gamma-\gamma^{\frac{2}{3}}<\gamma'\leqq\gamma-\eta}$$

$$=O\left(\frac{\gamma^{\frac{2}{3}}\log\gamma}{\gamma}+\frac{\gamma\log\gamma}{\gamma^{\frac{4}{3}}}+\frac{\gamma^{\frac{2}{3}}\log\gamma}{\gamma}\right)=O\left(\frac{\log\gamma}{\gamma^{\frac{1}{3}}}\right),$$

so daß die Reihe (4) konvergiert. Aus (3) folgt also

$$(5)\qquad \int_2^x\left(\frac{\psi(t)-t}{t}\right)^2dt=\log x\sum_\varrho\left|\frac{n_\varrho}{\varrho}\right|^2$$
$$+\sum\sum_{0<|\gamma+\gamma'|<\eta}\frac{n_\varrho n_{\varrho'}(x^{\varrho+\varrho'-1}-2^{\varrho+\varrho'-1})}{i\varrho\varrho'(\gamma+\gamma')}+O(1)$$

und es muß nur noch gezeigt werden, daß, wenn ε beliebig klein gegeben ist, η und x_0 so gewählt werden können, daß

$$\left|\sum_\varrho\sum_{0<|\gamma+\gamma'|<\eta}\frac{n_\varrho n_{\varrho'}(x^{i(\gamma+\gamma')}-2^{i(\gamma+\gamma')})}{\varrho\varrho'(\gamma+\gamma')}\right|<\varepsilon\log x$$

für $x>x_0$ gilt.

Da die Gliederzahl der inneren Summe endlich ist und in der Summe $\sum_\varrho'$ die Glieder nach wachsendem $|\gamma|$ geordnet sind, kann man immer zwei Glieder $(\varrho=\frac{1}{2}+i\gamma,\ \varrho'=\frac{1}{2}+i\gamma')$ und $(\varrho=\frac{1}{2}-i\gamma,\ \varrho'=\frac{1}{2}-i\gamma')$ zusammenfassen. Es genügt also, die Ungleichung

$$\left|\sum_\varrho\sum_{0<\gamma+\gamma'<\eta}\frac{n_\varrho n_{\varrho'}}{\gamma+\gamma'}\,\Im\,\frac{x^{i(\gamma+\gamma')}-2^{i(\gamma+\gamma')}}{\varrho\varrho'}\right|<\frac{1}{2}\varepsilon\log x$$

oder sogar

$$(6)\qquad \sum_\varrho\sum_{0<\gamma+\gamma'<\eta}\frac{n_\varrho n_{\varrho'}}{|\varrho|^2|\varrho'|^2(\gamma+\gamma')}\left|\Im\,\bar\varrho\,\bar\varrho'x^{i(\gamma+\gamma')}\right|<\frac{1}{4}\varepsilon\log x$$

für alle $x\geqq 2$, bei zweckmäßiger Wahl von η, zu beweisen. Es gilt nun

$$\left|\Im\,\bar\varrho\,\bar\varrho'x^{i(\gamma+\gamma')}\right|=\left|-\tfrac{1}{2}(\gamma+\gamma')\cos(\gamma+\gamma')\log x+(\tfrac{1}{4}-\gamma\gamma')\sin(\gamma+\gamma')\log x\right|$$
$$<\gamma+\gamma'+2\,|\gamma\gamma'\sin(\gamma+\gamma')\log x|<3\,|\gamma\gamma'|\,(\gamma+\gamma')\log x;$$

die linke Seite von (6) ist also kleiner als

$$(7)\qquad 3\log x\sum_\varrho\sum_{0<\gamma+\gamma'<\eta}\frac{n_\varrho n_{\varrho'}}{|\gamma\gamma'|}.$$

Es sei nun K eine solche Konstante, daß für alle $T>2$ die Ungleichung $T-1<\gamma<T+1$ von höchstens $K\log T$ Nullstellen (jede mit ihrer Vielfachheit gezählt) erfüllt ist, und es sei $A>0$ so gewählt, daß

$$\sum_{\gamma>A}n_\varrho\frac{\log\gamma}{\gamma(\gamma-1)}<\frac{\varepsilon}{24K}$$

gilt. Dann wählen wir η so, daß erstens $\eta < 1$ und zweitens η kleiner als die kleinste Differenz zweier verschiedener γ im Intervalle $0 < \gamma \leqq A + 1$ wird. Die Glieder der Doppelsumme in (7) können dann nur solche ϱ enthalten, für welche $|\gamma| > A$ gilt. Wir haben also

$$3 \log x \sum_{\varrho} \sum_{0<\gamma+\gamma'<\eta} \frac{n_\varrho n_{\varrho'}}{|\gamma\gamma'|} < 3 \log x \sum_{|\gamma|>A} \sum_{0<\gamma+\gamma'<\eta} \frac{n_\varrho n_{\varrho'}}{|\gamma|(|\gamma|-1)}$$

$$< 6 \log x \sum_{\gamma>A} \frac{K \log \gamma \cdot n_\varrho}{\gamma(\gamma-1)} < \frac{\varepsilon}{4} \log x .$$

Hierdurch ist (6) bewiesen, und aus (5) folgt dann der erste Teil unseres Satzes:

$$\int\limits_2^x \left(\frac{\psi(t)-t}{t}\right)^2 dt = \log x \sum_{\varrho} \left|\frac{n_\varrho}{\varrho}\right|^2 + o(\log x).$$

3. In gleicher Weise wie oben erhalten wir, wenn $\alpha < 1$ und von x unabhängig ist,

$$(8) \qquad \frac{1}{x^{2(1-\alpha)}} \int\limits_2^x \left(\frac{\psi(t)-t}{t^\alpha}\right)^2 dt = \sum_{\varrho} \frac{n_\varrho}{\varrho} \sum_{\varrho'} \frac{n_{\varrho'}}{\varrho'} \cdot \frac{x^{\varrho+\varrho'-1}}{\varrho+\varrho'+1-2\alpha} + o(1).$$

Die Hauptschwierigkeit bei dem soeben behandelten Falle $\alpha = 1$ lag nun darin, daß der in der Doppelreihe in (3) im Nenner auftretende Faktor $\varrho + \varrho' - 1$ beliebig kleine Werte annehmen konnte. Deshalb mußte diese Reihe in zwei Teile geteilt und die beiden Teile in verschiedener Weise abgeschätzt werden. Hier aber hat der entsprechende Faktor $\varrho + \varrho' + 1 - 2\alpha$ den reellen Teil $2(1-\alpha) > 0$, und man kann die ganze Doppelreihe in genau derselben Weise wie vorhin die Glieder mit $|\varrho + \varrho' - 1| \geqq \eta$ behandeln. Man zeigt in dieser Weise, daß die Doppelreihe unbedingt konvergent und von der Form $O(1)$ ist. Hierdurch ist schon bewiesen, daß die Unbestimmtheitsgrenzen von (2) *endlich* sind, und wir haben nur noch zu zeigen, daß sie *verschieden* sind.

Zuerst bemerken wir, daß die absolut konvergente Doppelreihe

$$(9) \qquad \sum_{\varrho} \frac{n_\varrho}{\varrho} \sum_{\varrho'} \frac{n_{\varrho'}}{\varrho'} \cdot \frac{x^{\varrho+\varrho'-1}}{\varrho+\varrho'+1-2\alpha}$$

jedenfalls nicht für alle $x > 2$ konstant sein kann. Denn in diesem Falle würde man durch Multiplikation mit $x^{2(1-\alpha)}$ und Differentiation

$$\left(\sum \frac{n_\varrho}{\varrho} x^\varrho\right)^2 = 2(1-\alpha) K x$$

für alle $x \neq p^m$ erhalten. (Für alle solche Werte von x darf man nach bekannten Sätzen gliedweise differentiieren.) Das ist aber z. B. mit dem oben zitierten Littlewoodschen Satze nicht verträglich.

Wenn wir jetzt noch beweisen können, daß die Doppelreihe (9) jedem Werte, den sie überhaupt für $x > 2$ annimmt, wieder für beliebig große Werte von x beliebig nahe kommt, so folgt daraus, daß die linke Seite von (8) keinen bestimmten Limes für $x \to \infty$ haben kann, und unser Satz ist dann vollständig bewiesen. Dies folgt aber leicht aus dem bekannten Dirichlet-Kroneckerschen Satz über diophantische Approximationen, der etwa folgendermaßen ausgesprochen werden kann[6]):

„Es seien n reelle Größen $\alpha_1, \alpha_2, \ldots, \alpha_n$ und zwei positive Zahlen t_0 (groß) und ε (klein) gegeben. Dann kann man n ganze Zahlen $\mu_1, \mu_2, \ldots, \mu_n$ und ein $t > t_0$ bestimmen, so daß die Ungleichungen

$$|t\alpha_1 - 2\pi\mu_1| < \varepsilon$$
$$\cdots\cdots\cdots$$
$$|t\alpha_n - 2\pi\mu_n| < \varepsilon$$

alle erfüllt sind."

Betrachten wir nun die Doppelreihe (9). Sie nimmt für $x = z > 2$ einen gewissen Wert A an. Wenn x_0 beliebig gegeben ist, so wollen wir zeigen, daß es oberhalb x_0 Werte von x gibt, wo die Doppelreihe dem Wert A beliebig nahe kommt. Da die Doppelreihe für alle $x > 2$ absolut und gleichmäßig konvergiert, können wir, wenn ε beliebig klein gegeben ist, n Glieder herausnehmen und dabei n so wählen, daß die Summe der übrigen Glieder für $x > 2$ absolut $< \varepsilon$ ist. Setzen wir in der Summe der n fraglichen Glieder $x = ze^t$ und wenden dann den obigen Satz an, wobei wir $t_0 = \log\frac{x_0}{z}$ nehmen und unter $\alpha_1, \ldots, \alpha_n$ die n entsprechenden Werte von $\gamma + \gamma'$ verstehen, so ergibt sich, daß wir ein $t > t_0$, d. h. ein $x > x_0$ derart finden können, daß

$$\left|\sum_{\varrho}\frac{n_\varrho}{\varrho}\sum_{\varrho'}\frac{n_{\varrho'}}{\varrho'}\cdot\frac{x^{\varrho+\varrho'-1}}{\varrho+\varrho'+1-2\alpha} - A\right|$$
$$< 2\varepsilon + \sum_{\nu=1}^{n}\left|\frac{n_\varrho n_{\varrho'}}{\varrho\varrho'(\varrho+\varrho'+1-2\alpha)}\left(e^{i(t\alpha_\nu - 2\pi\mu_\nu)} - 1\right)\right|$$
$$< \varepsilon\left(2 + \sum_{\varrho}\sum_{\varrho'}\left|\frac{n_\varrho n_{\varrho'}}{\varrho\varrho'(\varrho+\varrho'+1-2\alpha)}\right|\right).$$

Hierdurch ist die behauptete Eigenschaft unserer Doppelreihe bewiesen, und nach dem oben Gesagten folgt daraus der letzte Teil des Satzes.

4. *Wenn die Riemannsche Vermutung falsch ist, so kann weder* (1) *noch* (2) *für unendlich wachsendes x beschränkt bleiben. Aus der*

[6]) Vgl. z. B. H. Bohr und E. Landau, Über das Verhalten von $\zeta(s)$ und $\zeta_\varkappa(s)$ in der Nähe der Geraden $\sigma = 1$, Göttinger Nachrichten, 1910, S. 303–330.

Beschränktheit von (1) *oder* (2) *folgt also umgekehrt die Riemannsche Vermutung.*

Es genügt, dies für (1) zu beweisen. Es gilt nämlich, wenn wir den Ausdruck (2) gleich $f_\alpha(x)$ setzen,

$$\int_2^x \left(\frac{\psi(t)-t}{t}\right)^2 dt = f_\alpha(x) + 2(1-\alpha)\int_2^x \frac{f_\alpha(t)}{t}\, dt,$$

so daß aus der Beschränktheit von (2) auch die Beschränktheit von (1) folgt. — Aus der bekannten Schwarzschen Ungleichung erhalten wir

$$\left(\int_2^x \frac{\psi(t)-t}{t}\, dt\right)^2 \leqq (x-2)\int_2^x \left(\frac{\psi(t)-t}{t}\right)^2 dt,$$

also, wenn (1) beschränkt ist,

$$\int_2^x \frac{\psi(t)-t}{t}\, dt = O(\sqrt{x\log x}),$$

d. h.

$$\sum_{n<x} \Lambda(n)\log\frac{x}{n} = x + O(\sqrt{x\log x}),$$

wo wie gewöhnlich $\Lambda(n) = \psi(n) - \psi(n-1)$ gesetzt ist. Da nun

$$\sum_{n<x} \log\frac{x}{n} = x + O(\log x)$$

gilt, wäre also

$$\sum_{n<x} (\Lambda(n) - 1)\log\frac{x}{n} = O(\sqrt{x\log x}).$$

Die Reihe $\sum_1^\infty (\Lambda(n)-1)n^{-\sigma-it}$ wäre also für $\sigma > \frac{1}{2}$ summierbar[7]) $(\log n, 1)$ und daraus folgt bekanntlich die Regularität der Funktion von $s = \sigma + it$

$$\frac{\zeta'(s)}{\zeta(s)} + \zeta(s)$$

für $\sigma > \frac{1}{2}$, d. h. die Riemannsche Vermutung.

Mörby, Stocksund, den 7. Juni 1921.

---

[7]) Vgl. Hardy und Riesz, The general theory of Dirichlet's series, Cambridge 1915, Theorem **31**, S. 45.

(Eingegangen am 13. Juni 1921.)

15.

Über zwei Sätze des Herrn G. H. Hardy

Math. Z. **15**, 201–210 (1922)

Zwei der interessantesten Funktionen der analytischen Zahlentheorie, die besonders in der neuesten Zeit vielfach behandelt wurden, sind die Funktionen $D(x)$ und $R(x)$, welche in folgender Weise erklärt sind: Es bezeichne $d(n)$ die Anzahl der Teiler der positiven ganzen Zahl n, $r(n)$ die Anzahl ihrer Zerlegungen in zwei Quadrate; dann setzen wir für $x > 0$

$$D(x) = \sum_{n \leqq x} d(n), \qquad R(x) = \sum_{n \leqq x} r(n).$$

Die beiden Funktionen lassen sich bekanntlich auch in einfacher Weise als Gitterpunktanzahlen interpretieren. Die Hauptglieder der Funktionen für große Werte von x sind schon lange[1]) bekannt, sie gehen aus den Formeln

$$D(x) = x \log x + (2C - 1)x + \varDelta(x),$$
$$R(x) = \pi x + P(x)$$

hervor. Hier bedeutet C die Eulersche Konstante, und die Größenordnung der beiden Restglieder $P(x)$ und $\varDelta(x)$ ist jedenfalls kleiner als die der Hauptglieder. Die wirkliche Größenordnung dieser Restglieder zu finden, ist aber eins der schwierigsten noch ungelösten Probleme der analytischen Zahlentheorie. In dieser Hinsicht ist bewiesen[2]), daß

[1]) Vgl. Gauß, De nexu inter multitudinem classium etc., Werke 2, S. 269 und Dirichlet, Über die Bestimmung der mittleren Werte in der Zahlentheorie, Werke 2, S. 51.

[2]) Vgl. G. H. Hardy, On the expression of a number as the sum of two squares, Quarterly Journal of Mathematics **46** (1915), S. 263—283, und On Dirichlet's divisor problem, Proc. Lond. Math. Soc. (2) **15** (1916), S. 1—25. In diesen Arbeiten findet man zahlreiche Angaben über die Literatur unserer Probleme.

$$
(1)\qquad \begin{cases} \Delta(x) = O(x^{\frac{1}{3}} \log x), \\ P(x) = O(x^{\frac{1}{3}}) \end{cases}
$$

und

$$
(2)\qquad \begin{cases} \Delta(x) \neq o(x^{\frac{1}{4}} (\log x)^{\frac{1}{4}} \log\log x), \\ P(x) \neq o(x^{\frac{1}{4}} (\log x)^{\frac{1}{4}}) \end{cases}
$$

gilt. Bezeichnet also $\varkappa$ die untere Grenze der Zahlen k, für welche z. B. $\Delta(x) = O(x^k)$ ist, so muß $\frac{1}{4} \leqq \varkappa \leqq \frac{1}{3}$ sein; mehr weiß man nicht über $\varkappa$. Dagegen hat Hardy[3]) bewiesen, daß *die Mittelwerte* der beiden Restglieder von einer $x^{\frac{1}{4}}$ nicht wesentlich übersteigenden Größenordnung sind; es gilt nämlich für jedes $\varepsilon > 0$

$$
(3)\qquad \begin{cases} \dfrac{1}{x}\displaystyle\int_1^x |\Delta(t)|\, dt = O(x^{\frac{1}{4}+\varepsilon}), \\ \dfrac{1}{x}\displaystyle\int_1^x |P(t)|\, dt = O(x^{\frac{1}{4}+\varepsilon}). \end{cases}
$$

Diese Resultate ergeben sich mit Hilfe der Schwarzschen Ungleichung als unmittelbare Folgerungen aus

$$
(4)\qquad \begin{cases} \displaystyle\int_1^x (\Delta(t))^2\, dt = O(x^{\frac{3}{2}+\varepsilon}), \\ \displaystyle\int_1^x (P(t))^2\, dt = O(x^{\frac{3}{2}+\varepsilon}). \end{cases}
$$

Ich werde im folgenden einen neuen, kürzeren Beweis der beiden durch (3) oder (4) ausgedrückten Hardyschen Sätze mitteilen und dieselben gleichzeitig auch verschärfen. Es wird sich nämlich ergeben, daß (4) durch

$$
(5)\qquad \begin{cases} \displaystyle\int_1^x (\Delta(t))^2\, dt = \alpha x^{\frac{3}{2}} + O(x^{\frac{5}{4}+\varepsilon}), \\ \displaystyle\int_1^x (P(t))^2\, dt = \beta x^{\frac{3}{2}} + O(x^{\frac{5}{4}+\varepsilon}) \end{cases}
$$

[3]) The average order of the arithmetical functions $P(x)$ and $\Delta(x)$, Proc. Lond. Math. Soc. (2) **15** (1916), S. 192—213, und Additional note on two problems in the analytic theory of numbers, Ebenda (2) **18** (1918), S. 201—204.

mit konstanten α und β ersetzt werden kann. Übrigens ist[4])

$$(6)\qquad \begin{cases} \alpha = \dfrac{1}{6\pi^2} \displaystyle\sum_1^\infty \left(\dfrac{d(n)}{n^{\frac{3}{4}}}\right)^2, \\ \beta = \dfrac{1}{3\pi^2} \displaystyle\sum_1^\infty \left(\dfrac{r(n)}{n^{\frac{3}{4}}}\right)^2, \end{cases}$$

was mit den für nicht ganzzahlige x geltenden Formeln[2])

$$\Delta(x) = \frac{x^{\frac{1}{4}}}{\pi\sqrt{2}} \sum_1^\infty \frac{d(n)}{n^{\frac{3}{4}}} \cos\left(4\pi\sqrt{nx} - \frac{1}{4}\pi\right) + O(1),$$

$$P(x) = \frac{x^{\frac{1}{4}}}{\pi} \sum_1^\infty \frac{r(n)}{n^{\frac{3}{4}}} \sin\left(2\pi\sqrt{nx} - \frac{1}{4}\pi\right) + O(1)$$

in einem gewissen Einklang steht. Hierdurch wird natürlich auch (3) zu

$$(7)\qquad \begin{cases} \dfrac{1}{x}\displaystyle\int_1^x |\Delta(t)|\,dt = O(x^{\frac{1}{4}}) \\ \dfrac{1}{x}\displaystyle\int_1^x |P(t)|\,dt = O(x^{\frac{1}{4}}) \end{cases}$$

verschärft. Es ist interessant, dieses Resultat mit (2) zu vergleichen.

Die Beweise der beiden durch (5) ausgedrückten Sätze verlaufen sehr analog, und ich werde daher nur den Beweis des Satzes über $P(x)$ vollständig entwickeln, um dann einige kurze Andeutungen über $\Delta(x)$ zu geben.

§ 1.

Ich gehe von der leicht beweisbaren[5]) Gleichung

$$(8)\qquad \int_0^t R(u)\,du = \frac{\pi}{2}t^2 - t + \frac{1}{\pi}\sum_1^\infty \frac{r(n)}{n}\, t\, J_2(2\pi\sqrt{nt})$$

aus und erhalte unmittelbar, wenn noch allgemein

$$\delta f(x) = f(x+1) - f(x)$$

---

[4]) Die Konstante α kann auch durch die Riemannsche Zetafunktion ausgedrückt werden. Man hat nämlich $\alpha = \dfrac{\zeta^4\left(\frac{3}{2}\right)}{6\pi^2\zeta(3)}$.

[5]) Vgl. E. Landau, Über die Gitterpunkte in einem Kreise, Math. Zeitschr. 5 (1919), S. 319—320. Das dortige $A(x)$ ist gleich $R(x)+1$.

14*

gesetzt wird,

$$\int_t^{t+1} R(u)\,du = \pi t + \frac{1}{\pi}\sum_1^\infty \frac{r(n)}{n}\,\delta t\, J_2(2\pi\sqrt{nt}) + O(1).$$

Nun ist aber

$$\int_t^{t+1} R(u)\,du = R(t) + \int_t^{t+1}[R(u) - R(t)]\,dt,$$

also wegen $r(n) = O(n^\varepsilon)$

$$\int_t^{t+1} R(u)\,du = \pi t + P(t) + O(t^\varepsilon)$$

für beliebig kleines $\varepsilon > 0$. Hieraus folgt

$$P(t) = \frac{1}{\pi}\sum_1^\infty \frac{r(n)}{n}\,\delta t\, J_2(2\pi\sqrt{nt}) + O(t^\varepsilon),$$

und, wenn jetzt die asymptotische Entwicklung der Besselschen Funktion in der Gestalt

$$J_2(x) = -\sqrt{\frac{2}{\pi x}}\cos\left(x - \frac{1}{4}\pi\right) + \frac{15}{8}\sqrt{\frac{2}{\pi x^3}}\sin\left(x - \frac{1}{4}\pi\right) + O(x^{-\frac{5}{2}})$$

eingeführt wird,

$$P(t) = -\frac{1}{\pi^2}\sum_1^\infty \frac{r(n)}{n^{\frac{5}{4}}}\,\delta t^{\frac{3}{4}}\cos\left(2\pi\sqrt{nt} - \frac{1}{4}\pi\right)$$

$$+ \frac{15}{16\pi^3}\sum_1^\infty \frac{r(n)}{n^{\frac{7}{4}}}\,\delta t^{\frac{1}{4}}\sin\left(2\pi\sqrt{nt} - \frac{1}{4}\pi\right) + O(t^\varepsilon),$$

In dieser Gleichung ist aber die zweite Summe rechts von der Form $O(t^\varepsilon)$, denn es gilt

$$\delta t^{\frac{1}{4}}\sin\left(2\pi\sqrt{nt} - \frac{1}{4}\pi\right) = \int_t^{t+1}\frac{d}{dv}v^{\frac{1}{4}}\sin\left(2\pi\sqrt{nv} - \frac{1}{4}\pi\right)dv = O\left(n^{\frac{1}{2}}t^{-\frac{1}{4}}\right);$$

wir haben also

$$P(t) = -\frac{1}{\pi^2}\sum_1^\infty \frac{r(n)}{n^{\frac{5}{4}}}\,\delta t^{\frac{3}{4}}\cos\left(2\pi\sqrt{nt} - \frac{1}{4}\pi\right) + O(t^\varepsilon) = P_1(t) + O(t^\varepsilon).$$

Hieraus folgt nun

$$(9)\qquad \int_1^x (P(t))^2\,dt = \int_1^x (P_1(t))^2\,dt + \int_1^x (P_1(t))\cdot O(t^\varepsilon)\,dt + O(x^{1+\varepsilon}),$$

wo

$$(10)\qquad \int_1^x (P_1(t))^2\,dt$$

$$= \frac{1}{\pi^4} \sum_{m=1}^{\infty} \sum_{n=1}^{\infty} \frac{r(m)\,r(n)}{(mn)^{\frac{5}{4}}} \int_1^x \delta t^{\frac{3}{4}} \cos\left(2\pi\sqrt{mt} - \frac{1}{4}\pi\right) \cdot \delta t^{\frac{3}{4}} \cos\left(2\pi\sqrt{nt} - \frac{1}{4}\pi\right) dt$$

gilt. Wir wollen zuerst beweisen, daß in der hier auftretenden Doppelreihe die Summe aller Glieder mit $m \neq n$ von der Form $O(x^{\frac{5}{4}+\varepsilon})$ ist. Es genügt natürlich, nur die Glieder mit $m > n$ zu betrachten. Schreiben wir

$$\delta_m(t) = \delta t^{\frac{3}{4}} \cos\left(2\pi\sqrt{mt} - \frac{1}{4}\pi\right) = \frac{1}{2} e^{-\frac{1}{4}\pi i}\, \delta t^{\frac{3}{4}}\, e^{2\pi i\sqrt{mt}}$$
$$+ \frac{1}{2} e^{\frac{1}{4}\pi i}\, \delta t^{\frac{3}{4}}\, e^{-2\pi i\sqrt{mt}},$$

so folgt

$$(11)\qquad \delta_m(t)\,\delta_n(t) = \frac{1}{2}\Re\left(e^{-\frac{1}{2}\pi i} \cdot \delta t^{\frac{3}{4}} e^{2\pi i\sqrt{mt}} \cdot \delta t^{\frac{3}{4}} e^{2\pi i\sqrt{nt}}\right)$$
$$+ \frac{1}{2}\Re\left(\delta t^{\frac{3}{4}} e^{2\pi i\sqrt{mt}} \cdot \delta t^{\frac{3}{4}} e^{-2\pi i\sqrt{nt}}\right),$$

und hieraus

$$(12)\qquad \left|\int_1^x \delta_m(t)\,\delta_n(t)\,dt\right| < \left|\int_1^x \delta t^{\frac{3}{4}} e^{2\pi i\sqrt{mt}} \cdot \delta t^{\frac{3}{4}} e^{2\pi i\sqrt{nt}}\,dt\right|$$
$$+ \left|\int_1^x \delta t^{\frac{3}{4}} e^{2\pi i\sqrt{mt}} \cdot \delta t^{\frac{3}{4}} e^{-2\pi i\sqrt{nt}}\,\delta t\right|.$$

Das zweite Glied rechts in (12) ist nun für $m > n$ gleich

$$(13)\qquad \left|\int_1^x F(t)\,d\,\frac{e^{2\pi i(\sqrt{m}-\sqrt{n})\sqrt{t}}}{\sqrt{m}-\sqrt{n}}\right| < \frac{|F(x)| + |F(1)| + \int_1^x |F'(t)|\,dt}{\sqrt{m}-\sqrt{n}},$$

wo

$$(14)\qquad F(t) = \frac{t^2}{\pi i}\left[\left(1+\frac{1}{t}\right)^{\frac{3}{4}} e^{2\pi i\sqrt{m}(\sqrt{t+1}-\sqrt{t})} - 1\right] \cdot \left[\left(1+\frac{1}{t}\right)^{\frac{3}{4}} e^{-2\pi i\sqrt{n}(\sqrt{t+1}-\sqrt{t})} - \right.$$
$$= \frac{t^2}{\pi i} F_1(t)\,F_2(t).$$

Durch einfache Rechnung ergibt sich für $t \geqq 1$

$$|F_1(t)| < \begin{cases} 3, \\ 8\,m^{\frac{1}{2}}\,t^{-\frac{1}{2}} \end{cases}$$

$$|F_1'(t)| < 5\,m^{\frac{1}{2}}\,t^{-\frac{3}{2}}$$

und analoge Ungleichungen für $F_2(t)$ und $F_2(t)$, also folgt aus (14)

$$|F(t)| < \begin{cases} \frac{t^2}{\pi} \cdot 3 \cdot 3 & < 3t^2 \\ \frac{t^2}{\pi} \cdot 8 m^{\frac{1}{2}} t^{-\frac{1}{2}} \cdot 8 n^{\frac{1}{2}} t^{-\frac{1}{2}} & < 22 t m^{\frac{1}{2}} n^{\frac{1}{2}} \end{cases}$$

$$|F'(t)| < \begin{cases} \frac{2t}{\pi} \cdot 3 \cdot 3 + \frac{t^2}{\pi} \cdot 15 t^{-\frac{3}{2}} (m^{\frac{1}{2}} + n^{\frac{1}{2}}) < 6t + 10 t^{\frac{1}{2}} m^{\frac{1}{2}} \\ \frac{2t}{\pi} \cdot 8 m^{\frac{1}{2}} t^{-\frac{1}{2}} \cdot 8 n^{\frac{1}{2}} t^{-\frac{1}{2}} + \frac{t^2}{\pi} \cdot 8 m^{\frac{1}{2}} n^{\frac{1}{2}} t^{-2} < 70 m^{\frac{1}{2}} n^{\frac{1}{2}}. \end{cases}$$

Durch Einführung von diesen Ausdrücken in (13) ergeben sich für das zweite Glied rechts in (12) die beiden Majoranten

$$(15) \qquad \frac{7 x^{\frac{3}{2}} (\sqrt{x} + \sqrt{m})}{\sqrt{m} - \sqrt{n}} \quad \text{und} \quad \frac{120 x \sqrt{mn}}{\sqrt{m} - \sqrt{n}}.$$

Das erste Glied rechts in (12) läßt sich natürlich genau ebenso behandeln, jedoch tritt hier der Nenner $\sqrt{m} + \sqrt{n}$ auf, so daß die Abschätzung auch für Glieder mit $m = n$ brauchbar ist. Schließlich finden wir also für $m > n$

$$(16) \qquad \left| \int_1^x \delta_m(t) \delta_n(t) dt \right| < \begin{cases} \frac{14 x^{\frac{3}{2}} (\sqrt{x} + \sqrt{m})}{\sqrt{m} - \sqrt{n}} \\ \frac{240 x \sqrt{mn}}{\sqrt{m} - \sqrt{n}}. \end{cases}$$

Die Glieder mit $m > n$ in der Doppelsumme in (10) zerlegen wir nun in zwei Teile, indem wir schreiben

$$s_1 = \sum_{m \leqq x} \sum_{n < m} \frac{r(m) r(n)}{(mn)^{\frac{5}{4}}} \int_1^x \delta_m(t) \delta_n(t) dt,$$

$$s_2 = \sum_{m > x} \sum_{n < m} \frac{r(m) r(n)}{(mn)^{\frac{5}{4}}} \int_1^x \delta_m(t) \delta_n(t) dt.$$

Durch Anwendung von (16) erhalten wir für beliebig kleines $\varepsilon > 0$ (wir nehmen insbesondere $\varepsilon < \frac{1}{4}$ an)

$$s_1 = O\left(\sum_{m \leqq x} \sum_{n < m} (mn)^{-\frac{5}{4} + \varepsilon} \frac{x \sqrt{mn}}{\sqrt{m} - \sqrt{n}} \right) = O\left(x \sum_{m \leqq x} m^{-\frac{3}{4} + \varepsilon} \sum_{n < m} \frac{\sqrt{m} + \sqrt{n}}{n^{\frac{3}{4} + \varepsilon} (m - n)} \right)$$

$$= O\left(x \sum_{m \leqq x} m^{-\frac{1}{4} + \varepsilon} \sum_{n < m} \frac{1}{n^{\frac{3}{4} - \varepsilon} (m - n)} \right) = O\left(x \sum_{m \leqq x} m^{-1 + 2\varepsilon} \log m \right) = O(x^{1 + 3\varepsilon})$$

und

$$s_2 = O\left(\sum_{m>x}\sum_{n<m}(mn)^{-\frac{5}{4}+\varepsilon}\frac{x^{\frac{3}{2}}(\sqrt{x}+\sqrt{m})}{\sqrt{m}-\sqrt{n}}\right)$$
$$= O\left(x^{\frac{3}{2}}\sum_{m>x}m^{-\frac{1}{4}+\varepsilon}\sum_{n<m}\frac{1}{n^{\frac{5}{4}-\varepsilon}(m-n)}\right) = O\left(x^{\frac{3}{2}}\sum_{m>x}m^{-\frac{5}{4}+\varepsilon}\right) = O(x^{\frac{5}{4}+\varepsilon}).$$

Es folgt somit aus (10)

$$(17)\qquad \int_1^x (P_1(t))^2\,dt = \frac{1}{\pi^4}\sum_{n=1}^{\infty}\left(\frac{r(n)}{n^{\frac{5}{4}}}\right)^2\int_1^x(\delta_n(t))^2\,dt + O(x^{\frac{5}{4}+\varepsilon}).$$

Nach (11) haben wir

$$(\delta_n(t))^2 = \frac{1}{2}\Re\left[-i(\delta t^{\frac{3}{4}}e^{2\pi i\sqrt{nt}})^2\right] + \frac{1}{2}\delta t^{\frac{3}{4}}e^{2\pi i\sqrt{nt}}\cdot\delta t^{\frac{3}{4}}e^{-2\pi i\sqrt{nt}},$$

und das Integral des ersten Gliedes rechts kann nach der Bemerkung zu (15) durch den Ausdruck

$$\frac{120x\sqrt{n}\cdot n}{\sqrt{n}+\sqrt{n}} = 60x\sqrt{n}$$

majoriert werden. Der Beitrag der betreffenden Glieder in (17) ist also von der Form

$$O\left(x\sum_1^{\infty}n^{-\frac{5}{2}+\varepsilon}\cdot n^{\frac{1}{2}}\right) = O(x),$$

woraus weiter

$$(18)\int_1^x(P_1(t))^2\,dt = \frac{1}{2\pi^4}\sum_1^{\infty}\left(\frac{r(n)}{n^{\frac{5}{4}}}\right)^2\int_1^x\delta t^{\frac{3}{4}}e^{2\pi i\sqrt{nt}}\cdot\delta t^{\frac{3}{4}}e^{-2\pi i\sqrt{nt}}\,dt + O(x^{\frac{5}{4}+\varepsilon})$$

folgt.

Für $n \geqq x$ setzen wir hier

$$\delta t^{\frac{3}{4}}e^{+2\pi i\sqrt{nt}} = O(t^{\frac{3}{4}});$$

für $n < x$ dagegen wenden wir die Identität

$$\delta f(t) = f(t+1) - f(t) = f'(t) + \int_t^{t+1}du\int_t^u f''(v)\,dv$$

an und erhalten

$$\delta t^{\frac{3}{4}}e^{2\pi i\sqrt{nt}} = \left(\pi i n^{\frac{1}{2}}t^{\frac{1}{4}} + \frac{3}{4}t^{-\frac{1}{4}}\right)e^{2\pi i\sqrt{nt}} + O(nt^{-\frac{1}{4}})$$
$$= \pi i n^{\frac{1}{2}}t^{\frac{1}{4}}e^{2\pi i\sqrt{nt}} + O(nt^{-\frac{1}{4}}),$$

$$\delta t^{\frac{3}{4}}e^{2\pi i\sqrt{nt}}\cdot\delta t^{\frac{3}{4}}e^{-2\pi i\sqrt{nt}} = \pi^2 n t^{\frac{1}{2}} + O(n^{\frac{3}{2}} + n^2 t^{-\frac{1}{2}})$$
$$= \pi^2 n t^{\frac{1}{2}} + O(n^{\frac{3}{2}}x^{\frac{1}{2}}t^{-\frac{1}{2}}),$$

wo die Abschätzungen gleichmäßig für $n < x$, $1 \leqq t \leqq x$ gelten. Durch Einsetzung in (18) ergibt sich

$$\begin{aligned}(19)\quad \int_1^x (P_1(t))^2 dt &= \frac{1}{2\pi^4} \sum_{n<x} \left(\frac{r(n)}{n^{\frac{5}{4}}}\right)^2 \left(\frac{2}{3}\pi^2 n x^{\frac{3}{2}} + O(n^{\frac{3}{2}} x)\right) \\ &\qquad + O\Big(\sum_{n \geqq x} n^{-\frac{5}{2}+\varepsilon} \cdot x^{\frac{5}{2}}\Big) + O(x^{\frac{5}{4}+\varepsilon}) \\ &= \frac{x^{\frac{3}{2}}}{3\pi^2} \sum_{n<x} \left(\frac{r(n)}{n^{\frac{3}{4}}}\right)^2 + O(x^{\frac{5}{4}+\varepsilon}) \\ &= \frac{x^{\frac{3}{2}}}{3\pi^2} \sum_1^\infty \left(\frac{r(n)}{n^{\frac{3}{4}}}\right)^2 + O(x^{\frac{5}{4}+\varepsilon}).\end{aligned}$$

Endlich erhalten wir also aus (9)

$$\int_1^x (P(t))^2 dt = \beta x^{\frac{3}{2}} + \int_1^x P_1(t) \cdot O(t^\varepsilon) dt + O(x^{\frac{5}{4}+\varepsilon}),$$

wo β durch (6) gegeben ist. Aus (19) folgt aber nach der Schwarzschen Ungleichung

$$\int_1^x P_1(t) \cdot O(t^\varepsilon) dt = O\left(\sqrt{\int_1^x (P_1(t))^2 dt \cdot \int_1^x O(t^{2\varepsilon}) dt}\right) = O(x^{\frac{5}{4}+\varepsilon})$$

und hieraus

$$\int_1^x P(t)^2 dt = \beta x^{\frac{3}{2}} + O(x^{\frac{5}{4}+\varepsilon}).$$

§ 2.

Um die analogen Resultate für das Restglied $\Delta(x)$ in dem Dirichletschen Teilerproblem ableiten zu können, brauchen wir einen Ausdruck von der Form (8) für das Integral

$$\int_0^t D(u) du = \sum_{n \leqq t} (t-n) d(n),$$

woraus dann der Satz in derselben Weise wie oben durch asymptotische Entwicklung und Differenzbildung abgeleitet werden könnte.

Ein solcher Ausdruck ließe sich zwar ohne Schwierigkeit durch Benutzung der in den oben angeführten Abhandlungen dargestellten Hardyschen Methoden aufstellen; auch gibt E. Landau in einer Arbeit[6]) einige ähnliche Formeln. In keiner von diesen Arbeiten findet sich jedoch das

[6]) Über Dirichlets Teilerproblem, Münchener Sitzungsberichte 1915, S. 317–328.

genaue Gegenstück von (8), dagegen gibt Voronoï[7]) die asymptotische Entwicklung des gesuchten Ausdruckes in einer Form, die wir gerade anwenden können. Er beweist nämlich folgendes: Für jedes $k = 0, 1, 2, \ldots$ und $m = 1, 2, \ldots$ gilt

$$\sum_{n \leqq t} d(n) \frac{(t-n)^k}{k!} = \int_0^t \frac{(t-n)^k}{k!} (\log u + 2C)\, du + \sum_{\lambda=0}^{k} \zeta^2(-\lambda) \frac{(-1)^\lambda}{\lambda!} \frac{t^{k-\lambda}}{(k-\lambda)}$$

$$+ \frac{1}{2} 0^k d(t) + 2\sqrt{\pi} \sum_{\lambda=0}^{m-1} t^{\frac{k-\lambda}{2}+\frac{1}{4}} \frac{(2k+2\lambda+1)(2k+2\lambda-1)\ldots(2k+3-2\lambda)}{2^{4\lambda}\cdot\lambda!}$$

$$\cdot \sum_{n=1}^{\infty} d(n) \frac{\cos\left[4\pi\sqrt{nt} + \frac{\pi}{2}\left(\lambda - k - \frac{1}{2}\right)\right]}{(4\pi^2 n)^{\frac{k+\lambda}{2}+\frac{3}{4}}} + O\left(t^{\frac{k-m}{2}+\frac{1}{4}}\right),$$

wo $0^0 = 1$ ist und $d(t)$ für nicht ganzzahliges t Null bedeutet. Setzen wir in dieser Beziehung $k = 1$ und $m = 2$, so erhalten wir die gesuchte Formel

$$\int_0^t D(u)\, du = \int_0^t [u \log u + (2C-1)u]\, du + at + b$$

$$+ \frac{t^{\frac{3}{4}}}{2\sqrt{2}\pi^2} \sum_1^{\infty} \frac{d(n)}{n^{\frac{5}{4}}} \sin\left(4\pi\sqrt{nt} - \frac{1}{4}\pi\right) + \frac{15 t^{\frac{1}{4}}}{2^6\sqrt{2}\pi^3} \sum_1^{\infty} \frac{d(n)}{n^{\frac{7}{4}}} \cos\left(4\pi\sqrt{nt} - \frac{1}{4}\pi\right)$$

$$+ O(t^{-\frac{1}{4}}).$$

Von hier aus gelangen wir genau wie in § 1 zu der Gleichung

$$\Delta(t) = \frac{1}{2\sqrt{2}\pi^2} \sum_1^{\infty} \frac{d(n)}{n^{\frac{5}{4}}} \delta t^{\frac{3}{4}} \sin\left(4\pi\sqrt{nt} - \frac{1}{4}\pi\right) + O(t^{\varepsilon}) = \Delta_1(t) + O(t^{\varepsilon})$$

und es ergibt sich weiter

$$\int_1^x (\Delta(t))^2 dt = \int_1^x (\Delta(t))^2 dt + \int_1^x \Delta_1(t) \cdot O(t^{\varepsilon})\, dt + O(x^{1+\varepsilon}),$$

$$\int_1^x (\Delta_1(t))^2 dt = \frac{1}{8\pi^4} \sum_{m=1}^{\infty} \sum_{n=1}^{\infty} \frac{d(m)\, d(n)}{(mn)^{\frac{5}{4}}} \int_1^x \delta t^{\frac{3}{4}} \sin\left(4\pi\sqrt{mt} - \frac{1}{4}\pi\right)$$

$$\cdot \delta t^{\frac{3}{4}} \sin\left(4\pi\sqrt{nt} - \frac{1}{4}\pi\right) dt.$$

---

[7]) Sur une fonction transcendente et ses applications à la sommation de quelques séries, Ann. de l'Ecole Normale (3) 21 (1904), S. 207–268 und S. 459–534. Vgl. insbesondere S. 218, wo jedoch im Wortlaut des Satzes ein Druckfehler vorkommt, der bei dem Beweise (vgl. S. 499 und 514) berichtigt ist.

Die Abschätzungen des vorigen Paragraphen lassen sich wieder durchführen mit den trivialen Modifikationen, die durch die Verschiedenheit einiger Konstanten bedingt werden. Das Resultat wird

$$\int_1^x (\Delta(t))^2 dt = \frac{x^{\frac{3}{2}}}{6\pi^2} \sum_1^x \left(\frac{d(n)}{n^{\frac{3}{4}}}\right)^2 + O(x^{\frac{5}{4}+\varepsilon}).$$

Damit sind also die beiden Gleichungen (5) bewiesen; mit Hilfe der Schwarzschen Ungleichung folgen daraus unmittelbar die Gleichungen (7).

(Eingegangen am 18. Oktober 1921.)

16.

Contributions to the analytic theory of numbers

Proc. 5th Scand. Math. Congress, Helsingfors 1922, 266–272

1. Let us denote by $f(x)$ a real function of x, with the period 2π, the square of which is integrable in the sense given by LEBESGUE. In the corresponding FOURIER series

$$f(x) \sim \sum_{1}^{\infty} (a_n \cos nx + b_n \sin nx) \tag{1}$$

we suppose the constant term equal to zero. Then it is well known that we have

$$\frac{1}{2\pi} \int_{-\pi}^{\pi} f^2(t)\, dt = \frac{1}{2} \sum_{1}^{\infty} (a_n^2 + b_n^2),$$

so that the average value of the square of $f(x)$ in any interval of length 2π can be simply expressed in terms of the coefficients of the series (1).

If, on the other hand, we consider a function $f(x)$ represented by or at least formally connected with a trigonometric series of the more general type

$$\sum_{1}^{\infty} (a_n \cos \lambda_n x + b_n \sin \lambda_n x), \qquad (0 < \lambda_1 < \lambda_2 < \cdots,\ \lambda_n \to \infty), \tag{2}$$

we cannot in general expect to have an equally simple formula for a finite interval; it is, however, in many cases possible to prove the relation

$$\lim_{x \to \infty} \frac{1}{2x} \int_{-x}^{x} f^2(t)\, dt = \frac{1}{2} \sum_{1}^{\infty} (a_n^2 + b_n^2), \tag{3}$$

which may be said to give the average value of $f^2(x)$ for an infinite interval. This relation holds in particular whenever the series (2) is absolutely convergent and represents $f(x)$. The series (2) is the real part of a DIRICHLET's series

$$F(s) = \sum_{1}^{\infty} (a_n + i\, b_n)\, e^{-\lambda_n s},$$

if we put here $s = i\,x$. The corresponding mean value theorem for $F(s)$ is

$$\lim_{x \to \infty} \frac{1}{2x} \int_{-x}^{x} |F(i\,t)|^2\, dt = \sum_{1}^{\infty} (a_n^2 + b_n^2), \tag{4}$$

and it is well known that this is true whenever the DIRICHLET's series representing $F(s)$ is absolutely convergent on the imaginary axis. — A considerable amount of labour has been expended by different authors in order to find more general conditions for the validity of the relations (3) and (4). I am not going to speak here of the general theorems obtained in this way; my object is only to call attention to a certain type of problems occurring in the Analytic Theory of Numbers, where we are concerned with series of the type (2), and where relations more or less corresponding to (3) may be proved, though it does not seem possible to apply any of the general theorems just mentioned.

In some of the most important problems of the Analytic Theory of Numbers it is demanded to find asymptotic representations of certain arithmetical functions for large values of the variable. The most famous example is afforded by the „Prime Number Theorem", which asserts that the number $\pi(x)$ of primes less than x is asymptotically equal to $Li(x)$, *i. e.* that $Li(x)$ is the „dominating term" of the function $\pi(x)$. As soon as we are able to write down like this the dominating term of an arithmetical function, the question arises as to whether it is possible or not to assign an upper limit to the error committed by representing the function by its dominating term, and, in particular, to determine the „true maximum order" of the error. Problems of this type very often appear to be extremely difficult to deal with, and in many cases still remain unsolved. On the other hand, it is sometimes possible to deduce „explicit formulæ" for the error terms, and these formulæ contain as their essential part certain trigonometric expansions of the type (2). The error term being a function with an infinite number of discontinuities, the corresponding trigonometric series cannot, in general, be absolutely convergent (in some cases it is not even possible to prove that they are *convergent,* but only that they are *summable*), and this of course makes it a very difficult task to use them for determination of the maximum order of the error. If we are able to prove, for such a series, the relation corresponding to (3), this will give us at least some information as to the „average order" of the error, and this method often enables us to show that the error is *in general* of a lower order than what can be proved to be *universally* the case. Sometimes it is even possible to show that the „average order" is effectively smaller than the „maximum order"; in such cases we may infer that the graph of the absolute value of the error possesses an infinite number of very scarcely occurring maxima, where it raises much higher than usually above the axis of x.

In the sequel, we shall be concerned with three essentially different topics, *viz.* 1. The theory of the distribution of primes; 2. A problem concerning the divisors of integral numbers; 3. The theory of ideals in a quadratic field. In all these problems, we have to deal with trigonometric series of the type (2) and it is possible to deduce relations corresponding to (3) and, from them, to find new properties of the arithmetical functions. — It is, of course, possible to prove general theorems, from which these applications would follow as particular cases. It seems, however, that these general theorems cannot, so far, be made sufficiently general to deserve much interest for their own sake; this is the reason why I prefer to give only the applications.

2. The first example of an explicit formula for an arithmetical function was originally given by RIEMANN in his famous work on the distribution of primes, though a rigorous proof was not obtained until long afterwards, when the matter was brought into contact with an analogous formula due to VON MANGOLDT. We shall here principally consider the latter, which is formally simpler; the original RIEMANN formula may, however, be treated in the same way.

We shall assume the Riemann hypothesis concerning the roots of the zeta function throughout this paragraph[1]; then V. MANGOLDT's formula may be written

[1]) We shall assume also that $\zeta(s)$ has no multiple zeros. This is only a matter of formal simplification and makes no real difference.

$$\psi(x) = x - \sqrt{x} \sum_{\gamma > 0} \left(\frac{x^{i\gamma}}{\frac{1}{2} + i\gamma} + \frac{x^{-i\gamma}}{\frac{1}{2} - i\gamma} \right) + O(\log x),$$

where

$$\psi(x) = \sum_{p^m \leqq x} \log p \qquad (p \text{ prime}, \ m = 1, 2, \cdots)$$

denotes the arithmetical function first introduced by TSCHEBYSCHEF, and $\frac{1}{2} + i\gamma$ runs through all the zeros of the function $\zeta(s)$ belonging to the *upper half* of the plane of s. Thus the dominating term of the function $\psi(x)$ is x; as for the maximum order of the error the most that is known is that the error term is of the form $O(\sqrt{x} \log^2 x)$. The series occurring in the expression of the error term is readily seen to be of the type (2), with the variable $\log x$ in the place of x. It is possible to prove for this series a relation corresponding to (3), where the mean value is formed for the interval from $\log 2$ to $\log x$, and by re-introducing the variable x, we obtain the relation

$$\lim_{x \to \infty} \frac{1}{\log x} \int_2^x \left(\frac{\psi(t) - t}{t} \right)^2 dt = \sum_{\varrho} \frac{1}{|\varrho|^2},$$

where the last sum is extended over *all* the complex zeros ϱ of $\zeta(s)$. In the same way it may be deduced that

$$\frac{1}{x} \int_2^x |\psi(t) - t| \, dt = O(\sqrt{x}),$$

so that the error term is, *on the average*, of the form $O(\sqrt{x})$, whereas it has been proved by LITTLEWOOD that the relation

$$\psi(x) - x = O(\sqrt{x})$$

is *not* universally true. — For the function

$$f(x) = \sum_{p^m \leqq x} \frac{1}{m}$$

introduced by RIEMANN, we have the corresponding relation

$$\lim_{x \to \infty} \frac{1}{\log x} \int_2^x \left(\frac{f(t) - Li(t)}{Li(t)} \right)^2 dt = \sum_{\varrho} \frac{1}{|\varrho|^2}.$$

3. Denoting by $\sigma_a(n)$ the sum of the a:th powers of the divisors of an integer n, we have for all sufficiently large values of s

$$\sum_1^\infty \frac{\sigma_a(n)}{n^s} = \sum_1^\infty \frac{1}{n^s} \cdot \sum_1^\infty \frac{1}{n^{s-a}} = \zeta(s)\,\zeta(s-a).$$

In this paragraph we shall be concerned with the summatory function

$$S_a(x) = \sum_{n \leqq x} \sigma_a(n);$$

to avoid unnecessary complications we shall restrict ourselves to the case $|a| \leqq 1$. It is easy to find the dominating terms of this function; if we put

$$T_a(x) = \begin{cases} \dfrac{\zeta(1+a)}{1+a} x^{1+a} + \zeta(1-a)\,x - \dfrac{1}{2}\zeta(-a) & \text{for } -1 < a \leqq 1,\ a \neq 0 \\ x \log x + (2C-1)\,x + \dfrac{1}{4} & \text{,, } \quad a = 0, \\ \dfrac{\pi^2}{6} x - \dfrac{1}{2}\log x - \dfrac{1}{2}(C + \log 2\pi) & \text{,, } \quad a = -1, \end{cases}$$

and

$$S_a(x) = T_a(x) + R_a(x),$$

then $T_a(x)$ gives the dominating terms, and it can be shown that the error term $R_a(x)$ is at most of the order

$$O\left(x^{\frac{1}{3-2a}} (\log x)^{1+a}\right) \qquad \text{for } -1 \leqq a \leqq 0,$$

$$O\left(x^{\frac{1}{3+2a}+a} (\log x)^{1-a}\right) \qquad \text{for } \quad 0 \leqq a \leqq 1.$$

For $a = 0$ this was first proved by Voronoï, for $a = \pm 1$ more precise results have been found by Wigert, *viz.*

$$(5) \qquad \begin{cases} \limsup\limits_{x \to \infty} \dfrac{|R_{-1}(x)|}{\log x} \leqq \dfrac{1}{4}, \\ \limsup\limits_{x \to \infty} \dfrac{|R_1(x)|}{x \log x} \leqq \dfrac{1}{4}. \end{cases}$$

By means of the relation

$$\sum_1^\infty \frac{\sigma_a(n)}{n^{\frac{1+a}{2}}} e^{-s\sqrt{n}} = \frac{1}{2\pi i} \int_{3-i\infty}^{3+i\infty} \Gamma(z)\, s^{-z}\, \zeta\left(\frac{1+a+z}{2}\right) \zeta\left(\frac{1-a+z}{2}\right) dz,$$

we can deduce the following explicit formula for the error term:

$$R_a(x) = \frac{x^{\frac{2a+1}{4}}}{\pi\sqrt{2}} \sum_1^\infty \frac{\sigma_a(n)}{n^{\frac{2a+3}{4}}} \cos\left(4\pi\sqrt{nx} - \frac{\pi}{4}\right) -$$

$$-\frac{(2a+1)(2a+3)}{32\pi^2\sqrt{2}} x^{\frac{2a-1}{4}} \sum_1^\infty \frac{\sigma_a(n)}{n^{\frac{2a+5}{4}}} \sin\left(4\pi\sqrt{nx} - \frac{\pi}{4}\right) + O\left(x^{\frac{2a-3}{4}}\right).$$

This relation holds for all non-integral values of x; for integral x the value of the second member is $\frac{1}{2}\left(R_a(x+0) + R_a(x-0)\right)$. In the second member, the first series is simply convergent for $|a| < \frac{1}{2}$; for $\frac{1}{2} \leqq |a| \leqq 1$ we cannot prove that it is convergent, but only that it is summable by CESÀRO's means of the first order. The second series is absolutely convergent for $|a| < \frac{1}{2}$, simply convergent for $\frac{1}{2} \leqq |a| \leqq 1$. The additional term $O\left(x^{\frac{2a-3}{4}}\right)$ denotes a continuous function, which obviously tends to zero as x increases. — For $a = 0$ this expression was found independently by VORONOÏ and HARDY, for $a = \pm 1$ by WALFISZ. Some similar expressions for certain related functions have been given by WIGERT.

As long as $|a| < \frac{1}{2}$, we can obtain for the trigonometric series occurring here mean value theorems of the type characterized above. Thus we get for $|a| < \frac{1}{2}$

$$\int_1^x (R_a(t))^2\,dt \sim \beta x^{\frac{3}{2}+a} \tag{6}$$

with

$$\beta = \frac{1}{(4a+6)\pi^2} \sum_1^\infty \left(\frac{\sigma_a(n)}{n^{\frac{2a+3}{4}}}\right)^2.$$

This gives for $|a| < \frac{1}{2}$

$$\frac{1}{x}\int_1^x |R_a(t)|\,dt = O\left(x^{\frac{2a+1}{4}}\right),$$

so that the error term is, *on the average,* of the form $O\left(x^{\frac{2a+1}{4}}\right)$, while on the other hand it is possible to show that the relation

$$R_a(x) = O\left(x^{\frac{2a+1}{4}}\right)$$

is distinctly false.

For $|a| \geqq \frac{1}{2}$ the infinite series involved in the expression for β is readily seen to be divergent, so that (6) cannot be expected to hold in this case. It is, however, still possible to draw from the explicit formula some information as to the average order of the error; we can prove, *e. g.*, the relations

$$\frac{1}{x}\int_1^x |R_a(t)|\,dt = O\left((\log x)^{1+\frac{a}{2}}\right) \qquad \text{for } -1 \leqq a \leqq -\frac{1}{2},$$

$$\frac{1}{x}\int_1^x |R_a(t)|\,dt = O\left(x^a (\log x)^{1-\frac{a}{2}}\right) \qquad \text{for } \frac{1}{2} \leqq a \leqq 1,$$

which give, in particular, for the average order of $R_{\pm 1}(x)$ a lower limit than that obtained from (5).

4. Let K be a quadratic field of discriminant $\varDelta$; then it is well known that the number $H(x)$ of ideals in K with norm $\leqq x$ has for its dominant term kx, where k is a constant depending on the field. The error term is of the form $O\left(x^{\frac{1}{3}}\right)$, and there is an explicit formula for the error, the principal part of which is

$$\frac{(x|\varDelta|)^{\frac{1}{4}}}{\pi\sqrt{2}}\sum_1^\infty \frac{F(n)}{n^{\frac{3}{4}}}\sin\left(4\pi\sqrt{\frac{nx}{|\varDelta|}}+\frac{\varepsilon}{4}\pi\right),$$

where $F(n)$ denotes the number of representations of n as the norm of an ideal belonging to K, and ε is $+1$ for a real, -1 for an imaginary field. Hence the average value of the error term may be deduced on the same lines as before, and the result is

$$\int_1^x (H(t)-kt)^2\,dt \sim \gamma\, x^{\frac{3}{2}}$$

with

$$\gamma = \frac{\sqrt{|\varDelta|}}{6\pi^2}\sum_1^\infty \left(\frac{F(n)}{n^{\frac{3}{4}}}\right)^2.$$

An immediate corollary is

$$\frac{1}{x}\int_1^x |H(t)-kt|\,dt = O\left(x^{\frac{1}{4}}\right).$$

17.

Ein Satz über Dirichletsche Reihen

Ark. Mat. Astr. Fys. **18** (2), 1–7 (1923)

Mitgeteilt am 11. April 1923 durch I. Bendixson und H. von Koch.

Es sei eine Funktion $f(s)$ durch eine auf der imaginären Achse der s-Ebene unbedingt konvergente Dirichlet'sche Reihe

$$f(s) = \sum_{1}^{\infty} c_n e^{-\lambda_n s} \qquad (c_n = a_n + i b_n) \tag{1}$$

dargestellt. Dann gilt bekanntlich

$$\lim_{x \to \infty} \frac{1}{2x} \int_{-x}^{x} |f(it)|^2 \, dt = \sum_{1}^{\infty} (a_n^2 + b_n^2).$$

Ebenso gilt für den reellen Teil von $f(it)$:

$$\varphi(t) = \sum_{1}^{\infty} (a_n \cos \lambda_n t + b_n \sin \lambda_n t)$$

die folgende Gleichung

$$\lim_{x \to \infty} \frac{1}{x} \int_{-x}^{x} \varphi^2(t) \, dt = \sum_{1}^{\infty} (a_n^2 + b_n^2). \tag{2}$$

Bekanntlich ist eine grosse Arbeit darauf niedergelegt worden, die Voraussetzung der unbedingten Konvergenz von (1) durch allgemeinere Voraussetzungen zu ersetzen; ich erwähne hier nur die zusammenfassende Darstellung von LANDAU[1] sowie die neueren Arbeiten von CARLSON.[2] — Andererseits habe ich[3] für einige spezielle Reihen, die in der analytischen Zahlentheorie auftreten, gewisse zu (2) analoge Formeln abgeleitet. Die entsprechenden DIRICHLET'schen Reihen haben auf der imaginären Achse unendlich viele Singularitäten, und es erscheint überhaupt schwierig, die erwähnten allgemeinen Sätze hier zu benutzen; dafür kann man aber über die Eigenschaften der Randfunktion hier viel speziellere Voraussetzungen als im allgemeinen Falle machen. Ich habe bereits angedeutet[3], dass meine speziellen Überlegungen auch etwas allgemeiner formuliert werden könnten, und ich möchte nun diese Andeutung etwas näher ausführen, indem ich den folgenden Satz beweisen werde.

Es sei die Dirichlet'sche Reihe (1) *für* $\sigma > 0$ *konvergent, und es seien noch die folgenden Voraussetzungen erfüllt:*

a) *Der reelle Teil* $Rf(s)$ *sei in jedem endlichen Teil der Viertelebene* $\sigma > 0$, $t > 1$ *beschränkt. Daraus folgt schon* (*nach Fatou*) *die Existenz einer Randfunktion*

$$\varphi(t) = \lim_{\sigma \to 0} Rf(s)$$

für »fast alle» $t > 1$.

[1] E. LANDAU, Handbuch der Lehre von der Verteilung der Primzahlen, Leipzig und Berlin 1909, Band II.

[2] F. CARLSON, Sur les séries de Dirichlet, C. r. 172 (1921) p. 838; Contribution à la théorie des séries de Dirichlet, Note I, Arkiv för matematik, bd 16 n:o 18 (1922). — Ich benutze diese Gelegenheit, um die folgende Bemerkung zu machen. In einer in diesem Arkiv veröffentlichten Arbeit (Un théorème sur les séries de Dirichlet et son application, bd 13 n:o 22, 1918) habe ich einen Beweis des folgenden, wie ich damals glaubte, neuen Satzes gegeben: »Eine ganze Funktion $\varphi(z)$, die für $z = 1, 2, \ldots$ gleich Null wird, und die für hinreichend grosse $|z|$ der Bedingung $|\varphi(z)| < e^{k|z|}$ mit $k < \pi$ genügt, ist identisch gleich Null». Dieser Satz ist aber als Spezialfall in einem Satz enthalten, der auf p. 58 der Dissertation des Herrn CARLSON steht: Sur une classe de séries de Taylor, Upsala 1914.

[3] H. CRAMÉR, Contributions to the Analytic Theory of Numbers, Congrès des Mathématiques à Helsingfors 1922.

b) *Die Reihe* $\sum\limits_1^\infty (a_n^2 + b_n^2)$ *sei konvergent,* $\sum\limits_1^\infty \frac{|c_n|}{\lambda_n^k}$ *sei auch für hinreichend grosse* $k > 0$ *konvergent.*

c) *Zu jedem* $\varepsilon > 0$ *gebe es ein* $h > 0$ *und ein* $t_0 > 0$, *so dass für* $t > t_0$, $0 < u < h$,

$$|\varphi(t+u) - \varphi(t)| < \varepsilon$$

gilt.

Dann folgt

$$\lim_{x\to\infty} \frac{1}{x}\int_1^x \varphi^2(t)\,dt = \frac{1}{2}\sum_1^\infty (a_n^2 + b_n^2). \tag{3}$$

Es ist natürlich wesentlich, dass wir die Beschränktheit von $Rf(s)$ nur in jedem *endlichen* Teil von $\sigma > 0$, $t > 1$ verlangen. Beschränktheit in der *ganzen* Viertelebene würde ja eine viel engere Voraussetzung sein. Wird die Viertelebene durch die Halbebene $\sigma > 0$ ersetzt und die Voraussetzung c) entsprechend modifiziert, so kommt als Resultat genau (2) hervor; (3) ist aber für die obengenannten zahlentheoretischen Anwendungen wichtiger. Es muss bemerkt werden, dass jene Anwendungen nicht *unmittelbar* aus dem Satze folgen, da die Voraussetzung c) dort im Allgemeinen nicht genau in der obigen Fassung erfüllt ist, sondern durch die Einführung einer gewissen Abhängigkeit zwischen h und t abgeändert werden muss. — Die Notwendigkeit, c) oder eine ähnliche Voraussetzung einzuführen, vermindert natürlich den Wert des Satzes, da man im Allgemeinen die Randfunktion nicht hinreichend genau kennt. Es ist einleuchtend, dass durch c) die Stetigkeit der Randfunktion $\varphi(t)$ *nicht* gefordert wird.

Ich gebe unten den Beweis des Satzes für den Fall, dass $\sum \frac{|c_n|}{\lambda_n}$ konvergiert, so dass $k = 1$ gewählt werden kann. Im allgemeinen Falle verläuft der Beweis genau analog, worüber ich auch eine Andeutung geben werde.

Es sei also $F(s) = \sum \frac{c_n}{\lambda_n} e^{-\lambda_n s}$ für $\sigma \geqq 0$ unbedingt konvergent. Wird $IF(it) = \Phi(t)$ gesetzt, so folgt aus a) für alle $t > 1$, $h > 0$:

$$\int_t^{t+h} \varphi(u)\,du = -(\Phi(t+h) - \Phi(t))$$

und hieraus

$$(4) \qquad \varphi(t) = -\frac{\Phi(t+h) - \Phi(t)}{h} - \int_0^h \frac{\varphi(t+u) - \varphi(t)}{h}\,du$$

$$= \qquad A \qquad + \qquad B\,,$$

$$\frac{1}{x}\int_1^x \varphi^2(t)\,dt = \frac{1}{x}\int_1^x A^2 dt + \frac{2}{x}\int_1^x AB\,dt + \frac{1}{x}\int_1^x B^2 dt.$$

Aus der Voraussetzung c) folgt nun

$$\limsup_{x\to\infty} \frac{1}{x}\int_1^x B^2 dt < \varepsilon(h),$$

wo $\varepsilon(h)$ mit h gegen Null strebt. Wegen

$$\left(\int_1^x AB dt\right)^2 \leqq \int_1^x A^2 dt \,.\, \int_1^x B^2 dt$$

genügt es, noch folgendes zu beweisen:

Zu jedem $\varepsilon > 0$ *können* x_0 *und* h *so gefunden werden, dass* $x_0 > 0,\ 0 < h < \varepsilon$ *ist, und dass*

$$\left|\frac{1}{x}\int_1^x A^2 dt - \frac{1}{2}\sum_1^\infty (a_n^2 + b_n^2)\right| < \varepsilon$$

für alle $x > x_0$ *gilt.*

Offenbar ist

$$\int_1^x A^2 dt = J_1 + R(J_2),$$

wo

$$J_1 = \frac{1}{2h^2}\sum_{m,n}\frac{c_m \bar{c}_n}{\lambda_m \lambda_n}(e^{-\lambda_m ih}-1)(e^{\lambda_n ih}-1)\int_1^x e^{-(\lambda_m-\lambda_n)it}\,dt,$$

$$J_2 = -\frac{1}{2h^2}\sum_{m,n}\frac{c_m c_n}{\lambda_m \lambda_n}(e^{-\lambda_m ih}-1)(e^{-\lambda_n ih}-1)\int_1^x e^{-(\lambda_m+\lambda_n)it}\,dt$$

gesetzt ist. Weiter haben wir

$$J_1 = \frac{x-1}{2}\sum_1^\infty \frac{4\sin^2\frac{\lambda_n h}{2}}{\lambda_n^2 h^2}|c_n|^2 +$$

$$+\frac{1}{2h^2}\sum_{m \neq n}\frac{c_m \bar{c}_n}{\lambda_m \lambda_n}(e^{-\lambda_m ih}-1)(e^{\lambda_n ih}-1)\int_1^x e^{-(\lambda_m-\lambda_n)it}\,dt =$$

$$= \frac{x-1}{2}\sum_1^\infty \frac{\sin^2\frac{\lambda_n h}{2}}{\frac{1}{4}\lambda_n^2 h^2}|c_n|^2 + Q.$$

Für die Abschätzung von Q führen wir eine positive Zahl p ein, die mit h in unten näher anzugebenden Weise gegen Null strebt, und trennen die Glieder der Doppelsumme in zwei Gruppen, je nachdem $|\lambda_m - \lambda_n| < p$ oder $|\lambda_m - \lambda_n| \geqq p$ gilt. Hierdurch folgt

$$|Q| < \frac{4}{h^2 p}\sum_{m,n}\frac{|c_m c_n|}{\lambda_m \lambda_n} + \frac{2x}{h^2}\sum_{0<|\lambda_m-\lambda_n|<p}\frac{|c_m c_n|}{\lambda_m \lambda_n}$$

$$< \frac{K}{h^2 p} + \frac{\delta(p)}{h^2}x,$$

wo K (wie auch unten) eine von h, p und x unabhängige Konstante bezeichnet, und $\delta(p)$ mit p gegen Null strebt, von x aber nicht abhängt. In analoger — aber einfacherer — Weise findet man leicht

$$|J_2| < \frac{K}{h^2};$$

hieraus folgt schliesslich

$$\left|\frac{1}{x}\int_1^x A^2\,dt - \frac{1}{2}\sum_1^\infty (a_n^2 + b_n^2)\right| <$$

$$< \frac{1}{2}\sum_1^\infty \left(1 - \frac{\sin^2 \frac{1}{2}\lambda_n h}{\frac{1}{4}\lambda_n^2 h^2}\right)|c_n|^2 + \frac{K}{h^2 p x} + \frac{\delta(p)}{h^2}.$$

Es sei jetzt $\varepsilon > 0$ beliebig klein gegeben. Da nach Voraussetzung $\Sigma|c_n|^2$ konvergiert, kann ein positives $h < \varepsilon$ derart gewählt werden, dass das erste Glied rechts kleiner als $\frac{1}{3}\varepsilon$ wird. Dann wird p so klein gewählt, dass das dritte Glied kleiner $\frac{1}{3}\varepsilon$ wird; dies ist möglich, da ja $\lim\limits_{p\to 0}\delta(p) = 0$ ist. Schliesslich wählen wir x_0 so, dass $\frac{K}{h^2 p x} < \frac{1}{3}\varepsilon$ für $x > x_0$ wird; dann wird die ganze rechte Seite $< \varepsilon$, und nach dem oben gesagten folgt daraus unser Satz für den Fall $k = 1$.

Zuletzt machen wir einige Bemerkungen über den Fall eines beliebigen k. Es muss dann nur die Identität (4) in geeigneter Weise verallgemeinert werden, sonst geht alles genau wie oben. Nehmen wir z. B. $k = 2$ an; wir setzen dann $F(s) = \sum \frac{c_n}{\lambda_n^2} e^{-\lambda_n s}$, wo die Reihe für $\sigma \geqq 0$ unbedingt konvergiert. Setzen wir weiter $RF(it) = \Phi(t)$, so wird (4) durch die Identität

$$\varphi(t) = \frac{\Phi(t+2h) - 2\Phi(t+h) + \Phi(t)}{h^2} -$$

$$- \int_0^h \int_0^h \frac{\varphi(t+u+v) - \varphi(t)}{h^2} \, du \, dv$$

ersetzt, die sich in derselben Weise wie oben behandeln lässt. Der Fall eines beliebigen k bringt keine neue Schwierigkeiten.

Tryckt den 20 december 1923.

Uppsala 1923. Almqvist & Wiksells Boktryckeri-A.-B.

18.

Das Gesetz von Gauss und die Theorie des Risikos

Skand. Aktuarietidskr. 6, 209–237 (1923)

Einleitung.

1. Die Theorie des Risikos in der Lebensversicherung hat für die Praxis bisher wohl nie die Bedeutung erlangt, die man von verschiedenen Seiten für sie beansprucht hat. Die Beurteilung der Stabilität einer Versicherungsanstalt gegen Sterblichkeitsschwankungen, die Festsetznng von Höchstbeträgen auf eigenem Risiko, die Bildung und Verwaltung der Sicherheitsreserven — alles dies wird wohl im allgemeinen nach verhältnissmässig wenig durchgedachten praktischen Erwägungen geregelt, ohne dass dabei die Theorie zur Sprache kommen darf.

Es ist dies nach meiner Ansicht ein Übelstand, dessen Grund fast ausschliesslich in Mangeln der Theorie gesucht werden muss. Erstens hat die Theorie des Risikos die mannigfaltigen Verhältnisse der Praxis, die auf jene Probleme einwirken, meist nicht hinreichend berücksichtigt. Die hier vorliegenden Probleme können unmöglich durch allgemeine und einfache Formeln gelöst werden, die etwa nur von der Verteilung der Versicherungssummen im Bestande abhängen. Eine praktisch verwendbare Lösung muss notwendig alle eigenartige Züge des vorgelegten Falles berücksichtigen: die Rechnungsgrundlagen, die Art der Gewinnverteilung, die benutzte Rückversicherungsmethode u. s. w.

Zweitens aber, und das ist für meinen jetzigen Zweck wichtiger, lässt die Theorie des Risikos auch in rein mathematischer Hinsicht viel zu wünschen übrig. Ihre Entwicklung ist immer von aus der Fehlertheorie übernommenen Anschauungen stark beeinflusst gewesen, und es wurden dabei oft gewisse für die Fehlertheorie fundamentale Sätze und Begriffe ohne genügende Kritik auch in der Theorie des Risikos benutzt. Ich denke hier in erster Linie auf den Begriff des *mittleren Fehlers* sowie auf den Satz, dass die Summe einer grossen Anzahl unabhängiger Elementarfehler dem *Fehlergesetz von* Gauss folgt. Die ganze Theorie des Risikos, wie sie z. B. in den bekannten Bohlmann'schen Berichten[1] zusammengefasst vorliegt, ist in der Tat auf jenem Begriff und jenem Satz aufgebaut.[2] Die Rolle der Elementarfehler wird hierbei von den einzelnen Versicherungen übernommen, und für den Totalgewinn in dem ganzen Bestande, der sich aus den Gewinnen bez. Verlusten bei den einzelnen Versicherungen additiv zusammensetzt, wird dann die Gültigkeit des Gauss'schen Gesetzes behauptet. Das mittlere Risiko einer Versicherung, bez. eines Bestandes, in genauer Anschliessung an die Fehlertheorie zu definieren ist ein leichtes, und man glaubt dann aus dem Gesetz von Gauss ohne weiteres schliessen zu können, dass Abweichungen, die das drei- oder vierfache des mittleren Risikos übersteigen, bereits »sehr» selten sind. Diese Behauptung mag nun richtig sein oder nicht; est ist aber klar, dass die hier skizzierte Art der Begründung gänzlich ungenügend ist. Denn der herangezogene Satz der Fehlertheorie besagt nur, dass *in der Grenze* (für eine unendliche Anzahl Elementarfehler) das Gesetz von Gauss gilt; wenn man aber keine Vorstellung von der Grösse der Approximation bei einer endlichen Anzahl Elementarfehler hat, so darf man in einem vorgelegten Falle in der Theorie des Risikos keineswegs hieraus schliessen, dass die Wahrscheinlichkeit einer gegebenen Abweichung »sehr» klein ist, auch wenn das Gauss'sche Gesetz

[1] Lebensversicherungs-Mathematik, Encykl. d. matematischen Wissenschaften I D 4 b; Die Theorie des mittleren Risikos in der Lebensversicherung, Verhandlungen des 6. Kongresses für Versicherungs-Wissenschaft, Wien 1909.

[2] Von den Theorien, die z. B. den Begriff des *durchschnittlichen* Risikos benutzen, wird hier abgesehen. Sie werden von den gleichen Bemerkungen getroffen.

eine Zahl mit beliebig vielen Nullen hinter dem Dezimalkomma ergibt. Die Ableitungen des Gauss'schen Gesetzes läufen nun sämmtlich darauf hinaus, das Zustandekommen eben dieser Form der Fehlerkurve unter möglichst allgemeinen Voraussetzungen zu *erklären;* die Frage nach dem Grade der Approximation bleibt aber meistens unberücksichtigt. Ich möchte im Folgenden eben diese Frage in den Vordergrund stellen und so einen kleinen Beitrag zur Vertiefung der Theorie geben. Bei diesen Untersuchungen drängt sich die Überzeugung auf, dass der Begriff des mittleren Risikos und der Satz von der Gültigkeit des Gauss'schen Gesetzes als Fundamente für eine befriedigende Theorie des Risikos nicht ausreichen. Es erscheint vielmehr als ein selbständiges Problem, und zwar als eine der Hauptaufgaben jener Theorie, die Eigenschaften der auftretenden Wahrscheinlichkeitsfunktionen zu untersuchen. Auch hierüber werde ich im Folgenden etwas zu sagen haben. — In einer späteren Arbeit hoffe ich, einige Anwendungen der unten abgeleiteten Resultate zeigen zu können.

2. Es seien $x_1, x_2, \ldots x_n$ Grössen, die vom Zufall abhängen, von einander aber im gewöhnlichen Sinne des Wortes unabhängig sind. Ich werde immer voraussetzen, dass jedes x_ν den Mittelwert Null hat, was bekanntlich nur eine formelle Vereinfachung bedeutet. Wird der Mittelwert des Quadrates x_ν^2 durch ϱ_ν bezeichnet, so lässt sich *unter gewissen Voraussetzungen* die Wahrscheinlichkeit der Beziehung

$$(1) \qquad x_1 + x_2 + \cdots + x_n \leqq x \sqrt{2(\varrho_1 + \varrho_2 + \cdots + \varrho_n)}$$

approximativ durch das Fehlerintegral

$$\Phi(x) = \frac{1}{\sqrt{\pi}} \int_{-\infty}^{x} e^{-t^2} dt$$

ausdrücken. Dieser Satz, dessen Inhalt auf Laplace zurückgeht, nimmt in der Fehlertheorie eine zentrale Stellung ein; für die Theorie des Risikos ist er aber, auch wenn die nötigen weiteren Voraussetzungen präzisiert werden, noch nicht hinreichend genau formuliert. Ein strenger Beweis des Satzes ist wohl zuerst von TSCHEBYSCHEF gegeben worden; von neueren Bearbeitern nenne ich LIAPOUNOFF[1], v. MISES[2], PÓLYA[3] und LINDEBERG[4]; der letztgenannte hat unter sehr allgemeinen Voraussetzungen einen Beweis von überraschender Einfachheit gegeben. Unter den Methoden, welche diese Verfasser benutzt haben, dürfte diejenige von LIAPOUNOFF die genaueste Abschätzung des Approximationsgrades liefern, sie scheint aber verhältnissmässig wenig bekannt zu sein. Ich werde hier diese Methode in vereinfachter Form benutzen und weiterführen, zuerst aber möchte ich eine kurze Zusammenfassung meiner Resultate geben.

Es sei t eine beliebige reelle Zahl. Wenn die Wahrscheinlichkeit der Beziehung

$$x_\nu \leqq t$$

durch $V_\nu(t)$ gegeben wird, so soll $V_\nu(t)$ *Verteilungsfunktion* von x_ν genannt werden. Offenbar muss dann $V_\nu(t)$ eine nirgends abnehmende Funktion von t sein, die für $t \to -\infty$ den Grenzwert 0, für $t \to +\infty$ den Grenzwert 1 hat. Aus unseren Voraussetzungen folgt

$$\int_{-\infty}^{\infty} d\,V_\nu(t) = 1, \quad \int_{-\infty}^{\infty} t\,d\,V_\nu(t) = 0, \quad \int_{-\infty}^{\infty} t^2\,d\,V_\nu(t) = \varrho_\nu,$$

wo die Integrale in STIELTJES'schem Sinne genommen sind. Wir setzen weiter

---

[1] Sur une proposition de la théorie des probabilités, Bull. de l'Acad. St. Petersbourg vol. 13 (1900); Nouvelle forme du théoreme sur la limite de probabilité, Mémoires de d'Acad. St. Petersbourg, vol. 12 (1901).

[2] Fundamentalsätze der Wahrscheinlichkeitsrechnung, Math. Zeitschrift, Bd 4 (1919).

[3] Über den zentralen Grenzwertsatz der Wahrscheinlichkeitsrechnung und das Momentenproblem, Math. Zeitschr. Bd 8 (1920).

[4] Über des GAUSS'sche Fehlergesetz, Skandinavisk Aktuarietidskrift 1922; Eine neue Herleitung des Exponentialgesetzes in der Wahrscheinlichkeitsrechnung, Math. Zeitschr. Bd 15 (1922).

$$\int_{-\infty}^{\infty} t^3\, d\, V_\nu(t) = \sigma_\nu, \quad \int_{-\infty}^{\infty} |\, t\,|^3\, d\, V_\nu(t) = \bar{\sigma}_\nu, \quad \int_{-\infty}^{\infty} t^4\, d\, V_\nu(t) = \tau_\nu,$$

$$\varrho = \sum_1^n \varrho_\nu, \qquad \sigma = \sum_1^n \sigma_\nu,$$

$$\bar{\sigma} = \sum_1^n \bar{\sigma}_\nu, \qquad \tau = \sum_1^n \tau_\nu,$$

indem wir die Existenz jener Integrale voraussetzen. Unsere Überlegungen lassen sich zwar zum Teil auch unter allgemeineren Voraussetzungen durchführen, eine solche Allgemeinheit dürfte aber für die Theorie des Risikos ziemlich belanglos sein.

Die Verteilungsfunktion der Summe $x_1 + x_2 + \cdots + x_n$ sei durch $V(t)$ bezeichnet, und es sei

$$S(x) = V(x\sqrt{2\varrho}),$$

so dass $S(x)$ die Wahrscheinlichkeit der Beziehung (1) bedeutet. Es muss also untersucht werden, ob die Funktion $S(x)$ durch das Fehlerintegral $\Phi(x)$ angenähert dargestellt werden kann, und was wir über den Grad der Approximation aussagen können.

Bekanntlich lässt sich $S(x)$ formal in eine Reihe, die sog. BRUNSsche Reihe, entwickeln, die in folgender Weise anfängt:

$$S(x) = \Phi(x) - \frac{\sigma}{3!\,(2\varrho)^{\frac{3}{2}}}\, \Phi'''(x) + \frac{\tau - 3\,\Sigma\varrho_\nu^2}{4!\,(2\varrho)^2}\, \Phi''''(x) + \cdots$$

$$= \frac{1}{\sqrt{\pi}} \int_{-\infty}^{x} e^{-t^2} dt - \frac{\sigma}{3!\,(2\varrho)^{\frac{3}{2}}} \cdot \frac{4x^2 - 2}{\sqrt{\pi}}\, e^{-x^2} +$$

$$+ \frac{\tau - 3\,\Sigma\varrho_\nu^3}{4!\,(2\varrho)^2} \cdot \frac{-8x^3 + 12x}{\sqrt{\pi}}\, e^{-x^2} + \cdots.$$

Diese Reihe ist jedoch nicht unter hinreichend allgemeinen Bedingungen konvergent[1], und eine anwendbare Abschätzung des Restes liegt soweit mir bekannt nicht vor. Schreibt man aber die Reihe für den speziellen Fall auf, dass alle n Verteilungsfunktionen $V_\nu(t)$ einander gleich sind, so zeigt es sich dass die Koeffizienten der Reihenglieder, als Funktionen von n betrachtet, von abnehmender Grössenordnung sind. Es gilt z. B.

$$\frac{\sigma}{\varrho^{\frac{3}{2}}} = \frac{\sigma_1}{\varrho_1^{\frac{3}{2}}} \cdot \frac{1}{\sqrt{n}}, \qquad \frac{\tau}{\varrho^2} = \frac{\tau_1}{\varrho_1^2} \cdot \frac{1}{n}.$$

Man wird hiernach vermuten können, dass auch im allgemeinen Falle die Grössenordnung der Differenz $S(x) - \Phi(x)$ mit den ersten Koeffizienten der obigen Reihe in irgend einem Zusammenhang stehe. Diese Vermutung wird nun gewissermassen durch die beiden folgenden Sätze bestätigt.

Satz 1. *Es bezeichne* η *die grösste der beiden Zahlen 1 und* $\log \frac{\varrho^{\frac{3}{2}}}{\sigma}$. *Dann gilt für alle Werte von* x

$$|S(x) - \Phi(x)| < 6\eta \frac{\bar{\sigma}}{\varrho^{\frac{3}{2}}}.$$

Es wurde schon von Liapounoff gezeigt, dass seine Methode eine Schranke von der Ordnung[2] $\frac{\bar{\sigma}}{\varrho^{\frac{3}{2}}} \log \frac{\varrho^{\frac{3}{2}}}{\sigma}$ ergibt; er hat aber das nicht näher ausgeführt.

---

[1] Das Konvergenzproblem dieser und analoger Reihen wurde neulich von Herrn Phragmén in zwei Vorträgen in Svenska Aktuarieföreningen behandelt.

[2] Liapounoff behandelt auch den Fall, dass nicht die Existenz der Momente dritter Ordnung, sondern nur die der Momente von der Ordnung $2 + \alpha$, bei irgend einem $\alpha > 0$, vorausgesetzt wird.

Satz 2. *Es gibt Werte von x, für welche*

$$|S(x)-\Phi(x)| > \frac{1}{24\sqrt{\pi}} \cdot \frac{|\sigma|}{\varrho^{\frac{3}{2}}} - 2\frac{\tau}{\varrho^2}$$

gilt.

Wie man sofort sieht, ist dieser Satz unter gewissen Umständen trivial. Wendet man ihn z. B. auf den soeben besprochenen Spezialfall an, so erhält man in Falle $\sigma_1=0$ nur das triviale Ergebniss $|S(x)-\Phi(x)| > -2\frac{\tau}{\varrho^2}$; wenn aber $\sigma_1 \neq 0$ ist, so besagt unser Satz, dass der Grad der Approximation bei hinreichend grossen Werten von n nicht von »besserer» Grössenordnung als $\frac{1}{\sqrt{n}}$ sein kann, was offenbar ein nicht triviales Resultat ist.

Es wurde oben erwähnt, dass die Annäherung an das Gesetz von Gauss in der Theorie des Risikos meistens dazu benutzt wird, auf die Seltenheit der einer gewissen Grenze übersteigenden Abweichungen zu schliessen. Ohne die Benutzung des Gauss'schen Gesetzes liesse sich zwar mit Hilfe der sog. Tschebyschef'schen Ungleichung

$$S(x) < \frac{1}{2x^2} \quad \text{für } x<0$$

bez.

$$1-S(x) < \frac{1}{2x^2} \quad \text{für } x>0$$

schliessen; diese Beziehungen reichen aber für die Bedürfnisse der Theorie des Risikos noch lange nicht aus. Durch die folgenden Ungleichungen kann man in vielen Fällen bedeutend schärfere Ergebnisse erhalten.

Satz 3. *Es gilt für $x<0$*

$$S(x) < \frac{2}{\sqrt{\pi}} \int_{-\infty}^{\frac{x}{\sqrt{2}}} e^{-t^2}\,dt - \frac{\sigma}{\varrho^{\frac{3}{2}}} \cdot \frac{x^2-1}{12\sqrt{\pi}} e^{-\frac{x^2}{2}} + 4\frac{\tau}{\varrho^2}$$

und für $x > 0$

$$1 - S(x) < \frac{2}{\sqrt{\pi}} \int_{\frac{x}{\sqrt{2}}}^{\infty} e^{-t^2} dt + \frac{\sigma}{\varrho^{\frac{3}{2}}} \cdot \frac{x^2 - 1}{12\sqrt{\pi}} e^{-\frac{x^2}{2}} + 4 \frac{\tau}{\varrho^2}.$$

Um den Sinn unserer Sätze zu erläutern, wollen wir auch noch die Sätze 1 und 3 für den Fall spezialisieren, wo alle V_ν gleich sind. Es bedeutet dann $S(x)$ die Wahrscheinlichkeit der Beziehung

$$x_1 + x_2 + \cdots + x_n \leqq x\sqrt{2 n \varrho_1},$$

d. h. die Wahrscheinlichkeit, dass die Summe der in n Versuchen erhaltenen Zahlen nicht grösser als $x\sqrt{2 n \varrho_1}$ ist, wenn in jedem Versuch die Wahrscheinlichkeit $V_1(t)$ dafür besteht, eine t nicht übersteigende Zahl zu erhalten. Satz 1 gibt dann

$$|S(x) - \Phi(x)| < K \frac{\log n}{\sqrt{n}},$$

während Satz 3 z. B. für negatives x

$$S(x) < \frac{2}{\sqrt{\pi}} \int_{-\infty}^{\frac{x}{\sqrt{2}}} e^{-t^2} dt + \frac{K_1}{\sqrt{n}} (x^2 - 1) e^{-\frac{x^2}{2}} + \frac{K_2}{n}$$

ergibt, wo K, K_1 und K_2 von n und x unabhängig sind.

Einführung der adjungierten Funktionen. Umkehrungsformeln.

3. Jeder der n Verteilungsfunktionen $V_\nu(t)$ ordnen wir jetzt eine neue Funktion $v_\nu(t)$ zu, indem wir setzen

$$(2) \qquad v_\nu(t) = \int_{-\infty}^{\infty} e^{-\frac{itx}{\sqrt{2\varrho}}} d V_\nu(x),$$

wobei t reell ist. Analog setzen wir

(3) $$v(t) = \int_{-\infty}^{\infty} e^{-\frac{itx}{\sqrt{2\varrho}}} d\,V(x),$$

$$= \int_{-\infty}^{\infty} e^{-itx} d\,S(x).$$

Aus der Voraussetzung von der gegenseitigen Unabhängigkeit der x_ν folgt aber[1]

$$V(x) = \int_{-\infty}^{\infty} \cdots \int_{-\infty}^{\infty} V_n(x - t_1 - t_2 - \cdots - t_{n-1}) \cdot d\,V_1(t_1)\, d\,V_2(t_2) \cdots d\,V_{n-1}(t_{n-1}),$$

und hieraus ergibt sich ohne Schwierigkeit

(4) $$v(t) = v_1(t)\, v_2(t) \ldots v_n(t).$$

Durch formale Rechnung kann man ohne Schwierigkeit die Umkehrungsformel[2]

$$S(x) - \Phi(x) = \frac{1}{\pi} I \left[\int_0^{\infty} \frac{e^{itx}}{t} \left(v(t) - e^{-\frac{t^2}{4}} \right) dt \right]$$

erhalten; der Konvergenzbeweis bereitet aber Schwierigkeiten, und die Formel erscheint deshalb für eine Abschätzung des Fehlers wenig geeignet. Man kann sich dann des folgenden, von Liapounoff herrührenden, Kunstgriffs bedienen. — Es sei

(5) $$f(x) = \frac{1}{\sqrt{\pi\lambda}} \int_{-\infty}^{x} e^{-\frac{t^2}{\lambda}} dt,$$

[1] Nach der üblichen Definition eines Stieltjes'schen Integrals wird $V(x)$ durch die Formel des Textes in der Tat nur in allen Kontinuitätspunkten bestimmt. Unsere allgemeine Definition einer Verteilungsfunktion legt dann auch die Werte in den übrigen Punkten fest.

[2] Wenn a und b reell sind, setzen wir $I(a + bi) = b$.

wo λ eine von x unabhängige positive Grösse bezeichnet, über die wir später noch verfügen werden. Wir setzen[1]

$$(6)\qquad \overline{S}(x) = \int_{-\infty}^{\infty} f(x-t)\, dS(t) = \int_{-\infty}^{\infty} S(x-t)\, df(t)$$

und

$$(7)\qquad \overline{\Phi}(x) = \int_{-\infty}^{\infty} f(x-t)\, d\Phi(t) = \int_{-\infty}^{\infty} \Phi(x-t)\, df(t) =$$

$$= \frac{1}{\sqrt{\pi(1+\lambda)}} \int_{-\infty}^{x} e^{-\frac{t^2}{1+\lambda}}\, dt,$$

und behaupten nunmehr, dass die Umkehrungsformel

$$(8)\qquad \overline{S}(x) - \overline{\Phi}(x) = \frac{1}{\pi} I \left[\int_0^{\infty} \frac{e^{itx}}{t} \left(v(t) - e^{-\frac{1}{4}t^2} \right) e^{-\frac{1}{4}\lambda t^2}\, dt \right]$$

gilt, so dass wir durch die Einführung des Parameters λ ein absolut konvergentes Integral erhalten haben. In der Tat folgt aus der Formel

$$\int_0^{\infty} \frac{\sin \alpha t}{t} e^{-\beta t^2}\, dt = \frac{1}{2} \sqrt{\frac{\pi}{\beta}} \int_{-\infty}^{\alpha} e^{-\frac{t^2}{4\beta}}\, dt - \frac{\pi}{2},$$

da alles absolut konvergent ist:

---

[1] Offenbar ist dann $\overline{S}(x)$ die Verteilungsfunktion der Summe $\frac{x_1}{\sqrt{2\varrho}} + \cdots + \frac{x_n}{\sqrt{2\varrho}} + z$, wo z eine Hilfsvariable mit der Verteilungsfunktion $f(x)$ bezeichnet.

$$\frac{1}{\pi} I\left[\int_0^\infty \frac{e^{itx}}{t}\left(v(t)-e^{-\frac{1}{4}t^2}\right) e^{-\frac{1}{4}\lambda t^2}\, dt\right]=$$

$$=\frac{1}{\pi}\int_0^\infty \frac{1}{t} e^{-\frac{1}{4}\lambda t^2}\, dt\left(\int_{-\infty}^{\infty} \sin t(x-u)\, dS(u)-\sin t x\, e^{-\frac{1}{4}t^2}\right)$$

$$=\frac{1}{\pi}\int_{-\infty}^{\infty} dS(u)\int_0^\infty \frac{\sin t(x-u)}{t} e^{-\frac{1}{4}\lambda t^2}\, dt-$$

$$-\frac{1}{\pi}\int_0^\infty \frac{\sin t x}{t} e^{-\frac{1}{4}(1+\lambda)t^2}\, dt$$

$$=\int_{-\infty}^{\infty}\left[f(x-u)-\frac{1}{2}\right] dS(u)-\overline{\Phi}(x)+\frac{1}{2}$$

$$=\overline{S}(x)-\overline{\Phi}(x).$$

Die Formel (8) werden wir für den Beweis von Satz 1 in der Weise anwenden, dass wir zuerst die Differenz $\overline{S}-\overline{\Phi}$ abschätzen und nachträglich durch geeignete Bestimmung des Parameters λ zur Differenz $S-\Phi$ übergehen. Für den Beweis der Sätze 2 und 3 dagegen brauchen wir die folgende Formel

$$\overline{S}(x)-\overline{\Phi}(x)+\frac{\sigma}{3!\,(2\varrho)^{\frac{3}{2}}}\overline{\Phi}'''(x)=$$

(9)

$$=\frac{1}{\pi} I\left[\int_0^\infty \frac{e^{itx}}{t}\left(v(t)-e^{-\frac{1}{4}t^2}-\frac{\sigma}{3!\,(2\varrho)^{\frac{3}{2}}}\, i t^3 e^{-\frac{1}{4}t^2}\right) e^{-\frac{1}{4}\lambda t^2}\, dt\right],$$

die unmittelbar aus (8) und den Gleichungen

$$\frac{1}{\pi}\int_0^\infty \frac{\sin tx}{t} e^{-\frac{1+\lambda}{4}t^2} dt = \overline{\Phi}(x) - \frac{1}{2},$$

$$-\frac{1}{\pi}\int_0^\infty t^2 \cos tx\, e^{-\frac{1+\lambda}{4}t^2} dt = \overline{\Phi}'''(x)$$

folgt.

Beweis von Satz 1.

4. Aus (8) folgt

$$(10) \qquad |\overline{S}(x) - \overline{\Phi}(x)| < \frac{1}{\pi}\int_0^\infty \frac{1}{t} e^{-\frac{1}{4}\lambda t^2} \left| v(t) - e^{-\frac{1}{4}t^2} \right| dt,$$

und es muss zunächst das hier auftretende Integral abgeschätzt werden. Wir setzen

$$v(t) = r e^{i\omega}, \quad v_\nu(t) = r_\nu e^{i\omega_\nu},$$

wobei die Argumente so normiert werden, dass sie für $t=0$ verschwinden; es wird dann nach (4)

$$r = r_1 r_2 \ldots r_n,$$

$$\omega = \omega_1 + \omega_2 + \cdots + \omega_n.$$

Nach (2) gilt dann

$$r_\nu^2 = r_\nu e^{i\omega_\nu} . r_\nu e^{-i\omega_\nu} = \int_{-\infty}^\infty \int_{-\infty}^\infty e^{\frac{it(x-y)}{\sqrt{2\varrho}}} dV_\nu(x)\, dV_\nu(y)$$

$$= \int_{-\infty}^\infty \int_{-\infty}^\infty \cos\frac{t(x-y)}{\sqrt{2\varrho}} dV_\nu(x)\, dV_\nu(y).$$

Mit Hilfe der Ungleichungen

$$1-\frac{1}{2}x^2 \leqq \cos x \leqq 1-\frac{1}{2}x^2+\frac{1}{6}|x|^3$$

und

$$|x-y|^3 \leqq 4(|x|^3+|y|^3),$$

die für beliebige reelle x und y gelten, folgert man nun hieraus

$$(11) \qquad 1-\frac{\varrho_\nu}{2\varrho}t^2 \leqq r_\nu^2 \leqq 1-\frac{\varrho_\nu}{2\varrho}t^2+\frac{4\bar{\sigma}_\nu}{3(2\varrho)^{\frac{3}{2}}}t^3$$

Hieraus folgt nun einerseits, da für alle $x > 0$

$$\log x \leqq x-1$$

gilt,

$$2\log r_\nu \leqq -\frac{\varrho_\nu}{2\varrho}t^2+\frac{4\bar{\sigma}_\nu}{3(2\varrho)^{\frac{3}{2}}}t^3,$$

und weiter durch Addition

$$\log|v(t)| = \log r \leqq -\frac{1}{4}t^2+\frac{\bar{\sigma}}{3\sqrt{2}\varrho^{\frac{3}{2}}}t^3,$$

$$|v(t)| \leqq e^{-\frac{1}{4}t^2+\frac{\bar{\sigma}}{3\sqrt{2}\varrho^{\frac{3}{2}}}t^3}$$

für alle $t > 0$.

Andererseits folgt aus (11), so oft $\frac{\varrho_\nu t^2}{2\varrho} < 1$ ist,

$$(12) \qquad \log r_\nu \geqq \frac{1}{2}\log\left(1-\frac{\varrho_\nu t^2}{2\varrho}\right).$$

Wir setzen von nun an

$$\frac{\sigma}{\varrho^{\frac{3}{2}}} = \varepsilon^3,$$

und behaupten, dass im Intervalle

$$0 < t < \frac{1}{\varepsilon}$$

die Ungleichung

$$\frac{\varrho_\nu t^2}{\varrho} < 1$$

für jedes ν stattfindet, so dass a fortiori (12) für jeden Wert von ν dort gilt. In der Tat haben wir, da die binäre Form

$$F(x, y) = \int_{-\infty}^{\infty} (x\,|\,t\,|^\alpha + y\,|\,t\,|^\beta)^2\, d\,V_\nu(t)$$

für beliebige α und β definit positiv ist,

$$c_{\alpha+\beta}^2 \leqq c_{2\alpha} \cdot c_{2\beta},$$

wo

$$c_p = \int_{-\infty}^{\infty} |\,t\,|^p\, d\,V_\nu(t)$$

gesetzt ist. Insbesondere ist also.

$$c_1^2 \leqq c_0\, c_2, \quad c_2^2 \leqq c_1\, c_3,$$

und hieraus folgt

$$c_1\, c_2 \leqq c_0\, c_3, \quad c_2^3 \leqq c_0\, c_3^2,$$

d. h. schliesslich

$$\varrho_\nu^3 \leqq \bar{\sigma}_\nu^2.$$

Für $0 < t < \frac{1}{\varepsilon}$ gilt demnach

$$\frac{\varrho_\nu t^2}{\varrho} \leqq \left(\frac{\bar{\sigma}_\nu t^3}{\varrho^{\frac{3}{2}}}\right)^{\frac{2}{3}} < \left(\frac{\bar{\sigma} t^3}{\varrho^{\frac{3}{2}}}\right)^{\frac{2}{3}} = \varepsilon^2 t^2 < 1.$$

Für $0 < x < 1$ gilt

$$\log(1-x) > -x - \frac{x^2}{2(1-x)}; \tag{13}$$

unter Benutzung dieser Ungleichung folgt aus (12), wenn $0 < t < \frac{1}{\varepsilon}$ angenommen wird,

$$\log r_\nu > -\frac{\varrho_\nu}{4\varrho} t^2 - \frac{\varrho_\nu^2 t^4}{16\varrho^2\left(1 - \frac{\varrho_\nu t^2}{2\varrho}\right)}$$

$$> -\frac{\varrho_\nu}{4\varrho} t^2 - \frac{1}{8}\left(\frac{\varrho_\nu t^2}{\varrho}\right)^{\frac{1}{2}} \cdot \frac{\bar{\sigma}_\nu t^3}{\varrho^{\frac{3}{2}}}$$

$$> -\frac{\varrho_\nu}{4\varrho} t^2 - \frac{1}{8} \cdot \frac{\bar{\sigma}_\nu t^3}{\varrho^{\frac{3}{2}}},$$

woraus

$$\log |v(t)| = \log r > -\frac{1}{4} t^2 - \frac{1}{8}\varepsilon^3 t^3,$$

$$|v(t)| > e^{-\frac{1}{4}t^2 - \frac{1}{8}\varepsilon^3 t^3}.$$

Es muss jetzt ω, das Argument von $v(t)$, abgeschätzt werden, wobei wir uns auch im Intervall $0 < t < \frac{1}{\varepsilon}$ halten. Wir benutzen dann die Ungleichungen.

$$\cos x \geqq 1 - \frac{1}{2}x^2,$$

$$|\sin x - x| \leqq \frac{1}{6}|x|^3$$

und erhalten

$$r_\nu \cos \omega_\nu = \int_{-\infty}^{\infty} \cos \frac{tx}{\sqrt{2\varrho}}\, d\,V_\nu(x) \geqq 1 - \frac{\varrho_\nu}{4\varrho}\, t^2 > \frac{3}{4},$$

$$|r_\nu \sin \omega_\nu| = \left| \int_{-\infty}^{\infty} \left(\sin \frac{tx}{\sqrt{2\varrho}} - \frac{tx}{\sqrt{2\varrho}} \right) d\,V_\nu(x) \right|$$

$$\leqq \frac{1}{12\sqrt{2}} \cdot \frac{\bar{\sigma}_\nu}{\varrho^{\frac{3}{2}}}\, t^3,$$

und hieraus

$$|\omega_\nu| \leqq |\operatorname{tg} \omega_\nu| < \frac{1}{9\sqrt{2}} \cdot \frac{\sigma_\nu}{\varrho^{\frac{3}{2}}}\, t^3,$$

$$|\omega| \leqq |\omega_1| + \cdots + |\omega_n| < \frac{1}{9\sqrt{2}}\, \varepsilon^3 t^3.$$

5. In der vorigen Nr. haben wir die folgenden Eigenschaften von $v(t)$ bewiesen: erstens

$$|v(t)| \leqq e^{-\frac{1}{4}t^2 + \frac{1}{3\sqrt{2}}\varepsilon^3 t^3}$$

für alle $t > 0$, und zweitens die beiden Ungleichungen

$$|v(t)| > e^{-\frac{1}{4}t^2 - \frac{1}{8}\varepsilon^3 t^3},$$

$$|\arg v(t)| < \frac{1}{9\sqrt{2}}\, \varepsilon^3 t^3$$

für $0 < t < \frac{1}{\varepsilon}$. Hieraus folgt nun, dass in dem letztgenannten Intervalle

$$\left|v(t)-e^{-\frac{1}{4}t^2}\right| \leqq \left|v(t)-e^{-\frac{1}{4}t^2+i\omega}\right| + \left|e^{-\frac{1}{4}t^2}(e^{i\omega}-1)\right|$$

$$\leqq \left|r-e^{-\frac{1}{4}t^2}\right| + |\omega| \,.\, e^{-\frac{1}{4}t^2}$$

$$< \frac{1}{3\sqrt{2}}\varepsilon^3 t^3 e^{-\frac{1}{4}t^2+\frac{1}{3\sqrt{2}}\varepsilon^3 t^3} + \frac{1}{9\sqrt{2}}\varepsilon^3 t^3 e^{-\frac{1}{4}t^2}$$

$$< \left(\frac{1}{3\sqrt{2}}e^{\frac{1}{3\sqrt{2}}} + \frac{1}{9\sqrt{2}}\right)\varepsilon^3 t^3 e^{-\frac{1}{4}t^2}$$

$$< 0{,}4\,\varepsilon^3 t^3 e^{-\frac{1}{4}t^2}.$$

gilt, woraus weiter

$$\frac{1}{\pi}\int_0^{\frac{1}{\varepsilon}} \frac{1}{t}e^{-\frac{1}{4}\lambda t^2}\left|v(t)-e^{-\frac{1}{4}t^2}\right| dt < \frac{0{,}4}{\pi}\varepsilon^3 \int_0^\infty t^2 e^{-\frac{1}{4}t^2}\,dt < 0{,}5\,\varepsilon^3$$

geschlossen wird.

Für den Beweis von Satz 1 können wir ohne Beschränkung der Allgemeinheit $\varepsilon^3 < 0{,}1$ voraussetzen, da dieser Satz, wie eine leichte Ausrechnung ergibt, sonst trivial ist. Wir erhalten dann

$$\frac{1}{\pi}\int_{\frac{1}{\varepsilon}}^{\frac{3\sqrt{2}}{8}\cdot\frac{1}{\varepsilon^3}} \frac{1}{t}e^{-\frac{1}{4}\lambda t^2}\left|v(t)-e^{-\frac{1}{4}t^2}\right| dt <$$

$$< \frac{\varepsilon}{\pi}\int_{\frac{1}{\varepsilon}}^{\frac{3\sqrt{2}}{8}\cdot\frac{1}{\varepsilon^3}} \left(e^{-\frac{1}{4}t^2+\frac{1}{3\sqrt{2}}\varepsilon^3 t^3} + e^{-\frac{1}{4}t^2}\right) dt$$

$$< \frac{\varepsilon}{\pi}\int_{\frac{1}{\varepsilon}}^{\infty} e^{-\frac{1}{8}t^2}\,dt + \frac{\varepsilon}{\pi}\int_{\frac{1}{\varepsilon}}^{\frac{3\sqrt{2}}{8}\cdot\frac{1}{\varepsilon^3}} e^{-\frac{1}{4}t^2}\,dt$$

$$< \frac{4\,\varepsilon^2}{\pi} e^{-\frac{1}{8\varepsilon^2}} + \frac{\varepsilon}{\pi} \int\limits_{\frac{1}{\varepsilon}}^{\frac{3\sqrt{2}}{8}\cdot\frac{1}{\varepsilon^3}} e^{-\frac{1}{4}t^2}\, dt$$

$$< 1{,}6\,\varepsilon^3 + \frac{\varepsilon}{\pi} \int\limits_{\frac{1}{\varepsilon}}^{\frac{3\sqrt{2}}{8}\cdot\frac{1}{\varepsilon^3}} e^{-\frac{1}{4}t^2}\, dt.$$

Nach (3) ist $|v(t)| < 1$ für alle Werte von t, und hieraus folgt

$$\frac{1}{\pi} \int\limits_{\frac{3\sqrt{2}}{8}\cdot\frac{1}{\varepsilon^3}}^{\infty} \frac{1}{t} e^{-\frac{1}{4}\lambda t^2} \left| v(t) - e^{-\frac{1}{4}t^2} \right| dt <$$

$$< \frac{8}{3\pi\sqrt{2}}\varepsilon^3 \int\limits_{\frac{3\sqrt{2}}{8}\cdot\frac{1}{\varepsilon^3}}^{\infty} \left(e^{-\frac{1}{4}\lambda t^2} + e^{-\frac{1}{4}t^2} \right) dt$$

$$< \frac{64}{9\pi}\cdot\frac{\varepsilon^6}{\lambda} e^{-\frac{9}{128}\cdot\frac{\lambda}{\varepsilon^6}} + \frac{\varepsilon}{\pi} \int\limits_{\frac{3\sqrt{2}}{8}\cdot\frac{1}{\varepsilon^3}}^{\infty} e^{-\frac{1}{4}t^2}\, dt.$$

Aus den obigen folgt schliesslich

$$\frac{1}{\pi} \int\limits_{0}^{\infty} \frac{1}{t} e^{-\frac{1}{4}\lambda t^2} \left| v(t) - e^{-\frac{1}{4}t^2} \right| dt <$$

$$< 2{,}1\,\varepsilon^3 + \frac{\varepsilon}{\pi} \int\limits_{\frac{1}{\varepsilon}}^{\infty} e^{-\frac{1}{4}t^2}\, dt + \frac{64}{9\pi}\cdot\frac{\varepsilon^6}{\lambda} e^{-\frac{9}{128}\cdot\frac{\lambda}{\varepsilon^6}}$$

$$< 2{,}6\,\varepsilon^3 + \frac{64\,\varepsilon^6}{9\pi\,\lambda} e^{-\frac{9}{128}\cdot\frac{\lambda}{\varepsilon^6}},$$

d. h. nach (10)

$$|\bar{S}(x)-\bar{\Phi}(x)| < 2{,}6\,\varepsilon^3 + \frac{64}{9\pi}\cdot\frac{\varepsilon^6}{\lambda}\,e^{-\frac{9}{128}\cdot\frac{\lambda}{\varepsilon^6}}.$$

6. Es bleibt jetzt nur übrig, von der Differenz $\bar{S}-\bar{\Phi}$ zu $S-\Phi$ zu übergehen. — Wir müssen dann eine weitere von x unabhängige positive Hilfsgrösse μ einführen, die später genauer bestimmt werden soll. Nach (5) und (6) haben wir

$$\bar{S}(x) = \int_{-\infty}^{\infty} S(x-t)\,df(t) \geqq \int_{-\infty}^{\mu} S(x-t)\,df(t) \geqq$$

$$\geqq S(x-\mu)f(\mu) \geqq S(x-\mu)-f(-\mu),$$

$$\bar{S}(x) \leqq \int_{-\infty}^{-\mu} df(t) + \int_{-\mu}^{\infty} S(x-t)\,df(t) \leqq f(-\mu) + S(x+\mu).$$

Hieraus folgt offenbar

$$\bar{S}(x-\mu)-f(-\mu) \leqq S(x) \leqq \bar{S}(x+\mu)+f(-\mu)$$

und analog

$$\bar{\Phi}(x-\mu)-f(-\mu) \leqq \Phi(x) \leqq \bar{\Phi}(x+\mu)+f(-\mu).$$

Es gilt aber

$$f(-\mu) = \frac{1}{\sqrt{\pi\lambda}}\int_{-\infty}^{-\mu} e^{-\frac{t^2}{\lambda}}\,dt < \frac{1}{2}\sqrt{\frac{\lambda}{\pi\mu^2}}\,e^{-\frac{\mu^2}{\lambda}}$$

und

$$\bar{\Phi}(x+\mu)-\bar{\Phi}(x-\mu) < \frac{2\mu}{\sqrt{\pi}};$$

aus der Endgleichung der vorigen Nr. folgt dann

$$|S(x)-\Phi(x)| < 2{,}6\,\varepsilon^3 + \frac{64}{9\pi}\cdot\frac{\varepsilon^6}{\lambda}\,e^{-\frac{}{128}\cdot\frac{\lambda}{\varepsilon^6}} + \frac{2}{\sqrt{\pi}} + \sqrt{\frac{\lambda}{\pi\mu^2}}\,e^{-\frac{\mu^2}{\lambda}}.$$

Wir bestimmen jetzt die Parameter λ und μ durch die Gleichungen

$$\lambda = \frac{128}{3}\,\varepsilon^6 \log\frac{1}{\varepsilon},$$

$$\mu = 8\sqrt{2}\,\varepsilon^3 \log\frac{1}{\varepsilon},$$

wobei wir immer noch $\varepsilon^3 < 01$ voraussetzen. Durch Einsetzung erhalten wir

$$|S(x) - \Phi(x)| <$$

$$< 2{,}6\,\varepsilon^3 + \frac{1}{6\pi\log\frac{1}{\varepsilon}}\,e^{-3\log\frac{1}{\varepsilon}} + 16\sqrt{\frac{2}{\pi}}\,\varepsilon^3\log\frac{1}{\varepsilon} +$$

$$+ \sqrt{\frac{1}{3\pi\log\frac{1}{\varepsilon}}\,e^{-3\log\frac{1}{\varepsilon}}}$$

$$= \varepsilon^3\log\frac{1}{\varepsilon^3}\left(\frac{2{,}6}{\log\frac{1}{\varepsilon^3}} + \frac{1}{2\pi\log^2\frac{1}{\varepsilon^3}} + \frac{16}{3}\sqrt{\frac{2}{\pi}} + \frac{1}{\sqrt{\pi}\left(\log\frac{1}{\varepsilon^3}\right)^{\frac{3}{2}}}\right)$$

$$< \varepsilon^3\log\frac{1}{\varepsilon^3}(1{,}2 + 0{,}1 + 4{,}3 + 0{,}2)$$

$$< 6\,\varepsilon^3\log\frac{1}{\varepsilon^3}.$$

Satz 1 ist hierduch für den Fall $\varepsilon^3 < 0{,}1$ beweisen; im entgegengesetzten Falle ist der Satz aber, wie oben hervorgehoben wurde, trivial.

Beweis der Sätze 2 und 3.

7. In (9) setzen wir jetzt von vorn herein $\lambda = 1$. Es wird dann nach (7)

$$\bar{\Phi}(x) = \frac{1}{\sqrt{2\pi}} \int_{-\infty}^{x} e^{-\frac{1}{2}t^2} dt,$$

und aus (9) folgt

$$(14)\quad \begin{aligned} &\left| \bar{S}(x) - \bar{\Phi}(x) + \frac{\sigma}{3!\,(2\varrho)^{\frac{3}{2}}} \bar{\Phi}'''(x) \right| < \\ &< \frac{1}{\pi} \int_0^{\infty} \frac{1}{t} e^{-\frac{1}{4}t^2} \left| v(t) - e^{-\frac{1}{4}t^2} - \frac{\sigma}{3!\,(2\varrho)^{\frac{3}{2}}} i t^3 e^{-\frac{1}{4}t^2} \right| dt. \end{aligned}$$

Das hier auftretende Integral soll nun in ähnlicher Weise wie in Nr. 4 abgeschätzt werden. Wie dort setzen wir

$$v(t) = r e^{i\omega}, \quad v_\nu(t) = r_\nu e^{i\omega_\nu}$$

und erhalten

$$r_\nu^2 = \int_{-\infty}^{\infty} \int_{-\infty}^{\infty} \cos \frac{t(x-y)}{\sqrt{2\varrho}} \, d V_\nu(x) \, d V_\nu(y).$$

Wir benutzen jetzt die Ungleichungen

$$1 - \frac{1}{2} x^2 \leqq \cos x \leqq 1 - \frac{1}{2} x^2 + \frac{1}{24} x^4;$$

heraus folgt

$$\begin{aligned} 1 - \frac{\varrho_\nu}{2\varrho} t^2 \leqq r_\nu^2 &\leqq 1 - \frac{\varrho_\nu}{2\varrho} t^2 + \frac{\tau_\nu + 3\varrho_\nu^2}{48\varrho^2} t^4 \\ &\leqq 1 - \frac{\varrho_\nu}{2\varrho} t^2 + \frac{\tau_\nu}{12\varrho^2} t^4, \end{aligned}$$

da aus der Positivität der quadratischen Form

$$F(x, y) = \int_{-\infty}^{\infty} (x + y t^2)^2 \, d V_\nu(t)$$

unmittelbar

$$\varrho_\nu^2 \leqq \tau_\nu$$

folgt. Erstens haben wir nun wie in Nr. 4

$$2 \log r_\nu \leqq -\frac{\varrho_\nu}{2\varrho} t^2 + \frac{\tau_\nu}{12 \varrho^2} t^4,$$

und hieraus, wenn

$$\frac{\tau}{\varrho^2} = \zeta^4$$

gesetzt wird,

$$\log r \leqq -\frac{1}{4} t^2 + \frac{1}{24} \zeta^4 t^4,$$

$$r = |v(t)| \leqq e^{-\frac{1}{4} t^2 + \frac{1}{24} \zeta^4 t^4}$$

für alle $t > 0$.

Zweitens aber ist, wenn $0 < t < \frac{1}{\zeta}$ angenommen wird,

$$\frac{\varrho_\nu t^2}{\varrho} \leqq \left(\frac{\tau_\nu t^4}{\varrho^2}\right)^{\frac{1}{2}} < \left(\frac{\tau t^4}{\varrho^2}\right)^{\frac{1}{2}} = \zeta^2 t^2 < 1,$$

so dass für diese Werte von t

$$\log r_\nu \geqq \frac{1}{2} \log \left(1 - \frac{\varrho_\nu t^2}{2 \varrho}\right)$$

gilt. Aus (13) folgt dann

$$\log r_\nu > -\frac{\varrho_\nu}{4\varrho}t^2 - \frac{\varrho_\nu^2 t^4}{16\varrho^2\left(1-\frac{\varrho_\nu t^2}{2\varrho}\right)}$$

$$> -\frac{\varrho_\nu}{4\varrho}t^2 - \frac{\tau_\nu}{8\varrho^2}t^4,$$

$$\log r > -\frac{1}{4}t^2 - \frac{1}{8}\zeta^4 t^4,$$

$$r = |v(t)| > e^{-\frac{1}{4}t^2 - \frac{1}{8}\zeta^4 t^4}.$$

Es muss jetzt ω, das Argument von $v(t)$, abgeschätzt werden, wobei wir uns auch im Intervalle $0 < t < \frac{1}{\zeta}$ halten. Wir benutzen die Ungleichungen

$$1 - \cos x \leqq \frac{1}{2}x^2,$$

$$|\sin x - x| \leqq \frac{1}{6}|x|^3,$$

$$\left|\sin x - x + \frac{1}{6}x^3\right| \leqq \frac{1}{24}x^4,$$

$$|\operatorname{tg} x - x| \leqq 4x^2,$$

von denen die drei ersten für alle reelle x, die letzte für $|\operatorname{tg} x| < \frac{4}{3}$ und $|x| < \frac{\pi}{2}$ gelten. Wir haben nun

$$\left|\omega_\nu - \frac{\sigma_\nu}{3!(2\varrho)^{\frac{3}{2}}}t^3\right| \leqq$$

$$(15) \qquad \leqq |\omega_\nu - \operatorname{tg}\omega_\nu| + \frac{\left|r_\nu \sin\omega_\nu - \frac{\sigma_\nu}{3!(2\varrho)^{\frac{3}{2}}}t^3\right|}{r_\nu\cos\omega_\nu} +$$

$$+ \frac{|\sigma_\nu|}{3!(2\varrho)^{\frac{3}{2}}} \cdot \frac{1 - r_\nu\cos\omega_\nu}{r_\nu\cos\omega_\nu}t^3$$

$$= \quad A_1 + A_2 + A_3.$$

Im Intervalle $0 < t < \frac{1}{\zeta}$ gilt

$$r_\nu \cos \omega_\nu \geqq 1 - \frac{\varrho_\nu}{4\varrho} t^2 > \frac{3}{4},$$

$$|r_\nu \sin \omega_\nu| \leqq r_\nu \leqq 1,$$

d. h.

$$|\operatorname{tg} \omega_\nu| < \frac{4}{3}.$$

Es folgt also

$$A_1 \leqq 4\,\omega_\nu^2 \leqq 4 \operatorname{tg}^2 \omega_\nu < \frac{64}{9}(r_\nu \sin \omega_\nu)^2 \leqq \frac{64}{9}\left(\frac{\bar{\sigma}_\nu}{3!\,(2\varrho)^{\frac{3}{2}}} t^3\right)^2.$$

Im analoger Weise wie oben folgert man aber aus der Positivität gewisser quadratischer Formen

$$\bar{\sigma}_\nu^4 \leqq \tau_\nu^3,$$

und es ergibt sich

$$A_1 < \frac{2}{81} \cdot \left(\frac{\tau_\nu t^4}{\varrho^2}\right)^{\frac{3}{2}} < \frac{2}{81} \cdot \frac{\tau_\nu t^4}{\varrho^2}.$$

Für A_2 haben wir

$$A_2 < \frac{4}{3}\left| r_\nu \sin \omega_\nu - \frac{\sigma_\nu}{3!\,(2\varrho)^{\frac{3}{2}}} t^3 \right| \leqq$$

$$\leqq \frac{4}{3}\int_{-\infty}^{\infty}\left|\sin \frac{tx}{\sqrt{2\varrho}} - \frac{tx}{\sqrt{2\varrho}} + \frac{t^3 x^3}{3!\,(2\varrho)^{\frac{3}{2}}}\right| d\,V_\nu(x)$$

$$\leqq \frac{\tau_\nu t^4}{72\,\varrho^2},$$

und schliesslich für A_3

$$A_3 < \frac{|\sigma_\nu|}{3!\,(2\varrho)^{\frac{3}{2}}} \cdot \frac{4t^3}{3}(1 - r_\nu \cos\omega_\nu) \leq \frac{\bar{\sigma}_\nu \varrho_\nu}{36\sqrt{2}\,\varrho}\, t^5$$

$$\leq \frac{1}{36\sqrt{2}} \left(\frac{\tau_\nu t^4}{\varrho^2}\right)^{\frac{5}{4}} < \frac{1}{36\sqrt{2}} \cdot \frac{\tau_\nu t^4}{\varrho^2},$$

da aus $\varrho_\nu^2 \leqq \tau_\nu$ und $\bar{\sigma}_\nu^4 \leqq \tau_\nu^3$

$$(\bar{\sigma}_\nu \varrho_\nu)^4 \leqq \tau_\nu^5.$$

folgt. Aus (15) ergibt sich dann

$$\left| \omega_\nu - \frac{\sigma_\nu}{3!\,(2\varrho)^{\frac{3}{2}}}\, t^3 \right| < \left(\frac{2}{81} + \frac{1}{72} + \frac{1}{36\sqrt{2}}\right) \frac{\tau_\nu t^4}{\varrho^2} < 0{,}1\, \frac{\tau_\nu t^4}{\varrho^2}$$

und hieraus durch Addition

$$\left| \omega - \frac{\sigma}{3!\,(2\varrho)^{\frac{3}{2}}}\, t^3 \right| < 0{,}1\, \zeta^4 t^4.$$

8. Für die Funktion $v(t) = r\, e^{i\omega}$ gilt nach der vorigen Nr. erstens

$$r \leqq e^{-\frac{1}{4} t^2 + \frac{1}{24} \zeta^4 t^4}$$

für alle Werte von t, und zweitens

$$r > e^{-\frac{1}{4} t^2 - \frac{1}{8} \zeta^4 t^4},$$

$$\left| \omega - \frac{\sigma}{3!\,(2\varrho)^{\frac{3}{2}}}\, t^3 \right| < 0{,}1\, \zeta^4 t^4$$

für $0 < t < \frac{1}{\zeta}$. In dem letztgenannten Intervall gilt also

$$\left| v(t) - e^{-\frac{1}{4}t^2} - \frac{\sigma}{3!\,(2\varrho)^{\frac{3}{2}}}\, i\, t^3 e^{-\frac{1}{4}t^2} \right| \leqq$$

$$\leqq \left| r e^{i\omega} - e^{-\frac{1}{4}t^2 + i\omega} \right| + e^{-\frac{1}{4}t^2} \cdot \left| e^{i\omega} - e^{\frac{\sigma}{3!\,(2\varrho)^{\frac{3}{2}}}\, i\, t^3} \right| +$$

$$+ e^{-\frac{1}{4}t^2} \cdot \left| e^{\frac{\sigma}{3!\,(2\varrho)^{\frac{3}{2}}}\, i\, t^3} - 1 - \frac{\sigma}{3!\,(2\varrho)^{\frac{3}{2}}}\, i t^3 \right| \leqq$$

$$\leqq \left| r - e^{-\frac{1}{4}t^2} \right| + e^{-\frac{1}{4}t^2} \cdot \left| \omega - \frac{\sigma}{3!\,(2\varrho)^{\frac{3}{2}}}\, i t^3 \right| +$$

$$+ e^{-\frac{1}{4}t^2} \cdot \frac{1}{2} \left(\frac{\sigma t^3}{3!\,(2\varrho)^{\frac{3}{2}}} \right)^2$$

$$< \frac{1}{8} \zeta^4 t^4 e^{-\frac{1}{4}t^2} + \frac{1}{10} \zeta^4 t^4 e^{-\frac{1}{4}t^2} + \frac{1}{576} \zeta^4 t^6 e^{-\frac{1}{4}t^2}$$

$$< \left(\frac{1}{4} \zeta^4 t^4 + \frac{1}{576} \zeta^4 t^6 \right) e^{-\frac{1}{4}t^2}.$$

Wir haben hier die Ungleichung $\sigma^2 \leqq \varrho\tau$ benutzt, welche aus der Positivität der Form

$$F(x, y) = \int_{-\infty}^{\infty} (x t + y t^2)^2\, d\left(\sum_{1}^{n} V_\nu(x) \right)$$

folgt. Es ergibt sich weiter, wenn

$$h(t) = \left| v(t) - e^{-\frac{1}{4}t^2} - \frac{\sigma}{3!\,(2\varrho)^{\frac{3}{2}}}\, i\, t^3 e^{-\frac{1}{4}t^2} \right|$$

gesetzt wird,

$$(16)\qquad \frac{1}{\pi}\int_0^{\frac{1}{\zeta}} \frac{1}{t} e^{-\frac{1}{4}t^2} h(t)\,dt < \frac{1}{\pi}\zeta^4 \int_0^{\infty} \left(\frac{1}{4}t^3 + \frac{1}{576}t^5\right) e^{-\frac{1}{2}t^2}\,dt =$$

$$= \frac{1}{\pi}\left(\frac{1}{2} + \frac{1}{72}\right)\zeta^4 < 0{,}2\,\zeta^4$$

und

$$\frac{1}{\pi}\int_{\frac{1}{\zeta}}^{\infty} \frac{1}{t} e^{-\frac{1}{4}t^2} h(t)\,dt <$$

$$(17)\qquad < \frac{1}{\pi}\int_{\frac{1}{\zeta}}^{\infty} \left(\frac{1}{t} e^{-\frac{1}{4}t^2} + \frac{1}{t} e^{-\frac{1}{2}t^2} + \frac{|\sigma|}{3!\,(2\varrho)^{\frac{3}{2}}} t^2 e^{-\frac{1}{2}t^2}\right) dt$$

$$< \frac{1}{\pi}\left(2\zeta^2 e^{-\frac{1}{4\zeta^2}} + \zeta^2 e^{-\frac{1}{2\zeta^2}} + \frac{1}{12\sqrt{2}}\zeta^2 \left(\zeta + \frac{1}{\zeta}\right) e^{-\frac{1}{2\zeta^2}}\right).$$

Wenn wir jetzt vorläufig $\zeta < 1$ voraussetzen, so zeigt eine leichte Ausrechnung, dass die rechte Seite von (17) immer kleiner als $1{,}8\,\zeta^4$ ist; aus (16) und (17) folgt somit

$$\frac{1}{\pi}\int_0^{\infty} \frac{1}{t} e^{-\frac{1}{4}t^2} h(t)\,dt < 2\,\zeta^4,$$

d. h. nach (14)

$$(18)\qquad \left|\overline{S}(x) - \overline{\Phi}(x) + \frac{\sigma}{3!\,(2\varrho)^{\frac{3}{2}}}\overline{\Phi}'''(x)\right| < 2\zeta^4.$$

Dies gilt aber ganz allgemein ohne die Voraussetzung $\zeta < 1$, denn die linke Seite von (18) ist trivialerweise kleiner als

$$1 + \frac{|\sigma|}{3!\,(2\varrho)^{\frac{3}{2}}} |\overline{\Phi}'''(x)| \leqq 1 + \frac{1}{24\sqrt{\pi}} \cdot \frac{|\sigma|}{\varrho^{\frac{3}{2}}} \leqq$$

$$\leqq 1 + \frac{1}{24\sqrt{\pi}}\left(\frac{\tau}{\varrho^2}\right)^{\frac{1}{2}} = 1 + \frac{1}{24\sqrt{\pi}}\zeta^2,$$

und für $\zeta \geqq 1$ ist dieser Ausdruck kleiner als $2\,\zeta^4$.

9. Die Ungleichung (18) soll jetzt zum Beweis von Satz 2 benutzt werden. Aus (18) folgt für alle Werte von x

$$\frac{|\sigma|}{3!\,(2\varrho)^{\frac{3}{2}}}\,|\overline{\Phi}'''(x)| < |\overline{S}(x)-\overline{\Phi}(x)|+2\,\zeta^4.$$

Wenn M die obere Grenze der Werte $|S(x)-\Phi(x)|$ für alle x bedeutet, folgt hieraus unter Benutzung von (6) und (7)

$$\frac{|\sigma|}{3!\,(2\varrho)^{\frac{3}{2}}}\,|\overline{\Phi}'''(x)| < \int\limits_{-\infty}^{\infty} |S(x-t)-\Phi(x-t)|\,d f(t)+2\,\zeta^4$$

$$\leqq M+2\,\zeta^4,$$

also inbesondere für $x=0$

$$\frac{1}{24\sqrt{\pi}}\cdot\frac{|\sigma|}{\varrho^{\frac{3}{2}}} < M+2\,\zeta^4,$$

d. h.

$$M > \frac{1}{24\sqrt{\pi}}\cdot\frac{|\sigma|}{\varrho^{\frac{3}{2}}}-2\,\frac{\tau}{\varrho^2}.$$

10. Auch für den Beweis von Satz 3 werden wir die Ungleichung (18) benutzen. Aus (6) ergibt sich zuerst

$$\overline{S}(x)=\int\limits_{-\infty}^{\infty} S(x-t)\,d f(t) \geqq \int\limits_{-\infty}^{0} S(x-t)\,d f(t) \geqq$$

$$\geqq S(x)\int\limits_{-\infty}^{0} d f(t)=\frac{1}{2}\,S(x),$$

also

$$S(x) \leqq 2\,\overline{S}(x)$$

und in analoger Weise

$$1 - S(x) \leqq 2\,(1 - \overline{S}(x)\,.$$

Aus (18) folgt nun unmittelbar

$$S(x) < 2\left(\overline{\Phi}(x) - \frac{\sigma}{3!\,(2\varrho)^{\frac{3}{2}}}\,\overline{\Phi}'''(x)\right) + 4\,\frac{\tau}{\varrho^2}$$

$$= \frac{2}{\sqrt{\pi}}\int\limits_{-\infty}^{\frac{x}{\sqrt{2}}} e^{-t^2}\,dt - \frac{\sigma}{12\sqrt{\pi}\,\varrho^{\frac{3}{2}}}\,(x^2-1)\,e^{-\frac{1}{2}x^2} + 4\,\frac{\tau}{\varrho^2}$$

und

$$1 - S(x) < 2\left(1 - \overline{\Phi}(x) + \frac{\sigma}{3!\,(2\varrho)^{\frac{3}{2}}}\,\overline{\Phi}'''(x)\right) + 4\,\frac{\tau}{\varrho^2}$$

$$= \frac{2}{\sqrt{\pi}}\int\limits_{\frac{x}{\sqrt{2}}}^{\infty} e^{-t^2}\,dt + \frac{\sigma}{12\sqrt{\pi}\,\varrho^{\frac{3}{2}}}\,(x^2-1)\,e^{-\frac{1}{2}x^2} + 4\,\frac{\tau}{\varrho^2}\,.$$

19.

with H. Bohr

Die neuere Entwicklung der analytischen Zahlentheorie

Enzykl. d. Math. Wissensch. II, C 8, 722–849 (1923)

Dieser Artikel, welcher den 1900 abgeschlossenen *Bachmann*schen Artikel (I C 3) weiterführen soll, besteht aus zwei Teilen, von denen der erste, der von *Bohr* ausgearbeitet ist, insofern einen vorbereitenden Charakter trägt, als er sich ausschließlich mit den für die Behandlung der zahlentheoretischen Probleme nötigen funktionen- und reihentheoretischen Hilfsmitteln beschäftigt, während der zweite, welcher von *Cramér* herrührt, die betreffenden Probleme selbst behandelt.

Es wurde von den Verfassern zweckmäßig gefunden, dem Artikel, obwohl er sich nur in geringem Grade mit der älteren, in dem *Bachmann*schen Artikel behandelten Literatur befaßt, jedoch eine in sich abgerundete Form zu geben, so daß er gewissermaßen als ein selbständiges Ganzes hervortritt.*)

Inhaltsübersicht.

Erster Teil.

I. Allgemeine Theorie der Dirichletschen Reihen.

---

*) Bei der Ausarbeitung ist uns die von dem Meister des Gebietes, *J. Hadamard*, in der französischen Ausgabe der Encyklopädie gegebene Bearbeitung und Weiterführung des *Bachmann*schen Artikels von großer Bedeutung gewesen. Dasselbe gilt von dem klassischen Werk von *E. Landau*, Handbuch der Lehre von der Verteilung der Primzahlen, Bd. 1—2, Leipzig und Berlin 1909, welches wir im folgenden einfach mit „Handbuch" zitieren werden.

Erster Teil.

In diesem Teil, der, wie in den einleitenden Worten gesagt, einen rein analytischen Charakter trägt, d. h. von den zahlentheoretischen Anwendungen prinzipiell absieht, wird die Theorie der *Dirichletschen Reihen* besprochen, welche sich — obwohl ihre wesentliche Bedeutung in ihrer Stellung als besonders geeignetes Hilfsmittel zur funktionentheoretischen Behandlung von zahlentheoretischen Aufgaben zu ersehen ist, und sie immer noch ihre meisten Problemstellungen der analytischen Zahlentheorie verdankt — doch im Laufe der letzten Jahrzehnte zu einem selbständigen Abschnitt der allgemeinen Reihenlehre entwickelt hat. Das Referat ist in zwei Kapitel eingeteilt, von denen das erste die Theorie der allgemeinen *Dirichlet*schen Reihen behandelt, während das zweite der für das Studium der Primzahlen fundamentalen speziellen *Dirichlet*schen Reihe, welche die *Riemann*sche Zetafunktion darstellt, gewidmet ist. Bei der Abfassung ist mehr Gewicht auf eine bequeme Übersicht der wichtigeren Resultate als auf strenge Vollständigkeit gelegt.

I. Allgemeine Theorie der Dirichletschen Reihen.[1])

1. Definition einer Dirichletschen Reihe. Unter *einer allgemeinen Dirichletschen Reihe* wird eine unendliche Reihe der Form

$$(1) \qquad f(s) = \sum_{n=1}^{\infty} a_n e^{-\lambda_n s}$$

verstanden; hierbei bedeutet $s = \sigma + it$ eine komplexe, unabhängige Variable, die Koeffizienten a_n sind beliebige komplexe Zahlen, während die Exponentenfolge $\{\lambda_n\}$ eine reelle monoton wachsende Zahlenfolge mit $\lambda_n \to \infty$ bezeichnet.[2]) Für die folgende Darstellung wird es bequem sein, die (unwesentliche) Annahme $\lambda_1 \geq 0$ zu machen. Für $\lambda_n = n$ ist (1) eine Potenzreihe in der Variablen e^{-s}. In dem beson-

1) Betreffs vieler Einzelheiten in der Theorie sei der Leser auf *E. Landau,* Handbuch, und *G. H. Hardy-M. Riesz,* The general theory of Dirichlet's series, Cambridge tracts, Nr. 18 (1915), verwiesen.

2) *W. Schnee,* Über irreguläre Potenzreihen und Dirichletsche Reihen, Dissertation, Berlin 1908, und *K. Väisälä,* Verallgemeinerung des Begriffes der Dirichletschen Reihen, Acta Universitatis Dorpatensis (1921), betrachten auch Reihen mit *komplexen* Exponenten λ_n und untersuchen, unter welchen Bedingungen solche Reihen sich „ähnlich" benehmen wie Reihen mit reellen Exponenten.

ders wichtigen Spezialfall $\lambda_n = \log n$ erhalten wir die *gewöhnlichen Dirichletschen Reihen*

(2) $$\sum_{n=1}^{\infty} a_n e^{-s \log n} = \sum_{n=1}^{\infty} \frac{a_n}{n^s}.$$

Die spezielle Reihe (2), bei welcher $a_n = 1$ ist für alle n, also die Reihe

(3) $$\sum \frac{1}{n^s} = 1 + \frac{1}{2^s} + \frac{1}{3^s} + \frac{1}{4^s} + \cdots,$$

definiert die *Riemannsche Zetafunktion*, deren Theorie in einem besonderen Kapitel behandelt wird. Als ein anderes wichtiges Beispiel einer gewöhnlichen *Dirichlet*schen Reihe (2) sei eine solche erwähnt[3]), bei der die Koeffizienten a_n sich periodisch wiederholen (etwa mit der Periode k), und die Summe der Koeffizienten erstreckt über eine Periode gleich 0 ist, wo also

(4) $$a_n = a_m \quad \text{für} \quad m \equiv n \pmod{k}, \quad \sum_{n=1}^{k} a_n = 0.$$

Zu diesem Typus gehört z. B. die Zetareihe mit abwechselndem Vorzeichen

(5) $$\sum \frac{(-1)^{n+1}}{n^s} = 1 - \frac{1}{2^s} + \frac{1}{3^s} - \frac{1}{4^s} + \cdots,$$

welche durch formale Multiplikation der Zetareihe (3) mit dem Faktor $1 - 2^{1-s}$ entsteht. Andere wichtige Typen gewöhnlicher *Dirichlet*scher Reihen werden in Nr. 7 besprochen.

2. Die drei Konvergenzabszissen. Eine *Dirichlet*sche Reihe (1), die in einem Punkte $s_0 = \sigma_0 + it_0$ *absolut konvergiert*, wird offenbar in jedem Punkte $s = \sigma + it$ mit $\sigma \geqq \sigma_0$ absolut konvergieren; denn es ist ja, $s - s_0 = s'$ gesetzt,

(6) $$\sum a_n e^{-\lambda_n s} = \sum a_n e^{-\lambda_n s_0} \cdot e^{-\lambda_n s'}$$

und $|e^{-\lambda_n s'}| \leqq 1$ für $\Re(s') \geqq 0$. Jede Reihe (1) besitzt daher eine *absolute Konvergenzabszisse* σ_A derart, daß (1) für $\sigma > \sigma_A$ absolut konvergiert, für $\sigma < \sigma_A$ dagegen nicht; hierbei sind, den Werten $+\infty$ und $-\infty$ von σ_A entsprechend, diejenigen Fälle mit inbegriffen, wo die Reihe nirgends bzw. überall absolut konvergiert.

Tiefer liegt der Satz von *Jensen*[4]), daß, wenn die Reihe (1) im Punkte $s_0 = \sigma_0 + it_0$ *konvergiert*, sie dann auch in der ganzen Halbebene $\sigma > \sigma_0$ konvergiert. Diesen Hauptsatz der Theorie beweist *Jensen*

---

3) *G. Lejeune Dirichlet*, Recherches sur diverses applications de l'Analyse infinitésimale à la Théorie des Nombres, Crelles J. 19 (1839), p. 324—369 = Werke, Bd. 1, p. 411 u. f.

4) *J. L. W. V. Jensen*, Om Rækkers Konvergens, Tidsskr. for Math. (5) 2 (1884), p. 63—72.

48*

von (6) aus, indem er mit Hilfe partieller (*Abel*scher) Summation nachweist, daß bei festem s' mit $\Re(s') > 0$ die Zahlenfolge $\{e^{-\lambda_n s'}\}$ eine „konvergenzerhaltende" ist in dem Sinne, daß aus der Konvergenz einer Reihe $\sum b_n$ die Konvergenz der „multiplizierten" Reihe $\sum b_n e^{-\lambda_n s'}$ folgt. Es gibt also auch eine *Konvergenzabszisse* $\sigma_B (\leqq \sigma_A)$ derart, daß (1) für $\sigma > \sigma_B$ konvergiert, für $\sigma < \sigma_B$ divergiert.

Cahen[5]), der zuerst die *Dirichlet*schen Reihen einer systematischen Untersuchung unterworfen hat, zeigt, daß (1) in jedem Gebiete $\sigma > \sigma_B + \varepsilon$, $|s| < K$ *gleichmäßig konvergiert* und somit in der Konvergenzhalbebene $\sigma > \sigma_B$ eine *reguläre analytische Funktion* $f(s)$ darstellt. Im allgemeinen konvergiert aber eine Reihe (1) *nicht* gleichmäßig in der *ganzen* Halbebene $\sigma > \sigma_B + \varepsilon$, und *Bohr*[6]) hat daher die *gleichmäßige Konvergenzabszisse* σ_G eingeführt, welche definiert wird als die untere Grenze aller Abszissen σ_0, für die (1) in der ganzen Halbebene $\sigma > \sigma_0$ gleichmäßig konvergiert. Hierbei ist offenbar $-\infty \leqq \sigma_B \leqq \sigma_G \leqq \sigma_A \leqq +\infty$, und es können die drei Konvergenzabszissen alle Werte tatsächlich haben, welche mit diesen Ungleichungen verträglich sind.[7])

Die drei Konvergenzabszissen einer Reihe (1) können leicht aus den Koeffizienten und Exponenten der Reihe bestimmt werden. Für die Abszisse σ_B gilt nach *Cahen*[8]) der Satz: Falls $\sigma_B > 0$ ist[9]), wird

---

5) *E. Cahen,* Sur la fonction $\zeta(s)$ de Riemann et sur des fonctions analogues, Ann. Éc. Norm. (3) 11 (1894), p. 75—164.

6) *H. Bohr*, a) Sur la convergence des séries de Dirichlet, Paris C. R. 151 (1910), p. 375—377; b) Über die gleichmäßige Konvergenz Dirichletscher Reihen, Crelles J. 143 (1913), p. 204—211; c) Nogle Bemærkninger om de Dirichletske Rækkers ligelige Konvergens, Mat. Tidsskr. B 1921, p. 51—55.

7) *L. Neder*, Über die Lage der Konvergenzabszissen einer Dirichletschen Reihe zur Beschränktheitsabszisse ihrer Summe, Arkiv för Mat., Astr. och Fys. 16 (1922), No. 20.

8) *E. Cahen,* a. a. O. 5). Ein Teil des Satzes findet sich schon bei *J. L. W. V. Jensen,* Sur une généralisation d'une théorème de Cauchy, Paris C. R. 106 (1888), p. 833—836.

9) Die Bedingung $\sigma_B > 0$ bedeutet keine wesentliche Einschränkung der Allgemeinheit, weil ja die Konvergenzabszisse σ_B, falls sie $> -\infty$ ist, immer durch die einfache Transformation $s = s' - c$ um eine Konstante c vergrößert werden kann. Ausdrücke für σ_B, die im Falle $\sigma_B < 0$ oder sogar für jede Lage von σ_B gelten, sind gegeben von *S. Pincherle,* Alcune spigolature nel campo delle funzioni determinanti, Atti d. IV Congr. intern. d. Mat. 2 (Rom 1908), p. 44—48; *K. Knopp,* Über die Abszisse der Grenzgeraden einer Dirichletschen Reihe, Sitzungsber. Berl. Math. Ges. 10 (1910), p. 1—7; *W. Schnee,* Über die Koeffizientendarstellungsformel in der Theorie der Dirichletschen Reihen, Gött. Nachr. 1910, p. 1—42; *T. Kojima,* a) On the convergence-abscissa of general Dirichlet's series, Tôhoku J. 6 (1914), p. 134—139; b) Note on the convergence-abscissa of

sie durch den Ausdruck

(7) $$\sigma_B = \limsup_{n\to\infty} \frac{\log|S_n|}{\lambda_n} \qquad \left(S_n = \sum_1^n a_m\right)$$

gegeben; d. h. σ_B ist die untere Grenze aller positiven Zahlen α, für welche die „summatorische" Funktion S_n gleich $O(e^{\lambda_n \alpha})$ ist.[10])

Aus (7) ergibt sich sofort, daß im Falle $\sigma_A > 0$

$$\sigma_A = \limsup_{n\to\infty} \frac{\log R_n}{\lambda_n}. \qquad \left(R_n = \sum_1^n |a_m|\right)$$

Für die gleichmäßige Konvergenzabszisse σ_G gilt schließlich, falls $\sigma_G > 0$ ist, die entsprechende Formel[11]):

$$\sigma_G = \limsup_{n\to\infty} \frac{\log T_n}{\lambda_n},$$

wo T_n, bei festem n, die obere Grenze von $\left|\sum_1^n a_m e^{-\lambda_m i t}\right|$ für $-\infty < t < \infty$ bezeichnet.

Für Reihen (1), bei denen die Exponentenfolge $\{\lambda_n\}$ hinreichend schnell ins Unendliche wächst (z. B. für die Potenzreihen, wo $\lambda_n = n$ ist), gilt immer die Gleichung $\sigma_A = \sigma_B$ $(= \sigma_G)$, d. h. sie besitzen keinen bedingten Konvergenzstreifen. Die genaue notwendige und hinreichende Bedingung, die eine Exponentenfolge erfüllen muß, damit jede zu ihr gehörige *Dirichlet*sche Reihe der Bedingung $\sigma_A = \sigma_B$ genügt, ist

(8) $$\lim_{n\to\infty} \frac{\log n}{\lambda_n} = 0.$$

Dirichlet's series, Tôhoku J. 9 (1916), p. 28—37; *M. Fujiwara*, a) On the convergence-abscissa of general Dirichlet's series, Tôhoku J. 6 (1914), p. 140—142; b) Über Konvergenzabszisse der Dirichletschen Reihe, Tôhoku J. 17 (1920), p. 344—350; *E. Lindh* (bei *Mittag-Leffler*), Sur un nouveau théorème dans la théorie des séries de Dirichlet, Paris C. R. 160 (1915), p. 271—273; *B. Malmrot*, Sur une formule de M. Fujiwara, Arkiv för Mat., Astr. och Fys. 14 (1919), No. 4, p. 1—10.

10) Soll die Reihe (1) noch in Punkten *auf* der Konvergenzgeraden $\sigma = \sigma_B$ (> 0) konvergieren, ist es nach *Jensen*, a. a. O. 8), notwendig (aber nicht hinreichend, vgl. Nr. 5), daß die summatorische Funktion S_n der Bedingung $S_n = o(e^{\lambda_n \sigma_B})$ genügt.

11) Für *gewöhnliche Dirichlet*sche Reihen $(\lambda_n = \log n)$ bei *H. Bohr*, Darstellung der gleichmäßigen Konvergenzabszisse einer Dirichletschen Reihe $\sum_{n=1}^{\infty} \frac{a_n}{n^s}$ als Funktion der Koeffizienten der Reihe, Arch. Math. Phys. (3) 21 (1913), p. 326—330, für *beliebige Dirichlet*sche Reihen bei *M. Kuniyeda*, Uniform convergence-abscissa of general Dirichlet's series, Tôhoku J. 9 (1916), p. 7—27. In der letzten Arbeit sind auch Formeln für σ_G angegeben, die für jede Lage von σ_G gelten. (Vgl. Note 9).)

Allgemein gilt der Satz[13]), daß *die maximale Breite M des bedingten Konvergenzstreifens* $\sigma_B \leqq \sigma \leqq \sigma_A$ für alle zu einer gegebenen Exponentenfolge gehörigen Reihen (1) durch den Ausdruck

$$M = \limsup_{n \to \infty} \frac{\log n}{\lambda_n}$$

gegeben wird. Für die gewöhnlichen *Dirichlet*schen Reihen (2) ist somit die maximale Breite $M = 1$. Diese Breite 1 wird z. B. bei jeder Reihe (2), die den Bedingungen (4) genügt, erreicht; in der Tat ist hier $\sigma_A = 1$, $\sigma_B = 0$.

3. Der Eindeutigkeitssatz. Aus der einfachen Bemerkung, daß die Funktion $e^{-\lambda s} = e^{-\lambda(\sigma + it)}$ $(\lambda > 0)$ für $\sigma \to \infty$ um so schneller gegen 0 abnimmt, je größer der Exponent λ ist, ergibt sich leicht: falls eine *Dirichlet*sche Reihe (1) mit $\sigma_B < \infty$ die Bedingung $\sigma_A < \infty$ oder nur die Bedingung $\sigma_G < \infty$[6c]) erfüllt, dann *überwiegen für $\sigma \to \infty$ die Anfangsglieder der Reihe den Rest,* d. h. es gilt, bei jedem festen N, für $\sigma \to \infty$ *gleichmäßig in t* die Limesgleichung

$$\sum_{n=1}^{\infty} a_n e^{-\lambda_n s} = \sum_{n=1}^{N} a_n e^{-\lambda_n s} + o(e^{-\lambda_N \sigma}); \tag{9}$$

hieraus folgt sofort, daß, wenn nicht sämtliche Koeffizienten a_n gleich 0 sind, die Summe $f(s)$ bei hinreichend großem K in der ganzen Halbebene $\sigma > K$ von 0 verschieden sein wird. Für *Reihen* (1) *mit* $\sigma_G < \infty$ gilt daher der folgende *Eindeutigkeitssatz:* Sind zwei *Dirichlet*sche Reihen $f(s) = \sum a_n e^{-\lambda_n s}$ und $g(s) = \sum b_n e^{-\mu_n s}$ gleichgroß in allen Punkten einer Zahlenfolge $\{s_n = \sigma_n + it_n\}$ mit $\sigma_n \to \infty$, dann sind die beiden Reihen identisch; denn in der *Dirichlet*schen Reihe $\sum c_n e^{-\nu_n s}$, welche durch Subtraktion von $f(s)$ und $g(s)$ entsteht, müssen ja alle Koeffizienten c_n gleich 0 sein.

Für eine *beliebige Dirichlet*sche Reihe (1) mit $\sigma_B < \infty$ gilt die Limesgleichung (9) für $\sigma \to \infty$ im allgemeinen *nicht* gleichmäßig in t, wenn t das *ganze* Intervall $-\infty < t < \infty$ durchläuft. Dagegen gilt (9), wie von *Perron*[13]) bewiesen, gleichmäßig in t, wenn t durch eine Bedingung der Form $|t| < e^{k\sigma}$ beschränkt wird, wo k eine beliebige Konstante bedeutet. In diesem *allgemeinen Fall* finden wir daher den folgenden *Eindeutigkeitssatz:* Wenn zwei *Dirichlet*sche Reihen mit

---

12) *E. Cahen,* a. a. O. 5). Vgl. auch *Hardy-Riesz,* a. a. O. 1), p. 9.

13) *O. Perron,* Zur Theorie der Dirichletschen Reihen, Crelles J. 134 (1908), p. 95—143. Daß die Limesgleichung (9) für ein *festes* t gilt, steht schon bei *Dirichlet,* Vorlesungen über Zahlentheorie, herausgegeben von *Dedekind,* Braunschweig 1863, p. 410—414. Vgl. auch eine (in Math. Ztschr. bald erscheinende) Arbeit von *L. Neder,* Über Gebiete gleichmäßiger Konvergenz Dirichletscher Reihen.

$\sigma_B < \infty$ in den Punkten einer Zahlenfolge $\{s_n\}$ mit $\sigma_n \to \infty$ und $|t_n| < e^{k\sigma}$ gleichgroß sind, so sind die beiden Reihen identisch. Hier kann die Forderung $|t_n| < e^{k\sigma_n}$ nicht weggelassen werden, denn es existieren tatsächlich Reihen (1), deren Koeffizienten nicht alle 0 sind, die jedoch eine Folge von Nullstellen $\{s_n\}$ mit $\sigma_n \to \infty$ besitzen.[14])

4. Die Koeffizientendarstellungsformel. Aus dem Eindeutigkeitssatze in Nr. 3 folgt sofort: wenn eine in einer gewissen Halbebene $\sigma > \sigma_0$ reguläre analytische Funktion $f(s)$ durch eine konvergente *Dirichlet*sche Reihe darstellbar ist, dann müssen die Exponenten λ_n und die Koeffizienten a_n dieser Reihe aus der Funktion $f(s)$ eindeutig bestimmt werden können. Die tatsächliche Bestimmung dieser beiden Zahlenfolgen $\{\lambda_n\}$ und $\{a_n\}$ wird durch den unten folgenden Satz gegeben, dessen *formale* Herleitung[15]) sich aus der bekannten, für jedes positive c gültigen Formel

$$\frac{1}{2\pi i}\int_{c-i\infty}^{c+i\infty}\frac{e^{\alpha s}}{s}\,ds = \begin{cases}1 & \text{für } \alpha > 0\\ 0 & \text{für } \alpha < 0\end{cases}$$

ergibt, während seine strenge Begründung zuerst von *Hadamard* und *Perron*[16]) gegeben wurde. Dieser Satz lautet: *Es sei* (1) *eine beliebige Dirichletsche Reihe mit der Konvergenzabszisse* $\sigma_B < \infty$ *und* c *eine positive Zahl* $> \sigma_B$. *Dann gilt für jedes* x *im Intervalle* $\lambda_N < x < \lambda_{N+1}$ *die Formel*

$$\sum_1^N a_n = \frac{1}{2\pi i}\int_{c-i\infty}^{c+i\infty} f(s)\frac{e^{xs}}{s}\,ds. \tag{10}$$

Es ist also das auf der rechten Seite stehende Integral $J(x)$ streckenweise konstant (für $0 < x < \infty$) und die Exponenten λ_n sind die Unstetigkeitsstellen von $J(x)$, während die Koeffizienten a_n sich

---

14) *H. Bohr*, Beweis der Existenz Dirichletscher Reihen, die Nullstellen mit beliebig großer Abszisse besitzen, Palermo Rend. 31 (1911), p. 235–243.

15) Vgl. *L. Kronecker*, Notiz über Potenzreihen, Monatsber. Akad. Berlin (1878), p. 53—58, und *E. Cahen*, a. a. O. 5). Ein Spezialfall kommt schon bei *B. Riemann*, Ueber die Anzahl der Primzahlen unter einer gegebenen Grösse, Monatsber. Akad. Berlin 1859, p. 671—680 = Werke, p. 145—153, vor.

16) *J. Hadamard*, Sur les séries de Dirichlet, Palermo Rend. 25 (1908), p. 326–330, beweist den Satz unter der Annahme, daß die Reihe eine *absolute* Konvergenzhalbebene besitzt (also $\sigma_A < \infty$) und *O. Perron*, a. a. O. 13) für den allgemeinen Fall. Vgl. auch *E. Phragmén*, Über die Berechnung der einzelnen Glieder der Riemannschen Primzahlformel, Oefvers. af Kgl. Vetensk. Förh. 48 (Stockholm 1891), p. 721—744 und *H. v. Mangoldt*, Auszug aus einer Arbeit unter dem Titel: Zu Riemanns Abhandlung „Über die Anzahl der Primzahlen unter einer gegebenen Größe", Sitzungsber. Akad. Berlin 1894, p. 883–896.

als die Sprünge in den Punkten λ_n ergeben.[17]) In einer Unstetigkeitsstelle λ_n selbst ist das Integral $J(x)$ wohl nicht direkt konvergent, es hat aber einen Hauptwert, definiert durch $\lim\limits_{T\to\infty}\frac{1}{2\pi i}\int\limits_{c-iT}^{c+iT}$, und dieser Hauptwert ist gleich dem Mittelwert $\frac{1}{2}(J(\lambda_n+0)+J(\lambda_n-0))$.

Das Integral in (10) konvergiert im allgemeinen nur bedingt. Bei verschiedenen Untersuchungen ist es deshalb bequem, statt (10) die Formel

$$\frac{1}{2\pi i}\int\limits_{c-i\infty}^{c+i\infty} f(s)\frac{e^{xs}}{s^2}\,ds = \sum_{1}^{N} a_n(x-\lambda_n) \tag{11}$$

zu benutzen, wo das Integral (wenigstens im Falle $\sigma_G < \infty$, vgl. Nr. 6) absolut konvergiert. Die Formeln (10) und (11) sind übrigens Spezialfälle der allgemeinen Formel[18])

$$\frac{1}{2\pi i}\int\limits_{c-i\infty}^{c+i\infty} f(s)\frac{e^{xs}}{s^\alpha}\,ds = \frac{1}{\Gamma(\alpha)}\sum_{1}^{N} a_n(x-\lambda_n)^{\alpha-1}, \tag{12}$$

wo $\alpha \geqq 1$ ist.

5. Beziehung zwischen der Reihe auf der Konvergenzgeraden und der Funktion bei Annäherung an die Konvergenzgerade. In den Punkten der Konvergenzgeraden $\sigma = \sigma_B$ einer *Dirichlet*schen Reihe (1) kann das Verhalten der Reihe sehr verschiedenartig sein. Wie im Spezialfall einer Potenzreihe ($\lambda_n = n$) bestehen aber auch bei den allgemeinen *Dirichlet*schen Reihen wichtige Zusammenhänge zwischen dem Verhalten der *Reihe* in einem Punkte der Konvergenzgeraden und dem Verhalten der dargestellten *Funktion* $f(s)$, wenn die Variable s sich diesem Punkte nähert. Da dies Problem im Spezialfall $\lambda_n = n$ im Artikel II C 4 ausführlich besprochen ist, sollen hier nur einige Hauptresultate erwähnt werden. Zuerst nennen wir den Satz (Analogon zum *Abel-Stolz*schen Satze über Potenzreihen): wenn die Reihe (1) in einem Punkte s_0 der Konvergenzgeraden $\sigma = \sigma_B$ *konvergiert* mit der Summe A, dann existiert der *Grenzwert* $\lim f(s)$ und ist $= A$, wenn s sich von rechts längs einer horizontalen Geraden oder sogar

17) Eine andere, von *Hadamard* herrührende Methode, um die Koeffizienten a_n einer *Dirichlet*schen Reihe aus der durch die Reihe dargestellten Funktion zu bestimmen, wird in Nr. 9 besprochen; diese letzte Methode — und nicht die oben angegebene — ist übrigens als die unmittelbare Verallgemeinerung der *Cauchy*schen Methode zur Bestimmung der Koeffizienten einer Potenzreihe anzusehen.

18) *J. Hadamard,* Sur la distribution des zéros de la fonction $\zeta(s)$ et ses conséquences arithmétiques, Bull. Soc. math. France 24 (1896), p. 199—220. Wegen der strengen Begründung im Falle $\sigma_A = \infty$ vgl. *O. Perron,* a. a. O. 13).

in einem der Halbebene $\sigma > \sigma_B$ ganz angehörenden Winkelraum dem Punkte s_0 nähert.[19]) Dieser Satz läßt sich natürlich *nicht* ohne weiteres *umkehren*, d. h. aus der Existenz des Grenzwertes folgt nicht die Konvergenz der Reihe im Punkte s_0. Bedingungen, unter welchen die Umkehrung erlaubt ist, wurden von *Landau*, *Schnee*, *Littlewood* und *Hardy-Littlewood* gegeben.[20]) Hier sei nur der tiefliegende Satz von *Littlewood* erwähnt, wonach die Bedingung

$$a_n e^{-\lambda_n s_0} = O\left(\frac{\lambda_n - \lambda_{n-1}}{\lambda_n}\right) \qquad (\text{für } n \to \infty)$$

für die besprochene Umkehrung genügt.

Von etwas anderer Art — weil *Regularität* im Punkte s_0 statt Grenzwert für $s \to s_0$ vorausgesetzt wird — ist ein für verschiedene Anwendungen sehr wichtiger Satz von *M. Riesz*[21]), der als die Verallgemeinerung eines *Fatou*schen Satzes über Potenzreihen ($\lambda_n = n$) anzusehen ist, und der besagt, daß, falls eine *Dirichlet*sche Reihe (1) mit $\sigma_B > 0$ die Bedingung

$$(13) \qquad S_n = a_1 + \cdots a_n = o(e^{\lambda_n \sigma_B})$$

erfüllt, sie in jedem Punkte der Konvergenzgeraden $\sigma = \sigma_B$, in welchem die Funktion $f(s)$ *regulär* ist, *konvergiert*, und zwar gleichmäßig in jedem Regularitätsintervall. Die Bedeutung dieses Satzes zeigt sich

19) Für Annäherung längs einer horizontalen Geraden siehe *Dirichlet-Dedekind*, a. a. O. 13), p. 410—414; für Annäherung im Winkelraum *E. Cahen*, a. a. O. 5).

20) *E. Landau*, a) Über die Konvergenz einiger Klassen von unendlichen Reihen am Rande des Konvergenzgebietes, Monatsh. Math. Phys. 18 (1907), p. 8—28; b) Über einen Satz des Herrn Littlewood, Palermo Rend. 35 (1913), p. 265—276; *W. Schnee*, Über Dirichletsche Reihen, Palermo Rend. 27 (1909), p. 87—116; *J. Littlewood*, The converse of Abel's theorem on power series, Proc. London math. Soc. (2) 9 (1910), p. 434—448; *G. H. Hardy* u. *J. Littlewood*, Tauberian theorems concerning power series and Dirichlet's series whose coefficients are positive, Proc. London math. Soc. (2) 13 (1913), p 174—191.

21) *M. Riesz*, a) Sur les séries de Dirichlet et les séries entières, Paris C. R. 149 (1909), p. 309—312; b) Ein Konvergenzsatz für Dirichletsche Reihen, Acta Math. 40 (1916), p. 349—361. Ein Beweis des Spezialfalls $\lambda_n = \log n$ wurde schon früher (nach einer Mitteilung von *Riesz*) von *E. Landau*, Über die Bedeutung einiger neuer Grenzwertsätze der Herren Hardy und Axer, Prac. Mat. Fiz. 21 (1910), p. 97—177, veröffentlicht. Vgl. auch *D. Kojima*, On the double Dirichlet series, Reports Tôhoku University 9 (1920), p. 351—400.

Riesz hat bedeutende Verallgemeinerungen seines Satzes in Aussicht gestellt. Vgl. eine demnächst in den Acta Univ. hung. Francesco-Jos. erscheinende Arbeit. Eine besonders wichtige dieser Verallgemeinerungen — welche den Fall Summabilität statt Konvergenz behandelt, vgl. Nr. 13 — ist in der zahlentheoretischen Arbeit von *H. Cramér*, Über das Teilerproblem von Piltz, Ark. f. Mat., Astr. och Fys. 16 (1922), No. 21, nach einer Mitteilung von *Riesz* veröffentlicht. Vgl. auch *A. Walfisz*, Über die summatorischen Funktionen einiger Dirichletscher Reihen, Diss. Göttingen 1922, p. 1—56.

schon darin, daß die Bedingung (13), wie früher[10]) erwähnt, *notwendig* ist, damit die Gerade $\sigma = \sigma_B$ überhaupt eine Konvergenzstelle der Reihe enthalte.

An die erstgenannten Sätze schließt sich eine Reihe von weiteren Sätzen an, wo an Stelle der Konvergenz der Reihe im Punkte s_0 und der Existenz des Grenzwertes der Funktion bei Annäherung an diesen Punkt, bestimmte Art von (eigentlicher) *Divergenz* der Reihe im Punkte s_0 und entsprechende bestimmte Art von *Unendlichwerden* der Funktion bei Annäherung an den Punkt tritt. Solche Sätze, die durch Vergleich mit speziellen einfachen Typen *Dirichlet*scher Reihen abgeleitet werden, verdankt man besonders *Knopp*[22]) und *Schnee*[23]). Als ein einfaches Beispiel für eine gewöhnliche *Dirichlet*sche Reihe (2) sei der folgende Satz genannt (wo es sich um den Punkt $s_0 = 0$ handelt). Aus

$$\lim_{n\to\infty} \frac{a_1 + a_2 + \cdots a_n}{\log^\alpha n} = A \qquad (\alpha > 0)$$

folgt

$$\lim_{s\to 0} s^\alpha f(s) = A\,\Gamma(\alpha + 1),$$

wo s durch positive Werte gegen 0 strebt. Mit der viel schwierigeren Frage nach der *Umkehrung* solcher Sätze haben sich *Hardy* und *Littlewood*[24]) beschäftigt. So haben sie z. B. die Umkehrung des eben erwähnten Satzes in dem Falle bewiesen, wo die Koeffizienten a_n sämtlich *positiv* sind. Der weitestgehende von *Hardy* und *Littlewood* bewiesene Satz, welcher den allgemeinen Typus *Dirichlet*scher Reihen (1) betrifft (wo jedoch $\lambda_n : \lambda_{n+1} \to 1$ vorausgesetzt wird) besagt[24b]), daß, wenn eine Reihe (1) mit der Konvergenzabszisse $\sigma_B = 0$ die Limesgleichung

$$\lim_{s\to 0} s^\alpha f(s) = A \qquad (\alpha \geqq 0)$$

erfüllt, und ihre Koeffizienten a_n reell sind und der „einseitigen" Bedingung $a_n > - K\lambda_n^{\alpha-1}(\lambda_n - \lambda_{n-1})$ genügen[25]), die Gleichung gilt:

$$\lim_{n\to\infty} \frac{a_1 + a_2 + \cdots a_n}{\lambda_n^\alpha} = \frac{A}{\Gamma(\alpha + 1)}.$$

22) *K. Knopp,* a) Grenzwerte von Reihen bei der Annäherung an die Konvergenzgrenze, Diss. Berlin 1907; b) Divergenzcharaktere gewisser Dirichletscher Reihen, Acta Math. 34 (1911), p. 165—204; c) Grenzwerte von Dirichletschen Reihen bei der Annäherung an die Konvergenzgrenze, Crelles J. 138 (1910), p. 109—132.

23) *W. Schnee,* a) a. a. O. 2); b) a. a. O. 20). In der letzten Arbeit gibt *Schnee* einige interessante spezielle Typen *Dirichlet*scher Reihen an, die als „Vergleichsreihen" besonders geeignet sind.

24) Vgl. insbesondere *G. H. Hardy* u. *J. Littlewood,* a) a. a. O. 20); b) Some theorems concerning Dirichlet's series, Mess. of math. 43 (1914), p. 134—147.

25) Hieraus folgt sofort als Corollar, daß der Satz, im Falle komplexer Koeffizienten, gültig ist, falls die oben angegebene „einseitige" Bedingung durch

Mit den obigen Fragestellungen eng verwandt ist das Problem nach der Beziehung des Verhaltens der Funktion bei Annäherung an einen Punkt s_0 auf der Konvergenzgeraden $\sigma = \sigma_B$ und der Art der Divergenz der Reihe in einem Punkte s_1, welcher *links* von dieser Geraden in derselben Höhe wie s_0 gelegen ist; der einfachen Formulierung halber seien beide Punkte auf der reellen Achse angenommen, und zwar $s_1 = 0$ (also $s_0 = \sigma_B > 0$), so daß die Partialsummen im Punkte s_1 die Werte der summatorischen Funktion $S_n = a_1 + \cdots a_n$ ergeben. Hier ist vor allem ein Satz von *Dirichlet*[26]) über gewöhnliche *Dirichlet*sche Reihen (mit $\sigma_B = 1$) zu erwähnen, der besagt, daß aus

$$\frac{S_n}{n} \to A \qquad \text{(für } n \to \infty)$$

folgt

$$f(s)(s-1) \to A. \qquad \text{(für zu 1 abnehm. } s)$$

Auch dieser Satz läßt sich nicht ohne weiteres umkehren[27]), und zwar nicht einmal, wenn den Koeffizienten der Reihe Bedingungen der Art auferlegt werden (z. B. daß sie alle positiv sein sollen), welche beim vorhergehenden Problem für die Gültigkeit des Umkehrsatzes genügten; es läßt sich im allgemeinen nur behaupten[28]), daß aus $f(s)(s-1) \to A$ folgt

$$\limsup_{n\to\infty} \frac{S_n}{n} \geqq A \quad \text{und} \quad \liminf_{n\to\infty} \frac{S_n}{n} \leqq A.$$

Bei den obigen Sätzen, wo aus dem Verhalten der Funktion auf das Verhalten der Reihe geschlossen wurde, bezog sich die Annahme über die Funktion stets auf ihr Verhalten in der Nähe eines *einzigen* Punktes auf der Konvergenzgeraden. Von *Landau*[29]) rührt der folgende

---

die „allseitige" Bedingung $a_n = O(\lambda_n^{\alpha-1}(\lambda_n - \lambda_{n-1}))$ ersetzt wird. Im speziellen Falle $\alpha = 0$ reduziert sich dieser letzte Satz auf den oben erwähnten *Littlewood*schen Satz (über Konvergenz).

26) *G. Lejeune Dirichlet,* Sur un théorème relatif aux séries, J. de math. (2) 1 (1856), p. 80—81 = Werke, Bd. 2, p. 195—200. Verallgemeinerungen solcher Sätze finden sich z. B. bei *A. Pringsheim,* Zur Theorie der Dirichletschen Reihen, Math. Ann. 37 (1890), p. 38—60; *A. Berger,* Recherches sur les valeurs moyennes dans la théorie des nombres, Nova Acta Upsala (3) 14 (1891), Nr. 2; *J. Franel,* Sur la théorie des séries, Math. Ann. 52 (1899), p. 529—549.

27) Wäre dies der Fall, so „würde das ganze Gebäude der Primzahltheorie mit großer Geschwindigkeit errichtet werden können" (*Landau,* Handbuch, Bd. 1, p. 114).

28) *O. Hölder,* Grenzwerte von Reihen an der Convergenzgrenze, Math. Ann. 20 (1882), p. 535—549. Vgl. auch *E. Landau,* Über die zu einem algebraischen Zahlkörper gehörige Zetafunktion und die Ausdehnung der Tschebyschefschen Primzahlentheorie auf das Problem der Vertheilung der Primideale, Crelles J. 125 (1903), p. 64—188.

29) *E. Landau,* Beiträge zur analytischen Zahlentheorie, Palermo Rend. 26 (1908), p. 169—302. Eine Verschärfung seines Satzes gab *Landau* a. a. O. 21).

tiefliegende Satz her, in welchem Voraussetzungen über die Funktion bei Annäherung an *alle* Punkte der Konvergenzgeraden gemacht werden und daraus ein sehr genaues Resultat über das Verhalten der Reihe (nämlich Umkehrung des obigen *Dirichlet*schen Satzes) hergeleitet wird: Es sei eine gewöhnliche *Dirichlet*sche Reihe (2) mit positiven Koeffizienten (und $\sigma_B = 1$) in allen Punkten der Konvergenzgeraden $\sigma = 1$ regulär mit Ausnahme des Punktes $s = 1$, wo sie einen Pol erster Ordnung mit dem Residuum A besitzt; ferner sei für $\sigma \geqq 1$ (und $|t| \to \infty$) die Relation $f(s) = O(|t|^k)$ bei passender Wahl einer Konstanten k erfüllt. Dann ist

$$\lim_{n\to\infty} \frac{a_1 + a_2 + \cdots a_n}{n} = A.$$

Landau[30]) hat später diesen Satz auf beliebige *Dirichlet*sche Reihen (1) übertragen. Eine Verallgemeinerung dieses *Landau*schen Satzes und andere ähnliche Sätze haben auf anderem Wege *Hardy* und *Littlewood*[31]) gefunden.

6. Das Konvergenzproblem. In Nr. 2 wurde besprochen, wie die drei Konvergenzabszissen σ_A, σ_B, σ_G von den Koeffizienten und Exponenten der Reihe aus bestimmt werden können. Wir wenden uns nun zu einem viel schwierigeren Problem, dem sogenannten *Konvergenzproblem* der *Dirichlet*schen Reihen, nämlich zur Frage, ob und in welcher Weise die Lage dieser Abszissen (und vor allem der Konvergenzabszisse σ_B) mit einfachen analytischen Eigenschaften der durch die Reihe dargestellten Funktion $f(s)$ zusammenhängt. Im speziellen Fall $\lambda_n = n$ (Potenzreihe in e^{-s}) ist diese Frage ja einfach dahin zu beantworten, daß die Reihe genau so weit konvergiert, wie die Funktion $f(s)$ *regulär* bleibt; in der Tat, es liegt ja hier immer ein singulärer Punkt auf der Konvergenzgeraden $\sigma = \sigma_B$ $(= \sigma_A = \sigma_G)$. Es gilt aber nicht nur in dem ganz speziellen Fall $\lambda_n = n$, sondern für alle solche *Dirichlet*sche Reihen (1), deren Exponentenfolge die Bedingung

$$\lim_{n\to\infty} \frac{\log n}{\lambda_n} = 0 \tag{14}$$

erfüllt (wo also, nach Nr. 2, $\sigma_B = \sigma_A$ ist), daß das Konvergenzproblem in einfachster Weise zu lösen ist; die Funktion $f(s)$ braucht wohl hier *nicht* auf (oder in unendlicher Nähe links von) der Konvergenzgeraden

30) *E. Landau,* Handbuch, p. 874.

31) *G. H. Hardy* u. *J. Littlewood,* a) New proofs of the prime-number theorem and similar theorems, Quart. J. 46 (1915), p. 215—219; b) Contributions to the theory of the Riemann Zetafunction and the theory of the distributions of primes, Acta Math. 41 (1918), p. 119—196.

$\sigma = \sigma_B$ Singularitäten zu besitzen, es gilt aber der fast ebenso einfache Satz, daß die Reihe genau so weit konvergiert, wie die Funktion $f(s)$ *regulär und beschränkt* bleibt, d. h. es ist $\sigma_B (= \sigma_A) = \sigma_b$, wo σ_b (wie überall im folgenden) *die untere Grenze aller Zahlen* σ_0 *bezeichnet, für welche* $f(s)$ *in der Halbebene* $\sigma > \sigma_0$ *regulär ist und einer Ungleichung* $|f(s)| < K = K(\sigma_0)$ *genügt.*[32]) Für Reihen (1), deren Exponentenfolge „sehr" schnell ins Unendliche wächst, gilt übrigens, daß die Funktion $f(s)$ überhaupt nicht über die Konvergenzgerade hinaus fortgesetzt werden kann; es läßt sich nämlich, wie zuerst *Wennberg*[33]) und später allgemeiner *Carlson* und *Landau*[34]) und *Szász*[35]) gezeigt haben, der *Hadamard-Fabry*sche Lückensatz für Potenzreihen auf beliebige *Dirichlet*sche Reihen übertragen. Der Satz lautet hier, daß für jede zu einer Exponentenfolge mit $\lambda_n : n \to \infty$ und $\liminf (\lambda_{n+1} - \lambda_n) > 0$ gehörige Reihe (1) die Konvergenzgerade $\sigma = \sigma_B (= \sigma_A)$ eine *wesentlich singuläre Linie* ist.*)

In anderer Richtung — weil Voraussetzungen über die *Koeffizienten* und nicht über die Exponenten gemacht werden — liegt ein

---

32) Dieser Satz wurde zuerst von *H. Bohr*, a. a. O. 6 b) bewiesen. Einen äußerst einfachen Beweis gab *E. Landau*, Über die gleichmäßige Konvergenz Dirichletscher Reihen, Math. Ztschr. 11 (1921), p. 317—318. Der Satz umfaßt offenbar den für die Potenzreihen ($\lambda_n = n$) gültigen Satz als Spezialfall, denn im Falle $\lambda_n = n$ ist ja $f(s)$ *periodisch* mit der Periode $2\pi i$, und $f(s)$ wird daher von selbst in jeder Halbebene $\sigma \geqq \sigma_0$ beschränkt sein, wenn sie dort regulär ist.

Zur Definition der Abszisse σ_b vgl. auch die Arbeit von *H. Bohr*, Ein Satz über Dirichletsche Reihen, Münch. Sitzungsber. 1913, p. 557—562, worin bewiesen wird, daß, falls die durch eine beliebige *Dirichlet*sche Reihe (mit $\sigma_A < \infty$) definierte Funktion $f(s)$ nur in irgendeiner *Viertelebene* $\sigma > \sigma_0$, $t > t_0$ regulär und beschränkt ist, sie von selbst in der ganzen *Halbebene* $\sigma > \sigma_0$ regulär und beschränkt bleiben wird.

33) *S. Wennberg*, Zur Theorie der Dirichletschen Reihen, Diss. Upsala 1920.

34) *F. Carlson* u. *E. Landau*, Neuer Beweis und Verallgemeinerungen des Fabryschen Lückensatzes, Gött. Nachr. 1921, p. 184—188. Vgl. hierzu auch *L. Neder*, Über einen Lückensatz für Dirichletsche Reihen, Math. Ann. 85 (1922), p. 111—114.

35) *O. Szasz*, Über Singularitäten von Potenzreihen und Dirichletschen Reihen am Rande des Konvergenzbereiches, Math. Ann. 85 (1922), p. 99—110.

*) In einer soeben erschienenen interessanten Abhandlung von *A. Ostrowski*, Über vollständige Gebiete gleichmäßiger Konvergenz von Folgen analytischer Funktionen, Hamburger Seminar 1 (1922), p. 327—350, die sich allgemein mit den Abschnittsfolgen einer *Dirichlet*schen Reihe beschäftigt, wird u. a. auch ein *Lückensatz* bewiesen, wo die Exponentenfolge $\{\lambda_n\}$ nur „ab und zu" große Lücken aufweist; es wird gezeigt, daß die den Lücken entsprechende Abschnittsfolge so weit konvergiert, wie es von vornherein überhaupt gehofft werden konnte, d. h. so weit, wie die Funktion sich regulär verhält. Vgl. hierzu auch *H. Bohr*, a. a. O.[44])

wichtiger Satz von *Landau*[36]), der ebenfalls die Verallgemeinerung eines bekannten (*Vivanti*schen) Satzes über Potenzreihen darstellt und der besagt, daß, wenn alle Koeffizienten a_n *positiv* sind, der Punkt σ_B, worin die Konvergenzgerade durch die reelle Achse geschnitten wird, immer ein *singulärer* Punkt der Funktion ist.

Für solche *Dirichlet*sche Reihen (1), für welche die Exponentenfolge $\{\lambda_n\}$ die Bedingung (14) *nicht* erfüllt, z. B. für die *gewöhnlichen Dirichlet*schen Reihen (2), stellt sich das Konvergenzproblem (wenn keine besonderen Bedingungen über die Koeffizienten gemacht werden) viel schwieriger, und es scheint hier überhaupt zweifelhaft, ob es möglich ist, die Lage der Konvergenzgeraden $\sigma = \sigma_B$ durch „einfache" analytische Eigenschaften der dargestellten Funktion genau zu charakterisieren.[37]) Bevor wir über die vorliegenden Resultate berichten können, müssen einige charakteristische Eigenschaften erörtert werden, die einer jeden von einer *Dirichlet*schen Reihe (1) dargestellten Funktion zukommen, und die das Verhalten dieser Funktion $f(s)$ für ins Unendliche wachsende Werte der Ordinate t betreffen. Zuerst nennen wir den Satz, daß jede solche Funktion $f(s)$ in der Halbebene $\sigma > \sigma_B + \varepsilon$ die Limesgleichung

$$(15) \qquad f(s) = f(\sigma + it) = o(|t|) \qquad (\text{für } |t| \to \infty)$$

erfüllt, sogar gleichmäßig in σ.[38]) Es bezeichne nunmehr hier (und überall im folgenden) σ_e ($\leq \sigma_B$) *die untere Grenze aller Abszissen* σ_0,

---

36) *E. Landau*, Über einen Satz von Tschebyschef, Math. Ann. 61 (1905), p. 527—550. Verallgemeinerungen des *Landau*schen Satzes sind gegeben von *M. Fekete*, a) Sur les séries de Dirichlet, Paris C. R. 150 (1910), p. 1033—1036; b) Sur une théorème de M. Landau, Paris C. R. 151 (1910), p. 497—500.

Für die von *Landau* betrachteten Reihen mit $a_n > 0$ ist offenbar $\sigma_A = \sigma_B$; es sei beiläufig bemerkt, daß das bloße Bestehen dieser Gleichung $\sigma_A = \sigma_B$ *nicht* genügt um zu schließen, daß die Konvergenzgerade einen singulären Punkt enthält. *H. Bohr*, Über die Summabilität Dirichletscher Reihen, Gött. Nachr. 1909, p. 247—262.

37) So kennt man z. B. keinen allgemeinen Satz über gewöhnliche *Dirichlet*sche Reihen (2), der uns aus einfachen analytischen Eigenschaften der durch die Zetareihe mit abwechselndem Vorzeichen (5) definierten ganzen transzendenten Funktion $\zeta(s)(1 - 2^{1-s})$ darüber Aufschluß gibt, daß diese Reihe eben die Konvergenzabszisse $\sigma_B = 0$ besitzt. Anders verhält es sich, wie aus den späteren Ausführungen hervorgehen wird, mit der gleichmäßigen Konvergenzabszisse $\sigma_G = 1$ und der absoluten Konvergenzabszisse $\sigma_A = 1$ dieser Reihe.

38) *E. Landau*, Handbuch, Bd. 2, p. 824. Der Satz findet sich schon, wie von *Landau* angegeben, implizite bei *O. Perron*, a. a. O. 13). Wie von *H. Bohr*, Bidrag til de Dirichlet'ske Rækkers Theori, Habilitationsschrift, Kopenhagen 1910, p. 32, bewiesen, läßt sich die Gleichung $f(s) = o(|t|)$ durch keine Gleichung der Form $f(s) = o(|t|^\alpha)$ mit $\alpha < 1$ ersetzen.

für welche $f(s)$ in der Halbebene $\sigma > \sigma_0$ regulär und von endlicher Größenordnung in bezug auf t ist, d. h. gleich $O(|t|^k)$ bei passender Wahl von $k = k(\sigma_0)$. Für jedes feste $\sigma > \sigma_e$ definieren wir alsdann die „Größenordnung" $\mu = \mu(\sigma)$ von $f(s)$ auf der vertikalen Geraden mit der Abszisse σ als die untere Grenze aller Zahlen α, für die $f(\sigma + it) = O(|t|^\alpha)$ ist. Die somit für $\sigma > \sigma_e$ definierte *Funktion* $\mu(\sigma)$ ist nach (15) gewiß $\leqq 1$ für $\sigma > \sigma_B$, und sie ist ferner, wie leicht zu sehen[39]), immer $\geqq 0$ für $\sigma > \sigma_B$. Die genaue Bestimmung der zu einer gegebenen *Dirichlet*schen Reihe gehörigen μ-Funktion ist im allgemeinen ein sehr schwieriges Problem. Doch läßt sich mit Hilfe der bekannten allgemeinen Sätze von *Phragmén* und *Lindelöf* (Artikel II C 4, Nr. **10**) über das Verhalten analytischer Funktionen in der Nähe einer wesentlich singulären Stelle (hier des Punktes $s = \infty$) leicht zeigen, *daß $\mu(\sigma)$ im ganzen Definitionsintervall $\sigma > \sigma_e$ eine stetige konvexe Funktion ist, die überall $\geqq 0$ ist, und die mit abnehmendem σ niemals abnimmt.* Wenn nicht nur $\sigma_B < \infty$, sondern auch $\sigma_G < \infty$ ist (was ja z. B. für jede gewöhnliche *Dirichlet*sche Reihe mit $\sigma_B < \infty$ der Fall ist), wird übrigens $\mu(\sigma)$ gleich 0 sein für alle hinreichend großen σ, nämlich mindestens für $\sigma > \sigma_G$.[40])

Kehren wir jetzt zum Konvergenzproblem zurück. *Landau*[41]) war der erste, der mit Erfolg die Frage angegriffen hat, inwiefern man aus der Kenntnis der Größenordnung der durch eine *Dirichlet*sche Reihe dargestellten Funktion (d. h. aus ihrer μ-Funktion) Schlüsse über die Lage der Konvergenzgeraden $\sigma = \sigma_B$ ziehen kann. Das Problem wurde später von *Schnee*[42]) in einer bedeutsamen Arbeit und von *Landau*[43]) selbst weiter verfolgt. Die Untersuchungen umfassen

---

39) *K. Ananda-Rau,* Note on a property of Dirichlet's series, London math. Soc. (2) 19 (1920), p. 114—116; *T. Jansson,* Über die Größenordnung Dirichletscher Reihen, Arkiv f. Mat., Astr. och Fys. 15 (1920), No. 6.

40) Die angeführten Resultate über die μ-Funktion finden sich im wesentlichen implizite bei *E. Lindelöf,* Quelques remarques sur la croissance de la fonction $\zeta(s)$, Bull. de Soc. math. (2) 32 (1908), p. 341—356. Vgl. auch *H. Bohr,* a. a. O. 38), p. 28—36; *G. H. Hardy-M. Riesz,* a. a. O. 1), p. 16—18, und die a. a. O. 39) erwähnten Abhandlungen.

Eine sich auf das Verhalten der oberen Grenze $L(\sigma)$ der Funktion $|f(s)|$ im Intervall $\sigma > \sigma_G$ beziehende Ergänzung des *Lindelöf*schen Satzes über die Konvexität der μ-Funktion ist von *G. Doetsch,* Über die obere Grenze des absoluten Betrages einer analytischen Funktion auf Geraden, Math. Ztschr. 8 (1920), p. 237—240, gegeben.

41) *E. Landau,* a. a. O. 29).

42) *W. Schnee,* Zum Konvergenzproblem der Dirichletschen Reihen, Math. Ann. 66 (1909), p. 337—349.

43) *E. Landau,* a) Über das Konvergenzproblem der Dirichletschen Reihen,

nicht den allgemeinsten Typus *Dirichlet*scher Reihen, sondern es wird der Exponentenfolge $\{\lambda_n\}$ die (für $\lambda_n = \log n$ erfüllte) Bedingung

$$(16) \qquad \frac{1}{\lambda_{n+1} - \lambda_n} = O(e^{\lambda_n k}) \qquad (k > 0)$$

auferlegt, welche offenbar darauf hinausläuft, daß die Exponenten nirgends allzu dicht aufeinander folgen dürfen.[44]) Indem wir uns der Einfachheit halber auf die gewöhnlichen *Dirichlet*schen Reihen (2) beschränken, besagt das allgemeinste Resultat von *Landau* und *Schnee*: Es sei die Reihe (2) in einer gewissen Halbebene $\sigma > \sigma_0$ nicht nur absolut konvergent, sondern „so deutlich" absolut konvergent, daß $a_n n^{-\sigma_0}$ gleich $O(n^{-1+\varepsilon})$ bei jedem $\varepsilon > 0$ ist; es sei ferner die durch die Reihe dargestellte Funktion $f(s)$ für $\sigma > \sigma_0 - \alpha$ $(\alpha > 0)$ regulär und gleich $O(|t|^k)$. Dann konvergiert die Reihe jedenfalls für

$$\sigma > \sigma_0 - \frac{\alpha}{1+k}.$$

Hierin ist speziell das Resultat (von *Schnee*[42])) enthalten, daß, falls $f(s)$ für $\sigma > \sigma_1$ $(= \sigma_0 - \alpha)$ regulär und, bei jedem $\delta > 0$, gleich $O(|t|^\delta)$ ist, $\sigma_B \leqq \sigma_1$ ist, d. h. eine *Dirichlet*sche Reihe (2) ist mindestens so weit nach links konvergent, wie die zugehörige μ-Funktion gleich 0 ist. Die genannten Sätze geben, mit Hilfe der μ-Funktion, hinreichende Bedingungen für die Konvergenz der Reihe in einer gewissen Halbebene, aber keine Bedingungen, die zugleich notwendig und hinreichend sind. Solche Bedingungen gibt es aber überhaupt nicht, d. h. es ist nicht möglich, von der bloßen Kenntnis der μ-Funktion zu einer *genauen* Bestimmung der Konvergenzabszisse σ_B zu gelangen; in der Tat[45]), es existieren *Dirichlet*sche Reihen, sogar vom Typus (2), die dieselbe μ-Funktion, aber verschiedene Konvergenzabszissen σ_B besitzen.

Palermo Rend. 28 (1909), p. 113—151; b) Neuer Beweis eines Hauptsatzes aus der Theorie der Dirichletschen Reihen, Leipziger Ber. 69 (1917), p. 336—348.

44) Die Bedingung (16) ist übrigens nicht die von *Landau* und *Schnee* benutzte; sie wurde erst später von *H. Bohr,* Einige Bemerkungen über das Konvergenzproblem Dirichletscher Reihen, Palermo Rend. 37 (1914), p. 1—16, eingeführt, der zeigte, daß sie die für die betreffenden Untersuchungen „genau richtige" Bedingung ist, d. h. die für die Gültigkeit der *Landau-Schnee*schen Sätze notwendige und hinreichende.

Zur Orientierung sei bemerkt, daß eine Exponentenfolge $\{\lambda_n\}$, die der Bedingung (16) genügt, auch der Bedingung $\limsup \log n : \lambda_n < \infty$ genügt (aber nicht umgekehrt), so daß (nach Nr. 2) jede Reihe (1), die (16) erfüllt, gewiß ein *absolutes* Konvergenzgebiet besitzt, falls sie überhaupt ein Konvergenzgebiet besitzt.

45) *H. Bohr,* a. a. O. 38), p. 34.

Ganz anders verhält es sich mit dem Problem der Bestimmung der *gleichmäßigen* Konvergenzabszisse σ_G. Hier gilt nach *Bohr*[46]) der einfache Satz, daß jede *Dirichlet*sche Reihe (1), deren Exponentenfolge die Bedingung (16) erfüllt[47]), also speziell jede gewöhnliche *Dirichlet*sche Reihe (2), so weit nach links gleichmäßig konvergiert, wie von vornherein überhaupt gehofft werden konnte, d. h. es ist $\sigma_G = \sigma_b$, wo σ_b die oben definierte „Regularitäts- und Beschränktsheitsabszisse" bedeutet.

Es erübrigt die Frage nach dem Zusammenhang der Lage der *absoluten* Konvergenzgeraden $\sigma = \sigma_A$ mit den analytischen Eigenschaften der dargestellten Funktion zu erörtern. Diese Frage kann auch so gestellt werden, daß es sich um die Bestimmung der Breite des Streifens $\sigma_b \leqq \sigma \leqq \sigma_A$ handelt, in welchem die Funktion $f(s)$ über die absolute Konvergenzhalbebene hinaus regulär und beschränkt bleibt, und dann natürlich vor allem um den maximalen Wert dieser Breite bei gegebener Exponentenfolge $\{\lambda_n\}$. Diese letztere Frage, zu deren Behandlung Hilfsmittel ganz anderer Art herangezogen werden müssen als diejenigen, worauf die oben referierten Untersuchungen beruhen, wird am Ende der nächsten Nummer besprochen. Dabei werden wir uns wesentlich auf die gewöhnlichen *Dirichlet*schen Reihen (2) beschränken; bei diesen Reihen ist, nach dem obigen, $\sigma_b = \sigma_G$, und der besprochene Streifen $\sigma_b \leqq \sigma \leqq \sigma_A$ kann daher auch als derjenige Streifen charakterisiert werden, in welchem die Reihe gleichmäßig konvergiert ohne absolut zu konvergieren.

7. Anwendung der Theorie der diophantischen Approximationen. Die Rolle, welche die diophantischen Approximationen beim Studium der *Dirichlet*schen Reihen spielen, tritt am deutlichsten hervor bei der Aufgabe, die Menge der Werte zu bestimmen, welche eine gewöhnliche *Dirichlet*sche Reihe (2) annimmt, wenn die Variable s eine feste vertikale Gerade $\sigma = \sigma_0$ durchläuft. Hierbei umkreist offenbar jedes *einzelne* Glied, d. h. sein Bildpunkt in einer komplexen Ebene, einen festen *Kreis*; in der Tat, es ist, $a_n = \varrho_n e^{i\varphi_n}$ gesetzt,

$$\frac{a_n}{n^{\sigma_0+it}} = \frac{\varrho_n}{n^{\sigma_0}} \cdot e^{i(\varphi_n - t\log n)},$$

46) *H. Bohr,* a. a. O. 6a) und b).

47) Bei diesem Problem — im Gegensatz zu dem obigen — ist die Bedingung (16) übrigens *nicht* die „genau richtige", d. h. die für die Gültigkeit des Satzes notwendige und hinreichende. Eine wesentliche Erweiterung der Bedingung (16) ist von *E. Landau,* a. a. O. 32) gegeben. Vgl. hierzu auch *L. Neder,* a) a. a. O. 7); b) Zum Konvergenzproblem der Dirichletschen Reihen beschränkter Funktionen, Math. Ztschr. 14 (1922), p. 149—158.

wo der Modul $r_n = \varrho_n n^{-\sigma_0}$ nicht von t abhängt. Wie unmittelbar zu sehen, bewegen sich aber die Glieder nicht in der Weise „quasi unabhängig" voneinander jedes auf seinem Kreise, daß man bei passender Wahl der Variablen t erreichen kann, daß eine beliebig vorgegebene Anzahl N dieser Glieder beliebig nahe an N beliebig gegebene Punkte der entsprechenden N Kreisperipherien fallen; es ist ja dies z. B. für die drei Glieder $\frac{a_2}{2^s}, \frac{a_3}{3^s}, \frac{a_6}{6^s}$ gewiß nicht der Fall, denn aus der Gleichung $\frac{1}{2^s} \cdot \frac{1}{3^s} = \frac{1}{6^s}$ folgt sofort, daß, wenn die Bildpunkte der beiden ersten Glieder „sehr" nahe an zwei festen Punkten P_2 und P_3 auf ihren respektiven Kreisen liegen, der Bildpunkt des dritten Gliedes von selbst sehr nahe an einen festen, von P_2 und P_3 abhängigen, Punkt P_6 auf seiner Kreisperipherie fallen wird. Betrachten wir aber nicht die Größen $\frac{1}{n^s}$, wo n die sämtlichen Zahlen $1, 2, 3 \cdots$ durchläuft, sondern nur die Größen $\frac{1}{p_n^s}$, wo p_n die *Primzahlen* $2, 3, 5 \cdots$ durchläuft, so stellt die Sache sich ganz anders. Hier können wir nämlich, bei passender Wahl von t, erreichen, daß die Bildpunkte der N Größen $\frac{1}{2^s}, \frac{1}{3^s} \cdots \frac{1}{p_N^s}$ mit beliebig vorgegebener Genauigkeit in N beliebig gegebene Punkte ihrer N Kreisperipherien fallen; die Amplituden dieser Größen sind nämlich durch $-t\log 2, -t\log 3, \cdots -t\log p_N$ gegeben, und weil die Primzahllogarithmen — wegen der eindeutigen Zerlegbarkeit einer ganzen Zahl in Primfaktoren — im rationalen Körper *linear unabhängig* sind, können die genannten N Amplituden nach einem berühmten *Kronecker*schen Satz über diophantische Approximationen beliebig nahe (modulo 2π) an N beliebig gegebene Größen gebracht werden. Von dieser Bemerkung ausgehend hat *Bohr*[48]) die Bedeutung der diophantischen Approximationen für verschiedene Probleme in der Theorie der *Dirichlet*schen Reihen gezeigt; es sollen im folgenden die wesentlichsten Resultate dieser Untersuchung kurz angegeben werden.

Es bezeichne $p_{n_1}^{\nu_1} p_{n_2}^{\nu_2} \cdots p_{n_r}^{\nu_r}$ die Zerlegung der ganzen Zahl n in Primfaktoren, und es sei in der beliebig gegebenen gewöhnlichen *Dirichlet*schen Reihe

$$\sum \frac{a_n}{n^s} = \sum a_n \left(\frac{1}{p_{n_1}^s}\right)^{\nu_1} \left(\frac{1}{p_{n_2}^s}\right)^{\nu_2} \cdots \left(\frac{1}{p_{n_r}^s}\right)^{\nu_r}$$

48) Vgl. insb. *H. Bohr,* Über die Bedeutung der Potenzreihen unendlich vieler Variabeln in der Theorie der Dirichletschen Reihen $\sum \frac{a_n}{n^s}$, Gött. Nachr. 1913, p. 441—488.

$\frac{1}{p_1^s} = x_1, \frac{1}{p_2^s} = x_2, \cdots \frac{1}{p_m^s} = x_m \cdots$ gesetzt, wodurch die Reihe die Form annimmt:

$$P(x_1, x_2, \cdots x_m \cdots) = \sum_{n=1}^{\infty} a_n x_{n_1}^{\nu_1} x_{n_2}^{\nu_2} \cdots x_{n_r}^{\nu_r}$$
$$= c + \sum c_\alpha x_\alpha + \sum c_{\alpha,\beta} x_\alpha x_\beta + \sum c_{\alpha,\beta,\gamma} x_\alpha x_\beta x_\gamma + \cdots$$

wo $c = a_1$, $c_\alpha = a_{p_\alpha}$, $c_{\alpha,\beta} = a_{p_\alpha p_\beta}$, $\cdots$ ist. Hier sind vorläufig die Größen x_m alle Funktionen der einen Variablen s. Nun denken wir uns aber — weil ja oben gesehen wurde, daß die $x_m = p_m^{-s}$ sich in gewisser Beziehung „fast" so benehmen, als wären sie unabhängig voneinander — das Band zwischen den x_m ganz aufgelöst, d. h. wir fassen die x_m als *voneinander unabhängige* Variablen auf. Die obige Reihe $P(x_1, x_2, \cdots x_m \cdots)$ wird dann offenbar *eine Potenzreihe in den unendlich vielen Variabeln* $x_1, x_2, \cdots$, von der wir sagen werden, daß sie der gegebenen *Dirichlet*schen Reihe (2) entspricht. Betreffs der am Anfang des Paragraphen gestellten Frage nach *dem Verhalten der Reihe* (2) *auf einer vertikalen Geraden* $\sigma = \sigma_0$ ergibt sich dann der Satz: Es sei $\sigma_0 > \sigma_A$ (oder nur $\sigma_0 > \sigma_G$), und es bezeichne $U(\sigma_0)$ bzw. $W(\sigma_0)$ die Menge der Werte, welche die Reihe $f(s)$ auf bzw. in unendlicher Nähe[49]) der Geraden $\sigma = \sigma_0$ annimmt. Ferner bezeichne $M = M(\sigma_0)$ die Menge der Werte, welche die der *Dirichlet*schen Reihe entsprechende Potenzreihe $P(x_1, x_2 \cdots)$ annimmt, wenn die Variabeln $x_1, x_2 \cdots$ *unabhängig voneinander* die Kreise $|x_m| = p_m^{-\sigma_0}$ $(m = 1, 2 \cdots)$ durchlaufen. Dann gilt, 1. daß die Menge U in der Menge M *überall dicht* liegt, und 2. daß die Menge W mit der Menge M *identisch* ist. Die Wirkungsweise dieses Satzes wird durch seine später zu erwähnende Anwendung auf die Zetareihe deutlich hervorgehen.

Über die (in Nr. 6 erwähnte) Frage nach *der oberen Grenze* T *der Differenz* $\sigma_A - \sigma_b$ *für alle Dirichletschen Reihen* (2), findet man ferner mit Hilfe der Theorie der diophantischen Approximationen den Satz: Es ist
$$T = \frac{1}{S},$$
wo S die obere Grenze aller positiven Zahlen α mit der Eigenschaft bezeichnet, daß jede in einem Gebiete $|x_m| \leq G_m$ $(m = 1, 2 \cdots)$ *beschränkte*[50]) Potenzreihe $P(x_1, x_2 \cdots)$ im Gebiet $|x_m| \leq \varepsilon_m G_m (m = 1, 2 \ldots)$

49) Dies letzte so zu verstehen, daß eine Zahl w dann und nur dann zur Menge $W(\sigma_0)$ gehört, falls die Gleichung $f(s) = w$ in jedem Streifen $\sigma_0 - \varepsilon < \sigma < \sigma_0 + \varepsilon$ eine Lösung besitzt.

50) Eine Potenzreihe $P(x_1, x_2 \ldots)$ in unendlichvielen Variabeln heißt — nach *D. Hilbert*, Wesen und Ziele einer Analysis der unendlichvielen unabhängigen

absolut konvergiert, wenn nur $\Sigma \varepsilon_m^\alpha$ konvergiert (und $0 < \varepsilon_m < 1$). Es ist hierdurch die Bestimmung der „Maximalbreite" T auf die Bestimmung der (in der Theorie der Potenzreihen wesentlichen) Konstanten S zurückgeführt. Über diese Konstante S findet man sofort, daß sie $\geqq 2$ ist, woraus folgt, daß $T \leqq \frac{1}{2}$ ist.[51]) Die besonders wichtige Frage, ob nicht $T = 0$ ist (d. h. ob nicht immer $\sigma_A = \sigma_b$ ist), wurde von *Toeplitz*[52]) gelöst, der durch Untersuchungen über quadratische Formen mit unendlichvielen Variabeln zeigte, daß $S \leqq 4$, also $T \geqq \frac{1}{4}$ ist. Das Problem, S (und damit T) genau zu bestimmen, ist noch ungelöst.

Ein bemerkenswertes Resultat ergibt sich, wenn man den besprochenen Zusammenhang zwischen *Dirichlet*schen Reihen und Potenzreihen mit unendlichvielen Variabeln nicht auf die allgemeinen *Dirichlet*schen Reihen vom Typus (2), sondern auf zwei *spezielle Klassen* solcher Reihen anwendet, nämlich auf diejenigen Reihen (2), die formal eine Zerlegung in Addenden bzw. in Faktoren derart zulassen, daß dadurch *die einzelnen Primzahlen separiert werden*, d. h. deren Koeffizienten entweder die Bedingung: $a_n = 0$ für alle n, die mindestens zwei verschiedene Primzahlen enthalten, oder die Bedingung: $a_m a_l = a_{ml}$ für teilerfremde m und l erfüllen. Für diese beiden Typen *Dirichlet*scher Reihen — die übrigens fast alle in der analytischen Zahlentheorie vorkommenden Reihen (2) umfassen — gilt immer die Gleichung $\sigma_A = \sigma_b$, d. h. eine jede *Dirichlet*sche Reihe einer dieser Typen ist (im Gegensatz zu einer beliebigen Reihe (2)) *genau so weit*

Variabeln, Palermo Rend. 27 (1909), p. 59—74 — *beschränkt* in einem Gebiete $|x_m| \leqq G_m (m = 1, 2 \cdots)$, wenn 1. bei jedem festen m der m^{te} „Abschnitt" $P_m(x_1, \cdots x_m)$ im Gebiete $|x_1| \leqq G_1, \cdots |x_m| \leqq G_m$ absolut konvergiert, und 2. eine absolute Konstante K derart existiert, daß bei jedem m und $|x_1| \leqq G_1, \cdots |x_m| \leqq G_m$ die Ungleichung $|P_m(x_1, \cdots x_m)| < K$ besteht.

51) *H. Bohr*, a. a. O. 48). Dies spezielle Resultat $T \leqq \frac{1}{2}$ ist, wie *G. H. Hardy*, The application of Abel's method of summation to Dirichlet's series, Quart. J. of math. 47 (1916), p. 176—192 gezeigt hat, kein tiefliegendes, d. h. es läßt sich auch ohne Zurückgreifen auf die Theorie der Potenzreihen mit unendlichvielen Variabeln leicht herleiten. Vgl. auch eine interessante Note von *F. Carlson*, Sur les séries de Dirichlet, Paris C. R. 172 (1921), p. 838—840, und die eben erschienene Arbeit von *K. Grandjot*, Über das absolute Konvergenzproblem der Dirichletschen Reihen, Diss. Göttingen 1922, in welcher ein dem *Schnee-Landau*schen Satze über das Konvergenzproblem (vgl. Nr. 6) entsprechender Satz über das absolute Konvergenzproblem abgeleitet wird.

52) *O. Toeplitz*, Über eine bei den Dirichletschen Reihen auftretende Aufgabe aus der Theorie der Potenzreihen von unendlichvielen Veränderlichen, Gött. Nachr. 1913, p. 417—432.

absolut konvergent, wie die dargestellte Funktion regulär und beschränkt bleibt.[53])

Die oben erwähnten Untersuchungen können von den gewöhnlichen *Dirichlet*schen Reihen (2) auf den allgemeinen Typus (1) erweitert werden.[54]) Eine ähnliche Rolle, wie die von den Primzahllogarithmen gebildete Zahlenfolge für die spezielle Exponentenfolge $\{\lambda_n = \log n\}$, spielt im Falle einer *beliebigen Exponentenfolge* $\{\lambda_n\}$ eine sogenannte *Basis* dieser Folge $\{\lambda_n\}$, d. h. eine (aus endlich oder abzählbarvielen Zahlen bestehende) Folge von *linear unabhängigen* Zahlen $\beta_1, \beta_2, \ldots$ mit der Eigenschaft, daß jeder der Exponenten λ_n als lineare Funktion endlichvieler β mit rationalen Koeffizienten darstellbar ist. Ein besonders einfacher Fall liegt vor, wenn die Exponenten λ_n *selbst* linear unabhängig sind (also selbst eine Basis bilden). Hier gilt ganz allgemein der Satz, daß $\sigma_A = \sigma_b$ ist.[55]) Dies ist die Verallgemeinerung eines obigen Satzes über gewöhnliche *Dirichlet*sche Reihen (2), nach welchem die Gleichung $\sigma_A = \sigma_b$ immer gilt, wenn $a_n = 0$ ist für zusammengesetztes n.

8. Über die Darstellbarkeit einer Funktion durch eine Dirichletsche Reihe. Beim Konvergenzproblem in Nr. 6 (und Nr. 7) waren wir von einer Funktion $f(s)$ ausgegangen, von der *vorausgesetzt* wurde, daß sie in einer gewissen Halbebene durch eine *Dirichlet*sche Reihe dargestellt war, und es handelte sich darum, die Lage der Konvergenzabszissen dieser Reihe aus den analytischen Eigenschaften der Funktion zu bestimmen. Mit dieser Frage verwandt, aber davon wesentlich zu trennen, ist die Frage, welche Bedingungen eine in einer gewissen Halbebene $\sigma > \sigma_0$ *beliebig gegebene* analytische Funktion erfüllen muß, damit sie überhaupt in eine (dort konvergente) *Dirichlet*sche Reihe entwickelt werden kann. Es liegt hierbei nahe, von dem Satze über

---

53) Der „Grund", weshalb die Zetareihe mit abwechselndem Vorzeichen (die ja der Bedingung $a_m a_l = a_{ml}$ genügt) die absolute Konvergenzabszisse $\sigma_A = 1$ besitzt, ist also, daß die durch die Reihe dargestellte (ganze transzendente) Funktion $\zeta(s)(1-2^{1-s})$ nicht über die Gerade $\sigma = 1$ hinaus beschränkt bleibt.

54) *H. Bohr*, Zur Theorie der allgemeinen Dirichletschen Reihen, Math. Ann. 79 (1919), p. 136—156.

55) *H. Bohr*, Lösung des absoluten Konvergenzproblems einer allgemeinen Klasse Dirichletscher Reihen, Acta Math. 36 (1913), p. 197—240. Bei diesem Satze über die Bestimmung der *absoluten* Konvergenzabszisse σ_A ist bemerkenswert, daß — im Gegensatze zu den Sätzen in Nr. 6 über die Konvergenzabszisse σ_B und die gleichmäßige Konvergenzabszisse σ_G — überhaupt keine Bedingung über die „ungefähre" Lage der λ_n (z. B daß sie nicht allzu dicht aufeinander folgen dürfen) nötig ist, sondern nur die angegebene *arithmetische* Bedingung der linearen Unabhängigkeit, welche ja die „genaue" Lage der λ_n betrifft.

die Koeffizientendarstellung in Nr. 4 auszugehen, welcher die Koeffizienten und Exponenten der Reihe von der Funktion aus bestimmt, und zu untersuchen, ob nicht etwa die Konvergenz und streckenweise Konstanz des dort vorkommenden Integrals $J(x) = \frac{1}{2\pi i}\int\limits_{c-i\infty}^{c+i\infty} \frac{e^{xs} f(s)}{s} ds$ für die Entwickelbarkeit einer Funktion $f(s)$ in eine *Dirichlet*sche Reihe genügt. Es zeigt sich nun, daß eine solche unmittelbare Umkehrung des Satzes in Nr. 4 *nicht* gilt[56]), daß sie aber *unter gewissen einschränkenden Bedingungen* gelingt.[57]) Die hierdurch gewonnenen Resultate sind jedoch von einem etwas komplizierten Charakter, und es zeigen überhaupt viele Eigenschaften der *Dirichlet*schen Reihen, daß dieser Reihentypus zur Darstellung von Funktionen allgemeinen Charakters nicht geeignet ist. In diesem Zusammenhange ist vor allem eine schöne Arbeit von *Ostrowski*[58]) zu erwähnen, worin zunächst der Satz bewiesen wird, daß eine durch eine *Dirichlet*sche Reihe (1) dargestellte Funktion $f(s)$ nur in dem sehr speziellen Fall einer *algebraischen „Differenzendifferentialgleichung"* genügen kann, in welchem die Exponentenfolge $\{\lambda_n\}$ eine *endliche* lineare Basis besitzt.[59]) Bei den weiteren Untersuchungen von *Ostrowski* erweist es sich als bequem, die Transformation $e^{-s} = x$ auszuführen, also statt einer *Dirichlet*schen Reihe (1) die entsprechende „irreguläre" Potenzreihe

$$F(x) = \sum_{n=1}^{\infty} a_n x^{\lambda_n}$$

zu betrachten, die offenbar im Punkte $x = 0$ (welcher $\sigma = +\infty$ ent-

56) Vgl. *O. Perron,* a. a. O. 13) und *E. Landau,* Handbuch, p. 833.

57) Vgl. *J. Hadamard,* a) Sur les séries de Dirichlet, Palermo Rend. 25 (1908), p. 326—330; b) Rectification à la note „Sur les séries de Dirichlet", Palermo Rend. 25 (1908), p. 395—396 und insbesondere die Abhandlungen von *W. Schnee,* a. a. O. 9) und *M. Fujiwara,* Über Abelsche erzeugende Funktion und Darstellbarkeitsbedingung von Funktionen in Dirichletschen Reihen, Tôhoku J. 17 (1920), p. 363 bis 383. In anderer Richtung liegt eine Untersuchung von *J. Steffensen,* Eine notwendige und hinreichende Bedingung für die Darstellbarkeit einer Funktion als Dirichletsche Reihe, Nyt Tidskr. f. Mat. 1917, p. 9—11.

58) *A. Ostrowski,* Über Dirichletsche Reihen und algebraische Differentialgleichungen, Math. Ztschr. 8 (1920), p. 241—298.

59) Für den speziellen Fall der *Zetafunktion* war es schon durch *D. Hilbert,* Sur les problèmes futurs des Mathématiques, C. R. du 2 congr. intern. d. math. Paris 1902, p. 58—114, bekannt, daß sie keiner algebraischen Differentialgleichung genügt. Vgl. auch *V. Stadigh,* Ein Satz über Funktionen, die algebraische Differentialgleichungen befriedigen, und über die Eigenschaft der Funktion $\zeta(s)$ keiner solchen Gleichung zu genügen, Dissertation Helsingfors 1902.

spricht) einen Verzweigungspunkt unendlich hoher Ordnung besitzt. Die Frage nach den Funktionen $f(s)$, welche in eine *Dirichlet*sche Reihe entwickelt werden können, tritt dann hier in der Gestalt auf, welche Art von Singularitäten im Punkte $x = 0$ durch eine irreguläre Potenzreihe bewältigt werden können. *Ostrowski* zeigt nun u. a., daß nur in dem oben genannten speziellen Fall, wo die Exponentenfolge eine endliche lineare Basis besitzt, die durch eine solche irreguläre Potenzreihe dargestellte Funktion $F(x)$ einer an der Stelle $x = 0$ *analytischen* Differentialgleichung genügen kann. Durch diesen Satz tritt deutlich zutage, wie „schwer" die Singularität ist, die eine *Dirichlet*sche Reihe im unendlichfernen Punkte besitzt.

9. Der Mittelwertsatz. Aus der Gleichung

$$\lim_{T\to\infty}\frac{1}{2T}\int_{-T}^{T} e^{i\alpha t}dt = \begin{cases} 0 \text{ für reelles } \alpha \neq 0 \\ 1 \text{ für } \alpha = 0 \end{cases}$$

folgt sofort durch formales Rechnen, daß, wenn

$$f(s) = \sum a_n e^{-\lambda_n s}, \quad g(s) = \sum b_n e^{-\lambda_n s}$$

zwei beliebige (zur selben λ-Folge gehörige) *Dirichlet*sche Reihen sind, die Gleichung gilt

$$(17) \qquad \lim_{T\to\infty}\frac{1}{2T}\int_{-T}^{T} f(\sigma_1 + it)g(\sigma_2 - it)dt = \sum a_n b_n e^{-\lambda_n(\sigma_1+\sigma_2)},$$

worin speziell, $b_n = \bar{a}_n$ und $\sigma_1 = \sigma_2$ entsprechend, die Gleichung

$$\lim_{T\to\infty}\frac{1}{2T}\int_{-T}^{T} |f(\sigma_1 + it)|^2 dt = \sum |a_n|^2 e^{-2\lambda_n \sigma_1}$$

enthalten ist. *Hadamard*[60]), der zuerst auf die Gleichung (17) hingewiesen hat, hat ihre Gültigkeit für den Fall bewiesen, in dem die zwei Reihen auf den Geraden $\sigma = \sigma_1$ bzw. $\sigma = \sigma_2$ *absolut* konvergieren, und *Landau*[61]) und *Schnee*[62]) haben später (unter einer gewissen einschränkenden Bedingung über die Dichte der λ-Folge) bewiesen, daß die Formel auch in anderen allgemeinen Fällen gültig bleibt. Als ein für die Anwendungen (z. B. auf die Zetafunktion) besonders wichtiges

60) *J. Hadamard,* Théorème sur les séries entières, Acta Math. 22 (1899), p. 55—63.

61) *E. Landau,* a) a. a. O. 29); b) Neuer Beweis des Schneeschen Mittelwertsatzes über Dirichletsche Reihen, Tôhoku J. 20 (1922), p. 125—130.

62) *W. Schnee,* Über Mittelwertsformeln in der Theorie der Dirichletschen Reihen, Wiener Sitzungsber. (IIa) 118 (1909), p. 1439—1522.

Beispiel der *Landau-Schnee*schen Resultate nennen wir den sogenannten *Schneeschen Mittelwertsatz für gewöhnliche Dirichletsche Reihen* (2), der besagt, daß die Gleichung

$$\lim_{T\to\infty} \frac{1}{2T}\int_{-T}^{T} |f(\sigma_1 + it)|^2 dt = \sum \frac{|a_n|^2}{n^{2\sigma_1}}$$

für jedes $\sigma_1 > \frac{1}{2}(\sigma_A + \sigma_B)$ besteht (aber im allgemeinen *nicht* für $\sigma_1 \leqq \frac{1}{2}(\sigma_A + \sigma_B)$).

Aus der Gleichung (17) folgt ferner (indem $g(s)$ gleich $e^{-\lambda_n s}$ und $\sigma_2 = -\sigma_1$ gesetzt wird) die *Koeffizientendarstellungsformel*[63])

$$\lim_{T\to\infty} \frac{1}{2T}\int_{-T}^{T} f(\sigma_1 + it) e^{\lambda_n(\sigma_1 + it)} dt = a_n;$$

diese Formel gilt nach *Landau*[61a]) bei jedem $\sigma > \sigma_B$, und nach *Schnee*[62]) konvergiert der Ausdruck auf der linken Seite sogar *gleichmäßig* in n (unter der oben erwähnten einschränkenden Bedingung über $\{\lambda_n\}$).

10. Über die Nullstellen einer Dirichletschen Reihe. Schon bei Besprechung des Eindeutigkeitssatzes in Nr. 3 wurde die Frage nach der Verteilung der *Nullstellen* einer *Dirichlet*schen Reihe berührt, indem gezeigt wurde, daß gewisse Gebiete der Konvergenzhalbebene nullpunktsfrei sind. Die erste allgemeine Untersuchung des Problems, *wie viel* Nullstellen eine *Dirichlet*sche Reihe in einer Halbebene $\sigma > \sigma_0$ $(> \sigma_B)$ besitzen kann, rührt von *Landau*[64]) her, der mit Hilfe des bekannten *Jensen*schen Satzes bewies, daß für jede *gewöhnliche Dirichlet*sche Reihe (2) die Anzahl $n(\sigma_B + \varepsilon, T)$ der im Gebiete $\sigma > \sigma_B + \varepsilon$, $T < t < T + 1$ gelegenen Nullstellen gleich $O(\log T)$ und also die Anzahl $N(\sigma_B + \varepsilon, T)$ von Nullstellen im Gebiete $\sigma > \sigma_B + \varepsilon$, $0 < t < T$ gleich $O(T \log T)$ ist. Für *beliebige Dirichlet*sche Reihen (1) bewies *Landau*[64a]) einen entsprechenden Satz, wo nur $\log T$ durch $\log^2 T$ ersetzt ist; später hat *Landau*[64b]) gezeigt, daß in der Formel $n(\sigma_B + \varepsilon, T) = O(\log^2 T)$ der Buchstabe O durch o ersetzt werden kann, während *Wennberg*[33]) bewiesen hat, daß man in der *Landau*schen Formel $N(\sigma_B + \varepsilon, T) = O(T \log^2 T)$ ganz allgemein, d. h. für jede *Dirichlet*sche Reihe (1), $\log^2 T$ durch $\log T$ ersetzen kann, so daß wir also für $N(\sigma_B + \varepsilon, T)$ (aber nicht für $n(\sigma_B + \varepsilon, T)$) genau dieselbe Formel bekommen, wie für die gewöhnlichen Reihen (2).

---

63) Vgl. Note 17).

64) a) *E. Landau,* Über die Nullstellen Dirichletscher Reihen, Berliner Sitzungsber. 14 (1913), p. 897—907; b) Über die Nullstellen Dirichletscher Reihen, Math. Ztschr. 10 (1921), p. 128—129.

Tiefer — weil auf dem *Schnee*schen Mittelwertsatze beruhend — liegt ein Satz von *Bohr* und *Landau*[65a]) über *gewöhnliche Dirichlet*sche Reihen (2), welcher besagt, daß bei jedem $\sigma_1 > \sigma_B + \frac{1}{2}$ die Relation $N(\sigma_1, T) = O(T)$ besteht.[66]) Dieser Satz läßt sich nicht verbessern; wohl aber gilt[65b]) für gewisse *spezielle*, für die zahlentheoretischen Anwendungen besonders wichtige Reihen (2), daß der Ausdruck $O(T)$ durch $o(T)$ ersetzt werden kann. Im Anschluß an diese Untersuchungen hat *Carlson*[67]) einen allgemeinen Satz über die Anzahl der Nullstellen einer gewöhnlichen *Dirichlet*schen Reihe gefunden, von dem ein (wegen Anwendung auf die Zetafunktion) besonders wichtiger Spezialfall so lautet: In der Reihe $f(s) = \sum \frac{a_n}{n^s}$ sei $a_1 \neq 0$, und es sei σ_B etwa gleich 0; es mögen ferner die Koeffizienten b_n der (formal entwickelten) Reihe $1 : f(s) = \sum \frac{b_n}{n^s}$ die Bedingung $\lim |b_n| : \log n = 0$ erfüllen. Dann ist bei jedem $\varepsilon > 0$ die Anzahl $N(\frac{1}{2} + \varepsilon, T)$ nicht nur gleich $o(T)$, sondern sogar gleich $O(T^{1-4\varepsilon^2+\delta})$, wo δ beliebig klein ist.

Mit Hilfe von Sätzen aus dem *Picard-Landau*schen Satzkreis lassen sich ferner verschiedene interessante Resultate über den Wertvorrat einer *Dirichlet*schen Reihe (1) ableiten. So ergibt sich nach *Lindelöf*[68]), daß, falls $\sigma_b < \infty$ ist, und $f(s)$ für $\sigma > \sigma_b - \varepsilon$ regulär (also dann gewiß nicht beschränkt) bleibt, $f(s)$ in jedem Streifen um die Gerade $\sigma = \sigma_b$ *sämtliche Werte, höchstens mit einer einzigen Ausnahme* annimmt. Dasselbe Resultat gilt in jedem Streifen um die Gerade

---

65) *H. Bohr* und *E. Landau*, a) Ein Satz über Dirichletsche Reihen mit Anwendung auf die ζ-Funktion und die L-Funktionen, Palermo Rend. 37 (1914), p. 269—272; b) Sur les zéros de la fonction $\zeta(s)$ de Riemann, Paris C. R. 158 (1914), p. 106—110.

66) Dieselbe Relation $N(\sigma_1, T) = O(T)$ gilt nach *Wennberg* 33) für eine beliebige *Dirichlet*sche Reihe (1), wenn $\sigma_1 > \sigma_b$ angenommen wird, und sie ist hier (wie *Wennberg* mit Hilfe diophantischer Approximationen beweist) die bestmögliche in dem Sinne, daß, falls die Reihe in der Halbebene $\sigma > \sigma_b + \varepsilon$ überhaupt eine Nullstelle besitzt, die Anzahl $N(\sigma_b + \varepsilon, T) \neq o(T)$ ist.

67) *F. Carlson*, Über die Nullstellen der Dirichletschen Reihen und der Riemannschen ζ-Funktion, Arkiv für Mat., Ast. och Fys. 15 (1920), No. 20. Vgl. auch *E. Landau*, Über die Nullstellen der Dirichletschen Reihen und der Riemannschen ζ-Funktion, Arkiv för Mat., Ast. och Fys. 16 (1921), No. 7, der mit Hilfe einer neuen Beweismethode des *Schnee*schen Mittelwertsatzes (vgl. 61b) einen abgekürzten Beweis des *Carlson*schen Satzes gibt.

68) *E. Lindelöf*, Mémoire sur certaines inégalités dans la théorie des fonctions monogènes, et sur quelques propriétés nouvelles de ces fonctions dans le voisinage d'un point singulier essentiel, Acta soc. sc. Fenn. 35 (1908), No. 7. Vgl. auch *H. Bohr* und *E. Landau*, Über das Verhalten von $\zeta(s)$ und $\zeta_x(s)$ in der Nähe der Geraden $\sigma = 1$, Gött. Nachr. 1910. p. 303—330.

$\sigma = \sigma_e$, falls $f(s)$ für $\sigma > \sigma_e - \varepsilon$ regulär ist.[69]) Ferner wird, nach *Wennberg*[33]), jede *Dirichlet*sche Reihe mit $\sigma_b = \infty$, in jeder Halbebene $\sigma > \sigma_0$ sämtliche Werte, höchstens mit einer Ausnahme, annehmen. Schließlich sei noch erwähnt, daß jede *Dirichlet*sche Reihe mit *linear unabhängiger* Exponentenfolge (und also mit $\sigma_b = \sigma_A$), falls sie nicht in der ganzen Halbebene $\sigma > \sigma_b$ beschränkt ist, in dieser Halbebene überhaupt *jeden* Wert unendlich oft annimmt.[55])

11. Zusammenhang verschiedener Dirichletscher Reihen. Es seien

(18a) $$f(s) = \sum a_n e^{-\lambda_n s} \qquad (s = \sigma + it)$$

und

(18b) $$F(z) = \sum a_n e^{-\mu_n z} \qquad (z = x + iy)$$

zwei *Dirichlet*sche Reihen mit *denselben Koeffizienten* a_n, deren *Exponenten durch die Relation* $\mu_n = e^{\lambda_n}$ *verbunden sind.* Wie von *Cahen*[5]) gezeigt, besteht ein interessanter Zusammenhang zwischen den beiden Funktionen $f(s)$ und $F(z)$, indem jede von ihnen durch ein bestimmtes Integral dargestellt werden kann, dessen Integrand in einfacher Weise von der anderen der beiden Funktionen abhängt. Formal ergeben sich diese Darstellungen sehr leicht aus der Integraldarstellung der Γ-Funktion

$$\Gamma(s) = \int_0^\infty e^{-x} x^{s-1} dx \qquad (\sigma > 0)$$

und ihrer im *Mellin*schen Sinne „reziproken" Formel[70])

$$e^{-z} = \frac{1}{2\pi i} \int_{c-i\infty}^{c+i\infty} \Gamma(s) z^{-s} ds \qquad (c > 0,\ x > 0).$$

In der Tat, es lassen sich diese beiden Formeln (nach einer einfachen Transformation) so schreiben:

$$e^{-\lambda_n s}\Gamma(s) = \int_0^\infty x^{s-1} e^{-\mu_n x} dx, \quad e^{-\mu_n z} = \frac{1}{2\pi i}\int_{c-i\infty}^{c+i\infty} \Gamma(s) z^{-s} e^{-\lambda_n s} ds,$$

69) Ein Beweis findet sich (implizite) bei *H. Bohr,* Über die Summabilitätsgrenzgerade der Dirichletschen Reihen, Wiener Sitzungsber. (IIa) 119 (1910), p. 1391—1397.

70) Diese Formel, deren große Bedeutung sich in den Untersuchungen von *H. Mellin* gezeigt hat, ist (nach *Mellin,* Bemerkungen im Anschluß an den Beweis eines Satzes von *Hardy* über die Zetafunction, Ann. Acad. sc. Fenn. (A) 11 (1917), No. 3) schon von *S. Pincherle,* Sulle funzioni ipergeometriche generalizzate, Rend Ac. Linc. 4 (1888), p. 694—700 in etwas anderer Form angegeben. Auch in neueren Arbeiten von *Hardy* und *Littlewood* (vgl. z. B. a. a. O. 31) spielt diese „*Cahen-Mellin*sche Formel" eine wichtige Rolle.

und hieraus folgen sofort (durch Multiplikation mit a_n und Summation) die gesuchten Integraldarstellungen

$$f(s) = \frac{1}{\Gamma(s)} \int_0^\infty x^{s-1} F(x)\, dx \tag{19a}$$

und

$$F(z) = \frac{1}{2\pi i} \int_{c-i\infty}^{c+i\infty} \Gamma(s) z^{-s} f(s)\, ds. \tag{19b}$$

Bei *Cahen* waren die Konvergenzuntersuchungen noch nicht streng durchgeführt. Dies geschah erst durch *Perron*[71]), der den Satz bewies: Wenn die Reihe (18a) für $\sigma > \sigma_0 > 0$ konvergiert (woraus leicht folgt, daß (18b) mindestens für $x > 0$ konvergiert), so gilt die Formel (19a) für $\sigma > \sigma_0$, und die Formel (19b) bei festem $c > \sigma_0$ für $x > 0$. Im Spezialfalle $\lambda_n = \log n$, $\mu_n = n$ haben wir es mit einer *gewöhnlichen Dirichletschen Reihe* $f(s) = \sum \frac{a_n}{n^s}$ und einer einfachen *Potenzreihe* $F(z) = \sum a_n e^{-nz}$ zu tun.[72]) Ist außerdem noch $a_n = 1$ für alle n, wird $f(s) = \zeta(s)$ und $F(z) = \frac{1}{e^z - 1}$, und wir erhalten aus der obigen Formel (19a) die von *Riemann*[73]) benutzte wichtige Integraldarstellung der Zetafunktion

$$\zeta(s) = \frac{1}{\Gamma(s)} \int_0^\infty \frac{x^{s-1}}{e^x - 1}\, dx \qquad (\sigma > 1).^{74)}$$

---

71) *O. Perron*, a. a. O. 13). Vgl. auch *G. H. Hardy*, On a case of term-by-term integration of an infinite series, Mess. of Math. 39 (1910), p. 136—139.

72) Wie aus der Integraldarstellung (19a) ersichtlich ist, hängt das analytische Verhalten der *Dirichlet*schen Reihe $f(s) = \sum \frac{a_n}{n^s}$ mit dem Verhalten der, für $x > 0$ konvergenten, Potenzreihe $F(z) = \sum a_n e^{-nz}$ bei Annäherung an den Punkt $z = 0$, d. h. mit dem Verhalten der (für $|u| < 1$ konvergenten) Potenzreihe $\varphi(u) = \sum a_n u^n$ bei Annäherung an den Punkt $u = 1$, eng zusammen. Vgl. hierüber *G. H. Hardy*, The application to Dirichlet's series of Borel's exponential method of summation, Lond n math. Soc. (2) 8 (1909), p. 277—294 und *M. Fekete*, a. a O. 36) So besteht z. B. der Satz [*A. Hurwitz*, Über die Anwendung eines funktionentheoretischen Principes auf gewisse bestimmte Integrale, Math. Ann. 53 (1900), p. 220—224], daß, falls $\varphi(u)$ im Punkte $u = 1$ *regulär* ist, $f(s)$ gewiß eine *ganze* Funktion ist. Auch die später zu erwähnende Untersuchung von *Hardy* über *Abel*sche Summabilität *Dirichlet*scher Reihen (2) basiert auf der Verbindung zwischen $f(s)$ und $\varphi(u)$.

73) *B. Riemann*, Über die Anzahl der Primzahlen unter einer gegebenen Größe, Berliner Monatsber 1859, p. 671—680 = Werke (2. Aufl.), p. 145—153.

74) Eine von der Integraldarstellung (19a) wesentlich verschiedene Integraldarstellung einer allgemeinen *Dirichlet*schen Reihe $f(s) = \sum a_n e^{-\lambda_n s}$ ist von

Bei dem oben besprochenen Zusammenhang zweier *Dirichlet*scher Reihen handelte es sich um Reihen mit denselben Koeffizienten, aber verschiedenen Exponenten. Wie von *Cramér*[75]) gezeigt, besteht auch ein gewisser Zusammenhang zwischen zwei *Dirichlet*schen Reihen $f(s) = \sum a_n e^{-\lambda_n s}$ und $g(s) = \sum b_n e^{-\lambda_n s}$ mit *denselben Exponenten,* deren Koeffizienten a_n und b_n aber derart voneinander abhängen, *daß* $b_n = a_n \varphi(\lambda_n)$ *ist, wo* $\varphi(z)$ *eine ganze transzendente Funktion von* z *ist,* welche die Bedingung $|\varphi(z)| < e^{k|z|}$ für alle hinreichend großen $|z|$ erfüllt. *Cramér* beweist nämlich, daß, falls die Funktion $f(s)$, welche durch die erste Reihe definiert wird, in einem Gebiete G_1, das über die Konvergenzgerade $\sigma = \sigma_B$ dieser Reihe hinausreicht, regulär ist, die durch die zweite Reihe definierte Funktion $g(s)$ ebenfalls über die „entsprechende" Gerade $\sigma = \sigma_B + k$ analytisch fortsetzbar sein wird, und zwar auf ein Gebiet G_2, das vom Gebiete G_1 in einfach angebbarer Weise abhängt.

12. Multiplikation Dirichletscher Reihen. Wie leicht zu sehen, wird man durch „gewöhnliches" Rechnen mit *Dirichlet*schen Reihen wieder zu *Dirichlet*schen Reihen geführt; speziell entsteht durch *Multiplikation* zweier beliebiger *Dirichlet*scher Reihen

(20) $$f(s) = \sum a_n e^{-\lambda_n s} \quad \text{und} \quad g(s) = \sum b_n e^{-\mu_n s}$$

wiederum eine *Dirichlet*sche Reihe $\sum c_n e^{-\nu_n s}$, und zwar führt die Multiplikation zweier *gewöhnlicher Dirichlet*scher Reihen $\sum \frac{a_n}{n^s}$ und $\sum \frac{b_n}{n^s}$ wieder zu einer *gewöhnlichen Dirichlet*schen Reihe $\sum \frac{c_n}{n^s}$, deren Koeffizienten c_n durch die Formel $c_n = \sum a_m b_l$ bestimmt werden, wobei m und $l = n : m$ alle Teiler von n durchlaufen.[76])

J. Steffensen, Ein Satz über *Stieltjes*sche Integrale mit Anwendung auf *Dirichlet*sche Reihen, Palermo Rend. 36 (1913), p. 213—219, angegeben; die *Steffensen*sche Formel, die die absolute Konvergenz der Reihe für $\sigma > 0$ voraussetzt, lautet

$$f(s) = \frac{\sin \pi s}{\pi} \int_0^\infty x^{-s} B(x)\, dx \qquad (0 < \sigma < 1),$$

wo $B(x)$ die *Partialbruchreihe*

$$B(x) = \sum \frac{a_n}{x + \mu_n} \qquad (\mu_n = e^{\lambda_n})$$

bezeichnet.

75) *H. Cramér,* a) Sur une classe de séries de Dirichlet, Dissertation Upsala (Stockholm 1917); b) Un théorème sur les séries de Dirichlet et son application, Arkiv f. Mat., Astr. och Fys. 13 (1918), No. 22.

76) Aus diesem Bildungsgesetz der Koeffizienten c_n folgt z. B., wie von *H. Mellin,* Ein Satz über Dirichletsche Reihen, Ann. Ac. sc. Fenn. (A) 11 (1917), No. 1 hervorgehoben, daß die modulo q gebildeten „Partialreihen" der Produkt-

Sind die gegebenen Reihen (20) in einem Punkte s_0 beide *absolut konvergent*, so wird offenbar auch die durch Multiplikation entstandene Reihe im Punkte s_0 absolut konvergieren (und zwar mit der Summe $f(s_0) \cdot g(s_0)$). Einem bekannten *Mertens*schen Satze über Potenzreihen entsprechend (und ihn verallgemeinernd) gilt ferner nach *Stieltjes*[77]) der Satz, daß die Produktreihe konvergiert in jedem Punkt s_0 (mit der Summe $f(s_0) \cdot g(s_0)$), in welchem nur *eine* der Faktorenreihen absolut konvergiert, während die andere nur bedingt konvergiert. Dagegen braucht die Produktreihe in einem Punkte, worin *beide* Faktorenreihen bedingt konvergieren, *nicht* zu konvergieren, und dieses kann nicht nur am Rande der Konvergenzgebiete der Reihen der Fall sein; denn, wie *Landau*[29]) gezeigt hat, gibt es sogar zwei gewöhnliche *Dirichlet*sche Reihen (2), die beide in einer gewissen Halbebene $\sigma > \sigma_0$ konvergieren, deren Produktreihe aber *nicht* in der ganzen Halbebene $\sigma > \sigma_0$ konvergiert. Andererseits gibt es doch, nach *Stieltjes* und *Landau*[78]), wichtige Sätze, welche die Konvergenz der Produktreihe in Gebieten, in welchen die Faktorenreihen *beide nur bedingt konvergieren*, sichern. Als ein charakteristisches Beispiel nennen wir den Satz, daß, falls die Faktorenreihen beide für $\sigma > \alpha$ konvergieren und für $\sigma > \alpha + \beta$ absolut konvergieren, die Produktreihe mindestens für $\sigma > \alpha + \frac{\beta}{2}$ konvergiert. Hierbei läßt sich die Zahl $\alpha + \frac{\beta}{2}$ durch keine bessere (d. h. kleinere) ersetzen, denn wie *Bohr*[38]) gezeigt hat, gibt es eine *Dirichlet*sche Reihe (2) mit $\sigma_A = 1$, $\sigma_B = 0$, deren Quadratreihe die Konvergenzabszisse $\sigma_B = \frac{1}{2}$ besitzt.[79])

---

reihe $\sum \frac{c_n}{n^s}$ in einfacher Weise durch die „Partialreihen" der gegebenen Reihen $\sum \frac{a_n}{n^s}$ und $\sum \frac{b_n}{n^s}$ ausgedrückt werden können.

77) *T. Stieltjes*, Note sur la multiplication de deux séries, Nouv. Ann. de Math. (3) 6 (1887), p. 210—215.

78) *T. Stieltjes* a. a. O. 77) und Sur une loi asymptotique dans la théorie des nombres, Paris C. R. 101 (1885), p. 368—370 gibt ohne Beweise die wesentlichsten dieser Sätze an, doch nur für die gewöhnlichen *Dirichlet*schen Reihen (2). Verallgemeinerungen auf den Fall beliebiger *Dirichlet*scher Reihen (1) (sowie Verallgemeinerungen anderer Art) und Beweise der Sätze sind von *E. Landau*, Über die Multiplikation Dirichletscher Reihen, Palermo Rend. 24 (1907), p. 81—160 und Handbuch, p. 755—762 gegeben.

79) Von etwas anderer Art als die obigen Sätze ist ein Satz von *G. H. Hardy*, On the multiplication of Dirichlet's series, London math. Soc. (2) 10 (1911), p. 396—405, welcher besagt, daß, falls die Faktorenreihen beide im Punkte $s = 0$ konvergieren und $a_n = O\left(\frac{\lambda_n - \lambda_{n-1}}{\lambda_n}\right)$, $b_n = O\left(\frac{\mu_n - \mu_{n-1}}{\mu_n}\right)$, auch die Produktreihe im Punkte 0 konvergiert. Vgl. hierzu auch *A. Rosenblatt*, Über einen Satz

Der bekannte *Cesàro*sche Satz über Potenzreihen ($\lambda_n = n$), der ja besagt, daß, wenn $\sum a_n x^n$ und $\sum b_n x^n$ in einem Punkte, etwa $x = 1$, mit den Summen A und B konvergieren, die Produktreihe $\sum c_n x^n$ im Punkte $x = 1$ gewiß summabel $(C, 1)$ mit der Summe $A \cdot B$ ist (d. h. das arithmetische Mittel ihrer Partialsummen strebt gegen $A \cdot B$) wurde von *Phragmén, M. Riesz* und *Bohr*[80]) auf beliebige *Dirichlet*sche Reihen (1) übertragen; der Satz besagt hier, daß, falls $\sum a_n e^{-\lambda_n s}$ und $\sum b_n e^{-\mu_n s}$ in einem Punkte, etwa $s = 0$, mit den Summen A und B konvergieren, die Produktreihe $\sum c_n e^{-\nu_n s}$ im Punkte $s = 0$ in dem Sinne *summabel mit der Summe* AB ist, daß, $C_n = \sum_1^n c_n$ gesetzt,

$$\lim_{n \to \infty} \frac{C_1(\nu_2 - \nu_1) + C_2(\nu_3 - \nu_2) + \cdots + C_{n-1}(\nu_n - \nu_{n-1})}{\nu_n} = AB.$$

Im speziellen Falle *gewöhnlicher Dirichlet*scher Reihen

$$(\lambda_n = \mu_n = \nu_n = \log n)$$

lautet also die Gleichung:

$$\lim_{n \to \infty} \frac{C_1(\log 2 - \log 1) + \cdots + C_{n-1}(\log n - \log(n-1))}{\log n} = AB.$$ [81])

Diese Mittelwertbildung (mit Gewichten) bildet den Ausgangspunkt für die bekannte, von *M. Riesz* ausgearbeitete, allgemeine Summabilitätsmethode für beliebige *Dirichlet*sche Reihen, über die wir im nächsten Paragraphen näher berichten werden. Aus dem oben angegebenen Satze geht speziell hervor, daß, falls die Produktreihe

des Herrn Hardy, Jahresber. d. Deutsch. Math.-Ver. 23 (1914), p. 80—84 (welcher zeigt, daß eine bei *Hardy* der Exponentenfolge auferlegte Bedingung unnötig ist) und *E. Landau,* Über einen Satz des Herrn Rosenblatt, Jahresber. d. Deutsch. Math.-Ver. 29 (1920), p. 238. Ferner ist ein Satz von *Hardy* und *Littlewood,* a. a. O. 24 b) zu erwähnen, welcher aus der Voraussètzung der Konvergenz gewisser aus den beiden zu multiplizierenden Reihen gebildeten Hilfsreihen die Konvergenz der durch Multiplikation entstandenen Reihe folgert.

80) Der Beweis von *E. Phragmén* wurde brieflich *E. Landau* mitgeteilt und findet sich im Handbuch, p. 762—765. Vgl. auch *M. Riesz,* Sur la sommation des séries de Dirichlet. Paris C. R. 149 (1909), p. 18—21 und *H. Bohr,* a. a. O. 36).

81) Dagegen braucht das einfache [und, *Riesz,* a. a. O. 80), sogar auch das beliebig oft wiederholte] „arithmetische" Mittel $\frac{1}{n}(C_1 + C_2 + \cdots C_n)$ *nicht* für $n \to \infty$ zu konvergieren. [Es ist ein allgemeines Prinzip, daß eine Summabilitätsmethode durch Mittelwertbildungen der Form $\frac{\mu_1 C_1 + \cdots + \mu_n C_n}{\mu_1 + \cdots + \mu_n}$ um so kräftiger ist, je „langsamer" $\mu_1 + \cdots + \mu_n \to \infty$.] Über das nähere Verhältnis der „arithmetischen" Mittelwertbildung zu der „logarithmischen" Mittelwertbildung vgl. Nr. 13, Note 86.

konvergent (und nicht nur summabel) ist, sie gewiß die „richtige" Summe, d. h. die Summe $A \cdot B$ hat. Dieser letzte Satz war schon früher von *Landau*[78]) in dem speziellen Falle, wo mindestens eine der Faktorenreihen eine *absolute* Konvergenzhalbebene besitzt, durch funktionentheoretische Überlegungen bewiesen.

Wir verlassen hiermit die Konvergenztheorie der *Dirichlet*schen Reihen, um uns der Summabilitätstheorie dieser Reihen zuzuwenden. Hierbei werden wir sehen, daß die Erweiterungen des Konvergenzbegriffes für die Theorie der *Dirichlet*schen Reihen eine noch größere Rolle spielt, als es z. B. bei den Potenzreihen der Fall ist. In der Tat, bei den *Dirichlet*schen Reihen können schon die allereinfachsten Summabilitätsmethoden in ganzen *Gebieten* außerhalb der Konvergenzhalbebene verwendet werden, während solche Methoden bei den Potenzreihen nur *auf dem Rande* des Konvergenzgebietes von Bedeutung sind.

13. Summabilität Dirichletscher Reihen. Der in Nr. 2 erwähnte Hauptsatz, daß das Konvergenzgebiet einer *Dirichlet*schen Reihe (1) eine Halbebene $\sigma > \sigma_B$ ist, beruhte auf dem Satze, daß die Zahlenfolge $\{e^{-\lambda_n s}\}$ bei festem s mit $\sigma > 0$ eine „konvergenzerhaltende" war; es ist in derselben Weise klar, daß auch das Gebiet, in welchem eine *Dirichlet*sche Reihe (1) nach einer angegebenen Summabilitätsmethode *summabel* ist, ebenfalls eine *Halbebene* $\sigma > \sigma_0$ sein wird, sobald die betreffende Summabilitätsmethode die Eigenschaft besitzt, daß die Zahlenfolge $\{e^{-\lambda_n s}\}$ für $\sigma > 0$ eine „summabilitätserhaltende" ist. Dies ist nach *Bohr*[82]), der die Summabilität *Dirichlet*scher Reihen in Gebieten der komplexen Ebene zuerst untersucht hat[83]), für die *gewöhnlichen Dirichlet*schen Reihen (2) der Fall, wenn die benutzte Summabilitätsmethode die einfache *Cesàrosche Methode* (C, r) ist, wo r eine beliebige positive ganze Zahl bedeutet (Artikel II C 4, p. 477 u. f.). Es besitzt also jede *Dirichlet*sche Reihe (2) eine Folge von *Summabilitätsabszissen* $\sigma_B = \sigma^{(0)} \geqq \sigma^{(1)} \geqq \sigma^{(2)} \dots \geqq \sigma^{(r)} \dots$ derart, daß die Reihe für $\sigma > \sigma^{(r)}$ summabel (C, r) ist, für $\sigma < \sigma^{(r)}$ dagegen nicht. Bezeichnet $\Omega = \lim_{r \to \infty} \sigma^{(r)}$ die *Summabilitätsgrenzabszisse* der Reihe, so ergibt sich ferner, daß die „Summe" der Reihe *in der ganzen Halbebene* $\sigma > \Omega$ *eine reguläre analytische Funktion* darstellt, so daß wir

82) *H. Bohr,* a) Sur la série de Dirichlet, Paris C. R. 148 (1909), p. 75—80; b) a. a. O. 36); c) Habilitationsschrift, a. a. O. 38); in dieser letzten Arbeit wurde eine zusammenfassende Darstellung der Theorie gegeben.

83) Für *Dirichlet*sche Reihen als Funktionen einer *reellen* Variablen s war die Summabilität schon früher von *G. H. Hardy,* Generalisation of a theorem in the theory of divergent series, London math. Soc. (2) 6 (1908), p. 255—264 untersucht.

also durch die *Cesàro*sche Summabilität die *analytische Fortsetzung* der durch die Reihe in ihrer Konvergenzhalbebene $\sigma > \sigma_B$ bestimmten Funktion über die ganze Summabilitätshalbebene $\sigma > \Omega$ erhalten. Für die in Nr. 1 erwähnten speziellen Reihen (2), deren Koeffizienten den Bedingungen (4) genügen, findet man z. B. $\sigma^{(r)} = -r (r = 0, 1, 2 \ldots)$, also $\Omega = -\infty$; jede dieser Reihen ist also in der ganzen Ebene summabel und definiert somit (was übrigens auf anderem Wege schon bekannt war) eine ganze transzendente Funktion. *Bohr*[82c]) gab ferner explizite Ausdrücke der Summabilitätsabszissen $\sigma^{(r)}$ als Funktionen der Koeffizienten und zeigte, daß diese Abszissen den beiden folgenden Ungleichungen genügen

$$\sigma^{(r)} - \sigma^{(r+1)} \leqq 1, \quad \sigma^{(r)} - \sigma^{(r+1)} \leqq \sigma^{(r-1)} - \sigma^{(r)},$$

d. h. die Breite jedes Summabilitätsstreifens ist höchstens 1, und diese Breite kann mit wachsender Summabilitätsordnung r niemals zunehmen; diese beiden Ungleichungen sind ferner die für die Verteilung der Summabilitätsabszissen notwendigen und hinreichenden, in dem Sinne, daß es zu jeder monoton abnehmenden Zahlenfolge $\{\sigma^{(r)}\}$, die diesen Ungleichungen genügt, eine *Dirichlet*sche Reihe (2) gibt, die eben diese Zahlen $\sigma^{(r)}$ als Summabilitätsabszissen besitzt.[84])

M. Riesz[85]), der etwas später als *Bohr*, aber unabhängig von ihm,

---

84) In der bekannten Arbeit von *G. H. Hardy* u. *J. Littlewood*, Contributions to the arithmetic theory of series, London math. Soc. (2) 11 (1912), p. 411—478 wird u. a. die oben referierte Untersuchung über die Verteilung der Summabilitätsabszissen dadurch verfeinert, daß auch das Summabilitätsverhalten der Reihe in Punkten *auf* den Summabilitätsgeraden $\sigma = \sigma^{(r)}$ selbst betrachtet wird. Zur Charakterisierung der gewonnenen Resultate sei der Satz erwähnt, daß eine Reihe (2), falls sie in einem Punkte $s = \sigma_1$ summabel $(C, r + 1)$ und in einem Punkte $s = \sigma_2$ summabel $(C, r - 1)$ ist, im Mittelpunkte $s = \frac{1}{2}(\sigma_1 + \sigma_2)$ summabel (C, r) ist; in diesem Satze ist die obige Ungleichung $\sigma^{(r)} - \sigma^{(r+1)} \leqq \sigma^{(r-1)} - \sigma^{(r)}$ speziell enthalten. Ferner werden, unter gewissen spezielleren Annahmen über die Größenordnung der Koeffizienten, genauere Sätze über die Lage der Summabilitätsgeraden und das Verhalten der Reihe auf diesen Geraden bewiesen.

85) *M. Riesz*, a) Sur les séries de Dirichlet, Paris C. R. 148 (1909), p. 1658—1660; b) Sur la sommation des séries de Dirichlet, Paris C. R. 149 (1909), p. 18—21. Eine zusammenfassende Darstellung der *Riesz*schen Untersuchungen findet sich in dem a. a. O. 1) zitierten Cambridge tract von *G. H. Hardy* und *M. Riesz.* Vgl. auch die Arbeiten von *P. Nalli*, a) Sulle serie di Dirichlet, Palermo Rend. 40 (1915), p. 44—70; b) Aggiunta alla memoria: „Sulle serie di Dirichlet", Palermo Rend. 40 (1915), p. 167—168, und *M. Kuniyeda*, a) Note on Perron's integral and summability-abscissae of Dirichlet's series, Quart. J. 47 (1916), p. 193—219; b) On the abscissa of summability of general Dirichlet's series, Tôhoku J. 9 (1916), p. 245—262, welche sich nahe an die *Riesz*schen Arbeiten anschließen.

die *Cesàro*-Summabilität der *Dirichlet*schen Reihen untersucht hat, beschränkt sich nicht auf Summabilität ganzzahliger Ordnung und — was wesentlicher ist — betrachtet sogleich die *allgemeinen Dirichlet*schen Reihen. *Riesz* mußte daher zunächst das *Cesàro*sche Summabilitätsverfahren so verallgemeinern, daß es auf diesen allgemeineren Reihentypus (1) angewendet werden konnte, und er wurde hierbei auf eine neue bedeutsame Summabilitätsmethode geführt, die von ihm „*Summation nach typischen Mitteln*" genannt wurde. Schon bei dem Multiplikationssatz in Nr. 12 haben wir gesehen, daß es bei den gewöhnlichen *Dirichlet*schen Reihen (2) zweckmäßig sein kann, ein Summabilitätsverfahren zu benutzen, dessen erste Stufe darin besteht, das „logarithmische" Mittel

$$\frac{C_1(\log 2 - \log 1) + \cdots + C_{n-1}(\log n - \log(n-1))}{\log n}$$

statt des arithmetischen Mittels

$$\frac{C_1 + C_2 + \cdots + C_n}{n}$$

zu bilden. Indem *Riesz* diesen Gedanken ausführt und verallgemeinert, führt er, einer gegebenen *Dirichlet*schen Reihe (1) (oder vielmehr einer gegebenen Exponentenfolge $\{\lambda_n\}$) entsprechend, *zwei* verschiedene Summationsmethoden ein, deren eine der logarithmischen, die andere der arithmetischen Mittelwertbildung analog ist. Es sei

$$e^{\lambda_n} = l_n, \quad a_n e^{-\lambda_n s} = a_n l_n^{-s} = c_n,$$

$$C_\lambda(\tau) = \sum_{\lambda_n < \tau} c_n, \quad C_l(t) = \sum_{l_n < t} c_n$$

und

$$C_\lambda^{(k)}(\omega) = \sum_{\lambda_n < \omega} (\omega - \lambda_n)^k c_n = k\int_0^\omega C_\lambda(\tau)(\omega - \tau)^{k-1} d\tau,$$

$$C_l^{(k)}(w) = \sum_{l_n < w} (w - l_n)^k c_n = k\int_0^w C_l(t)(w - t)^{k-1} dt,$$

wobei k eine beliebige positive (ganze oder nicht ganze) Zahl bedeutet. Die Ausdrücke

$$\frac{C_\lambda^{(k)}(\omega)}{\omega^k} \quad \text{und} \quad \frac{C_l^{(k)}(w)}{w^k}$$

heißen dann nach *Riesz* die typischen Mittelwerte der k-Ordnung von der ersten bzw. zweiten Art, welche zu der gegebenen Reihe (1) gehören. Wenn nun

$$\frac{C_\lambda^{(k)}(\omega)}{\omega^k} \to C \quad \text{bzw.} \quad \frac{C_l^{(k)}(w)}{w^k} \to C$$

für $\omega \to \infty$ bzw. $w \to \infty$, wird die Reihe (1) *summabel* (λ, k) *bzw.* (l, k) *mit der Summe* C genannt. Die „Kraft" der Summabilitätsmethode steigt mit wachsender Summabilitätsordnung k, d. h. wenn eine Reihe summabel (λ, k) bzw. (l, k) ist, so ist sie a fortiori summabel (λ, k') bzw. (l, k') für $k' > k$.

Riesz zeigt nun für eine beliebige Exponentenfolge $\{\lambda_n\}$, daß die Zahlenfolge $\{e^{-\lambda_n s}\}$ bei festem s mit $\sigma > 0$ eine summabilitätserhaltende Faktorenfolge ist, sowohl für die Summabilitätsmethode (λ, k) als für die Methode (l, k), woraus folgt, daß der Gültigkeitsbereich der (einen oder anderen) Summabilitätsmethode eine Halbebene ist. Über die Tragweite der beiden Methoden (λ, k) und (l, k) gegeneinander gilt der Satz: in jedem Punkte s, wo die Reihe (1) summabel (l, k) ist, ist sie gewiß auch summabel (λ, k), so daß (λ, k) die „kräftigere" Methode ist; die Methode (l, k) ist aber „beinahe" ebenso stark, d. h. wenn die Reihe (1) in einem Punkte $s_0 = \sigma_0 + it_0$ summabel (λ, k) ist, braucht sie wohl nicht im Punkte s_0 selbst summabel (l, k) zu sein, ist es aber in jedem Punkte $s = \sigma + it$ mit $\sigma > \sigma_0$.

Aus diesen Sätzen folgt, daß zu jeder *Dirichlet*schen Reihe (1) eine *Summabilitätsabszissenfunktion* $\sigma^{(k)}$ $(0 < k < \infty)$ derart existiert, daß die Reihe bei jedem $k > 0$ für $\sigma > \sigma^{(k)}$ summabel (λ, k) und (l, k) ist, während sie für $\sigma < \sigma^{(k)}$ weder summabel (λ, k) noch (l, k) ist.[86])

Für die *gewöhnlichen Dirichlet*schen Reihen (2) $(\lambda_n = \log n)$ ist die *Riesz*sche Summabilität (l, k) mit der *Cesaro*schen Summabilität (C, k) inhaltsmäßig identisch[87]), und es sind somit für diese Reihen (2), und ganzzahlige k, die hier definierte Summabilitätsabszissen $\sigma^{(k)}$ mit den früher besprochenen identisch. Die dort angegebenen Ungleichungen über die Verteilung der Summabilitätsabszissen werden, sogar für den Fall einer *beliebigen Dirichlet*schen Reihe (1), von *Riesz* dahin verallgemeinert, daß die Summabilitätsabszissenfunktion $\sigma^{(k)}$ eine *konvexe* Funktion von k ist.[88]) Ferner verallgemeinert *Riesz* die expliziten Ausdrücke für die Summabilitätsabszissen $\sigma^{(k)}$ als Funktionen der Koeffizienten und Exponenten auf beliebige *Dirichlet*sche Reihen (1) und beliebiges nicht ganzzahliges k.

86) In Punkten *auf* der Summabilitätsgeraden $\sigma = \sigma^{(k)}$ selbst kann es vorkommen (vgl. eine Bemerkung oben), daß die Reihe summabel (λ, k) aber *nicht* (l, k) ist. So ist nach *Riesz*, a. a. O. 85a) und 85b) die Zetareihe $\sum \frac{1}{n^s}$, für welche $\sigma^{(k)} = 1$ für alle k ist, bei keinem k summabel (l, k) in irgendeinem Punkte der Geraden $\sigma = 1$, während sie in jedem Punkte $s \neq 1$ dieser Geraden summabel $(\lambda, 1)$ ist. (Vgl. Note 81.)

87) *M. Riesz,* Une méthode de sommation équivalente à la méthode des moyennes arithmétiques, Paris C. R. 152 (1911), p. 1651—1654. Die von *Riesz* angegebene Formulierung der *Cesàro*schen Summationsmethode hat sich bei verschiedenen Anwendungen als wesentlich bequemer als die ursprüngliche Formulierung gezeigt.

88) Der Beweis dieses Satzes wird demnächst in den Acta Univ. hung. Francesco-Jos. erscheinen.

Über den Zusammenhang der Summabilitätseigenschaften einer Reihe (1) mit den analytischen Eigenschaften der durch die Reihe dargestellten Funktion $f(s)$ gilt zunächst der folgende leicht beweisbare Satz: *Für* $\sigma > \sigma^{(k)} + \varepsilon$ *ist* $f(s) = O\,(|t|^{k+1})$. Die Frage nach der *Umkehrung* dieses Satzes ist (wie im Falle $k = 0$, d. h. Konvergenz) viel schwieriger; es zeigt sich, daß eine unmittelbare Umkehrung *nicht* gilt, dagegen eine solche, in welcher *der Exponent* $k + 1$ *durch* k *ersetzt wird,* d. h. wenn die durch eine *Dirichlet*sche Reihe definierte Funktion $f(s)$ für $\sigma > \sigma_0$ regulär und gleich $O(|t|^k)$ ist, so wird die Summabilitätsabszisse $\sigma^{(k)}$ gewiß $\leqq \sigma_0$ sein. Es geben diese *Riesz*schen Sätze einerseits notwendige und andererseits hinreichende Bedingungen für die Summabilität k^{ter} Ordnung, aber (ganz wie im Falle $k = 0$) keine Bedingungen, die zugleich notwendig und hinreichend sind. Betrachten wir aber den *Grenzwert* Ω der abnehmenden Funktion $\sigma^{(k)}$ (für $k \to \infty$), so können wir aus den obigen Sätzen den folgenden Hauptsatz über die funktionentheoretische Bestimmung dieser *Summabilitätsgrenzabszisse* Ω ableiten: Es ist die Reihe genau so weit summabel (von irgendeiner Ordnung) wie die dargestellte Funktion $f(s)$ regulär und von endlicher Ordnung in bezug auf t ist, d. h. *es ist* Ω *gleich der früher eingeführten Abszisse* σ_ε. Dieser Satz wurde, für die gewöhnlichen *Dirichlet*schen Reihen (2), zuerst von *Bohr*[36]) explizite aufgestellt, der ihn aus einigen, den *Riesz*schen ähnlichen, aber nicht so weitreichenden Sätzen herleitete.[89])

Bei einer näheren Untersuchung zeigt es sich, daß die Einführung der *Cesàro-Riesz*schen Summabilität für fast alle Probleme in der Theorie der *Dirichlet*schen Reihen von wesentlicher Bedeutung ist, weil dadurch frühere Resultate aus der Konvergenztheorie sich in wichtiger Weise verallgemeinern lassen. Wegen der allgemeinen Durchführung solcher Untersuchungen und der dabei erhaltenen Resultate sei der Leser auf das *Hardy-Riesz*sche Buch verwiesen.[1]) Hier soll nur noch ein besonders interessantes Resultat über die *Multiplikation Dirichlet*scher Reihen erwähnt werden, welches den klassischen Satz von *Cesàro* über Multiplikation von Potenzreihen ($\lambda_n = n$) auf den allgemeinsten Typus *Dirichlet*scher Reihen (1) verallgemeinert, und so lautet[90]): *Wenn* $\sum a_n e^{-\lambda_n s}$ *im Punkte* $s = s_0$ *summabel* (λ, α) *mit*

89) Vgl. auch eine Arbeit von *W. Schnee,* Über den Zusammenhang zwischen den Summabilitätseigenschaften Dirichletscher Reihen und ihrem funktionentheoretischen Charakter. Acta Math. 35 (1912), p. 357—398, worin der *Landau-Schnee*sche Satz über das Konvergenzproblem (vgl. Nr. 6) von Konvergenz auf Summabilität verallgemeinert wird.

90) *Hardy-Riesz,* a. a. O. 1), p. 64.

50*

der Summe A und $\sum b_n e^{-\mu_n s}$ *im selben Punkte* s_0 *summabel* (μ, β) *mit der Summe B ist, so ist die Produktreihe* $\sum c_n e^{-\nu_n s}$ *im Punkte* s_0 *summabel* $(\nu, \alpha + \beta + 1)$ *mit der Summe AB.* Im speziellen Fall $\alpha = \beta = 0$ erhalten wir den in Nr. 12 erwähnten Satz über die Multiplikation zweier *konvergenter Dirichlet*scher Reihen.

Außer den *Cesàro-Riesz*schen Methoden wurden auch *andere* Summationsmethoden auf die *Dirichlet*schen Reihen angewendet. So hat *Hardy*[91]) die Wirkung der *Borel*schen Summation auf die gewöhnlichen *Dirichlet*schen Reihen (2) geprüft. Auch bei dieser Summationsmethode ist die Zahlenfolge $\left\{\frac{1}{n^s}\right\}$ $(\sigma > 0)$ eine summabilitätserhaltende Faktorenfolge, und das Summabilitätsgebiet also eine *Halbebene* $\sigma > \sigma^{(B)}$. Diese Halbebene $\sigma > \sigma^{(B)}$ kann aber niemals über die *Cesàro-Riesz*sche Summabilitätshalbebne $\sigma > \Omega$ hinausreichen und braucht nicht immer so weit zu reichen. Anders verhält es sich mit einer anderen von *Hardy*[92]) untersuchten Summabilitätsmethode, der sogenannten *Abel*schen Methode, nach welcher eine Reihe $\sum a_n$ summabel mit der Summe A heißt, wenn die Potenzreihe $f(x) = \sum a_n x^n$ für $0 < x < 1$ konvergiert und die Bedingung $f(x) \to A$ für $x \to 1$ erfüllt. *Hardy* beweist, daß auch hier das Summabilitätsgebiet einer gewöhnlichen *Dirichlet*schen Reihe (2) eine *Halbebene* $\sigma > \sigma^{(A)}$ ist, und daß $\sigma^{(A)}$ einfach die untere Grenze aller Abszissen σ_0 ist, für welche die durch die Reihe dargestellte Funktion $f(s)$ in der Halbebene $\sigma > \sigma_0$ *regulär und gleich* $O(e^{k|t|})$ *mit* $k < \frac{\pi}{2}$ *ist.*[93])

Schließlich ist noch eine schöne Arbeit von *M. Riesz*[94]) zu erwähnen, in welcher es ihm gelungen ist, die bekannten *Mittag-Leffler*schen Resultate über die analytische Darstellung der durch eine

91) *G. H. Hardy*, a. a. O. 72). Vgl. auch *Fekete*, a. a. O. 72) und *G. H. Hardy-J. Littlewood*, The relations between Borel's and Cesàro's method of summation, Proc. London math. Soc. (2) 11 (1913), p. 1—16.

92) *G. H. Hardy*, a) Sur la sommation des séries de Dirichlet, Paris C. R. 162 (1916), p. 463—465; b) a. a. O. 51).

93) Einfache Beispiele *Dirichlet*scher Reihen, bei welchen erst die *Abel*sche Summabilität — also nicht die *Cesàro*sche — imstande ist, die durch die Reihe dargestellte Funktion über die Konvergenzhalbebene $\sigma > \sigma_B$ hinaus analytisch fortzusetzen $\Big($weil die Funktion für $\sigma < \sigma_B$ stärker als $|t|^k$, aber nicht so stark wie $e^{\frac{\pi}{2}|t|}$ wächst$\Big)$, wurden von *G. H. Hardy*, a) a. a. O. 92) und b) Example to illustrate a point in the theory of Dirichlet's series, Tôhoku J. 8 (1915), p. 59—66, angegeben.

94) *M. Riesz*, Sur la représentation analytique des fonctions définies par des séries de Dirichlet, Acta Math. 35 (1912), p. 253—270.

Potenzreihe ($\lambda_n = n$) definierten Funktion in ihrem Hauptstern auf den allgemeinen Reihentypus (1) zu übertragen. *Riesz* beweist u. a. den folgenden Satz: Es habe die *Dirichlet*sche Reihe (1) ein Konvergenzgebiet, und es sei H der *Hauptstern* der durch die Reihe definierten Funktion $f(s)$, d. h. das Gebiet, welches aus der s-Ebene entsteht, wenn alle mit der negativen reellen Achse parallelen Halbgeraden, die von den singulären Punkten von $f(s)$ ausgehen, entfernt werden. Dann gilt *im ganzen Hauptstern die Darstellung* $f(s) = \lim\limits_{\alpha \to 0} \varphi_\alpha(s)$, *wo* $\varphi_\alpha(s)$ *die (ganze transzendente) Funktion*

$$\varphi_\alpha(s) = \sum \frac{1}{\Gamma(\alpha\lambda_n + 1)} a_n e^{-\lambda_n s}$$

bezeichnet. Auch die von *Mittag-Leffler* benutzten *Integraldarstellungen* zur analytischen Fortsetzung einer durch eine gegebene Potenzreihe ($\lambda_n = n$) definierten Funktion wurden von *Riesz* auf die *Dirichlet*schen Reihen übertragen.

II. Die Riemannsche Zetafunktion.

14. Die Zetafunktion und ihre Funktionalgleichung. Die Zetafunktion wird (vgl. Nr. **1**) durch die *Dirichlet*sche Reihe

(21) $$\zeta(s) = \sum \frac{1}{n^s} = 1 + \frac{1}{2^s} + \frac{1}{3^s} + \cdots$$

definiert. Obwohl schon *Euler* diese Funktion betrachtet und ihre zahlentheoretische Bedeutung erkannt hat, wird sie doch gewöhnlich als die „*Riemann*sche“ Zetafunktion bezeichnet, weil *Riemann* sie zuerst in seiner berühmten Abhandlung über die Anzahl der Primzahlen[95]), welche auch für die Entwicklung der neueren Funktionentheorie von fundamentaler Bedeutung gewesen ist, einem tiefergehenden Studium unterworfen hat. Über die Bedeutung der Zetafunktion für das Primzahlproblem sei in diesem Kapitel, das sich ausschließlich mit den rein funktionentheoretischen Eigenschaften von $\zeta(s)$ beschäftigen soll, nur bemerkt, daß sie in der *Euler*schen *Identität*

(22) $$\zeta(s) = \sum n^{-s} = \prod \frac{1}{1 - p_m^{-s}},$$

wo p_m die Primzahlen durchläuft, wurzelt; diese *Euler*sche Produktdarstellung spielt übrigens auch (vgl. z. B. Nr. **17**) bei manchen funktionentheoretischen Untersuchungen von $\zeta(s)$ eine bedeutsame Rolle.

Die die Zetafunktion definierende Reihe (21) konvergiert nur in der Halbebene $\sigma > 1$, und auch das Produkt (22) ist für $\sigma < 1$ divergent und gibt somit keinen Aufschluß über die Möglichkeit analyti-

95) *B. Riemann*, Über die Anzahl der Primzahlen unter einer gegebenen Größe, Monatsber. Akad. Berlin 1859, p. 671—680 = Werke (2. Aufl.), p. 145—153.

scher Fortsetzung über die Gerade $\sigma = 1$ hinaus. Anders verhält es sich mit der in Nr. **11** erwähnten Integraldarstellung

$$\zeta(s) = \frac{1}{\Gamma(s)} \int_0^\infty \frac{x^{s-1}}{e^x - 1} dx;$$

in der Tat, es läßt sich dieses zunächst ebenfalls nur für $\sigma > 1$ brauchbare Integral als ein komplexes Kurvenintegral

$$(23) \qquad \zeta(s) = \frac{1}{\Gamma(s)} \frac{1}{e^{-\pi s i} - e^{\pi s i}} \int_W \frac{(-x)^{s-1}}{e^x - 1} dx = \frac{i\Gamma(1-s)}{2\pi} \int_W \frac{(-x)^{s-1}}{e^x - 1} dx$$

schreiben, wo der Integrationsweg W eine Schleife ist, die vom Punkte $x = +\infty$ ausgeht und nach einem einmaligen Umkreisen des Punktes $x = 0$ zum Punkte $x = +\infty$ zurückkehrt. Aus dieser Integraldarstellung, die offensichtlich für jedes s konvergiert, schloß *Riemann*, daß die Funktion $\zeta(s)\Gamma(s)\sin \pi s$ eine ganze Transzendente ist, und hieraus weiter, *daß $\zeta(s)$ in der ganzen Ebene als eine eindeutige Funktion existiert, die überall regulär ist mit Ausnahme des einzigen Punktes $s = 1$, wo sie einen Pol erster Ordnung (mit dem Residuum 1) besitzt.*

Aus der Darstellung (23) leitete *Riemann* des weiteren durch eine Deformation des Integrationsweges und Anwendung des *Cauchy*schen Satzes eine fundamentale Eigenschaft der Zetafunktion ab, nämlich daß sie der *Funktionalgleichung*

$$(24) \qquad \zeta(1-s) = \frac{2}{(2\pi)^s} \cos \frac{s\pi}{2} \Gamma(s)\zeta(s)$$

genügt[96]), oder anders ausgedrückt, daß die Funktion

$$\eta(s) = \zeta(s)\Gamma\left(\frac{s}{2}\right)\pi^{-\frac{s}{2}}$$

ungeändert bleibt, wenn die Variable s durch $1-s$ ersetzt wird. Für die Funktionalgleichung in dieser letzten Form: $\eta(s) = \eta(1-s)$ und gleichzeitig auch für die Existenz von $\zeta(s)$ in der ganzen Ebene[97])

96) Nach *E. Landau*, Euler und die Funktionalgleichung der Riemannschen Zetafunktion, Bibl. Math. (3) 7 (1906—7), p. 69—79 war diese Funktionalgleichung schon *Euler* bekannt.

97) Außer den beiden, von *Riemann* selbst herrührenden, Beweisen des Satzes, daß $\zeta(s)$ von der Definitionshalbebene $\sigma > 1$ aus in die ganze Ebene fortgesetzt werden kann, gibt es eine Menge anderer Beweise dieses Satzes. So hat z. B. *J. L. W. V. Jensen*, Interméd. math. 1 (1895), p. 346—347 verschiedene Integraldarstellungen für $\zeta(s)$ angegeben, aus welchen die Existenz von $\zeta(s)$ in der ganzen Ebene unmittelbar ersichtlich ist; vgl. hierzu auch *E. Lindelöf*, Le calcul des résidus et ses applications à la théorie des fonctions (Collection Borel), Paris 1905, p. 1—141. Ein anderer Beweis von *Jensen*, Sur la fonction $\zeta(s)$ de

hat *Riemann*[95]) auch einen anderen Beweis gegeben, welcher sich bei der Anwendung auf mit der Zetafunktion verwandte Funktionen als sehr verallgemeinerungsfähig erwiesen hat. *Riemann* geht hierbei vom Integral

$$\frac{1}{n^s}\Gamma\left(\frac{s}{2}\right)\pi^{-\frac{s}{2}} = \int_0^\infty e^{-n^2\pi x} x^{\frac{s}{2}-1} dx$$

aus und erhält durch Summation die Formel

$$(25)\qquad \zeta(s)\Gamma\left(\frac{s}{2}\right)\pi^{-\frac{s}{2}} = \int_0^\infty x^{\frac{s}{2}-1}\,\omega(x)\,dx, \qquad (\sigma > 1)$$

wo $\omega(x)$ die Reihe

$$\omega(x) = \sum_{n=1}^\infty e^{-n^2\pi x}$$

bezeichnet. Nun ist aber, nach einer bekannten Formel aus der Theorie der elliptischen Thetafunktionen,

$$1 + 2\omega(x) = \frac{1}{\sqrt{x}}\left(1 + 2\omega\left(\frac{1}{x}\right)\right), \qquad (x > 0)$$

woraus sich durch Einsetzen in (25) und eine leichte Rechnung die Formel

$$(26)\qquad \zeta(s)\Gamma\left(\frac{s}{2}\right)\pi^{-\frac{s}{2}} - \frac{1}{s(s-1)} = \int_1^\infty \left(x^{\frac{1-s}{2}-1} + x^{\frac{s}{2}-1}\right)\omega(x)\,dx$$

Riemann, Paris C. R. 104 (1887), p. 1156—1159 beruht auf einer Relation zwischen den unendlich vielen Gliedern der Folge $\zeta(s)$, $\zeta(s+1)$, $\zeta(s+2)$, ...; ähnliche Relationen, welche überdies die Eigenschaft besitzen, in sich als Definitionsgleichungen der Zetafunktion gelten zu können, wurden später von *J. Hadamard,* Sur une propriété fonctionelle de la fonction $\zeta(s)$ de Riemann, Bull. Soc. math. France 37 (1909), p. 59—60, angegeben. *Ch. de la Vallée Poussin,* Démonstration simplifiée du théorème de Dirichlet sur la progression arithmétique, Mém. Acad. Belgique 53 (1895—96), No. 6, p. 1—32, beweist den Satz durch Vergleich der Reihe $\sum \frac{1}{n^s} = \zeta(s)$ mit dem entsprechenden Integral $\int_1^\infty \frac{du}{u^s} = \frac{1}{s-1}$, indem er (durch partielle Integrationen) nachweist, daß die Differenz $\zeta(s) - \frac{1}{s-1}$ eine ganze Transzendente ist; die Idee dieser Beweismethode ist von *H. Cramér,* Sur une classe de séries de Dirichlet, Diss. Upsala (Stockholm 1917), p. 1—51, zur Untersuchung beliebiger *Dirichlet*scher Reihen verallgemeinert. Setzt man die Theorie der *Cesàro*schen Summabilität *Dirichlet*scher Reihen (vgl. Nr. 13) als bekannt voraus, dürfte der einfachste Beweis für die Existenz von $\zeta(s)$ in der ganzen Ebene wohl derjenige sein, daß man die Funktion $\zeta(s)(1 - 2^{1-s})$ betrachtet, welche durch die in der ganzen Ebene summable Reihe $\sum \frac{(-1)^{n+1}}{n^s}$ dargestellt wird und somit sich sofort als eine ganze Transzendente erweist.

ergibt, welche sofort erkennen läßt, daß die auf der linken Seite stehende Funktion eine ganze Transzendente ist, die ungeändert bleibt, wenn s durch $1 - s$ ersetzt wird.[98])

Die Funktionalgleichung (24) verbindet die Werte der Zetafunktion in zwei Punkten s und $1 - s$, welche in bezug auf den Punkt $\frac{1}{2}$ symmetrisch gelegen sind. Hieraus folgt, daß man das Studium der Zetafunktion wesentlich auf die Halbebene $\sigma \geqq \frac{1}{2}$ beschränken kann (übrigens sogar auf die Viertelebene $\sigma \geqq \frac{1}{2}$, $t \geqq 0$, da $\zeta(s)$ in konjugierten Punkten konjugierte Werte annimmt); denn die Funktionalgleichung erlaubt ja das Verhalten von $\zeta(s)$ in der Halbebene $\sigma < \frac{1}{2}$ aus dem Verhalten der Funktion für $\sigma > \frac{1}{2}$ abzulesen. So können wir z. B. aus der aus der *Euler*schen Identität (22) unmittelbar folgenden Tatsache, daß $\zeta(s)$ *in der Halbebene* $\sigma > 1$ *überall von* 0 *verschieden* ist, mittels der Funktionalgleichung (24) sofort die Nullstellen von $\zeta(s)$ in der Halbebene $\sigma < 0$ bestimmen; in der Tat, es folgt ja aus (24), daß diese Nullstellen mit den Nullstellen der Funktion $1 : \left\{\cos\frac{s\pi}{2}\Gamma(s)\right\}$ für $\sigma < 0$ übereinstimmen, d. h. daß $\zeta(s)$ *in der besprochenen Halbebene* $\sigma < 0$ *Nullstellen* (*einfache*) *in den Punkten* $s = -2, -4, -6, \ldots$ und nur in diesen Punkten besitzt. Es werden diese Nullstellen gewöhnlich als die „trivialen“ Nullstellen von $\zeta(s)$ bezeichnet, im Gegensatze zu den (in Nr. **15** zu erwähnenden) „nichttrivialen“ Nullstellen im Streifen $0 \leqq \sigma \leqq 1$. Diese letzten Nullstellen liegen übrigens alle im *Innern* des Streifens $0 < \sigma < 1$; denn wie *de la Vallée Poussin*[99]) und *Hadamard*[100]) unabhängig von ein-

98) *H. Hamburger*, a) Über die Riemannsche Funktionalgleichung der ζ-Funktion, Math. Ztschr. 10 (1921), p. 240—254; 11 (1922), p. 224—245; 13 (1922), p. 283—311; b) Über einige Beziehungen, die mit der Funktionalgleichung der Riemannschen ζ-Funktion äquivalent sind, Math. Ann. 85 (1922), p. 129—140, beweist, daß $\zeta(s)$ bis auf einen konstanten Faktor *eindeutig* bestimmt ist durch die folgenden Eigenschaften: sie ist eine meromorphe Funktion mit nur endlich vielen Polen, die 1. der Funktionalgleichung (24) genügt, 2. für $|s| \to \infty$ gleich $O(e^{|s|^k})$ und 3. für $\sigma > 1$ durch eine absolut konvergente *Dirichlet*sche Reihe $\sum \frac{a_n}{n^s}$ darstellbar ist. Im Laufe des Beweises dieses Satzes [durch welchen eine Fragestellung von *J. Hadamard*, a. a. O. 60) behandelt wird] gibt *Hamburger* einen neuen Beweis der *Riemann*schen Funktionalgleichung. Vgl. auch *C. Siegel*, Bemerkung zu einem Satz von Hamburger über die Funktionalgleichung der Riemannschen Zetafunktion, Math. Ann. 86 (1922), p. 276—279.

99) *Ch. de la Vallée Poussin*, Recherches analytiques sur la théorie des nombres premiers, Ann. soc. sc. Bruxelles 20² (1896), p. 183—256 und p. 281—397.

100) *J. Hadamard*, Sur la distribution des zéros de la fonction $\zeta(s)$ et ses conséquences arithmétiques, Bull. Soc. math. France 24 (1896), p. 199—220.

ander durch sinnreiche Überlegungen bewiesen haben, ist *die Gerade* $\sigma = 1$ (*und daher auch die Gerade* $\sigma = 0$) *nullpunktsfrei.* In diesem Zusammenhange sei noch erwähnt, daß *Mertens*[101]) schon früher durch eine interessante Abschätzung bewiesen hatte, daß das *Euler*sche Produkt (22) in jedem Punkte $s \neq 1$ auf der Geraden $\sigma = 1$, in welchem $\zeta(s) \neq 0$ ist (also nach dem obigen Satze in den sämtlichen Punkten $s \neq 1$) noch konvergiert; mit Hilfe des in Nr. 5 besprochenen *Riesz*schen Konvergenzsatzes (auf $\log \zeta(s)$ verwendet) läßt sich dieses Resultat[102]), oder was damit gleichbedeutend ist, die Konvergenz der Reihe $\sum \frac{1}{p_m^{1+it}}$ für alle $t \gtrless 0$, unmittelbar ohne jede spezielle Abschätzung aus dem Nichtverschwinden von $\zeta(s)$ auf der Geraden $\sigma = 1$ ableiten.

15. Die Riemann-Hadamardsche Produktentwicklung. Betrachten wir mit *Riemann* die Funktion[103])

$$\xi(s) = \frac{s(s-1)}{2} \Gamma\left(\frac{s}{2}\right) \pi^{-\frac{s}{2}} \zeta(s),$$

welche eine ganze Transzendente ist, deren Nullstellen mit den nichttrivialen Nullstellen von $\zeta(s)$ übereinstimmen (indem die trivialen Nullstellen weggeschafft sind), und die der Gleichung $\xi(s) = \xi(1-s)$ genügt. Die durch diese Funktionalgleichung ausgedrückte Eigenschaft kann auch dadurch zum Ausdruck gebracht werden, daß die Funktion $\Xi(z)$, welche aus $\xi(s)$ durch die Transformation $s = \frac{1}{2} + iz$ entsteht, eine *gerade* Funktion von z ist, d. h. eine Funktion von z^2, die wir mit $g(z^2)$ bezeichnen werden, wo also $g(x)$ eine ganze Funktion von x ist. Jedem Nullstellenpaar $\pm \lambda$ von $\Xi(z)$, d. h. jedem Nullstellenpaar ϱ und $1 - \varrho$ von $\xi(s)$ entspricht also nur eine einzige Nullstelle $\mu = \lambda^2$ von $g(x)$. Über diese Funktion $g(x)$ behauptete *Riemann*[95]), daß sie unendlich viele Nullstellen μ_n hat — also daß die Zetafunktion unendlich viele nichttriviale Nullstellen besitzt — und ferner, daß sie durch das Produkt

$$g(x) = g(0) \prod_{n=1}^{\infty} \left(1 - \frac{x}{\mu_n}\right) \tag{27}$$

101) *F. Mertens*, Über die Konvergenz einer aus Primzahlpotenzen gebildeten unendlichen Reihe, Gött. Nachr. 1887, p. 265—269.

102) Vgl. *E. Landau*, a. a. O. 21).

103) Die folgenden Bezeichnungen der Funktionen sind die von *E. Landau* benutzten (und jetzt üblichen), welche von den *Riemann*schen etwas abweichen. Die von *Riemann* mit ξ bezeichnete Funktion ist die unten erwähnte Funktion Ξ.

dargestellt werden kann. Die Richtigkeit dieser Behauptung wurde bekanntlich zuerst von *Hadamard*[104]) durch seine grundlegenden Untersuchungen über ganze transzendente Funktionen endlichen Geschlechtes bewiesen. Es ist nämlich, wie leicht zu zeigen,

$$g(x) = O\left(e^{|x|^{\frac{1}{2}+\varepsilon}}\right) \qquad (\varepsilon > 0)$$

und nach einem allgemeinen Satz der *Hadamard*schen Theorie folgt aus dieser Abschätzung sofort die Richtigkeit der obigen Behauptung.[105]) Wie oben erwähnt, entsprechen jeder Nullstelle μ_n von $g(x)$ zwei Nullstellen von $\zeta(s)$, nämlich $\frac{1}{2} \pm i\sqrt{\mu_n}$, welche symmetrisch in bezug auf den Punkt $\frac{1}{2}$ liegen. Wird die Produktentwicklung (27) von $g(x)$ zu einer Produktentwicklung der Funktion $\zeta(s)$ selbst umgeschrieben, so findet man — indem der Bequemlichkeit halber Konvergenzfaktoren hinzugefügt werden, die das Produkt von der Reihenfolge der Faktoren (d. h. von dem paarweisen Zusammennehmen zweier „entsprechender" Nullstellen ϱ und $1 - \varrho$) unabhängig machen — die grundlegende Formel

$$(28) \qquad (s-1)\zeta(s) = \frac{1}{2} e^{bs} \frac{1}{\Gamma\left(\frac{s}{2}+1\right)} \prod_{\varrho} \left(1 - \frac{s}{\varrho}\right) e^{\frac{s}{\varrho}},$$

wo ϱ die sämtlichen nichttrivialen Nullstellen durchläuft, und b eine Konstante ($b = \log 2\pi - 1 - \frac{1}{2}C$, wo C die *Euler*sche Konstante ist) bezeichnet. In der Primzahlentheorie kommt diese Formel (28) meistens in der Form

$$(29) \qquad \frac{\zeta'}{\zeta}(s) = b - \frac{1}{s-1} - \frac{1}{2}\frac{\Gamma'}{\Gamma}\left(\frac{s}{2}+1\right) + \sum_{\varrho}\left(\frac{1}{s-\varrho} + \frac{1}{\varrho}\right)$$

zur Anwendung.

Es sei schon hier erwähnt, daß *Riemann*[95]) des weiteren die Vermutung ausgesprochen hat — aber mit ausdrücklicher Hervorhebung, daß er diese Vermutung nicht beweisen konnte — daß die Nullstellen von $g(x)$ alle *reell* sind, d. h. *daß die nichttrivialen Nullstellen von $\zeta(s)$ alle auf der Geraden $\sigma = \frac{1}{2}$ liegen.* Ob diese berühmte „*Riemann*sche Vermutung" richtig ist oder nicht, ist bekanntlich noch heute unentschieden, und man weiß auch nicht, durch welche Ar-

104) *J. Hadamard,* Étude sur les propriétés des fonctions entières et en particulier d'une fonction considérée par Riemann, J. de math. (4) 9 (1893), p. 171—215.

105) In dem ursprünglichen *Hadamard*schen Beweise wird übrigens nicht die Größenordnung der Funktion $g(x)$ selbst, sondern — was nach *Hadamard* auf genau dasselbe hinauskommt — die Größenordnung der Koeffizienten a_n der Potenzreihe $g(x) = \Sigma a_n x^n$ abgeschätzt.

gumente (abgesehen von der *symmetrischen Lage der Nullstellen in bezug auf die Gerade* $\sigma = \frac{1}{2}$) *Riemann* auf diese Vermutung geführt worden ist.

16. Die Riemann-v. Mangoldtsche Formel für die Anzahl der Nullstellen. Über die nähere Verteilung der Ordinaten der nichttrivialen Nullstellen von $\zeta(s)$ hat *Riemann*[95]) ohne Beweis eine Formel angegeben, die viel präziser ist als diejenigen Resultate, welche man aus der *Hadamard*schen Theorie direkt entnehmen kann, nämlich die Formel

$$(30)\qquad N(T) = \frac{1}{2\pi} T \log T - \frac{1+\log 2\pi}{2\pi} T + O(\log T),$$

wo $N(T)$ die Anzahl der Nullstellen von $\zeta(s)$ im Rechteck $0 < \sigma < 1$, $0 < t \leqq T$ bezeichnet. Es gelang erst *v. Mangoldt*[106]), diese Formel streng zu beweisen. Betreffs des Beweises sei nur erwähnt, daß man von dem Ausdruck $N(T) = \frac{1}{2\pi i}\int \frac{\zeta'(s)}{\zeta(s)} ds$ ausgehend, wo das Integral längs des Randes eines Rechteckes mit den Eckpunkten 2, $2 + iT$, $-1 + iT$, -1 erstreckt ist, durch einfache Rechnungen (unter Benutzung der Funktionalgleichung der Zetafunktion und bekannter Eigenschaften der Gammafunktion) leicht findet, daß

$$(31)\qquad N(T) = \frac{1}{2\pi} T \log T - \frac{1+\log 2\pi}{2\pi} T + R(T)$$

ist, wo das Restglied $R(T) = O(1) + \frac{1}{2\pi i}\int\limits_{2+iT}^{-1+iT} \frac{\zeta'(s)}{\zeta(s)} ds$, und daß die ganze, erst von *v. Mangoldt* überwundene Schwierigkeit darin liegt, dies letzte Integral, welches den „kritischen" Streifen $0 < \sigma < 1$ durchsetzt, abzuschätzen. Der *v. Mangoldt*sche Beweis der Ungleichung $R(T) = O(\log T)$, wie auch ein später vereinfachter von *Landau*[107]), stützt sich wesentlich auf die *Hadamard*schen Resultate, d. h. auf die Produktentwicklung von $\zeta(s)$. Vor einigen Jahren wurde ein sehr eleganter Beweis dieser Ungleichung von *Backlund*[108]) gefunden, der

106) *H. v. Mangoldt,* Zur Verteilung der Nullstellen der Riemannschen Funktion $\xi(t)$, Math. Ann. 60 (1905), p. 1—19. Schon früher hatte *v. Mangoldt* [zu *Riemanns* Abhandlung „Über die Anzahl der Primzahlen unter einer gegebenen Größe", Crelles J. 114 (1895), p. 255—305] die Formel (30) mit einem Restgliede $O(\log^2 T)$ (statt $O(\log T)$) bewiesen.

107) *E. Landau,* Über die Verteilung der Nullstellen der Riemannschen Zetafunktion und einer Klasse verwandter Funktionen, Math. Ann. 66 (1909), p. 419—445.

108) *R. Backlund,* a) Sur les zéros de la fonction $\zeta(s)$ de Riemann, Paris C. R. 158 (1914), p. 1979—1981; b) Über die Nullstellen der Riemannschen Zeta-

nicht die Produktentwicklung (28), sondern nur eine ganz grobe Abschätzung von $\zeta(s)$ benutzt.

Ob die Abschätzung $R(T) = O(\log T)$ verbessert werden kann, weiß man nicht (vgl. jedoch Nr. 20, wo über Folgerungen der „*Riemann*schen Vermutung" berichtet wird); dagegen weiß man nach *Cramér*[109]), daß der *Mittelwert* $\frac{1}{T}\int_0^T R(t)\,dt$ beschränkt ist, sogar daß er für $T \to \infty$ einem bestimmten Grenzwert, nämlich dem Werte $\frac{7}{8}$, zustrebt.[110]) Verfeinerte Resultate dieser Art sind neulich von *Littlewood*[111]) angegeben.

17. Über die Werte von $\zeta(s)$ auf einer vertikalen Geraden $\sigma = \sigma_0 (> \frac{1}{2})$. Betrachten wir zunächst die Halbebene $\sigma > 1$; da $\zeta(s)$ hier $\neq 0$ ist, ist es ein natürlicheres (und allgemeineres) Problem, nach den Werten von $\log \zeta(s)$, statt nach den Werten von $\zeta(s)$ selbst zu fragen, wo $\log \zeta(s)$ z. B. denjenigen (für $\sigma > 1$ regulären) Zweig des Logarithmus der Zetafunktion bezeichnet, der für reelles $s > 1$ reell ist. Dieser Zweig ist, nach der *Euler*schen Identität (22) durch

$$\log \zeta(s) = -\sum_{m=1}^{\infty} \log(1 - p_m^{-s}) \qquad (\sigma > 1) \tag{32}$$

gegeben, also (wenn $\log(1 - p_m^{-s})$ in eine Potenzreihe entwickelt wird) durch eine *Dirichlet*sche Reihe, in welcher die einzelnen Primzahlen separiert sind. Durch diesen Umstand wird es möglich, das in Nr. 7 angegebene Verfahren, welches auf der Theorie diophantischer Approximationen beruht, in einfacher Weise durchzuführen, indem die dort mit $M(\sigma_0)$ bezeichnete Wertmenge explizite bestimmt werden

---

funktion, Dissertation Helsingfors 1916, p. 1—24; c) Über die Nullstellen der Riemannschen Zetafunktion, Acta Math. 41 (1918), p. 345—375.

109) *H. Cramér*, Studien über die Nullstellen der Riemannschen Zetafunktion, Math. Ztschr. 4 (1919), p. 104—130.

110) Daß der Grenzwert gerade den Wert $\frac{7}{8}$ hat, hängt damit zusammen, daß $R(T)$ auf die Form $R(T) = \frac{7}{8} + O\left(\frac{1}{T}\right) + \frac{1}{\pi}\Delta \arg \zeta(s)$ gebracht werden kann (*Backlund*, a. a. O. 108c), wo die Konstante $\frac{7}{8}$ von der Gammafunktion und den anderen in der Funktionalgleichung eingehenden elementaren Funktionen herrührt, während $\Delta \arg \zeta(s)$ den Zuwachs von $\arg \zeta(s)$ angibt, wenn s den gebrochenen Linienzug $2,\ 2 + iT,\ \frac{1}{2} + iT$ durchläuft.

111) *J. E. Littlewood*, Researches in the theory of the Riemann ζ-function, Proc. London math. Soc. 20 (1922), (Records et cet. p. XXII—XXVIII). In dieser kurzen Mitteilung wird, ohne Beweise, eine Reihe sehr tiefgehender Sätze über $\zeta(s)$ angegeben.

kann. Es ergibt sich, daß diese Menge $M(\sigma_0)$ ein endliches Gebiet (in der komplexen Ebene) ist, das je nach der Lage der Abszisse $\sigma_0(> 1)$ von einer oder von zwei konvexen Kurven begrenzt wird. Bei festem σ_0 läuft nun (nach Nr. 7) die von $\log \zeta(\sigma_0 + it)$ $(-\infty < t < \infty)$ beschriebene Kurve im Gebiete $M(\sigma_0)$ überall dicht herum, und es ist die Menge der Werte, welche $\log \zeta(s)$ in unendlicher Nähe der Geraden $\sigma = \sigma_0$ annimmt, mit $M(\sigma_0)$ identisch. Für σ_0 nahe an 1 ist $M(\sigma_0)$ ein einfach zusammenhängendes Gebiet (d. h. nur von einer Kurve begrenzt), das sich für $\sigma_0 \to 1$ nach und nach über die ganze Ebene ausbreitet[112]); hiermit ist speziell gefunden, daß $\log \zeta(s)$ in der Halbebene $\sigma > 1$ jeden Wert unendlich oft annimmt, also a fortiori, *daß* $\zeta(s)$ *in der Halbebene* $\sigma > 1$ (übrigens sogar in jedem Streifen $1 < \sigma < 1 + \delta$) *sämtliche Werte außer* 0 *unendlich oft annimmt.*[113])

Wesentlich schwieriger ist die Bestimmung der Werte von $\zeta(s)$ auf einer vertikalen Geraden $\sigma = \sigma_0$, welche im Streifen $\frac{1}{2} < \sigma \leqq 1$ liegt, weil das *Euler*sche Produkt, das ja die Quelle der obigen Untersuchung war, für $\sigma < 1$ divergiert (und für $\sigma = 1$, $t \neq 0$ nur bedingt konvergiert). Die auf der Theorie der diophantischen Approximationen beruhende Untersuchungsmethode läßt sich aber, obwohl in einer wesentlich modifizierten Form, auch hier verwenden[114]), und es ergibt sich, daß $\zeta(s)$ auf jeder festen Geraden $\sigma = \sigma_0$ im Streifen $\frac{1}{2} < \sigma \leqq 1$

112) *H. Bohr,* Sur la fonction $\zeta(s)$ dans le demi-plan $\sigma > 1$, Paris C. R. 154 (1912), p. 1078—1081. Die genaue Ausführung der betreffenden geometrischen Überlegungen findet sich in der Abhandlung: Om Addition af uendelig mange konvekse Kurve, Overs. Vidensk. Selsk. Köbenhavn 1913, p. 326—366. Die entsprechende Untersuchung der (bei den zahlentheoretischen Anwendungen wichtigen) Funktion $\frac{\zeta'}{\zeta}(s)$ findet sich bei *H. Bohr,* Über die Funktion $\frac{\zeta'}{\zeta}(s)$, Crelles J. 141 (1912), p. 217—234; die Untersuchung gestaltet sich hier wesentlich einfacher, weil die (konvexen) Begrenzungskurven der Gebiete $M(\sigma_0)$ einfach Kreise werden.

113) Über einen Beweis dieses letzteren (spezielleren) Resultates siehe *H. Bohr,* Über das Verhalten von $\zeta(s)$ in der Halbebene $\sigma > 1$, Gött. Nachr. 1911, p. 409—428. Schon früher hatten *H. Bohr* und *E. Landau,* Über das Verhalten von $\zeta(s)$ und $\zeta_\varkappa(s)$ in der Nähe der Geraden $\sigma = 1$, Gött. Nachr. 1910, p. 303—330 mit Hilfe allgemeiner funktionentheoretischer Methoden den weniger aussagenden Satz bewiesen, daß $\zeta(s)$ in jedem Streifen $1 - \delta < \sigma < 1 + \delta$ alle Werte, höchstens mit einer einzigen Ausnahme, annimmt.

114) Hierbei spielt eine von *H. Weyl* (Über die Gleichverteilung von Zahlen mod. Eins, Math. Ann. 77 (1916), p. 313—352) herrührende Verschärfung des in Nr. 7 erwähnten *Kronecker*schen Satzes über diophantische Approximationen eine wesentliche Rolle.

Werte annimmt, die *in der ganzen Ebene überall dicht* liegen[115]), und ferner, daß die Menge der Werte von $\zeta(s)$ in unendlicher Nähe einer solchen Geraden gewiß *sämtliche Werte, höchstens mit Ausnahme des einen Wertes* 0, *enthält.*[116]) Bei der besonders wichtigen Frage, ob auch der „kritische" Wert 0 angenommen wird oder nicht — also ob die „*Riemann*sche Vermutung" falsch oder richtig ist — versagt aber die Methode, und sie vermag nur (weil sie im Grunde eine Wahrscheinlichkeitsmethode ist) zu zeigen, daß, falls 0 in einem Streifen $\sigma_0 - \delta < \sigma < \sigma_0 + \delta$ überhaupt angenommen wird, 0 jedenfalls „unendlich seltener" angenommen wird als jeder andere Wert a, d. h. wenn $N_0(T)$ und $N_a(T)$ die Anzahl von 0-Stellen bzw. a-Stellen im Rechteck $\sigma_0 - \delta < \sigma < \sigma_0 + \delta$, $0 < t < T$ bezeichnen, so gilt für $T \to \infty$ die Gleichung $\lim N_0(T) : N_a(T) = 0$.[116])

18. Über die Größenordnung der Zetafunktion auf vertikalen Geraden. Man findet sehr leicht, daß $\zeta(s)$ in jeder Halbebene $\sigma > \sigma_0$ von endlicher Größenordnung in bezug auf t ist, und es läßt sich daher im ganzen Intervalle $-\infty < \sigma < \infty$ eine endliche *Größenordnungsfunktion* $\mu(\sigma)$ definieren (vgl. Nr. 6) als die untere Grenze aller Zahlen α, für welche $\zeta(\sigma + it)$ bei festem σ gleich $O(|t|^\alpha)$ ist. Die Funktionalgleichung (24) liefert die Relation $\mu(\sigma) = \mu(1-\sigma) + \frac{1}{2} - \sigma$, und es genügt somit $\mu(\sigma)$ für $\sigma \geqq \frac{1}{2}$ zu untersuchen. Für $\sigma > 1$ ist $\mu(\sigma) = 0$ (es ist sogar $|\zeta(s)|$ und $\frac{1}{|\zeta(s)|}$ beschränkt auf jeder vertikalen Geraden $\sigma = \sigma_0 > 1$). Die Schwierigkeit besteht darin, $\mu(\sigma)$ für $\frac{1}{2} \leqq \sigma \leqq 1$ zu bestimmen. Nachdem zuerst *Mellin* und später *Landau* gewisse Abschätzungen der μ-Funktion gewonnen hatten[117]), gelang es

115) *H. Bohr* und *R. Courant,* Neue Anwendungen der Theorie der diophantischen Approximationen auf die Riemannsche Zetafunktion, Crelles J. 144 (1914), p. 249-274.

116) *H. Bohr,* a) Sur la fonction $\zeta(s)$ de Riemann, Paris C. R. 158 (1914), p. 1986—1988; b) Zur Theorie der Riemannschen Zetafunktion im kritischen Streifen, Acta Math. 40 (1915), p. 67—100.

Schon früher hatten *H. Bohr* und *E. Landau,* Beiträge zur Theorie der Riemannschen Zetafunktion, Math. Ann. 74 (1913), p. 3—30 durch Überlegungen ganz anderer Art gezeigt, daß unter Annahme der Richtigkeit der „Riemannschen Vermutung" die Wertmenge von $\zeta(s)$ im Streifen $\sigma_0 - \delta < \sigma < \sigma_0 + \delta$ $(\frac{1}{2} < \sigma_0 < 1)$ alle Werte außer 0 enthält.

117) *H. Mellin,* Eine Formel für den Logarithmus transcendenter Funktionen von endlichem Geschlechte, Acta Soc. Sc. Fenn. 29 (1900), No. 4, p. 1—50, bewies, daß $\mu(\sigma) \leqq 1 - \sigma$ für $\frac{1}{2} \leqq \sigma \leqq 1$, und *E. Landau,* Sur quelques inégalités dans la théorie de la fonction $\zeta(s)$ de Riemann, Bull. Soc. math. France 33 (1905), p. 229—241 verschärfte dieses Resultat zu $\mu(\sigma) \leqq \frac{3}{4}(1-\sigma)$ für $\frac{1}{2} \leqq \sigma \leqq 1$.

Lindelöf[118]) durch allgemeine funktionentheoretische Betrachtungen zu beweisen (vgl Nr. 6), *daß $\mu(\sigma)$ eine stetige konvexe Funktion von σ ist*, woraus sofort folgt, wegen $\mu(\sigma) = 0$ für $\sigma > 1$ und $\mu(\sigma) = \frac{1}{2} - \sigma$ für $\sigma < 0$), daß die μ-Kurve für $0 \leqq \sigma \leqq 1$ im Dreieck mit den Endpunkten $(0, \frac{1}{2})$ $(\frac{1}{2}, 0)$ $(1, 0)$ verläuft, also speziell, daß $\mu(\sigma) \leqq \frac{1}{2}(1 - \sigma)$ für $\frac{1}{2} \leqq \sigma \leqq 1$. Ganz neuerdings ist es *Hardy* und *Littlewood*[119]) gelungen, über das *Lindelöf*sche Resultat hinauszukommen, und zwar u. a. zu beweisen, daß die μ-Kurve im Punkte $\sigma = 1$ die Abszissenachse berührt. Der genaue Verlauf der μ-Funktion für $0 < \sigma < 1$ ist noch heute unbekannt (vgl. jedoch Nr. 20).

Ein besonderes Interesse bietet die Untersuchung der Größenverhältnisse von $\zeta(s)$ $\left(\text{und } \frac{1}{\zeta(s)}\right)$ *auf der Geraden* $\sigma = 1$, die den kritischen Streifen von der „trivialen" Halbebene $\sigma > 1$ trennt, und wo $\zeta(s)$ (nach Nr. 17) „zum ersten Mal" Werte annimmt, die in der ganzen Ebene überall dicht liegen. Über das Resultat $\mu(1) = 0$, d. h. $\zeta(1 + it) = O(t^{\epsilon})$ hinaus, bewies *Mellin*[120]) das viel schärfere Resultat $\zeta(1 + it) = O(\log t)$, und mit Hilfe tiefgehender Untersuchungen über diophantische Approximationen haben später *Hardy-Littlewood*[121]) und *Weyl*[122]) die *Mellin*sche Abschätzung zu $\zeta(1 + it) = o(\log t)$ und *Weyl* sogar zu

$$\zeta(1 + it) = O\left(\frac{\log t}{\log\log t}\right) \tag{33}$$

verschärfen können. Andererseits haben *Bohr* und *Landau*[123]) ebenfalls mit Hilfe diophantischer Approximationen bewiesen, daß

$$\zeta(1 + it) \neq o(\log\log t) \tag{34}$$

ist. Die Frage nach der „wahren" Größenordnung von $\zeta(1 + it)$ ist aber hiermit noch lange nicht gelöst (vgl. jedoch Nr. 20), denn es besteht ja noch eine beträchtliche Lücke zwischen (33) und (34). Die entsprechende Frage über $1 : \zeta(1 + it)$ ist noch weniger aufgeklärt. Nachdem es zuerst *Mertens*[124]), durch eine neue Beweisanordnung des

118) *E. Lindelöf,* Quelques remarques sur la croissance de la fonction $\zeta(s)$, Bull. Sc. math. (2) 32 I (1908), p. 341—356.

119) Vgl. *J. Littlewood,* a. a. O. 111).

120) *H. Mellin,* a. a. O. 117).

121) *G. H. Hardy* und *J. Littlewood,* Some problems of diophantine approximation, Internat. Congr. of math. Cambridge 1 (1912), p. 223—229.

122) *H. Weyl,* Zur Abschätzung von $\zeta(1 + it)$, Math. Ztschr. 10 (1921), p. 88—100.

123) *H. Bohr* und *E. Landau,* a. a. O. 113).

124) *F. Mertens,* Über eine Eigenschaft der Riemannschen ζ-Funktion, Sitzungsber. Acad. Wien 107, II^a (1898), p. 1429—1434.

*Hadamard-de la Vallée Poussin*schen Satzes: $\zeta(1+it) \neq 0$, gelungen war, eine obere Grenze $\varphi(t)$ für $\frac{1}{|\zeta(1+it)|}$ explizite anzugeben, fand *Landau*[125]) die für seine zahlentheoretischen Zwecke wichtige Abschätzung $1:\zeta(1+it) = O\{(\log t)^c\}$, die von *Landau* selbst[29]) auf $O(\log t \cdot \log\log t)$, dann von *Gronwall*[126]) auf $1:\zeta(1+it) = O(\log t)$ und neuerdings von *Littlewood*[111]) auf

$$1:\zeta(1+it) = O\left(\frac{\log t}{\log\log t}\right)$$

verschärft wurde; andererseits weiß man aber nur, nach *Bohr*, daß

$$1:\zeta(1+it) \neq O(1)$$

ist, also daß $\zeta(s)$ auf der Geraden $\sigma = 1$ beliebig kleine Werte annimmt[127])), ohne daß man bis jetzt imstande gewesen ist, irgendeine mit t ins Unendliche wachsende Funktion $\psi(t)$ explizite anzugeben, für die $1:\zeta(1+it) \neq O(\psi(t))$ ist.

Obwohl $\zeta(s)$ auf keiner Geraden $\sigma = \sigma_0 \leqq 1$ beschränkt bleibt, ist sie doch bei jedem σ_0 im Intervalle $\frac{1}{2} < \sigma < 1$ *im Mittel* beschränkt, ja es ist sogar ihr Quadrat im Mittel beschränkt; denn aus dem *Schnee*schen Mittelwertsatze (Nr. 9) ergibt sich leicht[128]), daß

$$\lim_{T\to\infty} \frac{1}{2T}\int_{-T}^{T} |\zeta(\sigma+it)|^2 dt = \sum \frac{1}{n^{2\sigma}} = \zeta(2\sigma). \qquad (\tfrac{1}{2} < \sigma < 1)$$

Hieraus folgt mit Hilfe der Funktionalgleichung, daß für $\sigma < \frac{1}{2}$ der Mittelwert $\frac{1}{2T}\int_{-T}^{T} |\zeta(\sigma+it)|^2 dt$ sich asymptotisch wie

$$\tfrac{1}{2}(2\pi)^{2\sigma-1}\frac{\zeta(2-2\sigma)}{2-2\sigma}\cdot T^{1-2\sigma}$$

verhält. Viel schwieriger ist das Problem des Verhaltens von $\frac{1}{2T}\int_{-T}^{T} |\zeta(\sigma+it)|^2 dt$ für $\sigma = \frac{1}{2}$; dieses wurde von *Hardy* und *Little-*

125) *E. Landau*, Neuer Beweis des Primzahlsatzes und Beweis des Primidealsatzes, Math. Ann. 56 (1903), p. 645—670.

126) *T. Gronwall*, Sur la fonction $\zeta(s)$ de Riemann au voisinage de $\sigma = 1$, Palermo Rend. 35 (1913), p. 95—102.

127) Ein direkter Beweis dieses Satzes (der ja als Spezialfall in dem in Nr. 17 erwähnten Resultate über $\zeta(1+it)$ enthalten ist) findet sich bei *H. Bohr*, Sur l'existence de valeurs arbitrairement petites de la fonction $\zeta(s) = \zeta(\sigma+it)$ de Riemann pour $\sigma > 1$, Oversigt. Vidensk. Selsk. Köbenhavn 1911, p. 201—208. Vgl. auch *H. Bohr*, Note sur la fonction Zéta de Riemann $\zeta(s) = \zeta(\sigma+it)$ sur la droite $\sigma = 1$, Oversigt Vidensk. Selsk. Köbenhavn 1913, p. 3—11.

128) Vgl. z. B. *E. Landau*, Handbuch, a. a. O. 1), § 228.

wood gelöst[31b]), und zwar mit dem Ergebnis

$$\frac{1}{2T}\int_{-T}^{T}|\zeta(\tfrac{1}{2}+it)|^2 dt \sim \log T.$$

Neuerdings haben *Hardy* und *Littlewood*[129]) bewiesen, daß auch $\frac{1}{2T}\int_{-T}^{T}|\zeta(\sigma+it)|^4 dt$ bei festem σ im Intervalle $\frac{1}{2}<\sigma<1$ beschränkt bleibt und sogar für $T\to\infty$ einem bestimmten Grenzwerte zustrebt. Der Beweis basiert auf der von *Hardy* und *Littlewood*[129]) entdeckten sogenannten „*approximativen Funktionalgleichung*", welche besagt, daß für $|\sigma|<k$, $x>K$, $y>K$ und $2\pi xy=|t|$

$$\zeta(s)=\sum_{n<x}\frac{1}{n^s}+\chi\sum_{n<y}\frac{1}{n^{1-s}}+R,$$

wo $\chi=2(2\pi)^{s-1}\sin\frac{\pi s}{2}\Gamma(1-s)$ und $R=O(x^{-\sigma})+O(y^{\sigma-1}|t|^{\frac{1}{2}-\sigma})$. Diese Formel, welche bei Abschätzungen der Zetafunktion im kritischen Streifen $0<\sigma<1$ von der größten Bedeutung ist, ist[129]) eine Art „Kompromiß zwischen der für $\sigma>1$ gültigen Formel $\zeta(s)=\sum\frac{1}{n^s}$ und der für $\sigma<0$ gültigen Formel $\zeta(s)=\chi\sum\frac{1}{n^{1-s}}$".

Die oben erwähnten Mittelwertsformeln und andere ähnliche, die sich durch Anwendung des *Schnee*schen Mittelwertsatzes auf mit der Zetareihe beschlechtete *Dirichlet*sche Reihen abgeleitet werden, spielen bei neueren Untersuchungen über die Zetafunktion eine immer wichtigere Rolle.

19. Näheres über die Nullstellen im kritischen Streifen. Aus dem Satze (Nr. **14**), daß keine Nullstelle von $\zeta(s)$ *auf* der Geraden $\sigma=1$ gelegen ist, folgt sofort, daß für eine mit $t\to\infty$ „hinreichend" schnell zu 0 abnehmende Funktion $\varphi(t)$ der asymptotisch unendlich schmale Streifen $1\geqq\sigma>1-\varphi(t)$ ebenfalls nullpunktsfrei ist. Mit Hilfe der *Hadamard*schen Produktentwicklung von $\zeta(s)$ gelang es *de la Vallée Poussin*[130]), und später durch elementarere Mittel *Landau*[125]), eine solche Funktion $\varphi(t)$ explizite anzugeben; das *de la Vallée Poussin*-

129) *G. H. Hardy* und *J. Littlewood*, The approximate functional equation in the theory of the zeta-function, with applications to the divisor-problems of Dirichlet and Piltz, Proc. London math. Soc. 21 (1922), p. 39—74.

130) *Ch. de la Vallée Poussin*, Sur la fonction $\zeta(s)$ de Riemann et le nombre des nombres premiers inférieurs à une limite donnée, Mém. Acad. Belgique 59, No. 1 (1899—1900), p. 1—74.

sche Resultat, welches das genauere war, besagt, daß $\frac{k}{\log t}$ (bei passender Wahl von $k > 0$) eine zulässige Funktion $\varphi(t)$ ist. Ganz neuerdings hat *Littlewood*[111]) dieses dahin verschärft, daß $\varphi(t) \sim \frac{k \log\log t}{\log t}$ angenommen werden darf. Ob es aber eine Konstante $\sigma_0 < 1$ gibt mit der Eigenschaft, daß $\zeta(s) \neq 0$ für $\sigma > \sigma_0$, ist immer noch unentschieden.

Wie in Nr. **16** erwähnt, ist die Anzahl $N(T)$ von Nullstellen im Rechteck $0 < \sigma < 1$, $0 < t < T$ für $T \to \infty$ asymptotisch gleich $k \cdot T \log T$. Über die Verteilung dieser Nullstellen haben *Bohr* und *Landau*[65a]) bewiesen, daß *ihre Mehrzahl in nächster Nähe der Geraden* $\sigma = \frac{1}{2}$ *gelegen ist,* d. h. bei jedem festen $\delta > 0$ ist die Anzahl $N_1(T)$ von Nullstellen, welche innerhalb des obigen Rechtecks, aber außerhalb des dünnen Streifens $\frac{1}{2} - \delta < \sigma < \frac{1}{2} + \delta$ liegen, gleich $o(T \log T)$; dies folgt aus einem allgemeinen, in Nr. **10** erwähnten Satz über *Dirichlet*sche Reihen (auf die Zetareihe mit abwechselndem Vorzeichen angewendet), nach welchem die besprochene Anzahl $N_1(T)$ sogar gleich $O(T)$ ist. Durch eine weitergehende Untersuchung wurde dieses Resultat zuerst[65b]) zur Gleichung $N_1(T) = o(T)$ und später von *Carlson*[131]) mit Hilfe seines in Nr. **10** erwähnten Satzes über *Dirichlet*sche Reihen zu
$$N_1(T) = O(T^{1-4\delta^2+\varepsilon}) \qquad (\varepsilon > 0)$$
verschärft. Ferner gelang es *Littlewood*[111]) eine mit $t \to \infty$ zu 0 abnehmende Funktion $\varphi(t)$ explizite anzugeben mit der Eigenschaft, daß das Hauptresultat $N_1(T) = o(T \log T)$ noch gültig bleibt, wenn $N_1(T)$ die Anzahl der Nullstellen im Gebiete $\sigma > \frac{1}{2} + \varphi(t)$, $0 < t < T$ angibt.

Ein sehr bedeutsamer Fortschritt in den Untersuchungen über die Nullstellen von $\zeta(s)$ wurde von *G. H. Hardy*[132]) gemacht, dem es zuerst zu beweisen gelang, daß *auf der Geraden* $\sigma = \frac{1}{2}$ *tatsächlich unendlich viele Nullstellen liegen.* Daß es überhaupt auf dieser „kritischen" Geraden Nullstellen gibt, war schon früher durch numerische Untersuchungen festgestellt.[133]) Der ursprüngliche *Hardy*sche Beweis

---

131) *F. Carlson,* a. a. O. 67). Vgl. auch eine frühere Arbeit von *S. Wennberg,* a. a. O. 33), worin die weniger genaue Relation $N_1(T) = O\left(T : (\log T)^{\frac{\delta}{1-\delta}}\right)$ bewiesen wird.

132) *G. H. Hardy,* Sur les zéros de la fonction $\zeta(s)$ de Riemann, Paris C. R. 158 (1914), p. 1012—1014.

133) Vgl. *J. Gram,* a) Note sur le calcul de la fonction $\zeta(s)$ de Riemann, Oversigt Vidensk. Selsk. Köbenhavn 1895, p. 303—308; b) Note sur les zéros de la fonction $\zeta(s)$ de Riemann, Acta Math. 27 (1903), p. 289—304; *Ch. de la Vallée Poussin,* a. a. O. 130); *E. Lindelöf,* a) Quelques applications d'une formule som-

dieses Satzes nahm seinen Ausgangspunkt in der folgenden von *Mellin*[134]) herrührenden Integraldarstellung einer Thetareihe durch die Zetafunktion

$$\sum_{n=-\infty}^{n=\infty} e^{-n^2 y} = 1 + \sqrt{\frac{\pi}{y}} + \frac{1}{2\pi i}\int_{\frac{1}{2}-i\infty}^{\frac{1}{2}+i\infty} \Gamma\left(\frac{s}{2}\right) y^{-\frac{s}{2}} \zeta(s)\, ds, \quad (\Re(y) > 0)$$

welche zu den in Nr. **11** erwähnten Typen von Integraldarstellungen einer *Dirichlet*schen Reihe durch eine andere *Dirichlet*sche Reihe gehört. Wird unter dem Integralzeichen statt $\zeta(s)$ die in Nr. **14** erwähnte Funktion $\eta(s) = \Gamma\left(\frac{s}{2}\right)\pi^{-\frac{s}{2}}\zeta(s)$ eingeführt, und $\eta(\frac{1}{2} + it) = \varrho(t)$ gesetzt, so geht die *Mellin*sche Formel, wenn $y = \pi e^{2\alpha i}\left(-\frac{\pi}{4} < \alpha < \frac{\pi}{4}\right)$ gewählt wird, in die Formel

$$(35)\quad \int_0^\infty (e^{\alpha t} + e^{-\alpha t})\,\varrho(t)\,dt = -4\pi\cos\frac{\alpha}{2} + 2\pi e^{\frac{\alpha i}{2}} \sum_{n=-\infty}^{n=\infty} e^{-n^2\pi e^{2\alpha i}}$$

über, wobei noch benutzt ist, daß (wegen der Funktionalgleichung $\eta(s) = \eta(1-s)$) die Funktion $\varrho(t)$ eine gerade ist. Es handelt sich darum zu beweisen, daß $\varrho(t)$ unendlich viele *reelle* Nullstellen besitzt, also daß der Integrand in (35) in unendlich vielen Punkten des Integrationsintervalles $0 < t < \infty$ verschwindet. Der Beweis beruht nun darauf, daß die Funktion $\eta(s)$ (wie z. B. aus der *Riemann*schen Integraldarstellung (26) ersichtlich) auf der betrachteten Geraden $s = \frac{1}{2} + it$ *reell* ist, d. h. daß $\varrho(t)$ für reelles t reell ist, und daß daher, wenn $\varrho(t)$ nur *endlich viele* Nullstellen auf der reellen Achse besäße, für alle t von einer gewissen Stelle t_0 an durchweg die Gleichung $\varrho(t) = |\varrho(t)|$ oder durchweg die Gleichung $\varrho(t) = -|\varrho(t)|$ stattfände. Die ursprüng-

---

matoire générale, Acta soc. sc. Fenn. 31, No. 3 (1903), p. 1—46; b) Sur une formule sommatoire générale, Acta Math. 27 (1903), p. 305—311; *R. Backlund,* Einige numerische Rechnungen, die Nullpunkte der Riemann'schen ζ-Funktion betreffend, Öfversigt Finska Vetensk. Soc. (A) 54 (1911—12), No. 3, p. 1—7 und a. a. O. 108a), b). Das weitestgehende Resultat rührt von *Backlund* her, welcher in der letztzitierten Arbeit beweist, 1. daß auf der Strecke $\mathfrak{s} = \frac{1}{2}$, $0 < t < 200$ genau 79 Nullstellen von $\zeta(s)$ liegen, und 2. daß es außer diesen 79 Nullstellen keine einzige Nullstelle von $\zeta(s)$ im Rechtecke $0 < \sigma < 1$, $0 < t < 200$ gibt. Dies Resultat gehört zu den kräftigsten Argumenten für den Glauben an die Richtigkeit der „*Riemann*schen Vermutung".

134) *H. Mellin,* Über eine Verallgemeinerung der Riemannschen Funktion $\zeta(s)$, Acta soc. sc. Fenn. 24, No. 10 (1899), p. 1—50. Ein anderes Integral der Funktion $\zeta(\frac{1}{2} + it)$, welches (nach *Hardy*) ebenfalls zum Beweis des *Hardy*schen Satzes verwendbar ist, ist von *S. Ramanujan,* New expressions for Riemann's functions $\xi(s)$ and $\Xi(t)$, Quart. J. 183 (1915), p. 253—261, angegeben.

51*

liche Fassung des *Hardy*schen Beweises, daß diese Annahme mit der Gleichung (35) in Widerspruch steht, wurde bald von *Landau*[135]) etwas vereinfacht. In der vereinfachten Form kommt dieser Widerspruch einfach so heraus, daß einerseits aus (35), unter der (falschen) Annahme $\varrho(t) = |\varrho(t)|$ (oder $\varrho(t) = -|\varrho(t)|$), durch den Grenzübergang $\alpha \to \frac{\pi}{4}$ (bei welchem die Thetareihe verschwindet) die Konvergenz des Integrales

$$\int_0^\infty e^{\frac{\pi}{4}t} |\varrho(t)|\, dt$$

geschlossen wird, woraus, bei Wiedereinführung der Zetafunktion selbst, die *Konvergenz* von

$$(36) \qquad \int_1^\infty t^{-\frac{1}{4}} |\zeta(\tfrac{1}{2} + it)|\, dt$$

sich ergibt, während andererseits mit Hilfe des *Cauchy*schen Satzes leicht gezeigt wird, daß für hinreichend große T das Integral $|\int_1^T \zeta(\frac{1}{2} + it)\, dt|$, und also um so mehr das Integral $\int_1^T |\zeta(\frac{1}{2} + it)|\, dt$, größer als kT ist, woraus durch eine grobe Abschätzung die Ungleichung

$$\int_1^T t^{-\frac{1}{4}} |\zeta(\tfrac{1}{2} + it)|\, dt > kT^{\frac{3}{4}},$$

also gewiß die *Divergenz* von (36) erfolgt.

Der *Hardy*sche Beweis wurde bald so umgeformt, daß er nicht nur die Relation $M(T) \to \infty$, wo $M(T)$ die Anzahl der Nullstellen *auf* der Geraden $\sigma = \frac{1}{2}$ mit Ordinaten zwischen 0 und T bezeichnet, sondern zugleich auch eine untere Abschätzung von $M(T)$ liefern konnte. Nachdem zuerst *Landau*[135]) die Ungleichung $M(T) > K \log\log T$ bewiesen hatte, wurden wesentlich weitergehende Abschätzungen von *de la Vallée Poussin*[136]) und unabhängig davon von *Hardy* und *Littlewood*[31b]) gegeben; die letzteren, welche die genaueren Resultate erhielten, bewiesen u. a., daß $M(T) > KT^{\frac{3}{4}-\varepsilon}$. In den Beweisen dieser weitergehenden Sätze wurden verschiedene wesentliche Änderungen der ursprünglichen *Hardy*schen Beweismethode vorgenommen (vor allem konnten *Hardy* und *Littlewood* in ihrem Beweis der Ungleichung $M(T) > K \cdot T^{\frac{3}{4}-\varepsilon}$ den Gebrauch der *Mellin*schen Formel (35) und da-

135) *E. Landau,* Über die Hardysche Entdeckung unendlich vieler Nullstellen der Zetafunktion mit reellem Teil $\frac{1}{2}$, Math. Ann 76 (1915), p. 212—243.

136) *Ch. de la Vallée Poussin,* Sur les zéros de $\zeta(s)$ de Riemann, Paris C. R. 163 (1916), p. 418—421 und p. 471—473.

durch die Einführung der elliptischen Thetafunktionen gänzlich vermeiden); die wesentliche Idee der Beweismethode ist aber immer dieselbe geblieben. Neuerdings ist es *Hardy* und *Littlewood*[137]) durch eine sehr verfeinerte Analyse gelungen, sogar die Abschätzung

$$M(T) > KT$$

zu beweisen, und damit festzustellen, daß die Anzahl $M(T)$ von Nullstellen auf der Geraden $\sigma = \frac{1}{2}$ jedenfalls „fast" von derselben Größenordnung ist als die Anzahl $N(T) (\sim \frac{1}{2\pi} T \log T)$ von Nullstellen im ganzen Streifen $0 < \sigma < 1$.

20. Folgerungen aus der „Riemannschen Vermutung". Es wird in dieser Nummer über einige Untersuchungen referiert, deren Resultate nicht auf gesicherte Wahrheit Anspruch erheben dürfen, weil sie auf der *Annahme* der Richtigkeit der *Riemann*schen Vermutung, daß alle nicht-trivialen Nullstellen auf der Geraden $\sigma = \frac{1}{2}$ liegen, beruhen.[138]) Der Weg zu solchen Untersuchungen wurde von *Littlewood*[138]) geöffnet, der bei dem Problem der Bestimmung der μ-Funktion zuerst gezeigt hat, in welcher Weise die Annahme $\zeta(s) \neq 0$ für $\sigma > \frac{1}{2}$ für das funktionentheoretische Studium der Zetafunktion ausgenützt werden kann. Die *Littlewood*sche Methode, welche auf der Anwendung des sogenannten *Hadamard*schen Dreikreisensatzes (vgl. Art. II C 4, Nr. 62) auf die (unter Annahme der Richtigkeit der *Riemann*schen Vermutung) für $\sigma > \frac{1}{2}$, $t > 0$ *reguläre* Funktion $\log \zeta(s)$ beruht, lieferte die *genaue Bestimmung der μ-Funktion für alle σ*, und zwar mit dem Resultat $\mu(\sigma) = 0$ für $\sigma \geqq \frac{1}{2}$, also (vgl. Nr. 18) $\mu(\sigma) = \frac{1}{2} - \sigma$

137) *G. H. Hardy* und *J. Littlewood*, The zeros of Riemann's Zeta-Funktion on the critical line, Math. Ztschr. 10 (1921), p. 283—317. Die Verfasser beweisen übrigens noch mehr, nämlich daß bei jedem $a > 0$ und $U = T^a$ die Ungleichung $M(T + U) - M(T) > KU$ für alle hinreichend großen T besteht. Der Beweis dieses letzten Satzes basiert auf der „approximativen Funktionalgleichung" (Nr. 18), welche in dieser Abhandlung zum ersten Mal (obwohl nicht in ihrer weitestgehenden Form) bewiesen wird.

138) Im Laufe der vielen (bisher mißglückten) Versuche, die „*Riemann*sche Vermutung" zu beweisen, haben verschiedene Forscher das Problem in mannigfacher Weise umgeformt. Vor allem hat *J. Littlewood*, Quelques conséquences de l'hypothèse que la fonction $\zeta(s)$ de Riemann n'a pas de zéros dans le demiplan $R(s) > \frac{1}{2}$, Paris C. R. 154 (1912), p. 263—266, entdeckt, daß die *Riemann*sche Hypothese gleichwertig ist mit der Hypothese, daß die *Dirichlet*sche Reihe für $1 : \zeta(s)$ (siehe Nr. 22) die Konvergenzabszisse $\sigma_B = \frac{1}{2}$ besitze. Vgl. auch eine Arbeit von *M. Riesz*, Sur l'hypothèse de Riemann, Acta Math. 40 (1916), p. 185—190, in welcher die Umformung des Problems auf einer von *Riesz* gefundenen interessanten Integraldarstellung der Funktion $1 : \zeta(s)$ beruht.

für $\sigma \leqq \frac{1}{2}$.[139]) Über das Resultat $\mu(\sigma) = 0$ für $\sigma \geqq \frac{1}{2}$ hinaus bewies *Littlewood*[138]), daß bei jedem σ des Intervalles $\frac{1}{2} < \sigma < 1$ und jedes $a > 2$

(37) $$\log \zeta(s) = O((\log t)^{a(1-\sigma)}),$$

und er konnte ferner $\zeta(s)$ *auf der Geraden* $\sigma = 1$ viel genauer abschätzen, als es ohne Benutzung der „*Riemann*schen Vermutung" möglich gewesen war (vgl. Nr. 18). Neuerdings hat *Littlewood*[111]) seine Abschätzung von $\zeta(1 + it)$ noch verbessert, und zwar die Relation

$$\zeta(1 + it) = O(\log\log t)$$

bewiesen; hiermit ist das Problem, die Größenordnung von $\zeta(1 + it)$ zu bestimmen, zu einem gewissen Abschluß gebracht, weil ja andererseits bekannt ist (Nr. 18), daß $\zeta(1 + it) \neq o(\log\log t)$.

An die erste *Littlewood*sche Arbeit schloß sich eine Arbeit von *Bohr* und *Landau*[140]) an, worin (unter Annahme der *Riemann*schen Vermutung) die Relation

(38) $$\log \zeta(s) \neq O((\log t)^{b(1-\sigma)}) \qquad (\tfrac{1}{2} < \sigma < 1)$$

bei passender Wahl einer Konstanten $b > 0$ bewiesen wurde. Hiermit wurde (unter Berücksichtigung des *Littlewood*schen Resultates (37)) auch die Größenordnung von $\log \zeta(s)$ im kritischen Streifen einigermaßen genau bestimmt.

Mit der Frage nach der Größenordnung von $\zeta(s)$ eng verbunden ist die Frage nach der „feineren" Verteilung der Ordinaten der nichttrivialen Nullstellen von $\zeta(s)$, d. h. die Frage nach dem Verhalten des Restgliedes $R(T)$ in der *Riemann-v. Mangoldt*schen Formel (31)[141]), und auch bei diesem Problem ist es möglich gewesen, unter Heranziehen der *Riemann*schen Hypothese recht genaue Aufschlüsse zu erhalten. Einerseits hat *Landau*[142]) bewiesen, daß $R(T) \neq O(1)$ (also daß $R(T)$ nicht beschränkt bleibt), und später haben *Bohr* und *Lan-*

139) Nach *R. Backlund*, Über die Beziehung zwischen Anwachsen und Nullstellen der Zetafunktion, Öfversigt Finska Vetensk. Soc. 61 (1918—19), No. 9, ist es, um den Beweis der Gleichung $\mu(\sigma) = 0$ für $\sigma \geqq \frac{1}{2}$ zu führen, nicht nötig, die *Riemann*sche Vermutung in ihrem vollen Umfange zu benutzen. Vielmehr ist die Annahme: $\mu(\sigma) = 0$ für $\sigma \geqq \frac{1}{2}$ mit der Annahme, das bei jedem festen $\delta > 0$ die Anzahl $A(T)$ von Nullstellen im Rechtecke $\frac{1}{2} + \delta < \sigma < 1$, $T < t < T + 1$ gleich $o(\log T)$ ist, gleichwertig. Sichergestellt (d. h. ohne irgendeine Annahme bewiesen) ist nur die Abschätzung $A(T) = O(\log T)$.

140) *H. Bohr* und *E. Landau*, Beiträge zur Theorie der Riemannschen Zetafunktion, Math. Ann. 74 (1913), p. 3—30.

141) Der Zusammenhang dieser beiden Probleme ist neuerdings von *J. Littlewood*, a. a. O. 111) einem tiefgehenden Studium unterworfen worden.

142) *E. Landau*, Zur Theorie der Riemannschen Zetafunktion, Vierteljahrschr. Naturf. Ges. Zürich 56 (1911), p. 125—148.

dau[140]) aus der Ungleichung (38) gefolgert, daß sogar

$$R(T) \neq O((\log T)^c)$$

bei passender Wahl einer Konstanten $c > 0$. Andererseits verbesserte *Bohr*[143]) die *v. Mangoldt*sche Abschätzung $R(T) = O(\log T)$ zu $R(T) = o(\log T)$; dieses Resultat wurde dann von *Cramér*[144]), *Landau*[145]) und *Littlewood*[111]) noch etwas verschärft; letzterer bewies, daß

$$R(T) = O\left(\frac{\log T}{\log\log T}\right)$$

und außerdem (vgl. Nr. **16**), daß

$$\frac{1}{T}\int_0^T \left|R(t) - \tfrac{7}{8}\right| dt = O(\log\log T).$$

Schließlich sei noch ein interessanter Satz von *Littlewood*[143]) über das Verhalten von $\zeta(s)$ in der unmittelbaren Nähe der kritischen Geraden $\sigma = \frac{1}{2}$ erwähnt, welcher (unter Annahme der *Riemann*schen Vermutung) das Resultat (vgl. Nr. **17** und **19**), daß $\zeta(s)$ in jedem Streifen $\frac{1}{2} - \delta < \sigma < \frac{1}{2} + \delta$ sämtliche Werte unendlich oft annimmt, dahin verschärft, daß $\zeta(s)$ bei festem $K > 0$, $\delta > 0$ in jedem Kreise $|s - (\frac{1}{2} + i\tau)| < \delta$ $(\tau > \tau_0 = \tau_0(K, \delta))$ sämtliche Werte vom absoluten Betrage $< K$ annimmt.

21. Verallgemeinerte Zetafunktionen. An die *Riemann*sche Zetafunktion schließen sich mehrere Klassen anderer „Zetafunktionen" an, welche ebenfalls durch *Dirichlet*sche Reihen definiert werden und Eigenschaften besitzen, die in vielen Hinsichten mit denjenigen der *Riemann*schen Zetafunktion übereinstimmen. Die interessantesten Klassen solcher Funktionen werden, wegen des zahlentheorischen Charakters der Koeffizienten ihrer Reihenentwicklung, erst im zweiten Teil des Artikels besprochen, wo sie im Zusammenhange mit den zahlentheoretischen Problemen, für deren Behandlung sie erfunden sind, eingeführt werden. In diesem Paragraphen sollen nur von rein analytischem Gesichtspunkte aus gewisse „verallgemeinerte" Zetafunktionen

---

143) Vgl. *H. Bohr, E. Landau* und *J. Littlewood,* Sur la fonction $\zeta(s)$ dans le voisinage de la droite $\sigma = \frac{1}{2}$, Bull. Acad. Belgique 15 (1913), p. 1144—1175.

144) *H. Cramér,* Über die Nullstellen der Zetafunktion, Math. Ztschr. 2 (1918), p. 237—241. In dieser Abhandlung wird u. a. bewiesen, daß zur Herleitung der Abschätzung $R(T) = o(\log T)$ nicht die volle „*Riemann*sche Vermutung" nötig ist, sondern nur die (vgl. Note 139) weniger aussagende sogenannte „*Lindelöf*sche Vermutung" $\mu(\sigma) = 0$ für $\sigma \geqq \frac{1}{2}$.

145) *E. Landau,* Über die Nullstellen der Zetafunktion, Math. Ztschr. 6 (1920), p. 151—154. Vgl. hierzu auch *H. Cramér,* Bemerkung zu der vorstehenden Arbeit des Herrn E. Landau, Math. Ztschr. 6 (1920), p. 155—157.

kurz besprochen werden, deren Definitionen kein zahlentheoretisches Moment enthalten.

Betrachten wir zunächst die Reihe

(39) $$\zeta(w, s) = \sum_{n=0}^{\infty} \frac{1}{(w+n)^s}$$ [146])

oder allgemeiner die von *Lipschitz*[147]) und *Lerch*[148]) untersuchte Reihe

$$\sum_{n=0}^{\infty} \frac{e^{2\pi i n x}}{(w+n)^s},$$

wo x eine komplexe Zahl bedeutet, deren reeller Teil $\Re(x)$ etwa dem Intervalle $0 \leqq x < 1$ angehört, während ihr imaginärer Teil $\Im(x) \geqq 0$ ist. Diese Reihe, als Funktion von s betrachtet, ist offenbar im Falle $\Im(x) > 0$ in der ganzen s-Ebene konvergent und stellt eine ganze Transzendente dar; für $\Im(x) = 0$ ist sie, abgesehen vom Fall $x = 0$, in der Halbebene $\sigma > 0$ konvergent, definiert aber auch hier eine ganze Transzendente, und im speziellen Falle $x = 0$ (d. h. im Falle der Reihe (39)) konvergiert sie für $\sigma > 1$ und stellt, wie die Zetareihe selbst, eine meromorphe Funktion dar, die überall regulär ist mit Ausnahme des einzigen Punktes $s = 1$, wo sie einen Pol erster Ordnung besitzt. Dies ersieht man in ganz ähnlicher Weise, wie *Riemann* die Fortsetzbarkeit von $\zeta(s)$ bewies, d. h. es wird die Reihe zunächst

---

146) Diese Funktion ist besonders von *H. Mellin*, Über eine Verallgemeinerung der Riemannschen Funktion $\zeta(s)$, Acta soc. sc. Fenn. 24, No. 10 (1899), im Zusammenhange mit seinen Studien über Umkehrformeln (vgl. Note 70) näher untersucht. Vgl. auch *A. Piltz*, Über die Häufigkeit der Primzahlen in arithmethischen Progressionen und über verwandte Gesetze, Habilitationsschrift, Jena 1884, p. 1—48; *E. Lindelöf*, a. a. O. 97) und *E. W. Barnes*, a) The theory of the Gamma function, Mess. of Math. (2) 29 (1899), p. 64—128; b) The theory of the double Gamma function, London Phil. Trans. (A) 196 (1901), p. 265—387; c) On the theory of the multiple Gamma function, Cambridge Phil. Trans. 19 (1904), p. 374—425; *Barnes* untersucht auch Reihen der Form

$$\sum_{n_1=0}^{\infty} \cdots \sum_{n_p=0}^{\infty} \frac{1}{(w + n_1\omega_1 + \cdots n_p\omega_p)^s}.$$

147) *R. Lipschitz*, a) Untersuchung einer aus vier Elementen gebildeten Reihe, Crelles J. 54 (1857), p. 313—328; b) Untersuchung der Eigenschaften einer Gattung von unendlichen Reihen, Crelles J. 105 (1889), p. 127—156.

148) *M. Lerch*, Note sur la fonction $K(w, x, s) = \sum_{k=0}^{\infty} \frac{e^{2k\pi i x}}{(w+k)^s}$, Acta Math. 11 (1887), p. 19—24.

durch ein einfaches bestimmtes Integral dargestellt und dieses wieder in ein komplexes Kurvenintegral umgeformt. Aus dieser letzten Integraldarstellung folgt weiter, wie bei *Riemann*, durch Deformation des Integrationsweges und Anwendung des *Cauchy*schen Satzes[149]), daß unsere Funktion einer der *Riemann*schen ähnlichen *Funktionalgleichung* genügt, welche in der Form

$$\sum_{n=0}^{\infty} \frac{e^{2\pi i n x}}{(w+n)^s} = \frac{\Gamma(1-s)}{(2\pi i)^{1-s}} \sum_{m=-\infty}^{\infty} \frac{e^{2\pi i w(m-x)}}{(-x+m)^{1-s}}$$

geschrieben werden kann.

Eine wesentlich weitergehende Verallgemeinerung der *Riemann*schen Zetafunktion ist von *Epstein*[150]) gegeben, dessen Untersuchungen an den zweiten *Riemann*schen Beweis der Funktionalgleichung von $\zeta(s)$, d. h. an die Darstellung der Zetafunktion durch ein bestimmtes Integral mit Hilfe elliptischer Thetafunktionen anknüpfen. *Epstein* betrachtet Reihen der Form

$$(40) \qquad \sum_{m_1} \cdots \sum_{m_p} \frac{e^{2\pi i \sum_{\mu=1}^{p} m_\mu x_\mu}}{\{\varphi(y+m)\}^{\frac{s}{2}}},$$

wo $x_1, \ldots x_p, y_1, \ldots y_p$ Konstanten sind und $\varphi(\alpha+\beta)$ ein symbolischer Ausdruck für die quadratische Form $\sum_\mu \sum_\nu (\alpha_\mu + \beta_\mu)(\alpha_\nu + \beta_\nu)$ der $2p$ Variabeln $\alpha_1, \ldots \beta_p$ ist; die durch eine solche Reihe (40) definierte Funktion wird eine *Zetafunktion* p^{ter} *Ordnung* genannt. *Epstein* zeigt nun, daß die Reihe (40) durch ein bestimmtes Integral mit Hilfe allgemeiner Thetafunktionen dargestellt werden kann, und durch Anwendung von Transformationsformeln dieser Thetafunktionen beweist er, daß auch diese allgemeine Zetafunktion einer Funktionalgleichung von ähnlichem Charakter wie die *Riemann*sche für $\zeta(s)$ genügt.

---

149) Vgl. *M. Lerch*, a. a. O. 148); die Funktionalgleichung wurde zuerst von *R. Lipschitz*, a. a. O. 147a), gefunden, welcher sie mit Hilfe der Theorie der *Fourier*schen Integrale herleitete.

150) *P. Epstein*, a) Zur Theorie allgemeiner Zetafunktionen, Math. Ann. 56 (1903), p. 615—644; b) Zur Theorie allgemeiner Zetafunktionen II, Math. Ann. 63 (1907), p. 205—216.

Zweiter Teil.

22. Einleitung. Bezeichnungen. Dieser Teil handelt von den zahlentheoretischen Anwendungen der im Vorhergehenden entwickelten Theorien. Die *gewöhnlichen Dirichlet*schen Reihen $\sum a_n n^{-s}$ sind als Hilfsmittel für analytisch-zahlentheoretische Untersuchungen besonders wertvoll; einen „Grund" hierfür kann man in ihrer Multiplikationsregel (vgl. Nr. 12) sehen, wonach bei der Koeffizientenbildung die „multiplikativen" Eigenschaften der Zahlen zur Geltung kommen. Die wichtigsten zahlentheoretischen Funktionen treten als Koeffizienten gewisser *Dirichlet*scher Reihen auf, die mit der *Riemann*schen Zetafunktion in einfacher Weise zusammenhängen. Indem man auf diese Reihen die Sätze über Beziehungen zwischen den Koeffizienten einer *Dirichlet*schen Reihe und der von der Reihe dargestellten Funktion (vgl. Nr. 4 und 5) anwendet, gelangt man mit Hilfe der Theorie der Zetafunktion zu neuen Ergebnissen über die Natur der zahlentheoretischen Funktionen. Manche Probleme erfordern die Einführung neuer erzeugender Funktionen, die alle der *Riemann*schen $\zeta(s)$ mehr oder weniger ähnlich sind. Verschiedene Probleme lassen sich auch durch Methoden angreifen, die von der Theorie der *Dirichlet*schen Reihen gänzlich unabhängig sind.

Es dürfte zweckmäßig sein, einige der im folgenden gebrauchten Bezeichnungen hier zusammenzustellen; die unten gegebenen Definitionen werden also im Texte nicht wiederkehren.

A. Die folgenden zahlentheoretischen Funktionen seien für ganze $n \geqq 1$ definiert:

$$\Lambda(n) = \begin{cases} \log p & \text{für } n = p^m \text{ (p Primzahl, $m \geqq 1$ ganz)}, \\ 0 & \text{sonst;} \end{cases}$$

$$\mu(n) = \begin{cases} (-1)^\nu & \text{für } n = p_1 p_2 \ldots p_\nu \text{ (die p_i \textit{verschiedene} Primzahlen)}, \\ 1 & \text{für } n = 1, \\ 0 & \text{sonst;} \end{cases}$$

$$\lambda(n) = \begin{cases} (-1)^{\alpha_1 + \alpha_2 + \cdots \alpha_\nu} & \text{für } n = p_1^{\alpha_1} p_2^{\alpha_2} \cdots p_\nu^{\alpha_\nu}, \\ 1 & \text{für } n = 1; \end{cases}$$

$d(n) =$ Anzahl der Teiler von n;

$\sigma(n) =$ Summe der Teiler von n;

$\varphi(n) =$ Anzahl der zu n teilerfremden positiven ganzen Zahlen $\leqq n$.

Diese Funktionen sind alle mit $\zeta(s)$ nahe verbunden; es gilt in der Tat für hinreichend große σ

$$\sum_1^\infty \frac{\Lambda(n)}{n^s} = -\frac{\zeta'}{\zeta}(s), \qquad \sum_1^\infty \frac{\Lambda(n)}{\log n \cdot n^s} = \log \zeta(s),$$

$$\sum_1^\infty \frac{\mu(n)}{n^s} = \frac{1}{\zeta(s)}, \qquad \sum_1^\infty \frac{\lambda(n)}{n^s} = \frac{\zeta(2s)}{\zeta(s)},$$

$$\sum_1^\infty \frac{d(n)}{n^s} = (\zeta(s))^2, \qquad \sum_1^\infty \frac{\sigma(n)}{n^s} = \zeta(s)\zeta(s-1),$$

$$\sum_1^\infty \frac{\varphi(n)}{n^s} = \frac{\zeta(s-1)}{\zeta(s)}.$$

B. Die folgenden summatorischen Funktionen der obigen und einiger verwandten *Dirichlet*schen Reihen seien für jeden positiven Wert von x definiert:

$$\pi(x) = \text{Anzahl der Primzahlen} \leqq x$$
$$= \sum_{p \leqq x} 1,$$

$$\Pi(x) = \sum_{p^m \leqq x} \frac{1}{m}$$ (p durchläuft die Primzahlen, m die ganzen positiven Zahlen)

$$= \sum_{n=1}^x \frac{\Lambda(n)}{\log n}$$
$$= \pi(x) + \tfrac{1}{2}\pi(x^{\frac{1}{2}}) + \tfrac{1}{3}\pi(x^{\frac{1}{3}}) + \cdots,$$

$$\vartheta(x) = \sum_{p \leqq x} \log p,$$

$$\psi(x) = \sum_{p^m \leqq x} \log p,$$
$$= \sum_{n=1}^x \Lambda(n),$$
$$= \vartheta(x) + \vartheta(x^{\frac{1}{2}}) + \vartheta(x^{\frac{1}{3}}) + \cdots,$$

$$M(x) = \sum_{n=1}^x \mu(n), \qquad \Phi(x) = \sum_{n=1}^x \varphi(n),$$

$$D(x) = \sum_{n=1}^x d(n), \qquad S(x) = \sum_{n=1}^x \sigma(n).$$

C. Es sei $g(x)$ irgendeine der unter B. eingeführten summatorischen Funktionen. Aus $g(x)$ werde die Funktion $\bar{g}(x)$ dadurch abgeleitet, daß für alle $x > 0$

$$\bar{g}(x) = \lim_{\varepsilon \to 0} \frac{g(x+\varepsilon) + g(x-\varepsilon)}{2}$$

gesetzt wird. $\bar{g}(x)$ ist also nur in den Unstetigkeitspunkten (d. h. für gewisse ganzzahlige x) von $g(x)$ verschieden.

III. Die Verteilung der Primzahlen.

23. Der Primzahlsatz. Ältere Vermutungen und Beweisversuche. Schon früh entstand das Problem, die Anzahl der Primzahlen zwischen zwei gegebenen Grenzen zu bestimmen, also insbesondere für die Funktion $\pi(x)$ einen (angenäherten oder exakten) Ausdruck aufzustellen. Bei dem höchst unregelmäßigen Verlauf dieser Funktion schien es von vornherein unmöglich, sie durch eine einfache analytische Funktion genau darzustellen; man mußte also zunächst darauf ausgehen, ein *asymptotisches* Resultat, etwa von der Form $\pi(x) \sim f(x)$, zu erhalten. Hierdurch ist schon die Fragestellung angebahnt, die zu dem berühmten *Primzahlsatz*[151]) führte: es gilt für unendlich wachsendes x:

$$\pi(x) \sim Li(x), \tag{41}$$

wo

$$Li(x) = \lim_{\varepsilon \to 0} \left(\int_0^{1-\varepsilon} \frac{du}{\log u} + \int_{1+\varepsilon}^{x} \frac{du}{\log u} \right)$$

gesetzt ist. — Dieser Satz kann wohl als das wichtigste Ergebnis der analytischen Zahlentheorie bezeichnet werden; durch die Anstrengungen, ihn zu beweisen, wurden ihre feinsten Methoden geschaffen und ausgebildet.

In (41) kann man, ohne den Sinn der Formel zu verändern, $Li(x)$ durch jede der Bedingung $f(x) \sim Li(x)$ genügende Funktion, z. B. durch $\frac{x}{\log x}$, ersetzen. Eine zu (41) äquivalente Behauptung wurde zuerst von *Legendre*[152]) ohne Beweis ausgesprochen: es werde $\pi(x)$ angenähert durch die Funktion $\frac{x}{\log x - 1{,}08366}$ dargestellt. Schon vor *Legendre* war *Gauß*[153]), wie aus einem viel später geschriebenen Briefe ersichtlich ist, auf die Vermutung $\pi(x) \sim \int_2^x \frac{du}{\log u}$ gekommen. Von

151) Die Benennung rührt von *H. v. Schaper* her: Über die Theorie der *Hadamard*schen Funktionen und ihre Anwendung auf das Problem der Primzahlen, Diss. Göttingen 1898.

152) *A. M. Legendre,* a) Essai sur la théorie des nombres (2. Aufl.), Paris 1808, p. 394; b) Théorie des nombres (3. Aufl.), Paris 1830, Bd. 2, p. 65.

153) *C. F. Gauß,* Werke 2, 2. Aufl., p. 444—447.

Dirichlet[154]) wurde gelegentlich behauptet, $\sum_{n=1}^{x} \frac{1}{\log n}$ sei eine bessere Vergleichsfunktion als diejenige von *Legendre*.

Einen präzisen Sinn erhielten diese Andeutungen erst durch die Arbeiten von *Tschebyschef*.[155]) In moderner Ausdrucksweise können seine wichtigsten Resultate etwa folgendermaßen zusammengefaßt werden: Er betrachtet die Funktionen $\pi(x)$, $\vartheta(x)$ und $\psi(x)$; zwischen ihnen besteht ein Zusammenhang, der durch die Beziehungen[156])

$$\pi(x) = \frac{\vartheta(x)}{\log x} + \int_2^x \frac{\vartheta(u)}{u \log^2 u}\, du$$

$$(42) \qquad \vartheta(x) = \pi(x)\log x - \int_2^x \frac{\pi(u)}{u}\, du$$

$$\psi(x) = \vartheta(x) + O(\sqrt{x})$$

ausgedrückt wird (in den beiden ersten Gleichungen kann man übrigens π bzw. ϑ durch Π bzw. ψ ersetzen). Hieraus läßt sich unmittelbar ablesen, daß für unendlich wachsendes x alle drei Quotienten

$$(43) \qquad \frac{\pi(x)}{Li(x)}, \quad \frac{\vartheta(x)}{x}, \quad \frac{\psi(x)}{x}$$

dieselben oberen bzw. unteren Unbestimmtheitsgrenzen haben. Werden diese durch l (lim inf) und L (lim sup) bezeichnet, so findet *Tschebyschef*

$$a \leqq l \leqq 1 \leqq L \leqq \tfrac{6}{5} a,$$

mit $a = 0{,}92129$.

Insbesondere folgt hieraus[157]): existiert für irgendeinen der Quotienten (43) ein Grenzwert, so haben alle drei Quotienten den Grenzwert 1. Außer durch (41) läßt sich also der Primzahlsatz durch irgendeine der Gleichungen

$$(44) \qquad \vartheta(x) \sim x$$

$$(45) \qquad \psi(x) \sim x \qquad \text{ausdrücken.}$$

154) Vgl. *G. Lejeune-Dirichlet,* Werke 1, p. 372, Fußnote 2).

155) *P. Tschebyschef,* a) Sur la fonction qui détermine la totalité des nombres premiers inférieurs à une limite donnée, Mém. présentés Acad. Pétersb. 6 (1851), p. 141—157; J. math. pures appl. (1) 17 (1852), p. 341—365; Œuvres 1, St. Pétersbourg 1899, p. 27—48; b) Mémoire sur les nombres premiers, J. math. pures appl. (1) 17 (1852), p. 366—390; Mém. présentés Acad. Pétersb. 7 (1854), p. 15—33; Œuvres 1, p. 49—70.

156) Die beiden ersten Gleichungen werden einfach durch partielle Summation aus den Definitionsgleichungen für $\pi(x)$ und $\vartheta(x)$ abgeleitet.

157) Das kann jetzt unmittelbar aus elementaren Sätzen über *Dirichlet*sche Reihen gefolgert werden. Vgl. Nr. 5, insbesondere die Fußnoten 27) und 28), vgl. auch *E. Landau,* Handbuch, § 31.

Die *Tschebyschef*schen Resultate wurden teils durch Betrachtung der Funktionen $\zeta(s)$ und $\log \zeta(s)$ für reelle, gegen 1 abnehmende Werte von s, teils durch elementare Summenabschätzungen mit Hilfe der Identität[158])

$$\psi(x) + \psi\left(\frac{x}{2}\right) + \psi\left(\frac{x}{3}\right) + \cdots = \log([x]!)$$

abgeleitet. Die Schranken für l und L wurden später von anderen[159]) mit analogen Methoden verengert; es ist jedoch bisher niemand gelungen, auf diesem Wege die Existenz eines Grenzwertes, d. h. den Primzahlsatz, zu beweisen.

24. Die Beweise von Hadamard und de la Vallée Poussin. Der Weg, der zu einem strengen Beweis des Primzahlsatzes führen konnte, wurde erst geöffnet durch die Erscheinung der grundlegenden *Riemann*schen Arbeit[95]) vom Jahre 1859, wo zum ersten Male die komplexe Funktionentheorie auf das Problem angewandt und die Zetafunktion völlig allgemein untersucht wurde. Das Endziel dieser Arbeit war allerdings nicht der Beweis des Primzahlsatzes, doch findet man hier schon die Integralformel für die Koeffizientensumme einer *Dirichlet*schen Reihe (vgl. Nr. 4), auf

$$\log \zeta(s) = \sum_{p,\,m} \frac{1}{m p^{ms}} = \sum_{n=1}^{\infty} \frac{\Lambda(n)}{\log n \cdot n^s}$$

angewandt. *Halphen*[160]) und *Cahen*[161]) versuchten, diesen *Riemann*schen Ansatz für den Beweis des Primzahlsatzes zu benutzen, ein vollständiger Beweis wurde jedoch erst im Jahre 1896 gegeben, und zwar fast gleichzeitig von *Hadamard*[162]) und *de la Vallée Poussin.*[163])

Die früheren Versuche waren hauptsächlich an den folgenden zwei Schwierigkeiten gescheitert: 1. die Eigenschaften der komplexen Null-

158) Diese Identität wurde unabhängig von *Tschebyschef*[155]) und *de Polignac*, Recherches nouvelles sur les nombres premiers, Paris 1851, entdeckt.

159) Betreffs der an *Tschebyschef* in dieser Richtung anschließenden Arbeiten sei auf *G. Torelli*, Sulla totalità dei numeri primi fino a un limite assegnato, Neapel 1901 (Atti Accad. sc. fis. mat. (2) 11 No. 1), Cap. 4—5 verwiesen. In dieser Monographie wird die Geschichte des Gegenstandes ausführlich dargestellt.

160) *G. H. Halphen*, Sur l'approximation des sommes de fonctions numériques, Paris C. R. 96 (1883), p. 634—637. Auch *T. J. Stieltjes* gibt an, einen Beweis gefunden zu haben: Correspondance d'Hermite et de Stieltjes, Paris 1905, verschiedene Stellen, vgl. z. B. Lettre 314.

161) *E. Cahen*, Sur la somme des logarithmes des nombres premiers qui ne dépassent pas x, Paris C. R. 116 (1893), p. 85—88.

162) *J. Hadamard*, Sur la distribution des zéros de la fonction $\zeta(s)$ et ses conséquences arithmétiques, Bull. soc. math. France 24 (1896), p. 199—220.

163) *Ch. de la Vallée Poussin*, Recherches analytiques sur la théorie des nombres premiers, Première partie, Ann. soc. sc. Bruxelles 20:2 (1896), p. 183—256.

stellen von $\zeta(s)$ waren noch nicht hinreichend bekannt; 2. die Integrale

$$(46) \qquad -\frac{1}{2\pi i}\int_{a-i\infty}^{a+i\infty}\frac{x^s}{s}\frac{\zeta'}{\zeta}(s)\,ds$$

und

$$(47) \qquad \frac{1}{2\pi i}\int_{a-i\infty}^{a+i\infty}\frac{x^s}{s}\log\zeta(s)\,ds,$$

die für $a > 1$ die Funktionen $\overline{\psi}(x)$ bzw. $\overline{\Pi}(x)$ darstellen (vgl. Nr. **4**; in einem Unstetigkeitspunkt muß man nach den dortigen Ausführungen die Hauptwerte der Integrale nehmen), sind nur bedingt konvergent.

Die erste Schwierigkeit wurde von *Hadamard* und *de la Vallée Poussin* dadurch überwunden, daß sie zeigten: jede Nullstelle von $\zeta(s)$ liegt links von der Geraden $\sigma = 1$ (vgl. Nr. **14**). Dieser Satz wird bei allen bisher bekannten Beweisen des Primzahlsatzes als wesentliche Grundlage benutzt. — Um unbedingt konvergente Ausdrücke zu erhalten, benutzen die beiden Verfasser an der Stelle von (46) und (47) Integralausdrücke für gewisse mit $\overline{\psi}$ und $\overline{\Pi}$ zusammenhängende Funktionen.

Hadamard betrachtet das für $\mu > 1$ unbedingt konvergente Integral (vgl. (12) Nr. **4**).

$$(48) \qquad -\frac{1}{2\pi i}\int_{a-i\infty}^{a+i\infty}\frac{x^s}{s^\mu}\frac{\zeta'}{\zeta}(s)\,ds = \frac{1}{\Gamma(\mu)}\sum_{n=1}^{x}\Lambda(n)\log^{\mu-1}\frac{x}{n}$$

$$= \frac{1}{\Gamma(\mu-1)}\int_2^x\frac{\psi(t)}{t}\log^{\mu-2}\frac{x}{t}\,dt.$$

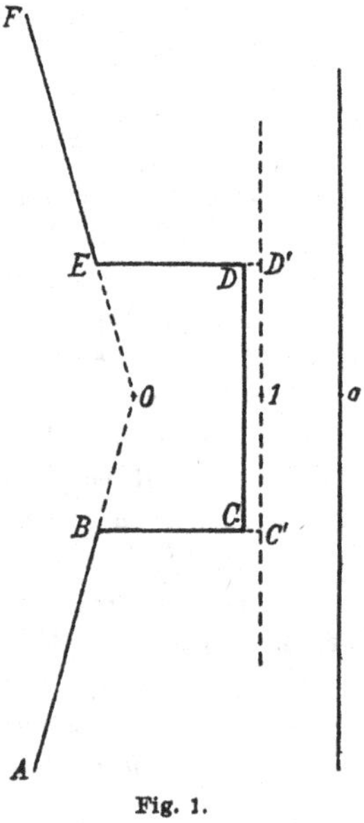

Fig. 1.

Durch eine Verschiebung des Integrationsweges folgert er, unter Benutzung der Tatsache, daß die ganze Funktion $(s-1)\zeta(s)$ vom Geschlechte 1 ist (vgl. Nr. **15**),

$$\frac{1}{\Gamma(\mu-1)}\int_2^x\frac{\psi(t)}{t}\log^{\mu-2}\frac{x}{t}\,dt$$

$$= x - \sum_\varrho{}'\frac{x^\varrho}{\varrho^\mu} - \frac{1}{2\pi i}\int_{ABCDEF}\frac{x^s}{s^\mu}\frac{\zeta'}{\zeta}(s)\,ds.$$

Rechts durchläuft ϱ in Σ' nur die oberhalb $D'E$ oder unterhalb BC' gelegenen Nullstellen von $\zeta(s)$; da auf $\sigma = 1$ keine Nullstellen liegen, kann DD' so klein gewählt werden, daß auch noch das Rechteck $CDD'C'$ nullstellenfrei wird (vgl. Figur 1).

Da der neue Integrationsweg ganz in der Halbebene $\sigma < 1$ verläuft, schließt man hieraus

$$\frac{1}{\Gamma(\mu-1)}\int_2^x \frac{\psi(t)}{t} \log^{\mu-2}\frac{x}{t}\,dt \sim x \tag{49}$$

und speziell für $\mu = 2$

$$\int_2^x \frac{\psi(t)}{t}\,dt = \sum_{n=1}^{x} \Lambda(n) \log\frac{x}{n} \sim x. \tag{50}$$

Hadamard zeigt, daß hieraus unmittelbar zu (44) oder (45) übergegangen werden kann (vgl. auch Nr. **25**), womit der Primzahlsatz bewiesen ist.

Auch *de la Vallée Poussin* nimmt als Ausgangspunkt ein unbedingt konvergentes Integral, nämlich

$$\int_{a-i\infty}^{a+i\infty} \frac{x^s}{(s-u)(s-v)} \frac{\zeta'}{\zeta}(s)\,ds.$$

Durch Anwendung der Gleichung (29), Nr. **15**, erhält er, da $\Re(\varrho) < 1$ ist,

$$\int_2^x \frac{\psi(t)}{t^2}\,dt = \sum_{n=1}^{x} \Lambda(n)\left(\frac{1}{n} - \frac{1}{x}\right) \tag{51}$$

$$= \log x + K - \sum_{\varrho} \frac{x^{\varrho-1}}{\varrho(\varrho-1)} + \frac{a}{x} + O\left(\frac{1}{x^3}\right) = \log x + K + o(1),$$

und zeigt, wie man hieraus zu (50) und (45) übergehen kann.

25. Die Beweismethoden von Landau. Bei den ersten Beweisen des Primzahlsatzes traten als wichtige Hilfsmittel Sätze auf, die durch die Anwendung der *Hadamard*schen Theorie der ganzen Funktionen auf $(s-1)\zeta(s)$ gefunden wurden und also die Existenz der Zetafunktion in der ganzen Ebene und gewisse Eigenschaften ihrer Nullstellen voraussetzen.

Landau hat aber gezeigt, daß der Beweis in weitgehendem Maße von diesen Voraussetzungen befreit werden kann, was für die Anwendung der Methode auf allgemeinere Fälle wichtig ist (vgl. Nr. **42**).

Durch Benutzung der elementar nachweisbaren Ungleichung

$$\left|\frac{\zeta'}{\zeta}(s)\right| < K(\log t)^A \quad \text{für} \quad \sigma > 1 - \frac{1}{(\log t)^B}, \quad t > t_0, \tag{52}$$

mit konstanten A, B, K, t_0, gelang es ihm[164]) den Primzahlsatz zu beweisen, indem er mit dem *Hadamard*schen Integral (48) für $\mu = 2$ und mit einem in jenem Gebiete verlaufenden Integrationsweg arbeitete.

164) *E. Landau*, a. a. O. 125) und Handbuch, § 51—54.

Später[165]) zeigte er, daß man die Voraussetzungen sogar noch mehr verringern kann: für den Beweis des Primzahlsatzes ist in der Tat nur wesentlich, daß $\frac{\zeta'}{\zeta}$ auf der Geraden $\sigma = 1$ (abgesehen vom Pole $s = 1$) regulär ist und für $\sigma \geqq 1$, $|t| \to \infty$ gleichmäßig von der Form $O(|t|^k)$ ist. Der am Ende von Nr. 5 genannte Satz von *Landau*[29]) über *Dirichlet*sche Reihen mit positiven Koeffizienten ist nämlich unmittelbar auf $-\frac{\zeta'}{\zeta}(s) = \sum \Lambda(n) n^{-s}$ anwendbar und liefert gerade die Beziehung (45). Für den Beweis dieses Satzes werden gewisse allgemeine Grenzwertsätze herangezogen, die speziell den Übergang von (50) oder (51) zu (45) ermöglichen (vgl. auch Nr. 33). Wenn beispielsweise die Funktion $f(t)$ für $t > a$ nirgends abnimmt, so kann man von

$$\int_a^x \frac{f(t)}{t}\, dt \sim x$$

auf die asymptotische Gleichheit der Ableitungen schließen: $f(x) \sim x$.

Durch Benutzung des Integranden $\frac{x^s}{s^2} \log \zeta(s)$ anstatt $\frac{x^s}{s^2} \frac{\zeta'}{\zeta}(s)$ kann man, wie *Landau*[166]) zeigt, den Satz (41) über $\pi(x)$ direkt, d. h. ohne den Umweg über $\psi(x)$ oder $\vartheta(x)$, beweisen; auch gelingt es ihm[167]) mit Hilfe des nur bedingt konvergenten Integrals (46) direkt zu (45) — und sogar zur Gleichung (53) von Nr. 27 — ohne den Umweg über (50) zu gelangen.

26. Andere Beweise. Der Beweis von *H. v. Koch*[168]) weicht von den vorhergehenden dadurch ab, daß er gar nicht mit Integralen von der Form $\int \frac{x^s}{s^\mu} f(s)\, ds$ arbeitet. Er gibt für die summatorischen Funktionen der *Dirichlet*schen Reihen für $\frac{\zeta'}{\zeta}$ und $\log \zeta$ unter Benutzung gewisser Diskontinuitätsfaktoren Ausdrücke, die in der folgenden Darstellungsformel für die Koeffizientensumme einer beliebigen *Dirichlet*schen Reihe $f(s) = \sum a_n e^{-\lambda_n s}$ (mit absolutem Konvergenzbereich) zu-

165) *E. Landau*, a. a. O. 29) und 21) sowie Zwei neue Herleitungen für die asymptotische Anzahl der Primzahlen unter einer gegebenen Grenze, Sitzungsber. Akad. Berlin 1908, p. 746—764 und Handbuch, § 66.

166) *E. Landau*, a. a. O. 165) (Zwei neue Herleitungen ...) und Handbuch, § 64.

167) *E. Landau*, Über den Gebrauch bedingt konvergenter Integrale in der Primzahltheorie, Math. Ann. 71 (1912), p. 368—379.

168) *H. v. Koch*, Sur la distribution des nombres premiers, Acta Math. 24 (1901), p. 159—182.

sammengefaßt werden können[169]):

$$\sum_{\lambda_n < x} a_n = \lim_{c \to \infty} \sum_{\nu=1}^{\infty} \frac{(-1)^{\nu-1}}{\nu!} e^{c\nu x} f(c\nu), \qquad (x \neq \lambda_n).$$

Diese Formel erscheint dadurch bemerkenswert, daß $f(s)$ darin nur mit einem reellen und positiven Argument auftritt.

Hardy und *Littlewood*[170]) beweisen, wie schon in Nr. **5** erwähnt wurde, mit Hilfe des „*Cahen-Mellin*schen Integrals" (vgl. Nr. **11**) einen Satz über *Dirichlet*sche Reihen, der den *Landau*schen, in Nr. **5** und **25** erwähnten, Satz — und damit den Primzahlsatz — als Spezialfall enthält.

Steffensen[171]) zeigt, daß eine von ihm und schon früher von *Mellin*[172]) gefundene Integraldarstellung für die Koeffizientensumme einer *Dirichlet*schen Reihe zum Beweis des Primzahlsatzes benutzt werden kann.

27. Die Restabschätzung. Schon durch die Resultate von *Tschebyschef*[155]) wurde die Vermutung nahe gelegt, daß unter allen asymptotisch gleichwertigen Funktionen, die man als Vergleichsfunktionen für $\pi(x)$ benutzt hatte, dem Integrallogarithmus eine besonders ausgezeichnete Stellung zukommt. Streng entschieden wurde diese Frage erst durch *de la Vallée Poussin*[173]), der aus seiner Gleichung (51) mit Hilfe seines Satzes (vgl. Nr. **19**)

$$\zeta(s) \neq 0 \quad \text{für} \quad \sigma > 1 - \frac{k}{\log t}, \quad t > t_0$$

die Folgerung

$$\pi(x) = Li(x) + O\left(x e^{-\alpha\sqrt{\log x}}\right) \tag{53}$$

für jedes $\alpha < \sqrt{k}$ zog. Gleichzeitig folgt, daß auch die Differenzen

$$\Pi(x) - Li(x), \quad \vartheta(x) - x, \quad \psi(x) - x$$

alle drei von der Größenordnung $O\left(x e^{-\alpha\sqrt{\log x}}\right)$ sind. Der Integral-

169) Vgl. auch *Hj. Mellin,* Die Dirichletschen Reihen, die zahlentheoretischen Funktionen und die unendlichen Produkte von endlichem Geschlecht, Acta Math. 28 (1904), p. 37—64.

170) *G. H. Hardy* und *J. E. Littlewood,* a. a. O. 31).

171) *J. F. Steffensen,* Analytiske studier med anvendelser paa taltheorien, Diss. Kopenhagen 1912; Über eine Klasse von ganzen Funktionen und ihre Anwendung auf die Zahlentheorie, Acta Math. 37 (1914), p. 75—112; vgl. auch: Über Potenzreihen, im besonderen solche, deren Koeffizienten zahlentheoretische Funktionen sind, Palermo Rend. 38 (1914), p. 376—386.

172) a. a. O. 169).

173) *Ch. de la Vallée Poussin,* Sur la fonction $\zeta(s)$ de *Riemann* et le nombre des nombres premiers inférieurs à une limite donnée, Mém. couronnés et autres mém. Acad. Belgique 59 (1899—1900), No. 1.

logarithmus stellt demnach $\pi(x)$ in einem ganz präzisen Sinne *besser* dar als $\frac{x}{\log x}$ oder irgendeine der Funktionen

$$f_q(x) = \frac{x}{\log x} + \frac{1!\,x}{\log^2 x} + \cdots \frac{(q-1)!\,x}{\log^q x} = Li(x) + O\left(\frac{x}{\log^{q+1} x}\right),$$

die bei der asymptotischen Entwicklung von $Li(x)$ auftreten.[174]) Nach (53) gilt nämlich für $q = 1, 2, \ldots$

(54) $$\pi(x) - f_q(x) \sim \frac{q!\,x}{\log^{q+1} x},$$

aber

(55) $$\pi(x) - Li(x) = o\left(\frac{x}{\log^{q+1} x}\right).$$

Bei *Landau*[164]) wird mit Hilfe von (52), also ohne Benutzung der Fortsetzbarkeit von $\zeta(s)$ oder der Existenz ihrer Nullstellen, die Abschätzung

(56) $$\pi(x) = Li(x) + O\left(xe^{-\sqrt[13]{\log x}}\right)$$

bewiesen; diese ist weniger scharf als (53), reicht aber doch für die Folgerungen (54) und (55) aus. *Landau*[175]) hat übrigens auch den Beweis von (53) wesentlich vereinfacht; diese Gleichung, mit dem von ihm angegebenen Werte von α, stellt die schärfste bisher mit Sicherheit bekannte Abschätzung von $\pi(x)$ dar.[176])

Nimmt man dagegen an, die Riemannsche Vermutung über die Nullstellen der Zetafunktion sei richtig (vgl. Nr. **20**), so erhält man noch schärfere Resultate, nämlich

(57) $$\left.\begin{matrix}\pi(x) - Li(x) \\ \Pi(x) - Li(x)\end{matrix}\right\} = O(\sqrt{x}\log x),$$

(58) $$\left.\begin{matrix}\vartheta(x) - x \\ \psi(x) - x\end{matrix}\right\} = O(\sqrt{x}\log^2 x).$$

174) Hieraus folgt die Richtigkeit einer von *Lionnet,* Question 1075, Nouv. ann. math. (2) 11 (1872), p. 190, ausgesprochenen Vermutung, daß für große x mehr Primzahlen im Intervalle $(1, x)$ als in $(x, 2x)$ liegen. Es gilt nämlich $2\pi(x) - \pi(2x) \sim 2\log 2 \frac{x}{\log^2 x}$; vgl. *E. Landau,* Solutions de questions proposées, 1075, Nouv. ann. math. (4) 1 (1901), p. 281—282.

175) *E. Landau,* a) a. a. O. 29); b) Neue Beiträge zur analytischen Zahlentheorie, Palermo Rend. 27 (1909), p. 46—58; c) Handbuch § 81; *Landau* zeigt, daß (53) für alle $\alpha < \frac{1}{\sqrt{18{,}53}}$, also z. B. für $\alpha = \frac{1}{5}$, gilt.

176) Der von *Littlewood,* a. a. O. 111) ohne Beweis ausgesprochene Satz $\zeta(s) \neq 0$ für $\sigma > 1 - \frac{c \operatorname{loglog} t}{\log t}$ würde eine Verbesserung von (53) zulassen, indem er ein Restglied von der Form $O\left(xe^{-\alpha\sqrt{\log x \operatorname{loglog} x}}\right)$ liefern würde.

Diese Gleichungen sind zuerst von *v. Koch*[168]) mit seiner in der vorigen Nummer erwähnten Methode bewiesen; sie können auch aus der *de la Vallée Poussin*schen Gleichung (51) erhalten werden, durch ein Verfahren, das von *Holmgren*[177]) und in einem analogen Fall von *Landau*[178]) benutzt wurde. *Landau*[179]) hat diese Abschätzungen auch auf anderem Wege bewiesen; die etwas unschärfere Abschätzung $O\left(x^{\frac{1}{2}+\varepsilon}\right)$ folgt nach den *Littlewood*schen Ergebnissen über die μ-Funktion (vgl. Nr. 20) direkt aus dem Konvergenzsatz von *Landau-Schnee* (vgl. Nr. 6).

Bezeichnet man allgemein durch Θ die obere Grenze der reellen Teile der Nullstellen von $\zeta(s)$, wobei also $\frac{1}{2} \leqq \Theta \leqq 1$ ist, so bleiben (57) und (58) jedenfalls richtig, wenn $\sqrt{x}$ durch x^{Θ} ersetzt wird.[179]) (Im Falle $\Theta = 1$ ist dies natürlich trivial.) Die *Dirichlet*sche Reihe

$$(59) \qquad \sum_{1}^{\infty} \frac{\Lambda(n)-1}{n^s} = -\left(\frac{\zeta'}{\zeta}(s) + \zeta(s)\right)$$

konvergiert also für $\sigma > \Theta$. Aus (53) folgt, daß sie jedenfalls auf der ganzen Geraden $\sigma = 1$ konvergiert.

Auch wenn die *Riemann*sche Vermutung bewiesen wird, kann man nicht hoffen, die durch (57) und (58) gegebenen Abschätzungen wesentlich zu verbessern. Jedenfalls kann für kein $\eta < \Theta$ z. B.

$$\psi(x) - x = O(x^{\eta})$$

sein[180]), denn daraus würde die Konvergenz der linken Seite von (59) — und also die Regularität der rechten Seite — für $\sigma > \eta$ folgen. Weitere Sätze in dieser Richtung gaben *Phragmén*[181]), *Schmidt*[182]) und *Landau*[183]), der die Frage in Beziehung zu seinem Satze über *Dirichlet*-

177) *E. Holmgren,* Om primtalens fördelning, Öfvers. af Kgl. Vetensk. Förh. 59, Stockholm 1902—1903, p. 221—225.

178) *E. Landau,* Über einige ältere Vermutungen und Behauptungen in der Primzahltheorie, Math. Ztschr. 1 (1918), p. 1—24.

179) *E. Landau,* a. a. O. 175b) und Handbuch, § 93—94.

180) Dies wurde schon von *A. Piltz* behauptet: Über die Häufigkeit der Primzahlen in arithmetischen Progressionen und über verwandte Gesetze, Habilitationsschrift, Jena 1884. Vgl. auch *T. J. Stieltjes,* a. a. O. 160), Lettre 299.

181) *E. Phragmén,* Sur le logarithme intégral et la fonction $f(x)$ de *Riemann,* Öfvers. af Kgl. Vetensk. Förh. Stockholm 48 (1891—1892), p. 599—616 und Sur une loi de symétrie relative à certaines formules asymptotiques, ibid. 58 (1901—1902), p. 189—202.

182) *E. Schmidt,* Über die Anzahl der Primzahlen unter gegebener Grenze, Math. Ann. 57 (1903), p. 195—204.

183) *E. Landau,* Über einen Satz von *Tschebyschef,* Math. Ann. 61 (1905), p. 527—550, und Handbuch, § 201—204.

sche Reihen mit positiven Koeffizienten (vgl. Nr. 6) setzte. Ein bedeutender Fortschritt wurde von *Littlewood*[184]) gemacht; durch eine Methode, auf die wir in der nächsten Nummer zurückkommen, bewies er die Existenz einer positiven Konstanten K derart, daß alle vier Ungleichungen

$$(60)\quad \begin{cases} \pi(x) - Li(x) > \quad K\frac{\sqrt{x}}{\log x}\operatorname{logloglog} x \\ \pi(x) - Li(x) < -K\frac{\sqrt{x}}{\log x}\operatorname{logloglog} x \\ \vartheta(x) - x \qquad > \quad K\sqrt{x}\operatorname{logloglog} x \\ \vartheta(x) - x \qquad < -K\sqrt{x}\operatorname{logloglog} x \end{cases}$$

beliebig große Lösungen besitzen. Das gleiche gilt für die entsprechenden Ungleichungen mit Π an der Stelle von π und ψ an der Stelle von ϑ. Dies ist das beste bisher bekannte Resultat über die wirklich stattfindenden Unregelmäßigkeiten der Primzahlfunktionen; wäre es aber gelungen, die *Falschheit* der *Riemann*schen Vermutung (d. h. $\Theta > \frac{1}{2}$) zu beweisen, so könnte nach *Schmidt*[182]) der Faktor von K in (60) sogar durch $x^{\Theta-\varepsilon}$ ersetzt werden. — Das Resultat von *Littlewood* ist besonders darum bemerkenswert, weil man früher die Beziehung

$$(61)\qquad \pi(x) < Li(x)$$

als höchst wahrscheinlich betrachtet hat[185]); diese Beziehung gilt insbesondere für alle $x < 10.000.000$. Nach (60) kann sie aber nicht allgemein gelten.

Nach (60) ist z. B. die Funktion $\frac{\psi(x)-x}{\sqrt{x}}$ sicher nicht beschränkt. Wenn die *Riemann*sche Vermutung richtig ist, so hat sie trotzdem, wie *Cramér*[186]) zeigt, einen beschränkten quadratischen Mittelwert, d. h.

$$\frac{1}{x}\int_2^x \left(\frac{\psi(t)-t}{\sqrt{t}}\right)^2 dt$$

ist beschränkt, strebt aber für $x \to \infty$ keinem bestimmten Grenzwert

184) *J. E. Littlewood,* Sur la distribution des nombres premiers, Paris C. R. 158 (1914), p. 1869—1872; *G. H. Hardy* und *J. E. Littlewood,* a. a. O. 31b).

185) Vgl. *Gauß,* a. a. O. 153), Bemerkung von *E. Schering* in *Gauß'* Werke 2, p. 520; *Phragmén,* a. a. O. 189); *Lehmer,* List of prime numbers from 1 to 10.006.721, Washington 1914.

186) *H. Cramér,* Some theorems concerning prime numbers, Arkiv f. Mat., Astr. och Fys. 15 (1920), No. 5.

zu, was dagegen von

$$\frac{1}{\log x}\int\limits_2^x \left(\frac{\psi(t)-t}{t}\right)^2 dt$$

gilt.[187])

28. Die Riemannsche Primzahlformel. Das Hauptziel der *Riemann*schen Primzahlarbeit[95]) war die Aufstellung eines *exakten* Ausdrucks für die Funktion $\overline{\Pi}(x)$; durch die Betrachtung des Integrals (47) wurde *Riemann* nämlich auf die Formel

$$(62)\quad \overline{\Pi}(x) = Li(x) - \sum_{\gamma>0} \left(Li(x^{\varrho}) + Li(x^{1-\varrho})\right) + \int\limits_x^{\infty} \frac{dt}{(t^2-1)t\log t} - \log 2$$

geführt. (Ein unwesentlicher Schreib- oder Rechenfehler im letzten, konstanten Gliede wurde von *Genocchi*[188]) berichtigt.) Die Summe ist hier über alle Nullstellen $\varrho = \beta + \gamma i$ von $\zeta(s)$ zu erstrecken, die der oberen Halbebene angehören, und es ist

$$Li(x^{a+bi}) = \int\limits_{-\infty + bi\log x}^{(a+bi)\log x} \frac{e^z}{z}\,dz \pm \pi i$$

gesetzt, je nachdem $b \log x \gtrless 0$ gilt.[189]) Über seinen Beweis der Konvergenz dieser Reihe gab *Riemann* nur eine unbestimmte Andeutung, und auch aus anderen Gründen war die Formel als nur heuristisch begründet anzusehen. Wegen der äußerst verwickelten Natur der auftretenden Funktionen wurde sogar an der Möglichkeit gezweifelt, die Formel überhaupt beweisen oder jedenfalls daraus irgendwelche Schlüsse ziehen zu können.[190]) Es hat auch lange gedauert, bis ein vollständiger Beweis gegeben wurde; nach verschiedenen Versuchen[191]) gelang dies zuerst *v. Mangoldt*[192]), der eine entsprechende Formel für die

---

187) *H. Cramér,* Ein Mittelwertsatz in der Primzahltheorie, Math. Ztschr. 12 (1922), p. 147—153; vgl. auch Sur un problème de *M. Phragmén,* Arkiv f. Mat., Astr. och Fys. 16 (1922), No. 27.

188) *A. Genocchi,* Formole per determinare quanti siano i numeri primi fino ad un dato limite, Ann. Mat. pura appl. (1) 3 (1860), p. 52—59.

189) Über den Sinn der Formel und ihre Verwendung für numerische Rechnungen vgl. *E. Phragmén,* Über die Berechnung der einzelnen Glieder der *Riemann*schen Primzahlformel, Öfvers. af Kgl. Vetensk. Förh. 48, Stockholm 1891—1892, p. 721—744.

190) Vgl. z. B. *Ch. de la Vallée Poussin,* a. a. O. 163), p. 252—256.

191) Vgl. z. B. *A. Piltz,* a. a. O. 180); *J. P. Gram,* Undersögelser angaaende Mængden af Primtal under en given Grænse, Kgl. Danske Vidensk. Selsk. Skrifter, naturv. og math. Afd. (6) 2 (1881—1886), p. 183—308.

192) *H. v. Mangoldt,* a. a. O. 16) und Zu *Riemanns* Abhandlung „Über die Anzahl der Primzahlen unter einer gegebenen Größe", Crelles J. 114 (1895), p. 255—305.

Funktion

$$F(x, r) = \sum_{n=1}^{x} \frac{\Lambda(n)}{n^r} - \frac{\Lambda(x)}{2x^r} \quad \text{(für nicht ganze } x \text{ bedeutet } \Lambda(x) \text{ Null)}$$

aufstellte, um dann durch Integration nach dem Parameter r zur *Riemann*schen Formel überzugehen.[193]) $F(x, 0)$ ist mit $\overline{\psi}(x)$ identisch, und in diesem Falle lautet die Formel

$$(63) \qquad \overline{\psi}(x) = x - \sum_{\rho} \frac{x^\rho}{\rho} - \frac{1}{2}\log\left(1 - \frac{1}{x^2}\right) - \log 2\pi,$$

wo jetzt die Summe über alle komplexen ϱ, nach absolut wachsenden Ordinaten geordnet, erstreckt wird. Diese Formel, deren einzelne Glieder elementare Funktionen sind, ist für die spätere Entwicklung sogar wichtiger als (62) geworden. Formal kommt sie bei der Betrachtung des Integrals (46) unmittelbar heraus, da rechts die Summe der Residuen des Integranden links vom Integrationswege steht.

Da $\overline{\psi}(x)$ in den Punkten $x = p^m$ unstetig ist, kann $\sum \frac{x^\varrho}{\varrho}$ dort nicht gleichmäßig konvergieren. *Landau*[194]), der den *v. Mangoldt*schen Beweis vereinfacht und auch (62) direkt aus (47) abgeleitet hat, zeigt aber, daß die Reihe in jedem Intervall, das rechts von $x = 1$ liegt und von den $x = p^m$ frei ist, gleichmäßig konvergiert. Er dehnt seine Untersuchungen auch auf die allgemeinere Reihe

$$(64) \qquad \sum_{\gamma > 0} \frac{x^\varrho}{\varrho^k} \qquad (0 < k \leqq 1)$$

aus[195]), welche analoge Konvergenzeigenschaften besitzt[196]), die auf das Verhalten der endlichen Summe $\sum\limits_{0 < \gamma \leqq T} x^\varrho$ zurückgeführt werden können. *Cramér*[197]) betrachtet diese Reihen auch für komplexe Werte der

193) Zu diesem Übergang vgl. *H. Cramér*, Über die Herleitung der *Riemann*schen Primzahlformel, Arkiv f. Mat., Astr. och Fys. 13 (1918), No. 24.

194) *E. Landau*, Neuer Beweis der Riemannschen Primzahlformel, Sitzungsber. Akad. Berlin 1908, p. 737—745; Nouvelle démonstration pour la formule de Riemann sur le nombre des nombres premiers inférieurs à une limite donnée, et démonstration d'une formule plus générale pour le cas des nombres premiers d'une progression arithmétique, Ann. Éc. Norm. (3) 25 (1908), p. 399—442.

195) *E. Landau*, Über die Nullstellen der Zetafunktion, Math. Ann. 71 (1912), p. 548—564.

196) In den Unstetigkeitspunkten $x = p^m$ ist jedoch (64) divergent, während $\sum\limits_{\varrho} \frac{x^\varrho}{\varrho}$ für alle $x > 0$ konvergiert.

197) *H. Cramér*, Studien über die Nullstellen der Riemannschen Zetafunktion, Math. Ztschr. 4 (1919), p. 104—130.

Veränderlichen, indem er insbesondere die Funktion

$$V(z) = \sum_{\gamma > 0} e^{\varrho z}$$

untersucht. Wird die z-Ebene längs der negativen imaginären Achse aufgeschnitten, so ist $V(z)$ im Innern der aufgeschnittenen Ebene meromorph und hat nur die singulären Stellen $z = \pm \log p^m$, welche Pole erster Ordnung sind. Hierdurch wird es möglich, auf die Reihe (64) den Konvergenzsatz von *M. Riesz* anzuwenden (vgl. Nr. 5). — Alle diese Erscheinungen deuten auf irgendeinen arithmetischen Zusammenhang zwischen den Nullstellen ϱ und den Primzahlen p hin.

Die Formeln (62) und (63) setzen die Hauptglieder der Funktionen $\overline{\Pi}(x)$ bzw. $\overline{\psi}(x)$ in Evidenz; wegen der nur bedingten Konvergenz der auftretenden Reihen läßt sich aus ihnen jedoch nicht einmal der Primzahlsatz *unmittelbar* erschließen. Zwar ist z. B. in $\sum \frac{x^\varrho}{\varrho}$ jedes Glied von der Form $o(x)$ — wenn die *Riemann*sche Vermutung wahr ist, sogar von der Form $O\left(x^{\frac{1}{2}}\right)$ — wegen der Divergenz von $\sum \left|\frac{x^\varrho}{\varrho}\right|$ ist es aber nicht zulässig, unmittelbar hieraus $\sum \frac{x^\varrho}{\varrho} = o(x)$ zu folgern.[198]) *v. Koch*[199]) hat diese Formeln dadurch für asymptotische Zwecke verwerten können, daß er in die unendlichen Reihen konvergenzerzeugende Faktoren einführt und die Reihen dann durch endliche Summen ersetzt. Auf diese Weise ist es ihm gelungen, $\Pi(x)$ als Summe einer absolut konvergenten Reihe und eines beschränkten Fehlergliedes darzustellen, für $\psi(x)$ erhält er z. B. den Ausdruck

$$(65) \qquad \psi(x) = x - \sum_{|\varrho| < \sqrt{x}} \frac{x^\varrho}{\varrho} \Gamma\left(1 - \frac{\varrho \log x}{3\sqrt{x}}\right) + O(\sqrt{x} \log^2 x).$$

Landau[200]) zeigt, daß diese Gleichung auch dann richtig bleibt, wenn

198) Die Ausführungen von *H. v. Mangoldt,* Über eine Anwendung der Riemannschen Formel für die Anzahl der Primzahlen unter einer gegebenen Grenze, Crelles J. 119 (1898), p. 65—71, enthalten nur einen Übergang von (45) zu (41).

199) *H. v. Koch,* Über die Riemannsche Primzahlfunction, Math. Ann. 55 (1902), p. 441—464; Contribution à la théorie des nombres premiers, Acta Math. 33 (1910), p. 293—320.

200) *E. Landau,* Über einige Summen, die von den Nullstellen der Riemannschen Zetafunktion abhängen, Acta Math. 35 (1911), p. 271—294. Vgl. auch *A. Hammerstein,* Zwei Beiträge zur Zahlentheorie, Diss., Göttingen 1919. — *Littlewood,* a. a. O. 111) hat sogar (ohne Beweis) die Formel

$$\psi(x) = x - \sum_{|\varrho| \leqq y} \frac{x^\varrho}{\varrho} + O(\sqrt{x} \log x),$$

man den Γ-Faktor wegläßt. *Cramér*[186]) gibt die Formel

$$(66) \qquad \psi(x) = x - \sum_{\varrho} \frac{x^{\varrho}}{\varrho} e^{-\frac{|\gamma|}{x^2}} + O(\log^2 x),$$

wo die Reihe absolut konvergiert.

Wenn die Riemannsche Vermutung richtig ist, so kann aus (65), der entsprechenden *Landau*schen Formel, oder (66) sofort (58) erhalten werden. Unter derselben Voraussetzung folgt aus der *v. Mangoldt*schen Formel (63)

$$\psi(x) = x - 2\sqrt{x} \sum_{\gamma > 0} \frac{\sin(\gamma \log x)}{\gamma} + O(\sqrt{x}).$$

Die hier auftretende Reihe stellt den „kritischen Teil" von $\psi(x)$ dar; wird jedes Glied mit dem entsprechenden $e^{-\gamma\sigma}$ multipliziert und $\log x = t$ gesetzt, so erhält man den imaginären Teil der Funktion von $s = \sigma + it$

$$\sum_{\gamma > 0} \frac{1}{\gamma} e^{-\gamma s}.$$

Durch Betrachtung dieser Funktion beweist *Littlewood*[184]) unter Benutzung eines Satzes über diophantische Approximationen sein oben erwähntes, durch (60) ausgedrücktes Resultat.

Aus (62) erhält man mit Hilfe der Beziehung (vgl. Nr. 32)

$$(67) \qquad \bar{\pi}(x) = \sum_{n=1}^{\infty} \frac{\mu(n)}{n} \overline{\Pi}\left(x^{\frac{1}{n}}\right)$$

eine explizite Formel für $\bar{\pi}(x)$. Diese Formel hat früher die theoretische Stütze der (falschen) Vermutung (61) geliefert[201]), es läßt sich jedoch zur Zeit daraus nicht wesentlich mehr über $\bar{\pi}(x)$ folgern, als schon aus der einfacheren Beziehung

$$\bar{\pi}(x) = \overline{\Pi}(x) + O\left(\frac{\sqrt{x}}{\log x}\right)$$

folgt.

29. Theorie der L-Funktionen. Es sei $k > 1$ eine gegebene ganze Zahl; dann muß jede Primzahl, mit Ausnahme der endlich vielen in k aufgehenden, irgendeiner der $\varphi(k)$ zu k teilerfremden Restklassen

---

gleichmäßig für $y \geq \sqrt{x}$, angegeben, deren Gültigkeit aber nur unter Voraussetzung der *Riemann*schen Vermutung behauptet wird.

201) *Riemann* (a. a. O. 95) sagt z. B. bei der Besprechung der Formel (67): „Die bekannte Näherungsformel $F(x) = Li(x)$ (sein $F(x)$ ist unser $\bar{\pi}(x)$) ist also nur bis auf Größen von der Ordnung $x^{\frac{1}{2}}$ richtig und gibt einen etwas zu großen Wert."

modulo k angehören. Schon von *Legendre*[202]) wurde (mit falschem Beweis) die Behauptung ausgesprochen, daß jede dieser Restklassen unendlich viele Primzahlen — und sogar asymptotisch gleich viele wie jede andere — enthält. Für die erste Behauptung gab *Dirichlet*[203]) einen strengen Beweis, die zweite aber wurde erst von *Hadamard*[162]) und *de la Vallée Poussin*[204]) bewiesen. Bei diesen Untersuchungen treten als Hilfsmittel gewisse Funktionen auf, die auch bei verschiedenen anderen Fragen der analytischen Zahlentheorie eine Rolle spielen (vgl. Nr. **35, 40, 41**), und die deshalb jetzt besprochen werden müssen.

Die obengenannten $\varphi(k)$ Restklassen bilden in bezug auf die gewöhnliche Multiplikation eine *Abel*sche Gruppe. Es sei $X(K)$ irgendeiner der $\varphi(k)$ Charaktere der Gruppe (vgl. I A 6, Nr. **20**); diese Funktion nimmt für jede der fraglichen Restklassen K einen bestimmten Wert an, der übrigens immer eine $\varphi(k)$-te Einheitswurzel ist. Es sei nun die zahlentheoretische Funktion $\chi(n)$ für $n = 0, \pm 1, \pm 2, \ldots$ folgendermaßen erklärt: für jedes n einer mit k gemeinteiligen Restklasse sei $\chi(n) = 0$; für jedes n einer zu k teilerfremden Restklasse K sei $\chi(n) = X(K)$. Unter den so eingeführten $\varphi(k)$ verschiedenen *Charakteren modulo* k zeichnet sich besonders der *Hauptcharakter* aus, der für jedes zu k teilerfremde n den Wert 1 hat. Um die verschiedenen Charaktere zu unterscheiden, bezeichnet man sie durch $\chi_1(n), \chi_2(n), \ldots \chi_{\varphi(k)}(n)$, wobei $\chi_1(n)$ immer der Hauptcharakter ist. Die Charaktere besitzen die folgenden vier Fundamentaleigenschaften:

$$\text{a) } \chi(n) = \chi(n') \quad \text{für} \quad n \equiv n' \pmod{k},$$

$$\text{b) } \chi(n) \cdot \chi(n') = \chi(nn'),$$

$$\text{c) } \sum_{n=1}^{k} \chi_\nu(n) = \begin{cases} \varphi(k) & \text{für } \nu = 1, \\ 0 & \text{sonst}, \end{cases}$$ [205]),

202) *A.-M. Legendre*, a. a. O. 152a) p. 404; 152b) p. 77 und 99. Vgl. auch *A. Dupré*, Examen d'une proposition de Legendre relative à la théorie des nombres, Paris 1859; *C. Moreau*, Extrait d'une lettre, Nouv. Ann. math. (2) 12 (1873), p. 322—324; *A. Piltz*, a. a. O. 180).

203) *G. Lejeune-Dirichlet*, Beweis des Satzes, daß jede unbegrenzte arithmetische Progression, deren erstes Glied und Differenz ganze Zahlen ohne gemeinschaftlichen Faktor sind, unendlich viele Primzahlen enthält, Abh. Akad. Berlin 1837, math. Abhandl., p. 45—71 und Werke 1, p. 313—342.

204) *Ch. de la Vallée Poussin*, Recherches analytiques sur la théorie des nombres premiers, Deuxième partie, Ann. soc. sc. Bruxelles 20: 2 (1896), p. 281—362.

205) Hieraus folgt speziell, daß $\left|\sum_{1}^{N} \chi(n)\right|$ für jeden Nicht-Hauptcharakter

d) $\sum_{\nu=1}^{\varphi(k)} \chi_\nu(n) = \begin{cases} \varphi(k) & \text{für} \quad n \equiv 1 \pmod{k}, \\ 0 & \text{sonst.} \end{cases}$

Die *Dirichlet*sche[203]) Reihe

$$(68) \qquad L_\nu(s) = L(s, \chi_\nu) = \sum_{n=1}^{\infty} \frac{\chi_\nu(n)}{n^s}$$

ist für $\sigma > 1$ absolut konvergent; wegen b) gilt auch dort

$$(69) \qquad L_\nu(s) = \prod_p \left(1 - \frac{\chi_\nu(p)}{p^s}\right)^{-1}.$$

Diese L-Funktionen können als Verallgemeinerungen von $\zeta(s)$ — die dem Falle $k = 1$ entspricht — angesehen werden und besitzen auch durchaus analoge Eigenschaften.[206]) Für den Fall des Hauptcharakters folgt unmittelbar

$$(70) \qquad L_1(s) = \sum_{n=1}^{\infty} \frac{\chi_1(n)}{n^s} = \zeta(s) \cdot \prod_{p|k} \left(1 - \frac{1}{p^s}\right).$$

($p|k$ bedeutet: p geht in k auf.) Die Funktion $L_1(s)$ läßt sich somit direkt auf $\zeta(s)$ zurückführen; sie besitzt wie diese in $s = 1$ einen Pol erster Ordnung und ist sonst überall im Endlichen regulär. Für $\nu > 1$ folgt dagegen aus c), daß (68) für $\sigma > 0$ konvergiert und sogar für jeden Wert von s durch die *Cesàro*sche Methode summabel ist (vgl. Nr. **13**); für jedes vom Hauptcharakter verschiedene χ ist also $L(s)$ eine ganze transzendente Funktion.[207])

Dirichlet untersuchte die L-Funktionen nur für reelle s; verschiedene andere Verfasser[208]) haben dann auch komplexe s berücksichtigt und

---

unter einer nur von k abhängenden Schranke liegt. In der Tat gilt sogar $\left|\sum_1^N \chi(n)\right| < c\sqrt{k}\log k$, wo c eine absolute Konstante bedeutet. Vgl. *G. Pólya,* Über die Verteilung der quadratischen Reste und Nichtreste, Göttinger Nachr. 1918, p. 21—29; *J. Schur,* Einige Bemerkungen zu der vorstehenden Arbeit des Herrn G. Pólya, Gött. Nachr. 1918, p. 30—36; *E. Landau,* Abschätzungen von Charaktersummen, Einheiten und Klassenzahlen, Gött. Nachr. 1918, p. 79—97.

206) Eine ausführliche Darstellung der Theorie gibt *E. Landau,* Handbuch, § 95—140.

207) Dies folgt auch aus der Identität

$$L(s) = \sum_{m=1}^{k-1} \chi(m) \sum_{n=0}^{\infty} \frac{1}{(m+nk)^s} = k^{-s} \sum_{m=1}^{k-1} \chi(m)\, \zeta\left(\frac{m}{k}, s\right)$$

und den in Nr. **21** erwähnten Untersuchungen über $\zeta(w, s)$.

208) Vgl. *C. J. Malmstén,* Specimen analyticum etc., Diss., Upsala 1842 und De integralibus quibusdam definitis, seriebusque infinitis, Crelles J. 38 (1849), p. 1—39; *R. Lipschitz,* a. a. O. 147); *H. Kinkelin,* Allgemeine Theorie der har-

die L-Funktionen auf die verallgemeinerten Zetafunktionen von Nr. **21** zurückgeführt. Aus diesen Untersuchungen geht vor allem hervor, daß jede L-Funktion eine Funktionalgleichung besitzt, die derjenigen von $\zeta(s)$ (vgl. Nr. **14**) analog gebaut ist. Wenn $\chi(n)$ einem sog. eigentlichen Charakter[209]) entspricht, so gilt in der Tat

(71) $$L(s) = \Theta \frac{\sqrt{k}}{\pi} \left(\frac{2\pi}{k}\right)^s \sin \frac{\pi(s+\alpha)}{2} \Gamma(1-s) \bar{L}(1-s),$$

wo $\bar{L}(s)$ mit dem konjugiert komplexen Charakter $\bar{\chi}(n)$ gebildet ist, Θ eine Konstante vom absoluten Betrage 1 und $\alpha = 0$ oder 1 ist.[210]) Dies wird z. B. dadurch bewiesen, daß $L(s)$ durch die Funktion $\sum \chi(n) e^{-n^2 z}$ (für $\alpha = 0$) oder durch $\sum \chi(n) n e^{-n^2 z}$ für ($\alpha = 1$) analog wie bei $\zeta(s)$ (vgl. Nr. **14**) ausgedrückt wird[211]), wonach die Funktionalgleichung aus der Transformationstheorie der Thetafunktionen folgt. — Gehört $L(s)$ dagegen zu einem uneigentlichen Charakter, so lassen sich immer ein echter Teiler k' von k und ein eigentlicher Charakter $\chi'(n)$ modulo k' derart angeben, daß für $\sigma > 1$

(72) $$L(s) = \sum_{n=1}^{\infty} \frac{\chi'(n)}{n^s} \cdot \prod_{p \mid k} \left(1 - \frac{\chi'(p)}{p^s}\right)$$

gilt.

In diesem Falle unterscheidet sich $L(s)$ also nur um einen trivialen Faktor von einer zu einem eigentlichen Charakter gehörigen L-Funktion (modulo k'); (70) stellt offenbar einen Spezialfall hiervon dar.

Jetzt können die Eigenschaften von $L(s)$ genau wie bei $\zeta(s)$ abgeleitet werden. In der Halbebene $\sigma > 1$ ist $L(s) \neq 0$, für $\sigma < 0$ gibt es nur die vom Faktor $\sin \frac{\pi(s+\alpha)}{2}$ in (71) herrührenden „tri-

monischen Reihen, mit Anwendung auf die Zahlentheorie, Progr. d. Gewerbeschule, Basel 1862; *A. Piltz*, a. a. O. 180); *A. Hurwitz*, Einige Eigenschaften der Dirichletschen Funktionen $F(s) = \sum \left(\frac{D}{n}\right) \frac{1}{n^s}$, die bei der Bestimmung der Klassenanzahl binärer quadratischer Formen auftreten; *M. Lerch*, a. a. O. 148). Bei *Hadamard*, a. a. O. 162) und *de la Vallée Poussin*, a. a. O. 204), werden die früheren Resultate zusammengestellt und die Hilfsmittel der modernen Funktionentheorie zum erstenmal auf die L-Funktionen angewandt.

209) Ein Charakter $\chi(n)$ modulo k heißt *uneigentlich*, wenn es einen echten Teiler k' von k und einen Charakter $\chi'(n)$ modulo k' gibt, so daß für jedes n entweder $\chi(n) = 0$ oder $\chi(n) = \chi'(n)$ gilt. Sonst heißt $\chi(n)$ *eigentlich*. Der Hauptcharakter ist für $k > 1$ immer uneigentlich. Vgl. z. B. *Landau*, Handbuch, Bd. 1, p. 478.

210) Nämlich $\alpha = 0$ im Falle $\chi(-1) = 1$, $\alpha = 1$ im Falle $\chi(-1) = -1$.

211) *de la Vallée Poussin*, a. a. O. 204). Seine Darstellung wurde von *Landau*, a. a. O. 206) vereinfacht.

vialen" Nullstellen, auch der Punkt $s = 0$ kann unter Umständen Nullstelle sein. Im Streifen $0 \leqq \sigma \leqq 1$ liegen unendlich viele von Null verschiedene Nullstellen $\varrho = \beta + \gamma i$, und die Anzahl $N(T)$ der ϱ, deren Ordinaten dem Intervall $0 < \gamma \leqq T$ angehören, ist gleich

$$N(T) = \frac{1}{2\pi} T \log T - cT + O(\log T),$$

wo c von k und χ abhängt.[212]) (Vgl. Nr. **16**.) Für jeden Nicht-Hauptcharakter gibt es eine Produktentwicklung[211])

$$(73) \qquad L(s) = a s^{\mu} e^{bs} \frac{1}{\Gamma\left(\frac{s+\alpha}{2}\right)} \prod_{\varrho} \left(1 - \frac{s}{\varrho}\right) e^{\frac{s}{\varrho}},$$

wo μ ganz und $\geqq 0$ ist; beim Hauptcharakter muß auf der linken Seite das Produkt $(s-1)L(s)$ stehen (vgl. Nr. **15**).

Der für die Primzahltheorie besonders wichtige Satz, daß der Punkt $s = 1$ bei keiner L-Funktion eine Nullstelle ist, wurde schon von *Dirichlet*[203]) gefunden. Der Beweis ist ganz verschieden, je nachdem der Charakter ein *reeller* (d. h. ein für alle n reeller) oder ein *komplexer* (d. h. ein für wenigstens ein n nicht-reeller) ist. Im letzteren Falle wäre gleichzeitig mit $L(1)$ auch $\bar{L}(1)$ gleich Null, und die Funktionen $L(s)$ und $\bar{L}(s)$ wären nicht identisch. Dies wäre aber nicht mit der Identität

$$\prod_{\nu=1}^{\varphi(k)} L_\nu(s) = e^{\varphi(k) \sum\limits_{p^m \equiv 1 \pmod{k}} \frac{1}{m p^{ms}}}, \qquad (\sigma > 1)$$

verträglich, da die linke Seite für $s = 1$ eine Nullstelle hätte, während die rechte Seite für reelle $s > 1$ immer $\geqq 1$ ist. Für jeden komplexen Charakter gilt sogar[213])

$$\frac{1}{|L(1)|} < M \log k \, (\log\log k)^{\frac{3}{8}}$$

---

212) *E. Landau*, a. a. O. 107).

213) Vgl. *H. Gronwall*, Sur les séries de Dirichlet correspondant à des caractères complexes, Palermo Rend. 35 (1913), p. 145—159; *E. Landau*, a) Über das Nichtverschwinden der Dirichletschen Reihen, welche komplexen Charakteren entsprechen, Math. Ann. 70 (1911), p. 69—78; b) Über die Klassenzahl imaginär-quadratischer Zahlkörper, Gött. Nachr. 1918, p. 285—295; c) Zur Theorie der Heckeschen Zetafunktionen, welche komplexen Charakteren entsprechen, Math. Ztschr. 4 (1919), p. 152—162. Bei diesen Abschätzungen von $\frac{1}{L(1)}$ als Funktion von k zeigt sich eine eigenartige Analogie mit der Abschätzung von $\zeta(1 + ti)$ als Funktion von t (vgl. Nr. **18**). Für die reellen Charaktere wird das entsprechende Ergebnis (mit $\frac{1}{2}$ statt $\frac{3}{8}$) nur unter einer gewissen unbewiesenen Voraussetzung erhalten (vgl. Nr. **40**).

mit absolut konstantem M. — Bei den *reellen* Charakteren war der Beweis viel schwieriger; es war eben die Hauptleistung von *Dirichlet*[203]), $L(1)$ als Produkt von einer positiven Konstanten und einer gewissen Klassenzahl quadratischer Formen darzustellen (vgl. Nr. **40**); eo ipso war $L(1) \neq 0$. Vereinfachte Beweisanordnungen gaben *Mertens*[214]), *de la Vallée Poussin*[215]), *Teege*[216]) und *Landau*[217]), die den Beweis durch reihen- oder funktionentheoretische Überlegungen, ohne Benutzung der Theorie der quadratischen Formen, führten.[218])

Für jeden von $s = 1$ verschiedenen Punkt der Geraden $\sigma = 1$ läßt sich wie bei $\zeta(s)$ (vgl. Nr. **14**) $L(s) \neq 0$ nachweisen.[219]) Auch die entsprechenden schärferen Sätze gelten hier, es gibt z. B. eine absolute Konstante $a > 0$ derart, daß im Gebiete $\sigma > 1 - \frac{a}{\log|t|}$, $|t| > t_0$ keine Nullstellen von $L(s)$ liegen (vgl. Nr. **19**).[220]) Das Gegenstück der *Riemann*schen Vermutung wurde für die L-Funktionen von *Piltz*[180]) ausgesprochen. Da man im allgemeinen nicht weiß, ob Nullstellen auf der Strecke $0 < s < 1$ der reellen Achse liegen, und

214) *F. Mertens*, Über Dirichletsche Reihen, Sitzungsber. Akad. Wien 104 Abt. 2a (1895), p. 1093—1153; Über das Nichtverschwinden Dirichletscher Reihen mit reellen Gliedern, ebenda 104 Abt. 2a, p. 1158—1166; Über Multiplikation und Nichtverschwinden Dirichletscher Reihen, Crelles J. 117 (1897), p. 169—184; Über Dirichlets Beweis usw. Sitzungsber. Akad. Wien 106 Abt. 2a (1897), p. 254—286; Eine asymptotische Aufgabe, ebenda 108, Abt. 2a (1899), p. 32—37.

215) *Ch. de la Vallée Poussin*, a. a. O. 204) und Démonstration simplifiée du théorème de Dirichlet sur la progression arithmétique, Mém. couronnés et autres mém. Acad. Belgique 53 (1895—1896), No. 6.

216) *H. Teege*, Beweis, daß die unendliche Reihe $\sum \left(\frac{P}{n}\right) \frac{1}{n}$ einen positiven von Null verschiedenen Wert hat, Mitt. math. Ges. Hamburg 4 (1901), p. 1—11.

217) *E. Landau*, a. a. O. 29) und Über das Nichtverschwinden einer Dirichletschen Reihe, Sitzungsber. Akad. Berlin 1906, p. 314—320.

218) Vgl. auch eine Bemerkung von *Remak* bei *E. Landau*, Über imaginärquadratische Zahlkörper mit gleicher Klassenzahl, Gött. Nachr. 1918, p. 277—284.

219) Hiermit hängt zusammen, daß die Reihe $\sum_p \frac{\chi(p)}{p^s}$ und das Produkt in (69) auch noch für $\sigma = 1$ konvergieren (beim Hauptcharakter jedoch nur für $t \neq 0$) und daß (69) auch hier richtig bleibt (vgl. Nr. **14**). Ob diese Ausdrücke in der Halbebene $\sigma < 1$ einen einzigen Konvergenzpunkt besitzen, ist noch nicht entschieden. Vgl. *E. Landau*, a) Über die Primzahlen einer arithmetischen Progression, Sitzungsber. Akad. Wien 112, Abt. 2a (1903), p. 493—535; b) Über einen Satz von Tschebyschef, Math. Ann. 61 (1905), p. 527—550, wo eine Reihe früherer Arbeiten über den Gegenstand kritisiert werden; c) a. a. O. 178).

220) *E. Landau*, Handbuch, § 131, wo frühere Resultate desselben Verfassers verschärft werden.

da ferner nach (72) die imaginäre Achse unter Umständen unendlich viele Nullstellen enthalten kann, muß die Vermutung etwa so ausgesprochen werden: „für $\sigma > \frac{1}{2}$ ist $L(s) \neq 0$." [221]) Die Sätze von der Existenz unendlich vieler Nullstellen [222]) auf $\sigma = \frac{1}{2}$ und von der Häufung der Nullstellen in der Nähe dieser Geraden [223]) (vgl. Nr. **19**) gelten auch für die L-Funktionen.

Das Produkt zweier L-Reihen ist, sofern keine von ihnen einem Hauptcharakter entspricht, nach dem Satze von *Stieltjes* (vgl. Nr. **12**) für $\sigma > \frac{1}{2}$ konvergent. *Landau* [224]) beweist den folgenden Satz, der als Spezialfall eine Verschärfung hiervon enthält: Es seien $\chi_1(n)$ und $\chi_2(n)$ zwei beliebige [225]) Charaktere modulo k_1 bzw. k_2. Dann gibt es zwei Konstanten A und B, so daß die *Dirichlet*sche Reihe

$$\sum \frac{a_n}{n^s} = \sum \frac{\chi_1(n)}{n^s} \cdot \sum \frac{\chi_2(n)}{n^s} + A \sum \frac{\log n}{n^s} + B \sum \frac{1}{n^s}$$
$$= L_1(s) L_2(s) - A\zeta'(s) + B\zeta(s)$$

für $\sigma > \frac{1}{3}$ konvergiert. Hierbei ist $A = B = 0$, wenn weder χ_1 noch χ_2 Hauptcharakter ist, und $A = 0$, wenn nur einer von den beiden Hauptcharakter ist. Dieser Satz hat wichtige Anwendungen auf verschiedene zahlentheoretische Probleme (vgl. Nr. **34** und **35**).

30. Die Verteilung der Primzahlen einer arithmetischen Reihe. Nach (69) gilt für $\sigma > 1$

$$(74)\quad \begin{cases} \log L(s) = \sum\limits_{p,\,m} \dfrac{\chi(p^m)}{m\,p^{ms}} = \sum\limits_{n=1}^{\infty} \dfrac{\chi(n)\Lambda(n)}{\log n \cdot n^s} \\ -\dfrac{L'}{L}(s) = \sum\limits_{p,\,m} \dfrac{\chi(p^m)\log p}{p^{ms}} = \sum\limits_{n=1}^{\infty} \dfrac{\chi(n)\Lambda(n)}{n^s}. \end{cases}$$

Hieraus folgt nach den Eigenschaften b) und d) der Charaktere, wenn

221) Eine notwendige und hinreichende Bedingung für die Wahrheit dieser Vermutung gab neuerdings *H. Bohr* mit Hilfe des von ihm eingeführten Begriffes „Quasiperiodizität einer *Dirichlet*schen Reihe": Über eine quasi-periodische Eigenschaft Dirichletscher Reihen mit Anwendung auf die Dirichletschen L-Funktionen, Math. Ann. 85 (1922), p. 115—122. — *J. Großmann* hat die Vermutung durch numerische Untersuchungen gestützt: Über die Nullstellen der Riemannschen ζ-Funktion und der Dirichletschen L-Funktionen, Diss., Göttingen 1913.

222) *E. Landau*, a. a. O. 135).

223) *H. Bohr* und *E. Landau*, a. a. O. 65).

224) *E. Landau*, Über die Anzahl der Gitterpunkte in gewissen Bereichen, Gött. Nachr. 1912, p. 687—771.

225) Hier soll also $\chi_1(n)$ nicht wie oben notwendig den Hauptcharakter bezeichnen.

l eine beliebige zu k teilerfremde ganze Zahl bedeutet,

$$(75)\quad \begin{cases} \displaystyle\sum_{n \equiv l \,(\mathrm{mod}\, k)} \frac{\Lambda(n)}{\log n \cdot n^s} = \frac{1}{\varphi(k)} \sum_{\nu=1}^{\varphi(k)} \frac{1}{\chi_\nu(l)} \log L_\nu(s) \\ \displaystyle\sum_{n \equiv l \,(\mathrm{mod}\, k)} \frac{\Lambda(n)}{n^s} = -\frac{1}{\varphi(k)} \sum_{\nu=1}^{\varphi(k)} \frac{1}{\chi_\nu(l)} \cdot \frac{L_\nu'}{L_\nu}(s). \end{cases}$$

In beiden Gleichungen (75) wird die rechte Seite bei Annäherung an $s = 1$ unendlich, da dieser Punkt für $L_1(s)$ ein Pol, für die übrigen $L_\nu(s)$ dagegen weder Pol noch Nullstelle ist. Daraus schloß *Dirichlet*[203]), daß in der arithmetischen Reihe $l, l+k, l+2k, \ldots$ unendlich viele Primzahlen vorkommen; sonst würden ja in der Tat die linken Seiten von (75) für alle s endlich bleiben.[226])

Mit Hilfe der tieferen Eigenschaften der L-Funktionen konnten *Hadamard*[162]) und *de la Vallée Poussin*[204]) die dem Primzahlsatz entsprechenden Sätze

$$(76)\quad \begin{cases} \displaystyle \pi_{k,l}(x) = \sum_{\substack{p \leqq x \\ p \equiv l \,(\mathrm{mod}\, k)}} 1 \sim \frac{1}{\varphi(k)} Li(x), \\ \displaystyle \psi_{k,l}(x) = \sum_{\substack{n \leqq x \\ n \equiv l \,(\mathrm{mod}\, k)}} \Lambda(n) \sim \frac{1}{\varphi(k)} x \end{cases}$$

beweisen. Das Hauptargument beim Beweise bildet der in der vorigen Nummer erwähnte Satz: $L_\nu(1+ti) \neq 0$ für alle ν und alle reellen t. Von *Landau*[227]) wurden die Beweise vereinfacht und die Resultate verschärft, so daß das beste mit Sicherheit bekannte Resultat[228]) so lautet:

$$(77)\quad \begin{cases} \displaystyle \pi_{k,l}(x) = \frac{1}{\varphi(k)} Li(x) + O\left(x e^{-\alpha\sqrt{\log x}}\right) \\ \displaystyle \psi_{k,l}(x) = \frac{1}{\varphi(k)} x + O\left(x e^{-\alpha\sqrt{\log x}}\right) \end{cases}$$

mit absolut konstantem, d. h. von k und l unabhängigen, α. Die

---

226) In mehreren speziellen Fällen läßt sich der *Dirichlet*sche Satz elementar beweisen. Vgl. z. B. *L. E. Dickson,* History of the theory of numbers, Bd. 1, Washington 1919, p. 419.

227) *E. Landau,* Über die Primzahlen in einer arithmetischen Progression und die Primideale in einer Idealklasse, Sitzungsber. Akad. Wien 117, Abt. 2a (1908), p. 1095—1107; a. a. O. 219a); a. a. O. 107); Handbuch, § 119—121, § 131—132.

228) Wäre die verallgemeinerte *Riemann*sche Vermutung für die L-Funktionen bewiesen, so würden natürlich für die Primzahlen einer arithmetischen Reihe zu (57) und (58) analoge Beziehungen gelten. Vgl. *E. Landau,* Handbuch, § 239 und a. a. O. 178).

Hauptglieder rühren natürlich von den Singularitäten von $\log L_1(s)$ bzw. $\frac{L_1'}{L_1}(s)$ in $s = 1$ her. Aus (76) folgt speziell, wenn l_1 und l_2 beide zu k teilerfremd sind,

$$\pi_{k,l_1}(x) \sim \pi_{k,l_2}(x),$$

d. h. die zweite in der vorigen Nummer genannte *Legendre*sche Behauptung. — Die *Riemann-v. Mangoldt*sche Primzahlformel (vgl. Nr. 28) läßt sich auch für die Primzahlen einer arithmetischen Reihe verallgemeinern.[229]) Zunächst gilt für einen beliebigen Charakter $\chi_\nu(n)$ (vgl. (73))[230])

$$\text{(78)}\quad \begin{aligned} \sum_{n\leqq x} \chi_\nu(n)\Lambda(n) - \frac{1}{2}\chi_\nu(x)\Lambda(x) &= -\frac{1}{2\pi i}\int_{2-i\infty}^{2+i\infty} \frac{x^s}{s}\cdot\frac{L_\nu'}{L_\nu}(s)\,ds \\ &= \varepsilon_\nu x - \sum_\varrho \frac{x^\varrho}{\varrho} - \sum_{n=1}^{\infty}\frac{x^{\alpha-2n}}{\alpha-2n} + a_0 + a_1\log x, \end{aligned}$$

wo a_0 und a_1 von x unabhängig sind. ε_ν bedeutet Eins für $\nu = 1$, sonst Null. Aus (75) kann man jetzt, nach dem Eindeutigkeitssatz der *Dirichlet*schen Reihen, (vgl. Nr. 3) eine explizite Formel für die Funktion

$$\overline{\psi}_{k,l}(x) = \tfrac{1}{2}(\psi_{k,l}(x+0) + \psi_{k,l}(x-0))$$

erhalten.

Die bisher erwähnten Resultate laufen alle darauf hinaus, daß die Primzahlen auf die $\varphi(k)$ zu k teilerfremden Restklassen *gleichmäßig* verteilt sind. Schon *Tschebyschef*[231]) behauptete — freilich nur für den Fall $k = 4$ — dies könne nur bis zu einer bestimmten Grenze gelten, indem die Reihe $4n + 3$ „viel mehr" Primzahlen als die Reihe $4n + 1$ $(n = 1, 2, \ldots)$ enthalte. Er sprach (ohne Beweis) den Satz aus: es gibt eine Folge $x_1, x_2, \ldots$ mit $x_\nu \to \infty$, derart, daß für wachsendes ν

$$\text{(79)}\qquad \pi_{4,3}(x_\nu) - \pi_{4,1}(x_\nu) \sim \frac{\sqrt{x_\nu}}{\log x_\nu}$$

gilt. Dies wurde zuerst von *Phragmén*[181]) und dann einfacher von

229) Vgl. *A. Piltz*, a. a. O. 180) und *G. Torelli*, a. a. O. 159), Nuove formole per calcolare la totalità dei numeri primi etc., Rend. Accad. sc. fis. mat. Napoli (3) 10 (1904), p. 350—362 und (3) 11 (1905), p. 101—109. Vollständig ausgeführt wurde der Beweis erst von *E. Landau*, a. a. O. 194) und Handbuch, § 133—138.

230) Für nicht ganze x bedeuten $\chi(x)$ und $\Lambda(x)$ Null.

231) *P. Tschebyschef*, Lettre à M. Fuss, Bull. cl. phys.-math. Acad. St. Petersburg 11 (1853), p. 208 und Œuvres 1, p. 697—698; Sur une transformation de séries numériques, Nouv. corr. math. 4 (1878), p. 305—308 und Œuvres 2, p. 705—707.

Landau[183]) bewiesen; aus den Untersuchungen von *Littlewood*[184]) folgt aber, daß die obige Differenz, bei zweckmäßiger Wahl von K, für beliebig große Werte von x sowohl $> K\frac{\sqrt{x}}{\log x}\log\log\log x$ als auch $< -K\frac{\sqrt{x}}{\log x}\log\log\log x$ wird. Der Zusammenhang wird gewissermaßen durch die aus (78) folgenden Gleichungen

$$\vartheta_{4,1}(x) = \tfrac{1}{2}x - x^{\frac{1}{2}} - \tfrac{1}{2}\Bigl(\sum_\varrho \frac{x^\varrho}{\varrho} + \sum_{\varrho'} \frac{x^{\varrho'}}{\varrho'}\Bigr) + O\bigl(x^{\frac{1}{3}}\bigr),$$

$$\vartheta_{4,3}(x) = \tfrac{1}{2}x \qquad - \tfrac{1}{2}\Bigl(\sum_\varrho \frac{x^\varrho}{\varrho} - \sum_{\varrho'} \frac{x^{\varrho'}}{\varrho'}\Bigr) + O\bigl(x^{\frac{1}{3}}\bigr)$$

aufgeklärt. (Es ist $\vartheta_{k,l} = \sum_{\substack{p \leqq x \\ p \equiv l \,(\mathrm{mod}\, k)}} \log p$ gesetzt; ϱ durchläuft die komplexen Nullstellen von $\zeta(s)$ und ϱ' diejenigen von

$$L(s) = \sum_0^\infty \frac{(-1)^n}{(2n+1)^s} = \sum_1^\infty \frac{\chi(n)}{n^s},$$

wo $\chi(n)$ den Nicht-Hauptcharakter modulo 4 bezeichnet.) Die oszillierenden Glieder sind hier zwar von höherer Größenordnung als $x^{\frac{1}{2}}$, in der ersten Gleichung tritt aber ein Glied $-x^{\frac{1}{2}}$ *von konstantem Vorzeichen* auf, was wiederum daraus folgt, daß alle *Primzahlquadrate* ($2^2 = 4$ ausgenommen) von der Form $4n+1$ sind.

Tschebyschef[231]) behauptete auch: „wenn c gegen Null abnimmt, so gilt

(80) $e^{-3c} - e^{-5c} + e^{-7c} + e^{-11c} - \cdots = -\sum_p \chi(p)e^{-pc} \to \infty$.“

Von *Hardy-Littlewood*[232]) und *Landau*[233]) wurde gezeigt, daß dieser Satz mit dem folgenden (unbewiesenen) Analogon der *Riemann*schen Vermutung äquivalent ist: „Die zum Nicht-Hauptcharakter modulo 4 gehörige L-Funktion ist für $\sigma > \frac{1}{2}$ von Null verschieden.“

Die allgemeine *Tschebyschef*sche Aussage: „es gibt viel mehr Primzahlen von der Form $4n+3$ als von der Form $4n+1$“ kann also jedenfalls nur in ziemlich beschränktem Maße wahr sein[234]) und

232) *G. H. Hardy* und *J. E. Littlewood*, a. a. O. 31b). (Aus $L(s) \neq 0$ für $\sigma > \frac{1}{2}$ folgt (80)).

233) *E. Landau*, a. a. O. 178). (Aus (80) folgt $L(s) \neq 0$ für $\sigma > \frac{1}{2}$) und Über einige ältere Vermutungen und Behauptungen in der Primzahltheorie, zweite Abhandl., Math. Ztschr. 1 (1918), p. 213—219.

234) *E. Landau*, a. a. O. 178), p. 6, bemerkt, daß aus der Behauptung (80) von *Tschebyschef*, $\pi_{4,3}(x) - \pi_{4,1}(x) = O\bigl(x^{\frac{1}{2}}\log x\bigr)$ folgt. Die Differenz $\pi_{4,3} - \pi_{4,1}$

ist z. B. in der Fassung (80), die wenigstens wahr sein *könnte*, noch nicht bewiesen.

Die Resultate von *Phragmén* und *Landau* betreffend die *Tschebyschef*sche Behauptung (79) wurden von *Landau*[235]) für beliebige Moduln k (an der Stelle von 4) verallgemeinert.

31. Andere Primzahlprobleme. *Summen über Primzahlen.* Daß unter den n ersten ganzen Zahlen annäherungsweise $Li(n)$ Primzahlen vorkommen, kann wegen

$$Li(n) = \sum_{2}^{n} \frac{1}{\log n} + O(1)$$

in ungenauer Weise so ausgedrückt werden: „die Wahrscheinlichkeit, daß die beliebig gewählte Zahl n Primzahl ist, ist gleich $\frac{1}{\log n}$." Man wird hiernach vermuten, daß die beiden Reihen

$$\sum_{p} F(p) \quad \text{und} \quad \sum_{n} \frac{F(n)}{\log n} \tag{81}$$

sich mehr oder weniger ähnlich verhalten müssen. In der Tat besagt ja der Primzahlsatz

$$\sum_{p \leqq x} 1 \sim \sum_{n=2}^{x} \frac{1}{\log n}, \qquad (F(t) = 1)$$

bzw.

$$\sum_{p \leqq x} \log p \sim \sum_{n=2}^{x} 1, \qquad (F(t) = \log t).$$

Nach *Tschebyschef*[236]) sind die Reihen (81) gleichzeitig konvergent oder divergent, sobald $\frac{F(n)}{\log n}$ für hinreichend große n positiv und nie zunehmend ist. *Mertens*[237]) beweist

$$\sum_{p \leqq x} \frac{\log p}{p} = \sum_{n=2}^{x} \frac{1}{n} + O(1) = \log x + O(1) \tag{82}$$

und

$$\sum_{p \leqq x} \frac{1}{p} = \sum_{n=2}^{x} \frac{1}{n \log n} + A + O\left(\frac{1}{\log x}\right) = \log\log x + B + O\left(\frac{1}{\log x}\right) \tag{83}$$

---

wäre also nach (80) absolut genommen kleiner, als bisher bekannt war — nämlich $O\left(xe^{-\alpha\sqrt{\log x}}\right)$ — eine Folgerung, die ja in der entgegengesetzten Richtung von *Tschebyschefs* Interpretation seiner Behauptung liegt.

235) *E. Landau*, Handbuch, § 200.

236) *P. Tschebyschef*, a. a. O. 155 b).

237) *F. Mertens*, Ein Beitrag zur analytischen Zahlentheorie, Crelles J. 78 (1874), p. 46—62.

53*

mit konstantem A und B. Von *de la Vallée Poussin*[173]) wurde (82) zu

$$\sum_{p \leqq x} \frac{\log p}{p} = \log x - C - \sum_{p} \frac{\log p}{p(p-1)} + O\left(e^{-\alpha\sqrt{\log x}}\right)$$

verschärft, wo C die *Euler*sche Konstante bezeichnet. Die Abschätzung des Restgliedes in (83) kann in ähnlicher Weise verschärft werden. Hieraus folgt speziell

$$\lim_{x \to \infty} \left(\log x - \sum_{p \leqq x} \frac{\log p}{p-1}\right) = \lim_{n \to \infty} \left(\sum_{n=1}^{x} \frac{1}{n} - \log x\right) = C$$

und (vgl. (59))
$$\sum_{n=1}^{\infty} \frac{\Lambda(n) - 1}{n} = -2C.$$

Landau[238]) gibt verschiedene Sätze über Summen der Gestalt $\sum\limits_{p \leqq x} F(p)$ und $\sum\limits_{p \leqq x} F(p, x)$ und bespricht insbesondere die Möglichkeit, von einer Formel elementar zu den andern zu gelangen, ohne jedes Mal die Theorie der Zetafunktion zu benutzen (vgl. hierzu Nr. **33**). *Mertens*[237]) hat (82) und (83) auch für die Primzahlen einer arithmetischen Reihe verallgemeinert.

Die Konvergenz von $\sum p^{-s}$ und $\sum \chi(p) p^{-s}$ auf der Geraden $\sigma = 1$ wurde schon oben besprochen (vgl. Nr. **14** und Nr. **29**, Fußnote 219). Diese Reihen stellen für $\sigma > 1$ die Funktionen dar

$$\sum_{1}^{\infty} \frac{\mu(n)}{n} \log \zeta(ns) \quad \text{bzw.} \quad \sum_{1}^{\infty} \frac{\mu(n)}{n} \log L(ns, \chi^n),$$

die über $\sigma = 1$ hinaus bis zu $\sigma = 0$, aber nicht weiter, analytisch fortgesetzt werden können.[239]) — Die Funktion

$$F(z) = \sum_{p} \frac{z^p}{p}$$

wird bei Annäherung an einen „rationalen" Punkt $z = e^{\frac{2m\pi i}{n}}$ des Einheitskreises unendlich groß, sofern n eine quadratfreie Zahl ist.

238) *E. Landau,* Sur quelques problèmes relatifs à la distribution des nombres premiers, Bull. Soc. math. France 28 (1900), p. 25—38; Handbuch § 55—56 (vgl. auch p. 889).

239) *E. Landau* und *A. Walfisz,* Über die Nichtfortsetzbarkeit einiger durch Dirichletsche Reihen definierter Funktionen, Palermo Rend. 44 (1920), p. 82—86.

Vgl. auch *J. C. Kluyver*, Benaderingsformules betreffende de priemgetallen beneden eene gegeven grens, Akad. Wetensk. Amsterdam, Verslagen 8 (1900), p. 672—682 und *E. Landau*, a. a. O. 78).

Fatou[240]) schließt hieraus, daß $F(z)$ und

$$zF'(z) = \sum_p z^p$$

nicht über den Einheitskreis fortgesetzt werden können. Nach einer Bemerkung von *Landau*[241]) folgt dies auch direkt aus neueren Sätzen über die *Taylor*sche Reihe.

Die n^{te} *Primzahl und die Differenz* $p_{n+1} - p_n$. Wenn p_n die n^{te} Primzahl bezeichnet, so folgt aus der Gleichung

$$n = \pi(p_n) = Li(p_n) + O\left(p_n e^{-\alpha\sqrt{\log p_n}}\right)$$

durch Inversion

$$p_n = Li^{-1}(n) + O\left(n \log^2 n\, e^{-\alpha\sqrt{\log n}}\right),$$

wo $Li^{-1}(x)$ die zu $Li(x)$ inverse Funktion bedeutet. Insbesondere ist[242]) also

$$p_n \sim n \log n.$$

Tschebyschef[236]) bewies den früher von *Bertrand*[243]) vermuteten und empirisch bestätigten Satz, daß von einer gewissen Stelle an zwischen x und $2x$ wenigstens eine Primzahl liegt, d. h. daß für große n immer $\frac{p_{n+1}}{p_n} < 2$ ist. Der Primzahlsatz gibt sogar[244])

$$\lim_{n\to\infty} \frac{p_{n+1}}{p_n} = 1$$

oder

$$p_{n+1} - p_n = o(p_n).$$

Aus der genauen Restabschätzung (53) zum Primzahlsatz folgt[245])

$$p_{n+1} - p_n = O\left(p_n e^{-\alpha\sqrt{\log p_n}}\right),$$

240) *P. Fatou*, Sur les séries entières à coefficients entiers, Paris C. R. 138 (1904), p. 342—344.

241) *E. Landau*, a. a. O.78). Dieselbe Bemerkung hat auch *F. Carlson* gemacht: Über Potenzreihen mit ganzzahligen Koeffizienten, Math. Ztschr. 9 (1921), p. 1—13.

242) Mit der asymptotischen Darstellung von p_n beschäftigten sich u. a. *M. Perwuschin*, Formule pour la détermination approximative des nombres premiers etc., Verhandl. Math.-Kongr. Zürich 1897, Leipzig 1898, p. 166—167; *E. Cesàro*, Sur une formule empirique de M. Pervouchine, Paris C. R. 119 (1894), p. 848—849; *M. Cipolla*, La determinazione assintotica dell' n^{imo} numero primo, Rend. Accad. Sc. Fis. Mat. Napoli (3) 8 (1902), p. 132—166. Vgl. auch *E. Landau*, Handbuch, § 57.

243) *J. Bertrand*, Mémoire sur le nombre de valeurs que peut prendre une fonction quand on y permute les lettres qu' elle renferme, J. Éc. Polyt. 18 (1845), p. 123—140.

244) Ein direkter Beweis dieser Tatsache, der nicht zugleich den Primzahlsatz liefert, scheint nicht bekannt zu sein. Vgl. *E. Landau*, Gelöste und ungelöste Probleme aus der Theorie der Primzahlverteilung und der Riemannschen Zetafunktion, Proc. Fifth. Intern. Congr. Math., Cambridge 1913, 1 p. 93—108.

245) Vgl. *Ch. de la Vallée Poussin*, a. a. O.173), p. 55.

was die beste mit Sicherheit bekannte Abschätzung darstellt. Wenn die *Riemann*sche Vermutung vorausgesetzt wird, folgt aus (57)

$$p_{n+1} - p_n = O(\sqrt{p_n}\log^2 p_n),$$

wo nach *Cramér*[246]) $\log^2 p_n$ durch $\log p_n$ ersetzt werden kann. Es gibt demnach, wenn die *Riemann*sche Vermutung richtig ist, eine Zahl c, so daß für $n = 2, 3, \ldots$ zwischen n^2 und $(n + c\log n)^2$ immer wenigstens eine Primzahl liegt. *Oppermann*[247]) behauptete, daß dasselbe von dem Intervall $(n^2, (n+1)^2)$ gilt; das ist aber bisher nicht entschieden. *Piltz*[248]) hat sogar die Behauptung

$$p_{n+1} - p_n = O(p_n^{\varepsilon})$$

für jedes $\varepsilon > 0$, ausgesprochen; in dieser Hinsicht ist nur bekannt[249]), daß die Anzahl der $p_n \leqq x$, die der Ungleichung

$$p_{n+1} - p_n > p_n^k, \qquad (0 < k \leqq \tfrac{1}{2})$$

genügen, unter Voraussetzung der *Riemann*schen Vermutung von der Form $O\left(x^{1-\frac{3}{2}k+\varepsilon}\right)$ ist. — Im Mittel muß die Differenz $\delta_n = p_{n+1} - p_n$ von der Ordnung $\log p_n$ sein, denn es gilt

$$\frac{1}{n}(\delta_1 + \delta_2 + \cdots \delta_n) = \frac{1}{n}(p_{n+1} - 2) \sim \log p_n.$$

Nach unten ist keine bessere Abschätzung als die triviale $\delta_n \geqq 2$ für $n > 1$ bekannt; verschiedene Verfasser[250]) vermuten, daß in der Tat

(84) $$p_{n+1} - p_n = 2$$

246) *H. Cramér,* a. a. O. 186).

247) *L. Oppermann,* Om vor Kundskab om Primtallenes Mœngde mellem givne Grœnser, Overs. Danske Vidensk. Selsk. Forh. 1882, p. 169—179.

248) *A. Piltz,* a. a. O. 180), p. 46.

249) *H. Cramér,* On the distribution of primes, Proc. Cambr. Phil. Soc. 20 (1921), p. 272—280.

250) Vgl. *J. J. Sylvester,* On the partition of an even number into two primes, Proc. London math. Soc. (1) 4 (1871), p. 4—6 und Collected Math. Pap. 2, p. 709—711; On the Goldbach-Euler Theorem regarding prime numbers, Nature 55 (1896—1897), p. 196—197, 269 und Pap. 4. p. 734—737; *P. Stäckel,* Über Goldbachs empirisches Theorem etc., Gött. Nachr. 1896, p. 292—299; Die Darstellung der geraden Zahlen als Summen von zwei Primzahlen, Sitzungsber. Akad. Heidelberg 1916; Die Lückenzahlen r^{ter} Stufe und die Darstellung der geraden Zahlen als Summen und Differenzen ungerader Primzahlen. I—III, Sitzungsber. Akad. Heidelberg 1917—1918; *J. Merlin,* Un travail sur les nombres premiers, Bull. sc. math. (2) 39 (1915), p. 121—136; *V. Brun,* Über das Goldbachsche Gesetz und die Anzahl der Primzahlpaare, Arch. for. Math. og Naturv., Kristiania 34, Nr. 8 (1915); Sur les nombres premiers de la forme $ap + b$, ebenda 34, Nr. 14 (1917), *G. H. Hardy* und *J. E. Littlewood,* Note on Messrs Shah and Wilson's paper entitled: On an empirical formula connected with Goldbach's

für unendlich viele n gilt, und sogar daß

$$h(x) \sim a\frac{x}{\log^2 x}$$

mit konstantem a ist, wenn $h(x)$ die Anzahl der $p_n \leqq x$ bedeutet, die (84) genügen. *Brun*[251]) beweist

$$h(x) = O\left(\frac{x}{\log^2 x}\right).$$

Der Satz von Goldbach und verwandte Fragen. Goldbach[252]) sprach im Jahre 1742 den bis jetzt unbewiesenen Satz aus: „Jede gerade Zahl kann als Summe von zwei Primzahlen dargestellt werden.“ Verschiedene Verfasser[250]) vermuteten, daß die Anzahl $G(n)$ solcher Darstellungen einer geraden Zahl n sogar mit n ins Unendliche wächst, und zwar so, daß für alle geraden n

$$G(n) > b\frac{n}{\log^2 n}$$

mit konstantem b gilt.[253]) *Hardy* und *Littlewood*[250]) verallgemeinern das Problem und greifen es zuerst mit analytischen Mitteln an, indem sie in der Potenzreihe

$$f_k(z) = \sum_{n=1}^{\infty} a_n^{(k)} z^n = \left(\sum_p \log p\, z^p\right)^k,$$

(die über den Einheitskreis nicht fortsetzbar ist) ein beliebiges $a_n^{(k)}$ durch das Integral

$$a_n^{(k)} = \frac{1}{2\pi i}\int\limits_{|z|=r<1} \frac{f_k(z)}{z^{n+1}}\,dz$$

ausdrücken, um dann das Verhalten von $a_n^{(k)}$ für große n zu untersuchen. (Auf diese Methode kommen wir in Nr. 38 zurück.) Die Be-

theorem, Proc. Cambr. Phil. Soc. 19 (1919), p. 245—254; Some problems of Partitio Numerorum; III: On the expression of a number as a sum of primes, Acta math. 44 (1922), p. 1—70.

251) *V. Brun*, Le crible d'Eratosthène et le théorème de Goldbach, Vidensk. selsk. Skrifter, Mat-naturv. Kl. Kristiania 1920, Nr. 3 und Paris C. R. 168 (1919), p. 544—546. Vgl. auch: La série $\frac{1}{5}+\frac{1}{7}+\cdots$ où les dénominateurs sont „nombres premiers jumeaux“ est convergente ou finie, Bull. sc. Math. (2) 43 (1919), p. 1—9.

252) Vgl. Briefwechsel zwischen *Euler* und *Goldbach* bei *P. H. Fuss*, Correspondance math. phys. 1, St. Petersbourg 1843, p. 127, 135. Vgl. in bezug auf die ältere Geschichte des Satzes *L. E. Dickson*, a. a. O. 226), p. 421—425. Über die numerische Prüfung des Satzes vgl. z. B. *P. Stäckel*, a. a. O. 250). — Für $n = 2$ kann der Satz offenbar nur richtig sein, wenn 1 als Primzahl mitgezählt wird.

253) *E. Landau*, Über die zahlentheoretische Funktion $\varphi(n)$ und ihre Beziehung zum Goldbachschen Satz, Gött. Nachr. 1900, p. 177—186, zeigt, daß $G(2) + \cdots G(n) \sim \frac{n^2}{2\log^2 n}$ ist. Hieraus folgt, daß eine früher von *Stäckel*, a. a. O. 250), vorgeschlagene Formel falsch ist.

hauptung von *Goldbach*, $a_n^{(2)} > 0$ für alle geraden $n > 2$, läßt sich zwar nicht beweisen, es wird aber die Formel[254])

$$G(n) \sim c \frac{n}{\log^2 n} \prod_q \frac{q-1}{q-2}, \qquad (n \text{ gerade})$$

als wahrscheinlich hingestellt. Hierin ist c konstant, und q durchläuft die ungeraden Primteiler von n. Wenn die (unbewiesene) Annahme gemacht wird, daß *die obere Grenze der reellen Teile der Nullstellen von* $\zeta(s)$ *und von allen L-Funktionen kleiner als* $\frac{3}{4}$ *ist*, so läßt sich der folgende Satz beweisen: „Jede hinreichend große *ungerade* Zahl kann als Summe von *drei* Primzahlen dargestellt werden." — *Brun*[251]) beweist durch Anwendung einer Modifikation des sog. Siebverfahrens von *Eratosthenes* den Satz: „Jede hinreichend große gerade Zahl kann als Summe von zwei ganzen Zahlen dargestellt werden, die höchstens je neun Primfaktoren enthalten." Die beiden letztgenannten Sätze sind offenbar direkte Folgerungen aus dem *Goldbach*schen.

Das Problem, die Bedingungen für die Lösbarkeit einer unbestimmten Gleichung $ax + by + c = 0$ mittelst zweier Primzahlen x und y zu finden, ist eine Verallgemeinerung des *Goldbach*schen; es wurde auch von den oben erwähnten Verfassern behandelt. Mit der *Hardy-Littlewood*schen Methode lassen sich endlich auch verschiedene Probleme der Art: „Gibt es unendlich viele Primzahlen von der Form $n^2 + 1$, von der Form $n'^3 + n''^3 + n'''^3$," usw., angreifen. Auch hier läßt sich nichts beweisen, die Methode führt aber auf gewisse asymptotische Formeln, die in mehreren Fällen mit gutem Erfolg numerisch geprüft wurden.

IV. Weitere zahlentheoretische Funktionen.[255])

32. Die Funktionen $\mu(n)$, $\lambda(n)$ und $\varphi(n)$. Für $\sigma > 1$ gilt (vgl. Nr. 22)

$$\sum_1^\infty \frac{\mu(n)}{n^s} = \frac{1}{\zeta(s)}. \tag{85}$$

Die für $\sigma > 1$ unbedingt konvergente Reihe $\sum \mu(n) n^{-s}$ stellt also eine für $\sigma \geqq 1$ reguläre Funktion dar; daß die Reihe auch noch für

254) Mehr oder weniger ähnliche Formeln waren von den oben erwähnten Verfassern schon früher vorgeschlagen worden. Die *Hardy-Littlewood*sche Formel wurde von *N. M. Shah* und *B. M. Wilson* numerisch geprüft: On an empirical formula connected with Goldbach's theorem, Proc. Cambr. Phil. Soc. 19 (1919), p. 238—244.

255) Betreffs älterer Untersuchungen zu diesem Kapitel sei auf I C 3 verwiesen.

$s = 1$ konvergiert, hat schon *Euler*[256]) vermutet. Dies wurde von *v. Mangoldt*[257]) unter Benutzung der *Hadamard*schen Sätze über die Produktzerlegung von $(s-1)\zeta(s)$ (vgl. Nr. 15) bewiesen; nach (85) ist dann

$$\sum_1^\infty \frac{\mu(n)}{n} = 0, \quad \text{d. h.} \quad g(x) = \sum_1^x \frac{\mu(n)}{n} = o(1).$$

Landau[258]) zeigt, daß dieses Resultat auch elementar aus dem Primzahlsatz abgeleitet werden kann (vgl. Nr. 33). Eine unmittelbare Folgerung ist

$$M(x) = \sum_1^x \mu(n) = o(x),$$

und man kann nun nach dem Konvergenzsatz von *M. Riesz* (vgl. Nr. 5) schließen, daß (85) auf der ganzen Geraden $\sigma = 1$ gültig bleibt.[259]) *Landau*[260]) hat sogar die Konvergenz von

$$\sum_1^\infty \frac{\mu(n)(\log n)^q}{n^{1+ti}}$$

für beliebige reelle q und t festgestellt. Er[261]) gab — mit seiner bei dem Primzahlsatz angewandten Methode — die Abschätzungen

$$(86)\quad \begin{cases} M(x) = O\left(x e^{-\alpha\sqrt{\log x}}\right) \\ g(x) = O\left(e^{-\alpha\sqrt{\log x}}\right) \\ \sum_1^x \frac{\mu(n)\log n}{n} = -1 + O\left(e^{-\alpha\sqrt{\log x}}\right) \end{cases}$$

---

256) *L. Euler,* Introductio in analysin infinitorum, 1, Lausanne 1748, p. 229.

257) *H. v. Mangoldt,* Beweis der Gleichung $\sum_1^\infty \frac{\mu(k)}{k} = 0$, Sitzungsber. Akad. Berlin 1897, p. 835—852.

258) *E. Landau,* Neuer Beweis der Gleichung $\sum_1^\infty \frac{\mu(k)}{k} = 0$, Diss. Berlin 1899.

259) Vgl. *Landau,* a. a. O. 21). Das war natürlich nicht der erste Beweis dieses Satzes.

260) *E. Landau,* Über die zahlentheoretische Funktion $\mu(k)$, Sitzungsber. Akad. Wien 112, Abt. 2a (1903), p. 537—570. Die Konvergenz von $\sum \frac{\mu(n)\log n}{n}$ wurde schon von *A. F. Möbius* vermutet: Über eine besondere Art von Umkehrung der Reihen, Crelles J. 9 (1832), p. 105—123 und Werke 4 (1887), p. 589—612.

261) *E. Landau,* a. a. O. 29) und Handbuch, § 163—164. *Ch. de la Vallée Poussin,* a. a. O. 173), hatte eine unschärfere Abschätzung gegeben.

und verallgemeinerte[262]) alle diese Resultate für den Fall, daß n nur die Zahlen einer arithmetischen Reihe durchläuft.

Wie bei dem Primzahlsatz, so ist bei den Gleichungen (86) die Frage nach der möglichen Verschärfung der Abschätzungen eng mit der *Riemann*schen Vermutung verbunden. Von *Stieltjes*[263]) und *Mertens*[264]) wurde

(87) $$M(x) = O(\sqrt{x})$$

vermutet; *Stieltjes* behauptete in der Tat auf diesem Wege die *Riemann*sche Vermutung bewiesen zu haben, denn aus (87) würde (vgl. Nr. 2) die Konvergenz von $\sum\limits_1^\infty \mu(n) n^{-s}$ für $\sigma > \frac{1}{2}$, und damit die *Riemann*sche Vermutung, folgen. — Daß auch umgekehrt aus der *Riemann*schen Vermutung die Konvergenz von $\sum\limits_1^\infty \mu(n) n^{-s}$ für $\sigma > \frac{1}{2}$ folgt, wurde zuerst von *Littlewood*[138]) im Laufe seiner in Nr. **20** besprochenen Untersuchungen über die Zetafunktion bewiesen. Demnach ist die *Riemann*sche Vermutung mit der Behauptung

(88) $$M(x) = O\left(x^{\frac{1}{2}+\varepsilon}\right) \quad \text{für jedes} \quad \varepsilon > 0$$

vollständig äquivalent.[265]) Die weitere Vermutung von *Stieltjes*[263]) daß $\sum\limits_1^\infty \mu(n) n^{-s}$ auch noch für $s = \frac{1}{2}$ konvergiert, ist aber nach *Landau*[178]) sicher nicht richtig. A fortiori kann also (88) für kein negatives ε gelten.

Aus (85) folgt $\sum\limits_1^\infty n^{-s} \cdot \sum\limits_1^\infty \mu(n) n^{-s} = 1$ und hieraus für jedes ganze $n > 1$

$$\sum_{d \mid n} \mu(d) = 0,$$

262) Vgl. *J. C. Kluyver*, Reeksen, afgeleid uit de reeks $\sum \frac{\mu(m)}{m}$, Akad. Wetensk. Amsterdam, Verslagen 12 (1904), p. 432—439, und *E. Landau,* Bemerkungen zu der Abhandlung von Herrn Kluyver etc., ebenda 13 (1905), p. 71—83, Handbuch, § 169—175.

263) *T. J. Stieltjes*, a. a. O.160), Lettre 79 und Sur une fonction uniforme, Paris C. R. 101 (1885), p. 153—154.

264) *F. Mertens*, Über eine zahlentheoretische Funktion, Sitzungsber. Akad. Wien 106, Abt. 2a (1897), p. 761—830. Über die numerische Prüfung dieser Vermutung vgl. etwa *R. D. v. Sterneck*, Sitzungsber. Akad. Wien 110, Abt. 2a (1901), p. 1053—1102.

265) Aus (87) würde dagegen mehr als die *Riemann*sche Vermutung folgen, z. B. daß alle Wurzeln von $\zeta(s)$ einfach sind. Vgl. auch *H. Cramér* und *E. Landau*, Über die Zetafunktion auf der Mittellinie des kritischen Streifens, Arkiv för Mat., Astr. och Fys. 15 (1921), Nr. 28.

sowie für jedes $x \geq 1$

$$\sum_{n=1}^{x} \mu(n)\left[\frac{x}{n}\right] = 1.$$

Auf diesen Eigenschaften von $\mu(n)$ beruhen die sog. zahlentheoretischen Umkehrungsformeln. Es läßt sich z. B. die Gleichung (67), Nr. 28, leicht aus ihnen ableiten.

Die Funktion $\lambda(n)$ ist durch

$$\sum_{1}^{\infty} \frac{\lambda(n)}{n^s} = \frac{\zeta(2s)}{\zeta(s)} = \sum_{1}^{\infty} \frac{1}{n^{2s}} \cdot \sum_{1}^{\infty} \frac{\mu(n)}{n^s}, \qquad (\sigma > 1)$$

definiert; es folgt hieraus

$$\sum_{1}^{x} \lambda(n) = \sum_{1}^{\sqrt{x}} M\left(\frac{x}{n^2}\right),$$

und mit Hilfe dieser Identität lassen sich alle obigen Ergebnisse für $\lambda(n)$ verallgemeinern.[266]) Insbesondere zeigt es sich, daß es unter den N ersten ganzen Zahlen asymptotisch ebenso viele gibt, die aus einer geraden, als solche, die aus einer ungeraden Anzahl von Primfaktoren bestehen.[267])

Für die *Euler*sche Funktion $\varphi(n)$ gilt offenbar immer $\varphi(n) \leq n-1$, und sobald n eine Primzahl ist, muß hier das Gleichheitszeichen benutzt werden. Andererseits beweist *Landau*[268])

$$\liminf_{n\to\infty} \frac{\varphi(n) \log\log n}{n} = e^{-C}.$$

Daß die summatorische Funktion $\Phi(x)$ asymptotisch gleich $\frac{3}{\pi^2}x^2$ ist, war schon *Dirichlet*[269]) bekannt; nach *Mertens*[270]) gilt sogar

$$\Phi(x) = \frac{3}{\pi^2}x^2 + O(x\log x),$$

was völlig elementar bewiesen werden kann. Merkwürdigerweise ist es bisher nicht gelungen, aus der Beziehung $\sum_{1}^{\infty} \varphi(n) n^{-s} = \frac{\zeta(s-1)}{\zeta(s)}$ mit analytischen Mitteln eine bessere Abschätzung des Restgliedes zu

266) *E. Landau*, Handbuch, § 166—167, 169—172.

267) *E. Landau*, Handbuch, p. 571.

268) *E. Landau*, Über den Verlauf der zahlentheoretischen Funktion $\varphi(x)$, Arch. Math. Phys. (3) 5 (1903), p. 86—91.

269) *P. G. Lejeune-Dirichlet*, Über die Bestimmung der mittleren Werte in der Zahlentheorie, Abhandl. Akad. Berlin 1849, math. Abhandl. p. 69—83 (Werke 2, p. 49—66).

270) *F. Mertens*, Über einige asymptotische Gesetze der Zahlentheorie, Crelles J. 77 (1874), p. 289—338.

erhalten.[271]) — Nach *Landau*[272]) gilt

$$\sum_{n=1}^{x} \frac{1}{\varphi(n)} \sim \frac{315\,\zeta(3)}{2\pi^4} \log x.$$

Landau[273]) beweist mit Hilfe der Theorie der Multiplikation *Dirichlet*scher Reihen (vgl. Nr. **12**) die Konvergenz verschiedener hierhergehöriger Reihen, z. B.

$$\sum_{1}^{\infty} \frac{\chi(n)\,\mu(n)}{n}, \quad \sum_{1}^{\infty} \frac{\chi(n)\,\lambda(n)}{n}, \quad \sum_{1}^{\infty} \frac{\chi(n)\varphi(n)}{n^2},$$

wo $\chi(n)$ ein beliebiger Charakter (bei der letztgenannten Reihe jedoch nicht der Hauptcharakter) nach einem beliebigen Modul ist.

33. Zusammenhangssätze. Die im vorhergehenden erwähnten tieferen Ergebnisse der analytischen Zahlentheorie waren alle mit Hilfe der Theorie der Zetafunktion, bzw. deren Verallgemeinerungen, erreicht. Für die systematische Darstellung der Theorie erscheint es wichtig, die verschiedenen Hauptresultate in bezug auf ihre „Tiefe" zu vergleichen und insbesondere die Möglichkeit zu untersuchen, aus einem von ihnen die anderen *elementar* abzuleiten, ohne nochmals die transzendenten Methoden zu benutzen.

Die wichtigsten in dieser Richtung durch *Landau*[274]) und *Axer*[275]) bekannten Tatsachen lassen sich dahin zusammenfassen, daß die vier Gleichungen

$$\psi(x) = \sum_{1}^{x} \Lambda(n) = x + o(x), \tag{89}$$

$$\sum_{1}^{\infty} \frac{\Lambda(n) - 1}{n} = -2C, \tag{90}$$

$$M(x) = \sum_{1}^{x} \mu(n) = o(x), \tag{91}$$

$$\sum_{1}^{\infty} \frac{\mu(n)}{n} = 0, \tag{92}$$

---

271) Da $\Phi(x)$ unendlich viele Sprünge von der Größenordnung x macht, kann das Restglied jedenfalls nicht von niedrigerer Größenordnung als $O(x)$ sein.

272) *E. Landau*, a. a. O. 253).

273) *E. Landau*, a. a. O. 78) und Handbuch, § 184—195.

274) *E. Landau*, a. a. O. 258) und 21), sowie Über die Äquivalenz zweier Hauptsätze der analytischen Zahlentheorie, Sitzungsber. Akad. Wien **120**, Abt. 2a (1911), p. 973—988.

275) *A. Axer*, Beitrag zur Kenntnis der zahlentheoretischen Funktionen $\mu(n)$ und $\lambda(n)$, Prace Mat. Fiz. 21 (1910), p. 65—95.

alle in dem Sinne äquivalent sind, daß aus irgendeiner von ihnen die drei übrigen *elementar* folgen. [Nach den Ergebnissen von Nr. **23** können natürlich auch die (89) entsprechenden Formeln mit $\Pi(x)$, $\pi(x)$ und $\vartheta(x)$ hinzugesetzt werden.] In (90) und (92) ist nur die *Konvergenz* der betreffenden Reihe wesentlich, ist diese einmal festgestellt, so folgt die Wertbestimmung aus einfachen Stetigkeitsbetrachtungen. — Etwas tiefer liegt der Satz

$$\sum_{1}^{\infty} \frac{\mu(n)\log n}{n} = -1,$$

der mit einer schärferen Form von (92), nämlich mit

$$\sum_{1}^{x} \frac{\mu(n)}{n} = o\left(\frac{1}{\log x}\right),$$

äquivalent ist.[276]) — Der Übergang (durch partielle Summation) von (90) zu (89), bzw. von (92) zu (91), ist trivial; die anderen Übergänge folgen aus gewissen allgemeinen Grenzwertsätzen. *Landau*[277]) gibt einen Satz, aus dem alle jene Übergänge durch Spezialisierung folgen. *Hardy* und *Littlewood*[278]) zeigen, daß die Übergänge auch mit Hilfe von „*Tauber*schen" Sätzen (vgl. Nr. **5**) über die „*Lambert*schen Reihen"

$$\sum_{1}^{\infty} \frac{a_n}{e^{ns}-1}$$

ausgeführt werden können.

34. Teilerprobleme. Die Funktionen $d(n)$ und $\sigma(n)$, die Anzahl und die Summe der Teiler von n, sind vielfach untersucht worden. Über die Größenordnung dieser Funktionen ist zunächst trivial, daß immer

$$d(n) \geqq 2, \quad \sigma(n) \geqq n+1$$

ist, sowie daß in beiden Beziehungen unendlich oft (nämlich für alle Primzahlen) das Gleichheitszeichen gilt. Andererseits beweisen *Wigert*[279]) und *Gronwall*[280])

276) *A. Axer*, Über einige Grenzwertsätze, Sitzungsber. Akad. Wien 120, Abt. 2a (1911), p. 1253—1298.

277) *E. Landau*, Über einige neuere Grenzwertsätze, Palermo Rend. 34 (1912), p. 121—131.

278) *G. H. Hardy* und *J. E. Littlewood*, On a Tauberian theorem for Lambert's series and some fundamental theorems in the analytic theory of numbers, Proc. London math. Soc. (2) 19 (1919), p. 21—29.

279) *S. Wigert*, a) Sur l'ordre de grandeur du nombre des diviseurs d'un entier, Arkiv för Mat., Astr och Fys. 3 (1906—1907), No. 18; b) Sur quelques fonctions arithmétiques, Acta Math. 37 (1914), p. 113—140.

280) *H. Gronwall*, Some asymptotic expressions in the theory of numbers, Trans. Amer. math. Soc. 14 (1913), p. 113—122.

$$\limsup_{n \to \infty} \frac{\log d(n) \cdot \log\log n}{\log n} = \log 2,$$

$$\limsup_{n \to \infty} \frac{\sigma(n)}{n \log\log n} = e^C.$$

Gronwall gibt auch entsprechende Beziehungen für $\sigma_\alpha(n)$, die Summe der α^{ten} Potenzen der Teiler von n. *Ramanujan*[281]) beweist viele ins einzelne gehende Sätze über den Verlauf der Funktion $d(n)$. Er zeigt insbesondere, daß $d(n)$, wenn die *Riemann*sche Vermutung richtig ist, die „maximale Größenordnung“

$$2^{Li(\log n) + O(\log^\alpha n)} \qquad (\alpha < 1)$$

hat. Er nennt eine Zahl n „highly composite“, wenn $d(n) > d(\nu)$ für $\nu = 1, 2, \ldots n-1$ ist, und zeigt, wie man mit elementaren Mitteln erstaunend genaue Resultate über die Reihe der Exponenten in der Darstellung einer solchen Zahl als Produkt von Primzahlpotenzen ableiten kann. Er findet auch bemerkenswerte Beziehungen zwischen der Funktion $\sigma_\alpha(n)$ und gewissen trigonometrischen Summen; ein spezieller Fall hiervon lautet

$$\sigma(n) = \frac{\pi^2 n}{6} \sum_{\nu=1}^{\infty} \frac{c_\nu(n)}{\nu^2},$$

wo $c_\nu(n) = \sum_\mu \cos \frac{2\pi\mu n}{\nu}$ ist, und μ die $\varphi(\nu)$ zu ν teilerfremden ganzen positiven Zahlen $\leqq \nu$ durchläuft.

Die summatorische Funktion $D(x)$ gibt offenbar die Anzahl der *Gitterpunkte* (Punkte mit ganzzahligen Koordinaten) an, die in der (u, v)-Ebene dem Gebiet

$$(93) \qquad u > 0, \quad v > 0, \quad uv \leqq x$$

angehören; hieraus folgt leicht

$$D(x) = \sum_{n=1}^{x} \left[\frac{x}{n}\right] = 2 \sum_{n=1}^{\sqrt{x}} \left[\frac{x}{n}\right] - [\sqrt{x}]^2,$$

woraus die von *Dirichlet*[269]) gegebene Formel

$$D(x) = x(\log x + 2C - 1) + O(\sqrt{x})$$

---

281) *S. Ramanujan,* Highly composite numbers, Proc. London math. Soc. (2) 14 (1915), p. 347—409; On certain trigonometrical sums and their applications in the theory of numbers, Trans. Cambr. Phil. Soc. 22 (1918), p. 259—276. Vgl. auch: On certain arithmetical functions, Trans. Cambr. Phil. Soc. 22 (1916), p. 159—184, wo gewisse, die Funktion $\sigma_\alpha(n)$ enthaltende Summen untersucht werden.

gefolgert werden kann. Dieses Resultat wurde erst von *Voronoï*[282]) verschärft; er zieht in zweckmäßig gewählten Punkten der Hyperbel $uv = x$ die Tangenten, zerlegt dadurch das Gebiet (93) in mehrere Teilgebiete, schätzt die Anzahl der Gitterpunkte in jedem Teilgebiet ab und erhält

$$(94) \qquad D(x) = x(\log x + 2C - 1) + O\left(x^{\frac{1}{3}} \log x\right).$$

Neuerdings ist es *van der Corput*[306]) gelungen, die Abschätzung des Restgliedes sogar zu $O(x^M)$ zu verbessern, wo $M < \frac{33}{100}$ ist.

Schon vor *Voronoï* hatte *Pfeiffer*[283]) einen vermeintlichen Beweis von (94) — mit $O\left(x^{\frac{1}{3}+\varepsilon}\right)$ anstatt $O\left(x^{\frac{1}{3}} \log x\right)$ — veröffentlicht; seine Methode war freilich nicht einwandfrei, wurde aber von *Landau*[284]) umgearbeitet und u. a. zum Beweis von (94) benutzt. Diese „*Pfeiffer*sche Methode", auf die wir in Nr. 36 zurückkommen, beruht auf „reell-analytischer" Grundlage. Andererseits ist[285]) (vgl. Nr. 4 und 22)

$$(95) \qquad \overline{D}(x) = \frac{1}{2\pi i} \int\limits_{2-i\infty}^{2+i\infty} \frac{x^s}{s} (\zeta(s))^2 \, ds;$$

dieser für die Primzahltheorie grundlegende „komplex-analytische" Ansatz schien lange auf das Teilerproblem nicht anwendbar zu sein, es gelang jedoch *Landau*[286]) ihn zum Beweis von (94) zu benutzen. In (95) tritt die Zetafunktion nicht im Nenner auf; die Schwierigkeiten rühren daher nicht wie bei den Primzahlproblemen von den komplexen ζ-Nullstellen her, sie sind hier von ganz anderer Natur und sind hauptsächlich mit dem Aufsuchen einer oberen Grenze für das Integral

$$\int\limits_{-\varepsilon - Ti}^{-\varepsilon + Ti} \frac{x^s}{s} (\zeta(s))^2 \, ds \qquad (\varepsilon > 0)$$

verbunden, wobei T eine Funktion von x ist. *Landau*[286]) zeigt, daß

282) *G. Voronoï*, Sur un problème du calcul des fonctions asymptotiques, Crelles J. 126 (1903), p. 241—282.

283) *E. Pfeiffer*, Über die Periodizität in der Teilbarkeit der Zahlen und über die Verteilung der Klassen positiver quadratischer Formen auf ihre Determinanten, Jahresber. d. Pfeifferschen Lehr- und Erzieh.-Anstalt. Jena 1886, p. 1—21.

284) *E. Landau*, Die Bedeutung der Pfeifferschen Methode für die analytische Zahlentheorie, Sitzungsb. Akad. Wien 121, Abt. 2a (1912), p. 2195—2332.

285) Für ganzzahlige x muß wie oben der Hauptwert des Integrals genommen werden.

286) *E. Landau*, a) a. a. O. 224); b) Über die Anzahl der Gitterpunkte in gewissen Bereichen, zweite Abhandl., Gött. Nachr. 1915, p. 209—243; c) Über Dirichlets Teilerproblem, Sitzungsb. Akad. München 1915, p. 317—328.

diese Schwierigkeit bei einer ausgedehnten Klasse von Problemen überwunden werden kann, wo an der Stelle von $(\zeta(s))^2$ eine Funktion steht, die eine Funktionalgleichung vom Typus der *Riemann*schen besitzt und gewissen anderen Bedingungen genügt. Mit dieser Methode wurde z. B. der in Nr. 29 erwähnte Satz über die Multiplikation zweier L-Reihen bewiesen, der übrigens (94) — mit der Fehlerabschätzung $O\left(x^{\frac{1}{3}+\varepsilon}\right)$ — als Spezialfall enthält, da die Konvergenz von

$$(96) \qquad \sum_{1}^{\infty} \frac{d(n) - \log n - 2C}{n^s} = (\zeta(s))^2 + \zeta'(s) - 2C\zeta(s)$$

für $\sigma > \frac{1}{3}$ daraus folgt.[287])

Das Problem, die untere Grenze γ derjenigen α zu bestimmen, für welche

$$\Delta(x) = D(x) - x(\log x + 2C - 1) = O(x^\alpha)$$

gilt (γ ist also die Konvergenzabszisse von (96)), wird als „*Dirichlets* Teilerproblem" bezeichnet. Nach dem Obigen ist jedenfalls $\gamma < \frac{33}{100}$. Eine nicht triviale *untere* Abschätzung von γ hat *Hardy*[288]) gegeben, er beweist nämlich $\gamma \geqq \frac{1}{4}$. Er untersucht die Funktion

$$f(s) = \sum_{1}^{\infty} d(n) e^{-s\sqrt{n}} = \frac{1}{\pi i} \int_{2-i\infty}^{2+i\infty} \Gamma(2z) s^{-2z} \zeta^2(z)\, dz,$$

die in allen Punkten $s = \pm 4\pi i \sqrt{q}$ $(q = 1, 2, \ldots)$ algebraische Unendlichkeitsstellen von der Ordnung $\frac{3}{2}$ aufweist, während

$$f(s) + \frac{4(\log s - 1)}{s^2}$$

für $s = 0$ regulär ist. Hieraus folgt nach *Hardy* $\gamma \geqq \frac{1}{4}$ und sogar der schärfere Satz, daß bei zweckmäßiger Wahl einer positiven Konstanten K die Ungleichungen

$$(97) \qquad \begin{cases} \Delta(x) > K x^{\frac{1}{4}} \\ \Delta(x) < -K x^{\frac{1}{4}} \end{cases}$$

beide beliebig große Lösungen besitzen. *Hardy* deutet auch an, wie man durch die Anwendung der von *Littlewood* (vgl. Nr. 27 und 28) für die entsprechenden Probleme der Primzahltheorie geschaffenen

287) *Landau* gibt auch einen Beweis von (94) mit einer arithmetischen Methode, deren Grundgedanke von *Piltz* herrührt: Über Dirichlets Teilerproblem, Gött. Nachr. 1920, p. 13—32. Er hat auch (94) für den Fall verallgemeinert, daß nur solche Teiler, die einer gegebenen arithmetischen Reihe angehören, mitgezählt werden; vgl. a. a. O. 224) und 284).

288) *G. H. Hardy,* On Dirichlets Divisor Problem, Proc. London math. Soc. (2) 15 (1916), p. 1—25.

Methode in (97) sogar $x^{\frac{1}{4}}$ durch $(x \log x)^{\frac{1}{4}} \log\log x$ ersetzen kann. *Landau*[289]) beweist mit der vorhin erwähnten komplex-analytischen Methode einen allgemeinen Satz, der insbesondere $\gamma \geqq \frac{1}{4}$ ergibt.

Über γ ist also bis jetzt nur $\frac{1}{4} \leqq \gamma < \frac{33}{100}$ bekannt. *Im Mittel* ist aber $|\Delta(x)|$ von der Ordnung $x^{\frac{1}{4}}$; *Cramér*[290]) beweist nämlich (in Verschärfung früherer Resultate von *Hardy*[291]))

$$(98)\qquad \int_1^x (\Delta(t))^2\, dt = \frac{x^{\frac{3}{2}}}{6\pi^2} \sum_1^\infty \left(\frac{d(n)}{n^{\frac{3}{4}}}\right)^2 + O\left(x^{\frac{5}{4}+\varepsilon}\right)$$

und folgert hieraus

$$(99)\qquad \frac{1}{x}\int_1^x |\Delta(t)|\, dt = O\left(x^{\frac{1}{4}}\right).$$

Schließlich kennt man auch eine *explizite Formel* für die Funktion $\overline{D}(x)$; nach *Voronoï*[292]) gilt nämlich[293])

$$(100)\qquad \overline{D}(x) = x(\log x + 2C - 1) + \frac{1}{4} + \sqrt{x} \sum_1^\infty \frac{d(n)}{\sqrt{n}} \left(Y_1(4\pi\sqrt{nx}) - H_1(4\pi\sqrt{nx})\right),$$

289) *E. Landau*, a) Über die Anzahl der Gitterpunkte in gewissen Bereichen, dritte Abhandl., Gött. Nachr. 1917, p. 96—101; vgl. auch: b) Über die Heckesche Funktionalgleichung, ebenda 1917, p. 102—111.

290) *H. Cramér*, Über zwei Sätze des Herrn G. H. Hardy, Math. Ztschr. 15 (1922), p. 201—210.

291) *G. H. Hardy*, The average order of the arithmetical functions $P(x)$ and $\Delta(x)$, Proc. London math. Soc. (2) 15 (1916), p. 192—213; Additional note on two problems in the analytic theory of numbers, ebenda (2) 18 (1918), p. 201—204.

292) *G. Voronoï*, Sur une fonction transcendante et ses applications à la sommation de quelques séries, Ann. Éc. Norm. (3) 21 (1904), p. 207—268, 459—534.

293) In der folgenden Nummer machen wir über Formeln dieser Art einige allgemeine Bemerkungen. Eine Formel, die im wesentlichen mit (101) übereinstimmt, wurde schon 1891 — mit ungenügendem Beweis — von *L. Lorenz* gegeben: Analytiske Undersøgelser over Primtalmœngderne, Kgl. Danske Vidensk. Selsk. Skrifter, naturv. og math. Afd. (6) 5 (1889—1891), p. 427—450. Er entwickelt diese und sogar die entsprechenden Formeln für das *Piltz*sche Teilerproblem (s. u.) nach einer Methode, die im Grunde mit der — unstreng angewandten — *Pfeiffer*schen Methode identisch ist. Später wurde (100) von *Hardy* a. a. O. 288) unabhängig wiedergefunden. Einen Beweis von (100) mit der *Pfeiffer*schen Methode gab *W. Rogosinski*, Neue Anwendung der Pfeifferschen Methode bei Dirichlets Teilerproblem, Diss. Göttingen 1922. Vgl. auch *E. Landau*, Über Dirichlets Teilerproblem, zweite Mtlg., Gött. Nachr. 1922, p. 8—16 und *A. Walfisz*, a. a. O. 297).

wo $Y_1(v)$ die gewöhnliche „zweite Lösung" der *Bessel*schen Differentialgleichung bezeichnet und

$$H_1(v) = \frac{2}{\pi}\int_1^\infty \frac{te^{-vt}}{\sqrt{t^2-1}}\,dt \qquad (= O(e^{-v}))$$

eine *Hankel*sche Zylinderfunktion ist. Nach der bekannten asymptotischen Entwicklung von Y_1 hat man

$$(101)\qquad \overline{D}(x) = x(\log x + 2C - 1)$$
$$+ \frac{x^{\frac{1}{4}}}{\pi\sqrt{2}}\sum_1^\infty \frac{d(n)}{n^{\frac{3}{4}}}\cos\left(4\pi\sqrt{nx} - \tfrac{1}{4}\pi\right) + \tfrac{1}{4} + O\left(x^{-\frac{1}{4}}\right),$$

wo das Glied $O\left(x^{-\frac{1}{4}}\right)$ eine für $x \geqq 1$ stetige Funktion ist. Die Sprünge der Funktion $\overline{D}(x)$ rühren also von der in (101) auftretenden unendlichen Reihe her, die das „kritische Glied" von $\overline{D}(x)$ darstellt. *Voronoï*[292]) gibt auch analoge Formeln für $\sum_1^x d(n)(x-n)^k$, $k = 1, 2, \ldots$.[294])

Wigert[295]) untersucht die summatorischen Funktionen $S(x)$ und $\sum_1^x \frac{\sigma(n)}{n}$; er beweist

$$S(x) = \frac{\pi^2}{12}x^2 + x\Theta_1(x),$$

$$\sum_1^x \frac{\sigma(n)}{n} = \frac{\pi^2}{6}x - \frac{1}{2}\log x + \Theta_2(x),$$

wo für $\nu = 1, 2$

$$\limsup_{x\to\infty} \frac{|\Theta_\nu(x)|}{\log x} \leqq \frac{1}{4}$$

aber jedenfalls *nicht*

$$\Theta_\nu(x) = o(\log\log x)$$

294) *S. Wigert,* Sur la série de Lambert et son application à la théorie des nombres, Acta Math. 41 (1917), p. 197—218, und *E. Landau,* Gött. gel. Anz. 1915, p. 377—414, gaben einfachere Beweise für einen Teil der *Voronoï*schen Resultate. *Wigert* benutzt hierfür eine von ihm gefundene *asymptotische Funktionalgleichung* für die *Lambert*sche Reihe $\sum_1^\infty \frac{1}{e^{ns}-1} = \sum_1^\infty d(n)e^{-ns}$, für welche *Landau* einen vereinfachten Beweis gibt: Über die Wigertsche asymptotische Funktionalgleichung für die Lambertsche Reihe, Arch. Math. Phys. (3) 27 (1918), p. 144—146. Vgl. auch *S. Wigert,* Sur une équation fonctionnelle et ses conséquences arithmétiques, Arkiv för Mat., Astr. och Fys. 13 (1918), Nr. 16.

295) *S. Wigert,* a. a. O. 279 b).

gilt. Für die Funktion

$$\sum_1^x \frac{\sigma(n)}{n}(x-n)^k, \quad k=1, 2, \ldots,$$

gibt er erstens entsprechende asymptotische Formeln, die zum Teil von *Landau*[296]) verschärft wurden, und zweitens explizite Formeln, welche unendliche Reihen mit *Bessel*schen Funktionen enthalten. *Walfisz*[297]) zeigt, daß eine solche Formel auch für den Fall $k=0$ aufgestellt werden kann, und gibt für $\overline{S}(x)$ die entsprechende Entwicklung[298])

$$\overline{S}(x) = \frac{\pi^2}{12}x^2 - \frac{1}{2}x + \frac{1}{24} - x\sum_1^\infty \frac{\sigma(n)}{n} J_2(4\pi\sqrt{nx})$$

$$= \frac{\pi^2}{12}x^2 - \frac{1}{2}x + \frac{x^{\frac{3}{4}}}{\pi\sqrt{2}}\sum_1^\infty \frac{\sigma(n)}{n^{\frac{5}{4}}}\cos\left(4\pi\sqrt{nx} - \tfrac{1}{4}\pi\right) + O\left(x^{\frac{2}{3}}\right),$$

wobei jedoch die unendlichen Reihen mit *Cesàro*schen Mitteln von der ersten Ordnung summiert werden müssen, da ihre Konvergenz bisher nicht bewiesen werden konnte.

Ramanujan[299]) findet die Beziehung

$$\sum_1^\infty \frac{(d(n))^2}{n^s} = \frac{(\zeta(s))^4}{\zeta(2s)}$$

und schließt daraus $\displaystyle\sum_1^x (d(n))^2 \sim \frac{1}{\pi^2}x\log^3 x.$

Piltz[300]) verallgemeinert das *Dirichlet*sche Teilerproblem, indem er für $k=2, 3, \ldots$ die Funktion

$$D_k(x) = \sum_1^x d_k(n)$$

296) *E. Landau,* Gött. gel. Anz. 1915, p. 377—414.

297) *A. Walfisz,* Über die summatorischen Funktionen einiger Dirichletscher Reihen, Diss. Göttingen 1922

298) Die zweite Zeile der Formel ergibt sich durch Zusammenstellung der Ergebnisse von *Walfisz* mit denjenigen von *Wigert* a. a. O. 279 b) und *Landau,* a. a. O. 296).

299) *S. Ramanujan,* Some formulae in the analytic theory of numbers, Mess. of Math. 45 (1916), p. 81—84. Vgl. auch *B. M. Wilson,* Proofs of some formulae enunciated by Ramanujan, Proc. London math. Soc. (2) 21 (1922), p. 235—255.

300) *A. Piltz,* Über das Gesetz, nach welchem die mittlere Darstellbarkeit der natürlichen Zahlen als Produkte einer gegebenen Anzahl Faktoren mit der Größe der Zahlen wächst, Diss. Berlin 1881. Vgl. auch *E. Landau,* Über eine idealtheoretische Funktion, Trans. Amer. math. Soc. 13 (1912), p. 1—21.

betrachtet, wobei $$\sum_1^\infty \frac{d_k(n)}{n^s} = (\zeta(s))^k$$

gilt und $d_k(n)$ also die Anzahl der Zerlegungen von n in k Faktoren bezeichnet; insbesondere ist $d_2(n) = d(n)$. Er zeigte, daß — analog wie bei $k = 2$ — die Hauptglieder von $D_k(x)$ von dem Pol $s = 1$ der Funktion $\frac{x^s}{s}(\zeta(s))^k$ herrühren. Wird das dortige Residuum durch $x p_{k-1}(\log x)$ bezeichnet, wobei also p_{k-1} ein Polynom $(k-1)^{\text{ten}}$ Grades ist, und wird
$$D_k(x) = x p_{k-1}(\log x) + \Delta_k(x)$$
gesetzt, so weiß man nach *Hardy* und *Littlewood*[301]), daß
$$\Delta_k(x) = O\left(x^{\frac{k-2}{k}+\varepsilon}\right)$$
für jedes $\varepsilon > 0$ und alle $k \geqq 4$ ist. Für $k = 3$ wurde das schärfste Resultat von *Landau*[286]) gegeben, indem er
$$\Delta_k(x) = O\left(x^{\frac{k-1}{k+1}} \log^{k-1} x\right)$$
für alle $k \geqq 2$ beweist.[302]) — *Hardy*[288]) hat die durch (97) ausgedrückte Eigenschaft von $\Delta_2(x)$ für beliebige k verallgemeinert, wobei der Exponent $\frac{1}{4}$ durch $\frac{k-1}{2k}$ zu ersetzen ist. Die expliziten Formeln (100) und (101) wurden von *Walfisz*[297]) und *Cramér*[303]) verallgemeinert; das „kritische Glied" von (101) wird durch
$$\frac{x^{\frac{k-1}{2k}}}{\pi\sqrt{k}} \sum_1^\infty \frac{d_k(n)}{n^{\frac{k+1}{2k}}} \cos\left(2k\pi\sqrt[k]{nx} + \frac{k-3}{4}\pi\right)$$
ersetzt, wo von der unendlichen Reihe nur bekannt ist, daß sie durch *Cesàro*sche Mittel von der Ordnung $\left[\frac{k-1}{2}\right]$ summierbar ist. Der Fall $k = 2$ ist somit der einzige, wo die Konvergenz der auftretenden Reihen festgestellt ist.

---

301) *G. H. Hardy* und *J. E. Littlewood*, a. a. O. 129).

302) *Landau*, a. a. O. 224), bemerkt, daß aus der *Riemann*schen Vermutung $\Delta_k(x) = O\left(x^{\frac{1}{2}+\varepsilon}\right)$ für jedes $\varepsilon > 0$ folgen würde. Die Behauptung $\frac{1}{x}\int_1^x |\Delta_k(t)|\,dt = O\left(x^{\frac{1}{2}+\varepsilon}\right)$ für $k = 2, 3, \ldots$ ist nach *Hardy* und *Littlewood*, a. a. O. 301), der „*Lindelöf*schen Vermutung" $\zeta(\frac{1}{2}+it) = O(t^\varepsilon)$ äquivalent.

303) *H. Cramér*, Über das Teilerproblem von Piltz, Arkiv för Mat., Astr. och Fys. 16 (1922), No. 21.

35. Ellipsoidprobleme. Wenn $r(n)$ für $n \geqq 0$ die Anzahl der additiven Zerlegungen von n in zwei Quadrate bedeutet, gibt die summatorische Funktion $R(x) = \sum_0^x r(n)$ die Anzahl der Gitterpunkte (u, v) an, die der Kreisfläche $u^2 + v^2 \leqq x$ angehören. *Gauß*[304]) bewies durch eine einfache geometrische Überlegung

$$R(x) = \pi x + O(\sqrt{x});$$

der Flächeninhalt des Kreises stellt somit einen Annäherungswert für $R(x)$ dar. Die folgenden Hauptsätze über $R(x)$ entsprechen genau denjenigen über $D(x)$ und werden auch durch analoge Methoden bewiesen:

1. Nach *Sierpiński*[305]), der die *Voronoï*sche[282]) Methode benutzte, gilt

(102) $$R(x) = \pi x + O\left(x^{\frac{1}{3}}\right);$$

diese Abschätzung wurde neuerdings von *van der Corput*[306]) zu $O(x^M)$ mit $M < \frac{1}{3}$ verschärft.

2. Nach *Hardy*[307]) und *Landau*[308]) kann das Restglied für kein $h < \frac{1}{4}$ von der Form $O(x^h)$ sein.

3. *Im Mittel* ist das Restglied von der Größenordnung $x^{\frac{1}{4}}$; für die Funktion $R(x) - \pi x$ gelten nämlich Formeln, die zu (98) und (99) analog sind.

4. Die explizite Formel für $\bar{R}(x) = \frac{1}{2}(R(x+0) + R(x-0))$

---

304) *C. F. Gauss,* De nexu inter multitudinem classium etc., Werke 2 (1863), p. 269—291.

305) *W. Sierpiński,* O pewnem zagadnieniu z rachunku funkcyj asymptotycznych, Prace Mat.-Fiz. 17 (1906), p. 77—118. Vgl. auch *E. Landau,* a. a. O. 224), 286 b), 284), Über die Zerlegung der Zahlen in zwei Quadrate, Ann. Mat. pura ed appl. (3) 20 (1913), p. 1—28; Über einen Satz des Herrn Sierpiński, Giorn. di Mat. di Battaglini 51 (1913), p. 73—81; Über die Gitterpunkte in einem Kreise, erste Mtlg., Gött. Nachr. 1915, p. 148—160; Über die Gitterpunkte in einem Kreise, Math. Ztschr. 5 (1919), p. 319—320; *S. Wigert,* Über das Problem der Gitterpunkte in einem Kreise, Math. Ztschr. 5 (1919), p. 310—318.

306) *J. G. van der Corput,* a) Verschärfung der Abschätzung beim Teilerproblem, Math. Ann. 87 (1922), p. 39—65; b) Sur quelques approximations nouvelles, Paris C. R. 175 (1922), p. 856—859.

307) *G. H. Hardy,* On the expression of a number as the sum of two squares, Quart. J. 46 (1915), p. 263—283.

308) *E. Landau,* Über die Gitterpunkte in einem Kreise, zweite Mtlg., Gött. Nachr. 1915, p. 161—171; Neue Untersuchungen über die *Pfeiffer*sche Methode zur Abschätzung von Gitterpunktanzahlen, Sitzungsb. Akad. Wien 124, Abt. 2a (1915), p. 469—505.

lautet[309]) nach *Hardy*[307])

$$(103)\quad \begin{aligned} \bar{R}(x) &= \pi x + \sqrt{x}\sum_{1}^{\infty}\frac{r(n)}{\sqrt{n}}J_1(2\pi\sqrt{nx}) \\ &= \pi x + \frac{x^{\frac{1}{4}}}{\pi}\sum_{1}^{\infty}\frac{r(n)}{n^{\frac{3}{4}}}\sin\left(2\pi\sqrt{nx}-\tfrac{1}{4}\pi\right) + O\left(x^{-\frac{1}{4}}\right). \end{aligned}$$

Für $n \geqq 1$ ist bekanntlich $r(n) = 4(d_1(n) - d_3(n))$, (vgl. I C 2, c), wo $d_\nu(n)$ die Anzahl der Divisoren von n von der Form $4k + \nu$ bedeutet[310]); hieraus folgt für $\sigma > 1$

$$\sum_{1}^{\infty}\frac{r(n)}{n^s} = 4\sum_{1}^{\infty}\frac{1}{n^s}\cdot\sum_{1}^{\infty}\frac{\chi(n)}{n^s} = 4\zeta(s)L(s),$$

wenn $\chi(n)$ der Nicht-Hauptcharakter modulo 4 ist. Der Satz von *Landau* (vgl. Nr. 29) ergibt die Konvergenz von

$$\sum_{1}^{\infty}\frac{r(n)-\pi}{n^s} = 4\zeta(s)L(s) - \pi\zeta(s)$$

für $\sigma > \frac{1}{3}$; nach *van der Corput*[306]) ist diese Reihe sogar über $\sigma = \frac{1}{3}$ hinaus konvergent, für $\sigma < \frac{1}{4}$ ist sie aber jedenfalls divergent.

Das obige „Problem der Gitterpunkte in einem Kreise" ist als Spezialfall in dem Problem enthalten, die Anzahl der Gitterpunkte in dem k-dimensionalen ($k \geqq 2$) Ellipsoid

$$F(u_1, u_2, \ldots u_k) = \sum_{\mu,\nu=1}^{k} a_{\mu\nu}u_\mu u_\nu \leqq x \qquad (a_{\mu\nu} = a_{\nu\mu})$$

abzuschätzen, wenn F eine positiv-definite quadratische Form ist. Diese Anzahl ist nach *Landau*[311]) gleich

309) Einen Beweis dieser Formel mit der *Pfeiffer*schen Methode gab *E. Landau,* Über die Gitterpunkte in einem Kreise, dritte Mtlg., Gött. Nachr. 1920, p. 109—134. Vgl. auch *G. Voronoï,* Sur le développement, à l'aide des fonctions cylindriques, des sommes doubles $\sum f(pm^2 + 2qmn + rn^2)$ où $pm^2 + 2qmn + rn^2$ est une forme positive à coefficients entiers, Verhandl. des dritten intern. Math.-Kongr. Heidelberg 1904, p. 241—245.

310) Hieraus folgt insbesondere $r(n) \leqq 4d(n)$ und somit nach der vorigen Nummer eine obere Abschätzung für $r(n)$.

311) *E. Landau,* a. a. O. 224), 286 b), Zur analytischen Zahlentheorie der quadratischen Formen, (über die Gitterpunkte in einem mehrdimensionalen Ellipsoid) Sitzungsb. Akad. Berlin 1915, p. 458—476; Über eine Aufgabe aus der Theorie der quadratischen Formen, Sitzungsb. Akad. Wien, 124 Abt. 2a (1915), p. 445—468. Vgl. auch *J. G. van der Corput,* Over definiete kwadratische vormen, Nieuw Arch. voor Wisk. 13 (1919), p. 125—140. — Bei diesen Untersuchungen wird teils die *Pfeiffer*sche Methode benutzt, teils analytische Methoden, wobei die verallgemeinerten Zetafunktionen von *Epstein* (vgl. Nr. 21) zur Anwendung gelangen.

$$\frac{\pi^{\frac{k}{2}}}{\sqrt{\Delta}\,\Gamma\left(\frac{k}{2}+1\right)}\,x^{\frac{k}{2}}+O\left(x^{\frac{k(k-1)}{2(k+1)}}\right),$$

wo Δ die Determinante $\begin{vmatrix} a_{11} & \dots & a_{1k} \\ \cdot & \cdot & \cdot \\ \cdot & \cdot & \cdot \\ a_{k1} & \dots & a_{kk} \end{vmatrix}$

bezeichnet. Insbesondere ergibt sich für die dreidimensionale Kugel $u^2+v^2+w^2 \leqq x$ als Anzahl der Gitterpunkte

$$\tfrac{4}{3}\pi x^{\frac{3}{2}}+O\left(x^{\frac{3}{4}}\right),$$

was schon von *Cauer*[312]) gefunden war. *Landau*[311]) verallgemeinert seine Sätze nach verschiedenen Richtungen und gibt auch[313]) Verallgemeinerungen der Eigenschaft 2. von $R(x)$. Die *Hardy*sche Formel (103) wird von *Walfisz*[297]) für ein k-dimensionales Ellipsoid (mit ganzen $a_{\mu\nu}$) verallgemeinert[314]); für $k>2$ kann hierbei nur Summabilität, nicht Konvergenz der auftretenden Reihen bewiesen werden.

Die „expliziten Formeln", die in dieser und der vorhergehenden Nummer erwähnt sind, besitzen alle Eigenschaften, die denjenigen der *Riemann*schen Primzahlformel (vgl. Nr. 28) entsprechen. Da die auftretenden Reihen unstetige Funktionen darstellen, können sie jedenfalls nicht für alle x gleichmäßig konvergieren (bzw. summierbar sein); in jedem Intervall, das von Unstetigkeitspunkten frei ist, sind sie zwar gleichmäßig konvergent (bzw. summierbar), in keinem Falle jedoch unbedingt konvergent. In einigen Fällen ist es gelungen, derartige Formeln mit der „*Pfeiffer*schen Methode" zu beweisen[315]); im allgemeinen war es jedoch notwendig, die komplexe Funktionentheorie zu benutzen. Durch formale gliedweise Integration[316]) erhält man zunächst Formeln, die unbedingt konvergente Reihen enthalten und deshalb leicht bewiesen werden können. Es gilt z. B.[317])

(104) $$\int_0^x \overline{R}(t)\,dt=\frac{1}{2}\pi x^2+\frac{x}{\pi}\sum_1^{\infty}\frac{r(n)}{n}J_2(2\pi\sqrt{nx});$$

312) *D. Cauer,* Neue Anwendungen der Pfeifferschen Methode zur Abschätzung zahlentheoretischer Funktionen, Diss. Göttingen 1914.

313) *E. Landau*, a. a. O. 289) und 308).

314) *Hardy* hatte schon früher die Formel für eine Ellipse aufgestellt (a. a. O. 307)); vgl. auch *G. Voronoï* a. a. O. 309).

315) Vgl. 293) und 309).

316) Die in jedem Falle hinreichend oft auszuführen ist.

317) *E. Landau,* Über die Gitterpunkte in einem Kreise, Math. Ztschr. 5 (1919), p. 319—320.

die Zulässigkeit der gliedweisen Differentiation, die auf (103) führt, folgt nun aus dem Konvergenzsatz von *M. Riesz* (vgl. Nr. 5), der hier auf die *Dirichlet*sche Reihe $\sum_1^\infty \frac{r(n)}{n^{\frac{3}{4}}} e^{-s\sqrt{n}}$ anzuwenden ist. Die zahlentheoretischen Funktionen erscheinen hierbei gewissermaßen als Randwerte von analytischen Funktionen. *Steffensen*[171]) entwickelt eine ganz verschiedene Auffassungsweise; wenn eine zahlentheoretische Funktion $f(n)$ gegeben ist, interpoliert er nämlich die Folge $f(1)$, $f(2), \ldots$ durch eine ganze Funktion $f(z)$. Es sei z. B. $\varphi(s) = \sum f(n) n^{-s}$ für $\sigma \geqq 2$ unbedingt konvergent; dann definiert

$$f(z) = -\frac{\sin 2\pi z}{2\pi} \sum_1^\infty \varphi(n+1) z^n$$

für $|z| < 1$ eine ganze Funktion der gewünschten Art. *Steffensen* gibt verschiedene in der ganzen Ebene geltenden Darstellungen der Interpolationsfunktionen und wendet sie zur asymptotischen Untersuchung der zahlentheoretischen Funktionen an (vgl. Nr. 26).

Aus (104) ergibt sich leicht ein Beweis von (102), indem man $\int_x^{x+h} \overline{R}(t)\,dt$ bildet und dabei $h = x^{\frac{1}{3}}$ nimmt.[317]) Diese Differenzenbildung stellt einen sehr allgemein verwendbaren Kunstgriff dar.

36. Allgemeinere Gitterpunktprobleme. In den beiden vorhergehenden Nummern wurden verschiedene Spezialfälle der folgenden Aufgabe behandelt: Ein Gebiet G in der (u, v)-Ebene ist gegeben; man soll die Anzahl der in G oder auf der Begrenzung liegenden Gitterpunkte bestimmen. In allen jenen Spezialfällen konnte eine Annäherung an die gesuchte Anzahl sowie eine grobe Abschätzung des Fehlers durch triviale Mittel erhalten werden, und diese Abschätzung konnte durch neuere Methoden verschärft werden; die Aufgabe, die beste mögliche Abschätzung zu finden, war aber noch nicht gelöst. — Es gelingt nun, entsprechende Resultate auch bei viel allgemeineren Gebieten zu erhalten, und zwar gibt es hierfür mehrere verschiedene Methoden.

Die erste Methode, die auf solche allgemeinere Gebiete angewandt wurde, war die sog. *Pfeiffer*sche, die von *Landau* (vgl. Nr. 34) streng gemacht wurde. Wenn kein Gitterpunkt auf dem Rande von G liegt, und wenn außerdem gewisse Voraussetzungen über die Beschaffenheit des Randes gemacht werden, so kann die gesuchte Gitterpunktanzahl, wie *Landau* zeigt, durch

$$\lim_{m \to \infty} \iint_G \varphi_m(u) \varphi_m(v) du\, dv$$

ausgedrückt werden, wo

$$\varphi_m(u) = \sum_{-m}^{m} \cos 2n\pi u$$

gesetzt ist. *Landau*[318]), *Cauer*[319]) und *Hammerstein*[320]) benutzten diesen Ansatz, um bei verschiedenen speziellen Gebieten, von denen die wichtigsten in den vorhergehenden Nummern erwähnt wurden, die Gitterpunktanzahl abzuschätzen. *Van der Corput*[321]) faßt alle diese Ergebnisse in einem allgemeinen Satz zusammen, bei dem über den Rand von G nur sehr allgemeine Voraussetzungen gemacht werden. Er beweist diesen Satz auch mit der geometrischen *Voronoï*schen Methode (vgl. Nr. 34). *Landau* und *van der Corput*[322]) geben verschiedene analoge Sätze und Vereinfachungen der Beweise, wobei u. a. die arithmetische „*Piltz*sche Methode"[287]) zum Beweis von allgemeinen Gitterpunktsabschätzungen benutzt wird.

37. Verteilung von Zahlen, deren Primfaktoren vorgeschriebenen Bedingungen genügen. Es sei

(105) $$n = p_1^{\alpha_1} p_2^{\alpha_2} \ldots p_\nu^{\alpha_\nu}$$

die Darstellung von n als Produkt von Primzahlpotenzen. Die α sollen stets positiv und die p alle verschieden sein; p_μ soll *nicht* notwendig die μ^{te} Primzahl bezeichnen. Es liegt nahe, nach der Verteilung derjenigen Zahlen n zu fragen, deren Exponenten $\alpha_1 \ldots \alpha_\nu$ gegebenen Bedingungen genügen. Soll z. B. stets $\nu = 1$, $\alpha_1 = 1$ sein, so deckt sich diese Aufgabe offenbar mit derjenigen, die Verteilung der Primzahlen zu untersuchen. Als Verallgemeinerung hiervon kann das Problem aufgefaßt werden, die Verteilung der h Primfaktoren enthaltenden Zahlen zu bestimmen. Dies kann wiederum auf drei verschiedene Weisen aufgefaßt werden, die zu den folgenden Bedingungen führen:

318) *E. Landau*, a. a. O. 284), 293), 305), 308), 309).

319) *D. Cauer*, a. a. O. 312) und Über die Pfeiffersche Methode, Math. Abhandl., H. A. Schwarz zu seinem fünfzigjähr. Doktorjubiläum gewidmet, Berlin 1914, p. 432—447.

320) *A. Hammerstein*, a. a. O. 200).

321) *J. G. van der Corput*, Over roosterpunten in het platte vlak (De beteekenis van de methoden van Voronoï en Pfeiffer), Diss. Leiden 1919; Über Gitterpunkte in der Ebene, Math. Ann. 81 (1920), p. 1—20.

322) *E. Landau* und *J. G. van der Corput*, Über Gitterpunkte in ebenen Bereichen, Gött. Nachr. 1920, p. 135—171; *J. G. van der Corput*, Zahlentheoretische Abschätzungen nach der Piltzschen Methode, Math. Ztschr. 10 (1921), p. 105—120; Zahlentheoretische Abschätzungen, Math. Ann. 84 (1921), p. 53—79.

1) $\nu = h, \; \alpha_1 = \alpha_2 = \cdots = \alpha_h = 1,$

2) $\nu = h,$

3) $\alpha_1 + \alpha_2 + \cdots \alpha_\nu = h.$

Landau[323]) zeigt, daß die Anzahl der diesen Bedingungen genügenden Zahlen unterhalb x in jedem der drei Fälle asymptotisch gleich

$$\frac{1}{(h-1)!} \cdot \frac{x(\log\log x)^{h-1}}{\log x}$$

ist; für den Fall 1. war dies schon von *Gauß*[324]) vermutet worden. *Landau* gibt auch genauere Ausdrücke für jene Anzahlen. *Van der Corput*[325]) untersucht verschiedene allgemeinere Probleme dieser Art.

Läßt man ν unbestimmt, schreibt aber $\alpha_1 = \alpha_2 = \cdots = \alpha_\nu = 1$ vor, so bekommt man die sog. quadratfreien Zahlen. Bedeutet $Q(x)$ die Anzahl der quadratfreien Zahlen $\leq x$, so beweist man leicht[326])

$$Q(x) \sim \frac{6}{\pi^2} x.$$

Für $\sigma > 1$ gilt offenbar (wenn auch 1 als quadratfreie Zahl mitgerechnet wird)

$$\sum_1^\infty \frac{Q(n) - Q(n-1)}{n^s} = \prod_p \left(1 + \frac{1}{p^s}\right) = \frac{\zeta(s)}{\zeta(2s)} = \sum_1^\infty \frac{1}{n^s} \cdot \sum_1^\infty \frac{\mu(n)}{n^{2s}},$$

und die aus der Primzahltheorie geläufigen Methoden geben hier[327])

$$Q(x) = \frac{6}{\pi^2} x + O\left(\sqrt{x}\, e^{-\alpha\sqrt{\log x}}\right)$$

mit konstantem α. Wenn unter den $Q(x)$ quadratfreien Zahlen $\leq x$ $Q_1(x)$ aus einer ungeraden, $Q_2(x)$ aus einer geraden Anzahl von Primfaktoren besteht, so folgt aus (86)

$$\frac{Q_1(x)}{Q_2(x)} = 1 + O\left(e^{-\alpha\sqrt{\log x}}\right).$$

Hardy und *Ramanujan*[328]) lassen in (105) p_μ die μ^{te} Primzahl be-

323) *E. Landau*, Über die Verteilung der Zahlen, welche aus ν Primfaktoren zusammengesetzt sind, Gött. Nachr. 1911, p. 362—381; vgl. auch a. a. O. 238).

324) Vgl. *F. Klein*, Bericht über den Stand der Herausgabe von Gauß' Werken, neunter Bericht, Gött. Nachr. 1911, Geschäftl. Mitt., p. 26—32.

325) *J. G. van der Corput*, On an arithmetical function connected with the decomposition of the positive integers into prime factors, Proceed. Akad. Amsterdam 19 (1916), p. 826—855.

326) *L. Gegenbauer*, Asymptotische Gesetze der Zahlentheorie, Denkschriften Akad. Wien, 49 : 1 (1885), p. 37—80. Es werden hier auch analoge Beziehungen für „h^{te} potenzfreie Zahlen" bewiesen.

327) *A. Axer*, a. a. O. 276).

328) *G. H. Hardy* und *S. Ramanujan*, Asymptotic formulae for the distribution of integers of various types, Proc. London math. Soc. (2) 16 (1917),

zeichnen und führen die Bedingung $\alpha_1 \geqq \alpha_2 \geqq \alpha_3 \geqq \cdots$ ein. Für die Anzahl $A(x)$ der $n \leqq x$, die dieser Bedingung genügen, wird

$$\log A(x) \sim 2\pi \sqrt{\frac{\log x}{3 \log\log x}}$$

bewiesen.

Werden λ verschiedene zu k teilerfremde Restklassen mod. k gegeben, und wird vorgeschrieben, daß in (105) jedes p_μ einer von diesen Restklassen angehören muß, so ist nach *Landau*[329]) die Anzahl der $n \leqq x$ asymptotisch gleich

$$a x (\log x)^{\frac{\lambda}{\varphi(k)} - 1}, \quad (a > 0).$$

Die Summe $\sum 2^\nu$, über dieselben $n \leqq x$ erstreckt, ist dagegen asymptotisch gleich

$$b x (\log x)^{\frac{2\lambda}{\varphi(k)} - 1}, \quad (b > 0),$$

was schon *Lehmer*[330]) in einem speziellen Fall bewiesen hatte. Ein ähnlicher Satz von *Landau*[329]) enthält insbesondere das Resultat[331]): es gibt unterhalb x asymptotisch

$$c \frac{x}{\sqrt{\log x}}, \quad (c > 0),$$

ganze Zahlen, die als Summen von zwei Quadraten darstellbar sind. Hieraus folgt, wenn $B_\mu(x)$ die Anzahl der ganzen Zahlen $\leqq x$ bezeichnet, zu deren additiven Darstellung *genau* μ Quadrate erforderlich sind (bekanntlich ist $B_\mu(x) = 0$ für $\mu > 4$):

$$B_1(x) \sim \sqrt{x}, \quad B_2(x) \sim c \frac{x}{\sqrt{\log x}}, \quad B_3(x) \sim \frac{5}{6} x, \quad B_4(x) \sim \frac{1}{6} x.$$

38. Neuere Methoden der additiven Zahlentheorie. Als der Abschnitt über additive Zahlentheorie in I C 3 geschrieben wurde, war vor allem das große „*Waring*sche Problem" noch ungelöst. *Waring*[332]) vermutete 1782, daß jede ganze Zahl $n \geqq 0$ als Summe

---

p. 112—132. In der Abhandlung: The normal number of prime factors of n, Quart. J. 48 (1917), p. 76—92, beschäftigen sich die beiden Verfasser mit Problemen, die zu den in dieser Nummer behandelten Fragestellungen in einer gewissen Beziehung stehen.

329) *E. Landau*, a. a. O. 78); Bemerkungen zu Herrn D. N. Lehmers Abhandlung in Bd. 22 dieses Journals, Amer. J. of math. 26 (1904), p. 209—222; Lösung des Lehmerschen Problems, ebenda 31 (1909), p. 86—102.

330) *D. N. Lehmer*, Asymptotic Evaluation of certain Totient Sums, Amer. J. of math. 22 (1900), p. 293—335.

331) *E. Landau*, Über die Einteilung der positiven ganzen Zahlen in vier Klassen nach der Mindestzahl der zu ihrer additiven Zusammensetzung erforderlichen Quadrate, Arch. Math. Phys. (3) 13 (1908), p. 305—312.

332) *E. Waring*, Meditationes Algebraicae, 3. Aufl. Cambridge 1782, p. 349—350.

einer festen (d. h. nur von k, nicht von n, abhängenden) Anzahl von positiven k^{ten} Potenzen dargestellt werden konnte, und zwar für $k = 1, 2, \ldots$. Bis 1909 war dies nur für einige spezielle Werte von k bewiesen; es gelang aber *Hilbert*[333]) einen allgemeinen Beweis zu finden. Dieser Beweis benutzt zwar die Hilfsmittel der Integralrechnung; durch spätere Vereinfachungen[334]) ist aber gezeigt worden, daß dies gänzlich vermieden werden kann, so daß die Methode im Grunde eine rein algebraische ist und deshalb hier nicht eingehender besprochen werden soll.

Rein analytisch ist dagegen die Methode, welche neuerdings von *Hardy* und *Littlewood*[335]) auf das Problem angewandt worden ist. Es sei $k > 2$, und es werde für $|x| < 1$

---

333) *D. Hilbert*, Beweis für die Darstellbarkeit der ganzen Zahlen durch eine feste Anzahl n^{ter} Potenzen (Waringsches Problem), Gött. Nachr. 1909, p. 17—36 und Math. Ann. 67 (1909), p. 281—300.

334) Vgl. z. B. *F. Hausdorff*, Zur Hilbertschen Lösung des Waringschen Problems, Math. Ann. 67 (1909), p. 301—305; *E. Stridsberg*, Sur la démonstration de M. Hilbert du théorème de Waring, Math. Ann. 72 (1912), p. 145—152; Några elementära undersökningar rörande fakulteter och deras aritmetiska egenskaper, Arkiv för Mat., Astr. och Fys. 11 (1917), No. 25; *R. Remak*, Bemerkung zu Herrn Stridsbergs Beweis des Waringschen Theorems, Math. Ann. 72 (1912), p. 153—156. Für die ältere Literatur zum *Waring*schen Problem vgl. die Göttinger Dissertationen von *A. J. Kempner*, Über das Waringsche Problem und einige Verallgemeinerungen, 1912, und *W. S. Baer*, Beiträge zum Waringschen Problem, 1913.

335) *G. H. Hardy* und *J. E. Littlewood*, A new solution of Waring's problem, Quart. J. 48 (1919), p. 272—293; Some problems of Partitio numerorum, I: A new solution of Waring's problem, Gött. Nachr. 1920, p. 33—54, II: Proof that any large number is the sum of at most 21 biquadrates, Math. Ztschr. 9 (1921), p. 14—27, (III: a. a. O. 250)), IV: The singular series in Waring's problem and the value of the number $G(k)$, Math. Ztschr. 12 (1922), p. 161—188; *G. H. Hardy*, Some famous problems of the Theory of Numbers, and in particular Waring's problem, Inaugural lecture, Oxford 1920. Vgl. auch *E. Landau*, a) Zur Hardy-Littlewoodschen Lösung des Waringschen Problems, Gött. Nachr. 1921, p. 88—92; b) Zum Waringschen Problem, Math. Ztschr. 12 (1922), p. 219—247; c) Über die Hardy-Littlewoodschen Arbeiten zur additiven Zahlentheorie, Jahresb. d. deutschen Math.-Ver. 30 (1921), p. 179—185; *H. Weyl*, Bemerkungen über die Hardy-Littlewoodschen Untersuchungen zum Waringschen Problem, Gött. Nachr. 1921, p. 189—192; *A. Ostrowski*, Bemerkungen zur Hardy-Littlewoodschen Lösung des Waringschen Problems, Math. Ztschr. 9 (1921), p. 28—34. *E. Landau* (b.) berücksichtigt auch gewisse Verallgemeinerungen, die zuerst von *Kamke* mit der *Hilbert*schen Methode behandelt wurden: Verallgemeinerungen des Waring-Hilbertschen Satzes, Math. Ann. 83 (1921), p. 85—112. — Die im Texte gewählte Bezeichnungsweise weicht etwas von der *Hardy-Littlewood*schen ab und schließt sich an *Landau* (b.) an.

$$f(x)=\sum_{n=0}^{\infty}x^{n^k},\quad f^s(x)=\sum_{n=0}^{\infty}r(n)x^n$$

gesetzt, wo also $r(n)$ von s und k abhängt. Um die *Waring*sche Vermutung, $r(n)>0$ für $s>s_0=s_0(k)$, zu beweisen, setzen *Hardy* und *Littlewood*

$$r(n)=\frac{1}{2\pi i}\int\limits_{|x|=1-\frac{1}{n}}\frac{f^s(x)}{x^{n+1}}\,dx.$$

Bei dem Versuch, aus dieser Integraldarstellung ein asymptotisches Ergebnis über $r(n)$ zu gewinnen, stößt man auf ungeheure Schwierigkeiten, da die Funktion unter dem Integralzeichen nicht über den Einheitskreis fortgesetzt werden kann. *Hardy* und *Littlewood* bemerken nun, daß die Einheitswurzeln $\varrho=e^{\frac{2p\pi i}{q}}$ gewissermaßen die „schwersten" Singularitäten von $f(x)$ sind; bei radialer Annäherung an den Punkt $x=\varrho$ wird $f^s(x)$ asymptotisch gleich einer Hilfsfunktion, die durch eine Potenzreihe von der einfachen Form[336]) $\sum\nu^\alpha\left(\frac{x}{\varrho}\right)^\nu$, mit einer Konstanten multipliziert, dargestellt wird. Der Hauptgedanke der Methode ist nun, $f^s(x)$ durch eine Summe solcher Hilfsfunktionen, d. h. $r(n)$ durch die entsprechende Summe der Koeffizienten von x^n, zu approximieren. Die Durchführung dieses Ansatzes gelingt natürlich nur durch ziemlich verwickelte Überlegungen, wobei die Untersuchungen von *Weyl*[114]) über Diophantische Approximationen eine wichtige Rolle spielen. Das folgende Hauptresultat wird erhalten: Für alle $s\geqq s_0(k)$ ist bei unendlich wachsendem n

$$r(n)\sim\frac{\left(\Gamma\left(1+\frac{1}{k}\right)\right)^s}{\Gamma\left(\frac{s}{k}\right)}n^{\frac{s}{k}-1}S, \tag{106}$$

wo S die sog. „singuläre Reihe"

$$S=\sum_{q=1}^{\infty}\sum_{\substack{p=0\\(p,q)=1}}^{q-1}\left(\frac{S_{pq}}{q}\right)^s e^{-\frac{2\pi i n p}{q}},$$

mit

$$S_{pq}=\sum_{h=1}^{q}e^{\frac{2\pi i h^k p}{q}}$$

bezeichnet. Die Reihe S ist für $s\geqq s_0(k)$ konvergent und $>\sigma$, wo $\sigma=\sigma(k,s)$ nicht von n abhängt. Für s_0 ist insbesondere die Zahl $s_0=(k-2)2^{k-1}+5$ wählbar.

336) *Landau*, a. a. O. 335b), zeigt, daß man sogar eine noch einfachere, durch eine Binomialreihe dargestellte Hilfsfunktion benutzen kann.

Die *Hardy-Littlewood*sche Methode ergibt also wesentlich mehr als die *Hilbert*sche, welche nur einen Existenzsatz lieferte. Insbesondere folgt, daß es zu jedem k eine *kleinste* Zahl $G(k)$ gibt, so daß alle *hinreichend großen* n als Summen von höchstens je $G(k)$ k^{ten} Potenzen darstellbar sind, und daß $G(k) \leqq (k-2)2^{k-1}+5$ ist. Jede hinreichend große Zahl ist also die Summe von höchstens 21 Biquadraten; für den Fall $k=3$ liefert der Satz aber nur $G(k) \leqq 9$, während schon früher durch *Landau*[337]) bekannt war, daß jede hinreichend große Zahl die Summe von höchstens 8 Kuben ist. Andererseits ist bekannt[338]), daß immer $G(k) \geqq k+1$, und im Falle $k=2^m$ sogar $G(k) \geqq 4k$ ist.

Im Falle $k=2$ wird die erzeugende Funktion $f(x)$ durch eine Thetareihe ausgedrückt:

$$f(x)=\tfrac{1}{2}\left(\sum_{-\infty}^{\infty} x^{n^2}-1\right),$$

und die Transformationstheorie der Thetafunktionen gestattet nun, viel genauere Resultate als im allgemeinen Falle zu erhalten.[339]) Die asymptotische Gleichung (106) bleibt auch hier für $s \geqq 4$ richtig; es kann sogar für $3 \leqq s \leqq 8$ in (106) das Zeichen $\sim$ durch $=$ ersetzt werden. Im Falle $s=3$ ist $S=0$ für unendlich viele n (nämlich für $n=4^a(8b+7)$). Durch Umformung der so erhaltenen Ausdrücke erhält man neue Beweise der klassischen Formeln für die Anzahl der Darstellungen einer Zahl als Summe von Quadraten. Besonders wichtig für diese Untersuchungen waren einige neuere Arbeiten von *Mordell*[340]), der die Darstellung von Zahlen durch Quadratsummen mit Hilfe der Theorie der Modulfunktionen systematisch untersuchte.

337) *E. Landau,* Über eine Anwendung der Primzahltheorie auf das Waringsche Problem in der elementaren Zahlentheorie, Math. Ann. 66 (1909), p. 102—105. Bei dem Beweis wird ein Satz über Primzahlen in arithmetischen Reihen benutzt. Nach *Wieferich* ist *jede* Zahl die Summe von höchstens 9 Kuben; es gibt auch tatsächlich Zahlen (23, 239), die 9 Kuben erfordern: Math. Ann. 66 (1909), p. 95—101.

338) Außerdem kennt man z. B. $G(6) \geqq 9$. Eine Zusammenstellung der bekannten Resultate geben *Hardy* und *Littlewood*, Partitio numerorum IV (a. a. O. 335). Auf die Funktion $g(k)$, die man erhält, wenn man in der Definition von $G(k)$ die Wörter „hinreichend große" ausläßt, gehen wir hier nicht ein; es sei nur bemerkt, daß aus der Existenz von $G(k)$ unmittelbar die Existenz von $g(k)$ folgt.

339) *G. H. Hardy,* On the representation of a number as the sum of any number of squares, and in particular of five, Trans. Amer. math. Soc. 21 (1920), p. 255—284. Vgl. hierzu *S. Ramanujan,* On certain trigonometrical sums and their applications in the theory of numbers, Trans. Cambridge Phil. Soc. 22 (1919), p. 259—276.

340) *L. J. Mordell,* On the representations of numbers as a sum of $2r$

Über die Anwendung der Methode auf den „*Goldbach*schen Satz" und verwandte Primzahlprobleme wurde schon in Nr. **31** berichtet. — Zum ersten Male wurde die Methode nicht auf *Warings* Satz angewendet, sondern auf das Problem der Abschätzung der Funktion $p(n)$, welche die Anzahl der „unbeschränkten Partitionen" von n, d. h. die Anzahl der positiven ganzzahligen Lösungen von

$$n = x + 2y + 3z + 4u + \cdots,$$

angibt. Als erzeugende Funktion tritt hier

$$f(x) = 1 + \sum_1^\infty p(n) x^n = \prod_1^\infty \frac{1}{1 - x^n}$$

auf. *Hardy* und *Ramanujan*[341]) beweisen über $p(n)$ sehr genaue asymptotische Sätze, aus denen insbesondere

(107) $$p(n) \sim \frac{1}{4n\sqrt{3}} e^{\frac{1}{3}\pi\sqrt{6n}}$$

folgt. Der Hauptgedanke ist wieder derselbe: die nicht fortsetzbare Funktion $f(x)$ wird durch eine Summe fortsetzbarer Funktionen approximiert, die in je einer Einheitswurzel singulär sind. Die rechte Seite in (107) rührt übrigens von der „schwersten" Singularität $x = 1$ her.

39. Diophantische Approximationen. Durch die in den Nummern **7** und **17** besprochenen Anwendungen der Theorie der Diophantischen Approximationen wurde ein lebhaftes Interesse für diese Theorie erweckt. Jene Anwendungen gingen von dem grundlegenden *Kronecker*schen[342]) Satze aus, der in moderner Ausdrucksweise so lautet: Es seien $1, \alpha_1, \alpha_2, \ldots \alpha_k$ $(k \geq 1)$ *linear unabhängige* Zahlen, und es sei

$$(x) = x - [x]$$

gesetzt; dann liegen die Punkte mit den Koordinaten

(108) $$x_1 = (n\alpha_1),\ x_2 = (n\alpha_2), \ldots x_k = (n\alpha_k), \quad (n = 1, 2, \ldots)$$

im k-dimensionalen Einheitswürfel überall dicht. *Weyl*[114]) gibt eine

---

squares, Quart. J. 48 (1917), p. 93—104; On the representations of numbers as the sum of an odd number of squares, Trans. Cambridge Phil. Soc. 22 (1919), p. 361—372.

341) *G. H. Hardy* und *S. Ramanujan*, Une formule asymptotique pour le nombre des partitions de n, Paris C. R. 164 (1917), p. 35—38; Asymptotic formulae in combinatory analysis, Proc. London math. Soc. (2) 17 (1918), p. 75—115.

342) *L. Kronecker*, Die Periodensysteme von Funktionen reeller Variabeln, Sitzungsb. Akad. Berlin 1884, p. 31—46, Werke 3 : 1, p. 1071—1080; Näherungsweise ganzzahlige Auflösung linearer Gleichungen, Sitzungsb. Akad. Berlin 1884, p. 1179—1193, Werke 3 : 1, p. 47—110.

für die genannten Anwendungen wesentliche Vertiefung dieses Satzes, indem er zeigt, daß jene Punkte sogar in jedem Teile des Einheitswürfels *asymptotisch gleich dicht* liegen. Die Anzahl derjenigen unter den N ersten Punkten (108), die einem Teilgebiet vom Inhalt δ angehören, ist also asymptotisch gleich δN. *Weyl* beweist dies durch systematische Benutzung der „analytischen Invariante der Zahlklassen mod. 1“, der Funktion $e^{2\pi i s}$.

Hardy-Littlewood[343]) und *Weyl*[114]) geben auch wichtige Verallgemeinerungen auf den Fall, wo in (108) n durch n^q oder durch ein Polynom ersetzt wird; die hierbei von *Weyl* eingeführten, eleganten Methoden zur Transformation und Abschätzung von Summen mit dem allgemeinen Gliede $e^{2\pi i p(n)}$ (p ein Polynom) waren für die in der vorhergehenden Nummer besprochenen Untersuchungen über *Warings* Problem von grundlegender Bedeutung und haben auch zu neuen Resultaten über die Größenordnung von $\zeta(s)$ auf vertikalen Geraden geführt (vgl. Nr. **18**). *Hardy* und *Littlewood* haben insbesondere Summen der Gestalt

$$\sum_1^n e^{\left(\nu-\frac{1}{2}\right)^2 \alpha \pi i}, \quad \sum_1^n e^{\nu^2 \alpha \pi i}, \quad \sum_1^n (-1)^{\nu-1} e^{\nu^2 \alpha \pi i}$$

untersucht, die mit dem Verhalten der Thetareihen bei Annäherung an die Konvergenzgrenze zusammenhängen. Wenn α irrational ist, sind alle drei Summen von der Form $o(n)$. Auch über die Verteilung der Zahlen $(\lambda_n \alpha)$, wo $\lambda_1, \lambda_2, \ldots$ eine unbegrenzt und monoton wachsende Zahlfolge ist, gibt es Sätze, die den vorhergehenden entsprechen.[344]) Für $\lambda_n = a^n$ hängen diese Sätze mit der Verteilung der Ziffern in (verallgemeinerten) Dezimalbrüchen zusammen.[345])

Für die Summe $\sum(\nu\alpha)$ gilt bei irrationalem α immer

$$\sum_1^n (\nu\alpha) = \tfrac{1}{2}n + o(n).$$

Wird in dieser Formel das Restglied durch ein „besseres“ ersetzt, so

343) *G. H. Hardy* und *J. E. Littlewood,* Some problems of diophantine approximation, Intern. Congr. of math. Cambridge 1912, p. 223—229; Acta Math. 37 (1914), p. 155—238. Vgl. auch *J. G. van der Corput,* Über Summen, die mit den elliptischen Θ-Funktionen zusammenhängen, Math. Ann. 87 (1922), p. 66—77.

344) Vgl. hierzu auch *R. H. Fowler,* On the distribution of the set of points $(\lambda_n \Theta)$, Proc. London math. Soc. (2) 14 (1914), p. 189—206.

345) *E. Borel,* Les probabilités dénombrables et leurs applications arithmétiques, Palermo Rend. 27 (1909), p. 247—271 (vgl. auch Leçons sur la théorie des fonctions, deuxième éd., Paris 1914). Weitergehende Sätze geben *Hardy* und *Littlewood,* a. a. O. 343).

kann die neue Formel nicht für *alle* irrationalen α gelten; beschränkt man sich dagegen auf spezielle Klassen von Irrationalitäten, so kann die Abschätzung erheblich verschärft werden. Es gilt z. B. für ein α, bei dessen Entwicklung in einen gewöhnlichen Kettenbruch die auftretenden Nenner beschränkt sind[346])

$$\sum_1^n (\nu\alpha) = \tfrac{1}{2}n + O(\log n)$$

Ostrowski[346]) zeigt, daß bei *keinem* irrationalen α hier das O gegen o vertauscht werden kann. *Hardy* und *Littlewood*[346]) zeigen, daß das Problem der Abschätzung von $\sum(\nu\alpha)$ mit der Bestimmung der Gitterpunktanzahl in einem gewissen rechtwinkligen Dreieck nahe verbunden ist. *Hecke*[347]) hat jene Summen für quadratisch irrationale α näher untersucht. Ist insbesondere $\alpha = \sqrt{D}$, wo $4D$ eine positive Fundamentaldiskriminante ist (vgl. Nr. **40**), so konvergiert die *Dirichlet*sche Reihe

$$\sum_1^\infty \frac{(n\alpha) - \frac{1}{2}}{n^s}$$

für $\sigma > 0$ und stellt eine überall meromorphe Funktion dar, die in der Halbebene $\sigma \leqq 0$ unendlich viele Pole besitzt.

Es sei
$$\alpha = a_0 + \cfrac{1}{a_1 + \cfrac{1}{a_2 + \cdots}}$$
die gewöhnliche Kettenbruchentwicklung einer irrationalen Zahl α; hinsichtlich der Größenordnung der a_n sind u. a. folgende Sätze bekannt:[348])

1. Die Menge der α, für die von einer gewissen Stelle an $a_n > 1$ gilt, hat das Maß Null.

2. Es seien $d_1, d_2, \ldots$ und $k_1, k_2, \ldots$ monoton wachsende Folgen positiver Zahlen, und es sei $\sum \frac{1}{d_n}$ divergent, $\sum \frac{1}{k_n}$ konvergent. „Fast überall" ist dann von einer gewissen Stelle an $a_n < k_n$, und „fast überall" ist $a_n > d_n$ für unendlich viele n.

---

346) *M. Lerch*, L'interméd. des Math. 11 (1904), p. 145; *G. H. Hardy* und *J. E. Littlewood*, Some Problems of diophantine approximation: the lattice-points of a right-angled triangle, Proc. London math. Soc. (2) 20 (1921), p. 15—36; *A. Ostrowski*, Bemerkungen zur Theorie der Diophantischen Approximationen, Abh. Math. Seminar Hamburg 1 (1921), p. 77—98.

347) *E. Hecke*, Über analytische Funktionen und die Verteilung von Zahlen mod. eins, Abh. Math. Seminar Hamburg 1 (1921), p. 54—76.

348) *E. Borel*, a. a. O. 345); *F. Bernstein*, Über eine Anwendung der Mengenlehre auf ein aus der Theorie der säkularen Störungen herrührendes Problem, Math. Ann. 71 (1911), p. 417—439.

Hiermit hängen die Fragen nach der Approximation irrationaler Zahlen durch rationale nahe zusammen.[349]) Zu jedem irrationalen α gibt es unendlich viele rationale $\frac{p}{q}$, so daß

$$\left|\alpha - \frac{p}{q}\right| < \frac{1}{\sqrt{5}\, q^2}$$

gilt. Wenn k_n die obige Bedeutung hat, so bilden diejenigen α, die sich durch unendlich viele $\frac{p}{q}$ mit der Genauigkeit

$$\left|\alpha - \frac{p}{q}\right| < \frac{1}{q k_q}$$

approximieren lassen, eine Menge vom Maß Null. Wenn α eine algebraische Zahl vom Grade n ist, so gilt nach einem wichtigen Satze von *Siegel*[350]) für jedes rationale $\frac{p}{q}$

$$\left|\alpha - \frac{p}{q}\right| > \frac{\gamma}{q^{2\sqrt{n}}},$$

wo γ nur von α abhängt.

V. Algebraische Zahlen und Formen.

40. Quadratische Formen und Körper.[351]) Die nach *Dirichlet* benannten Reihen wurden von ihm gebraucht, um die Anzahl der verschiedenen Klassen binärer quadratischer Formen einer gegebenen Diskriminante zu berechnen;[352]) die Lösung dieser Aufgabe setzte ihn in den Stand, seinen Satz über die Primzahlen einer arithmetischen Reihe zu beweisen (vgl. Nr. **30**). Über die Berechnung jener Klassenanzahl und ihre Beziehung zu den *Gauß*schen Summen ist in I C 3, Nr. 2, über die analogen Probleme bei Formen mit mehr als zwei Veränderlichen in I C 2, d und e, berichtet worden; es seien hier nur einige Ergänzungen für die binären Formen gegeben.

---

349) Vgl. *A. Hurwitz,* Über die angenäherte Darstellung der Irrationalzahlen durch rationale Brüche, Math. Ann. 39 (1891), p. 279—284; *E. Borel,* Contribution à l'analyse arithmétique du continu, J. math. pures appl. (5) 9 (1903), p. 329—375; *O. Perron,* Irrationalzahlen, Berlin und Leipzig (Ver. wiss. Verleger) 1921.

350) *C. Siegel,* Approximation algebraischer Zahlen, Math. Ztschr. 10 (1921), p. 173—213.

351) Was sich unmittelbar durch Spezialisierung ($n = 2$) aus Formeln der beiden folgenden Nummern ergibt, wird hier nicht erwähnt.

352) *G. Lejeune-Dirichlet,* Recherches sur diverses applications de l'analyse infinitésimale à la théorie des nombres, Crelles J. 19 (1839), p. 324—369 und 21 (1840), p. 1—12, 134—155, Werke 1, p. 411—496.

Die quadratische Form werde in *Kronecker*scher Bezeichnungsweise[353])
$$f(xy) = ax^2 + bxy + cy^2 = (a, b, c)$$
geschrieben, ihre Diskriminante sei
$$b^2 - 4ac = D = Q^2 D_0,$$
wo D_0 eine sog. *Fundamentaldiskriminante*[354]) ist. Es sei
$$(a_1, b_1, c_1), \ldots (a_h, b_h, c_h)$$
ein Repräsentantensystem der *primitiven* — und im Falle $D < 0$ *positiven* — zu D gehörigen Klassen; die Koeffizienten a können dabei immer positiv und die b negativ vorausgesetzt werden. Dann ist für $\sigma > 1$

$$(109) \qquad \sum_{\nu=1}^{h} \sum_{x,y} (a_\nu x^2 + b_\nu xy + c_\nu y^2)^{-s} = \tau \sum_{n=1}^{\infty} \left(\frac{D}{n}\right) \frac{1}{n^s} \cdot \sum_{n}{}' \frac{1}{n^s},$$

wo links x und y alle die zugehörige quadratische Form zu Q teilerfremd machenden Paare ganzer Zahlen exkl. (0, 0) durchlaufen; im Falle $D > 0$ tritt jedoch die Beschränkung
$$0 \leqq y < \frac{2a_\nu U}{T - b_\nu U} x$$
hinzu, wo (T, U) die „Fundamentallösung" der Gleichung $t^2 - Du^2 = 4$ bezeichnet (vgl. I C 2 c, 2). Rechts durchläuft n in $\sum'$ alle zu Q teilerfremden positiven ganzen Zahlen, und es ist
$$\tau = \begin{cases} 2 & \text{für } D < -4 \\ 4 & \text{„} \quad D = -4 \\ 6 & \text{„} \quad D = -3 \\ 1 & \text{„} \quad D > 0. \end{cases}$$
Das *Kronecker*sche Symbol $\left(\frac{D}{n}\right)$ ist für $n > 0$ ein reeller Charakter mod. $|D|$, und die Reihe $\sum_{1}^{\infty} \left(\frac{D}{n}\right) n^{-s}$ ist deshalb eine von den in Nr. 29 untersuchten L-Reihen. Jede der h Doppelsummen auf der linken Seite von (109) hat für $s = 1$ einen Pol erster Ordnung mit einem von ν unabhängigen Residuum, und man findet durch Ver-

353) *L. Kronecker*, Zur Theorie der elliptischen Funktionen, Sitzungsb. Akad. Berlin 1885, p. 770; vgl. auch die ausführliche Darstellung von *J. de Séguier*, Formes quadratiques et multiplication complexe, Berlin 1894.

354) Wenn m eine quadratfreie Zahl bedeutet, so ist *entweder* $D_0 = m$, $m \equiv 1 \pmod{4}$, *oder* $D_0 = 4m$, $m \equiv 2$ oder $3 \pmod{4}$. Es wird immer $D_0 \neq 1$ vorausgesetzt, so daß D keine Quadratzahl ist.

gleichung der Residuen

$$(110)\qquad h = \begin{cases} \tau \dfrac{\sqrt{-D}}{2\pi} L(1) & \text{für} \quad D < 0 \\[2ex] \dfrac{\sqrt{D}}{\log \dfrac{T + U\sqrt{D}}{2}} L(1) & \text{für} \quad D > 0. \end{cases}$$

Die Summierung der unendlichen Reihen gestaltet sich am einfachsten, wenn D eine Fundamentaldiskriminante und daher $Q = 1$ ist; der allgemeine Fall läßt sich hierauf zurückführen. Für diesen Fall gilt

$$(111)\qquad h = \begin{cases} \dfrac{\tau}{2D} \displaystyle\sum_{n=1}^{|D|-1} \left(\frac{D}{n}\right) n & \text{für} \quad D < 0 \\[2ex] \dfrac{1}{\log \dfrac{T + U\sqrt{D}}{2}} \displaystyle\sum_{n=1}^{D-1} \left(\frac{D}{n}\right) \log \sin \frac{n\pi}{D} & \text{für} \quad D > 0. \end{cases}$$

Jede Fundamentaldiskriminante D ist die Grundzahl des durch $\sqrt{D}$ erzeugten quadratischen Zahlkörpers (vgl. I C 4a), und den Klassen quadratischer Formen der Diskriminante D entsprechen umkehrbar eindeutig die Idealklassen des Körpers $k(\sqrt{D})$.[355]) Die Anzahl der Idealklassen wird also auch durch (111) gegeben; diese Anzahl hängt nach I C 4a, Nr. 9 mit dem Residuum im Punkte $s = 1$ der zum Körper gehörigen Zetafunktion $\zeta_k(s)$ zusammen. In der Tat ist $\zeta_k(s)$ gleich der rechten Seite von (109), dividiert durch τ,

$$\zeta_k(s) = \sum_{1}^{\infty} \left(\frac{D}{n}\right) \frac{1}{n^s} \cdot \zeta(s) = L(s)\,\zeta(s),$$

so daß $\zeta_k(s)$ mit der Theorie von $\zeta(s)$ und den L-Funktionen schon erledigt ist.

Die Ausdrücke (111) können auch nach einer weniger analytischen Methode abgeleitet werden, wobei die *Dirichlet*schen Reihen nicht auftreten.[356]) Verschiedene Transformationen von (111), sowie andere Ausdrücke für Klassenzahlen sind von *Lerch*[357]) gegeben. Er gibt

355) Hierbei sind jedoch zwei Ideale nur dann zur selben Klasse gehörig, wenn ihr Quotient eine Zahl *positiver Norm* ist. Die Anzahl der so definierten Klassen ist entweder gleich der gewöhnlichen Klassenzahl (vgl. I C 4a, Nr. 8) oder doppelt so groß.

356) Vgl. z. B. *Ch. Hermite,* Paris C. R. 55 (1862), p. 684, Oeuvres 2, p. 255.

357) *M. Lerch,* Essais sur le calcul du nombre des classes de formes quadratiques binaires aux coefficients entiers, Mém. présentés Acad. sc. Paris (2) 33 (1906), No. 2; Paris C. R. 135 (1902), p. 1314—1315; Acta Math. 29 (1905), p. 333; 30 (1906), p. 203—293; Sur le nombre des classes de formes quadratiques binaires d'un discriminant positif fondamental, J. math. pures appl. (5) 9 (1903), p. 377—401.

insbesondere Formeln, welche für numerische Berechnung geeignet sind, z. B.

$$h = 2 \operatorname{sgn} D_2 \cdot \sum_{\mu=1}^{\frac{1}{2}|D_1|} \left(\frac{D_1}{\mu}\right) \sum_{\nu=1}^{\mu \left|\frac{D_2}{D_1}\right|} \left(\frac{D_2}{\nu}\right),$$

wo D_1 und D_2 Fundamentaldiskriminanten bedeuten, $D_1 D_2 < 0$ und h die Klassenzahl für die Diskriminante $D_1 D_2$ ist.

Für beliebige Diskriminanten gilt[358])

$$h = O\left(\sqrt{|D|} \log |D|\right)$$

und für $D > 0$ sogar $h = O(\sqrt{D})$.

Es gibt unendlich viele positive Diskriminanten mit gleicher Klassenzahl[359]), für negative D ist dies dagegen wahrscheinlich nicht der Fall — in der Tat gilt[360]) für negative Fundamentaldiskriminanten, wenn über die Nullstellen der obigen Funktion $\zeta_k(s)$ eine gewisse unbewiesene Annahme (die insbesondere aus der Richtigkeit der „verallgemeinerten *Riemann*schen Vermutung" für die L-Funktionen folgen würde) gemacht wird,

$$h > c \frac{\sqrt{|D|}}{\log |D|}.$$

Über die „mittlere Anzahl" der Klassen gab schon *Gauß*[361]) (ohne Beweis) einige Sätze; es gilt z. B. nach *Landau*[362]), wenn h_n die Klassenanzahl primitiver positiv-definiter Formen der Diskriminante $-n$ bedeutet,

$$\sum_1^x h_n = \frac{\pi}{18 \zeta(3)} x^{\frac{3}{2}} - \frac{3}{2\pi^2} x + O\left(x^{\frac{5}{8}} \log x\right).$$

Die Klassenzahl für negative Diskriminanten, insbesondere die Lehre von den sog. Klassenzahlrelationen, steht zu der Theorie der elliptischen Funktionen und deren komplexer Multiplikation in naher Beziehung[363]), darauf kann hier jedoch nicht eingegangen werden.

---

358) Vgl. *G. Pólya, J. Schur* und *E. Landau,* a. a. O. 205). *E. Landau* gibt auch analoge Resultate für beliebige algebraische Zahlkörper.

359) *G. Lejeune-Dirichlet,* Über eine Eigenschaft der quadratischen Formen von positiver Determinante, Ber. Verhandl. Akad. Berlin 1855, p. 493—495; Werke 2, p. 185—187.

360) *E. Landau,* Über imaginär-quadratische Zahlkörper mit gleicher Klassenzahl; Über die Klassenzahl imaginär-quadratischer Zahlkörper, Gött. Nachr. 1918, p. 277—295. Vgl. auch *T. Nagel,* Über die Klassenzahl imaginär-quadratischer Zahlkörper, Abhandl. Math. Seminar Hamburg 1 (1922), p. 140—150.

361) *C. F. Gauss,* Disquisitiones arithmeticae, Art. 302—304.

362) *E. Landau,* a. a. O. 284). Vgl. auch *F. Mertens,* a. a. O. 270).

363) Vgl. I C 6, Nr. 12 sowie II B 3, Nr. 75. Eine Darstellung der Lehre

Dirichlet[364]) stellte den Satz auf, daß durch jede primitive — und im Falle $D < 0$ positive — binäre quadratische Form einer nichtquadratischen Diskriminante unendlich viele Primzahlen dargestellt werden können. Er gab auch Andeutungen für den Beweis, der später von *Schering*[365]) und *Weber*[366]) vollständig ausgeführt wurde. *Mertens*[367]), *de la Vallée Poussin*[368]), *Bernays*[369]) und *Landau*[370]) gaben für die Anzahl der darstellbaren Primzahlen $\leqq x$ und für gewisse damit zusammenhängende Summen asymptotische Ausdrücke, die als Spezialfälle in den Sätzen von *Landau*[370]) über Primideale in Idealklassen (vgl. Nr. 42) enthalten sind. Jene Anzahl ergibt sich gleich

$$\frac{1}{h_0} Li(x) + O\left(x e^{-\alpha \sqrt{\log x}}\right)$$

mit konstantem α, wobei $h_0 = 2h$ oder $= h$ ist, je nachdem die betreffende Form einer zweiseitigen Klasse angehört oder nicht. Bei dem Beweis wird die Lehre von der Komposition der Klassen (vgl. I C 2c, 11) benutzt. Diese liefert bekanntlich eine *Abel*sche Gruppe, und indem man die Charaktere dieser Gruppe auf der linken Seite von (109) einführt, gewinnt man neue Funktionen, die den L-Funktionen (vgl. Nr. 29) entsprechen. Der Beweis verläuft dann ähnlich wie bei den Primzahlen in einer arithmetischen Reihe. Die hierbei auftretenden Summen

$$\sum_{x,y} (ax^2 + bxy + cy^2)^{-s},$$

---

von den Klassenzahlrelationen gab neuerdings *L. J. Mordell,* On class relation formulae, Messenger of Math. 46 (1916), p. 113—135.

364) *G. Lejeune-Dirichlet,* Über eine Eigenschaft der quadratischen Formen, Ber. Verhandl. Akad. Berlin 1840, p. 49—52; Werke 1, p. 497—502.

365) *E. Schering,* Beweis des Dirichletschen Satzes etc., Ges. math. Werke 2, p. 357—365.

366) *H. Weber,* Beweis des Satzes etc., Math. Ann. 20 (1882), p. 301—329; Über Zahlengruppen in algebraischen Körpern, Math. Ann. 48 (1897), p. 433—473; 49 (1897), p. 83—100; 50 (1898), p. 1—26.

367) a. a. O. 270).

368) *Ch. de la Vallée Poussin,* Recherches analytiques sur la théorie des nombres premiers, Troisième partie, Ann. soc. sc. Bruxelles 20 : 2 (1896), p. 363—397; Quatrième partie, ibid. 21 : 2 (1897), p. 251—342.

369) *P. Bernays,* Über die Darstellung von positiven, ganzen Zahlen durch die primitiven, binären quadratischen Formen einer nicht-quadratischen Diskriminante, Diss. Göttingen 1912.

370) *E. Landau,* a) Über die Verteilung der Primideale in den Idealklassen eines algebraischen Zahlkörpers, Math. Ann. 63 (1907), p. 145—204; b) Über die Primzahlen in definiten quadratischen Formen und die Zetafunktion reiner kubischer Körper, Math. Abhandl., H. Schwarz . . . gewidmet, Berlin (Springer) 1914, p. 244—273.

mit verschiedenen Bedingungen für die Summationsvariablen x und y, definieren in der ganzen Ebene eindeutige Funktionen, die nur für $s = 1$ einen Pol haben.[371]) Der Beweis dieses Satzes für positive Diskriminanten, der von *de la Vallée Poussin*[368]) gefunden wurde, war sehr kompliziert und wurde später von *Landau*[372]) vereinfacht. *Dirichlet*[373]) hat auch behauptet, daß unter den durch eine quadratische Form dargestellten Primzahlen unendlich viele vorkommen, die einer gegebenen arithmetischen Reihe angehören, vorausgesetzt, daß die Form überhaupt fähig ist, Zahlen von dieser Reihe darzustellen. *Meyer*[374]) hat diesen Satz bewiesen, *de la Vallée Poussin*[375]) und *Landau*[370]) haben ihn durch Angabe asymptotischer Formeln verschärft.

Aus den neueren Untersuchungen von *Hecke*[376]) geht hervor, daß die Form $ax^2 + bxy + cy^2$, wenn D Fundamentaldiskriminante ist, auch noch dann unendlich viele Primzahlen darstellt, wenn man die ganzzahligen Veränderlichen x und y auf einen beliebigen Winkelraum einschränkt.

Bernays[369]) zeigt, daß die Anzahl der ganzen Zahlen $n \leqq x$, die durch eine quadratische Form darstellbar sind, asymptotisch gleich

$$A \frac{x}{\sqrt{\log x}}$$

mit konstantem A ist (vgl. Nr. **37**, am Ende).

---

371) Vgl. Nr. **21**. Vgl. ferner *M. Lerch,* Základové theorie Malmsténovskych řad, Rozpravy české akad., 2. Kl., 1 (1892), Nr. **27**; Studie v oboru Malmsténovskych řad a invariantu forem kvadratickych, ibid. 2 (1893), Nr. **4**; *G. Herglotz,* Über die analytische Fortsetzung gewisser Dirichletscher Reihen, Math. Ann. 61 (1905), p. 551—560.

372) *E. Landau,* Neuer Beweis eines analytischen Satzes des Herrn de la Vallée Poussin, Jahresb. d. deutschen Math.-Ver. 24 (1915), p. 250—278.

373) *G. Lejeune-Dirichlet,* Extrait d'une lettre etc., Paris C. R. 10 (1840), p. 285—288; J. Math. pures appl. (1) 5 (1840), p. 72—74; Sur la théorie des nombres, Werke 1, p. 619—623.

374) *A. Meyer,* Über einen Satz von Dirichlet, Crelles J. 103 (1888), p. 98 bis 117. Vgl. auch *P. Bachmann,* Die analytische Zahlentheorie, Leipzig (Teubner) 1894, insbes. Abschn. 10.

375) *Ch. de la Vallée Poussin,* Recherches analytiques sur la théorie des nombres premiers; Cinquième partie, Ann. Soc. sc. Bruxelles **21**: 2 (1897), p. 343 bis 368.

376) *E. Hecke,* Eine neue Art von Zetafunktionen und ihre Beziehungen zur Verteilung der Primzahlen, Math. Ztschr. 1 (1918), p. 357—376; 6 (1920), p. 11—51.

41. Die Zetafunktionen von Dedekind und Hecke.[377]) *Dedekind*[378]) hat die *Riemann*sche Zetafunktion für einen beliebigen algebraischen Zahlkörper k vom Grade n verallgemeinert. Er setzt[379]) (vgl. I C 4a, Nr. 9)

$$\zeta_k(s) = \sum_{\mathfrak{a}} \frac{1}{N\mathfrak{a}^s} = \prod_{\mathfrak{p}} \left(1 - \frac{1}{N\mathfrak{p}^s}\right)^{-1} = \sum_{1}^{\infty} \frac{F(m)}{m^s},$$

wo $\mathfrak{a}$ die Ideale, $\mathfrak{p}$ die Primideale von k durchläuft und $F(m)$ die Anzahl der Darstellungen von m als Norm eines Ideals von k bedeutet. Alle drei Ausdrücke sind für $\sigma > 1$ absolut konvergent; $\zeta_k(s)$ ist demnach in dieser Halbebene regulär und $\neq 0$. Für den Körper der rationalen Zahlen ist $\zeta_k(s)$ mit $\zeta(s)$ identisch. Mit Hilfe einer *Weber*schen[380]) Verschärfung der *Dedekind*schen[378]) Abschätzung der Anzahl aller Ideale von k mit Norm $\leqq x$ zeigt *Landau*[381]), daß $\zeta_k(s)$ auch noch für $\sigma > 1 - \frac{1}{n}$ regulär ist, mit Ausnahme des Punktes $s = 1$, wo sie einen Pol erster Ordnung mit dem schon von *Dedekind*[378]) angegebenen Residuum

$$\frac{2^{r_1 + r_2} \pi^{r_2} R h}{w\sqrt{|d|}}$$

besitzt. Hier bedeutet (vgl. I C 4a) d die Grundzahl, R den Regulator, w die Anzahl der Einheitswurzeln und h die Anzahl der Idealklassen[382]) von k. Ferner bezeichnet r_1 die Anzahl der reellen und $2r_2 = n - r_1$ die Anzahl der nicht-reellen Wurzeln der irreduziblen Gleichung, die von einer den Körper k erzeugenden Zahl befriedigt wird.

Kann nun das Residuum von $\zeta_k(s)$ anderweitig bestimmt werden, so erhält man offenbar einen Ausdruck für die Klassenzahl h. (Im Falle eines quadratischen Körpers ist dieses Verfahren im wesent-

377) Vgl. hierzu II B 7, Nr. 129, wo die Beziehungen zu der allgemeinen Theorie der Thetafunktionen behandelt werden.

378) *G. Lejeune-Dirichlet,* Vorlesungen über Zahlentheorie, herausgeg. und m. Zusätzen versehen von *R. Dedekind,* 4. Aufl. 1894, p. 610.

379) Die kleinen deutschen Buchstaben bezeichnen Ideale, $N\mathfrak{a}$ ist die Norm des Ideals $\mathfrak{a}$, $N\mathfrak{a}^s$ bedeutet $(N\mathfrak{a})^s$.

380) *H. Weber,* Über einen in der Zahlentheorie angewandten Satz der Integralrechnung, Gött. Nachr. 1896, p. 275—281; Lehrbuch der Algebra 2, 2. Aufl. 1899, p. 672—678, 712.

381) *E. Landau,* a. a. O. 28).

382) Wo nichts anderes gesagt wird, ist der Begriff „Idealklasse" im „weitesten Sinne" genommen, d. h. zwei Ideale sind zur gleichen Klasse gehörig, sobald ihr Quotient eine Körperzahl ist.

lichen mit dem *Dirichlet*schen — vgl. Nr. 40 — identisch.) Dies läßt sich aber nur in wenigen Fällen durchführen[383]), und zwar:

a) Für die Kreiskörper und deren Unterkörper (vgl. I C 4b[384])). Im Körper der ν^{ten} Einheitswurzeln ist

$$\zeta_k(s) = L_1(s)\, L_2(s) \ldots L_{\varphi(\nu)}(s) \cdot \prod_{p \mid \nu} (1 - p^{-fs})^{-q},$$

wo alle $\varphi(\nu)$ L-Funktionen mod. ν (vgl. Nr. 29) multipliziert werden, und f und q gewisse von ν und p abhängige positive ganze Zahlen sind. Die Formel für die Klassenzahl gibt hier also einen neuen Beweis für das Nichtverschwinden sämtlicher Reihen $L(1)$.[385])

b) Für die Klassenkörper der komplexen Multiplikation[386]).

c) Für solche Körper 4. Grades, die durch eine Zahl von der Form $\sqrt{a + b\sqrt{c}}$ mit rationalen a, b, c, sowie $c > 0$, $a \pm b\sqrt{c} < 0$ erzeugt werden. *Hecke*[387]) beweist unter Anwendung der von ihm untersuchten *Gauß*schen Summen in algebraischen Zahlkörpern einen Satz über gewisse Relativklassenzahlen, aus dem insbesondere ein Ausdruck für die Klassenzahl der genannten Körper folgt. Es zeigen sich hierbei eigentümliche Zusammenhänge mit tiefliegenden Fragen aus der Theorie der Thetafunktionen.

Die analytische Fortsetzung von $\zeta_k(s)$ über $\sigma = 1 - \frac{1}{n}$ hinaus konnte lange nur bei speziellen Körpern ausgeführt werden. Ein sehr bedeutender Fortschritt wurde nun von *Hecke*[388]) gemacht, indem es

383) *E. Landau*, Über eine Darstellung der Anzahl der Idealklassen eines algebraischen Körpers durch eine unendliche Reihe, Crelles J. 127 (1904), p. 167 bis 174 (vgl. auch a. a. O. 78)), zeigt, daß die *Dirichlet*sche Reihe für $\zeta_k(s) \cdot \frac{1}{\zeta(s)}$ im Punkte $s = 1$ konvergiert, so daß die Klassenzahl immer durch eine konvergente unendliche Reihe dargestellt werden kann.

384) Vgl. auch die neuere Darstellung von *R. Fueter*, Synthetische Zahlentheorie, Berlin u. Leipzig (Göschen) 1917.

385) *Dirichlet-Dedekind*, a. a. O. 378), p. 625.

386) Jedes Eingehen auf die Theorie der Klassenkörper mußte aus diesem Bericht ausgeschlossen werden. Vgl. hierzu I C 6.

387) *E. Hecke*, Bestimmung der Klassenzahl einer neuen Reihe von algebraischen Zahlkörpern, Gött. Nachr. 1921, p. 1—23; Reziprozitätsgesetz und Gaußsche Summen in quadratischen Zahlkörpern, ibid. 1919, p. 265—278. Vgl. auch *L. J. Mordell*, On the reciprocity formula for the Gauss's sums in the quadratic field, Proc. London math. Soc. (2) 20 (1920), p. 289—296; *K. Reidemeister*. Über die Relativklassenzahl gewisser relativquadratischer Zahlkörper, Abhandl. Math. Seminar Hamburg 1 (1922), p. 27—48.

388) *E. Hecke*, Über die Zetafunktion beliebiger algebraischer Zahlkörper, Gött. Nachr. 1917, p. 77—89; eine Anwendung der Entdeckung auf die Theorie der Klassenkörper gibt die Arbeit: Über eine neue Anwendung der Zetafunktionen auf die Arithmetik der Zahlkörper, Gött. Nachr. 1917, p. 90—95.

ihm gelang, den zweiten *Riemann*schen Beweis für $\zeta(s)$ (vgl. Nr. **14**) zu verallgemeinern und dadurch nicht nur die Existenz von $\zeta_k(s)$ in der ganzen Ebene nachzuweisen, sondern auch eine der *Riemann*schen analoge Funktionalgleichung für $\zeta_k(s)$ aufzustellen und mit ihrer Hilfe die *Hadamard*schen Sätze über das Geschlecht und die Produktentwicklung der ganzen Funktion $(s-1)\zeta(s)$ (vgl. Nr. **15**) zu verallgemeinern.[389]) Es läßt sich nämlich, wenn das Ideal $\mathfrak{a}$ gegeben ist, die Summe[390])

$$\zeta(s, \mathfrak{a}) = \sum_{\mathfrak{a}\mathfrak{j} \sim 1} \frac{1}{N\mathfrak{j}^s}, \tag{112}$$

die über alle Ideale $\mathfrak{j}$ der Klasse $\mathfrak{a}^{-1}$ erstreckt ist, durch eine Thetareihe

$$\vartheta(\tau_1, \ldots \tau_n; \mathfrak{a}) = \sum_{\mu \equiv 0 \pmod{\mathfrak{a}}} e^{-\frac{\pi}{\sqrt[n]{N\mathfrak{a}^2|d|}} \sum_{p=1}^{n} \tau_p |\mu^{(p)}|^2} \tag{113}$$

ausdrücken, wobei μ alle Zahlen von $\mathfrak{a}$ durchläuft, $\mu^{(1)}, \ldots \mu^{(n)}$ die konjugierten Zahlen (inkl. μ selbst), in bestimmter Reihenfolge wie in I C 4a, Nr. 7, geordnet, und diejenigen τ_p, welche konjugiert imaginären Körpern entsprechen, einander gleich sind. *Hecke* findet in der Tat durch sinnreiche Überlegungen

$$\Phi(s, \mathfrak{a}) = A^s \left(\Gamma\left(\frac{s}{2}\right)\right)^{r_1} (\Gamma(s))^{r_2} \zeta(s, \mathfrak{a}) \tag{114}$$

$$= \frac{2^{r_1 - 1} n R}{w} \int_{-\frac{1}{2}}^{\frac{1}{2}} dx_1 \ldots \int_{-\frac{1}{2}}^{\frac{1}{2}} dx_r \int_0^\infty u^{\frac{ns}{2} - 1} (\vartheta(\tau_1, \ldots \tau_n; \mathfrak{a}) - 1)\, du,$$

wo

$$A = 2^{-r_2} \pi^{-\frac{n}{2}} \sqrt{|d|},$$

$$\tau_p = u\, e^{\frac{2}{n} \sum_{q=1}^{r} x_q \log |\varepsilon_q^{(p)}|}$$

gesetzt ist und $\varepsilon_1 \ldots \varepsilon_r$ ein System von Grundeinheiten des Körpers bezeichnet. (Wie üblich ist $r = r_1 + r_2 - 1$.) Die Thetareihe (113) genügt aber der Funktionalgleichung

$$\vartheta(\tau_1, \ldots \tau_n; \mathfrak{a}) = \frac{1}{\sqrt{\tau_1 \tau_2 \ldots \tau_n}} \vartheta\left(\frac{1}{\tau_1} \cdots \frac{1}{\tau_n}; \mathfrak{a}^{-1} \mathfrak{d}^{-1}\right),$$

389) Neuerdings wurde der *erste Riemann*sche Beweis von *C. Siegel* für $\zeta_k(s)$ verallgemeinert: Neuer Beweis für die Funktionalgleichung der Dedekindschen Zetafunktion, Math. Ann. 85 (1922), p. 123—128.

390) Die Ausdrücke (112), (113) und (114) sind nicht von dem Ideal $\mathfrak{a}$ selbst, sondern nur von der Klasse von $\mathfrak{a}$ abhängig.

wo $\mathfrak{d}$ das Grundideal von k ist (vgl I C.4a, Nr. 5); hieraus folgt

$$\Phi(s, \mathfrak{a}) - \frac{2^{r_1} R}{w\, s\,(s-1)} =$$

$$= \frac{2^{r_1-1} n R}{w} \int_{-\frac{1}{2}}^{\frac{1}{2}} d x_1 \ldots \int_{-\frac{1}{2}}^{\frac{1}{2}} d x_r \int_1^{\infty} \Big[u^{\frac{n s}{2}} (\vartheta(\tau_1 \ldots \tau_n;\, \mathfrak{a}) - 1) +$$

$$+ u^{\frac{n(1-s)}{2}} (\vartheta(\tau_1 \ldots \tau_n;\, \mathfrak{a}^{-1}\mathfrak{d}^{-1}) - 1) \Big] \frac{du}{u},$$

was der Gleichung (26) von Nr. **14** entspricht. Da $\mathfrak{a}^{-1}\mathfrak{d}^{-1}$ gleichzeitig mit $\mathfrak{a}$ alle h Idealklassen durchläuft, folgt weiter:

$$Z(s) = s\,(1-s)\, A^s \left(\Gamma\left(\frac{s}{2}\right)\right)^{r_1} (\Gamma(s))^{r_2}\, \zeta_k(s)$$

ist eine ganze Funktion, die sich bei Vertauschung von s mit $1-s$ nicht ändert. $\zeta_k(s)$ ist also, bis auf den Pol bei $s = 1$, in der ganzen Ebene regulär und besitzt die Funktionalgleichung

$$\zeta_k(1-s) = \left(\frac{2}{(2\pi)^s}\right)^n |d|^{s-\frac{1}{2}} \left(\cos\frac{s\pi}{2}\right)^{r_1+r_2} \left(\sin\frac{s\pi}{2}\right)^{r_1} \left(\Gamma(s)\right)^n \zeta_k(s).$$

(Vgl. [24] Nr. **14**.) Der Punkt $s = 0$ ist Nullstelle r^{ter} Ordnung, $s = -2, -4, \ldots$ Nullstellen $(r+1)^{\text{ter}}$ Ordnung, $s = -1, -3, \ldots$ Nullstellen $r_2{}^{\text{ter}}$ Ordnung; außerdem gibt es unendlich viele Nullstellen ϱ, die sämtlich dem Streifen $0 \leqq \sigma \leqq 1$ angehören, und es kann wie bei $\zeta(s)$ die Produktentwicklung

$$(s-1)\,\zeta_k(s) = a e^{b s} \frac{1}{s\left(\Gamma\left(\frac{s}{2}\right)\right)^{r_1} \left(\Gamma(s)\right)^{r_2}} \prod_{\varrho} \left(1 - \frac{s}{\varrho}\right) e^{\frac{s}{\varrho}}$$

(vgl. [28], Nr. **15**) abgeleitet werden.

Landau[391]) gibt eine Zusammenfassung der Theorie von $\zeta_k(s)$ und zeigt dabei, daß alle ϱ dem *Innern* des „kritischen Streifens" angehören[392]), und daß sogar das Gebiet (vgl. Nr. **19**)

$$\sigma > 1 - \frac{k}{\log t}, \quad t > t_0$$

391) *E. Landau*, Einführung in die elementare und analytische Theorie der algebraischen Zahlen und der Ideale, Leipzig u. Berlin (Teubner) 1918.

392) *Landau*, a. a. O. 28), hat schon vor *Heckes* Entdeckung gezeigt, daß keine Nullstelle auf $\sigma = 1$ liegt, und sogar daß $\left|\frac{\zeta_k'}{\zeta_k}\right| < c \log^9 t$ im Gebiete $\sigma > 1 - \frac{k}{\log^7 t}$, $t > t_0$ gilt, was für die Verallgemeinerung des Primzahlsatzes wichtig war (vgl. Nr. **42**). Vgl. hierzu auch *Landau*, Über die Wurzeln der Zetafunktion eines algebraischen Zahlkörpers, Math. Ann. 79 (1919), p. 388—401.

bei passender Wahl von k und t_0 von Nullstellen frei ist. Die Anzahl der ϱ im Rechteck $0 < \sigma < 1$, $0 < t \leqq T$ ist nach *Landau*[391])

$$N(T) = \frac{n}{2\pi} T \log T + \frac{\log|d| - n - n\log 2\pi}{2\pi} T + O(\log T).^{393)}$$

(Vgl. [30], Nr. **16**.) Ob bei der allgemeinen $\zeta_k(s)$ unendlich viele ϱ auf der Geraden $\sigma = \frac{1}{2}$ liegen (vgl. Nr. **19**), ist bisher nicht entschieden.

Die Sätze über die Werte von $\zeta(s)$ auf einer vertikalen Geraden[394]) und über die Größenordnung von $\zeta(s)$ wurden zum Teil auch für $\zeta_k(s)$ verallgemeinert. Insbesondere ist über die μ-Funktion von $\zeta_k(s)$ (vgl. Nr. **18**) bekannt, daß $\mu(\sigma) = 0$ für $\sigma > 1$ und $\mu(\sigma) = n\left(\frac{1}{2} - \sigma\right)$ für $\sigma < 0$ ist[391]), während für $0 < \sigma < 1$ die μ-Kurve im Dreieck mit den Eckpunkten $\left(0, \frac{n}{2}\right)$, $\left(\frac{1}{2}, 0\right)$, $(1, 0)$ verläuft.

Bei Untersuchungen über die Verteilung der Primideale in den verschiedenen Idealklassen des Körpers, bzw. in den Idealklassen mod. $\mathfrak{f}$ ($\mathfrak{f}$ ein ganzes Ideal), treten gewisse Funktionen auf, die den L-Funktionen des rationalen Körpers entsprechen (vgl. Nr. **29** und für den quadratischen Körper Nr. **40**). Sie werden für $\sigma > 1$ durch die Gleichung

$$\zeta_k(s, \chi) = \sum_{\mathfrak{a}} \frac{\chi(\mathfrak{a})}{N\mathfrak{a}^s} = \prod_{\mathfrak{p}} \left(1 - \frac{\chi(\mathfrak{p})}{N\mathfrak{p}^s}\right)^{-1}$$

definiert, wobei die idealtheoretische Funktion $\chi(\mathfrak{a})$ durch einen Charakter der betreffenden *Abel*schen Gruppe von Idealklassen in analoger Weise wie die zahlentheoretische Funktion $\chi(n)$ bei den L-Funktionen (vgl. Nr. **29**) bestimmt ist. Die analytische Fortsetzung wurde von *Landau*[370a]) bis zu $\sigma = 1 - \frac{1}{n}$, von *Hecke*[395]) über die ganze Ebene ausgeführt und die entsprechenden Funktionalgleichungen wurden von *Hecke*[395]) und *Landau*[396]) aufgestellt. Bei jedem vom Hauptcharakter verschiedenen χ ist $\zeta_k(s, \chi)$, bei dem Hauptcharakter aber $(s - 1) \cdot \zeta_k(s, \chi)$, eine ganze Funktion von Geschlechte Eins, die im Streifen

393) Einen Satz über den Mittelwert des Restgliedes in dieser Formel gibt *H. Cramér*, a. a. O. 186).

394) Vgl. *H. Bohr* und *E. Landau*, a. a. O. 113).

395) *E. Hecke*, Über die L-Funktionen und den Dirichletschen Primzahlsatz für einen beliebigen Zahlkörper, Gött. Nachr. 1917, p. 299—318.

396) *E. Landau*, Über Ideale und Primideale in Idealklassen, Math. Ztschr. 2 (1918), p. 52—154. Es werden hier auch andere Äquivalenzbegriffe, d. h. andere Definitionen des Begriffs „Idealklasse" berücksichtigt.

$0 < \sigma < 1$ unendlich viele Nullstellen besitzt[397]), aber für $\sigma \geqq 1$ durchweg von Null verschieden ist.[398])

Um ein genaueres Studium der Verteilung der Primideale von k zu ermöglichen, führt *Hecke*[376]) eine Klasse *verallgemeinerter Zetafunktionen*

$$\zeta(s, \lambda) = \sum_{\mathfrak{a}} \frac{\lambda(\mathfrak{a})}{\mathfrak{a}^s} = \prod_{\mathfrak{p}} \left(1 - \frac{\lambda(\mathfrak{p})}{N\mathfrak{p}^s}\right)^{-1}$$

ein, wo die $\lambda(\mathfrak{a})$ gewisse der Multiplikationsregel $\lambda(\mathfrak{a})\,\lambda(\mathfrak{b}) = \lambda(\mathfrak{a}\mathfrak{b})$ genügende „Größencharaktere" von $\mathfrak{a}$ sind, die von $n - 1$ „Grundcharakteren" abhängen. Durch die Angabe der Grundcharaktere und der Norm ist das Ideal $\mathfrak{a}$ eindeutig bestimmt. Diese $\zeta(s, \lambda)$ lassen sich wie $\zeta_k(s)$ durch Thetareihen ausdrücken und besitzen auch ähnliche Funktionalgleichungen. Durch Kombination dieser Resultate mit einem Satz von *Weyl* (vgl. Nr. 39) über diophantische Approximationen erhält *Hecke* neue Sätze über die Verteilung der Primideale und der Primzahlen in gewissen zerlegbaren Formen (vgl. Nr. 40). — Endlich sei auch noch erwähnt, daß *Hecke*[399]) neuerdings verschiedene einem Zahlkörper zugeordnete analytische Funktionen *mehrerer* Variablen eingeführt hat, wobei die $\zeta(s, \lambda)$ als Hilfsmittel dienen.

42. Die Verteilung der Ideale und der Primideale. Es war für die Verallgemeinerung des Primzahlsatzes wesentlich, daß die von *Landau* eingeführten Methoden (vgl. Nr. 25) nur das Verhalten von $\zeta(s)$ in der Nähe von $\sigma = 1$ benutzten. Ohne die Existenz der *Dedekind*schen $\zeta_k(s)$ in der ganzen Ebene — die damals nicht bekannt war — vorauszusetzen, gelang es ihm nämlich, den sog. *Primidealsatz*[400]) zu beweisen: Für jeden Körper k vom Grade n ist die Anzahl $\pi_k(x)$ der Primideale mit Norm $\leqq x$ asymptotisch gleich $Li(x)$. Unter Benutzung der Gleichung

$$\log \zeta_k(s) = \sum_{\mathfrak{p}, m} \frac{1}{m\, N\mathfrak{p}^{ms}} \qquad (\sigma > 1)$$

---

397) Außerdem gibt es nur „triviale" Nullstellen bei $s = 0$ und auf der negativen reellen Achse sowie — im Falle eines uneigentlichen Charakters — auf der imaginären Achse.

398) Vgl. *E. Hecke*, a. a. O. 395); *E. Landau*, a. a. O. 370a) und 396), Zur Theorie der Heckeschen Zetafunktionen, welche komplexen Charakteren entsprechen, Math. Ztschr. 4 (1919), p. 152—162.

399) *E. Hecke*, Analytische Funktionen und algebraische Zahlen, I. Teil Abhandl. Math. Seminar Hamburg 1 (1922), p. 102—126.

400) *E. Landau*, Neuer Beweis des Primzahlsatzes und Beweis des Primidealsatzes, Math. Ann. 56 (1903), p. 645—670.

ergibt sich dies genau wie bei dem Primzahlsatz. Es gilt sogar nach *Landau*[401])

$$\pi_k(x) = Li(x) + O\left(x e^{-\frac{\alpha}{\sqrt{n}}\sqrt{\log x}}\right)$$

$$\vartheta_k(x) = \sum_{N\mathfrak{p} \leqq x} \log N\mathfrak{p} = x + O\left(x e^{-\frac{\alpha}{\sqrt{n}}\sqrt{\log x}}\right)$$

mit absolut konstantem α, was jedoch erst nach der Entdeckung der Fortsetzbarkeit von $\zeta_k(s)$ bewiesen werden konnte. Hierin ist als Spezialfall die Abschätzung von $\pi(x)$ (vgl. Nr. 27) enthalten. Für die Primideale einer Idealklasse (bzw. einer Idealklasse mod. $\mathfrak{f}$, wo $\mathfrak{f}$ ein ganzes Ideal ist) gelten entsprechende Gleichungen[402]), und zwar gilt dies auch bei engerer Auffassung des Klassenbegriffs. Dies enthält wiederum als Spezialfall *Dirichlets* Satz von der arithmetischen Reihe (vgl. Nr. 30). Die *Littlewood*schen Beziehungen (60) von Nr. 27 wurden von *Landau*[396]) für einen beliebigen Körper k und für eine beliebige Idealklasse von k verallgemeinert. Insbesondere ist also

$$\liminf_{x \to \infty} \frac{\pi_k(x) - Li(x)}{\frac{\sqrt{x}}{\log x} \log\log\log x} < 0 < \limsup_{x \to \infty} \frac{\pi_k(x) - Li(x)}{\frac{\sqrt{x}}{\log x} \log\log\log x}.$$

Bei dem Beweis dieses Satzes dient als Hilfsmittel die Verallgemeinerung der *Riemann-v. Mangoldt*schen Primzahlformel[403]); über die Konvergenzeigenschaften von $\sum \frac{x^\varrho}{\varrho}$ gibt es auch hier ähnliche Sätze wie bei $\zeta(s)$ (vgl. Nr. 28).

Wird das Residuum von $\zeta_k(s)$ für $s = 1$ durch $h\lambda$ bezeichnet, so gilt nach *Landau*[404]) für die Anzahl $H(x, K)$ der Ideale mit Norm $\leqq x$, die einer Klasse K von k angehören,

$$H(x, K) = \lambda x + O\left(x^{1 - \frac{2}{n+1}}\right)$$

und für die Anzahl $H(x)$ aller Ideale des Körpers mit Norm $\leqq x$

$$H(x) = h\lambda x + O\left(x^{1 - \frac{2}{n+1}}\right).$$

---

401) *E. Landau*, a. a. O. 391). Die in der vorigen Fußnote erwähnte Arbeit gibt eine weniger gute Abschätzung.

402) *E. Hecke*, a. a. O. 395); *E. Landau*, a. a. O. 227), 370a und 396).

403) *E. Landau*, a. a. O. 391) und 396). Für einige Anwendungen einer analogen Formel vgl. *H. Cramér*, a. a. O. 186).

404) *E. Landau*, a. a. O. 289b), 391) und 396).

Andererseits zeigt *Walfisz*[297]), der die von *Hardy* bei den Teilerproblemen (vgl. Nr. **34**) angewandte Methode benutzt,

$$\liminf_{x \to \infty} \frac{H(x) - h\lambda x}{x^{\frac{n-1}{2n}}} < 0 < \limsup_{x \to \infty} \frac{H(x) - h\lambda x}{x^{\frac{n-1}{2n}}}.$$

In diesen Beziehungen sind als Spezialfälle verschiedene der in Nr. **35** erwähnten Resultate bei den Kreis- und Ellipsoidproblemen enthalten.[405]) *Walfisz*[297]) hat auch eine explizite Formel für $H(x)$ aufgestellt, die als Spezialfälle die entsprechenden Formeln bei jenen Problemen enthält.

Setzt man, analog wie bei $\zeta(s)$, für $\sigma > 1$

$$\frac{1}{\zeta_k(s)} = \prod_{\mathfrak{p}} \left(1 - \frac{1}{N\mathfrak{p}^s}\right) = \sum_{\mathfrak{a}} \frac{\mu(\mathfrak{a})}{N\mathfrak{a}^s},$$

so läßt sich über die idealtheoretische Funktion $\mu(\mathfrak{a})$ z. B.

$$\sum_{N\mathfrak{a} \leqq x} \mu(\mathfrak{a}) = o(x), \qquad \sum_{\mathfrak{a}} \frac{\mu(\mathfrak{a})}{N\mathfrak{a}} = 0,$$

mit entsprechenden schärferen Abschätzungen, beweisen.[406]) Auch die Zusammenhangssätze der rationalen Zahlentheorie (vgl. Nr. **33**) lassen sich für einen beliebigen Körper k verallgemeinern[407]), sowie auch verschiedene andere der in den Nummern **32**—**37** erwähnten Sätze über zahlentheoretische Funktionen.[408])

405) Die entsprechenden oberen Abschätzungen waren überhaupt für quadratische Körper schon früher bekannt. Vgl. z. B. *E. Landau*, a. a. O. 284); *A. Hammerstein*, a. a. O. 200).

406) *E. Landau*, Über die zahlentheoretische Funktion $\mu(k)$, Sitzungsber. Akad. Wien **112** (1903), Abt. 2a, p. 537—570.

407) *E. Landau*, a. a. O. 21); *A. Axer*, a. a. O. 275).

408) Vgl. z. B. *E. Landau*, a. a. O. 300) und 370a); *A. Axer*, Przyczynek do charakterystyki funkcyi idealowej $\varphi(\mathfrak{x})$ [Sur la fonction $\varphi(\mathfrak{x})$ dans la théorie des idéaux], Prace Math.-Fiz. 21 (1910), p. 37—41.

(Abgeschlossen im Mai 1922.)

20.

Remarks on correlation

Skand. Aktuarietidskr. 7, 220–240 (1924)

1. The fundamental problem of the mathematical theory of correlation is this: given two statistical variables x and y, we want to know if there may be said to exist some kind of *dependence* between them, and we want a method permitting us to give a simple and quantitative description of the formal nature of this dependence. This problem once solved in a particular case, another question immediately arises: to study the *real* nature of the connection between the variables — this, however, generally falls quite outside the scope of the mathematical theory of correlation. Of course our problem can be generalized to any number of variables; as far as concerns the questions dealt with in this paper, it will be sufficient to consider the case of two variables.

A good deal of the work laid down on this theory has centred round the problem of finding a numerical *measure* of the *degree of dependence* which may exist between two statistical variables. Quite in the same way as in so many other branches of science there has been a tendency to assume *a priori* that a complex and vague conception like that of »dependence» is capable of being adequately measured by a single number. That this is not so, seems to be in principle recognized by most modern authors on the subject; yet we can find numerous instances showing that the less critical view still holds a strong position.

Placing ourselves on the point of view of the theory of probability, and assuming the *a priori* probabilities for the different possible values of the variables to be known, we may ask if it is possible to give a reasonable meaning to the term »degree of dependence», permitting us to discuss the question of its characterization by a number.

2. It is convenient to start with a definition of »independence»; in this respect, we shall adopt the following definition, which seems to be the only one of sufficient generality among those hitherto proposed: *A statistical variable y is said to be independent of another variable x, if the probability that y takes any given value, or falls within any given limits, is in no way affected by our knowledge of the value which x may have taken.* If y is independent of x, it is easily proved that conversely x is independent of y; hence in this case the two variables may be said to be *independent of one another.*

We turn now to the case of two variables between which there exists some kind of dependence. In order to get a complete knowledge of the formal nature of this dependence, we must know the corresponding *probability function*, which may be generally defined in the following way:

$$F(u, v)$$

denotes the probability that we have simultaneously

$$x \leqq u, \; y \leqq v.$$

The knowledge of this function for all values of u and v gives the full answer to the question, but generally this answer does not satisfy us — because it contains too much. What we really want is a systematic method of characterizing by a small number of parameters the essential features of the result, so as to enable us to compare different cases without difficulty. In the present paper, this question will be discussed as far as the »degree of dependence» is concerned. Here, we want to find a definition which with each pair of statistical variables associates a number, in such a way that this number increases or decreases with the degree of dependence. In an absolutely strict sense, there is obviously no solution of this problem, but we may set out to look for the most *natural* definition we can possibly find. We shall not, however, consider an absolutely general probability function, but only two special cases, which we denote by A and B.

A. There is only a finite number of possible values for each variable. x may have one of the m values $x_1, x_2, \ldots x_m$ and y one of the n values $y_1, y_2, \ldots y_n$. The probability that we have simultaneously $x = x_i$ and $y = y_k$ is denoted by p_{ik}. Hence we have $\sum_{i,k} p_{ik} = 1$.

B. The probability that we have simultaneously $a < x < b$ and $c < y < d$ is for all values of a, b, c, d given by

$$\int_a^b \int_c^d f(u, v)\, du\, dv,$$

where $f(u, v)$ denotes a continuous and never negative function of its arguments, which are allowed to vary from $-\infty$ to $+\infty$. We have

$$\int_{-\infty}^{\infty} \int_{-\infty}^{\infty} f(u, v)\, du\, dv = 1.$$

3. In many of the cases which originally presented themselves for scientific treatment, the variables may be said to be, roughly speaking, linear functions of one another. It was through the study of such cases that Galton was led to the introduction of his famous *coefficient of correlation*, which is still the most frequently used measure of the degree of dependence between two statistical variables. Denoting generally by $E(\alpha)$ the »mathematical expectation» of the statistical variable α, and putting

$$\xi = x - E(x),$$
$$\eta = y - E(y), \tag{1}$$

the coefficient of correlation, r, is defined by the relation

$$r = \frac{E(\xi\eta)}{\sqrt{E(\xi^2)\,E(\eta^2)}}.$$

If x and y are independent according to the definition given above, we have

$$E(\xi\eta) = E(\xi)\,E(\eta) = 0$$

and hence $r = 0$. The converse, however, is not true, though this has often been asserted. Thus $r = 0$ is a necessary, but not a sufficient, condition for independence, and in the general case r does not at all give a satisfactory measure of the degree of dependence, although it may render excellent service in many particular cases, where the connection between the variables is more or less a linear one. Among these particular cases, perhaps the most important is represented by the so called *normal correlation function*

$$f(u, v) = \frac{\sqrt{ac - b^2}}{\pi} e^{-(au^2 + 2buv + cv^2)},$$

where a, b, c are constants such that $ac - b^2 > 0$. In this case we have

$$r = \frac{b}{\sqrt{ac}}$$

and hence $0 < r^2 < 1$. For small values of r^2, the variables are »almost» independent, whereas for r^2 just below 1 the dependence is almost complete.

4. Let us first consider case A. If the variables are independent, it is readily seen that p_{ik} is of the form

$$p_{ik} = q_i r_k, \tag{2}$$

where

$$q_i = \sum_k p_{ik},$$
(3)
$$r_k = \sum_i p_{ik},$$

so that q_i denotes the total probability that x takes the value x_i, while r_k has the corresponding significance for y. If the matrix formed by the numbers p_{ik} cannot be represented in the form (2), the variables are not independent.

This suggests that we should adopt as a measure of the degree of dependence the »deviation» of the table formed by the numbers p_{ik} *from the form* (2). *A natural way of measuring this deviation seems to be by calculating the minimum of the expression*

$$\sum_{i,k} (p_{ik} - u_i v_k)^2$$

for all real values of the $m + n$ *variables* u_i *and* v_k.

It will be shown in the following article how this minimum value can be calculated. For two independent variables it is equal to zero; in all other cases it is a positive quantity which is never greater than $\frac{1}{4}$. Obviously it is invariant for any permutation of rows and columns in the rectangular matrix formed by the numbers p_{ik}. — This measure bears a certain resemblance to the *mean square contingency* of K. Pearson,

$$\sum_{i,k} \frac{(p_{ik} - q_i r_k)^2}{q_i r_k},$$

which will also be discussed below.

While the above case can be treated by quite elementary means, case B requires a little more refined analysis. For two independent variables, we plainly have

$$f(u, v) = q(u)\, r(v),$$

where $\int_a^b q(u)\,d(u)$ denotes the probability of $a < x < b$, while $\int_c^d r(v)\,dv$ has the corresponding significance for y.

Thus in this case we are led by the same reasoning as before to adopt as a measure of the degree of dependence the minimum value of the expression

$$(4) \qquad \int_{-\infty}^{\infty}\int_{-\infty}^{\infty}[f(u, v) - g(u)\,h(v)]^2\,du\,dv$$

for all continuous functions $g(u)$ and $h(v)$.

The calculation of this minimum obviously forms a problem belonging to the calculus of variations. For the case of a finite domain of integration it has been solved by E. SCHMIDT[1], who treats it by means of the theory of FREDHOLM's integral equation. Owing to the assumptions that can be made as to the behaviour of our function $f(u, v)$ at infinity, his solution holds good even for infinite limits of integration, as will be shown in art. 7. Here again, the minimum is zero for two independent variables, and otherwise a positive quantity. There is, however, one additional remark to be made in this case. Suppose we have found the minimum value m for two variables x and y. Now if we change the units of measurement and consider the variables ax and by, we get the minimum value $\frac{m}{ab}$. Thus we ought to normalize the expression (4) by multiplying it with some convenient factor. It seems natural to choose for this purpose the factor $\sqrt{E(\xi^2)\,E(\eta^2)}$, ξ and η being given by (1), assuming that the infinite integrals representing $E(\xi^2)$ and $E(\eta^2)$ are convergent. This means simply that we choose the »standard deviations» of the variables as units of measurement.

[1] Zur Theorie der linearen und nichtlinearen Integralgleichungen. I. Teil, Math. Annalen 63, p. 433. Cf. in particular the fourth chapter.

As already stated above, there is no absolutely general measure of the degree of dependence. Every attempt to measure a conception like this by a single number must necessarily contain a certain amount of arbitrariness and suffer from certain inconveniences. Of course our definitions do not form an exception from this rule; they seem, however, in a way rather natural, and may deserve some attention, even if their practical value is somewhat limited. — In art. 8, the normal correlation function will be treated as an example.

5. In this article, we shall consider the minimum of the expression

$$\sum_{i,k} (p_{ik} - u_i v_k)^2, \tag{5}$$

associated with case A of art. 2. Putting

$$a_{ij} = \sum_{k=1}^{n} p_{ik} p_{jk}, \qquad (i, j = 1, 2, \ldots m),$$

$$b_{kl} = \sum_{i=1}^{m} p_{ik} p_{il}, \qquad (k, l = 1, 2, \ldots n),$$

we shall begin by proving the following theorem:

The minimum of (5) *is equal to*

$$\sum_{i,k} p_{ik}^2 - \varkappa_0,$$

where $\varkappa_0$ *denotes the greatest root of any of the equations*

$$\begin{vmatrix} a_{11} - \varkappa & a_{12} & \ldots & a_{1m} \\ a_{21} & a_{22} - \varkappa & \ldots & a_{2m} \\ \cdot & \cdot & \cdot & \cdot \\ a_{m1} & a_{m2} & \ldots & a_{mm} - \varkappa \end{vmatrix} = 0, \tag{6}$$

$$(7)\qquad \begin{vmatrix} b_{11}-\varkappa & b_{12} & \dots & b_{1n} \\ b_{21} & b_{22}-\varkappa & \dots & b_{2n} \\ \dots & \dots & \dots & \dots \\ b_{n1} & b_{n2} & \dots & b_{nn}-\varkappa \end{vmatrix} = 0,$$

the left members of which only differ by the factor $(-\varkappa)^{m-n}$. *Plainly this minimum is zero if and only if* x *and* y *are independent variables; otherwise it is positive but* $\leqq \frac{1}{4}$.

We note first that we may suppose

$$(8)\qquad \sum_i u_i^2 = 1$$

without altering the value of the minimum, since we have

$$\left(p_{ik} - cu_i \cdot \frac{v_k}{c}\right)^2 = (p_{ik} - u_i v_k)^2.$$

For a fixed system of values $u_1, u_2, \dots u_m$ satisfying (8), the expression (5) is easily found to take its smallest value if $v_k = \sum_i p_{ik} u_i$, and the value of this minimum is

$$(9)\qquad \sum_{i,k} p_{ik}^2 - \sum_k \Big(\sum_i p_{ik} u_i\Big)^2 = \sum_{i,k} p_{ik}^2 - \sum_{i,j} a_{ij} u_i u_j.$$

Hence our problem is reduced to that of finding the minimum of the last expression for all u_i satisfying (8), and it follows from a known theorem concerning the orthogonal transformation of quadratic forms[1] that this minimum value is

$$\mu = \sum_{i,k} p_{ik}^2 - \varkappa_0,$$

where $\varkappa_0$ denotes the greatest root of (6). In the same manner it may be proved, by starting from the assumption $\sum_k v_k^2 = 1$, that $\varkappa_0$ is also equal to the greatest root of (7). It may be

[1] Compare *e. g.* R. COURANT und D. HILBERT, Die Methoden der mathematischen Physik, Erster Band, Berlin (Springer) 1924, Chapter 1, § 3.

deduced from the elements of the theory of determinants that the left members of (6) and (7) are identical except for the factor $(-x)^{m-n}$; as this is not essential for our present purpose, we prefer, however, not to go into any details about this matter.

Since the quadratic form in the variables u_i occurring in (9) is a positive one, the equations (6) and (7) have no negative roots; the sum of all the roots is in both equations equal to $\sum_{i,k} p_{ik}^2$, which furnishes a verification of the result $\mu \geqq 0$. It will now be proved that we always have $\mu \leqq \frac{1}{4}$, this upper limit being attained *e. g.* for the particular system

$$p_{11} = p_{22} = \frac{1}{2}, \quad p_{ik} = 0 \text{ otherwise.}$$

This follows immediately from the inequality

$$\sum_{i,k} (p_{ik} - q_i r_k)^2 \leqq \frac{1}{4}, \tag{10}$$

(q_i and r_k being defined by the equations (3)), which we are now going to prove.

We begin by proving that, starting from any given rectangular table

$$\begin{matrix} p_{11} & \cdots\cdots & p_{1n} \\ \cdot & \cdots\cdots & \cdot \\ p_{m1} & \cdots\cdots & p_{mn} \end{matrix} \tag{11}$$

with $\sum_{i,k} p_{ik} = 1$, we can form another table of not-negative elements with the same number of rows and columns, such that no row and no column contains more than one element different from zero, without altering the value of the sum of all the mn elements and *without diminishing the value of the left hand side of* (10). — Consider two arbitrary elements belonging to the same row of (11)! Keeping all other elements fixed, we allow these two to vary in such a way that their sum remains constant. Denoting the two elements by z and $s-z$, and allowing z to vary from 0 to s, the left hand side of (10) takes the form

$$az^2 + bz + c \tag{12}$$

with $a \geqq 0$; a, b and c being independent of z. This expression takes its greatest value in one of the both extremes $z = 0$ or $z = s$, hence we can always put one of the two elements equal to zero without diminishing (12). By repeating the same argument we can obviously prove our assertion. Since the left hand side of (10) is invariant for any permutation of rows and columns, we may suppose the new matrix to have the form

$$\begin{matrix} \alpha_1 & 0 & 0 & \dots \\ 0 & \alpha_2 & 0 & \dots \\ 0 & 0 & \alpha_3 & \dots \\ \dots & & & \end{matrix}$$

Assuming, for instance, $m \leqq n$, the left hand side of (10), formed with the above elements, takes the value

$$\left(\sum_{i=1}^{m} \alpha_i^2\right)^2 + \sum_{i=1}^{m} \alpha_i^2 - 2\sum_{i=1}^{m} \alpha_i^3 = P,$$

the α_i being positive numbers subject to the condition

$$\sum_{i=1}^{m} \alpha_i = 1.$$

This gives us

$$\Sigma \alpha_i^2 \leqq \sqrt{\Sigma \alpha_i \cdot \Sigma \alpha_i^3} = \sqrt{\Sigma \alpha_i^3},$$

$$(\Sigma \alpha_i^2)^2 \leqq \Sigma \alpha_i^3,$$

hence we have

$$P \leqq \sqrt{\Sigma \alpha_i^3} - \Sigma \alpha_i^3 = \sqrt{\Sigma \alpha_i^3}\,(1 - \sqrt{\Sigma \alpha_i^3}) \leqq \frac{1}{4}$$

and the proof of the theorem is completed.

6. In the particular case when, for every value of x, there is only one possible value of y, the correspondence between the variables ceases to be a statistical one. If the same value of y is never associated with different values of x, the variables are uniform functions of one another. In this case we have $m=n$, and the numbers p_{ik} form a quadratic matrix, which in every row and every column has only one element different from zero. It must be regarded as a defect of our method that it does not yield the same numerical value of the degree of dependence in all these cases, where the dependence is complete. As a matter of fact, in the case

$$p_{ii}=\alpha_i, \ p_{ik}=0 \quad \text{for } i \neq k,$$

the minimum of the preceding article has the value

$$\sum_{i \neq i_0} \alpha_i^2 ,$$

where i_0 corresponds to the greatest α_i.

This inconvenience disappears if we consider the *mean square contingency* mentioned in art. 4:

$$\sum_{i,k} \frac{(p_{ik} - q_i r_k)^2}{q_i r_k}. \tag{13}$$

This quantity is zero if and only if the variables are independent, and in the above defined case of complete dependence it becomes equal to $n-1$. On the other hand, from our theoretical point of view there are other serious objections against (13). In the first place, it is not clear why we should choose to form the »deviation» of the table $|p_{ik}|$ from the particular table $|q_i r_k|$, since there may be other tables of the form $|u_i v_k|$ which give »better» representations of $|p_{ik}|$. Secondly, there seems to be no sufficient reason for introducing the particular »weights» $\frac{1}{q_i r_k}$, apart from the desire to obtain simple results.

Of course we may calculate the minimum of the expression

$$\sum_{i,k} \frac{(p_{ik} - u_i v_k)^2}{q_i r_k},$$

and we shall generally find something smaller than (13). This minimum obviously coincides with the minimum of

$$\sum_{i,k} \left(\frac{p_{ik}}{\sqrt{q_i r_k}} - w_i z_k \right)^2$$

and hence can be calculated by the method of the preceding article. In the case of complete dependence it is still equal to $n-1$, otherwise it is smaller. There is, however, still the objection as to the arbitrariness of the weights, which cannot be removed in the same way.

7. In this article, we are going to deal with case B of art. 2. The »frequency-function» of our two statistical variables being denoted by $f(x, y)$, we want to find the minimum of the expression

$$(14) \qquad \int_{-\infty}^{\infty} \int_{-\infty}^{\infty} [f(x, y) - g(x)\, h(y)]^2 \, dx \, dy$$

for all continuous $g(x)$ and $h(y)$. This minimum, multiplied by the factor

$$(15) \qquad \sqrt{E(\xi^2)\, E(\eta^2)},$$

(cf. art. 4) will constitute our measure of the degree of dependence.

In a paper quoted above, E. Schmidt has proved the following theorem, where a and b are supposed to be *finite:*

Given a function $f(x, y)$, *continuous for* $a \leqq x \leqq b$, $a \leqq y \leqq b$, *the minimum of the expression*

$$\int_a^b \int_a^b [f(x, y) - g(x) h(y)]^2 \, dx \, dy,$$

for all continuous $g(x)$ *and* $h(y)$, *is equal to*

$$\int_a^b \int_a^b [f(x, y)]^2 \, dx \, dy - \frac{1}{\lambda_0}, \tag{16}$$

where λ_0 *denotes the smallest characteristic number associated with any of the homogeneous integral equations*

$$\begin{aligned} \varphi(s) &= \lambda \int_a^b \overline{K}(s, t) \varphi(t) \, dt, \\ \psi(s) &= \lambda \int_a^b \underline{K}(s, t) \varphi(t) \, dt. \end{aligned} \tag{17}$$

Here, $\overline{K}(s, t)$ *and* $\underline{K}(s, t)$ *are symmetric kernels defined by the equations*

$$\begin{aligned} \overline{K}(s, t) &= \int_a^b f(s, r) f(t, r) \, dr, \\ \underline{K}(s, t) &= \int_a^b f(r, s) f(r, t) \, dr. \end{aligned} \tag{18}$$

The characteristic numbers, which are all real and positive, are the same for both kernels. Besides, we have

$$\sum \frac{1}{\lambda} = \int_a^b \int_a^b [f(x, y)]^2 \, dx \, dy,$$

when the sum is extended over all the characteristic numbers.

All this is, of course, strictly analogous to what we have proved by elementary means in art. 5. The relations between the two theorems are the same as those between the fundamental theorems on FREDHOLM's integral equation and the corresponding theorems on linear equations. — Obviously SCHMIDT's theorem cannot be extended to the case of infinite limits of integration without making any further hypotheses concerning $f(x, y)$. It would be sufficient to suppose that the repeated integrals

$$\int_{-\infty}^{\infty} dx \int_{-\infty}^{\infty} [f(x, y)]^2 \, dy \quad \text{and} \quad \int_{-\infty}^{\infty} dy \int_{-\infty}^{\infty} [f(x, y)]^2 \, dx$$

are convergent; we shall, however, suppose a little more, *viz.* that there are positive constants C and α such that

$$|f(x, y)| < \frac{C}{(x^2 + y^2)^{1+\alpha}} \tag{19}$$

as soon as $x^2 + y^2 > 1$. In the case of a statistical frequency-function, usually much more than this is true, and we have also the relation $f(x, y) \geqq 0$, which is, however, not essential for the extension of SCHMIDT's theorem.

In order to find the minimum of (14), we introduce the substitution

$$\begin{cases} x = \dfrac{1}{1-u} - \dfrac{1}{u}, \\[2ex] y = \dfrac{1}{1-v} - \dfrac{1}{v}, \end{cases}$$

$$\begin{cases} u = \dfrac{x - 2 + \sqrt{x^2 + 4}}{2x}, \\[2ex] v = \dfrac{y - 2 + \sqrt{y^2 + 4}}{2y}, \end{cases}$$

(the roots being taken positively) which replaces the domain of integration $-\infty < x < \infty$, $-\infty < y < \infty$ by $0 < u < 1$, $0 < v < 1$. Thus we have to calculate the minimum of

$$\int_0^1 \int_0^1 [F(u, v) - G(u)\, H(v)]^2\, du\, dv;$$

where

$$F(u, v) = \sqrt{\left(\frac{1}{(1-u)^2} + \frac{1}{u^2}\right)\left(\frac{1}{(1-v)^2} + \frac{1}{v^2}\right)} \cdot f\left(\frac{1}{1-u} - \frac{1}{u},\ \frac{1}{1-v} - \frac{1}{v}\right),$$

and this minimum, for all continuous $G(u)$ and $H(v)$, will be equal to the minimum of (14) for all continuous $g(x)$ and $h(y)$, such that $x g(x)$ and $y h(y)$ are continuous at infinity. Now, by the inequality (19), it is easily proved that $F(u, v)$ is continuous for $0 \leqq u \leqq 1$, $0 \leqq v \leqq 1$, so that SCHMIDT's theorem can be applied, and the minimum is given by the formula corresponding to (16). By re-introducing the variables x and y, this expression takes the form

$$\int_{-\infty}^{\infty} \int_{-\infty}^{\infty} [f(x, y)]^2\, dx\, dy - \frac{1}{\lambda_0}, \tag{20}$$

where λ_0 denotes the smallest characteristic number of any of the integral equations, formed with the function $F(u, v)$ and the domain $0 \leqq u \leqq 1$, $0 \leqq v \leqq 1$, according to the formulae (17) and (18). By re-introducing in these integral equations the original variables, we see that they can be put in the form

$$\text{(21)} \qquad \begin{aligned} \varphi(s) &= \lambda \int_{-\infty}^{\infty} \overline{K}(s, t)\, \varphi(t)\, dt, \\ \psi(s) &= \lambda \int_{-\infty}^{\infty} \underline{K}(s, t)\, \psi(t)\, dt, \end{aligned}$$

where

$$\overline{K}(s, t) = \int_{-\infty}^{\infty} f(s, r) f(t, r)\, dr,$$

$$\underline{K}(s, t) = \int_{-\infty}^{\infty} f(r, s) f(r, t)\, dr,$$

and conversely the equations (21) can always be transformed by the above substitutions into equations with continuous and symmetric kernels, defined in the domain $0 \leqq u \leqq 1$, $0 \leqq v \leqq 1$. Hence the ordinary theory of characteristic numbers holds for these equations in spite of their infinite limits of integration, and λ_0 in (20) may be defined also as the smallest characteristic number of any of the equations (21).

So far we have proved that, on the hypothesis (19), SCHMIDT's theorem holds even in the case of infinite limits of integration, as long as we consider the minimum of (14) for all continuous functions $g(x)$ and $h(y)$, *such that* $x g(x)$ *and* $y h(y)$ *are continuous at infinity.* We get, however, the same minimum even if we drop the last restriction. If $g(x)$ and $h(y)$ denote any continuous functions such that (14) is convergent, it is obvious that, given any positive ε, we can find continuous functions $\bar{g}(x)$ and $\bar{h}(y)$, such that $x\bar{g}(x)$ and $y\bar{h}(y)$ are continuous at infinity, and such that

$$\int_{-\infty}^{\infty}\int_{-\infty}^{\infty} \left| [f(x, y) - g(x)\, h(y)]^2 - [f(x, y) - \bar{g}(x)\, \bar{h}(y)]^2 \right| dx\, dy < \varepsilon .$$

Hence the minimum of (14) is not diminished by allowing $g(x)$ and $h(y)$ to be any continuous functions, *and thus the inequality* (19) *is sufficient to assure, without any further restriction, the validity of* SCHMIDT's *theorem for infinite limits of integration.*

8. What is, according to our definition, the degree of dependence shown by the normal correlation function? We have

$$f(x, y) = \frac{\sqrt{ac-b^2}}{\pi} e^{-(ax^2+2bxy+cy^2)},$$

and we are now going to show that the minimum of

$$\int_{-\infty}^{\infty}\int_{-\infty}^{\infty} [f(x, y) - g(x)\,h(y)]^2\,dx\,dy, \tag{22}$$

multiplied by the factor (15), is equal to

$$\frac{\sqrt{ac}}{4\pi\sqrt{ac-b^2}} \cdot \frac{\sqrt{ac} - \sqrt{ac-b^2}}{\sqrt{ac} + \sqrt{ac-b^2}}. \tag{23}$$

Expressed in terms of the corresponding correlation coefficient (cf. art. 3), this expression becomes

$$\frac{1}{4\pi\sqrt{1-r^2}} \cdot \frac{1-\sqrt{1-r^2}}{1+\sqrt{1-r^2}}.$$

When r^2 increases from 0 to 1, this quantity steadily increases from 0 to ∞, so that in this particular case r^2 may be said to give a satisfactory measure for the degree of dependence, as already stated in art. 3.

By the preceding article, the minimum of (22) is equal to

$$\int_{-\infty}^{\infty}\int_{-\infty}^{\infty} [f(x, y)]^2\,dx\,dy - \frac{1}{\lambda_0} = \frac{\sqrt{ac-b^2}}{2\pi} - \frac{1}{\lambda_0} \tag{24}$$

where λ_0 denotes the smallest characteristic number belonging to the symmetric kernel

$$\begin{aligned} K(s, t) &= \int_{-\infty}^{\infty} f(s, v)\,f(t, v)\,dv \\ &= \frac{ac-b^2}{\pi\sqrt{2c\pi}} e^{-\frac{2ac-b^2}{2c}(s^2+t^2)+\frac{b^2}{c}st}. \end{aligned} \tag{25}$$

Since $f(x, y)$ satisfies the condition (19) of the preceding article, it follows from the proof given there that the kernel $K(s, t)$ can always be transformed by a simple substitution into another, continuous and symmetric kernel defined in a finite domain. By this substitution, all the formulae used in the following proof are transformed into the corresponding formulae associated with the finite domain, which justifies our use of them. — If we denote by $D(\lambda)$ the FREDHOLM determinant associated with $K(s, t)$, it is known[1] that we have

$$-\frac{D'(\lambda)}{D(\lambda)} = A_1 + A_2\lambda + A_3\lambda^2 + \cdots, \tag{26}$$

where

$$A_n = \int_{-\infty}^{\infty} K_n(s, s)\, ds,$$

$K_1(s, t)$, $K_2(s, t)$ ... being the so called iterated kernels defined by the relations

$$K_1(s, t) = K(s, t),$$

$$K_2(s, t) = \int_{-\infty}^{\infty} K(s, v)\, K_1(v, t)\, dv,$$

$$\cdots\cdots\cdots\cdots\cdots$$

$$K_{n+1}(s, t) = \int_{-\infty}^{\infty} K(s, v)\, K_n(v, t)\, dv.$$

These are all symmetric functions. The series (26), regarded as a power series in λ, has as its cercle of convergence the cercle $|\lambda| \leqq |\lambda_1|$, where λ_1 denotes the characteristic number nearest to $\lambda = 0$. By the preceding article, the characteristic numbers of $K(s, t)$ are all real and positive, and thus we have $\lambda_1 = \lambda_0$, so that λ_0 ist equal to the radius of convergence of the series (26).

[1] Compare *e. g.* GOURSAT, Cours d'analyse, Tome III, p. 374.

From the expression (25) of $K(s, t)$, it is evident that $K_n(s, t)$ has the form

$$K_n(s,\, t)=k_n e^{-\alpha_n (s^2+t^2)+\beta_n st}. \tag{27}$$

(It follows from the symmetry of K_n that the coefficients of s^2 and t^2 in the exponent must be equal.) From the equation

$$K_{n+1}(s,\, t)=\int_{-\infty}^{\infty} K(s,\, v)\, K_n(v,\, t)\, dv,$$

it is easily deduced that we have

$$2\, c\alpha_{n+1}=2\, ac-b^2-\frac{b^4}{2\, ac-b^2+2\, c\alpha_n}, \tag{28}$$

$$\alpha_{n+1}=\alpha_n-\frac{c\beta_n^2}{2\,(2\, ac-b^2+2\, c\alpha_n)}, \tag{29}$$

$$k_{n+1}=\frac{ac-b^2}{\pi\sqrt{2\, ac-b^2+2\, c\alpha_n}}\, k_n. \tag{30}$$

Putting

$$2\, ac-b^2=p>0,$$

$$2\, c\alpha_n=\gamma_n,$$

it follows from (25) and (28) that we have

$$\gamma_{n+1}=p-\frac{b^4}{p+\gamma_n},$$

$$\gamma_1=p.$$

Hence we deduce

$$\gamma_1-\gamma_2=\frac{b^4}{2p}>0,$$

$$\gamma_n-\gamma_{n+1}=\frac{b^4(\gamma_{n-1}-\gamma_n)}{(p+\gamma_{n-1})(p+\gamma_n)}.$$

Observing that we always have $\gamma_n > 0$, since $K_n(s, t)$ is integrable all over the plane, we conclude tkat $\gamma_n - \gamma_{n+1}$ has the same sign as $\gamma_{n-1} - \gamma_n$, and hence

$$\gamma_n - \gamma_{n+1} > 0$$

for all n. Thus the positive numbers γ_n form a steadily decreasing sequence; it follows that they must tend to a limit γ which satisfies the equation

$$\gamma = p - \frac{b^4}{p+\gamma},$$

$$\gamma = \sqrt{p^2 - b^4} = 2\sqrt{ac(ac-b^2)}.$$

Hence we have

$$\lim_{n\to\infty} \alpha_n = \sqrt{\frac{a(ac-b^2)}{c}}$$

and from (29) and (30)

$$\lim_{n\to\infty} \beta_n = 0,$$

$$\lim_{n\to\infty} \frac{k_{n+1}}{k_n} = \frac{ac-b^2}{\pi\sqrt{2ac-b^2+2\sqrt{ac(ac-b^2)}}} = \frac{ac-b^2}{\pi(\sqrt{ac}+\sqrt{ac-b^2})}.$$

From (27) we obtain

$$A_n = \int_{-\infty}^{\infty} K_n(s, s)\, ds = k_n \sqrt{\frac{\pi}{2\alpha_n - \beta_n}},$$

$$\lim_{n\to\infty} \frac{A_{n+1}}{A_n} = \lim_{n\to\infty} \frac{k_{n+1}}{k_n} = \frac{ac-b^2}{\pi(\sqrt{ac}+\sqrt{ac-b^2})}.$$

This gives us for the radius of convergence of (26)

$$\lambda_0 = \frac{\pi(\sqrt{ac}+\sqrt{ac-b^2})}{ac-b^2},$$

and from (24) we find for the minimum of the expression (22)

$$(31)\qquad \frac{\sqrt{ac-b^2}}{2\pi}-\frac{ac-b^2}{\pi(\sqrt{ac}+\sqrt{ac-b^2})}=$$

$$=\frac{\sqrt{ac-b^2}}{2\pi}\cdot\frac{\sqrt{ac}-\sqrt{ac-b^2}}{\sqrt{ac}+\sqrt{ac-b^2}}\,.$$

Finally, we have for the normal correlation function

$$E(\xi^2)=\int_{-\infty}^{\infty}\int_{-\infty}^{\infty}x^2 f(x,y)\,dx\,dy=\frac{c}{2(ac-b^2)}\,,$$

$$E(\eta^2)=\int_{-\infty}^{\infty}\int_{-\infty}^{\infty}y^2 f(x,y)\,dx\,dy=\frac{a}{2(ac-b^2)}\,,$$

$$\sqrt{E(\xi^2)\,E(\eta^2)}=\frac{\sqrt{ac}}{2(ac-b^2)}\,.$$

Multiplying this by (31), we obtain the expression (23), and our proof is completed.

21.

On some classes of series used in mathematical statistics

Proc. 6th Scand. Math. Congr. Copenhagen 1925, 399–425

1. An investigator working in any of those branches of Science, where nowadays the statistical methods are being used, is confronted with many problems, the solution of which requires rather profound mathematical knowledge. Time and ability for mathematical investigations are, however, not always available, but in every actual case something has to be done, and consequently we often see that statistical practice is far in advance of theory. Thus certain methods are already in current use, although their mathematical theory is still far from complete. This is the case *e. g.* with certain classes of infinite series which hold a prominent position in the literature of mathematical statistics, particularly in Scandinavia and Germany. The theory of these series contains many points of purely mathematical interest, and I am going to deal with some of them in this lecture; I shall, however, mostly dwell upon such questions as are by the same time important from the point of view of statistical applications.

I will begin with the most important of the types of series to be dealt with. In statistical theory, these series occur in connection with the problem of finding a suitable analytical representation of the deviation of a given frequency curve from the curve representing the normal frequency function:

$$\varphi_{a,b}(x) = \frac{1}{b\sqrt{2\pi}}\, e^{-\frac{(x-a)^2}{2b^2}}. \tag{1}$$

This has always been regarded as one of the main problems of mathematical statistics (the importance of which has sometimes been slightly exaggerated) and many different solutions have been proposed. In the method of solution with which we are here concerned, the normal function itself is regarded as the first term of a series representing the given frequency function, and it is supposed that already the first three or four terms give a representation which is in general sufficiently accurate for practical purposes. This series is formed simply by taking the successive derivatives of the normal function and multiplying them with constant coefficients. We are thus led to consider the following type of expansions:

$$F(x) = \sum_{\nu=0}^{\infty} \frac{k_\nu}{\nu!} \varphi_{a,b}^{(\nu)}(x) . \tag{2}$$

With a view to statistical applications this series has been treated by a great number of authors, among whom I may mention *Gram*, *Thiele*, *Bruns*, *Charlier*, *Edgeworth*.

It is not necessary to deal throughout with the general form (1) of the generating function, with the parameters a and b. As a matter of fact, we may always, by a linear transformation, pass to the simple case $a = 0$, $b = 1$. Putting

$$\varphi(x) = \frac{1}{\sqrt{2\pi}} e^{-\frac{x^2}{2}} ,$$

$$F(a + bx) = f(x) ,$$

$$c_\nu = \frac{k_\nu}{b^{\nu+1}} ,$$

(2) is transformed into

$$f(x) = \sum_{0}^{\infty} \frac{c_\nu}{\nu!} \varphi^{(\nu)}(x) . \tag{3}$$

Thus it will be sufficient to consider expansions of the type (3). The integrals representing the coefficients of (2) and (3), which will be given below, are transformed into one another by the same substitution. For a series of the type (3), we shall use in the sequel the name exponential series.

The first question that would naturally be asked by a mathematician facing a new kind of series is concerned with the convergence properties of that series. In the case of the exponential series, and as far as its statistical applications are concerned, I hold, however, the convergence theory to be a matter of secondary interest, while certain other points of the theory are far more important. The reasons of this will be fully discussed below.

In spite of this I wish, however, to begin this lecture with an account of the convergence theory of the exponential series, The following remarks will explain why I have considered this desirable. — In a lecture delivered in 1921 before the Swedish Actuarial Society, prof. *Phragmén* pointed out that several of the authors dealing with the exponential series appear to hold quite wrong views about its convergence properties. Thus we can find in the literature a number of false proofs and false theorems concerning the exponential series. As for as I know, the most recent instance is furnished by *Arne Fisher*, who on p. 203 of the second edition of his book »The mathematical theory of probabilities« (New York 1922) enunciates a theorem, which he ascribes to the Russian mathematician *Vera Myller-Lebedeff.*[1]) This theorem, however, is false and does not coincide with the true theorem proved by *Vera Myller-Lebedeff.* As the fundamental mistake seems to be more or less the same for all these false theorems, I shall return to the matter below and give an explanation of the probable causes of the fallacy.

I now pass on to a brief survey of the convergence theory of the exponential series. The following sections of the paper deal with certain asymptotic properties of this series, which are of a fundamental importance for the statistical applications, and with certain other types of series used in mathematical statistics.

Convergence properties of the exponential series.

2. The polynomials of *Hermite* are defined by the relations

$$\frac{d^\nu}{dx^\nu} e^{-\frac{x^2}{2}} = (-1)^\nu H_\nu(x)\, e^{-\frac{x^2}{2}}, \quad (\nu = 0, 1, 2, \ldots).$$

$H_\nu(x)$ is a polynomial of degree ν, and we have

[1]) Die Theorie der Integralgleichungen in Anwendung auf einige Reihenentwicklungen. Math. Ann. 64 (1907), p. 388.

$$H_0(x) = 1,$$
$$H_1(x) = x,$$
$$H_2(x) = x^2 - 1,$$
$$H_3(x) = x^3 - 3x,$$
$$\text{— — — — — — —}$$
$$H_\nu(x) = x^\nu - \frac{\nu(\nu-1)}{2 \cdot 1!} x^{\nu-2} + \frac{\nu(\nu-1)(\nu-2)(\nu-3)}{2^2 \cdot 2!} x^{\nu-4} - \cdots.$$

These polynomials satisfy certain important relations, among which we quote the following

$$H_{\nu+1}(x) = x H_\nu(x) - \nu H_{\nu-1}(x),$$
$$H'_\nu(x) = \nu H_{\nu-1}(x),$$
$$\sum_{\nu=0}^{\infty} \frac{H_\nu(x)}{\nu!} t^\nu = e^{-\frac{1}{2}t^2 + tx},$$

$$\sum_{\nu=0}^{\infty} \frac{H_\nu(x) H_\nu(y)}{\nu!} t^\nu = \frac{1}{\sqrt{1-t^2}} e^{-\frac{t^2x^2 + t^2y^2 - 2txy}{2(1-t^2)}}, \qquad (|t| < 1).$$

From these relations, we can deduce the inequalities

$$(4) \qquad |H_\nu(x)| < K \sqrt{\nu!}\, e^{\frac{x^2}{4}}$$

for all values of x and ν, and

$$(5) \qquad |H_\nu(x)| < K \nu^{-\frac{1}{4}} \sqrt{\nu!}\, e^{\frac{x^2}{4}}$$

for $|x| \leq \sqrt{\nu}$, K being independent of x and ν. We note further the equation

$$(6) \qquad n! \sum_{\nu=0}^{n} \frac{H_\nu(x) H_\nu(y)}{\nu!} = \frac{H_{n+1}(x) H_n(y) - H_n(x) H_{n+1}(y)}{x - y}$$

and the important orthogonality relations

$$(7) \qquad \int_{-\infty}^{\infty} H_\mu(x) H_\nu(x) e^{-\frac{x^2}{2}} dx = \begin{cases} 0 & \text{for } \mu \neq \nu, \\ \nu! \sqrt{2\pi} & \text{» } \mu = \nu. \end{cases}$$

3. Writing as before

$$\varphi(x) = \frac{1}{\sqrt{2\pi}} e^{-\frac{x^2}{2}},$$

we have

$$\varphi^{(\nu)}(x) = (-1)^{\nu} H_{\nu}(x \ \varphi(x)$$

By means of the ortogonality relations (7), we obtain by formal calculation for the coefficients of the exponential series (3) expressions, analogous to the integrals giving the coefficients of an ordinary *Fourier* series. Using a notation well-known from the theory of these series, we may write

$$f(x) \sim \sum_{0}^{\infty} \frac{c_\nu}{\nu!} \varphi^{(\nu)}(x), \tag{8}$$

meaning simply that the c_ν are determined by the relations

$$c_\nu = (-1)^{\nu} \int_{-\infty}^{\infty} f(x) H_\nu(x)\, dx .$$

Let $f(x)$ be a given function, such that c_ν is finite for all values of ν; then we may ask if the series (8) is convergent and represents $f(x)$. This is the convergence problem of the exponential series.

For the more general expansion (2), we have the corresponding relations

$$F(x) \sim \sum_{0}^{\infty} \frac{k_\nu}{\nu!} \varphi_{a,b}^{(\nu)}(x) \tag{9}$$

and

$$k_\nu = (-b)^{\nu} \int_{-\infty}^{\infty} F(x) H_\nu \left(\frac{x-a}{b}\right) dx .$$

From the inequality (4), it is possible to deduce the following lemma, which is due to *Weyl.*[1])

Let $f(x)$ and $g(x)$ be two continuous functions, such that the integrals

---

[1]) Singuläre Integralgleichungen, Math. Anm. *66* (1909), p. 273. *Weyl* is working with the set of orthogonal functions $K_\nu(x) = \frac{H_\nu(x)\, e^{-\frac{x^2}{4}}}{\sqrt[4]{2\pi}\sqrt{\nu!}}$ and with the corresponding *Fourier* expansion. The lemma quoted here constitutes a »Parseval's theorem« for this orthogonal system. We are transforming *Weyl's* results to the exponential series.

$$\int_{-\infty}^{\infty} f^2(x)\, e^{\frac{x^2}{2}}\, dx\,, \quad \int_{-\infty}^{\infty} g^2(x)\, e^{\frac{x^2}{2}}\, dx$$

exist. Then, if in the above notation

$$f(x) \sim \sum_0^{\infty} \frac{c_\nu}{\nu!}\, \varphi^{(\nu)}(x)\,,$$

$$g(x) \sim \sum_0^{\infty} \frac{\gamma_\nu}{\nu!}\, \varphi^{(\nu)}(x)\,,$$

we have

$$\int_{-\infty}^{\infty} f(x)\; g(x)\; e^{\frac{x^2}{2}}\, dx = \frac{1}{\sqrt{2\pi}} \sum_0^{\infty} \frac{c_\nu\, \gamma_\nu}{\nu!}\,. \tag{10}$$

This lemma is proved with the aid of Weierstrass's well-known approximation theorem. It is easily shown that the conclusion of the lemma holds good, even when $f(x)$ and $g(x)$ need not be continuous, as soon as it is possible to find continuous functions $f_1(x)$ and $g_1(x)$, corresponding to any given $\varepsilon > 0$, such that

$$\int_{-\infty}^{\infty} \Big(f(x) - f_1(x)\Big)^2 e^{\frac{x^2}{2}}\, dx < \varepsilon\,,$$

$$\int_{-\infty}^{\infty} \Big(g(x) - g_1(x)\Big)^2 e^{\frac{x^2}{2}}\, dx < \varepsilon\,.$$

From this lemma a convergence theorem for the expansion (8) may be deduced. It was pointed out by *M. Riesz* that the original proof of *Weyl* may be simplified so as to avoid every reference to the theory of integral equations. We introduce in (10) the particular function $g(t)$ defined by the relations

$$g(t) = e^{-\frac{t^2}{2}} \quad \text{for } t \leq x\,,$$

$$g(t) = 0 \qquad \text{»} \quad t > x\,,$$

and we get

$$\int_{-\infty}^{x} f(t)\, dt - \frac{c_0}{\sqrt{2\pi}} \int_{-\infty}^{x} e^{-\frac{t^2}{2}}\, dt = \sum_0^{\infty} \frac{c_{\nu+1}}{(\nu+1)!}\, \varphi^{(\nu)}(x)\,.$$

It is readily seen that the right hand side is nothing but the exponential series corresponding to the function on the left hand side. As soon as a given function may be represented in this form, with an $f(x)$ satisfying the conditions of the lemma, we thus know that its exponential series is convergent, and even absolutely and uniformly convergent for all values of x. In this way we obtain the following theorem.

If $f(x)$ is a function which has a continuous[1]) *derivative such that the integral*

$$\int_{-\infty}^{\infty} (f'(x))^2 \, e^{\frac{x^2}{2}} \, dx \tag{11}$$

exists, and if $f(x)$ tends to zero as $|x|$ tends to infinity, then $f(x)$ may be developped in an exponential series

$$f(x) = \sum_{0}^{\infty} \frac{c_\nu}{\nu!} \varphi^{(\nu)}(x),$$

absolutely and uniformly convergent for $-\infty \leqq x \leqq \infty$.

4. The convergence problem may be treated also by another method, which is perfectly analogous to the classic method known from the theory of *Fourier* series. From the addition formula (6) we obtain

$$f(x) - \sum_{0}^{n} \frac{c_\nu}{\nu!} \varphi^{(\nu)}(x) =$$
$$\int_{-\infty}^{\infty} \Big(f(x)\,\varphi(t) - f(t)\,\varphi(x)\Big) \frac{H_{n+1}(x)\,H_n(t) - H_{n+1}(t)\,H_n(x)}{n!\,(x-t)} \, dt.$$

It has been proved by *Galbrun*[2]) that the properties of the »kernel«

$$\frac{H_{n+1}(x)\,H_n(t) - H_{n+1}(t)\,H_n(x)}{n!\,(x-t)}$$

are very similar to those of the corresponding kernel

---

[1]) Or »almost« continuous in the sense of the above remark concerning Weyl's lemma.

[2]) Sur un développement d'une fonction à variable réelle en séries de polynomes, Bulletin de la soc. math. de France, *41* (1913) p. 24. My attention was kindly drawn to this paper by Dr. *F. Nevanlinna* and Prof. *Lindelöf.*

$$\frac{\sin\,(2n+1)\frac{x-t}{2}}{\sin\frac{x-t}{2}}$$

occurring in the theory of *Fourier* series. *Galbrun* is mainly concerned with the case of a function $f(x)$ defined in a finite interval, and it follows from his investigations that the integral

$$\int_b^a \left(f(x)\,\varphi(t - f(t)\,\varphi(x)\right)\frac{H_{n+1}(x)\,H_n(t) - H_{n+1}(t)\,H_n(x)}{n!\,(x-t)}\,dt$$

tends to zero as n tends to infinity, as soon as $f\,x)$ is of bounded variation[1]) in (a, b) and x is a point of continuity. In a point of discontinuity, the same conclusion holds if $f(x)$ is replaced by

$$\frac{f(x+0)+f(x-0)}{2}.$$

For the general case *Galbrun* only gives rather incomplete results, but his method is capable of being considerably improved. For $b > x + 1$ and all sufficiently large values of n we obtain from the inequalities (4) and (5)

$$\left|\int_b^\infty \left(f(x)\,\varphi(t) - f(t)\,\varphi(x)\right)\frac{H_{n+1}(x)\,H_n(t) - H_{n+1}(t)\,H_n(x)}{n!\,(x-t)}\,dt\right| <$$

$$< K e^{\frac{x^2}{4}} \int_b^\infty (\,|\,f(x)\,|\,\varphi(t) + \varphi(x)\,|\,f(t)\,|\,)\,e^{\frac{t^2}{4}}\,dt,$$

K being an absolute constant. Now this may clearly be made as small as we please by choosing b sufficiently large, as soon as the integral

$$\int_{-\infty}^{\infty} |\,f(t)\,|\,e^{\frac{t^2}{4}}\,dt$$

is supposed to converge. The integral from $-\infty$ to a may be treated in the same way, and the integral from a to b has already been discussed. By paying some further attention to questions of uniformity, we obtain the following theorem.[2])

[1]) In reality *Galbrun* is working with the so called »*Dirichlet* conditions«.

[2]) A somewhat less general theorem was proved by a quite different method by *Phragmén* in a lecture given before the Swedish Actuarial Society in 1923.

If $f(x)$ is of bounded variation in every finite interval, and if the integral

$$(12) \qquad \int_{-\infty}^{\infty} |f(x)|\, e^{\frac{x^2}{4}}\, dx$$

exists, then the expansion of $f(x)$ in an exponential series

$$\sum_{0}^{\infty} \frac{c_\nu}{\nu!} \varphi^{(\nu)}(x)$$

converges everywhere to the sum $\frac{1}{2}\,(f(x+0)+f(x-0))$. *The convergence is uniform in every finite interval of continuity.*

If a function $f(x)$ satisfies the conditions of *Weyl*'s theorem, it is obviously of bounded variation in every finite interval, and it is easily proved that we have for $2 < \alpha < \beta$

$$\int_{\alpha}^{\beta} |f(x)|\, e^{\frac{x^2}{4}}\, dx < 4 \sqrt{\frac{1}{\alpha} \int_{\alpha}^{\infty} (f'(x))^2\, e^{\frac{x^2}{2}}\, dx},$$

so that the conditions of the last theorem are also satisfied. Thus the last theorem applies to a more general class of functions; on the other hand, its conclusion is less far-reaching, since it does not allow us to infer the absolute convergence of the series.

5. Anyone who is acquainted with the literature of mathematical statistics, will at once see that the above theorems assert considerably less than what has been assumed by several writers on the subject. Thus e. g. *Fisher* in the work quoted above states that the exponential series is convergent for every function $f(x)$ which has two continuous derivatives such that $f(x)$, $f'(x)$ and $f''(x)$ all tend to zero as x tends to infinity. In spite of this, we can show by examples that our theorems are not capable of being essentially improved as regards the behaviour of $f(x)$ at infinity.

Let us consider the function

$$f(x) = e^{-\lambda x^2}, \quad (\lambda > 0).$$

From the theorem of Art. 3, it follows that the corresponding exponential series is absolutely and uniformly convergent if $\lambda > \frac{1}{4}$. This series can without difficulty be explicitly formed, and if we put $x = 0$, we get the series

$$\frac{1}{\sqrt{2\lambda}} \sum_0^\infty \frac{(2\nu)!}{2^{2\nu}(\nu!)^2}\left(1 - \frac{1}{2\lambda}\right)^\nu,$$

which is absolutely convergent for $\lambda > \frac{1}{4}$, simply convergent for $\lambda = \frac{1}{4}$ and divergent for $\lambda < \frac{1}{4}$.

It is even possible to show that the exponential series may be divergent for a function which, for all values of x, differs from the normal function $\varphi(x)$ by less than any given ε and which has an arbitrary number of moments in common with $\varphi(x)$. Taking for instance

$$f(x) = \varphi(x) + \varepsilon\,(4\,\lambda^2 x^4 - 12\,\lambda\, x^2 + 3)\, e^{-\lambda x^2},$$

we have

$$\int_{-\infty}^{\infty} x^k f(x)\, dx = \int_{-\infty}^{\infty} x^k \varphi(x)\, dx$$

for $k = 0, 1, 2, 3$, and $f(x)$ tends uniformly towards $\varphi(x)$ as ε tends to zero. For $x = 0$, the corresponding exponential series becomes

$$\frac{1}{\sqrt{2\pi}} + \frac{\varepsilon}{\lambda^2 \sqrt{2\lambda}} \sum_2^\infty \nu(\nu - 1)\frac{(2\nu)!}{2^{2\nu}(\nu!)^2}\left(1 - \frac{1}{2\lambda}\right)^{\nu - 2},$$

which is absolutely convergent for $\lambda > \frac{1}{4}$, divergent for $\lambda \leq \frac{1}{4}$.

6. We have seen above that the question of the convergence of the exponential series has not always been treated with proper accuracy. One of the main reasons of the mistakes seems to lie in the fact that different authors define the *Hermite* polynomials and the corresponding series in different ways. In this paper, we have taken $e^{-\frac{x^2}{2}}$ as the generating function of the polynomials, but we might as well have taken e^{-x^2} or generally e^{-kx^2}. Of the authors quoted above, *Weyl* uses the same definition of $H_\nu(x)$ as we have done here, but he is working with the set of orthogonal functions

$$K_\nu x = \frac{H_\nu(x)\, e^{-\frac{x^2}{4}}}{\sqrt[4]{2\pi}\, \sqrt{\nu!}}$$

and with the corresponding *Fourier* expansions, which are not identical with our »exponential series«, though the relation between the two developments is of a very simple character.[1] — *Galbrun* and *Myller-Lebedeff* take e^{-x^2} as the generating function of the polynomials, and the former considers a series of polynomials

$$\Sigma\, \alpha_\nu P_\nu(x), \quad \frac{d^\nu}{dx^\nu} e^{-x^2} = P_\nu(x)\, e^{-x^2},$$

while the latter works with the corresponding set of orthogonal functions

$$Q_\nu(x) = \frac{P_\nu(x)\, e^{-\frac{x^2}{2}}}{2^{\frac{\nu}{4}} \sqrt[4]{\pi}\, \sqrt{\nu!}}$$

and the *Fourier* development $\Sigma\, \beta_\nu\, Q_\nu(x)$. It is obvious that we shall easily be led into mistakes if we do not take care to transform a theorem belonging to one set of definitions into the language of another set. Let us consider in particular the last-mentioned *Fourier* series. Its terms are polynomials multiplied by $e^{-\frac{x^2}{2}}$, exactly as is the case with our exponential series. But the two series must not be confused. The *Q*-series exists and converges under far more general conditions than the exponential series.

7. The convergence theorems of Art. 3 and 4 may be easily transformed to the more general expansion (2), with the generating function

[1]) The exponential series corresponding to $f(x)$ is identical with the »Fourier expansion« of $f(x)\, e^{\frac{x^2}{4}}$, multiplied with the factor $e^{-\frac{x^2}{4}}$.

$$\varphi_{a,\,b}(x) = \frac{1}{b\sqrt{2\pi}}\, e^{-\frac{(x-a)^2}{2b^2}}\,.$$

It follows from the substitution given in the introduction that both theorems hold good even for this expansion, as soon as we replace the integrals (11) and (12) by

$$\int_{-\infty}^{\infty} (F'(x))^2\, e^{\frac{(x-a)^2}{2b^2}}\, dx$$

and

$$\int_{-\infty}^{\infty} |\, F(x) \,|\, e^{\frac{(x-a)^2}{4b^2}}\, dx$$

respectively.

If $F(x)$ is a statistical frequency-function satisfying the relations

$$\int_{-\infty}^{\infty} F(x)\, dx = 1\,, \quad \int_{-\infty}^{\infty} x\, F(x)\, dx = m\,, \quad \int_{-\infty}^{\infty} (x - m)^2\, F(x)\, dx = \sigma^2\,,$$

it is customary to choose $a = m$, $b = \sigma$. In this case, the coefficients of the first and second derivatives disappear, and we get

$$F(x) \sim \varphi_{m,\,\sigma}(x) + \frac{c_3}{3!}\, \varphi'''_{m,\,\sigma}(x) + \ldots$$

If $F(x)$ is continuous and the integral

$$\int_{-\infty}^{\infty} F(x)\, e^{\frac{(x-m)^2}{4\sigma^2}}\, dx$$

converges, the series converges everywhere and represents $F(x)$.

It has sometimes been supposed that the question of convergence or divergence of the series is independent of the parameters a and b of the generating function. This is, however, quite wrong. Consider the development of the function $e^{-\lambda x^2}$ with $0 < \lambda < \frac{1}{4}$. As soon as we choose $b^2 > \frac{1}{4\lambda}$, our theorems show that the series is everywhere absolutely convergent, whereas we have seen in Art. 5 that, for the particular set of values $a = 0$, $b = 1$, $x = 0$, the series diverges.

Asymptotic properties of the exponential series. The hypothesis of elementary errors.

8. Perhaps it might be thought that the mathematical treatment of the exponential series would now be rather complete. We have proved certain convergence theorems, and we have shown by examples that they are not capable of being essentially improved.

But if we look closer into the way the series are actually used in statistical works and try to find out what is there demanded of them, then I think we shall find that the question of convergence is in reality rather inferior and that quite different problems stand in the foreground.

We find to begin with that in practical applications only a very limited number of terms (at most four or five) will ever be considered. When a statistical frequency-function is expanded in an exponential series, the latter is generally normalized by putting the coefficients of the first and second derivatives equal to zero (cf. Art 7) and it is then expected that the first term, the normal function, will already give a reasonable first approximation to the developed function; further, that the next term will furnish a real second approximation and finally that the following terms — two or at most three — will still more improve the approximation and render it fully satisfactory for practical purposes. This way of looking at the matter, which dominates the literature of the subject, has obviously very little to do with the convergence or divergence of the infinite series. The question is rather this: Is there a general class of statistical frequency-functions, which are able to be approximately represented by the first few terms of an exponential series? and will this approximation be considerably better with two terms than with one, better with three than with two, and so forth? Thus we are confronted with a question concerning the *asymptotic properties* of the series instead of its convergence properties.

To be able to say anything at all about a problem of this kind, it is necessary first to examine the nature of the statistical frequency-functions a little closer. So long as a frequency-function is considered merely as an analytical instrument for the graduation and interpolation of empirically given frequency-distributions, so long the theory offers nothing of deeper interest, and a question like the above mentioned has at most the significance of urging to further consideration. The theory will get a deeper meaning first when, as far as possible, it is brought into connection with the very causal structure of the phenomena that are to be investigated, so that a frequency function is something organically connected with a certain group of phenomena and not only an arbitrarily constructed interpolation function.

At this point it becomes natural to introduce the terminology of the calculus of probability, which has hitherto not been necessary. The calculus of probability furnishes, by its simple rules of combination, certain ideal schemes, which we compare with the observed phenomena. If, for some particular group of phenomena, we find a remarkable agreement between our observations on the one hand and a certain probability scheme on the other, then we feel, rightly or wrongly, inclined to form certain corresponding ideas as to the causal structure of the phenomena.

The particular probability scheme, with which we shall be principally concerned here, was first used in connection with the theory of errors of observation, where it served to »explain« the appearance of the normal error function. It is found already in the work of *Laplace*; by later authors it has been named the hypothesis of elementary errors.

We shall say in the following that a statistical variable is generated by elementary components, if it can be regarded as the sum of a great number of mutually independent variables, each of which has its own probability function. In statistical practice, we often meet with cases when it seems reasonable to make this assumption, and, as a matter of fact, it is only for variables of this kind that there has been any attempt towards giving a theoretical justification of the above statement concerning the asymptotic properties of the exponential series. From the same hypothesis, also other types of series have been deduced, for which analogous asymptotic properties are claimed.

The mathematical foundations of this theory seem, however, to be rather weak. Indeed it can be shown by easy formal calculation that we can form an exponential series corresponding to any variable which is generated by elementary components, and that the first terms of the series are generally rapidly decreasing, if the number of the components is large. But this kind of argument is obviously quite insufficient for the proof of anything about the asymptotic character of the series. As far as I can see, it has not yet been proved that there is anything more than a purely formal relation between the series and the corresponding probability function. The attempts hitherto made in this direction must be regarded as unsuccessful.

On the other hand, practical experience shows that in a great number of cases given frequency-distributions are remarkably well represented by the sum of the first two or three terms of the corresponding exponential series. But if, in a particular case, we want to regard this fact as supporting the hypothesis of elementary components, and if we want to claim for the coefficients of the series some particular significance as statistical characteristics of the frequency-distribution (e. g. measures of »skewness« and »excess«), then it becomes necessary to go much further into the mathematical theory of the subject. Above all, we must try to find out whether it is really possible to prove anything about the asymptotic properties of the series.

9. We shall begin by giving some formal definitions, which will enable us to give a precise mathematical form to the problem. We call a function $V(t)$ the *probability function* of a statistical variable x if, for all values of t,

$$V(t) = \text{the probability of } x \leqq t.$$

Obviously $V(t)$ is a never decreasing function, such that

$$V(-\infty) = 0, \quad V(\infty) = 1.$$

The mathematical expectation of a quantity $f(x)$ is expressed by a *Stieltjes'* integral

$$\int_{-\infty}^{\infty} f(t)\, d\, V(t).$$

Let $V_1(t)$, $V_2(t), \ldots V_n(t)$ denote the probability functions of the mutually independent variables x_1, $x_2, \ldots x_n$. Then the sum

$$x = x_1 + x_2 + \ldots + x_n$$

has the probability function

$$V(t) = \int_{-\infty}^{\infty} \ldots \int_{-\infty}^{\infty} V_n(t - t_1 - \ldots - t_{n-1})\, d V_1(t_1) \ldots d V_{n-1}(t_{n-1}).$$

It is true that, by the ordinary definition of *Stieltjes'* integral, this expression only gives the value of $V(t)$ in all its points of continuity. In a point of discontinuity, however, according to our general definition of a probability function $V(t)$ always assumes the upper limiting value.

Without restraining the generality we may suppose [1])

$$\int_{-\infty}^{\infty} t\, d V_k(t) = 0, \quad (k = 1, 2, \ldots n).$$

We put

$$\int_{-\infty}^{\infty} t^2\, d V_k(t) = \sigma_k^2,$$

$$\int_{-\infty}^{\infty} t^2\, d V(t) = \sigma_1^2 + \ldots + \sigma_n^2 = \sigma^2,$$

and

$$W(t) = V(\sigma t).$$

Then $W(t)$ will be the probability function of the variable

$$\frac{x}{\sigma} = \frac{x_1 + x_2 + + \ldots + x_n}{\sigma}$$

and it is readily seen that we have

$$(13) \qquad \int_{-\infty}^{\infty} t\, d W(t) = 0, \quad \int_{-\infty}^{\infty} t^2\, d W(t) = 1.$$

It is known from the theory of errors of observation that, under certain conditions, the function $W(x)$ is approximately equal to

$$\Phi(x) = \int_{-\infty}^{x} \varphi(t)\, dt = \frac{1}{\sqrt{2\pi}} \int_{-\infty}^{x} e^{-\frac{t^2}{2}}\, dt,$$

as soon as the number n of the elementary components is sufficiently large.

If we form, as in Art. 3, the exponential series corresponding to the difference $W(x) - \Phi(x)$, the coefficients of $\varphi(x)$ and

[1]) We shall always tacitly assume that the integrals occurring in our arguments are convergent.

$\varphi'(x)$ will disappear, and the series may be written, with a slightly altered notation,

$$W(x) \sim \Phi(x) + \sum_{3}^{\infty} \frac{c_\nu}{\nu!} \Phi^{(\nu)}(x), \tag{14}$$

where

$$c_\nu = \frac{(-1)^\nu}{\nu} \int_{-\infty}^{\infty} H_\nu(x)\, d\, W(x). \tag{15}$$

By means of the theorem of Art. 3, (cf. also Art. 7), it can be shown that the series is absolutely convergent and represents $W(x)$ if all the V_k have continuous derivatives such that

$$\int_{-\infty}^{\infty} (V'_k(x))^2\, e^{\lambda x^2}\, dx$$

exists for some $\lambda > \frac{1}{2\sigma_k^2}$.

We shall not, however, restrict ourselves to cases when all the coefficients (15) exist. It is readily seen that the c_ν may be expressed in terms of the moments

$$\int_{-\infty}^{\infty} x^i\, d\, V_k(x)$$

of the component functions, and that the existence of the ν: th moment for all these functions is sufficient to warrant the existence of $c_3, \ldots . c_\nu$.

10. We are now in a position to give a precise mathematical form to the problem discussed in Art. 8. The variable

$$x = x_1 + x_2 + \ldots . + x_n$$

considered in Art. 9 is a statistical variable, generated by the elementary components x_k, which are all measured from their respective means as origin. Choosing the standard deviation σ as our unit of measurement, we obtain the variable $\frac{x}{\sigma}$ with the probability function $W(t)$. We then ask: Under what circumstances is it true, that for large values of n

1) $W(x) - \Phi(x)$ is small,

2) $W(x) - \Phi(x)$ is asymptotically (for increasing n) represented by the first terms of the series $\Sigma \frac{c_\nu}{\nu!} \Phi^{(\nu)}(x)$?

Besides their general statistical importance, these questions have a direct bearing upon the problem of risk of an insurance company. In this case, the elementary components are formed by the gain or loss of the company on each single assurance, and the fundamental variable is the sum of all these: the total gain or loss of the company. In works on this theory, it is more or less customary to replace simply $W(x)$ by $\Phi(x)$, a procedure which leads, however, to results of a sometimes doubtful value. I hope to be able to discuss the applications to this theory in another paper.

11. The first question enunciated in the preceding Article has been treated by many different authors, and it appears from their work that, under very general conditions, $W(x)$ does tend uniformly towards $\Phi(x)$ as n tends to infinity. We shall mention only one particular case. Let us suppose that

$$\tau_k = \int_{-\infty}^{\infty} |t|^3 \, d\, V_k(t)$$

exist for $k = 1, 2, \ldots. n$, and put

$$\tau = \tau_1 + \tau_2 + \ldots + \tau_n .$$

Then if

$$\frac{\tau}{\sigma^3} \to 0$$

as $n \to \infty$, we have

$$W(x) \to \Phi(x)$$

uniformly for all values of x. In the particular case when all the probability functions V_k of the components are equal, we can even show that

$$| W(x) - \Phi(x) | < K \frac{\log n}{\sqrt{n}},$$

where K is independent of n and x.

More general theorems have been proved i. a. by *Liapounof*, *Lindeberg* and *Lévy*. It appears that, under certain conditions, we may assert the convergence of $W(x)$ towards $\Phi(x)$ even in cases when the moments

$$\int_{-\infty}^{\infty} t^i \, d\, V_k(t)$$

do not exist for any $i > 2$. In such a case, no term in the series $\Sigma \frac{c_\nu}{\nu!} \Phi^{(\nu)}(x)$ has a finite coefficient.

12. We pass now to the second question of Art. 10. Thus we assume the existence of a certain number of moments of the $V_k(t)$, so that at least some terms of the series $\Sigma \frac{c_\nu}{\nu!} \Phi^{(\nu)}(x)$ have finite coefficients, and we ask if these terms have any asymptotic significance when n is large.

In the particular case when all the V_k are equal, it is easily proved that we have for every fixed ν

$$c_\nu = O\left(n^{\left[\frac{\nu}{3}\right] - \frac{\nu}{2}}\right),$$

so that every particular term of the series tends to zero with increasing n. It follows that, with respect to their order of magnitude, the terms of the series can be arranged in groups in the following way:

Order of term.	Order of magnitude.
3	$n^{-\frac{1}{2}}$
4, 6	n^{-1}
5, 7, 9	$n^{-\frac{3}{2}}$
8, 10, 12	n^{-2}
11, 13, 15	$n^{-\frac{5}{2}}$
— — — — — —	— — — — — —

If the V_k are not actually equal, but satisfy certain conditions of regularity with respect to their moments, the order of magnitude of the terms will still be the same.

It would now be rather tempting to presume that in such cases the terms of the series would give an asymptotic expansion of $W(x) - \Phi(x)$, with an error indicated by the order of magnitude of the terms. This can, however, not be true. Let us consider the particularly simple case which corresponds to the well-known theorem of *Bernoulli.* — All the V_k are equal, and there are only two possible values for the elementary components x_k. If p and q are two positive numbers such that $p + q = 1$, then

$p =$ the probability that x_k takes the valne q,
$q =$ » » » » » » » $-p$,

The normalized probability function $W(x)$, which corresponds to the sum of n such components, expresses the probability of »succeeding« at most

$$np + x\sqrt{npq}$$

times in a series of n mutually independent throws, when the chance of succeeding in every throw is p. The graph representing $W(x)$ has the form of a »staircase« with $n + 1$ steps; in each of the points

$$x = s_\mu = \frac{\mu - np}{\sqrt{npq}}, \qquad (\mu = 0, 1, \ldots n),$$

falls a step of the height

$$\varrho_\mu = \binom{n}{\mu} p^\mu q^{n-\mu}.$$

If n tends to infinity while s_μ remains bounded, it is well known that ϱ_μ will be of the order $n^{-\frac{1}{2}}$. Thus every result of the type

$$\left| W(x) - \Phi(x) - \sum_3^h \frac{c_\nu}{\nu!} \Phi^{(\nu)}(x) \right| < \frac{K}{n^\alpha},$$

with $\alpha > \frac{1}{2}$ and K independent of x and n, is necessarily false. On the other hand, it can be proved that we have in this case

$$| W(x) - \Phi(x) | < \frac{K}{n^{\frac{1}{2}}},$$

and this shows that the asymptotic representation of $W(x)$ is not improved by considering the terms containing derivatives of Φ.

The case $p = q = \frac{1}{2}$ is particularly interesting. Then $c_3 = 0$ and no term of the series $\Sigma \frac{c_\nu}{\nu!} \overset{(\nu)}{\Phi}(x)$ is of an order greater than $\frac{1}{n}$. But still the difference

$$W(x) - \Phi(x) - \sum_{3}^{h} \frac{c_\nu}{\nu!} \overset{(\nu)}{\Phi}(x) \tag{16}$$

is, for any fixed h, sometimes of the order $n^{-\frac{1}{2}}$.

It might be thought that this negative result is essentially due to the fact that we have been working with discontinuous probability functions. It can, however, be shown by another example that this is not so. Let the probability function of x_k be

$$V_k(x) = p\ \Phi(\alpha k^2 (x - q)) + q\ \Phi(\alpha k^2 (x + p)),$$

p and q having the same meaning as before and α denoting a constant. $V_k(x)$ is continuous and, for $k \to \infty$, tends towards the elementary probability function considered in the preceding example. If we consider the sum $x_1 + x_2 + \ldots + x_n$ and form, in the same manner as before, its normalized probability function $W(x)$, the corresponding exponential series will obviously be everywhere absolutely and uniformly convergent. Its terms are, for increasing n, of the same orders of magnitude as those given above for the case when all V_k are equal. But in spite of all this it can be proved that, as soon as α is sufficiently large, the difference (16) is, for any fixed h, sometimes of the order $n^{-\frac{1}{2}}$. Thus we are led to the same conclusions as in the preceding example.

13. It follows from the above examples that, even in very regular cases, the order of the asymptotic representation of $W(x)$ is not always improved by taking into consideration the derivated terms of the exponential series. Thus we cannot assert i. e. that $\Phi(x) + \frac{c_3}{3!} \Phi'''(x)$ gives an »asymptotically better« approximation to $W(x)$ than that already furnished by $\Phi(x)$.

If we vant to get a positive result, it will obviously be necessary to modify our problem. We shall discuss two possible

27*

ways of doing this. In this Article, we shall modify the conception of an »asymptotitic representation« and show that some results may be obtained in this way. In the following Article, it will be shown that, if the V_k statisfy certain conditions of regularity, we shall really have an asymptotic expansion in the strict sense.

It can be proved that, under fairly general conditions, a sum of the type

$$\Phi(x) + \sum_{3}^{h} \frac{c_\nu}{\nu!} \Phi^{(\nu)}(x)$$

gives, *on the average*, a real asymptotic expansion of $W(x)$. In the further discussion of this point we shall restrict ourselves to the particular case when all n components $V_k(x)$ are equal. This will enable us to avoid the introduction of a complicated system of notations, without causing any essential modification in the general character of the result. In this case, we have the following theorem.

Let β denote a positive constant, which is arbitrary but fixed, and let h be a positive integer. Putting

$$z = x + t_1 + t_2 + \ldots + t_h,$$

we have

$$(17) \quad \int_{-\beta}^{\beta} \cdots \int_{-\beta}^{\beta} \left(W(z) - \Phi(z) - \sum_{3}^{3h} \frac{c_\nu}{\nu!} \Phi^{(\nu)}(z) \right) dt_1 \ldots dt_h = O\left(\frac{\log^2 n}{n^{\frac{h+1}{2}}} \right),$$

uniformly for all values of x. In cases when only a finite number of the c_ν exist, this relation holds as soon as c_{3h+1} has a finite value.

Thus we see that the mean value of the difference between $W(x)$ and a sum of terms of the exponential series, taken over an arbitrarily small interval, has an order of magnitude which decreases when more terms are taken into consideration. It is readily seen that, every time the number of integrations is increased by one, a group of three more terms will get asymptotic significance. It may be noted that in the sum $\sum_{3}^{3h}$ there is one term, viz. the term corresponding to $\nu = 3h - 1$, which has a smaller order of magnitude than the error term indicated by the theorem.

Roughly speaking, we might interprete this theorem by saying that the curve

$$y = \Phi(x) + \sum_{3}^{3h} \frac{c_\nu}{\nu!} \Phi^{(\nu)}(x)$$

furnishes a mechanical graduation of the curve $y = W(x)$. This result, and the more general theorem of which the above is a particular case, seem to be of a certain importance for the statistical applications of the exponential series. I am giving below the outlines of a proof, reserving further details for publication elsewhere.

We first introduce the adjuncts of the probality functions of Art. 9 by putting

$$v_k(t) = \int_{-\infty}^{\infty} e^{-itx} \, d\, V_k(x),$$

$$v(t) = \int_{-\infty}^{\infty} e^{-itx} d\, V(x),$$

$$w(t) = \int_{-\infty}^{\infty} e^{-itx} \, d\, W(x).$$

Then it is easily proved that we have

$$v(t) = v_1(t)\, v_2(t) \ldots . v_n(t),$$

$$w(t) = v\left(\frac{t}{\sigma}\right),$$

σ being defined as in Art. 9. Further, if we have

$$W(x) \sim \Phi(x) + \sum_{3}^{\infty} \frac{c_\nu}{\nu!} \Phi^{(\nu)}(x),$$

then we have also formally

$$e^{\frac{t^2}{2}} w(t) \sim 1 + \sum_{3}^{\infty} \frac{c_\nu}{\nu!} (it)^\nu .$$

By means of these formulae, we can prove the fundamental identity

$$\left(\int_{-\infty}^{x} dx\right)^{h} \left(W(x) - \Phi(x) - \sum_{3}^{3h} \frac{c_\nu}{\nu!} \Phi^{(\nu)}(x)\right) =$$

$$= \frac{1}{2\pi} \int_{-\infty}^{\infty} \frac{e^{-\frac{t^2}{2} + itx}}{(it)^{h+1}} \left(e^{\frac{t^2}{2}} w(t) - 1 - \sum_{3}^{3h} \frac{c_\nu}{\nu!} (it)^\nu\right) dt\,,$$

h being a positive integer. Corresponding to $h = 0$, there is an analogous formula for $W x) - \Phi(x)$, which contains a generally non-absolutely convergent integral. For $h > 0$, however, the integrals are absolutely convergent, and by a somewhat laborious evaluation of the integral on the right hand side we can prove our theorem. The theorem is a »best possible« one in the sense that h integrations are necessary in order to get an asymptotic expansion of $W(x)$ with an error of the order indicated by (17).

14. If the components $V_k(x)$ satisfy certain conditions, we can obtain an asymptotic expansion of $W(x)$ in the ordinary sense. As in the preceding Article, we shall restrict ourselves here to the case of equal components. Thus we have

$$V_1(x) = V_2(x) = \ldots = V_n(x).$$

Putting

$$V_0(x) = V_1(\sigma_1 x),$$

we have the following theorem, which is only a particular case of a more general one, the proof of which will be published at another occasion.

If the second derivative $V''_0(x)$ exists and satisfies the relation

$$\left| V''_0(x) + \frac{x}{\sqrt{2\pi}} e^{-\frac{x^2}{2}} \right| \leq \frac{1}{2\sqrt{\pi}} e^{-\frac{x^2}{4}}, \tag{18}$$

then we have, for any positive integer h,

$$\left| W(x) - \Phi(x) - \sum_{3}^{3h} \frac{c_\nu}{\nu!} \Phi^{(\nu)}(x) \right| < K n^{-\frac{h+1}{2}} e^{-\frac{x^2}{4}}$$

where K is independent of n and x.

The proof of this theorem is based on the inequality

$$\left| \frac{c_\nu}{\nu!} \Phi^{(\nu)}(x) \right| < K n^{-\frac{\nu}{2}} e^{-\frac{x^2}{4}} \sum_{i=1}^{\mu} \binom{n}{i} i^{\frac{\nu+i}{2}} \sqrt{\frac{\nu!}{(\nu+i)!}},$$

with

$$\mu = \mathrm{Min}\left(\left[\frac{\nu}{3}\right], n\right),$$

which may be deduced from (18).

15. From the hypothesis of elementary components, prof. *Charlier* has deduced another type of series, which he calls the B-series. We shall present this deduction here in a somewhat generalized form.

Let us consider a variable x generated by n elementary components. We introduce the fundamental assumption that the probability functions of the components depend on n in the following particular way. If λ is a positive constant and $E(t)$ denotes the probability function[1]) of a »variable« which is always $= 0$, then we put

$$V_1(t) = \ldots. = V_n(t) = \left(1 - \frac{\lambda}{n}\right) E(t) + \frac{\lambda}{n} U(t),$$

$U(t)$ being an arbitrary probability function. Now, if $V(t)$ denotes the probability function of x, and $v_1(t)$, $v(t)$ and $u(t)$ are the adjuncts of $V_1(t)$, $V(t)$ and $U(t)$ respectively, it is obvious that we have

$$v_1(t) = 1 + \frac{\lambda}{n}\left(u(t) - 1\right),$$

$$\lim_{n \to \infty} v(t) = \lim_{n \to \infty} \left(v_1(t)\right)^n = e^{\lambda(u(t)-1)}.$$

Putting

$$u(t) = \alpha(t) + i\beta(t),$$

it follows that we have,[2]) if certain convergence conditions are satisfied,

$$\lim_{n \to \infty} V(x) = \frac{1}{2} + \frac{1}{\pi} \int_0^\infty e^{\lambda(\alpha(t)-1)} \sin\left(tx - \lambda\beta(t)\right) \frac{dt}{t}.$$

[1]) Hence we have $E(x) = 0$ for $x < 0$, $E(x) = 1$ for $x \geqq 0$.

[2]) The expressions given in this Article represent, as a rule, the limit of $V(x)$ only in all points of continuity.

1. Taking, as a first example,

$$U(x) = \Phi(x)$$

we have

$$\lim_{n \to \infty} V(x) = e^{-\lambda} \sum_{0}^{\infty} \frac{\lambda^{\nu}}{\nu!} \Phi\left(\frac{x}{\sqrt{\nu}}\right),$$

where $\Phi\left(\frac{x}{0}\right)$ is to be interpreted as $E(x)$.

2. If we put

$$U'(x) = \frac{1}{2} \quad \text{for} \quad -1 \leqq x \leqq 1 ,$$

$$U'(x) = 0 \quad \text{for} \quad |x| > 1 ,$$

we have

$$u(t) = \frac{\sin t}{t},$$

$$\lim_{n \to \infty} V(x) = \frac{1}{2} + \frac{1}{\pi} e^{-\lambda} \int_{0}^{\infty} e^{\lambda \frac{\sin t}{t}} \frac{\sin tx}{t} dt .$$

3. Choosing with prof. *Charlier*

$$U(x) = E(x - 1),$$

we have

$$u(t) = e^{-it},$$

$$\lim_{n \to \infty} V(x) = \frac{1}{2} + \frac{1}{\pi} e^{-\lambda} \int_{0}^{\infty} e^{\lambda \cos t} \frac{\sin (tx - \lambda \sin t)}{t} dt .$$

The limit of $V(x)$ is a »staircase« with steps in the points $x = 0, 1, 2, \ldots$. The height of the step at $x = \nu$ is

$$\psi_{\lambda}(\nu) = \frac{\lambda^{\nu}}{\nu!} e^{-\lambda} = \frac{1}{\pi} e^{-\lambda} \int_{0}^{\pi} e^{\lambda \cos t} \cos (t\nu - \lambda \sin t) \, dt .$$

The function $V(x)$ itself is also represented by a staircase, and for the height of the step at $x = \nu$ there is an asymptotic expansion

$$\psi_{\lambda}(\nu) + \frac{\alpha_0 \psi_{\lambda}(\nu) + \alpha_1 \psi_{\lambda}(\nu - 1) + \alpha_2 \psi_{\lambda}(\nu - 2)}{n} +$$

$$+ \frac{\beta_0 \psi_{\lambda}(\nu) + \ldots \beta_4 \psi_{\lambda}(\nu - 4)}{n^2} + \ldots .$$

If, in this expansion, we consider ν as a continuous variable and use the second (integral) expression of $\psi_{\lambda}(\nu)$, we get the *Charlier* B-series. In practical applications, it has proved very useful. From a theoretical point of view, however, it is obvious that it has only an artificial connection with the hypothesis of elementary components. Further, as prof. *Steffensen* has shown, it suffers from the inconvenience that the integrals expressing the moments of the function $\psi^{\lambda}(x)$ are divergent.

22.

Sur quelques points du calcul des probabilités

Proc. London Math. Soc. **23**, LVIII–LXIII (1925)

Extracted from Records of Proceedings at the Meeting of the London Mathematical Society, June, 1924.

1. Faisons n tirages successives d'une urne renfermant des boules blanches et noires, la probabilité d'amener une boule blanche étant toujours égale à p. Quelle est la probabilité $\alpha_{n,\mu}$ d'avoir dans le cours de ces n tirages une suite de μ tirages ne donnant que des boules blanches?

Ce problème a été traité par De Moivre, Condorcet et Laplace, qui ont trouvé la formule de récurrence

$$\alpha_{n,\mu} = \alpha_{n-1,\mu} + p^\mu q(1-\alpha_{n-\mu-1,\mu}), \qquad (q = 1-p) \tag{1}$$

et la fonction génératrice

$$F_\mu(x) = \sum_{n=0}^{\infty} \beta_{n,\mu} x^n = \frac{1-p^\mu x^\mu}{1-x+p^\mu q x^{\mu+1}} \quad (\beta_{n,\mu} = 1-\alpha_{n,\mu}).$$

Plus récemment, notre problème a été longuement discuté par divers savants (on peut consulter sur ce sujet un ouvrage de M. Bortkiewicz: *Die Iterationen*, Berlin 1917); cependant les théorèmes suivants ne semblent pas avoir été observés.

Quelque petit que soit $\epsilon > 0$, *on peut choisir* n_0 *tel que, pour tout* $n > n_0$ *et pour toute valeur de* μ, *l'expression*

$$1-e^{-np^\mu q}$$

représente la probabilité $\alpha_{n,\mu}$ *avec une erreur dont la valeur absolue ne surpasse pas* ϵ.

En partant de l'égalité

$$F_\mu(x) = (1-p^\mu x^\mu) \sum_{n=0}^{\infty} \gamma_{n,\mu} x^n,$$

où
$$\gamma_{n,\mu} = \sum_{\nu=0}^{n/(\mu+1)} \frac{(n-\mu\nu)!}{\nu!\,(n-\mu\nu-\nu)!} (-p^\mu q)^\nu$$

$$= \sum_{\nu=0}^{n/(\mu+1)} \frac{(-np^\mu q)^\nu}{\nu!} \left(1-\frac{\mu\nu}{n}\right) \dots \left(1-\frac{\mu\nu+\nu-1}{n}\right),$$

on obtient d'abord $\qquad |\beta_{n,\mu} - e^{-np^\mu q}| < \epsilon$

pour tout n assez grand et pour toute valeur de μ telle que

$$np^{\mu}q < \tfrac{1}{2}\log n.$$

Pour achever la démonstration, il suffit de remarquer que, n restant fixe, $\beta_{n,\mu}$ doit être une fonction non décroissante de μ. Notre théorème est donc démontré; en voici quelques conséquences immédiates.

La probabilité pour que, dans n tirages, la plus longue des suites ne donnant que des boules blanches comprend précisément μ tirages est pour tout n suffisamment grand, à moins que ϵ près égale à $e^{-np^{\mu+1}q} - e^{-np^{\mu}q}$.

On peut déterminer un nombre h dépendant uniquement de ϵ de manière que la probabilité des inégalités

$$\frac{\log n}{\log(1/p)} - h < \mu < \frac{\log n}{\log(1/p)} + h$$

sera, pour tout n assez grand, supérieure à $1-\epsilon$.

Ainsi on aura, par exemple, dans le cas $p = q = \frac{1}{2}$, $n = 2^{\nu}$,

$$\nu - 3 \leqslant \mu \leqslant \nu + 3$$

avec une probabilité qui finit par rester supérieure à 0·95, lorsque ν tend vers l'infini.

2. Parmi les nombres entiers de 1 jusqu'à x, il y a comme on sait asymptotiquement $\int_2^x \frac{dt}{\log t}$ nombres premiers, distribués d'une manière très irrégulière. On pourrait interpréter ce fait en disant que, pour un nombre n pris au hasard, la probabilité d'être premier est égale à $1/\log n$. Considérons donc le schéma suivant en quelque sorte analogue: les urnes $U_1, U_2, \ldots, U_n, \ldots$ renferment des boules blanches et noires, la probabilité d'extraire une boule blanche de U_n étant $1/\log n$. Pour U_1 et U_2 cette définition doit être modifiée; on peut supposer, par exemple, que U_1 contient seulement des boules noires, U_2 seulement des boules blanches. En tirant une boule de chacune des n premières urnes, désignons le nombre de boules blanches obtenues par

$$\varpi(n) = \int_2^n \frac{dt}{\log t} + \lambda \sqrt{\left(\frac{n}{\log n}\right)}.$$

Alors, la probabilité pour que $|\lambda| < h$ tendra, pour n infini, vers la limite

$$\sqrt{\left(\frac{2}{\pi}\right)} \int_0^h e^{-\frac{1}{2}t^2}\, dt.$$

Rapprochons cela des théorèmes connus sur la distribution des nombres premiers. En désignant par $\pi(n)$ le nombre des nombres premiers inférieurs à n, on a, la célèbre hypothèse de Riemann étant admise,

$$\pi(n) = \int_2^n \frac{dt}{\log t} + \theta\sqrt{n}\log n,$$

tandis qu'on n'a pas

$$\pi(n) = \int_2^n \frac{dt}{\log t} + \theta\frac{\sqrt{n}}{\log n},$$

θ désignant d'une manière générale une quantité qui reste bornée lorsque n tend vers l'infini.

Considérons maintenant la différence la plus grande entre deux nombres premiers inférieurs à n. Même en admettant l'hypothèse de Riemann, on sait seulement que cette différence n'est pas d'un ordre supérieur à $\sqrt{n}\log n$, ce qui peut s'exprimer par la relation

$$p_{\nu+1} - p_\nu = \theta\sqrt{p_\nu}\log p_\nu,$$

p_ν désignant le ν-ième nombre premier. D'autre part, en étudiant dans notre schéma les suites de boules noires, nous aurons d'abord une formule de récurrence analogue à (1); par cette formule nous pouvons prouver que, avec une probabilité tendant vers l'unité quand n tend vers l'infini, le nombre des boules noires dans la plus longue suite obtenue dans les n premiers tirages sera inférieur à $(\log n)^2$.

Printed by C. F. Hodgson & Son, Newton Street, Kingsway, W.C.2.

23.

Some notes on recent mortality investigations

Skand. Aktuarietidskr. 9, 73–99 (1926)

1. It seems to be generally accepted that the phenomenon of human mortality may be adequately described by the language of the calculus of probability. The primary notion of the corresponding mathematical theory is the *force of mortality*, commonly denoted by μ_x. In the sequel, we shall always write $\mu(x)$, $l(x)$ etc. instead of the ordinary actuarial symbols μ_x, l_x etc.

The theory starts from the fundamental hypothesis that the probability of dying before attaining the age $x+dx$ is, for a person at age x, equal to $\mu(x)\,dx$. A hypothesis of this kind is not, of course, open to immediate statistical verification. The application of the general theorems of the calculus of probability leads, however, to the introduction of the well-known auxiliary functions

$$l(x)=l(0)\,e^{-\int_0^x \mu(t)\,dt},$$

$$d(x)=l(x)-l(x+1)=\int_x^{x+1}\mu(t)\,l(t)\,dt,$$

$$q(x)=\frac{d(x)}{l(x)},$$

etc. By the aid of these functions, it is possible to test the fundamental hypothesis and to obtain a numerical determination of the function $\mu(x)$ from statistical experience.

A prominent feature of modern mortality research is the attempt to penetrate deeper into the nature of the phenomenon by observing separately the effects of different causes of death or groups of causes of death.

If we want to base such an analysis of the mortality phenomenon on the calculus of probability, it is very important that our fundamental definitions should be appropriately chosen. According to the above, the most natural way of doing this is to work with the previously defined function $\mu(x)$ and to divide it into components.

Let us consider a distribution of all causes of death into n groups $G_1, G_2, \ldots G_n$. We introduce a force of mortality corresponding to the group G_i by putting the probability, for a person at age x, of dying before attaining the age $x+dx$ from a cause of death belonging to G_i equal to $\mu_i(x)\,dx$. By a well-known theorem in the calculus of probability we have

$$\mu(x)=\sum_{i=1}^{n}\mu_i(x).$$

The probable number of deaths in the age interval x to $x+1$ from causes of death belonging to G_i is, for a group of $l(x)$ persons at age x, equal to

$$d_i(x)=\int_x^{x+1}\mu_i(t)\,l(t)\,dt,$$

and the corresponding probability of dying within one year is

$$q_i(x)=\frac{d_i(x)}{l(x)}.$$

It is obvious that we have

$$d(x)=\sum_i d_i(x), \qquad q(x)=\sum_i q_i(x),$$

and that, in a stationary population, $\mu_i(x)\, l(x)$ would be *the frequency function for the age at death* corresponding to the group G_i.

While the force of mortality $\mu_i(x)$ is exclusively connected with the causes of death belonging to G_i, the expressions just given for $d_i(x)$ and $q_i(x)$ show that these functions contain the *total force of mortality* $\mu(x)$, which prevents them from being, from a theoretical point of view, of the same fundamental importance as $\mu_i(x)$. — As far as $q_i(x)$ is concerned, it may, however, be remarked that we have for most ages *approximately*

$$q_i(x) = \mu_i\left(x + \frac{1}{2}\right),$$

whereas for $d_i(x)$ there is no relation of this kind: $d_i(x)$ depends essentially on all components $\mu_1(x), \ldots \mu_n(x)$.

The probability of dying (at any time) from a cause of death belonging to G_i is, for a person at age x, equal to

$$(1) \qquad s_i(x) = \frac{1}{l(x)} \int_x^\infty \mu_i(t)\, l(t)\, d\,t.$$

2. At a mortality investigation, by far the greater part of the labour is required for the computation of the »exposed to risk», whereas the computation of the numbers of deaths is a much simpler task. This circumstance has given rise to a number of attempts to obtain some information as to mortality solely from numbers of deaths, classified according to ages and causes of death. The earliest attempts of this kind seem to have been done in connection with investigations concerning the problem of life insurance for sub-standard risks. I shall return below to these investigations, where the interest is principally directed towards the relation between the mortality among a certain class of sub-standard risks on one side and the corresponding class of normal risks on the other. I shall first, however, discuss a method for the construction of

mortality tables solely from data concerning deaths, which has been proposed by Mr. ARNE FISHER.[1]

This method is based on a hypothesis concerning the nature of the above-mentioned frequency functions $\mu_i(x)\,l(x)$, that has been stated in slightly different forms in the different publications of Mr. FISHER.[2] As some points of the theory are explained by the author in rather vague terms, it seems necessary to begin by giving a brief review of the theory, as I have understood it. Afterwards, I shall proceed to some critical remarks.

I shall first consider the original form of the method, as expounded in the book on *Frequency Curves*. It is assumed that *all causes of death can be distributed into a limited number of groups* G_i *— generally eight — such that the corresponding frequency functions for the age at death are of the form*

$$\mu_i(x)\,l(x)=\alpha_i\,F_i(x),$$

where the functions $F_i(x)$ *can be a priori fixed, while the constants* α_i *vary from one mortality table to another.*

Assuming this to be true, we have

$$\mu(x)\,l(x)=\sum_{i=1}^{n}\alpha_i\,F_i(x),$$

n denoting as before the total number of groups. The functions $F_i(x)$ being regarded as a priori known for all values of x, the only things that must be determined from statistical experience are the values of the parameters α_i. Since the function $l(x)$ contains an arbitrary factor, it is even sufficient

[1] Proceedings of the Casualty Actuarial and Statistical Society of America, Vol. IV, p. 65, 1917; Frequency curves and their applications etc., New York (Macmillan) 1922; Skandinavisk Aktuarietidskrift 1925, p. 163.

[2] In reality FISHER is working with $d_i(x)$ instead of $\mu_i(x)\,l(x)$. This is practically the same thing, but the introduction of $\mu_i(x)\,l(x)$ causes some formal simplification.

to determine the values of the ratios $\alpha_i : \alpha_n$. These ratios being known, we obtain by integration

$$(2) \qquad l(x) = \sum_i \alpha_i \int_x^\infty F_i(t)\, dt,$$

and consequently the mortality table is completely known.

The determination of the parameters may be performed e. g. in the following way, which is slightly different from the method given by FISHER. Suppose we are concerned with certain data of deaths, classified according to causes of death in the above n groups and according to age in m groups, the central age of the k:th group being x_k. Let the given number of deaths in the k:th age group from causes belonging to G_i be denoted by δ_{ik}, and the *unknown* corresponding number of the »exposed to risk» by λ_k. According to the ordinary method, we may write

$$\mu_i(x_k) = \frac{\delta_{ik}}{\lambda_k},$$

the fraction on the right hand side giving at least an approximate measure of $\mu_i(x_k)$. On the other hand, we have by the fundamental hypothesis

$$\mu_i(x_k)\, l(x_k) = \alpha_i F_i(x_k),$$

and hence

$$\frac{\alpha_i F_i(x_k)}{\delta_{ik}} = \frac{l(x_k)}{\lambda_k},$$

$$(3) \qquad \frac{\alpha_1 F_1(x_k)}{\delta_{1k}} = \frac{\alpha_2 F_2(x_k)}{\delta_{2k}} = \cdots = \frac{\alpha_n F_n(x_k)}{\delta_{nk}}.$$

For each of the m age groups, there is one such set of equations. The δ_{ik} and the F_i being known, it is possible to determine, e. g. by the method of least squares, the »best» values of the ratios $\alpha_i : \alpha_n$.

3. It is seen from the preceding Article that the principle of the method is a very simple one. Everything depends on the hypothesis; if that is accepted, there can be no fundamental objection against the method. Thus it will be convenient to direct our attention towards the nature of the hypothesis.

It does not seem quite clear, what FISHER wishes to be understood by the phrase that the functions $F_i(x)$ can be »a priori fixed». From the examples given in his own book, it follows that he is not attempting to use the same groups and the same functions for all mortality tables; on the other hand, it is essential that we should be able to assign *some* degree of invariableness to the F_i. It seems probable that the standpoint of FISHER in his book on *Frequency Curves* could be roughly expressed by the following modified form of the hypothesis: *If two mortality tables do not differ very much as far as race and epoch are concerned, the groups G_i can be formed in such a way that the corresponding functions $\mu_i(x)\, l(x)$ only differ by constant factors, i. e. we have*

$$\bar{\mu}_i(x)\, \bar{l}(x) = \alpha_i\, \mu_i(x)\, l(x). \qquad (i = 1, 2, \ldots n.)$$

We shall now develop some formal consequences of this hypothesis. It is obvious that, by the hypothesis, we have adopted the functions $\mu_i(x)\, l(x)$ as »characteristic functions» and the parameters α_i as »relative weights» of the causes of death belonging to G_i. As $\mu_i(x)\, l(x)$ contains, however, in its expression the total force of mortality $\mu(x)$, it can be shown that this choice leads to rather absurd consequences.

Let us consider two mortality tables, connected according to our hypothesis, and suppose that $\alpha_i = 1$ for $i = 2, 3, \ldots n$, while α_1 differs from 1. This would seem to indicate that the causes of death belonging to G_1 had changed their effects in the $\bar{\mu}$-table, while the others remain invariable. *We can, however,* *easily show that in this case no two of the functions $\overline{\mu}_i$ and μ_i can be equal.* If for instance $\overline{\mu}_2=\mu_2$, it follows from the supposition $\alpha_2=1$ that we have

$$\overline{l}(x)=l(x) \tag{4}$$

and hence

$$\overline{\mu}_i(x)=\mu^i(x)$$

for $i=2, 3, \ldots n$. Thus we have

$$\mu(x)=\mu(x)+(\alpha_1-1)\,\mu_1(x),$$

which is inconsistent with (4).

On the other hand, let us suppose that $\overline{\mu}_i(x)=\mu_i(x)$ for $i=2, 3, \ldots n$, while $\overline{\mu}_1(x)$ and $\mu_1(x)$ are not identically equal. This assumption was used by Pedersen[1] in his investigations concerning the mortality of sub-standard risks, and has been used with remarkable success also in other investigations, some of which will be discussed below. Further, it would be natural to make this assumption when comparing the tables of mortality for a certain population during two adjacent periods, an epidemic having increased the effect of a certain group of causes of death during one period. *Nevertheless, it can be shown that this assumption is inconsistent with the above hypothesis.* We can even show that, as soon as $\overline{\mu}_i(x)=\mu_i(x)$ for one single value of i, it follows from the hypothesis that we have

$$\sum_i \overline{\mu}_i(x)=\sum_i \mu_i(x),$$

which is clearly inconsistent with the assumption. — This follows immediately from the relations

$$\overline{\mu}_i(x)\,\overline{l}(x)=\alpha_i\,\mu_i(x)\,l(x),$$

$$\overline{\mu}_i(x)=\mu_i(x),$$

from which we conclude

[1] Über die Versicherung minderwertiger Leben. Jena 1906.

$$\bar{l}(x) = \alpha_i\, l(x),$$

$$\int_0^x \bar{\mu}(t)\, dt = \int_0^x \mu(t)\, dt + C,$$

$$\bar{\mu}(x) = \mu(x).$$

In a review in the Journal of the Institute of Actuaries (Vol. 54), signed W.P.E., the following question is suggested: »If two insurance offices have the same number insured, but one has double the rates of mortality of the other at each age, how can Mr. FISHER's method (or any other) get correct results from deaths alone?» The answer, from the point of view of FISHER's hypothesis, is simply that *the assumptions are inconsistent with the hypothesis.* As a matter of fact, for two mortality tables connected according to our hypothesis, we cannot have

$$\bar{\mu}_i(x) = c\, \mu_i(x)$$

for $i = 1, 2, \ldots n$, unless $c = 1$. For otherwise we should have

$$c\, \bar{l}(x) = \alpha_i\, l(x),$$

$$\frac{\bar{l}(x)}{l(x)} = \text{const.},$$

which is plainly impossible.

4. On p. 108—111 of his book, FISHER states his reasons for regarding the function $\mu(x)\, l(x)$ (or $d(x)$, which is practically equivalent) as fundamental instead of $\mu(x)$ or $q(x)$. They seem to me but little convincing. He contrasts the ordinary »empirical» method, starting from $\mu(x)$ or $q(x)$, with his own »rationalistic» method, that »starts with the d_x column and terminates with q_x as the by-product». — It is difficult to see, why a »rationalistic» method should not be able to start with $\mu(x)$ as the fundamental function. In the preceding Articles of this

paper it has been argued that, from the point of view of the calculus of probalility, it is $\mu(x)$ that represents the primary notion of the theory, and it has been shown that we shall be easily led into contradiction, if we try to base the theory on $d(x)$ and $d_i(x)$ (or $\mu_i(x)\,l(x)$).

In order to be able to construct a mortality table from deaths alone it is, however, essential that we should work with a hypothesis of the form

$$\mu(x)\,l(x)=\sum a_i F_i(x) \tag{5}$$

and *not* for instance

$$\mu(x)=\sum a_i F_i(x), \tag{6}$$

although theoretically the latter hypothesis seems preferable. This is due to the fact that, as we have seen in Article 2, it is sufficient to know the ratios $a_i : a_n$ when starting from (5), whereas in (6) we must know the absolute values of the parameters a_i. Starting from the hypothesis (6), we get exactly the same equations (3), which enable us to calculate the ratios $a_i : a_n$. Thus one more equation is wanted. It has been pointed out by PALMQUIST[1] in an analogous case, that such an equation can be formed e. g. if we know the total number of the exposed the risk for all groups taken as a whole. Denoting this number by L, and using otherwise the same notations as in Art. 2, we have in this case

$$\mu(x_k)=\frac{\sum\limits_i \delta_{ik}}{\lambda_k},$$

$$\mu(x_k)=\sum_i a_i F_i(x_k).$$

This gives us

[1] »On Investigations into Mortality based solely on Numbers of Deaths», in the forthcoming Proceedings of the 6:th Scandinavian Mathematical Congress at Copenhaguen 1925.

$$L = \sum_k \lambda_k = \sum_k \frac{\sum_i \delta_{ik}}{\sum_i \alpha_i F_i(x_k)}.$$

Of course we can also form an equation of this kind as soon as we know one of the quantities λ_k.

In many cases it is, however, impossible to obtain any information at all about the exposed to risk, and it might then be asked, if FISHER's method cannot be reasonably expected to give us at least an approximate idea of the true result. It seems not improbable that this is so. If we restrict ourselves to mortality tables, for which the μ-functions do not differ much from one another, the hypotheses (5) and (6) must lead to approximately identical results. But as we shall see below from an example, it is not at all sure that these results will be remarkably good.

5. There is another point in the method, which must be discussed here. Working with FISHER's hypothesis, it follows from (5) and (2) that we have

$$\mu(x) = \frac{\sum \alpha_i F_i(x)}{\sum \alpha_i \int_x^\infty F_i(t)\, dt};$$

hence it is essential that the F_i should be regarded as known for arbitrary large values of x. Now practical experience shows that it is very difficult to obtain sufficiently detailed statistical records concerning deaths at high ages. Further, it appears that, from a medical point of view, it is not always possible accurately to distinguish the different causes of death, and that these difficulties are particularly great at high ages. Thus it seems desirable to avoid the necessity of introducing the values of the F_i above a certain value of x. But in this case, we shall again find it necessary to have access to one more equation, which requires at least some information as to the exposed to risk. As a matter of fact, if the F_i are given for all values of $x<x_0$, we can determine the ratios $\alpha_i:\alpha_n$ in the same way as before, and we get

$$\mu(x)=\frac{\sum \alpha_i F_i(x)}{\sum \alpha_i \int\limits_x^{x_0} F_i(t)\,dt + l(x_0)},$$

and one more equation is needed because of the occurrence of the unknown $l(x_0)$. It is easily seen, that such an additional equation can be formed if we know the number L introduced in the preceding Article.

6. In the above quoted paper published in this journal, FISHER has given a modified form of the fundamental hypothesis. He puts

$$\mu_i(x)\,l(x)=\sum_{j=0}^{4} c_{ij}\,\varphi^{(j)}(z_i),$$

where

$$\varphi(z)=\frac{1}{\sqrt{2\pi}}e^{-\frac{z^2}{2}},$$

$$z_i=\alpha_i \log(110-x)-\beta_i.$$

The α_i and β_i are regarded as a priori fixed, while the c_{ij} are to be determined from experience, which can be done by methods similar to those above described.

Theoretically this form of the hypothesis is open to the same objections as the former one. On the other hand, it must be admitted that the new hypothesis seems more likely to give us at least an approximate idea as to the whereabouts of the mortality curves. In cases where it is impossible to get any information about the exposed to risk, it may thus render valuable service.

FISHER seems to attach a particular importance to the fact that the frequency functions representing his $d_i(x)$ curves are given by the well-known developments connected with the normal error function. He claims for this fact a definite theoretical foundation. Thus he begins his paper in this journal in the following way (the italics are mine):

»It has been known for some time that the d_x column in a mortality table can be considered as a compound frequency curve with a limited number of maxima and minima. From a theoretical point of view this is of course a self evident conclusion which follows directly from the so-called genetic theory of frequency originally introduced by Laplace. *He showed that any frequency distribution can be considered or generated as the sum of a very large number of elementary errors, referable to several sources of error, each group or error having its own peculiar law of error.*»

I hope to be pardoned for doubting the correctness of this statement.

7. In connection with the discussion of the functions $\mu_i(x)$ and $\mu_i(x)\,l(x)$, I shall now give an account of an investigation into Swedish mortality. From the official records of deaths and population[1], I have calculated mortality tables for Swedish males for the two periods 1916—17 and 1918—19. It was to be expected that these tables would differ considerably from one another owing to the epidemics during the latter period, and that this difference would affect only a very little number of causes of death, all other causes being almost unchanged. The μ-functions were graduated by MAKEHAM's formula

$$\mu(x)=\alpha+\beta\,c^x, \qquad ({}^{10}\log c=0{,}045)$$

for the period 1916—17 from $x=35$, for 1918—19 from $x=45$ upwards. The following values of the constants were found:

[1] Some of the yet unpublished data concerning the population were kindly put at my disposal by Mr. E. AROSENIUS of the Central Statistical Bureau of Stockholm.

1916—17	*1918—19*
$10^3\,\alpha = 4{,}42$	$10^3\,\alpha = 6{,}17$
$10^5\,\beta = 2{,}88$	$10^5\,\beta = 2{,}82$.

For lower ages the curves were graphically graduated. The resulting curves are shown in Figure 1, while the corresponding values of $\mu(x)$ are contained in the last columns of Tables I and II.

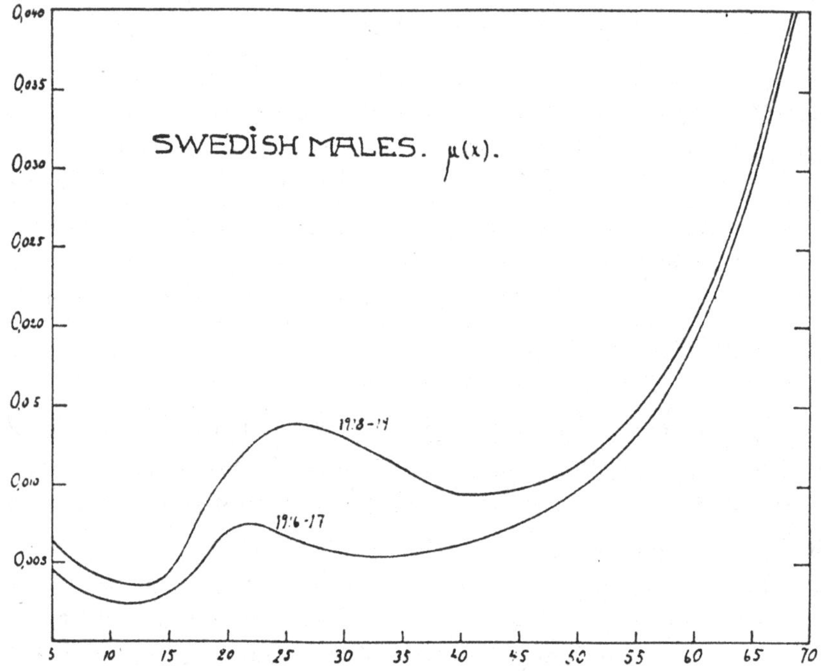

Fig. 1.

From the official publications on »Dödsorsaker» (Causes of Death), the numbers of deaths for both periods were calculated and classified into certain age groups (cf. Tables III and IV) and the following twelve groups of causes of death:

Group 1. Diseases typical of childhood (measles, scarlet fever, diphtheria, whooping cough etc.).
» 2. Influenza.
» 3. Pneumonia.
» 4. Tuberculosis.

Table I.

Swedish Males 1916—17.

x	$10^3\mu_i(x)$												$10^3\mu(x)$
	$i=1$	2	3	4	5	6	7	8	9	10	11	12	
5	1.25	0.02	0.23	1.02	0.20	0.12	0.19	0.23	0.08	0.04	0.48	0.61	4.47
10	0.45	0.01	0.13	0.62	0.14	0.17	0.05	0.13	0.07	0.03	0.33	0.39	2.52
15	0.18	0.01	0.08	1.23	0.15	0.23	0.04	0.13	0.07	0.05	0.45	0.42	3.04
20	0.11	0.01	0.31	3.62	0.15	0.24	0.09	0.24	0.13	0.09	1.03	1.00	7.02
25	0.09	0.01	0.35	3.38	0.19	0.33	0.10	0.24	0.18	0.10	0.91	0.80	6.68
30	0.04	0.01	0.35	2.65	0.21	0.36	0.10	0.25	0.20	0.11	0.74	0.61	5.63
35	0.01	0.01	0.35	2.19	0.21	0.38	0.10	0.28	0.24	0.18	0.72	0.84	5.51
40	0.01	0.02	0.41	1.95	0.26	0.47	0.12	0.38	0.34	0.47	0.69	1.12	6.24
45		0.02	0.65	1.82	0.42	0.66	0.22	0.53	0.46	0.84	0.60	1.26	7.48
50		0.03	1.00	1.80	0.67	1.03	0.32	0.63	0.59	1.44	0.61	1.43	9.55
55		0.04	1.35	1.93	1.15	1.76	0.52	0.86	0.82	2.52	0.64	1.44	13.03
60		0.06	1.71	2.08	1.85	3.21	0.89	1.17	1.17	4.06	0.66	2.01	18.87
65		0.11	2.38	2.04	3.01	5.85	1.48	1.58	1.92	5.98	0.66	3.67	28.68
70		0.16	3.31	1.94	4.58	10.03	2.20	2.04	2.93	8.38	0.66	8.91	45.14

Group 5. Diseases of the nervous system and organs of sense.
» 6. » » » circulatory system.
» 7. » » » respirat. syst. other than pneumonia.
» 8. » » » digestive system.
» 9. » » » urinary system and sex organs.
» 10. Tumours (cancer etc.).
» 11. Accidents.
» 12. Other causes.

For every age group, the ratios between the numbers of deaths in each of these twelve groups and the total number of deaths were calculated. For every group of causes of death, the diagram constructed by means of these ratios was submitted to a graphical graduation, and finally the total force of mortality was multiplied by the graduated ratios. The resulting values of the functions $\mu_i(x)$ are, for every fifth year of age, contained

Table II.

Swedish Males 1918—19.

x	$10^3\mu_i(x)$												$10^3\mu(x)$
	$i=1$	2	3	4	5	6	7	8	9	10	11	12	
5	1.76	1.25	0.40	0.80	0.13	0.13	0.29	0.15	0.09	0.03	0.54	0.83	6.40
10	0.84	0.84	0.17	0.50	0.10	0.14	0.08	0.15	0.08	0.04	0.40	0.42	3.76
15	0.25	1.40	0.17	0.90	0.14	0.18	0.06	0.20	0.08	0.04	0.45	0.28	4.15
20	0.13	4.72	0.57	2.73	0.16	0.24	0.09	0.22	0.13	0.05	0.94	0.94	10.92
25	0.06	6.97	0.72	3.01	0.17	0.28	0.11	0.25	0.17	0.09	0.98	0.91	13.72
30	0.03	7.23	0.64	2.69	0.18	0.32	0.11	0.27	0.19	0.12	0.82	0.46	13.06
35	0.01	5.06	0.63	2.29	0.20	0.34	0.12	0.32	0.24	0.20	0.72	0.97	11.10
40	0.01	3.16	0.64	1.91	0.26	0.41	0.12	0.42	0.32	0.47	0.66	1.14	9.52
45	0.01	2.30	0.76	1.80	0.37	0.57	0.17	0.52	0.42	0.84	0.61	1.42	9.79
50	0.01	1.90	1.01	1.75	0.59	0.97	0.27	0.63	0.55	1.44	0.56	1.51	11.19
55		1.60	1.42	1.71	0.96	1.77	0.50	0.90	0.78	2.52	0.53	1.90	14.59
60		1.83	2.02	1.76	1.77	3.21	0.94	1.20	1.07	4.04	0.57	1.90	20.31
65		1.87	2.91	1.90	2.87	5.80	1.59	1.57	1.58	5.86	0.76	3.19	29.90
70		1.87	4.11	1.92	4.41	9.95	2.39	2.01	2.56	8.16	0.78	7.86	46.02

in the Tables I and II, and the numbers of observed and calculated deaths are shown in Tables III and IV. Since in the official statistics on causes of death all ages above 70 form one single group, the results for these ages are very uncertain, and are therefore omitted.[1]

It is seen from these tables, that in the groups 1—4 there is a marked difference between the two periods, whereas the other groups are more or less invariable. The groups 2 and 3 are directly connected with the epidemic, and the increase of the force of mortality in the lower ages of group 1 may perhaps as well be ascribed to this. The decrease in the force of tuberculosis for ages below 30 seems a little more difficult to explain. Perhaps it might be suggested that young people

[1] Of course it must be remembered that the introduction of a constant force of mortality e. g. for influenza during the period 1918—19 depends on a fiction, since we know that the epidemic covered only a part of the period.

Table III. Calculated and

Group	5—10		10—15		15—20		20—30	
	Calc.	Obs.	Calc.	Obs.	Calc.	Obs.	Calc.	Obs.
1	476.7	466	180.1	181	73.4	71	76.3	82
2	7.9	9	5.8	4	5.5	3	9.1	10
3	102.7	102	61.5	50	97.1	106	316.1	315
4	480.4	484	439.3	430	1 262.6	1 259	3 077.2	3 094
5	88.0	86	84.8	94	82.4	72	168.6	202
6	88.6	89	120.3	121	129.0	129	284.7	296
7	62.3	64	26.3	27	37.6	39	89.4	105
8	96.5	100	76.0	76	104.3	105	221.1	222
9	45.8	46	40.9	41	53.5	55	155.4	164
10	15.3	17	23.3	24	36.0	36	90.5	91
11	247.6	250	205.6	209	385.8	385	848.5	845
12	297.0	313	228.4	232	350.0	334	773.5	722
Total	2 008.8	2 026	1 492.3	1 489	2 617.2	2 594	6 110.4	6 148

Table IV. Calculated and

1	752.3	757	311.8	328	103.6	106	62.3	55
2	608.8	610	511.6	508	1 730.7	1 735	6 113.8	6 114
3	156.1	156	89.2	90	198.4	195	629.7	628
4	365.6	360	350.8	350	961.9	947	2 793.2	2 777
5	65.8	66	68.8	70	85.1	86	156.7	156
6	82.8	86	92.6	85	123.1	125	258.3	263
7	92.1	93	39.3	40	36.3	35	100.2	103
8	82.2	82	103.9	105	118.8	119	228.8	232
9	53.7	67	47.6	47	58.7	61	153.1	148
10	18.7	19	23.8	23	23.0	23	81.0	82
11	281.2	278	240.7	257	357.3	365	903.4	905
12	345.4	288	278.4	301	466.4	350	672.3	797
Total	2 904.7	2 862	2 158.5	2 204	4 263.3	4 147	12 152.8	12 260

observed deaths. 1916—17.

30—40		40—50		50—60		60—70		Total	
Calc.	Obs.	Calc.	Obs.	Calc.	Obs.	Calc.	Obs.	Calc.	Obs.
12.8	9	—	4	—	2	—	1	819.3	816
10.4	11	11.4	10	18.5	20	38.5	40	107.1	107
259.7	255	361.8	383	650.2	634	824.8	821	2 673.9	2 666
1 625.8	1 625	1 031.9	1 020	935.5	938	707.1	707	9 559.8	9 557
158.8	148	235.3	233	563.0	569	1 035.3	1 021	2 416.2	2 425
281.2	278	376.8	369	895.1	904	2 041.7	2 039	4 217.4	4 225
74.9	73	119.5	118	260.2	262	505.2	514	1 175.4	1 202
211.4	206	283.4	280	418.5	438	542.1	547	1 953.3	1 974
182.5	170	252.9	256	401.8	406	661.5	661	1 794.3	1 799
157.0	172	476.2	473	1 233.3	1 283	2 052.5	2 040	4 084.1	4 136
522.3	521	345.7	338	309.0	311	228.7	228	3 093.2	3 087
619.2	625	708.1	673	759.0	891	1 393.9	1 325	5 129.1	5 115
4 116.0	4 093	4 203.0	4 157	6 444.1	6 658	10 031.3	9 944	37 023.1	37 109

observed deaths. 1918—19.

12.3	12	5.9	6	—	2	—	1	1 248.2	1 267
3 881.9	3 865	1 422.0	1 457	829.4	804	666.5	668	15 764.7	15 761
473.0	473	454.8	457	696.2	697	1 033.6	1 029	3 731.0	3 725
1 712.9	1 746	1 068.9	1 068	833.8	831	667.1	678	8 754.2	8 757
153.9	152	225.6	220	499.0	487	1 018.6	1 017	2 273.5	2 254
254.9	255	354.7	345	894.7	898	2 089.2	2 062	4 150.3	4 119
85.8	87	104.9	105	256.7	250	560.5	580	1 275.8	1 293
242.4	254	303.2	298	433.2	455	555.1	533	2 067.6	2 078
180.5	185	246.3	233	379.1	377	576.5	582	1 695.5	1 700
170.8	182	502.5	508	1 237.4	1 242	2 075.9	2 039	4 133.1	4 118
544.7	550	361.1	356	259.9	261	255.9	269	3 204.2	3 241
622.6	724	800.6	746	869.9	814	1 284.8	1 254	5 340.4	5 274
8 335.7	8 485	5 850.5	5 799	7 189.3	7 118	10 783.7	10 712	53 638.5	53 587

Table V.

x	Group 1—4				Group 5—12			
	$10^3 \mu_i(x)$		$\mu_i(x)\, l(x)$		$10^3 \mu_i(x)$		$\mu_i(x)\, l(x)$	
	16—17	18—19	16—17	18—19	16—17	18—19	16—17	18—19
5	2.52	4.21	252	421	1.95	2.19	195	219
10	1.21	2.85	119	229	1.31	1.41	129	138
15	1.50	2.72	146	261	1.54	1.43	150	137
20	4.05	8.15	384	752	2.97	2.77	282	256
25	3.83	10.76	350	932	2.85	2.96	261	256
30	3.05	10.59	271	857	2.58	2.47	229	200
35	2.56	7.99	221	609	2.95	3.11	255	237
40	2.39	5.72	200	414	3.85	3.80	323	275
45	2.49	4.87	202	336	4.99	4.92	404	340
50	2.83	4.67	220	306	6.72	6.52	522	427
55	3.32	4.73	244	291	9.71	9.86	713	606
60	3.85	5.61	261	317	15.02	14.70	1 020	829
65	4.53	6.68	274	333	24.15	23.22	1 459	1 158
70	5.41	7.90	273	327	39.73	38.12	2 003	1 578

suffering from tuberculosis were more easily infected by the epidemic diseases.

The difference in the behaviour of the various groups will become particularly clear if we put together the groups 1—4 on the one side and the remaining groups on the other, as it is seen from Table V and Figures 2 and 3. The μ-curve corresponding to groups 1—4 has suffered a violent change, while the other groups give practically the same curve for both periods. The table gives also the values of the functions $\mu_i(x)\, l(x)$, calculated from the initial value $l(5)=100\,000$. For the groups 5—12, the same thing is shown in graphical form in Figure 4. This function has changed considerably from 1916—17 to 1918—19, thus once more indicating the fact that $\mu_i(x)\, l(x)$ *depends on all causes of death.*

It is interesting to see, from the 1916—17 curve in Fig. 3, that the maximum of the total μ-curve about the age of

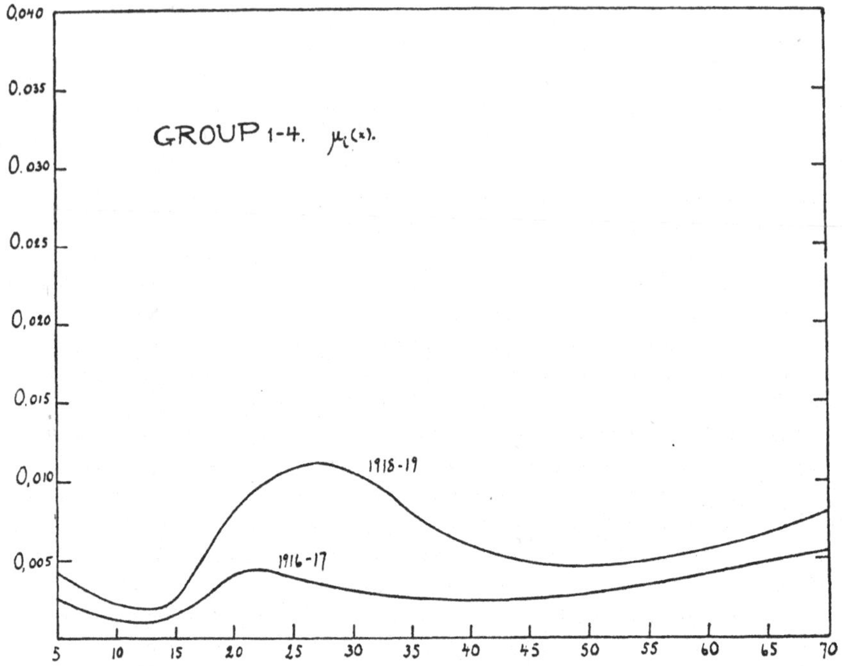
GROUP 1-4. $\mu_i(x)$.
1918-19
1916-17
0,040
0.035
0.030
0.025
0,020
0.015
0,010
0,005
5
10
15
20
25
30
35
40
45
50
55
60
65
70

Fig. 2.

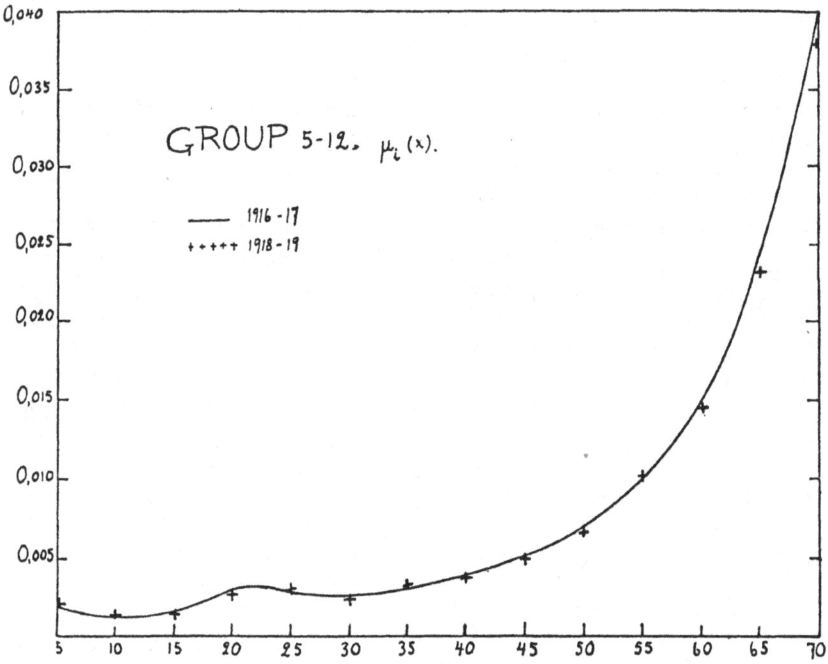
GROUP 5-12. $\mu_i(x)$.
1916-17
+++++ 1918-19
0,040
0,035
0,030
0,025
0,020
0,015
0,010
0,005
5
10
15
20
25
30
35
40
45
50
55
60
65
70

Fig. 3.

22 has almost, though not completely, disappeared by detaching tuberculosis.[1] The remaining maximum is, as may be seen from Table I, due to the groups 11 (accidents) and 12 (other causes). As for the last group, which includes the »unknown causes», it might perhaps be suspected that it contains a certain number of deaths, that in reality ought to be ascribed to tuberculosis. The mortality from accidents seems, however, to have a real maximum about $x=22$ for both periods.

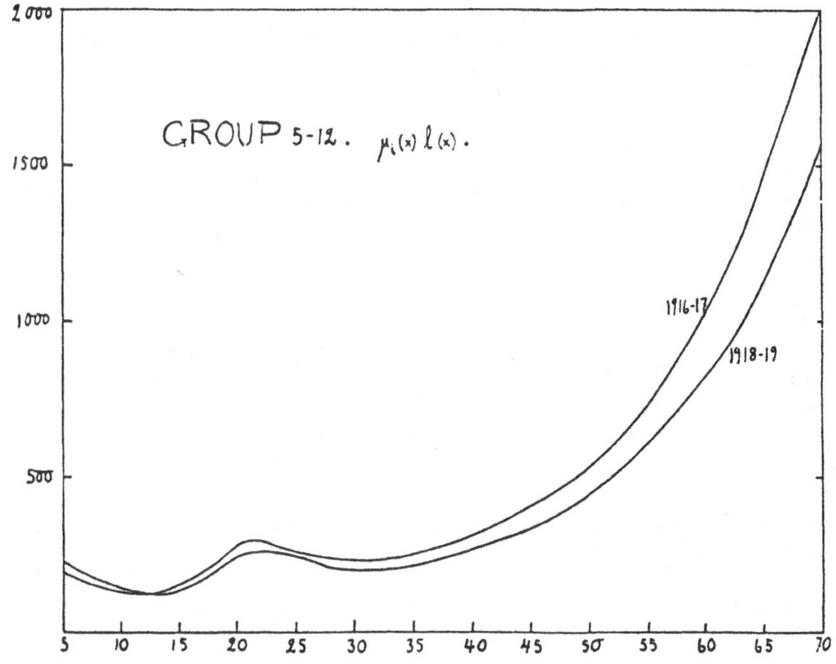

Fig. 4.

8. In the important jubilee volume recently published by Allmänna Pensionsförsäkringsbolaget[2] (in the sequel denoted by »A.P.»), Esscher has given a graduation of the company's mortality experience for insurances without medical examina-

[1] Hagström, in a paper quoted below, has observed the same phenomenon using the Swedish mortality for the period 1901—10 and the causes of death for 1912—16. His curves are, however, not exactly equivalent to ours, since he considers the force of mortality among persons which are sure to die from other diseases than tuberculosis.

[2] Allmänna Pensionsförsäkringsbolaget 1898—1922. Teknisk undersökning av tjugufem års verksamhet. Stockholm 1925.

Table VI.

A.P. Deaths from other causes than influenza.
Males 1900—1923.

Ages	1	2	3	4	5	6	Total
15—19	76	7	11	1	21	22	138
20—24	171	11	24	4	77	89	376
25—29	181	20	29	14	65	102	411
30—34	161	24	42	17	68	91	403
35—39	99	25	35	26	32	107	324
40—44	57	19	35	39	29	79	258
45—49	33	13	22	47	16	50	181
50—54	19	12	15	33	7	39	125
55—59	13	23	8	20	2	24	90
60—64	5	7	2	11	3	14	42
65—69		7	4	2	1	5	19
70—74		2				3	5
75—79						1	1
Total	815	170	227	214	321	626	2 373

tion (males) during the period (1900—$^{30}/_{6}$ 1918, $^{1}/_{7}$ 1919—1923). The statistics on causes of death contained in the same volume show, as to the relative proportions of the various groups of causes of death, a marked similarity to the corresponding proportions for the general population. This suggests the supposition that it would be possible to obtain an approximate reconstruction of the A.P. mortality table by some of the methods discussed above, using the 1916—17 population table as starting point. For this purpose, the causes of death were divided into the following six groups:

Group 1. Tuberculosis.
» 2. Diseases of the circulatory system.
» 3. » » » respiratory » .
» 4. Tumours.
» 5. Accidents.
» 6. Other causes.

After excluding all deaths from influenza (cf. p. 147 of the jubilee volume), the deaths of A.P. were distributed according to the following Table VI.

From these data, I have constructed a table of the function $\mu(x)$ by two different hypotheses, viz.

a) that the μ_i-functions of the six groups are equal to constant multiples of the corresponding μ_i-functions according to Table I (Swedish males 1916—17), and

b) the analogous hypothesis concerning the functions $\mu_i(x)\, l(x)$.

Table I giving the μ_i-functions only for ages not exceeding 70, it follows from the Arts. 4 and 5 that we shall need the total number of the exposed to risk for all ages, which is given on p. 86 of the work and amounts to 403 504. It must, however, be remembered that in Table VI the numbers of the deceased *persons* are counted, whereas the mortality investigation is based on the number of *insurances*. Thus we have first to determine the ratios $\alpha_i : \alpha_n$ according to the method given in Art. 2 and using exclusively the distribution of deaths. These ratios will have the same values on both hypotheses a) and b). Taking into account the approximate mean errors of the δ_{ik}, I have minimized the expression

$$\sum_k \sum_i \frac{[\alpha_i \delta_{nk} \mu_i(x_k) - \alpha_n \delta_{ik} \mu_n(x_k)]^2}{\delta_{nk} \mu_i^2(x_k) + \delta_{ik} \mu_n^2(x_k)},$$

the δ_{ik} being taken from Table VI and the μ_i from Table I. Thus I have obtained the following values:

$$\frac{\alpha_1}{\alpha_6} = 0{,}9391,$$

$$\frac{\alpha_2}{\alpha_6} = 0{,}9826,$$

$$\frac{\alpha_3}{\alpha_6} = 1{,}1142,$$

$$\frac{\alpha_4}{\alpha_6} = 1{,}5891,$$

$$\frac{\alpha_5}{\alpha_6} = 1{,}1991.$$

Table VII.

Age	A.P. graduation $10^3 \mu(x)$	Reconstruction from $\mu_i(x)$ (a)	Reconstruction from $\mu_i(x)\, l(x)$ (b)
15	2.58	2.10	2.17
20	4.15	4.81	4.94
25	4.44	4.58	4.65
30	4.03	3.88	3.90
35	3.72	3.84	3.83
40	4.43	4.46	4.41
45	5.87	5.47	5.36
50	8.10	7.15	6.91
55	11.56	9.96	9.49
60	16.91	14.57	13.59
65	25.19	22.05	19.88
70	38.02	34.24	29.09

Afterwards, the additional equation required has been formed according to the suggestion of Palmquist, using the total number of exposed to risk for the period (1900—$^{30}/_6$ 1918, $^1/_7$ 1919—1923) — 403 504 — and the corresponding distribution of deaths given on p. 97 of the work. The resulting μ-values are contained in Table VII; the corresponding curves are shown by Fig. 5. Up to the age of 45, the agreement is not unsatisfactory; for higher ages, however, there is a considerable discrepance.[1] For most ages, hypothesis a) gives the best results.

Thus by all the previous discussion the conclusion is strongly suggested, that simple hypotheses like the above discussed will, as a rule, only give us a rough approximation as to the true form of the mortality curves.

---

[1] Of course we ought to make allowance for the fact that the deaths in table VI include a small number of cases with medical examination, as well as deaths from other causes then influenza during the time $^1/_7$ 1918—$^{30}/_6$ 1919. None of these circumstances seems, however, likely to have a considerable influence on the result.

9. A hypothesis of a somewhat different character is the one made by Pedersen and alluded to in Art. 3. *It is assumed that, for a certain class of sub-standard risks, the increase of the force of mortality is limited to a certain number of causes of death, directly connected with the particular kind of disposition characterizing the class, whereas the force of mortality of every other disease is the same as for normal risks.*

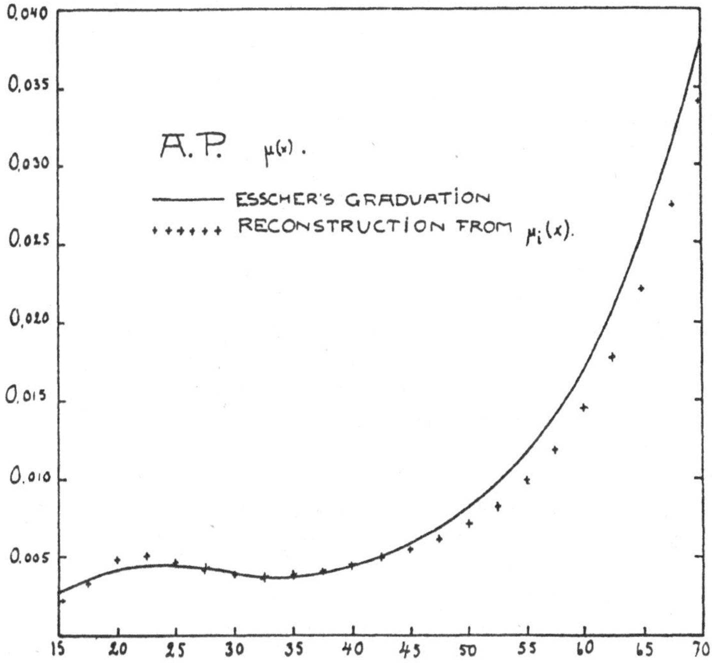

Fig. 5.

According to Art. 3, this hypothesis is inconsistent with Fisher's hypothesis (even in its modified form, as can be easily shown). The Pedersen hypothesis has been used e. g. in an investigation made by the Swedish Life Insurance Companies[1] and in other investigations by Palmquist[2], who acted as secretary of the committee for the investigation of the companies.

[1] Undersökning av dödligheten bland tuberkulöst belastade och med vissa andra sjukdomsanlag behäftade försäkrade. Stockholm 1921.

[2] See footnote p. 81. Cf also Some remarks on the Effect of Selection upon Mortality with regard to different Causes of Death. This journal, 1922, p. 115.

Table VIII.

$\bar{\mu}_r(x) : \mu_r(x)$. Ultimate table.

Age	T:1	T:2	T:3	T:4	T:5	T:6
27.5	1.20	1.62	1.45	1.10	1.09	3.38
32.5	1.36	1.61	1.51	1.10	1.18	4.09
37.5	1.49	1.61	1.54	1.13	1.27	4.61
42.5	1.53	1.60	1.51	1.15	1.33	4.40
47.5	1.55	1.56	1.50	1.14	1.36	3.56
52.5	1.55	1.55	1.47	1.14	1.34	2.97
57.5	1.49	1.53	1.37	1.14	1.26	2.71
62.5	1.44	1.49	1.29	1.16	1.20	2.74
67.5	1.40	1.43	1.18	1.16	1.14	2.98

From the tables of the committee, we can get the ratios

$$v(x) = \frac{\bar{\mu}(x)}{\mu(x)},$$

where $\bar{\mu}(x)$ denotes the force of mortality for a certain class of sub-standard risks, and $\mu(x)$ the same quantity for standard risks. Suppose now

$$\bar{\mu}(x) = \bar{\mu}_i(x) + \bar{\mu}_r(x),$$

$$\mu(x) = \mu_i(x) + \mu_r(x),$$

$$\bar{\mu}_i(x) = \mu_i(x),$$

where $\mu_i(x)$ denotes the force of mortality for all diseases not directly connected with the disposition characterizing the particular class that we consider. *Let us further suppose that, in the class of standard risks, which have been insured for at least 10 years,*

$$k(x) = \frac{\mu_r(x)}{\mu(x)}$$

has approximately the same value as in the general population for 1916—1917. Then we have

$$\frac{\bar{\mu}_r(x)}{\mu_r(x)} = 1 + \frac{v(x)-1}{k(x)}.$$

For the different tubercular classes of the committee, the values of this ratio are given in Table VIII, $\bar{\mu}_r$ and μ_r referring here to tuberculosis. The classes are characterized in the following way:

T: 1. The father of the insured has suffered from tuberculosis.

T: 2. The mother of the insured has suffered from tuberculosis.

T: 3. Brothers or sisters of the insured have suffered from tuberculosis.

T: 4. The insured has suffered from some none tubercular disease of the lungs.

T: 5. The insured has suffered from pleuritis.

T: 6. » » » » » tuberculosis (other than pleuritis).

It follows from the table that the ratios $\bar{\mu}_r : \mu_r$ are, in some of the classes, remarkably constant.

Hagström[1] introduces *the class of persons sure to die from tuberculosis.* Denoting by $p(x)$ the probability, for a person of age x, to belong to this class, he puts

$$\mu(x) = p(x)\, P(x) + (1 - p(x))\, \bar{P}(x),$$

P and $\bar{P}$ being the forces of mortality corresponding to the »purely tubercular» and »non-tubercular» classes respectively. Further, he attempts to characterize a tubercular class of substandard risks by putting

$$\bar{\mu}(x) = \alpha\, p(x)\, P(x) + (1 - \alpha\, p(x))\, \bar{P}(x),$$

thus indicating that *the probability to die from tuberculosis is equal to the corresponding probability for a normal risk multiplied*

[1] Zur Versicherung anormaler Leben. This journal, 1922, p. 159, 1923, p. 238.

by α. Using the tables of the committee, he calculates the values of α, which prove to be fairly constant for each of the classes $\mathrm{T}:\nu$.

It seems to me, however, that this assumption is not consistent with the hypothesis of the committee. For according to this hypothesis, the force of mortality for all non-tubercular diseases would be the same for a class $\mathrm{T}:\nu$ as for the class of normal risks, whereas it clearly follows from the above formulae that it has changed from $(1-p(x))\,\bar{P}(x)$ to $(1-\alpha\, p(x))\,\bar{P}(x)$.

In the above formulae, $P(x)\,d\,x$ denotes *the probability for a person at age* x, *of whom it is known that he is going to die from tuberculosis, to die before attaining the age* $x+d\,x$. I am not sure whether this is in all respects a legitimate concept; we get, however, by formal calculation

$$P(x)=\frac{\mu_r(x)}{s_r(x)},$$

$\mu_r(x)$ being the ordinary force of mortality corresponding to tuberculosis, and $s_r(x)$ the probability for a person at age x to die (at any time) from tuberculosis. According to (1) we have

$$s_r(x)=\frac{1}{l(x)}\int\limits_x^\infty \mu_r(t)\,l(t)\,d\,t,$$

and thus we may conclude that $P(x)$, like $\mu_r(x)\,l(x)$, depends on all causes of death.

24.

On an asymptotic expansion occurring in the theory of probability

J. Lond. Math. Soc. 2, 262–265 (1927)

[*Extracted from the Journal of the London Mathematical Society, Vol. 2, Part 4.*]

LET $F(x)$ be a never decreasing function of the real variable x, such that $F(-\infty)=0$ and $F(+\infty)=1$. The moments

$$a_\mu = \int_{-\infty}^{\infty} x^\mu \, dF(x), \quad \beta_\mu = \int_{-\infty}^{\infty} |x|^\mu \, dF(x)$$

are supposed to exist for $\mu = 1, 2, \ldots, k$ $(k \geqslant 2)$, and, for the sake of formal simplicity, we shall suppose that

$$a_1 = 0, \quad a_2 = \beta_2 = 1.$$

We then form two sequences of functions $F_n(x)$ and $G_n(x)$ by putting

$$F_1(x) = F(x),$$

$$F_2(x) = \int_{-\infty}^{\infty} F_1(x-t)\, dF(t),$$

$$\ldots \quad \ldots \quad \ldots \quad \ldots$$

$$F_n(x) = \int_{-\infty}^{\infty} F_{n-1}(x-t)\, dF(t),$$

* Received 21 September, 1927; read 10 November, 1927.

and* $$G_n(x) = F_n(x\sqrt{n}).$$

If x is a point of discontinuity, it will be understood that we have

$$F(x) = \tfrac{1}{2}[F(x+0)+F(x-0)],$$

and similarly for F_n and G_n. Then F_n and G_n are never decreasing functions of x, tending to the same limits as $F(x)$ when x tends to $\pm\infty$.

Putting $$\Phi(x) = \frac{1}{\sqrt{(2\pi)}}\int_{-\infty}^{x} e^{-t^2/2}\,dt,$$

it is known† that $$G_n(x) \to \Phi(x),$$

uniformly for all values of x, as n tends to infinity. This is a fact which is closely connected with the so-called Law of Large Numbers in the Theory of Probability‡. If we assume a certain additional hypothesis about $F(x)$, however, we can prove rather more, as the following theorem shows§.

$F(x)$ being a never decreasing function, it is possible to write

(1) $$F(x) = U(x)+V(x),$$

where U and V are both never decreasing and never negative functions, such that $U(x)$ is the indefinite integral of its derivative, while $V(x)$ has almost everywhere a derivative zero. Let us suppose that $U(x)$ is not identically zero.

Then we have, if $k > 2$, and ϵ denotes any positive number,

(2) $$G_n(x) = \Phi(x)+\sum_{\nu=1}^{k-3}\frac{p_{3\nu-1}(x)}{n^{\nu/2}}e^{-x^2/2}+O(n^{-[(k-2)/2]+\epsilon}),$$

uniformly for all values of x. Here $p_{3\nu-1}(x)$ denotes a polynomial of degree $3\nu-1$ in x, the coefficients of which depend only on the moments $\alpha_3, \ldots, \alpha_{\nu+2}$.

---

* Generally $G_n(x) = F_n[n\alpha_1 + x\sqrt{(n\alpha_2)}]$.

† See, *e.g.*, P. Lévy, *Calcul des probabilités* (Paris, 1925), 233.

‡ If $F(x)$ denotes the probability that a certain variable z takes a value less than x, then $G_n(x)$ will be the probability that $z_1+z_2+\ldots+z_n < x\sqrt{n}$, where $z_1, \ldots, z_n$ are the values of z obtained in a series of n mutually independent trials.

§ I have stated some preliminary theorems on the same subject in my paper "On some classes of series used in mathematical statistics", in the *Transactions of the Sixth Scandinavian Congress of Mathematics* (Copenhagen, 1925), 399.

*Further, it is possible to show by examples that, for $k > 3$, this theorem is not true if the condition $U(x) \not\equiv 0$ is omitted**.

In order to prove the last part of the theorem, take

$$F(x) = \begin{cases} 0 & \text{for } x < -1, \\ \frac{1}{2} & \text{,, } -1 < x < 1, \\ 1 & \text{,, } x > 1. \end{cases}$$

Then it is easily found that, at $x = 0$, $G_{2n}(x)$ has a discontinuity with the saltus

$$\binom{2n}{n} \cdot \left(\frac{1}{2}\right)^{2n} \sim \frac{1}{\sqrt{(\pi n)}}.$$

Thus there can be no expansion of the type (2) if k is greater than 3.

In order to prove the first part of the theorem we put

$$f(t) = \int_{-\infty}^{\infty} e^{-itx}\, dF(x), \quad f_n(t) = \int_{-\infty}^{\infty} e^{-itx}\, dF_n(x),$$

$$g_n(t) = \int_{-\infty}^{\infty} e^{-itx}\, dG_n(t),$$

when
$$f_n(t) = \{f(t)\}^n,$$

$$g_n(t) = f_n(t/\sqrt{n}) = \{f(t/\sqrt{n})\}^n.$$

From these relations it is possible to deduce by somewhat lengthy calculations that we have, for $|t| < n^{-\frac{1}{8}} \beta_k^{-1/k}$,

$$\text{(3)} \quad g_n(t) = e^{-t^2/2} + \sum_{\nu=1}^{k-3} \frac{P_{3\nu}(it)}{n^{\nu/2}} e^{-t^2/2} + O\left[n^{-(k-2)/2}(|t|^k + |t|^{3(k-2)})\, e^{-t^2/2}\right],$$

$P_{3\nu}(it)$ being a polynomial of degree 3ν, divisible by $(it)^{\nu+2}$. If

$$n^{-\frac{1}{8}} \beta_k^{-1/k} < |t| < \sqrt{n}/(4\beta_3),$$

it can be shown that we have

$$\text{(4)} \quad |g_n(t)| < e^{-t^2/3}.$$

Finally, it follows from (1) that we have

$$|f(t)| \leqslant \left|\int_{-\infty}^{\infty} e^{-itx}\, dU(x)\right| + \int_{-\infty}^{\infty} dV(x),$$

---

* For $k = 3$, no hypothesis on $U(x)$ is required. Cf. my paper: "Das Gesetz von Gauss und die Theorie des Risikos", *Skandinavisk Aktuarietidsskrift* (1923), 209.

and by means of our assumptions concerning $U(x)$ we can easily deduce that, for any given $c>0$, there is a $q<1$ such that $|f(t)|<q$ for $|t|>c$. Hence, taking $c=1/(4\beta_3)$, we obtain

$$|g_n(t)|<q^n \quad \left(|t|>\sqrt{n}/(4\beta_3)\right). \tag{5}$$

Now, denoting by $p_{3\nu-1}(x)$ the polynomial in x obtained by replacing in $P_{3\nu}(\Phi)$ every power Φ^j by the derivative $\Phi^{(j)}(x)$ and then multiplying by $e^{x^2/2}$, we put

$$R_n(x)=G_n(x)-\Phi(x)-\sum_{\nu=1}^{k-3}\frac{p_{3\nu-1}(x)}{n^{\nu/2}}e^{-x^2/2},$$

$$r_n(t)=g_n(t)-e^{-t^2/2}-\sum_{\nu=1}^{k-3}\frac{P_{3\nu}(it)}{n^{\nu/2}}e^{-t^2/2}$$

$$=\int_{-\infty}^{\infty}e^{-itx}\,dR_n(x).$$

We then have the fundamental identity

$$\frac{1}{\Gamma(\lambda)}\int_{-\infty}^{x}(x-t)^{\lambda-1}R_n(t)\,dt=\frac{1}{2\pi}\int_{-\infty}^{\infty}\frac{e^{itx}}{(it)^{1+\lambda}}r_n(t)\,dt$$

for $0<\lambda\leqslant 1$. By means of (3), (4), and (5) it follows that

$$\frac{1}{\Gamma(\lambda)}\int_{-\infty}^{x}(x-t)^{\lambda-1}R_n(t)\,dt=O(n^{-(k-2)/2})$$

uniformly in x. A theorem due to M. Riesz* now shows that we have, uniformly for all $h>0$,

$$\frac{1}{\Gamma(\lambda)}\int_{x-h}^{x}(x-t)^{\lambda-1}R_n(t)\,dt=O(n^{-(k-2)/2}).$$

Taking here $h=n^{-(k-2)/2}$ and $\lambda=2\epsilon/(k-2)$, we can easily prove that

$$R_n(x)=O(n^{-[(k-2)/2]+\epsilon})$$

uniformly in x, which completes the proof of our theorem†.

---

* M. Riesz, "Sur un théorème de la moyenne et ses applications", *Acta Univ. Hungaricae Franc.-Jos.*, 1 (1923), 114. *Cf.* also G. H. Hardy and M. Riesz, "The general theory of Dirichlet's series", *Cambridge Tracts*, No. 18 (1915), Lemma 7, 28.

† The theorem may be generalized in various directions. A full proof of some of these generalizations, which have important applications to certain actuarial problems, will appear shortly in the *Skandinavisk Aktuarietidsskrift*.

25.

On the composition of elementary errors

Skand. Aktuarietidskr. **11**, 13–74, 141–180 (1928)

First paper: Mathematical deductions.

Introduction.

1. By a *variable* in the sense of the Theory of Probability we mean a quantity z, which may assume certain real values with certain probabilities. We shall call $V(t)$ the *probability function* of z if, for every real t, $V(t)$ is equal to the probability that z has a value $< t$, increased by half the probability that z has exactly the value t.

Then $V(t)$ is a never decreasing function of t, such that $V(-\infty)=0$, $V(+\infty)=1$. For all values of t we have

$$V(t)=\frac{V(t-0)+V(t+0)}{2}.$$

In the particular case when a finite derivative $V'(t)$ exists for all values of t, $V'(t)$ will be called the *frequency function* of z. The probability that z takes a value in the infinitesimal interval $(t, t+dt)$ is, in this case, equal to $V'(t)dt$.

We shall be concerned here with the following classical problem, which is of a fundamental importance for various applications of the theory: *Let $z_1, z_2, \ldots, z_n$ be a set of variables with the probability functions $V_1(t), V_2(t), \ldots, V_n(t)$, and let us form the sum*

$$x_n = z_1 + z_2 + \cdots + z_n.$$

What can be said about the probability function $W_n(t)$ of x_n?

This problem was first treated by LAPLACE, in connection with the theory of errors of observation. According to the *hypothesis of elementary errors*, introduced by HAGEN and BESSEL, the error committed at a physical or astronomical measurement can be regarded as the sum of a great number of *mutually independent elementary errors*. If $z_1, z_2, \ldots, z_n$ are the elementary errors, it can be shown that, under certain supplementary conditions, the probability function W_n of the total error x_n will approximatively satisfy the relation

$$W_n(x) = \Phi\left(\frac{x-m}{\sigma}\right),$$

where m and σ are constants, and $\Phi(x)$ denotes the *normal probability function*

$$\Phi(x) = \frac{1}{\sqrt{2\pi}} \int_{-\infty}^{x} e^{-\frac{t^2}{2}} dt.$$

Thus the hypothesis of elementary errors may, to a certain extent, give an explanation of the fact that, as a rule, errors of measurement are found to be nearly normally distributed.

The same point of view can, however, be successfully applied also in many other branches of the theory of probability and its applications. The theorems of BERNOULLI and POISSON, e. g., can be expressed as particular cases of the theorem just alluded to concerning the sum x_n. Further, in certain statistical investigations concerning frequency-distributions and correlation, it seems not unnatural to regard the quantity measured as being the total effect of a large number of mutually independent causes, which sum up their effects. It is often necessary also to consider cases, where a similar hypothesis can be made, but where the causes must be regarded as more or less inter-dependent. Finally, we might take as an example also the problem of risk of an insurance company. The total gain or loss of the company during a certain period is equal to the sum of all the gains and losses on each single insurance, which shows that this question is closely related to our general problem.

In all these cases, a primary result is given by the theorem concerning the relation between the function $W_n(x)$ and the normal function. This is, however, not at all sufficient. Statistical frequency-curves often show considerable deviations from the normal form, even in cases when it seems plausible that the quantity measured can be considered as generated by independent elementary errors. Thus the problem arises to give an explanation of such deviations without possessing more than a general knowledge of the probability functions of the elementary errors. — In the theory of risk, on the other hand, the probability functions of the elementary errors are regarded as known, and we want a precise numerical knowledge of the corresponding function for the total gain or loss, under different assumptions concerning distribution of profits, reinsurance etc.

In this first paper, we shall restrict ourselves to a mathematical investigation concerning the composition of the probability functions of elementary errors.[1] The second paper in the series will contain a discussion of some statistical applications, and in the third the problem of risk of an insurance company will be considered.

Part I. Formal developments and summary of results from the point of view of the Theory of Probability.

Probability functions and their composition.

2. In all investigations connected with the Theory of Probability, it is very desirable to make a clear distinction between the purely mathematical deductions and those parts of the investigation which rest on the idea of probability.

[1] Some preliminary results of my investigations have been published in the papers: »On some classes of series used in Mathematical Statistics», Den sjette skandinaviske Matematikerkongress, Köbenhavn 1925, p. 399—425, and »On an asymptotic expansion occurring in the Theory of Probability», Journal of the London Mathematical Society, Vol. 2, p. 262—265.

The main contents of this first paper being purely mathematical, it is only in this introductory Part I that we shall use the conceptions and the terminology of the theory of probability. In the rest of the paper, we shall talk simply of never decreasing functions without even mentioning the word »probability» more than quite occasionally.

If z is a variable with the probability function $V(t)$, the *mathematical expectation* or *mean value* of a given function $h(z)$ is, by definition, equal to the Stieltjes' integral

$$\int_{-\infty}^{\infty} h(t)\, d\, V(t).$$

For the mean values of the powers z^ν and $|z|^\nu$, $(\nu=0, 1, 2, \ldots)$ we shall employ the notations

$$\alpha_\nu = \int_{-\infty}^{\infty} t^\nu d\, V(t), \qquad \beta_\nu = \int_{-\infty}^{\infty} |t|^\nu d\, V(t),$$

so that we have always $\alpha_{2\nu} = \beta_{2\nu}$. As a rule, we shall consider only probability functions, such that α_2 has a finite value, but shall not suppose that all the »moments» α_ν and β_ν are necessarily finite. — The square of the *standard deviation* of z is

$$\int_{-\infty}^{\infty} (t-\alpha_1)^2 d\, V(t) = \alpha_2 - \alpha_1^2.$$

Let $z_1, z_2, \ldots$ be mutually independent variables with the probability functions $V_1, V_2, \ldots$, the moments of V_i being denoted by $\alpha_\nu^{(i)}$ and $\beta_\nu^{(i)}$. Denoting generally by W_ν the probability function of the sum $x_\nu = z_1 + z_2 + \cdots + z_\nu$, we have[1]

[1] For the theory of probability functions and their composition, cf. R. v. Mises, Grundlagen der Wahrscheinlichkeitsrechnung, Math. Zeitschr. 5 (1919), p. 52, and Paul Lévy, Calcul des probabilités, Paris (Gauthier-Villars) 1925.

$$W_1(x) = V_1(x),$$

$$W_2(x) = \int_{-\infty}^{\infty} W_1(x-t)\, d\, V_2(t),$$

(1) $$W_3(x) = \int_{-\infty}^{\infty} W_2(x-t)\, d\, V_3(t),$$

. . . .

$$W_n(x) = \int_{-\infty}^{\infty} W_{n-1}(x-t)\, d\, V_n(t).$$

By the second equation, $W_2(x)$ is uniquely determined in all its points of continuity according to the ordinary definition of STIELTJES' integral. In a point of discontinuity, $W_2(x)$ is determined by the condition

$$W_2(x) = \frac{W_2(x-0) + W_2(x+0)}{2}.$$

The same is true for W_3, W_4, ... and generally for W_n. We shall say that W_n is *composed* of the *components* V_1, V_2, ..., V_n. We put

(2) $$m = \sum_{i=1}^{n} \alpha_1^{(i)},$$

$$\sigma^2 = \sum_{i=1}^{n} \left[\alpha_2^{(i)} - (\alpha_1^{(i)})^2\right],$$

so that m is the mean value and σ the standard deviation of the sum x_n. (Strictly speaking, we ought to write m_n and σ_n in the place of m and σ; for the sake of convenience we shall, however, adopt the simpler notation.) Further, if we put

(3) $$F_n(x) = W_n(m + \sigma x),$$

$F_n(x)$ is the probability function of the variable $\frac{x_n - m}{\sigma}$, which has the mean value 0 and the standard deviation 1. A variable satisfying these two conditions will be called a *normalized variable.* Similarly, $F_n(x)$ is a *normalized probability function,* satisfying the relations

$$(4) \qquad \int_{-\infty}^{\infty} x\, d F_n(x) = 0, \quad \int_{-\infty}^{\infty} x^2\, d F_n(x) = 1.$$

3. It is convenient to observe that no restriction is implied by assuming from the beginning that the mean value $\alpha_1^{(i)}$ is equal to zero for all the variables z_i. As a matter of fact, in the general case we need only consider the variables $z_i - \alpha_1^{(i)}$ with the mean values 0 and the sum $x_n - m$. The probability function of this sum evidently is $W_n(x+m)$, and if we know this function, we also know the function $W_n(x)$ itself.

Thus it will be assumed throughout the rest of the paper that we have

$$(5) \qquad \alpha_1^{(i)} = \int_{-\infty}^{\infty} t\, d\, V_i(t) = 0$$

for all the components, which signifies that the mean is taken as origin of measurement for each elementary error. The relations (2) then reduce to

$$(2\text{ a}) \qquad m = 0, \quad \sigma^2 = \sum_{i=1}^{n} \alpha_2^{(i)} = \sum_{i=1}^{n} \beta_2^{(i)}.$$

The normal function as a limiting form.

4. Under certain conditions concerning the components V_i it is possible to assert that, for large values of n, $F_n(x)$ is approximatively equal to the normal function $\Phi(x)$. This theorem, which must be regarded as the mathematical foundation of a considerable part of the statistical applications, goes back to LAPLACE, but a full proof under fairly general conditions was first given by TCHEBYCHEF and generalized by other Russian mathematicians, especially MARKOFF and LIAPOUNOF. Among the more recent authors who have treated the subject I mention v. MISES[1], PÓLYA[2], LINDEBERG[3], LÉVY[4] and CANTELLI[5].

I shall here only quote two particular cases of the general theorem. In the first place, consider the case in which all the components $V_1, V_2, \ldots$ are equal. Then it can be shown (cf. Art. 14) that, as soon as the moment of the second order $\alpha_2^{(1)} = \beta_2^{(1)}$ is finite, the normalized function $F_n(x)$ tends, uniformly for all values of x, towards $\Phi(x)$ as n tends to infinity. This seems to have been first proved by LINDEBERG.[6]

Secondly, if in the general case of unequal components we suppose the third moments $\beta_3^{(i)}$ to be finite (which of course implies the existence of $\alpha_2^{(i)} = \beta_2^{(i)}$), and if we put

$$B_2 = \frac{\beta_2^{(1)} + \beta_2^{(2)} + \cdots + \beta_2^{(n)}}{n}, \qquad B_3 = \frac{\beta_3^{(1)} + \beta_3^{(2)} + \cdots + \beta_3^{(n)}}{n}$$

then we have[7]

$$|F_n(x) - \Phi(x)| < 3\frac{\log n}{\sqrt{n}} \cdot \frac{B_3}{B_2^{3/2}} \tag{6}$$

for $n > 1$ and for all values of x. In the particular case when all the components are equal, B_2 and B_3 are independent of n, and it is obvious that, even if the components are not equal,

---

[1] Fundamentalsätze der Wahrscheinlichkeitsrechnung, Math. Zeitschrift, Bd 4 (1919).

[2] Über den zentralen Grenzwertsatz der Wahrscheinlichkeitsrechnung und das Momentenproblem, Math. Zeitschr. Bd 8 (1920).

[3] Über das GAUSS'sche Fehlergesetz, Skandinavisk Aktuarietidskrift 1922; Eine neuė Herleitung des Exponentialgesetzes in der Wahrscheinlichkeitsrechnung, Math. Zeitschr. Bd 15 (1922).

[4] Cf. the above quoted treatise by LÉVY and the references there given.

[5] Una nuova dimostrazione del secondo teorema-limite del calcolo delle probabilità, Rendic. del Circ. Matemat. di Palermo, 52 (1928).

[6] The theorem stated in the text is a particular case of Satz V in LINDEBERG's paper in the Math. Zeitschr. Cf. also p. 233 of LÉVY's treatise.

[7] This may be easily deduced from Satz 1 of my paper: »Das Gesetz von Gauss und die Theorie des Risikos», Skandinavisk Aktuarietidskrift 1923.

the factor $B_3 B_2^{-\frac{3}{2}}$ will in many cases remain bounded for large values of n.

If the derivative $F'_n(x)$ exists for all values of x, we have under certain additional conditions (cf. Art. 22)

$$\lim_{n \to \infty} F'_n(x) = \varphi(x)$$

with

$$\varphi(x) = \Phi'(x) = \frac{1}{\sqrt{2\pi}} e^{-\frac{x^2}{2}}.$$

Asymptotic expansions of probability functions.

5. As already pointed out in the Introduction, in certain applications we want a more precise knowledge of the probability function than that afforded by the theorem $\lim F_n = \Phi$. Such is, in particular, the case in the statistical applications of the hypothesis of elementary errors. According to Charlier[1] and Edgeworth[2], our function $F_n(x)$ may be formally developped in a series. Thus we may write, *formally*,

$$F_n(x) = \Phi(x) + \frac{c_3}{3!}\Phi'''(x) + \frac{c_4}{4!}\Phi''''(x) + \cdots, \tag{7}$$

where[3]

---

[1] »Über das Fehlergesetz», Arkiv för matematik etc., B. 2, n:o 8 (1905); »Über die Darstellung willkürlicher Funktionen», Arkiv, B. 2, n:o 20 (1905); »Researches into the Theory of Probability», Kongl. Fysiografiska Sällsk. Handl., B. 16 (1906); »Die strenge Form des Bernoullischen Theorems», Arkiv, B. 5, n:o 15 (1909); »Contributions to the mathematical Theory of Statistics 5», Arkiv, B. 9, n:o 25 (1914).

[2] »The law of error», Cambridge Philos. Trans. XX (1905); »The generalised law of errors», Journal of the R. Statistical Soc. LXIX (1906); »On the representation of statistical frequency by a series», Journal of the R. Statistical Soc. LXX (1907). — Series of the same type have, with a view to statistical applications, been considered also by various other authors, e.g. by Gram, Thiele, Bruns, Hausdorff.

[3] The relations (8) will be proved later in this paper. In a purely formal way, they are most easily deduced by means of the well known orthogonality relations satisfied by the polynomials H_ν. I take this opportunity of correcting an error in the expression for c_ν, given on p. 415 of my paper presented to the Copenhaguen Congress, where the integral is multiplied by $\frac{(-1)^\nu}{\nu}$ instead of $(-1)^\nu$.

$$c_\nu = (-1)^\nu \int_{-\infty}^{\infty} H_\nu(x)\, d F_n(x), \tag{8}$$

$H_\nu(x)$ being the ν:th polynomial of HERMITE, defined by the relation[1]

$$\frac{d^\nu}{dx^\nu} e^{-\frac{x^2}{2}} = (-1)^\nu H_\nu(x) e^{-\frac{x^2}{2}}.$$

It must be borne in mind that the coefficients c_ν, like the above introduced σ, depend on n. Since $F_n(x)$ is a normalized probability function, satisfying (4), it is readily seen that we have $c_1 = c_2 = 0$.

If we assume that the *frequency function* $F'_n(x)$ exists, we get by a formal differentiation of (7) the corresponding development

$$F'_n(x) = \varphi(x) + \frac{c_3}{3!}\varphi'''(x) + \frac{c_4}{4!}\varphi''''(x) + \cdots \tag{9}$$

This development is called by CHARLIER an *A-series*, and it is well known that it has been widely used as a general expression for statistical frequency-functions. In these practical applications, only a very limited number of terms of the series (generally two or three) are considered, and it is expected that the sum of these terms will give an approximation fully satisfactory for all practical purposes.

On the other hand, this use of the *A*-series has been severely criticised by several authors[2], who have objected to it on mathematical as well as on statistical grounds. Thus LINDEBERG expresses a very skeptical opinion as to the possibility of giving a theoretical justification (»— — — dürfte

[1] Thus we have $H_0(x)=1$, $H_1(x)=x$, $H_2(x)=x^2-1$, $H_3(x)=x^3-3x$, $H_4(x)=x^4-6x^2+3$; $H_5(x)=x^5-10x^3+15x$, $H_6(x)=x^6-15x^4+45x^2-15$.

[2] Cf. W. P. ELDERTON, »Frequency-curves and correlation», London 1906, p. 160—162 and a Notice in Biometrika, V, p. 206; J. F. STEFFENSEN, »Matematisk Iagttagelseslære», København 1923, p. 70—71; J. W. LINDEBERG, »Über die Begriffe Schiefheit und Exzess in der mathematischen Statistik», Skand. Aktuarietidskr. 1925, p. 125—127.

es aber nicht zu kühn sein, jeden Versuch in dieser Richtung als hoffnungslos zu bezeichnen»). As a matter of fact, it must be admitted that the mathematical foundations of the theory hitherto laid down seem to be rather weak.[1] — The present paper will be confined to a discussion of these questions from a mathematical point of view; in the following papers, we shall enter also upon the other sides of the subject. — Mathematically speaking, the problem is this:

Given a set of elementary errors $z_1, z_2, \ldots$ *with their probability functions* $V_1, V_2, \ldots$. *Is it true that, for large values of* n, *the probability function* $F_n(x)$ *and the frequency function* $F'_n(x)$ *(assuming that the latter exists) will be approximatively represented by the first few terms of* (7) *and* (9)? *And will this approximation be (on the whole) better with two terms than with one, better with three than with two, and so forth?*

Thus we are confronted with a question concerning the *asymptotic properties* of the series.[2] And this question has to be answered before we shall be theoretically justified in regarding the *A*-series as something more than an analytical instrument for the graduation and interpolation of given frequency-distributions, and in particular before it will be possible to claim for the coefficients of the series any special significance as statistical characteristics of the frequency-distribution (e.g. measures of »skewness» and »excess»).

6. From the hypothesis of elementary errors, CHARLIER[3] has deduced also another type of series, that he calls the *B-series*. This series has proved very useful in the practical

---

[1] As CHARLIER points out in Arkiv 9, n:o 25 (p. 2, footnote), his first proof, published in Arkiv 2, n:o 8, was vitiated by an error. His later deductions, and the deductions given by EDGEWORTH are, as far as I can see, of an essentially formal character. The error in CHARLIER's first proof has passed over into JÖRGENSEN's dissertation on the theory of correlation (»Undersögelser over Frekvensflader og Korrelation, Köbenhavn 1916, cf. p. 52—53).

[2] It must be remarked that our problem has only very little to do with the question of convergence or divergence of the *infinite* series (7) and (9). The latter question will, however, also be shortly discussed below.

[3] »Die zweite Form des Fehlergesetzes», Arkiv, B. 2, n:o 15 (1905); »Weiteres über das Fehlergesetz», Arkiv, B. 4, n:o 13 (1908).

applications and is generally regarded as possessing asymptotic properties, similar to those of the *A*-series.

We are brought to this new type of development if we consider a particular case of elementary errors, each of which has only a very small probability of deviating from a fixed value. Thus if we make a set of n drawings from an urn containing black and white balls, the chance of drawing a white ball being each time equal to $\frac{\lambda}{n}$, (λ constant), it is well known that the probability of getting a total number of ν white balls will, for increasing n, tend to the limit

$$\psi_\lambda(\nu) = \frac{\lambda^\nu}{\nu!} e^{-\lambda} = \frac{1}{\pi} e^{-\lambda} \int_0^\pi e^{\lambda \cos t} \cos(\lambda \sin t - \nu t)\, dt.$$

The number ν may be regarded as the sum of n mutually independent variables, one for each drawing, each of which has the only possible values 0 and 1, the corresponding probabilities being $1 - \frac{\lambda}{n}$ and $\frac{\lambda}{n}$. If in the last (integral) expression of $\psi_\lambda(\nu)$ we replace the integral variable ν by a continuous variable x, we get a function $\psi_\lambda(x)$, which is the generating function of the Charlier *B*-series:

$$k_0 \psi_\lambda(x) + k_1 \Delta \psi_\lambda(x) + k_2 \Delta^2 \psi_\lambda(x) + \cdots,$$

the Δ:s being the successive differences of $\psi_\lambda(x)$.

In this case, the probability function of each of the elementary errors which lead to $\psi_\lambda(x)$ may be written

$$\left(1 - \frac{\lambda}{n}\right) E(x) + \frac{\lambda}{n} E(x-1),$$

$E(x)$ being the particular probability function defined by

$$E(x) = \begin{cases} 0 & \text{for} \quad x < 0 \\ \frac{1}{2} & \text{»} \quad x = 0 \\ 1 & \text{»} \quad x > 0. \end{cases}$$

If we consider the more general case of a sum of n elementary errors, each of which has the probability function

$$\left(1 - \frac{\lambda}{n}\right) E(x) + \frac{\lambda}{n} U(x),$$

where $U(x)$ denotes an arbitrary probability function, we shall also get a definite limit for the probability function of the sum as n tends to infinity, but in the general case this limiting function has nothing[1] whatever to do with $\psi_\lambda(x)$.

Thus we see that $\psi_\lambda(x)$, considered as a generating function from the point of view of the theory of elementary errors, has not at all the same high degree of generality[2] as $\Phi(x)$. We shall not dwell further upon the B-series in the present paper, but shall return to the question in the following paper on statistical applications.

The coefficients of the A-series.

7. We proceed to a preliminary examination of the coefficients c_ν of the A-series. Putting for $\nu = 1, 2, \ldots$

$$\mu_\nu = \int_{-\infty}^{\infty} x^\nu \, d\, W_n(x),$$

(observe that μ_ν, like σ and c_ν, depends on n) we have by (3) and (2 a)

[1] For a particular form of $U(x)$, this has been pointed out by CHARLIER in Arkiv 2, n:o 15, p. 8. Cf. also my Copenhagen lecture, p. 423.

[2] Cf. EDGEWORTH, l. c., J. of the R. Stat. Soc., LXIX, p. 520, footnote.

$$\frac{\mu_\nu}{\sigma^\nu} = \int_{-\infty}^{\infty} x^\nu \, d F_n(x). \tag{10}$$

It is possible to express μ_ν in terms of the moments $\alpha_\nu^{(i)}$ of the components V_i. From the relations (1), we get without difficulty the symbolical expression

$$\mu_\nu = (\alpha^{(1)} + \alpha^{(2)} + \cdots + \alpha^{(n)})^\nu \tag{11}$$

where, after the expansion of the polynomial, each term $K(\alpha^{(1)})^h (\alpha^{(2)})^i \ldots (\alpha^{(n)})^l$ must be replaced by $K\alpha_h^{(1)} \alpha_i^{(2)} \ldots \alpha_l^{(n)}$. Thus if all the components have finite moments of the ν:th order, μ_ν and consequently c_ν are also finite. For the lowest values of ν, we have by (5)

$$\begin{aligned}
\mu_1 &= 0, \\
\mu_2 &= \sigma^2 = \sum_i \alpha_2^{(i)}, \\
\mu_3 &= \sum_i \alpha_3^{(i)}, \\
\mu_4 &= \sum_i \alpha_4^{(i)} + 6 \sum_{i<j} \alpha_2^{(i)} \alpha_2^{(j)}, \\
\mu_5 &= \sum_i \alpha_5^{(i)} + 10 \sum_{i \gtrless j} \alpha_2^{(i)} \alpha_3^{(j)}, \\
\mu_6 &= \sum_i \alpha_6^{(i)} + 15 \sum_{i \gtrless j} \alpha_2^{(i)} \alpha_4^{(j)} + 20 \sum_{i<j} \alpha_3^{(i)} \alpha_3^{(j)} + \\
&\qquad + 90 \sum_{i<j<k} \alpha_2^{(i)} \alpha_2^{(j)} \alpha_2^{(k)},
\end{aligned} \tag{12}$$

i, j and k passing through the values $1, 2, \ldots, n$. In the particular case when all the components V_i are equal, $\alpha_\nu^{(i)}$ is independent of i, and the expressions (12) become, if we write α_ν in the place of $\alpha_\nu^{(i)}$,

$$
\begin{aligned}
&\mu_2 = \sigma^2 = n\alpha_2, \\
&\mu_3 = n\alpha_3, \\
(13)\quad &\mu_4 = n\alpha_4 + 3n(n-1)\alpha_2^2, \\
&\mu_5 = n\alpha_5 + 10n(n-1)\alpha_2\alpha_3, \\
&\mu_6 = n\alpha_6 + 15n(n-1)\alpha_2\alpha_4 + 10n(n-1)\alpha_3^2 + \\
&\qquad\qquad + 15n(n-1)(n-2)\alpha_2^3.
\end{aligned}
$$

By the general formula (8) for the coefficients c_ν we get, with regard to (10),

$$
(14)\qquad
\begin{aligned}
c_3 &= -\frac{\mu_3}{\sigma^3}, \\
c_4 &= \frac{\mu_4}{\sigma^4} - 3, \\
c_5 &= -\frac{\mu_5}{\sigma^5} + 10\frac{\mu_3}{\sigma^3}, \\
c_6 &= \frac{\mu_6}{\sigma^6} - 15\frac{\mu_4}{\sigma^4} + 30, \\
&\quad \cdots\cdots\cdots\cdots
\end{aligned}
$$

and hence by (13) for the case of equal components

$$
(15)\qquad
\begin{aligned}
c_3 &= -\frac{\alpha_3}{\alpha_2^{3/2}} \cdot \frac{1}{n^{1/2}}, \\
c_4 &= \frac{\alpha_4 - 3\alpha_2^2}{\alpha_2^2} \cdot \frac{1}{n}, \\
c_5 &= -\frac{\alpha_5 - 10\alpha_2\alpha_3}{\alpha_2^{5/2}} \cdot \frac{1}{n^{3/2}}, \\
c_6 &= \frac{10\alpha_3^2}{\alpha_2^3} \cdot \frac{1}{n} + \frac{\alpha_6 - 15\alpha_2\alpha_4 - 10\alpha_3^2 + 30\alpha_2^3}{\alpha_2^3} \cdot \frac{1}{n^2}.
\end{aligned}
$$

Thus in the case of equal components the first coefficients of the A-series are polynomials in $n^{-1/2}$. It will appear below (as of course has been known long before) that this is true also for $\nu > 6$. If α_ν is finite, c_ν is also finite and is of the order n^{-q}, where[1] $q = \frac{\nu}{2} - \left[\frac{\nu}{3}\right]$.

It follows that, with respect to their order of magnitude, the terms of the A-series can be arranged in groups in the following way:

Order of term (ν):	Order of magnitude:
3	$n^{-1/2}$
4, 6	n^{-1}
5, 7, 9	$n^{-3/2}$
8, 10, 12	n^{-2}
11, 13, 15	$n^{-5/2}$
. . . .	. .

If the probability functions of the elementary errors are not actually equal, but satisfy certain conditions of regularity with respect to their moments, the order of magnitude of the terms will still be the same.

8. This behaviour of the coefficients c_ν for large values of the number of elementary errors seems to be the main ground for the general belief in the asymptotic properties of the A-series. It is obvious that, from a mathematical point of view, this is far from sufficient. In the general case, when only a limited number of moments of the elementary errors are finite, the A-series cannot be extended beyond a certain term, since the following coefficients are infinite. In such a case, it is evident that nothing can be inferred from the fact that the finite coefficients of the series tend to zero as n tends to infinity. And even if we suppose that all the moments are finite and that the A-series (7) or (9) is convergent for all values of x, it does not follow from the asymptotic behaviour of the individual terms that the partial sums

[1] $\left[\frac{\nu}{3}\right]$ denotes, as usual, the greatest integer $\leqq \frac{\nu}{3}$.

of the series will, for increasing n, furnish a true asymptotic expansion of the probability function $F_n(x)$ or the frequency function $F'_n(x)$. As a matter of fact, it will be shown below by examples that, unless certain restrictions of a quite different character are imposed upon the probability functions of the elementary errors, there are cases where *no asymptotic expansion exists*.

Thus we must conclude that the problem calls for fresh mathematical investigation.

A modification of the A-series.

9. As will be shown in the following, it is really possible to assign certain conditions for the components V_i, sufficient to establish the existence of an asymptotic expansion in the strict sense by the partial sums of the A-series. The defenders of the A-series will thus receive a certain amount of support from our considerations.

The expansion in A-series suffers, however, from certain formal inconveniences which have been in part pointed out by Edgeworth.[1] Thus it follows from the results mentioned in Art. 7 concerning the order of magnitude of the coefficients c_ν, that the terms of the series do not occur in the natural order. If, e. g., in the case of equal components we want an expression of $F_n(x)$, correct up to terms of the order $\frac{1}{n}$, we shall have to consider the terms with $\nu=3$, 4 and 6. According to (14) and (15) this requires the knowledge of the moments $\alpha_2, \alpha_3, \alpha_4, \alpha_6$ or $\mu_2, \mu_3, \mu_4, \mu_6$. But from (15) we see that c_6 contains also terms of the order $\frac{1}{n^2}$, and that the knowledge of the sixth moment is in reality unnecessary, since the only term of order $\frac{1}{n}$ in c_6 is

$$\frac{10\,\alpha_3^2}{\alpha_2^3}\cdot\frac{1}{n}=\frac{10\,\mu_3^2}{\mu_2^3}.$$

[1] l. c., J. of the R. Stat. Soc. LXX.

If we want to take into account also terms of the order $n^{-3/2}$, we shall have to introduce c_5, c_7 and c_9, which requires the knowledge of moments up to the ninth order, whereas in reality (as will be shown below) the fifth order is sufficient.

Thus we might ask for a modified type of expansion, not introducing moments of an order higher than is necessary to obtain any prescribed degree of approximation. In the case of equal components, the expansion ought to reduce directly to a development in powers of $n^{-1/2}$. The formal definition of an expansion of this type has, in reality, been given by EDGEWORTH[1], though his mathematical developments are far from rigorous.

In order to get this expansion, we shall use the *semi-invariants* introduced by THIELE. The semi-invariants γ_ν of a probability function $V(x)$ with the moments α_ν $(\alpha_1=0)$ are defined by the identity

$$(16) \qquad e^{\sum_2^\infty \frac{\gamma_\nu}{\nu!} z^\nu} = 1 + \sum_2^\infty \frac{\alpha_\nu}{\nu!} z^\nu,$$

which is to be interpreted in a purely formal sense, as a mere substitute for the system of linear relations between the γ_ν and the α_ν, to which it leads. For the lowest values of ν we get

$$(17) \qquad \begin{aligned} \gamma_2 &= \alpha_2, \\ \gamma_3 &= \alpha_3, \\ \gamma_4 &= \alpha_4 - 3\alpha_2^2, \\ \gamma_5 &= \alpha_5 - 10\alpha_2\alpha_3, \\ \gamma_6 &= \alpha_6 - 15\alpha_2\alpha_4 - 10\alpha_3^2 + 30\alpha_2^3. \end{aligned}$$

It is readily seen that the expression of γ_ν only contains moments up to the ν:th order, so that γ_ν is finite if α_ν is finite.

For our n components $V_i(x)$, we denote the semi-invariants by $\gamma_\nu^{(i)}$, and for the composed function $W_n(x)$ by λ_ν (observe

[1] l. c. Cambr. Phil. Trans.

that λ_ν, like μ_ν, c_ν and σ, depends on n), so that the λ_ν are connected with the moments μ_ν by relations analogous to (17) and summed up by the formal identity

$$e^{\sum_{2}^{\infty} \frac{\lambda_\nu}{\nu!} z^\nu} = 1 + \sum_{2}^{\infty} \frac{\mu_\nu}{\nu!} z^\nu. \tag{18}$$

Then it is known[1] that we have

$$\lambda_\nu = \sum_{i=1}^{n} \gamma_\nu^{(i)} \tag{19}$$

and hence in particular

$$\lambda_2 = \sum_i \alpha_2^{(i)} = \sigma^2. \tag{20}$$

Further, we consider the following symbolical identity

$$e^{\sum_{1}^{\infty} \frac{\lambda_{\nu+2} (-\Phi)^{\nu+2}}{\sigma^{\nu+2} (\nu+2)!} z^\nu} = 1 + \sum_{1}^{\infty} P_\nu(\Phi) z^\nu, \tag{21}$$

which defines $P_\nu(\Phi)$ as a polynomial in Φ, where after the expansion we replace every power Φ^j by the derivative $\Phi^{(j)}(x)$. It is readily seen that $P_\nu(\Phi)$ contains the derivatives $\Phi^{(\nu+2)}(x)$, $\Phi^{(\nu+4)}(x)$, $\cdots$, $\Phi^{(3\nu)}(x)$, and that P_ν is a polynomial in the arguments $\frac{\lambda_3}{\sigma^3}, \frac{\lambda_4}{\sigma^4}, \ldots, \frac{\lambda_{\nu+2}}{\sigma^{\nu+2}}$ and Φ, such that we have

$$\begin{aligned} P_\nu(\Phi) &= P_\nu\left(\frac{\lambda_3}{\sigma^3}, \frac{\lambda_4}{\sigma^4}, \ldots, \frac{\lambda_{\nu+2}}{\sigma^{\nu+2}}; \Phi\right) \\ &= \sigma^{-\nu} P_\nu\left(\frac{\lambda_3}{\sigma^2}, \frac{\lambda_4}{\sigma^2}, \ldots, \frac{\lambda_{\nu+2}}{\sigma^2}; \Phi\right), \end{aligned} \tag{22}$$

where of course Φ^j always must be replaced by $\Phi^{(j)}(x)$.

[1] Cf. STEFFENSEN, Matematisk Iagttagelseslære p. 77.

If all the components V_i have finite moments of the ν:th order, it follows from our formulae that $\lambda_2, \lambda_3, \cdots, \lambda_\nu$ and $P_1, P_2, \cdots, P_{\nu-2}$ are finite and can be expressed in terms of the moments.

For the lowest values of ν we get

$$P_1(\Phi) = -\frac{\lambda_3}{3!\,\sigma^3}\,\Phi^{(3)}(x),$$

$$(23)\quad P_2(\Phi) = \frac{\lambda_4}{4!\,\sigma^4}\,\Phi^{(4)}(x) + \frac{10\,\lambda_3^2}{6!\,\sigma^6}\,\Phi^{(6)}(x),$$

$$P_3(\Phi) = -\frac{\lambda_5}{5!\,\sigma^5}\,\Phi^{(5)}(x) - \frac{35\,\lambda_3\lambda_4}{7!\,\sigma^7}\,\Phi^{(7)}(x) - \frac{280\,\lambda_3^3}{9!\,\sigma^9}\,\Phi^{(9)}(x).$$

Expressed in terms of the moments μ_ν, these relations become

$$P_1(\Phi) = -\frac{1}{3!}\cdot\frac{\mu_3}{\sigma^3}\,\Phi^{(3)}(x),$$

$$(24)\quad P_2(\Phi) = \frac{1}{4!}\left(\frac{\mu_4}{\sigma^4} - 3\right)\Phi^{(4)}(x) + \frac{10}{6!}\left(\frac{\mu_3}{\sigma^3}\right)^2\Phi^{(6)}(x),$$

$$P_3(\Phi) = -\frac{1}{5!}\left(\frac{\mu_5}{\sigma^5} - 10\frac{\mu_3}{\sigma^3}\right)\Phi^{(5)}(x) - \frac{35}{7!}\cdot\frac{\mu_3}{\sigma^3}\left(\frac{\mu_4}{\sigma^4} - 3\right)\Phi^{(7)}(x)$$
$$- \frac{280}{9!}\left(\frac{\mu_3}{\sigma^3}\right)^3\Phi^{(9)}(x).$$

In the case of equal components, we have by (19) and (20) $\lambda_\nu = n\gamma_\nu$ and $\sigma^2 = n\alpha_2$. According to (16), γ_ν may be expressed by a polynomial in $\alpha_2, \alpha_3, \cdots, \alpha_\nu$, where all the terms are of the weight ν, and thus (22) gives in this case

$$(25)\qquad P_\nu(\Phi) = n^{-\frac{\nu}{2}}\,Q_\nu\left(\frac{\alpha_3}{\alpha_2^{3/2}}, \frac{\alpha_4}{\alpha_2^2}, \cdots, \frac{\alpha_{\nu+2}}{\alpha_2^{\frac{\nu+2}{2}}};\ \Phi\right),$$

where Q_ν is a polynomial in all its arguments, *independent of* n. For $\nu = 1, 2, 3$ we get in this case from (23)

$$P_1(\Phi) = -\frac{1}{n^{1/2}} \cdot \frac{1}{3!} \cdot \frac{\alpha_3}{\alpha_2^{3/2}} \Phi^{(3)}(x),$$

$$P_2(\Phi) = \frac{1}{n} \cdot \left[\frac{1}{4!}\left(\frac{\alpha_4}{\alpha_2^2} - 3\right) \Phi^{(4)}(x) + \frac{10}{6!} \cdot \frac{\alpha_3^2}{\alpha_2^3} \Phi^{(6)}(x)\right],$$

$$P_3(\Phi) = -\frac{1}{n^{3/2}} \cdot \left[\frac{1}{5!}\left(\frac{\alpha_5}{\alpha_2^{5/2}} - 10\frac{\alpha_3}{\alpha_2^{3/2}}\right) \Phi^{(5)}(x) + \frac{35}{7!} \cdot \frac{\alpha_3}{\alpha_2^{3/2}}\left(\frac{\alpha_4}{\alpha_2^2} - 3\right) \Phi^{(7)}(x) + \frac{280}{9!}\left(\frac{\alpha_3}{\alpha_2^{3/2}}\right)^3 \Phi^{(9)}(x)\right].$$

It is plain that we always have

$$\frac{d}{dx} P_\nu(\Phi) = P_\nu(\varphi),$$

where as before $\varphi(x) = \Phi'(x)$ and the expression $P_\nu(\varphi)$ is to be interpreted symbolically in the same way as $P_\nu(\Phi)$.

10. According to (14) and (24), the two developments

(26) $$\Phi(x) + P_1(\Phi) + P_2(\Phi) + \cdots$$

and

(27) $$\Phi(x) + \frac{c_3}{3!} \Phi^{(3)}(x) + \frac{c_4}{4!} \Phi^{(4)}(x) + \cdots,$$

as well as the developments in $\varphi(x)$ obtained by differentiation, coincide as to their two first terms, but differ from the third term onwards. It follows from (25) that, in the case of equal components, the P-development (26) reduces directly to a power series in $n^{-\frac{1}{2}}$, whereas the terms of the A-series (27) become polynomials in $n^{-\frac{1}{2}}$.

We shall consider the following mathematical problem: *A certain number of moments of the components V_i being regarded as known, it is required to determine the normalized probability function $F_n(x)$ with the greatest possible accuracy. — It will be shown that, under certain conditions, both developments* (26) *and* (27) *furnish asymptotic expansions of $F_n(x)$ and that,* *generally speaking, the P-development* (26) *gives the greatest accuracy that can be obtained with any given number of moments.*

In Part II, we shall deal thoroughly with the case of equal components, where results of a particularly simple type can be obtained. In Part III we shall give a brief account of various generalizations of the results of Part II.

It will follow from our results that, under fairly general conditions, the curve

$$y = \Phi(x) + P_1(\Phi) + \cdots + P_{k-3}(\Phi)$$

can be regarded as giving at least a sort of *mechanical graduation* of the curve $y = F_n(x)$. A similar graduation is furnished by the curve

$$y = \Phi(x) + \frac{c_3}{3!}\Phi^{(3)}(x) + \cdots + \frac{c_{k-1}}{(k-1)!}\Phi^{(k-1)}(x).$$

Here the moments are regarded as known up to the order $k-1$, and the moments of order k are assumed to be finite. Then the degree of approximation obtained by the P-series is, roughly speaking, of the order $n^{-\frac{k-2}{2}}$, whereas the A-series gives the order $n^{-\frac{\varkappa}{2}}$ with $\varkappa = \left[\frac{k+2}{3}\right]$. Thus for $k=3$ and $k=4$ we get the same result with both developments, but for $k>4$ the A-series does not give the best result that can be obtained with the available information. — This is shown by Theorem 1, generalized in Art. 25.

If the components satisfy certain conditions of regularity, we obtain real asymptotic expansions for $F_n(x)$ and even for the frequency function $F'_n(x)$. In the case of equal components, such conditions can be assigned in a very simple way (Theorems 2 and 3), but in the general case we cannot, so far, avoid a certain degree of complication (cf. Art. 26). The difference as to the order of approximation furnished by the two different expansions appears in all these theorems. Conditions for convergence of the A-series are given by Theorems 4 and 5. In Art. 24 we shall give some examples relating to cases where no asymptotic expansions exist.

Part II. The case of equal components.

Notations.

11. We are given a never decreasing function $V(x)$, such that $V(-\infty)=0$, $V(+\infty)=1$, and that $V(x)=\frac{1}{2}[V(x-0)+V(x+0)]$ for all real values of x. We suppose that the moments

$$\alpha_\nu=\int_{-\infty}^{\infty} x^\nu\, d\, V(x), \quad \beta_\nu=\int_{-\infty}^{\infty} |x|^\nu\, d\, V(x)$$

are finite for $\nu=1, 2, \cdots, k$ $(k\geqq 2)$, and that $\alpha_1=0$. We shall put

$$\varrho_\nu=\frac{\beta_\nu^{\frac{1}{\nu}}}{\beta_2^{\frac{1}{2}}}. \tag{28}$$

About the values of α_ν and β_ν for $\nu>k$ we shall not make any hypothesis — these moments may be finite or not.

We now form a sequence of functions $W_1(x)$, $W_2(x)$, ..., by putting

$$W_1(x)=V(x),$$

$$W_2(x)=\int_{-\infty}^{\infty} W_1(x-t)\, d\, V(t),$$

$$\cdots\cdots\cdots\cdots\cdots$$

$$W_n(x)=\int_{-\infty}^{\infty} W_{n-1}(x-t)\, d\, V(t),$$

the value of $W_n(x)$ in any point of discontinuity being determined by the condition $W_n(x)=\frac{1}{2}[W_n(x-0)+W_n(x+0)]$.

The moments

$$\mu_\nu = \int_{-\infty}^{\infty} x^\nu \, d\, W_n(x)$$

are finite for $\nu \leqq k$ and can be expressed by $\alpha_2, \alpha_3, \cdots, \alpha_\nu$ according to (11). We have $\mu_1 = 0$; for μ_2 we introduce the particular notation

$$\mu_2 = n\alpha_2 = \sigma^2.$$

The semi-invariants γ_ν and λ_ν are, according to (16) and (18), defined for $\nu = 2, 3, \cdots, k$ by the developments

(29) $$\log\left(1 + \sum_2^k \frac{\alpha_\nu}{\nu!} z^\nu\right) = \sum_2^k \frac{\gamma_\nu}{\nu!} z^\nu + L z^{k+1} + \cdots,$$

(30) $$\log\left(1 + \sum_2^k \frac{\mu_\nu}{\nu!} z^\nu\right) = \sum_2^k \frac{\lambda_\nu}{\nu!} z^\nu + M z^{k+1} + \cdots,$$

which are convergent for sufficiently small values of z. Similarly, the symbolical polynomial $P_\nu(\Phi)$ is defined for $\nu = 1, 2, \cdots, k-2$ by the development

(31) $$e^{\sum\limits_1^{k-2} \frac{\lambda_{\nu+2}(-\Phi)^{\nu+2}}{\sigma^{\nu+2}(\nu+2)!} z^\nu} = 1 + \sum_1^{k-2} P_\nu(\Phi) z^\nu + N z^{k-1} + \cdots$$

It follows directly from the definition that $P_\nu(\Phi)$ satisfies the relation (22). On the contrary, the relation (19) between the semi-invariants and the relation (25) for $P_\nu(\Phi)$ will not be regarded as previously known, but will be proved below.

$P_\nu(\Phi)$ has been defined as a *symbolical* polynomial, which in reality is a linear expression in the derivatives $\Phi^{(j)}(x)$. We shall also have occasion to use P_ν as a polynomial *in the ordinary sense*, with a purely imaginary argument. At these occasions, we shall use the notation $P_\nu(it)$, which is thus to be taken as a true polynomial in the argument it. We have, e. g., by (24)

$$P_1(it) = -\frac{\mu_3}{6\sigma^3}(it)^3, \quad P_1(\Phi) = -\frac{\mu_3}{6\sigma^3}\Phi^{(3)}(x).$$

Finally, we introduce the new sequence of functions $F_1(x)$, $F_2(x)$, $\cdots$ by putting

$$F_n(x) = W_n(\sigma x).$$

For $\nu = 1, 2, \cdots, k$ we then have

$$\frac{\mu_\nu}{\sigma^\nu} = \int_{-\infty}^{\infty} x^\nu \, dF_n(x).$$

Putting, for $\nu = 1, 2, \cdots, k$,

$$c_\nu = (-1)^\nu \int_{-\infty}^{\infty} H_\nu(x)\, dF_n(x), \tag{32}$$

we have $c_1 = c_2 = 0$.

It is to be noted that, of the sequences of numbers and functions above introduced, α_ν, β_ν, ϱ_ν and γ_ν are *independent of* n and only depend on the given function $V(x)$. μ_ν, λ_ν and c_ν, on the other hand, depend on the moments $\alpha_2, \alpha_3, \cdots, \alpha_\nu$ *and on* n. Finally P_ν depends on n, on the moments $\alpha_2, \alpha_3, \cdots, \alpha_{\nu+2}$ and on the variable x.

12. The *adjuncts* of the functions $V(x)$, $W_n(x)$ and $F_n(x)$ are complex functions, defined for all values of the real variable t by the relations

$$v(t) = \int_{-\infty}^{\infty} e^{-itx}\, dV(x),$$

$$w_n(t) = \int_{-\infty}^{\infty} e^{-itx}\, d\, W_n(x),$$

$$f_n(t) = \int_{-\infty}^{\infty} e^{-itx}\, dF_n(x).$$

It follows from the existence of the k:th moment β_k that all three functions $v(t)$, $w_n(t)$ and $f_n(t)$ as well as their k first derivatives are finite and continuous for all real values of t. The moduli of the three functions can never exceed unity; for $t=0$ we have $v(0)=w_n(0)=f_n(0)=1$.

From the relations between $V(x)$, $W_n(x)$ and $F_n(x)$, it readily follows that

$$\begin{aligned} w_n(t) &= (v(t))^n, \\ f_n(t) &= w_n\left(\frac{t}{\sigma}\right) = \left[v\left(\frac{t}{\sigma}\right)\right]^n. \end{aligned} \tag{33}$$

The adjunct of the particular function $\Phi(x)$ is $e^{-\frac{t^2}{2}}$. Actually, we get by a well-known formula in the Integral Calculus

$$\int_{-\infty}^{\infty} e^{-itx}\, d\,\Phi(x) = e^{-\frac{t^2}{2}}, \tag{34}$$

and hence by partial integration for $\nu=1, 2, \ldots$

$$\int_{-\infty}^{\infty} e^{-itx}\, d\,\Phi^{(\nu)}(x) = (it)^\nu e^{-\frac{t^2}{2}}. \tag{35}$$

Conversely, we have

$$\Phi^{(\nu)}(x) = \frac{1}{2\pi} \int_{-\infty}^{\infty} (it)^{\nu-1} e^{itx - \frac{t^2}{2}} dt \tag{36}$$

and hence

$$P_\nu(\Phi) = \frac{1}{2\pi} \int_{-\infty}^{\infty} \frac{e^{itx}}{it} P_\nu(it) e^{-\frac{t^2}{2}} dt. \tag{37}$$

13. The number k, the greatest index for which the moments α_ν and β_ν are regarded as known, plays an important part in the sequel. With respect to this number, we shall adopt the following convention.

We shall use the letter Θ as a general symbol denoting a real or complex quantity which may depend on n, x, t or any other variables, *but which is always less in absolute value than a constant depending only on k.*

Similarly, we shall denote by Δ a quantity which is *less in absolute value than a constant depending only on k and on the given function V*, but independent of n, x, t etc.

These symbols will thus be used in a sense which is closely similar to the well-known symbols O and o introduced by LANDAU.

A convergence theorem.

14. As a preparation for the following complicated developments, we shall now give a proof[1] of the theorem mentioned in Art. 4, that $F_n(x)$ tends to $\Phi(x)$ for $n \to \infty$, as soon as the second moment of $V(x)$ is finite.

Thus we take $k=2$, and we begin by proving

$$\lim_{t \to 0} \frac{v(t)-1}{t^2} = -\frac{1}{2}\alpha_2.$$

[1] The proof is similar to that given by LÉVY, l. c., but avoids the introduction of conditionally convergent integrals by the use of an argument due to PÓLYA, l. c. Another similar proof was given by PHRAGMÉN in a lecture in the Swedish Actuarial Society.

Given any $\varepsilon > 0$, we can choose M such that

$$0 \leqq \alpha_2 - \int_{-M}^{M} x^2 d\, V(x) < \varepsilon.$$

Since $\alpha_1 = 0$, we have

$$\frac{v(t)-1}{t^2} + \frac{1}{2}\alpha_2 = \int_{-\infty}^{\infty} \frac{e^{-itx} - 1 + itx + \frac{1}{2}t^2x^2}{t^2}\, d\, V(x).$$

Now we have for any real a

$$|e^{ia} - 1 - ia| \leqq \frac{1}{2}a^2,$$

$$\left|e^{ia} - 1 - ia + \frac{1}{2}a^2\right| \leqq \frac{1}{6}|a|^3,$$

and thus we find for the modulus of the fraction under the integral the upper limits x^2 and $\frac{1}{6}|tx^3|$. Using the first limit for $|x| > M$, the second for $|x| < M$, we obtain

$$\left|\frac{v(t)-1}{t^2} + \frac{1}{2}\alpha_2\right| < \frac{1}{6}M^3|t| + \varepsilon,$$

and this can be made $< 2\varepsilon$ by taking $|t|$ sufficiently small. Thus we have

$$v(t) = 1 - \frac{\alpha_2 t^2}{2}[1 + \omega(t)]$$

where $\omega(t)$ tends to zero with t. Hence it follows

$$v\left(\frac{t}{\sqrt{n\alpha_2}}\right) = 1 - \frac{t^2}{2n}\left[1 + \omega\left(\frac{t}{\sqrt{n\alpha_2}}\right)\right]$$

where, now, $\omega\left(\frac{t}{\sqrt{n\alpha_2}}\right)$ tends to zero as n tends to infinity, uniformly in every finite interval with respect to t. This gives us by (33)

$$\lim_{n\to\infty} f_n(t) = \lim_{n\to\infty}\left[v\left(\frac{t}{\sqrt{n\alpha_2}}\right)\right]^n = e^{-\frac{t^2}{2}}$$

uniformly in every finite interval. Thus the adjunct of $F_n(x)$ tends, for infinitely increasing n, uniformly towards the adjunct of $\Phi(x)$. As we shall now see, this makes it possible to prove also that $F_n(x)$ tends to $\Phi(x)$. — Consider the absolutely convergent integral

$$J_n(x) = \frac{1}{2\pi}\int_{-\infty}^{\infty}\frac{e^{itx}}{(it)^2}\left[f_n(t) - e^{-\frac{t^2}{2}}\right]dt,$$

which obviously tends to zero as $n\to\infty$, uniformly for all real values of x. Owing to the absolute convergence, we may write[1]

$$J_n(x) = -\frac{1}{2\pi}\int_{-\infty}^{\infty}\frac{dt}{t^2}\int_{-\infty}^{\infty} e^{it(x-y)}\, d[F_n(y) - \Phi(y)]$$

$$= \frac{1}{\pi}\int_{0}^{\infty}\frac{dt}{t^2}\int_{-\infty}^{\infty}[1-\cos t(x-y)]\, d[F_n(y) - \Phi(y)]$$

$$= \frac{1}{\pi}\int_{-\infty}^{\infty} d[F_n(y) - \Phi(y)]\int_{0}^{\infty}\frac{1-\cos t(x-y)}{t^2}\, dt.$$

But we have

$$\int_{0}^{\infty}\frac{1-\cos tz}{t^2}\, dt = \frac{\pi}{2}|z|,$$

[1] The inversion of the order of integration is legitimate, since the last repeated integral is absolutely convergent. Cf. Art. 18.

and hence

$$J_n(x) = \frac{1}{2}\int_{-\infty}^{x} (x-y)\, d\,[F_n(y) - \Phi(y)] + \frac{1}{2}\int_{x}^{\infty} (y-x)\, d\,[F_n(y) - \Phi(y)]$$

$$= \int_{-\infty}^{x} (x-y)\, d\,[F_n(y) - \Phi(y)]$$

$$= \int_{-\infty}^{x} [F_n(y) - \Phi(y)]\, dy,$$

so that

$$\int_a^b F_n(t)\, dt \to \int_a^b \Phi(t)\, dt$$

as $n \to \infty$, uniformly for all real a and b. If we put here first $a=x$, $b=x+h$ and then $a=x-h$, $b=x$, we get, since F_n and Φ are both never decreasing functions,

$$h F_n(x) < h\Phi(x+h) + \varepsilon_n,$$

$$h F_n(x) > h\Phi(x-h) - \varepsilon_n,$$

where ε_n tends to zero with $\frac{1}{n}$, uniformly in x and h. Hence finally, since $\Phi'(x)$ is always < 1,

$$F_n(x) < \Phi(x) + h + \frac{\varepsilon_n}{h},$$

$$F_n(x) > \Phi(x) - h - \frac{\varepsilon_n}{h}.$$

If we put here $h = \sqrt{\varepsilon_n}$, the theorem is proved.

Auxiliary theorems.

15. We begin with some simple Lemmas concerning the moments and semi-invariants.

Lemma 1. *We have* $\beta_1 \leqq \beta_2^{\frac{1}{2}} \leqq \beta_3^{\frac{1}{3}} \leqq \cdots \leqq \beta_k^{\frac{1}{k}}$.

The quadratic form in x and y

$$\int_{-\infty}^{\infty} \left(x\,|t|^{\frac{\nu-1}{2}} + y\,|t|^{\frac{\nu+1}{2}}\right)^2 d\,V(t) = \beta_{\nu-1}x^2 + 2\,\beta_\nu xy + \beta_{\nu+1}y^2$$

is never negative. This gives $\beta_\nu^2 \leqq \beta_{\nu-1}\,\beta_{\nu+1}$ and

$$\beta_\nu^{2\nu} \leqq \beta_{\nu-1}^{\nu}\,\beta_{\nu+1}^{\nu}.$$

Writing this inequality for $\nu=1, 2, \ldots, \nu$ and multiplying, we get $\beta_\nu^{\nu+1} \leqq \beta_{\nu+1}^{\nu}$, which proves the lemma. — Thus according to (28) we have $\varrho_\nu \leqq \varrho_k$ for $\nu < k$.

Lemma 2. *For* $\nu=2, 3, \ldots, k$ *is* $\gamma_\nu = \Theta\beta_\nu$. (For the signification of Θ cf. Art. 13.)

This follows immediately from the fact that, according to (29), γ_ν is equal to a polynomial in $\alpha_2, \ldots, \alpha_\nu$, where the coefficients are absolute constants and each term has the weight ν. By Lemma 1, each term in this polynomial is of the form $\Theta\beta_\nu$.

Lemma 3. *For* $\nu=2, 3, \ldots, k$ *is* $\lambda_\nu = n\gamma_\nu$.

In Art. 14, we have proved that

$$(38) \qquad v(t) = 1 + \frac{\alpha_2}{2!}(-it)^2 + o(t^2),$$

where, as usual, $o(t^2)$ denotes a quantity which, divided by t^2, tends to zero with t. In exactly the same way we get

$$
(39) \qquad
\begin{aligned}
v(t) &= 1 + \sum_2^k \frac{\alpha_\nu}{\nu!}(-it)^\nu + o(t^k), \\
w_n(t) &= 1 + \sum_2^k \frac{\mu_\nu}{\nu!}(-it)^\nu + o(t^k),
\end{aligned}
$$

and hence

$$
\log v(t) = \log\left(1 + \sum_2^k \frac{\alpha_\nu}{\nu!}(-it)^\nu\right) + o(t^k),
$$

$$
\log w_n(t) = \log\left(1 + \sum_2^k \frac{\mu_\nu}{\nu!}(-it)^\nu\right) + o(t^k),
$$

the logarithms being certainly finite for sufficiently small values of t. According to (29) and (30) this gives

$$
(40) \qquad
\begin{aligned}
\log v(t) &= \sum_2^k \frac{\gamma_\nu}{\nu!}(-it)^\nu + o(t^k), \\
\log w_n(t) &= \sum_2^k \frac{\lambda_\nu}{\nu!}(-it)^\nu + o(t^k).
\end{aligned}
$$

Since by (33) we have $\log w_n = n \log v$, the truth of the lemma follows. This proves, as shown in Art. 9, that $P_\nu(\Phi)$ satisfies the important relation (25).

Lemma 4. *If we put, according to* (31),

$$
P_\nu(\Phi) = \sum_{j=1}^{\nu} H_{\nu j} \Phi^{(\nu+2j)}(x),
$$

we have for $\nu = 1, 2, \ldots, k-2$

$$
H_{\nu j} = \Theta \varrho_k^{\nu+2j} n^{-\frac{\nu}{2}},
$$

ϱ_k *being defined by* (28).

By (31), $H_{\nu j}$ is a sum of less than k terms of the form

$$A \frac{\lambda_{\nu_1+2} \lambda_{\nu_2+2} \ldots \lambda_{\nu_j+2}}{\sigma^{\nu+2j}},$$

where $\nu_1 + \nu_2 + \cdots + \nu_j = \nu$ and A is an absolute constant. The truth of the lemma now follows immediately from the three preceding lemmas, if we remember that $\sigma = \sqrt{n \beta_2}$.

Lemma 5. *For $t \to 0$ we have*

$$e^{\frac{t^2}{2}} f_n(t) = 1 + \sum_{3}^{k} \frac{c_\nu}{\nu!} (i t)^\nu + o(t^k),$$

c_ν being given by (32).

This may be deduced from the relation

$$e^{\frac{t^2}{2}} f_n(t) = \int_{-\infty}^{\infty} e^{\frac{t^2}{2} - itx} \, d F_n(x)$$

and the development

$$e^{\frac{t^2}{2} - itx} = \sum_{0}^{\infty} \frac{H_\nu(x)}{\nu!} (-i t)^\nu$$

by a reasoning perfectly analogous to that which leads to (38) and (39).

Expansion of the adjunct.

16. Lemma 5 gives an asymptotic expansion of the adjunct $f_n(t)$ for small values of t. We shall, however, require the much more precise information furnished by the two following Lemmas. *We shall suppose henceforth that $k \geqq 3$, and put*

$$\varkappa = \left[\frac{k+2}{3} \right].$$

Lemma 6.[1] *For* $|t| < \frac{1}{\varrho_k}\sqrt[6]{n}$ *we have*

$$\text{(41 a)}\quad e^{\frac{t^2}{2}} f_n(t) = 1 + \sum_{1}^{k-3} P_\nu(it) + \Theta n^{-\frac{k-2}{2}}\left(|\varrho_k t|^k + |\varrho_k t|^{3(k-2)}\right)$$

and

$$\text{(41 b)}\quad e^{\frac{t^2}{2}} f_n(t) = 1 + \sum_{3}^{k-1} \frac{c_\nu}{\nu!}(it)^\nu + \Theta n^{-\frac{x}{2}}\left(|\varrho_k t|^k + |\varrho_k t|^{3(k-2)}\right).$$

This lemma shows how the difference between the »*P*-series» and the »*A*-series» arises. (41 a) gives, according to (25), an expansion in powers of $n^{-1/2}$, while (41 b) gives an expansion in powers of t.

For all real values of a we have

$$e^{ia} = \sum_{0}^{k-1} \frac{(ia)^\nu}{\nu!} + \vartheta \frac{a^k}{k!}$$

where $|\vartheta| < 1$. Hence

$$v\left(\frac{t}{\sigma}\right) = \int_{-\infty}^{\infty} e^{-\frac{itx}{\sigma}}\, dV(x)$$

$$= 1 + \sum_{2}^{k-1} \frac{\alpha_\nu}{\nu!}\left(-\frac{it}{\sigma}\right)^\nu + \vartheta \frac{\beta_k}{k!}\left(\frac{t}{\sigma}\right)^k.$$

If we put the last expression $= 1 + z$ and suppose $|t| < \frac{1}{\varrho_k}\sqrt[6]{n}$, we have by Lemma 1

$$\text{(42)}\qquad \left|\frac{\alpha_\nu t^\nu}{\sigma^\nu}\right| \leq \left(\frac{\beta_k^{\frac{1}{k}}|t|}{\sigma}\right)^\nu < n^{-\frac{\nu}{3}} \leq 1,$$

$$|z| < \sum_{2}^{\infty} \frac{1}{\nu!}\left(\frac{\beta_k^{\frac{1}{k}}|t|}{\sigma}\right)^\nu < (e-2)\left(\frac{\beta_k^{\frac{1}{k}}|t|}{\sigma}\right)^2 < 0{,}8.$$

[1] In my note in the J. of the Lond. Math. Soc. quoted above, $n^{-\frac{1}{6}}$ is misprinted for $n^{\frac{1}{6}}$.

For $|z|<0{,}8$ we have

$$\log(1+z) = \sum_{1 \leq j < \frac{k}{2}} \frac{(-1)^{j+1} z^j}{j} + \Theta z^{\frac{k}{2}},$$

and thus

(43) $$\log v\left(\frac{t}{\sigma}\right) = \sum_{1 \leq j < \frac{k}{2}} \frac{(-1)^{j+1} z^j}{j} + \Theta \frac{\beta_k t^k}{\sigma^k}.$$

Since by (42) the series $\sum_{2}^{\infty} \frac{1}{\nu!} \left(\frac{\beta_k^{\frac{1}{k}} |t|}{\sigma}\right)^\nu$ is a majorating expression for z, we have

$$z^j = \sum_{2j}^{k-1} \delta_{\nu j} \left(-\frac{it}{\sigma}\right)^\nu + \Theta \sum_{k}^{\infty} \frac{1}{\nu!} \left(\frac{j\beta_k^{\frac{1}{k}} |t|}{\sigma}\right)^\nu$$

$$= \sum_{2j}^{k-1} \delta_{\nu j} \left(-\frac{it}{\sigma}\right)^\nu + \Theta \frac{\beta_k t^k}{\sigma^k}$$

the coefficients $\delta_{\nu j}$ being independent of t. Introducing this in (43) and observing (40), we obtain

(44) $$\log v\left(\frac{t}{\sigma}\right) = \sum_{2}^{k-1} \frac{\gamma_\nu}{\nu!} \left(-\frac{it}{\sigma}\right)^\nu + \Theta \frac{\beta_k t^k}{\sigma^k},$$

and by (33) and Lemma 3

(45) $$\log f_n(t) = \sum_{2}^{k-1} \frac{\lambda_\nu}{\nu!} \left(-\frac{it}{\sigma}\right)^\nu + \Theta \frac{n \beta_k t^k}{\sigma^k}$$

$$= -\frac{t^2}{2} + \sum_{3}^{k-1} \frac{\lambda_\nu}{\nu!} \left(-\frac{it}{\sigma}\right)^\nu + \Theta n \left(\frac{\varrho_k t}{\sqrt{n}}\right)^k,$$

$$\log\left[e^{\frac{t^2}{2}} f_n(t)\right] = \sum_{3}^{k-1} \frac{\lambda_\nu}{\nu!}\left(-\frac{it}{\sigma}\right)^\nu + \Theta n \left(\frac{\varrho_k t}{\sqrt{n}}\right)^k.$$

Now by Lemmas 1—3 we have for the values of t here considered

$$(46)\qquad \left|\sum_{3}^{k-1} \frac{\lambda_\nu}{\nu!}\left(-\frac{it}{\sigma}\right)^\nu\right| < \Theta n \sum_{3}^{\infty} \frac{1}{\nu!}\left(\frac{\varrho_k |t|}{\sqrt{n}}\right)^\nu$$

$$< \Theta n \left(\frac{\varrho_k |t|}{\sqrt{n}}\right)^3 = \Theta(\varrho_k t)^2 \frac{\varrho_k |t|}{\sqrt{n}} < \Theta,$$

$$(47)\qquad \left|\Theta n \left(\frac{\varrho_k t}{\sqrt{n}}\right)^k\right| < \Theta n^{1-\frac{k}{3}} < \Theta,$$

and thus

$$(48)\qquad e^{\frac{t^2}{2}} f_n(t) = e^{\sum\limits_{1}^{k-3} \frac{\lambda_{\nu+2}(-it)^{\nu+2}}{\sigma^{\nu+2}(\nu+2)!}} + \Theta n \left(\frac{\varrho_k t}{\sqrt{n}}\right)^k.$$

If we denote by z a variable, such that $|z| \leqq 1$, and put

$$\zeta = \sum_{1}^{k-3} \frac{\lambda_{\nu+2}(-it)^{\nu+2}}{\sigma^{\nu+2}(\nu+2)!} z^\nu,$$

we have by (46)

$$(49)\qquad \begin{aligned} e^\zeta &= \sum_{0}^{k-3} \frac{\zeta^j}{j!} + \Theta \zeta^{k-2} \\ &= \sum_{0}^{k-3} \frac{\zeta^j}{j!} + \Theta(\varrho_k t)^{2(k-2)} \left(\frac{\varrho_k |tz|}{\sqrt{n}}\right)^{k-2}. \end{aligned}$$

A majorating expression for ζ is, by Lemmas 1—3,

$$\Theta \sum_{1}^{\infty} \frac{n}{\nu!}\left(\frac{\varrho_k |t|}{\sqrt{n}}\right)^{\nu+2} |z|^\nu = \Theta(\varrho_k t)^2 \sum_{1}^{\infty} \frac{1}{\nu!}\left(\frac{\varrho_k |tz|}{\sqrt{n}}\right)^\nu,$$

and consequently we have for $j=1, 2, \ldots, k-3$

$$\zeta^j = \sum_j^{k-3} \varepsilon_{\nu j} z^\nu + \Theta(\varrho_k t)^{2j} \sum_{k-2}^{\infty} \frac{1}{\nu!}\left(\frac{j\varrho_k |tz|}{\sqrt{n}}\right)^\nu$$

$$= \sum_j^{k-3} \varepsilon_{\nu j} z^\nu + \Theta(\varrho_k t)^{2j} \left(\frac{\varrho_k |tz|}{\sqrt{n}}\right)^{k-2},$$

the coefficients $\varepsilon_{\nu j}$ being independent of z. Introducing this in (49) and observing (31), we obtain

$$e^{\zeta} = 1 + \sum_1^{k-3} P_\nu(it) z^\nu + \Theta\left(\frac{\varrho_k |tz|}{\sqrt{n}}\right)^{k-2} [(\varrho_k t)^2 + (\varrho_k t)^{2(k-2)}].$$

Putting here $z=1$ and introducing in (48), we get the expansion (41 a), which for $k=3$ and $k=4$ coincides with (41 b).

In order to prove (41 b) also for $k>4$, we put according to Lemma 4

$$\sum_1^{k-3} P_\nu(it) = \sum_{\nu=1}^{k-3} \sum_{j=1}^{\nu} H_{\nu j}(it)^{\nu+2j}$$

(50)

$$= \sum_{h=3}^{k-1} (it)_h \sum_{j=1}^{\left[\frac{h}{3}\right]} H_{h-2j,j} + \sum_{h=k}^{3(k-3)} (it)^h \sum_{j=\left[\frac{h-k}{2}\right]+2}^{\left[\frac{h}{3}\right]} H_{h-2j,j}.$$

By Lemma 4, the second term in the last expression is of the form

$$\Theta \sum_{h=k}^{3(k-3)} |t|^h \sum_{j=1}^{\left[\frac{h}{3}\right]} \varrho_k^h n^{-\frac{h-2j}{2}} = \Theta \sum_{h=k}^{3(k-3)} |\varrho_k t|^h n^{-\frac{h}{2}+\left[\frac{h}{3}\right]}.$$

It is easily shown that we have for $h=k, k+1, \ldots$

$$\frac{h}{2} - \left[\frac{h}{3}\right] \geqq \frac{1}{2}\left[\frac{k+2}{3}\right] = \frac{\varkappa}{2}, \tag{51}$$

and that the sign $=$ holds in this relation, if we put h equal to the smallest multiple of 3 which is $\geqq k$. Thus (50) gives

$$\sum_{1}^{k-3} P_\nu(it) = \sum_{\nu=3}^{k-1} (it)^\nu \sum_{j=1}^{\left[\frac{\nu}{3}\right]} H_{\nu-2j,j} + \Theta n^{-\frac{\varkappa}{2}} (|\varrho_k t|^k + |\varrho_k t|^{3(k-2)}).$$

If we introduce this expression in (41 a) and use Lemma 5, we get the expansion (41 b), and the proof of Lemma 6 is completed.

At the same time, we obtain the relation

$$\frac{c_\nu}{\nu!} = \sum_{j=1}^{\left[\frac{\nu}{3}\right]} H_{\nu-2j,j},$$

and consequently by Lemma 4

$$|c_\nu| < \Theta \varrho_k^\nu n^{-\frac{\nu}{2}+\left[\frac{\nu}{3}\right]} < M n^{-\frac{\nu}{2}+\left[\frac{\nu}{3}\right]}$$

where M is independent of n, as already stated in Art. 7.

17. In Lemma 6, t is confined to an interval of the order $\sqrt[6]{n}$, and an inspection of the proof shows that full use of this condition has been made at the deduction of the relations (46) and (47). The following Lemma gives a less precise conclusion, which holds over an interval of the order $\sqrt{n}$.

Lemma 7. *For* $|t| < \frac{1}{4\varrho_k^3}\sqrt{n}$ *we have*

$$|f_n(t)| < e^{-\frac{t^2}{3}}.$$

We have

$$|v(t)|^2 = \int_{-\infty}^{\infty}\int_{-\infty}^{\infty} e^{it(x-y)}\, dV(x)\, dV(y) = \int_{-\infty}^{\infty}\int_{-\infty}^{\infty} \cos t(x-y)\, dV(x)\, dV(y),$$

but

$$\cos t(x-y) \leqq 1 - \frac{1}{2} t^2 (x-y)^2 + \frac{1}{6} |t|^3 \cdot |x-y|^3$$

$$\leqq 1 - \frac{1}{2} t^2 (x^2 - 2xy + y^2) + \frac{2}{3} |t|^3 (|x|^3 + |y|^3),$$

and hence

$$|v(t)|^2 \leqq 1 - \beta_2 t^2 + \frac{4}{3} \beta_3 |t|^3 \leqq e^{-\beta_2 t^2 + \frac{4}{3} \beta_3 |t|^3},$$

$$|f_n(t)| = \left| v\left(\frac{t}{\sigma}\right) \right|^n \leqq e^{-\frac{t^2}{2}\left(1 - \frac{4\varrho_3^3 |t|}{3\sqrt{n}}\right)}.$$

For $|t| < \frac{1}{4\varrho_k^3}\sqrt{n}$ we have by Lemma 1

$$1 - \frac{4\varrho_3^3 |t|}{3\sqrt{n}} > 1 - \frac{1}{3}\left(\frac{\varrho_3}{\varrho_k}\right)^3 \geqq \frac{2}{3},$$

and the lemma is proved.

Inversion formulae.

18. We shall write in the following

(52)
$$R_n(x) = F_n(x) - \Phi(x) - \sum_1^{k-3} P_\nu(\Phi),$$

$$r_n(t) = f_n(t) - e^{-\frac{t^2}{2}} - \sum_1^{k-3} P_\nu(it) e^{-\frac{t^2}{2}},$$

and

(53)
$$Q_n(x) = F_n(x) - \Phi(x) - \sum_3^{k-1} \frac{c_\nu}{\nu!} \Phi^{(\nu)}(x),$$

$$q_n(t) = f_n(t) - e^{-\frac{t^2}{2}} - \sum_3^{k-1} \frac{c_\nu}{\nu!} (it)^\nu e^{-\frac{t^2}{2}}.$$

By (34) and (35) we then have

(54)
$$r_n(t) = \int_{-\infty}^{\infty} e^{-itx}\, dR_n(x),$$
$$q_n(t) = \int_{-\infty}^{\infty} e^{-itx}\, dQ_n(x).$$

Conversely, it can be shown that we have

$$R_n(x) = \lim_{T\to\infty} \frac{1}{2\pi} \int_{-T}^{T} \frac{e^{itx}}{it}\, r_n(t)\, dt,$$

and a similar relation for $Q_n(x)$. The integrals occurring in these inversion formulae are, however, in general only conditionally convergent, and we shall find it convenient to introduce certain modifications so as to obtain absolutely convergent integrals.

For that purpose, we shall consider certain generalized (»RIEMANN-LIOUVILLE») integrals.[1] The *integral of order* $\omega > 0$ of a function $g(x)$ will be defined by the equation

$$I^{(\omega)} g(x) = \frac{1}{\Gamma(\omega)} \int_{-\infty}^{x} (x-t)^{\omega-1} g(t)\, dt.$$

The integrals of order ω of the functions $F_n(x)$, $R_n(x)$ and $Q_n(x)$ are finite for $0 < \omega \leqq k-1$, as may be easily deduced from the fact that β_k is finite.

Lemma 8.[2] *For* $0 < \omega \leqq k-1$ *we have*

[1] Reference may be made to a recent paper by HARDY and LITTLEWOOD: »Some properties of fractional integrals», Math. Zeitschr. 27 (1928).

[2] A particular case of this has been proved in Art. 14. The argument of it is taken as $+\frac{\pi}{2}$ for $t > 0$, $-\frac{\pi}{2}$ for $t < 0$.

$$\text{(55 a)} \qquad I^{(\omega)} R_n(x) = \frac{1}{2\pi} \int_{-\infty}^{\infty} \frac{e^{itx}}{(it)^{\omega+1}} r_n(t)\, dt,$$

$$\text{(55 b)} \qquad I^{(\omega)} Q_n(x) = \frac{1}{2\pi} \int_{-\infty}^{\infty} \frac{e^{itx}}{(it)^{\omega+1}} q_n(t)\, dt.$$

Obviously the integrals on the right are absolutely convergent. It is only necessary to prove (55 a), as (55 b) may be proved in exactly the same way. Further, it is sufficient to consider non-integral values of ω, the result being extended to integral values by continuity. — By Lemma 6 we have as $t \to 0$

$$r_n(t) = O(t^k), \qquad q_n(t) = O(t^k),$$

and hence by (54)

$$\int_{-\infty}^{\infty} x^\nu \, d R_n(x) = \int_{-\infty}^{\infty} x^\nu \, d Q_n(x) = 0$$

for $\nu = 0, 1, \ldots, k-1$. Thus we have for non-integral ω between 0 and $k-1$

$$\frac{1}{2\pi} \int_{-\infty}^{\infty} \frac{e^{itx}}{(it)^{\omega+1}} r_n(t)\, dt = \frac{1}{2\pi} \int_{-\infty}^{\infty} dt \int_{-\infty}^{\infty} \frac{e^{it(x-y)} - \sum\limits_{0}^{[\omega]} \frac{(it)^\nu (x-y)^\nu}{\nu!}}{(it)^{\omega+1}} \, d R_n(y)$$

$$= \frac{1}{\pi} \int_{0}^{\infty} dt \int_{-\infty}^{\infty} \Re \left(\frac{e^{it(x-y)} - \sum\limits_{0}^{[\omega]} \frac{(it)^\nu (x-y)^\nu}{\nu!}}{(it)^{\omega+1}} \right) d R_n(y) =$$

$$= \frac{1}{\pi} \int_{-\infty}^{\infty} d R_n(y) \int_{0}^{\infty} \Re \left(\frac{e^{it(x-y)} - \sum\limits_{0}^{[\omega]} \frac{(it)^\nu (x-y)^\nu}{\nu!}}{(it)^{\omega+1}} \right) dt.$$

$\Re(a)$ denotes the real part of a, and the reversion of the order of integration is permitted, since the last repeated integral is easily seen to be absolutely convergent.[1] It may, however, be deduced without difficulty that we have[2]

$$\frac{1}{\pi}\int_0^\infty \Re\left(\frac{e^{it(x-y)}-\sum_0^{[\omega]}\frac{(it)^\nu(x-y)^\nu}{\nu!}}{(it)^{\omega+1}}\right)dt=\begin{cases}\frac{(x-y)^\omega}{\Gamma(\omega+1)} & \text{for } x>y\\ 0 & \text{for } x\leqq y\end{cases}$$

and thus

$$\frac{1}{2\pi}\int_{-\infty}^{\infty}\frac{e^{itx}}{(it)^{\omega+1}}r_n(t)\,dt=\frac{1}{\Gamma(\omega+1)}\int_{-\infty}^{x}(x-y)^\omega\,dR_n(y).$$

An integration by parts (which is clearly permitted) is enough to complete the proof of Lemma 8.

19. By a combination of all the results previously obtained, we get the following lemma, which is fundamental for the proofs of our theorems.

Lemma 9. *For* $0<\omega\leqq k-1$ *we have*

$$I^{(\omega)}R_n(x)=\frac{1}{\pi}\Re\int_{\frac{\sqrt{n}}{4\varrho_k^3}}^{\infty}\frac{e^{itx}}{(it)^{\omega+1}}f_n(t)\,dt+\Theta\varrho_k^{3(k-2)}n^{-\frac{k-2}{2}},\tag{56 a}$$

$$I^{(\omega)}Q_n(x)=\frac{1}{\pi}\Re\int_{\frac{\sqrt{n}}{4\varrho_k^3}}^{\infty}\frac{e^{itx}}{(it)^{\omega+1}}f_n(t)\,dt+\Theta\varrho_k^{3(k-2)}n^{-\frac{\varkappa}{2}}.\tag{56 b}$$

The terms containing the factor Θ have continuous first and

[1] Cf. Hobson, The theory of functions of a real variable, 2d ed. (1926) Vol. II, p. 347.

[2] This relation is not true for integral values of ω. Cf. Art. 14.

second derivatives with respect to x, *both of which are of the form* $\Theta \varrho_k^{3(k-2)} n^{-\frac{k-2}{2}}$ *in* (56 a) *and* $\Theta \varrho_k^{3(k-2)} n^{-\frac{\varkappa}{2}}$ *in* (56 b).

Also in this case, we need only give the proof of the relation for $R_n(x)$. By Lemma 8 we have

$$I^{(\omega)} R_n(x) = \frac{1}{\pi} \Re \int_0^\infty \frac{e^{itx}}{(it)^{\omega+1}} r_n(t)\, dt \tag{57}$$

$$= A_1 + A_2 + A_3 + A_4,$$

where[1]

$$A_1 = \frac{1}{\pi} \Re \int_0^{\frac{\sqrt[6]{n}}{\varrho_k}} \frac{e^{itx}}{(it)^{\omega+1}} r_n(t)\, dt,$$

$$A_2 = \frac{1}{\pi} \Re \int_{\frac{\sqrt[6]{n}}{\varrho_k}}^\infty \frac{e^{itx}}{(it)^{\omega+1}} (r_n(t) - f_n(t))\, dt,$$

$$A_3 = \frac{1}{\pi} \Re \int_{\frac{\sqrt[6]{n}}{\varrho_k}}^{\frac{\sqrt{n}}{4\varrho_k^3}} \frac{e^{itx}}{(it)^{\omega+1}} f_n(t)\, dt,$$

$$A_4 = \frac{1}{\pi} \Re \int_{\frac{\sqrt{n}}{4\varrho_k^3}}^\infty \frac{e^{itx}}{(it)^{\omega+1}} f_n(t)\, dt.$$

---

[1] It is assumed here that we have $\frac{\sqrt[6]{n}}{\varrho_k} < \frac{\sqrt{n}}{4\varrho_k^3}$, which is obviously the case for all $n > 64\varrho^6$. For smaller values of n, only a slight modification of the proof is required.

By Lemma 6 we have

$$|A_1| < \Theta n^{-\frac{k-2}{2}} \int_0^\infty \left((\varrho_k t)^k + (\varrho_k t)^{3(k-2)}\right) e^{-\frac{t^2}{2}} \frac{dt}{t^{\omega+1}} < \Theta \varrho_k^{3(k-2)} n^{-\frac{k-2}{2}},$$

and by Lemma 7, putting $\frac{\sqrt[6]{n}}{\varrho_k} = \tau$,

$$|A_3| < \frac{1}{\pi} \int_\tau^\infty e^{-\frac{t^2}{3}} \frac{dt}{t^{\omega+1}}.$$

It will be found by some easy calculations that the function of τ and ω

$$\tau^{3(k-2)} \int_\tau^\infty e^{-\frac{t^2}{3}} \frac{dt}{t^{\omega+1}}$$

has, for $\tau > 0$ and $0 < \omega \leqq k-1$, a finite maximum which depends only on k (always assuming $k \geqq 3$). Thus

$$|A_3| < \frac{\Theta}{\tau^{3(k-2)}} = \Theta \varrho_k^{3(k-2)} n^{-\frac{k-2}{2}}.$$

Further, by Lemma 4 we have

$$|r_n(t) - f_n(t)| < e^{-\frac{t^2}{2}} \left(1 + \sum_{\nu=1}^{k-3} \sum_{j=1}^{\nu} |H_{\nu j} t^{\nu+2j}|\right)$$

$$< \Theta e^{-\frac{t^2}{2}} \left(1 + \sum_{\nu=1}^{k-3} n^{-\frac{\nu}{2}} (1 + \varrho_k^{3\nu} t^{3\nu})\right)$$

$$< \Theta e^{-\frac{t^2}{2}} \sum_0^{k-3} \left(\frac{t}{\tau}\right)^{3\nu} < \Theta e^{-\frac{t^2}{2}} \left(1 + \left(\frac{t}{\tau}\right)^{3(k-3)}\right),$$

and thus

$$|A_2| < \Theta \int_{\tau}^{\infty} e^{-\frac{t^2}{2}} \left(1 + \left(\frac{t}{\tau}\right)^{3(k-3)}\right) \frac{dt}{t^{\omega+1}}.$$

The function of τ and ω occurring here may be treated in the same way as above, and we obtain for $|A_2|$ an upper limit of the same form as for $|A_1|$ and $|A_3|$. Introducing this in (57), we obtain (56 a). It is readily seen that A_1, A_2 and A_3 all have two continuous derivatives with respect to x, and that an evaluation of the derivatives gives a result of the same form.

First theorem on asymptotic expansion.

20. We have assumed the existence of a certain number of moments of the given function $V(x)$. Without any further hypothesis on this function it is, as will be shown later, impossible to obtain a real asymptotic expansion of $F_n(x)$. Such an expansion can, however, be obtained for *the integral of a certain order of* $F_n(x)$, and the number of terms of the expansion depends on the number of moments that are brought into consideration, as shown by the following theorem.

Theorem 1. *If* $V(x)$ *has finite moments* $(\alpha_2 = \beta_2)$ *of the second order* $(k=2)$, *we have as* n *tends to infinity*

$$F_n(x) - \Phi(x) \to 0.$$

If β_3 *is also finite* $(k=3)$, *we have even*

$$|F_n(x) - \Phi(x)| < 3\varrho_3^3 \log n \,.\, n^{-\frac{1}{2}} \tag{58}$$

for $n > 1$ *and for all values of* x.

Finally, if β_k *is finite* $(k > 3)$, *we have for* $n > 1$ *and for all* x

$$I^{(k-3)}\left[F_n(x) - \Phi(x) - \sum_{1}^{k-3} P_\nu(\Phi)\right] = \Theta \varrho_k^{3(k-2)} (\log n)^2 n^{-\frac{k-2}{2}} \tag{59 a}$$

and

$$\text{(59 b)}\quad I^{(\varkappa-1)}\left[F_n(x)-\Phi(x)-\sum_{3}^{k-1}\frac{c_\nu}{\nu!}\Phi^{(\nu)}(x)\right]=\Theta\varrho_k^{3(k-2)}(\log n)^2 n^{-\frac{\varkappa}{2}}.$$

The cases $k=2$ and $k=3$ are previously known (cf. Arts 4 and 14) and are repeated here for the sake of completeness only. The relations (59 a) and (59 b), with slightly less precise error terms, follow directly from Lemma 9. In order to get the error terms stated in the theorem we proceed in the following way.

If ω is a positive integer $\leqq k-1$, we have

$$Z=\Re\int_{\frac{\sqrt{n}}{4\varrho_k^3}}^{\infty}\frac{e^{itx}}{(it)^{\omega+1}}f_n(t)\,dt=\int_{-\infty}^{\infty}dF_n(y)\int_{\frac{\sqrt{n}}{4\varrho_k^3}}^{\infty}\Re\frac{e^{it(x-y)}}{(it)^{\omega+1}}\,dt.$$

It is, however, easily shown that

$$\left|\int_{\frac{\sqrt{n}}{4\varrho_k^3}}^{\infty}\Re\frac{e^{it(x-y)}}{(it)^{\omega+1}}\,dt\right|<\frac{4^\omega}{\omega}\varrho_k^{3\omega}n^{-\frac{\omega}{2}}\operatorname{Min}\left(1,\frac{8\varrho_k^3\omega}{|x-y|\sqrt{n}}\right),$$

and consequently

$$\text{(60)}\qquad |Z|<\Theta\varrho_k^{3\omega}n^{-\frac{\omega}{2}}\int_{-\infty}^{\infty}\operatorname{Min}\left(1,\frac{\varrho_k^3}{|x-y|\sqrt{n}}\right)dF_n(y)$$

$$=\Theta\varrho_k^{3\omega}n^{-\frac{\omega}{2}}\sum_{1}^{5}D_\nu$$

where $D_1,\ldots,D_5$ are the integrals over the five intervals obtained by marking the points $x\pm\frac{\varrho_k^3}{\sqrt{n}}$ and $x\pm1$ on the axis of y. The sum of the integrals D_1 and D_5 over the two extreme intervals is less than $\frac{\varrho_k^3}{\sqrt{n}}$. For the evaluation of D_2, D_3 and D_4 we remark that, according to (58), we have for any real a and $b > a$

$$0 \leqq F_n(b) - F_n(a) < \Phi(b) - \Phi(a) + 6\varrho_3^3 \frac{\log n}{\sqrt{n}}$$

$$< b - a + 6\varrho_k^3 \frac{\log n}{\sqrt{n}}.$$

Hence we have

$$D_3 = \int_{x-\frac{\varrho_k^3}{\sqrt{n}}}^{x+\frac{\varrho_k^3}{\sqrt{n}}} dF_n(y) < \Theta\varrho_k^3 \frac{\log n}{\sqrt{n}},$$

$$D_4 < \sum_{\nu=1}^{\frac{\sqrt{n}}{\varrho_k^3}} \frac{\varrho_k^3}{\sqrt{n}} \int_{x+\nu\frac{\varrho_k^3}{\sqrt{n}}}^{x+(\nu+1)\frac{\varrho_k^3}{\sqrt{n}}} \frac{dF_n(y)}{y-x} < \sum_{1}^{\frac{\sqrt{n}}{\varrho_k^3}} \frac{1}{\nu} \int_{x+\nu\frac{\varrho_k^3}{\sqrt{n}}}^{x+(\nu+1)\frac{\varrho_k^3}{\sqrt{n}}} dF_n(y)$$

$$< \Theta\varrho_k^3 \frac{\log n}{\sqrt{n}} \sum_{1}^{\frac{\sqrt{n}}{\varrho_k^3}} \frac{1}{\nu} < \Theta\varrho_k^3 \frac{(\log n)^2}{\sqrt{n}}.$$

D_2 may be treated in the same way as D_4, and we obtain from (60)

$$Z = \Theta\varrho_k^{3(\omega+1)} (\log n)^2 n^{-\frac{\omega+1}{2}}.$$

If we introduce this result in (56 a) and put $\omega = k-3$, we obtain (59 a). If we introduce in (56 b) and put $\omega = \varkappa - 1$, we obtain (59 b).

Second theorem on asymptotic expansion.

21. If the given function $V(x)$ satisfies a certain additional condition, an asymptotic expansion in the ordinary sense can be obtained for $F_n(x)$, as we are now going to show.

$V(x)$ being a never decreasing function, it is known[1] that $V(x)$ can be uniquely represented as the sum of three never decreasing functions

$$V(x) = \xi(x) + \eta(x) + \zeta(x), \tag{61}$$

such that $\xi(-\infty) = \eta(-\infty) = \zeta(-\infty) = 0$ and that:

1. $\xi(x) = \int_{-\infty}^{x} \xi'(t)\,dt = \int_{-\infty}^{x} V'(t)\,dt$ for all values of x.
2. $\eta(x)$, the »saltus function», is equal to the sum of the saltuses of $V(x)$ at all the points of discontinuity which are less than x. If x coincides with a point of discontinuity, half the saltus in x should be included in this sum.
3. $\zeta(x)$, the »singular function», is a continuous function which has »almost everywhere» a derivative equal to zero.

We shall say that $V(x)$ satisfies the condition ξ, if the term $\xi(x)$ in the decomposition (61) *is not identically zero.*[2]

Theorem 2. *If $V(x)$ satisfies the condition ξ, and if the moment β_k is finite $(k \geqq 3)$, we have*

$$F_n(x) = \Phi(x) + \sum_{1}^{k-3} P_\nu(\Phi) + \Delta n^{-\frac{k-2}{2}}, \tag{62 a}$$

and

$$F_n(x) = \Phi(x) + \sum_{3}^{k-1} \frac{c_\nu}{\nu!} \Phi^{(\nu)}(x) + \Delta n^{-\frac{\varkappa}{2}}, \tag{62 b}$$

for $n > 1$ and all values of x, Δ being defined as in Art. 3.

[1] Cf. Hobson, l. c. Vol. I, p. 317 and p. 555.

[2] In the applications, the »singular» term $\zeta(x)$ generally does not occur. In this case the condition ξ may be simply expressed thus: *The sum of the saltuses at all the discontinuities of $V(x)$ is less than* 1.

If, for the particular case $k=3$, we compare this result with the relation (58), it is seen that the condition ξ has enabled us to omit the factor $\log n$ from the error term.

In order to prove the theorem we observe that, as x tends to $+\infty$, $\xi(x)$ tends to a limit which will be denoted by ξ^*. We then have

$$0<\xi^*\leqq 1,\quad \eta(\infty)+\zeta(\infty)=1-\xi^*,$$

and

$$v(t)=\int_{-\infty}^{\infty} e^{-itx}d(\xi(x)+\eta(x)+\zeta(x)),$$

$$|v(t)|\leqq\left|\int_{-\infty}^{\infty} e^{-itx}\xi'(x)\,dx\right|+1-\xi^*.$$

The first term on the right hand side of the last inequality tends to zero as t tends to infinity[1], and is for every $t\neq 0$ less than

$$\int_{-\infty}^{\infty}\xi'(x)\,dx=\xi^*.$$

Thus it follows that, in every finite or infinite interval which does not contain the point $t=0$, $|v(t)|$ has an upper limit *less than* 1. In particular, we can determine a number $b>0$, such that

$$|v(t)|<e^{-b}$$

for all $t\geqq\frac{1}{4}\beta_2\beta_k^{-\frac{3}{k}}$. Hence we conclude

$$(63)\qquad |f_n(t)|=\left|v\left(\frac{t}{\sigma}\right)\right|^n<e^{-bn}$$

[1] This is a simple consequence of a well-known theorem on FOURIER coefficients. Cf. HOBSON, l. c. Vol. II, p. 515.

for $t \geqq \frac{1}{4}\sigma\beta_2\beta_k^{-\frac{3}{k}} = \frac{\sqrt{n}}{4\varrho_k^3}$. According to Lemma 9 we then have

$$I^{(\omega)}R_n(x) = \frac{1}{\Gamma(\omega)}\int_{-\infty}^{x}(x-t)^{\omega-1}R_n(t)\,dt = \varDelta\left(\frac{e^{-bn}}{\omega n^{\frac{\omega}{2}}} + n^{-\frac{k-2}{2}}\right)$$

for $0<\omega<1$. It is essential to observe that, in accordance with Art. 13, the modulus of $\varDelta$ is less than a constant independent of n, x and ω.

By a theorem due to M. Riesz[1] we have, for every positive h

$$\frac{1}{\Gamma(\omega)}\int_{-\infty}^{x-h}(x-t)^{\omega-1}R_n(t)\,dt = \frac{1}{\Gamma(\omega)}\int_{-\infty}^{y}(y-t)^{\omega-1}R_n(t)\,dt$$

for some $y<x-h$, and thus

$$\int_{x-h}^{x}(x-t)^{\omega-1}R_n(t)\,dt = \frac{\varDelta}{\omega}\left(\frac{e^{-bn}}{\omega n^{\frac{\omega}{2}}} + n^{-\frac{k-2}{2}}\right). \tag{64}$$

This may be written

$$\int_{x-h}^{x}(x-t)^{\omega-1}F_n(t)\,dt = \int_{x-h}^{x}(x-t)^{\omega-1}\left(\Phi(t) + \sum_{1}^{k-3}P_\nu(\Phi)\right)dt + \\ + \frac{\varDelta}{\omega}\left(\frac{e^{-bn}}{\omega n^{\frac{\omega}{2}}} + n^{-\frac{k-2}{2}}\right).$$

[1] »Sur un théorème de la moyenne et ses applications», Acta Univ. Hungaricae Franc.-Jos., Vol. I (1923) p. 114. Cf. also Hardy and Riesz, »The general theory of Dirichlet s series», Cambridge Tracts, No. 18 (1915), Lemma 7, p. 28. — I owe to Prof. E. Phragmén the remark that an inversion formula can be constructed which leads directly to (64).

Since $F_n(x)$ is a never decreasing function, and the derivative of $\Phi + \Sigma P_\nu(\Phi)$ is uniformly bounded for all values of n and x, this gives us (cf. the corresponding proof in Art. 14)

$$\frac{h^\omega}{\omega} F_n(x) > \frac{h^\omega}{\omega}\left(\Phi(x) + \sum_1^{k-3} P_\nu(\Phi) - \varDelta h\right) - \frac{\varDelta}{\omega}\left(\frac{e^{-bn}}{\omega n^{\frac{\omega}{2}}} + n^{-\frac{k-2}{2}}\right).$$

In the same way we obtain, substituting in (64) $x+h$ for x,

$$\frac{h^\omega}{\omega} F_n(x) < \frac{h^\omega}{\omega}\left(\Phi(x) + \sum_1^{k-3} P_\nu(\Phi) + \varDelta h\right) + \frac{\varDelta}{\omega}\left(\frac{e^{-bn}}{\omega n^{\frac{\omega}{2}}} + n^{-\frac{k-2}{2}}\right),$$

and thus generally

$$F_n(x) = \Phi(x) + \sum_1^{k-3} P_\nu(\Phi) + \varDelta\left(h + \frac{h^{-\omega} e^{-bn}}{\omega n^{\frac{\omega}{2}}} + h^{-\omega} n^{-\frac{k-2}{2}}\right).$$

Putting here $h = n^{-\frac{k-2}{2}}$, $\omega = \frac{1}{\log n}$, we obtain (62 a). (62 b) is proved in a perfectly similar way.

Expansion of the »frequency function«.

22. If, in the decomposition (61), $\eta(x)$ and $\zeta(x)$ both vanish identically, we have for all x

$$V(x) = \int_{-\infty}^{x} V'(t)\,dt. \tag{65}$$

In this case also $F_n(x)$ has a unique derivative $F'_n(x)$, the »frequency function«. We put as before

$$\varphi(x) = \Phi'(x) = \frac{1}{\sqrt{2\pi}} e^{-\frac{x^2}{2}}$$

$P_\nu(\varphi)$ denotes the derivative of $P_\nu(\Phi)$.

Theorem 3. *If a) (65) holds for all values of x, b) the derivative $V'(x)$ is of bounded variation in the whole domain $-\infty < x < +\infty$, and c) the moment β_k is finite ($k \geqq 3$), then*

(66 a)
$$F'_n(x) = \varphi(x) + \sum_{1}^{k-3} P_\nu(\varphi) + A n^{-\frac{k-2}{2}},$$

and

(66 b)
$$F'_n(x) = \varphi(x) + \sum_{3}^{k-1} \frac{c_\nu}{\nu!} \varphi^{(\nu)}(x) + A n^{-\frac{\varkappa}{2}}$$

for $n > 1$ and all values of x.

It follows in particular that, as soon as V' is of bounded variation and β_3 is finite, $F'_n(x)$ tends towards $\varphi(x)$ as n tends to infinity.

Obviously (63) holds under the present assumptions. Further, we have

$$|v(t)| = \left| \frac{1}{it} \int_{-\infty}^{\infty} e^{-itx} d\, V'(x) \right| < \frac{1}{|t|} \int_{-\infty}^{\infty} |d\, V'(x)| = \frac{M}{\sqrt{\beta_2}\, |t|},$$

where by hypothesis M is finite. Consequently

(67)
$$|f_n(t)| = \left| v\left(\frac{t}{\sigma}\right) \right|^n < \left(\frac{M \sqrt{n}}{|t|} \right)^n.$$

If in Lemma 9 we put $\omega = 1$ we thus see that, since $n > 1$, the integrals may be twice differentiated, so that we get

(68)
$$R'_n(x) = \frac{1}{\pi} \Re \int_{\frac{\sqrt{n}}{4 \varrho_k^3}}^{\infty} e^{itx} f_n(t)\, dt + \Theta \varrho_k^{3(k-2)} n^{-\frac{k-2}{2}},$$

$$Q'_n(x) = \frac{1}{\pi} \Re \int_{\frac{\sqrt{n}}{4 \varrho_k^3}}^{\infty} e^{itx} f_n(t)\, dt + \Theta \varrho_k^{3(k-2)} n^{-\frac{\varkappa}{2}},$$

where the integrals are still absolutely convergent. Now we have by (63) and (67)

$$\left|\int\limits_{\frac{\sqrt{n}}{4\varrho_k^3}}^{\infty} e^{itx} f_n(t)\,dt\right| < \int\limits_{\frac{\sqrt{n}}{4\varrho_k^3}}^{Me^b\sqrt{n}} e^{-bn}\,dt + \int\limits_{Me^b\sqrt{n}}^{\infty}\left(\frac{M\sqrt{n}}{t}\right)^n dt$$

$$< A\left(\sqrt{n}\, e^{-bn} + \frac{1}{\sqrt{n}} e^{-bn}\right).$$

(If $Me^b < \frac{1}{4\varrho_k^3}$, the introduction of the first integral is superfluous.) Hence the theorem follows immediately by (68).

Convergent expansions.

23. It follows from the above theorems that an asymptotic expansion for $F_n(x)$ or $F'_n(x)$ may exist even in cases where only a limited number of terms of the A-series are finite. On the other hand, it will be shown later that there are cases where the A-series converges to the sum $F_n(x)$ (or $F'_n(x)$), but where no asymptotic expansion exists. As to the conditions of convergence of the A-series[1], I shall here only state (without proof) the following theorems.

Theorem 4.[2] *If the integral*

---

[1] The convergence problem of this and related series has been treated by many different authors. I quote among recent papers HILLE, A class of reciprocal functions, Annals of Math. Vol. 27 (1926) p. 427, which contains i. a. a useful Bibliography; SZEGÖ, Beiträge zur Theorie der Laguerreschen Polynome I, Math. Zeitschr. Vol. 25 (1926) p. 87; USPENSKY, On the development of arbitrary functions in series of Hermite's and Laguerre's polynomials, Annals of Math. Vol. 28 (1927) p. 593.

[2] KAMEDA, in an interesting paper (Theory of generating functions and its application to the Theory of Probability, Journ. of the Fac. Sc., Imper. Univ. of Tokyo, Vol. I (1925) p. 1) treats the convergence problem of the A-series from the point of view of the composition of elementary errors. His paper has several points of contact with the present paper, though he considers only the convergence properties of the expansions. My Theorem 4 is a generalization of his Theorem XXV.

$$\int_{-\infty}^{\infty} e^{\frac{x^2}{4\alpha_2}} d V(x) \tag{69}$$

converges, then we have

$$F_n(x) = \Phi(x) + \sum_{3}^{\infty} \frac{c_\nu}{\nu!} \Phi^{(\nu)}(x). \tag{70}$$

If, in addition, we have $V(x) = \int_{-\infty}^{x} V'(t)\,dt$ *for all values of* x, *and* $V'(x)$ *is of bounded variation in every finite interval, then we have also*

$$F'_n(x) = \varphi(x) + \sum_{3}^{\infty} \frac{c_\nu}{\nu!} \varphi^{(\nu)}(x). \tag{71}$$

The convergence of (70) *and* (71) *is uniform in every finite interval of continuity.*

This may be deduced without difficulty from a theorem given in my Copenhaguen lecture.[1] — The following theorem gives an expansion which is convergent and asymptotic, with an error term tending to zero as x tends to infinity. The proof requires somewhat intricate calculations and may be given at another occasion. We put

$$U(x) = V(x\sqrt{\alpha_2}).$$

$U(x)$ is a *normalized function* according to Art. 2.

Theorem 5. *If* $U(x)$ *has, for all values of* x, *a second derivative* $U''(x)$ *such that*

[1] l. c., p. 407. — The theorem given in that place is a generalization of a theorem due to GALBRUN: Sur un dévelóppement d'une fonction à variable réelle en série de polynomes, Bulletin de la Soc. Math. de France, Vol. 41 (1913), p. 24.

$$|U''(x)-\Phi''(x)| < \frac{1}{2\sqrt{\pi}} e^{-\frac{x^2}{4}},$$

the developments (70) *and* (71) *of Theorem 4 hold, and we have for any positive integer k*

$$F_n(x) = \Phi(x) + \sum_{3}^{k-1} \frac{c_\nu}{\nu!} \Phi^{(\nu)}(x) + \Theta n^{-\frac{k}{2}} e^{-\frac{x^2}{4}},$$

$$F'_n(x) = \varphi(x) + \sum_{3}^{k-1} \frac{c_\nu}{\nu!} \varphi^{(\nu)}(x) + \Theta n^{-\frac{k}{2}} e^{-\frac{x^2}{4}}.$$

Examples.

24. We shall now give two examples relating to our Theorems 2 and 3. In both cases we shall define a function $V(x)$, such that all the moments α_ν and β_ν are finite and that, according to Theorem 4, the A-series (70) (in the second example even (71)) converges.

A. We put

$$V(x) = \begin{cases} 0 & \text{for} \quad x < -1, \\ \frac{1}{2} & \text{»} \quad -1 < x < 1, \\ 1 & \text{»} \quad x > 1. \end{cases}$$

This function does not satisfy the »condition ξ» of Theorem 2 and we are going to show that in this case *no asymptotic expansion of $F_n(x)$ according to* (62 a) *or* (62 b) *exists for any* $k > 3$. As a matter of fact, the »probability distribution» is in this case symmetrical and we have $P_1(\Phi) = c_3 = 0$, so that (62 a) or (62 b) would give e.g. for $k=4$

$$F_n(x) = \Phi(x) + \frac{A}{n}$$

with Δ bounded for all values of n and x. But this is impossible, since it is easily seen that, at the point $x=0$, $F_{2n}(x)$ has a discontinuity with the saltus

$$\binom{2n}{n}\left(\frac{1}{2}\right)^{2n} \sim \frac{1}{\sqrt{n\pi}}.$$

This example corresponds (in the current notation) to the case $p=q=\frac{1}{2}$ of »BERNOULLI's theorem» in probability. Even in the case of an arbitrary p there can be no asymptotic expansion, since $F_n(x)$ has still discontinuities of the order $n^{-\frac{1}{2}}$.

B. Let $a_0, a_1, \ldots$ and $b_0=1, b_1, \ldots$ be two decreasing sequences of positive numbers, while $m_0=0, m_1, m_2, \ldots$ is an increasing sequence tending to infinity. Put

$$a_{-\nu}=a_\nu, \quad b_{-\nu}=b_\nu, \quad m_{-\nu}=-m_\nu,$$

and

$$V(x)=\sum_{-\infty}^{\infty} a_\nu \Phi\left(\frac{x-m_\nu}{b_\nu}\right).$$

If the equations

(72) $$\sum_{-\infty}^{\infty} a_\nu = 1, \quad \sum_{-\infty}^{\infty} a_\nu (b_\nu^2+m_\nu^2)=1$$

are satisfied, $V(x)$ is a »probability function» with $\alpha_1=0$, $\alpha_2=1$, such that all moments α_ν and β_ν are finite. The condition ξ is clearly satisfied, and thus the expansions (62 a) and (62 b) of Theorem 2 hold for arbitrarily large values of k.

The derivative $V'(x)$ exists for all values of x and is of bounded variation in every finite interval. If the series

(73) $$\sum_{-\infty}^{\infty} a_\nu e^{\frac{1}{2}m_\nu^2}$$

is convergent, it can be shown that the integral $\int\limits_{-\infty}^{\infty} e^{\frac{x^2}{4}}\, d\,V(x)$ converges, so that according to Theorem 4

$$F'_n(x) = \varphi(x) + \sum_{3}^{\infty} \frac{c_\nu}{\nu!}\varphi^{(\nu)}(x) \tag{74}$$

for all values of x. But if we do not know that $V'(x)$ is of bounded variation in the *infinite* interval $-\infty < x < \infty$, we cannot conclude from Theorem 3 that the last development is also asymptotic. Actually, it will be shown that *the a_ν, b_ν and m_ν can be so determined that the conditions* (72) *and* (73) *are satisfied, but that — in spite of* (74) *— already the relation*

$$\lim_{n\to\infty} F'_n(x) = \varphi(x)$$

is false. — We have in this case

$$W_n(x) = \sum_{\nu_1,\ldots,\nu_n} a_{\nu_1}, \ldots, a_{\nu_n}\, \Phi\left(\frac{x - m_{\nu_1} - \cdots - m_{\nu_n}}{\sqrt{b_{\nu_1}^2 + \cdots + b_{\nu_n}^2}}\right),$$

$$F_n(x) = \sum_{\nu_1,\ldots,\nu_n} a_{\nu_1}, \ldots, a_{\nu_n}\, \Phi\left(\frac{x\sqrt{n} - m_{\nu_1} - \cdots - m_{\nu_n}}{\sqrt{b_{\nu_1}^2 + \cdots + b_{\nu_n}^2}}\right),$$

$$F'_n(x) = \sqrt{n} \sum_{\nu_1,\ldots,\nu_n} \frac{a_{\nu_1}, \ldots, a_{\nu_n}}{\sqrt{b_{\nu_1}^2 + \cdots + b_{\nu_n}^2}}\, \varphi\left(\frac{x\sqrt{n} - m_{\nu_1} - \cdots - m_{\nu_n}}{\sqrt{b_{\nu_1}^2 + \cdots + b_{\nu_n}^2}}\right),$$

where $\nu_1, \ldots, \nu_n$ pass through all positive and negative integral values. If we substitute here $2n$ for n and group together the $\binom{2n}{n}$ terms where n of the indices have the particular value ν ($\nu = 1, 2, \ldots$), while the other n indices are equal to $-\nu$, we obtain

$$F'_{2n}(x) > \binom{2n}{n} \sum_{1}^{\infty} \frac{a_\nu^{2n}}{b_\nu} \varphi\left(\frac{x}{b_\nu}\right),$$

(75) $$F'_{2n}(0) > \frac{2^n a_n^{2n}}{b_n} \varphi(0).$$

If, for all values of n above a certain limit, we take

$$a_n = \frac{1}{n^2}, \quad b_n = \frac{1}{n^{4n}}, \quad m_n = \sqrt{\log n},$$

(73) is convergent and we get from (75)

$$F'_{2n}(0) > 2^n \varphi(0)$$

which shows that, in spite of the convergence of the A-series, $F'_n(x)$ does not tend to $\varphi(x)$ with increasing n. Obviously it is possible to give to the first a:s, b:s and m:s such values that (72) is satisfied.

Part III. Generalizations.

Unequal components.

25. In the general case we are given a sequence of components $V_1(x)$, $V_2(x)$, ... and consider the function $W_n(x)$ composed according to (1). We suppose that the moments $\alpha_\nu^{(j)}$ and $\beta_\nu^{(j)}$ of $V_j(x)$ are finite for $\nu = 1, 2, \ldots, k$ and that $\alpha_1^{(j)} = 0$. The normalized function $F_n(x)$ is equal to $W_n(\sigma x)$, σ being defined by (2 a). The semi-invariants $\gamma_\nu^{(j)}$ and λ_ν, as well as the polynomials $P_\nu(\Phi)$ and the coefficients c_ν, are defined according to (29); (30), (31) and (32). Further, we put as in Art. 4

$$B_\nu = \frac{\beta_\nu^{(1)} + \beta_\nu^{(2)} + \cdots + \beta_\nu^{(n)}}{n},$$

and

$$\varrho_\nu = \frac{B_\nu^{\frac{1}{\nu}}}{B_2^{\frac{1}{2}}}. \tag{76}$$

In the case of equal components, we have $B_\nu = \beta_\nu$ and the definition of ϱ_ν agrees with the one given by (28). For $\nu=2$, we have $\sigma^2 = n B_2$. The *adjuncts* are defined as in Art. 12, and (33) is here replaced by

$$f_n(t) = w_n\left(\frac{t}{\sigma}\right) = v_1\left(\frac{t}{\sigma}\right) v_2\left(\frac{t}{\sigma}\right), \ldots, v_n\left(\frac{t}{\sigma}\right).$$

Lemmas 1 and 2 remain true for each component, and we obtain further, by considering the moments of the function $\frac{1}{n}[V_1(x) + \cdots + V_n(x)]$

$$B_1 \leqq B_2^{\frac{1}{2}} \leqq B_3^{\frac{1}{3}} \leqq \cdots \leqq B_k^{\frac{1}{k}},$$

so that the relation $\varrho_\nu \leqq \varrho_k$ still holds. In the place of Lemma 3 we have here $\lambda_\nu = \gamma_\nu^{(1)} + \cdots + \gamma_\nu^{(n)}$ and thus by Lemma 2

$$\lambda_\nu = \Theta n B_\nu.$$

All the following lemmas 4—9 remain true without modification, and this is quite obvious in all cases with the only exception of Lemma 6. The proof of this lemma consists of two parts separated by the equation (44). Up to this equation we only have to do with one single component, and the corresponding equation

$$\log v_j\left(\frac{t}{\sigma}\right) = \sum_{2}^{k-1} \frac{\gamma_\nu^{(j)}}{\nu!}\left(-\frac{i\,t}{\sigma}\right)^\nu + \Theta \frac{\beta_k^{(j)} t^k}{\sigma^k} \tag{77}$$

can be proved in the same way for each component. There is, however, one delicate point in this stage of the proof, viz. the inequality (42) which must be replaced by

$$\left|\frac{\alpha_\nu^{(j)} t^\nu}{\sigma^\nu}\right| \leqq \frac{(\beta_k^{(j)})^{\frac{\nu}{k}} |t|^\nu}{\sigma^\nu} < \left(\frac{(nB_k)^{\frac{1}{k}} |t|}{\sqrt{nB_2}}\right)^\nu < n^{\nu\left(\frac{1}{k}-\frac{1}{3}\right)} \leqq 1.$$

By adding (77) for $j=1, 2, \ldots, n$ we obtain (45), and the rest of the proof of Lemma 6 requires no modification whatever.

Consequently, the cases $k=3$ and $k>3$ of Theorem 1 remain true also for the general case of unequal components, ϱ_k being determined by the generalized definition (76).

Actually, the proof of the case $k>3$ rests only on the case $k=3$, which is previously known (cf. Art. 4) and on Lemma 9. — The fundamental difference between the general case and the case of equal components is, of course, that in the former the quantity ϱ_k depends on n. Thus it is only in cases where ϱ_k does not increase too fast as n tends to infinity, that we can obtain any definite conclusion from Theorem 1.

26. The generalization of Theorems 2 and 3 is not quite so straightforward as in the case of Theorem 1. The proofs given in Part II rest on Lemma 9, that remains true with the new definition of ϱ_k, and on certain other conditions. Generally, it can be said that if these other conditions are *uniformly satisfied by all the components V_j*, we can still get asymptotic expansions. But the theorems obtained in this way are not very simple.

Consider first Theorem 2. If we have for all $n>1$

$$\frac{1}{4} B_2 B_k^{-\frac{3}{k}} \geqq h > 0,$$

where h is independent of n, and if, for $t>h$, we have

$$|v_j(t)| < e^{-b},$$

where $b>0$ is independent of j, we can still prove the inequality (63). The rest of the proof requires no modification, and we obtain the expansion (62 a) and (62 b), the upper limit of Δ depending only on k, b and ϱ_k.

If, in addition, we suppose $\sqrt{B_2} < g$ for all n, and if all the V_j have derivatives V'_j, such that the total variation of V'_j over the whole interval $-\infty < x < \infty$ is less than $\frac{M}{g}$, we can still prove the inequality (67) and so obtain the expansions for the »frequency function» $F'_n(x)$ given by Theorem 3. In this case, the upper limit of Δ depends also on M.

There is, however, one particular case in which these generalized theorems take a simpler form, viz. if the components $V_j(x)$ are of the form $V\left(\frac{x}{s_j}\right)$, where the function $V(x)$ is independent of j. If $V(x)$ satisfies e.g. the condition ξ of Theorem 2, and if the constants s_j are such that

$$s_j > h \frac{\left[\frac{1}{n}(s_1^k + \cdots + s_n^k)\right]^{\frac{3}{k}}}{\frac{1}{n}(s_1^2 + \cdots + s_n^2)},$$

where $h > 0$ is independent of j and n, then it can be shown that the expansions of Theorem 2 still hold, the upper limit of Δ depending on k, h and V. This case occurs in the theory of risk and will be subjected to further consideration in the third paper.

27. The convergence theorem 4 may be immediately generalized, so that if the conditions hold for each component, the developments (70) and (71) are convergent. There is also a generalized form of Theorem 5 which will, however, not be considered here.

Further generalizations.

28. In the preceding investigations, we have restricted ourselves to the consideration of sums of *mutually independent variables.* As already pointed out in the Introduction, it would be important to know how far this restriction can be removed.

As far as the convergence of the probability function $F_n(x)$ towards the normal function $\Phi(x)$ is concerned, this question has been treated by several Russian mathematicians, and particularly by S. Bernstein.

Let us consider a sum of *inter-dependent variables*

$$x_1 + x_2 + \ldots + x_n,$$

each of which has the mean value (considered *a priori*) zero, while the mean value of the product $x_i x_j$ is b_{ij}. If, in a particular case, the values of the variables $x_1, x_2, \ldots, x_r$ are known ($r < i$, $r < j$), the mean values of x_i and $x_i x_j$ will generally suffer certain variations and assume the new values δ_i and $b_{ij} + \varepsilon_{ij}$. Bernstein, in a recent important paper[1], proves several theorems which show that the relation

$$\lim F_n(x) = \Phi(x)$$

still holds under conditions, similar to those known for independent variables, if (generally speaking) the degree of dependence between two terms of Σx_ν decreases sufficiently rapidly with increasing difference between the indices. Thus in one of his theorems it is required that the above defined quantities δ_i and ε_{ij} decrease with a certain rapidity, as the differences $i - r$ and $j - r$ increase.

By a combination of Bernstein's methods with the methods used in the present paper, it is possible to investigate in detail also the problem of asymptotic expansion for the case of interdependent variables. Thus it can be shown, e. g., that the above Theorem 1 still holds in this case, if certain conditions of the same general character as those of Bernstein are satisfied. Considerations of space prohibit, however, a further entering upon the question in this paper.

29. Finally, we may also consider *errors and probability functions in several variables* without introducing any new difficulties of principle. This has already been done in many of

[1] »Sur l'extension du théorème limite du calcul des probabilités aux sommes de quantités dépendantes«, Math. Annalen, 97 (1926), p. 1.

the papers quoted above[1], and it is well known that we are here confronted with the *normal correlation function*, which is a generalization of the normal probability function. Subject to certain conditions, it can be shown e. g. that a two-dimensional probability distribution, generated by elementary errors in two variables, converges towards a normal correlation function of two variables. For the representation of such distributions, A-series in two (or more) variables have been formed, containing the partial derivatives of the normal correlation function. Concerning the asymptotic properties of these series, it is possible to prove theorems corresponding to those proved above for the case of one variable.

[1] Cf. also CHARLIER, »Contributions to the mathematical theory of statistics», 6, Arkiv för matematik etc., Vol. 9 (1914), n:o 26; JÖRGENSEN, »Undersögelser over Frekvensflader og Korrelation», Diss. Köbenhavn 1916; WICKSELL, »The correlation function of type A and the regression of its characteristics», K. Svenska Vet.-Akad:s Handl. 58 (1917), n:o 3.

Second paper: Statistical applications.

Analysis of statistical distributions.

1. Let m and σ denote the mean and the standard deviation of a statistical variable X, and let $W(x)$ be the probability function of that variable as defined in the first paper[1], Art. 1. If we put (cf. I, formula (3))

$$F(x) = W(m + \sigma x),$$

$F(x)$ is the probability function of the variable $\frac{X - m}{\sigma}$, with the mean value 0 and the standard deviation 1. Denoting by $\mu_2, \mu_3, \ldots$ the moments of $W(x)$, taken about the mean (cf. I, Art. 7, where m is supposed to be zero), we put, following CHARLIER,

$$S = -\frac{\mu_3}{2\sigma^3}, \qquad E = \frac{1}{8}\left(\frac{\mu_4}{\sigma^4} - 3\right),$$

[1] This journal, 1928, p. 13. We shall refer to that paper by the letter I. — The sense in which the words *probability function* and *frequency function* are used here must be carefully observed. If the probability that a certain variable lies between x and $x + dx$ is $f(x)\,dx$, then $f(x)$ is the *frequency function* of the variable. The *probability function* is, in cases where a finite frequency function exists, equal to the integral of the latter, taken over the interval from $-\infty$ to x. — The notations of the present paper will, as a rule, correspond to those of I, the most important exception being the symbol n, which will here always denote the number of observations in a statistical series and not, as in I, the number of elementary components.

and further

$$T = -\frac{1}{120}\left(\frac{\mu_5}{\sigma^5} - 10\frac{\mu_3}{\sigma^3}\right).$$

It has been proved in I that, if our variable can be regarded as the sum of a large number of mutually independent elementary components, we have under certain further conditions an *asymptotic expansion*

$$F(x) = \Phi(x) + P_1(\Phi) + P_2(\Phi) + \cdots, \tag{1}$$

where

$$\Phi(x) = \frac{1}{\sqrt{2\pi}}\int_{-\infty}^{x} e^{-\frac{t^2}{2}}\,dt$$

while, according to I, (24), the P_i are linear expressions in the derivatives of $\Phi(x)$:

$$\text{(2)}\quad \begin{cases} P_1(\Phi) = \frac{1}{3} S\,\Phi^{(3)}(x), \\ P_2(\Phi) = \frac{1}{3} E\,\Phi^{(4)}(x) + \frac{1}{18} S^2\,\Phi^{(6)}(x), \\ P_3(\Phi) = \quad T\,\Phi^{(5)}(x) + \frac{1}{9} ES\,\Phi^{(7)}(x) + \frac{1}{162} S^3\,\Phi^{(9)}(x), \\ \cdots\cdots\cdots\cdots\cdots\cdots \end{cases}$$

In the particular case when all the elementary components have identical probability functions, the term $P_i(\Phi)$ is of the order $N^{-i/2}$, where N is the number of the components (cf. I, (25)). Even in more general cases, the orders of magnitude of the terms $P_1(\Phi)$, $P_2(\Phi)$, ... will, generally speaking, form a decreasing sequence, and according to the expansion (1) the curves

$$\begin{aligned} y &= \Phi(x), \\ y &= \Phi(x) + P_1(\Phi), \\ y &= \Phi(x) + P_1(\Phi) + P_2(\Phi), \\ &\cdots\cdots\cdots\cdots \end{aligned}$$

will thus furnish a set of successive approximations to the curve $y = F(x)$. If, in such a case, we draw the curve

$$y = F(x) - \varPhi(x),$$

it will accordingly be found to lie close to the curve

$$y = P_1(\varPhi)$$

and the same thing will hold for the pairs of curves

$$\begin{cases} y = F(x) - \varPhi(x) - P_1(\varPhi), \\ y = P_2(\varPhi), \end{cases}$$

$$\begin{cases} y = F(x) - \varPhi(x) - P_1(\varPhi) - P_2(\varPhi), \\ y = P_3(\varPhi), \end{cases}$$

. . . .

It is the purpose of the present paper to discuss some statistical applications of asymptotic expansions of the type (1). We shall deal exclusively with distributions in *one* single variable, reserving the applications to correlation for a future occasion. — We shall first consider the question if, in statistical practice, probability functions conforming to (1) really do occur.

2. As already pointed out in I, Art. 5, similar expansions have been treated by Charlier, Edgeworth and various other writers, and have been widely used in statistical practice. Thus the three first terms of Charlier's A-series give the following expression for the probability function $F(x)$:

$$F(x) = \varPhi(x) + \frac{1}{3} S \varPhi^{(3)}(x) + \frac{1}{3} E \varPhi^{(4)}(x), \tag{3}$$

while the corresponding terms of (1) give, in accordance with the expansions used by Edgeworth,

$$F(x) = \varPhi(x) + \frac{1}{3} S \varPhi^{(3)}(x) + \frac{1}{3} E \varPhi^{(4)}(x) + \frac{1}{18} S^2 \varPhi^{(6)}(x). \tag{4}$$

I shall begin by making, from the point of view of the expansions, a somewhat more detailed analysis of statistical material than has hitherto, as far as I know, been published. The analysis will be based on a comparison between the curves $y = F(x) - \Phi(x)$, $y = P_1(\Phi)$ etc.

Let us first suppose that we know the »true values» of $m, \sigma, \mu_3, \mu_4, \ldots$ and that, in a series of n mutually independent observations, we have found the values $x_1, x_2, \ldots x_n$ of the variable X. Let us put

(5) $\nu =$ the number of those x_i which are $< m + \sigma x$,

where x is a given quantity such that no x_i is exactly equal to $m + \sigma x$. Denoting generally by $M[z]$ and $\varepsilon[z]$ the mathematical expectation (mean value) and the standard deviation (mean error) of a variable z, we have

$$M\left[\frac{\nu}{n}\right] = F(x), \quad \varepsilon\left[\frac{\nu}{n}\right] = \sqrt{\frac{F(x)\,[1 - F(x)]}{n}}.$$

Hence if n is large and if we mark, for a set of values of x, the corresponding points $y = \frac{\nu}{n}$ in a diagram, these points will give us an idea of the shape of the curve $y = F(x)$, which rises from 0 to 1 as x increases from $-\infty$ to $+\infty$. In the case of a statistical variable generated by elementary components, this curve will more or less approach to the normal curve $y = \Phi(x)$ and we are particularly interested in any systematic *deviations* shown from this form. These deviations are often very small compared with the whole range of the ordinates of the curves, and so an inconveniently large figure is required if we want to make a detailed examination of the deviations.

In order to simplify this examination, we form the differences

$$\Delta_1(x) = \frac{\nu}{n} - \Phi(x)$$

and mark the corresponding points $y = \Delta_1(x)$ in our diagram where, now, the scale of the ordinates may be taken much larger without causing any inconvenient dimensions of the figure. If the expansion (1) holds, the curve $y = P_1(\Phi)$ ought to show a tolerably good fit to the points $y = \Delta_1(x)$.

A further test on the expansion is supplied by the differences

$$\Delta_2(x) = \Delta_1(x) - P_1(\Phi)$$
$$= \frac{\nu}{n} - \Phi(x) - P_1(\Phi)$$

and the corresponding points $y = \Delta_2(x)$ which ought to agree with the curve $y = P_2(\Phi)$. — If there are still systematic deviations, we may obviously continue in the same way and form the differences

$$\Delta_3(x) = \Delta_2(x) - P_2(\Phi)$$
$$= \frac{\nu}{n} - \Phi(x) - P_1(\Phi) - P_2(\Phi)$$

that may be compared with the curve $y = P_3(\Phi)$, and so on.

If the expansion (1) holds, the fit to the observed data must be improved as we pass from the normal curve to the curves $y = \Phi + P_1$, $y = \Phi + P_1 + P_2, \ldots$, and so the differences $\Delta_1(x)$, $\Delta_2(x), \ldots$ will, on the average, form a decreasing sequence. As a standard for judging the values of the differences found in an actual case we shall choose the absolutely convergent integrals

$$J_i = \int_{-\infty}^{\infty} [\Delta_i(x)]^2 \, dx \tag{6}$$

that can be approximately calculated if the values of x for which $\Delta_i(x)$ is known lie sufficiently close together. Denoting the mathematical expectation $M[J_i]$ by M_i we have

$$M_1 = M\left[\int_{-\infty}^{\infty}\left(\frac{\nu}{n} - \Phi\right)^2 dx\right]$$

$$= M\left[\int_{-\infty}^{\infty}\left(\frac{\nu}{n} - F\right)^2 dx\right] + \int_{-\infty}^{\infty}(F - \Phi)^2\, dx$$

$$= \int_{-\infty}^{\infty}\frac{F(1-F)}{n}\, dx + \int_{-\infty}^{\infty}(F-\Phi)^2\, dx.$$

If we consider only distributions that do not deviate considerably from the normal form, and if we are contented with a rough approximation, we may here, according to the expansion (1), replace F by Φ in the first integral and $F - \Phi$ by P_1 in the second. Thus we get approximately

$$M_1 = \frac{1}{n}\int_{-\infty}^{\infty}\Phi(1-\Phi)\, dx + \int_{-\infty}^{\infty} P_1^2\, dx$$

and in a similar way

$$M_i = \frac{1}{n}\int_{-\infty}^{\infty}\Phi(1-\Phi)\, dx + \int_{-\infty}^{\infty} P_i^2\, dx. \tag{7}$$

Now, it can be shown without difficulty that we have

$$\int_{-\infty}^{\infty}\Phi(x)\,[1-\Phi(x)]\, dx = \frac{1}{\sqrt{\pi}}$$

and

$$\int_{-\infty}^{\infty}\Phi^{(r)}(x)\,\Phi^{(r+2s)}(x)\, dx = \frac{(-1)^s}{\sqrt{\pi}}\cdot\frac{(2r+2s-2)!}{2^{2r+2s-1}\,(r+s-1)!}$$

for every positive integer r and for $s=0, 1, 2, \ldots$ Hence we find, using the expressions (2),

$$
(8) \quad \begin{cases}
M_1 = \dfrac{1}{n\sqrt{\pi}} + \dfrac{S^2}{24\sqrt{\pi}}, \\[2ex]
M_2 = \dfrac{1}{n\sqrt{\pi}} + \dfrac{5}{2\,304\sqrt{\pi}}(48\,E^2 - 56\,ES^2 + 21\,S^4). \\[2ex]
M_3 = \dfrac{1}{n\sqrt{\pi}} + \dfrac{35}{165\,888\sqrt{\pi}}(15\,552\,T^2 + 4\,752\,E^2\,S^2 + \\[1ex]
\qquad + 715\,S^6 - 15\,552\,EST + 4\,752\,S^3T - 3\,432\,ES^4).
\end{cases}
$$

By comparing M_1, M_2 and M_3 with the values of J_1, J_2 and J_3 actually found, we thus have a further test on our expansion. Owing to the improvement of the fit, it is to be expected that we shall have, e.g., $J_2 < J_1$, and a comparison with the mean values M_1 and M_2 then will show if the actual improvement is as great as might be reasonably expected if the expansion (1) holds.

3. In ordinary practice we do not, however, possess any *a priori* knowledge as to the values of m, σ, etc., and so have to form estimates of these values from the observations x_1, $x_2, \ldots x_n$ themselves.[1] We put

$$
\begin{aligned}
m^* &= \frac{1}{n}\Sigma x_i, \\
\sigma^* &= \sqrt{\mu_2^*} = \sqrt{\frac{1}{n}\Sigma(x_i - m^*)^2}, \\
\mu_3^* &= \frac{1}{n}\Sigma(x_i - m^*)^3, \\
\mu_4^* &= \frac{1}{n}\Sigma(x_i - m^*)^4, \\
&\ldots\ldots\ldots\ldots
\end{aligned}
$$

[1] Of course this does not imply that we assume the actual *existence* of the »true values» which are a mere hypothetical construction.

all the sums being extended from $i=1$ to n. Further, we put

$$S^* = -\frac{\mu_3^*}{2\sigma^{*3}}, \qquad E^* = \frac{1}{8}\left(\frac{\mu_4^*}{\sigma^{*4}} - 3\right),$$

$$T^* = -\frac{1}{120}\left(\frac{\mu_5^*}{\sigma^{*5}} - 10\frac{\mu_3^*}{\sigma^{*3}}\right),$$

and

$$P_1^*(\Phi) = \frac{1}{3} S^* \Phi^{(3)}(x),$$

$$P_2^*(\Phi) = \frac{1}{3} E^* \Phi^{(4)}(x) + \frac{1}{18} S^{*2} \Phi^{(6)}(x),$$

. . . .

All these quantities can be empirically determined, and in particular the curves $y = P_1^*(\Phi)$, $y = P_2^*(\Phi), \ldots$ can be drawn as soon as the empirical moments have been computed from the statistics.

Since the true values m and σ are unknown, we cannot find the exact value ν corresponding to a given x according to (5). We put

ν^* = the number of those x_i which are $< m^* + \sigma^* x$,

so that ν^* can be empirically determined.

The mathematical expectations of the quantities $\frac{\nu^*}{n}$, $P_1^*(\Phi)$, $P_2^*(\Phi), \ldots$ are not actually equal to $F(x)$, $P_1(\Phi)$, $P_2(\Phi), \ldots$ When we regard the empirically determined quantities as estimates of the corresponding »true values», affected only by sampling errors, a certain *systematic error* is thus introduced. The latter being, however, of the same order of magnitude as $\frac{1}{n}$, while the errors of random sampling are of the order $\frac{1}{\sqrt{n}}$, the systematic errors will be comparatively insignificant when n is large.[1] In the statistical distributions considered in the present paper, n will never fall below 12 000 and the systematic errors will be neglected.

If, as is usually the case, the statistics are arranged in certain groups, the number of observations belonging to each group being the only data at our disposal, we thus have to determine the value $x = \frac{X - m^*}{\sigma^*}$ which corresponds to the upper extreme X of each group. The corresponding value ν^* is equal to the number of those observations which are $< X$, and is thus immediately given by the statistics. Thus we can form the differences

$$\Delta_1^*(x) = \frac{\nu^*}{n} - \Phi(x),$$

$$\Delta_2^*(x) = \frac{\nu^*}{n} - \Phi(x) - P_1^*(\Phi),$$

$$\Delta_3^*(x) = \frac{\nu^*}{n} - \Phi(x) - P_1^*(\Phi) - P_2^*(\Phi),$$

$$\cdots\cdots\cdots\cdots\cdots\cdots\cdots$$

and mark the corresponding points $y = \Delta_1^*(x)$, $y = \Delta_2^*(x)$, $y = \Delta_3(x)$ in our diagram. If n is large and if the expansion (1) holds, these points ought to agree fairly well with the curves $y = P_1^*(\Phi)$, $y = P_2^*(\Phi)$ and $y = P_3^*(\Phi)$ respectively.

It is obvious that, in making such comparisons, we must always pay regard to the influence of sampling errors upon the quantities used for the comparisons. Clearly we cannot reasonably expect, e.g., that the fit to the observed data will be improved by considering the term $P_1^*(\Phi)$ if the quantity S^* has not a value that differs *significantly* from zero with regard to its mean error. On the other hand, a bad agreement found in an actual case cannot be regarded as a piece

[1] Cf. A. A. Tschuprow, Grundbegriffe und Grundprobleme der Korrelationstheorie, Leipzig—Berlin 1925, p. 101.

of conclusive evidence against the expansion if the quantities S^*, etc., used for the comparison have not significant values.

For the mean errors of the quantities S^* and E^*, CHARLIER[1] gives the expressions

$$\varepsilon[S^*] = \sqrt{\frac{3}{2n}}, \quad \varepsilon[E^*] = \frac{1}{2}\sqrt{\frac{3}{2n}},$$

which are deduced under the hypothesis that the distribution is approximately normal. By the same argument we obtain the expression

$$\varepsilon[T^*] = \frac{1}{\sqrt{120\,n}}.$$

It has been pointed out by LINDEBERG[2] that this method of deducing mean error formulae is not in all respects satisfactory, and accordingly we must not rely too absolutely on the values calculated from the formulae.

As a standard for judging the gradual improvement of the fit as we pass from the normal curve to the curves $y = \Phi + P_1^*$, etc., we shall use the integrals

$$J_i^* = \int_{-\infty}^{\infty} [\Delta_i^*(x)]^2\,dx$$

analogous to (6), and the values M_i^* obtained by replacing in (8) the parameters S, E, T by the empirically determined S^*, E^* and T^*. Owing to the uncertainty of these quantities, obviously we cannot expect to find a very close agreement between the values of J_i^* and M_i^*. We shall have to be contented if the order of magnitude is the same and if the values J_i^* decrease in a way that shows some conformity to the values M_i^*.

[1] Vorlesungen über die Grundzüge der mathematischen Statistik, Lund 1920, p. 74.

[2] Über die Begriffe Schiefheit und Exzess in der mathematischen Statistik, Skand. Aktuarietidskrift 1925, p. 106.

4. By the method indicated in the preceding Articles, we shall now test the following five statistical materials (ordered according to increasing numerical value of the quantity S^*).

A. The stature of 46 981 Swedish conscripts and regular soldiers as published by the Swedish State Institute for Race Biology.[1]

B. The leg length of 46 982 Swedish conscripts and regular soldiers, taken from the same source as **A**.

C. The face height of 46 982 Swedish conscripts and regular soldiers, taken from the same source as **A**.

D. The breadth of 12 000 beans *(Phaseolus vulgaris)* according to JOHANNSEN.[2]

E. The barometric heights on 22 188 winter days at Den Helder according to KAPTEYN.[3]

The values of the arithmetical characteristics defined in the preceding Article are put together in the table on the following page. The moments have been adjusted according to the well-known formulae due to SHEPPARD.[4]

Thus in all cases the quantity S^* differs significantly from zero, while for E^* this takes place only in the materials **C**, **D** and **E**, and for T^* only in **E**.

For each of these five statistical series, the values of x, $\frac{\nu^*}{n}$, $\Phi(x)$, $P_1^*(\Phi)$, $P_2^*(\Phi)$ and $P_3^*(\Phi)$ have been calculated[5] for all the classes into which the material was subdivided, and then the differences $\Delta_1^*(x), \ldots, \Delta_4^*(x)$ have been found by subtraction. All these quantities are given in Tables **A**—**E** at the end of the paper.

An inspection of the tables shows that, in all five cases, the average order of magnitude of the final difference Δ_4^* is considerably smaller than that of the difference Δ_1^*, thus in-

---

[1] The Racial Characters of the Swedish Nation, Anthropologia Suecica MCMXXVI, edited by H. LUNDBORG and F. J. LINDERS, Uppsala 1926.

[2] Cf. CHARLIER, l. c., p. 73.

[3] KAPTEYN and VAN UVEN, Skew frequency curves in Biology and Statistics, 2nd paper, Groningen 1916, p. 57.

[4] This accounts for some slight deviations from the values given elsewhere.

[5] By means of JÖRGENSEN's tables.

	A	B	C	D	E
n	46 981	46 982	46 982	12 000	22 188
m^* in cm.	172.228 $\pm$0.028	92.020 $\pm$0.020	12.6574 $\pm$0.0032	0.85117 $\pm$0.00056	76.0318 $\pm$0.0066
σ^* in cm.	5.931 $\pm$0.019	4.301 $\pm$0.014	0.6937 $\pm$0.0023	0.06163 $\pm$0.00040	0.9854 $\pm$0.0047
S^*	−0.0598 $\pm$0.0057	−0.0681 $\pm$0.0057	−0.0829 $\pm$0.0057	−0.1439 $\pm$0.0112	+0.2080 $\pm$0.0082
E^*	−0.0011 $\pm$0.0028	+0.0047 $\pm$0.0028	+0.0104 $\pm$0.0028	+0.0244 $\pm$0.0056	−0.0137 $\pm$0.0041
T^*	−0.00062 $\pm$0.00042	−0.00066 $\pm$0.00042	+0.00005 $\pm$0.00042	+0.00109 $\pm$0.00083	−0.00602 $\pm$0.00061

dicating that the fit to the observed data has been improved by adding the terms P_1^*, P_2^* and P_3^* to the normal function Φ. In order to illustrate the gradual improvement of the fit, approximate values of the integrals

$$J_i^* = \int_{-\infty}^{\infty} (\Delta_i^*)^2 dx$$

have been calculated and compared with the corresponding M_i^* as shown by the following table. For the calculation of M_4^* it has been assumed that P_4^* can be neglected, so that M_4^* reduces to the first term on the right hand side of (7). In this way, we get a too small value of M_4^*, and thus are on the »safe side» for our comparisons.

i	A		B		C		D		E	
	$10^6 J_i^*$	$10^6 M_i^*$	$10^6 J_i^*$	$10^6 M_i^*$	$10^6 J_i^*$	$10^6 M_i^*$	$10^6 J_i^*$	$10^6 M_i^*$	$10^6 J_i^*$	$10^6 M_i^*$
1	73	96	113	121	231	174	544	534	1670	1043
2	7	13	6	12	30	15	33	58	230	125
3	4	13	7	12	27	12	14	58	87	52
4	3	12	7	12	24	12	31	47	38	25

It follows from this table that the average order of magnitude of the deviations in no case exceeds that which might have been reasonably expected to arise from random sampling. In all cases, the fit is considerably improved by passing from $i=1$ to $i=2$, i. e. by the addition of $P_1^*(\Phi) = \frac{1}{3} S^* \Phi^{(3)}(x)$ to the normal function. In the cases **A**, **C**, **D** and **E**, a further improvement is caused by adding the term P_2^*, while the consideration of P_3^* still reduces the deviations in **A**, **C** and **E**. In the cases **B** and **D**, where the quantities J_2^*, J_3^*, J_4^* are not steadily decreasing, it is seen that these quantities are, without any exception, smaller than the corresponding M_i^*, so that the increase may well be due to errors of sampling. This explanation is also indicated by the fact that, as shown above, the characteristics E^* and T^* in the material **B**, as well as T^* in the material **D**, do not differ significantly from zero. In **E**, where all our characteristics have significant values, the fit is steadily improved from $i=1$ to $i=4$.

The comparison between the points $y=\Delta_i^*(x)$ and the corresponding curves $y=P_i^*(\Phi)$ is illustrated by the figures 1—9, which have been drawn for those cases only, where the characteristics involved have significant values. The agreement is generally quite as good as might have been expected, and in some cases even striking. Especially the comparison between Δ_1^* and P_1^*, as shown by figures 1, 2, 3, 5 and 7, gives very good results. At the inspection of the figures, it must be borne in mind that, when comparing Δ_i^* and P_i^*, there are always systematic deviations due to the influence of the terms $P_{i+1}^*, P_{i+2}^*, \ldots$ that have been neglected. Our hypothesis is precisely that, on the average, these deviations are small as compared with P_i^*. In fig. 7, the reduction of the deviations caused by the consideration of the two first neglected terms is shown by the dotted curve. — Regard should also be paid to the large scale of the ordinates used in drawing the figures.

I think it will be admitted that the preceding analysis of our statistical material justifies the conclusion that, in each of the five investigated cases, all the evidence that can be legitimately drawn from the material supports the hypothesis that the corresponding probability function admits an expansion of the form (1).

5. The term $P_1^*(\Phi)$ is identical with the term $\frac{1}{3}S^*\Phi^{(3)}(x)$ furnished by the A-series. As far as moments up to the third order only are considered, our statistical material thus speaks in equal favour of both developments, the A-series as well as the P-series. But from the next term onwards the developments differ, as shown by the expressions (3) and (4) which arise when the fourth moment is also brought into consideration. If the fifth moment is also available, the P-series supplies

the term $P_3^*(\Phi)$ which is also different from the corresponding term of the A-series.

In order to decide between both developments we thus have, in the first place, to compare the fit to the points $y = \Delta_2^*(x)$ which is given by the two curves

$$(9) \qquad y = P_2^*(\Phi) = \frac{1}{3} E^* \Phi^{(4)}(x) + \frac{1}{18} S^{*2} \Phi^{(6)}(x)$$

and

$$(10) \qquad y = \frac{1}{3} E^* \Phi^{(4)}(x).$$

If S^* is small, the difference between these curves is insignificant, and accordingly we find that in our materials **A**, **B** and **C** the difference between the A-series and the P-series has not much practical importance. For **D** and **E**, the fit is found to be slightly less good with (10) than with (9). Thus if we put

$$\delta_3^*(x) = \Delta_2^*(x) - \frac{1}{3} E^* \Phi^{(4)}(x)$$

$$= \frac{\nu^*}{n} - \Phi(x) - \frac{1}{3} S^* \Phi^{(3)}(x) - \frac{1}{3} E^* \Phi^{(4)}(x)$$

and calculate the integral

$$j_3^* = \int_{-\infty}^{\infty} (\delta_3^*)^2 dx$$

we find the values

$$10^6 j_3^* = 21 \quad \text{in } \mathbf{D},$$

$$10^6 j_3^* = 171 \quad \text{in } \mathbf{E},$$

corresponding to the values

$$10^6 J_3^* = 14 \quad \text{in} \quad \mathbf{D},$$

$$10^6 J_3^* = 87 \quad \text{in} \quad \mathbf{E}.$$

In fig. 8, the curves (9) and (10) are shown for the case **E**, and it is obvious that, on the whole, it is (9) that gives the best fit.[1] The consideration of the fifth moment and the term P_3^* still improves this fit.

On the scope of the method.

6. By means of certain hypotheses — in the first place the »hypothesis of elementary errors» — concerning the genetic character of a given frequency-distribution, the validity of an asymptotic expansion for the corresponding probability function has been mathematically proved. A detailed analysis of statistical material has shown that, as far as might be reasonably expected, this expansion seems to be realized in all the cases here brought under examination. In some cases, the agreement between the observed values and the theoretical curves is even so striking that it strongly suggests the conjecture that the fundamental hypothesis may contain at least something which resembles the actual truth.

According to I, theorem 1, and the generalization of this theorem proved in I, Art. 25, the asymptotic expansion may be expected to hold, at least *on the average*, under fairly general conditions for a variable generated by mutually independent components. As a matter of fact, this theorem states that the integral of $F(x)$ of a certain order is asymptotically represented by the corresponding integral of the first terms of the expansion. This is equivalent to the assertion that, *on the average*, the curves $y = \Phi(x)$, $y = \Phi(x) + P_1(\Phi)$, ... furnish a set of successive approximations to the probability curve $y = F(x)$, which is of course exactly the kind of agreement we should expect to find when working with statistical

[1] Cf. S. Wicksell, The correlation function of type *A* and the regression of its characteristics, Kungl. Svenska Vet.-Ak. Handl., B. 58 (1917), p. 41, where some similar comparisons are made.

material. It has been indicated in I, Art. 28, that the validity of theorem I can be extended also to certain cases where the elementary components cannot be regarded as independent. Finally, theorems 2 and 3 of I show that, under certain more restricted conditions, even asymptotic expansions in the ordinary sense exist.

It is obvious that, in ordinary statistical practice, there will be no possibility of deciding *a priori* whether the conditions stated in our theorems are satisfied or not. We shall have to be contented if, in a general way, these conditions agree with our knowledge concerning the causal relations in the matter under investigation. It seems to be generally assumed that such an agreement is present in quite a number of biological and other cases. In such cases, I think it may be concluded from the preceding developments that we shall be justified in using for the probability functions an expansion of the type here discussed.

On the other hand, there are also plenty of cases occurring in practice where it is quite obvious that our conditions are not satisfied. Thus, to give only one single example, if we consider Head Length and Head Breadth as generated by the addition of elementary components and approximatively normally distributed, the preceding investigations do not give any justification for the application of our expansions to the distribution of Head Index.[1] In cases where no such justification is available, it has been customary to appeal to the properties of the A-series as a development of arbitrary functions, analogous to the Fourier series. But if we use a series simply as a development for arbitrary functions, without knowing anything about its asymptotic character, it is impossible to claim that two or three terms will be enough to give a good approximation to the sum, which is essential for the statistical applications. Thus in such cases the use of the method seems to be open to serious objections.

[1] It is, of course, not impossible that in this particular case such justification can be given by some other method.

7. For our analysis of statistical distributions in Arts 1—5 of the present paper, we have used the *probability function* $F(x)$ and not the corresponding *frequency function*[1] $f(x) = F'(x)$.

From a practical point of view, the law of distribution of a statistical variable should be regarded as known if the probability function is known with a certain approximation. As a matter of fact, it is this function and not the frequency function that is immediately furnished by the statistical experience.[2] Even if we form a class frequency, it is a difference between two values of $F(x)$ that we observe, not the frequency function $f(x)$ itself. The latter is a purely theoretical construction which has, however, the advantage of giving, to most minds, a more easily surveyable picture of the law of distribution of the variable.

Let us now suppose that, in a particular case, we know e.g. that the expression

$$\Phi(x) + P_1(\Phi) = \Phi(x) + \frac{1}{3} S \Phi^{(3)}(x)$$

approximates $F(x)$ with an error which is numerically less than ε. By differentiation we form the expression

$$\varphi(x) + P_1(\varphi) = \varphi(x) + \frac{1}{3} S \varphi^{(3)}(x)$$

with

$$\varphi(x) = \Phi'(x) = \frac{1}{\sqrt{2\pi}} e^{-\frac{x^2}{2}},$$

[1] Observe that f is here used to denote a *frequency function* and not, as in the paper I, the adjunct of F.

[2] Cf. P. Lévy, Calcul des probabilités, Paris 1925, p. 193 footnote, where the same view is expressed. — The choice of the probability function instead of the frequency function as basis for the comparison brings also the advantage that it will be possible to obtain some results even from statistical series consisting of a small number of observations. Cf. my treatise »Sannolikhetskalkylen och några av dess användningar», Stockholm (Gjallarhornets förlag) 1926, p. 121 and 132.

and we regard this as an approximation to the frequency function $f(x)$.

$F(x)$ is a never decreasing function. But obviously the relation

$$|F(x) - \Phi(x) - P_1(\Phi)| < \varepsilon$$

does not prevent that the approximating expression $\Phi(x) + P_1(\Phi)$ may, for certain values of x, be decreasing. For such a value, the derivative $\varphi(x) + P_1(\varphi)$ is negative, and thus it is seen that the occurrence of *negative frequencies* cannot always be avoided. This has always been one of the chief objections against the statistical use of the A-series[1], and it is obvious that the modification of the A-series which is principally used here is open to the same objections.

I cannot, however, consider this objection as conclusive. Even if the expression used as an approximation to the frequency function becomes negative for certain values of x, it may well occur that its integral gives a good approximation to a steadily increasing function. If it is admitted that it is the probability function, rather than the frequency function, that represents the primary notion of the theory, and if we remember that the expansions only give *approximations* to these functions, the question of the negative frequencies becomes less serious.

If the use of the expansions is carefully restricted to cases where, according to the theoretical conditions, they can be legitimately applied, as a rule only practically insignificant negative frequencies will occur. In fig. 10, the approximating frequency-curve

$$y = \varphi(x) + P_1^*(\varphi) + P_2^*(\varphi) + P_3^*(\varphi)$$

has been drawn for the above material **E** (Barometric Heights). The curve shows a considerable skewness, but the negative

---

[1] Cf. the authors quoted in I, p. 21, footnote 2. Reference should be made also to the second edition (1927) of Mr. Elderton's book on »Frequency-curves and correlation», which deals more fully than the first edition with the developments of frequency-functions in series.

frequencies are very small. They begin exactly at the point ($x = +2.6$) where the actual material ceases to give positive frequencies.

Measures of skewness and excess.

8. The quantities S and E (or rather the corresponding empirical estimates S^* and E^*) have been proposed by CHARLIER[1] as measures of the *skewness* and *excess* of a frequency distribution. The skewness is then geometrically interpreted as the abscissa of the maximum (the »mode») of the frequency-curve, when the origin is placed at the mean and the standard deviation is taken as unit of x. (If the sign is inverted, this agrees with the definition adopted by PEARSON.) CHARLIER considers the approximating frequency-curve

$$y = \varphi(x) + \frac{1}{3} S \varphi^{(3)}(x) + \frac{1}{3} E \varphi^{(4)}(x) \tag{11}$$

and shows that, in the first approximation, the mode is situated at the point $x = S$. — The excess, on the other hand, is defined as the relative excess, at the mean, of the actual frequency-curve over the normal curve $y = \varphi(x)$. Thus the excess is equal to the value of the expression

$$\frac{y - \varphi(x)}{\varphi(x)} \tag{12}$$

for $x = 0$. From (11) we get in this way the value E for the excess.

If we look at the matter from the point of view of the theory of elementary errors, we know that S, E and T are of the orders $N^{-1/2}$, N^{-1} and $N^{-3/2}$ respectively, where N denotes

[1] Researches into the Theory of Probability, Kongl. Fysiografiska Sällsk. Handl. B. 16 (1906), p. 10—11; Contributions to the mathematical Theory of Statistics, 5, Arkiv för matematik etc. B. 9 n:o 25 (1914), p. 12—13; Vorlesungen über die Grundzüge der mathematischen Statistik, Lund 1920, p. 70—72.

the number of the elementary errors.[1] In order to get expressions for the skewness and the excess, interpreted in the above geometrical sense, which are correct up to terms of the order N^{-1} we must, according to (4), add the term $\frac{1}{18}S^2\varphi^{(6)}(x)$ on the right hand side of (11). Neglecting terms of the order $N^{-3/2}$, we have still the expression $x=S$ for the abscissa of the mode, while the value of (12) for $x=0$ is changed into $E-\frac{5}{6}S^2$. Thus we have the following corrected empirical estimates for the Charlier skewness and excess:

$$\text{Skewness} = S^*$$

$$\text{Excess} \quad = E^* - \frac{5}{6}S^{*2}.$$

The ordinary formulae for the mean errors of the characteristics have been given in Art. 3. Calculating the mean error of the above expression for the excess as if S^* and E^* were independent, we get the approximative expression

$$\varepsilon\left[E^* - \frac{5}{6}S^{*2}\right] = (1 + 6S^{*2})\,\varepsilon\,[E^*]. \tag{13}$$

This is, of course, not strictly correct, but in view of the uncertainty of all these mean error formulae I have not considered it worth while to construct a more elaborate expression.

If we are dealing with a curve of considerable skewness, the omittance of the term $\frac{5}{6}S^{*2}$ may lead to quite false ideas as to the excess of the curve. Thus in the above material **D** (Beans) we have

$$E^* \qquad = +\,0.0244 \pm 0.0056,$$

$$E^* - \frac{5}{6}S^{*2} = +\,0.0071 \pm 0.0063.$$

[1] This does not imply that we should necessarily have e.g. $|E| < |S|$. The only thing which is proved is that S is *at most* of the order $N^{-1/2}$, E at most of the order N^{-1} etc. For all symmetrical distributions, we have of course $S=0$, but not necessarily $E=0$.

Judging only from the value of E^*, we should feel inclined to believe that this frequency curve shows a significant positive excess over the normal curve in the vicinity of the mean, while according to the corrected expression no such excess appears. — In the material E (Barometric Heights), the correction works in the opposite direction. We have here

$$E^* \qquad\qquad = -\,0.0137 \pm 0.0041,$$

$$E^* - \frac{5}{6}S^{*2} = -\,0.0498 \pm 0.0052.$$

If the fifth moment is also available, we can obtain expressions for the skewness and excess which are correct up to terms of the order $N^{-3/2}$. For this purpose, we have to use the equation

$$y = \varphi(x) + P_1(\varphi) + P_2(\varphi) + P_3(\varphi)$$

for the frequency curve and we find, neglecting terms of the order N^{-2}, that the above expression for the excess does not change, while the expression for the skewness becomes

$$S^* - 2S^{*3} + \frac{20}{3}E^* S^* - 15\,T^*.$$

9. It follows from the above that the CHARLIER expressions for skewness and excess are essentially connected with the expansion of the frequency-function that is obtained by the aid of the hypothesis of elementary errors. In cases where no such expansions hold, we have no evidence that the characteristics S and E will conserve their geometrical meaning as measures of skewness and excess.

In an interesting paper[1], LINDEBERG has introduced new arithmetical measures for skewness and excess. If $x_1, x_2, \ldots, x_n$

[1] Über die Begriffe Schiefheit und Exzess in der mathematischen Statistik, Skand. Aktuarietidskr. 1925, p. 106. Cf. also A. WALTHER, Zum Lindebergschen Exzess in der mathematischen Statistik, Skand. Aktuarietidskr. 1927, p. 246.

are the observed values in our statistical series, and if r denotes the number of those x_i which are greater than the empirical mean m^*, the LINDEBERG skewness is

$$\mathfrak{S}^* = 100\frac{r}{n} - 50. \tag{14}$$

Further, if ϱ denotes the number of those x_i which fall between the limits $m^* \pm \frac{1}{2}\sigma^*$, the quantity

$$\begin{aligned} \mathfrak{E}^* &= 100\frac{\varrho}{n} - \frac{100}{\sqrt{2\pi}}\int\limits_{-1/2}^{1/2} e^{-\frac{t^2}{2}}\,dt \\ &= 100\frac{\varrho}{n} - 38.29 \end{aligned} \tag{15}$$

is proposed by LINDEBERG as measure of the excess. For the mean errors of these quantities, LINDEBERG gives the expressions

$$\varepsilon[\mathfrak{S}^*] = \frac{30}{\sqrt{n}}, \qquad \varepsilon[\mathfrak{E}^*] = \frac{42}{\sqrt{n}}. \tag{16}$$

The simplicity and generality of these definitions represent undeniable advantages, and the numerical calculations required for the applications are much less complicated than in the case of the characteristics S^* and E^*.

$\mathfrak{S}^*$ and $\mathfrak{E}^*$ are empirically determined quantities. We denote by $\mathfrak{S}$ and $\mathfrak{E}$ the corresponding theoretical values, obtained by replacing in (14) and (15) the frequencies $\frac{r}{n}$ and $\frac{\varrho}{n}$ by the corresponding probabilities that the variable is greater than m, respectively falls between the limits $m \pm \frac{1}{2}\sigma$.

In a case where the probability function admits an expansion of the type studied above, it is easily seen that certain relations exist between S and E on the one hand, $\mathfrak{S}$ and $\mathfrak{E}$ on the other. Thus if we assume that we have approximately

$$F(x) = \Phi(x) + \frac{1}{3} S \Phi^{(3)}(x) + \frac{1}{3} E \Phi^{(4)}(x) + \frac{1}{18} S^2 \Phi^{(6)}(x),$$

the probability that an observed value of the variable will be greater than the theoretical mean m is equal to

$$1 - F(0) = \frac{1}{2} - \frac{1}{3} S \Phi^{(3)}(0)$$

$$= 0.5 + 0.133\, S,$$

and accordingly we have

$$\mathfrak{S} = 13.3\, S.$$

For large values of n the empirical estimates $\mathfrak{S}^*$ and S^* ought to agree well with the corresponding theoretical values, so that we should have approximately

$$\mathfrak{S}^* = 13.3\, S^*. \tag{17}$$

The probability that the variable falls between $m \pm \frac{1}{2}\sigma$ is equal to

$$F\left(\frac{1}{2}\right) - F\left(-\frac{1}{2}\right) = \Phi\left(\frac{1}{2}\right) - \Phi\left(-\frac{1}{2}\right) + \frac{2}{3} E \Phi^{(4)}\left(\frac{1}{2}\right) + \frac{1}{9} S^2 \Phi^{(6)}\left(\frac{1}{2}\right)$$

$$= 0.3829 + 0.323\, E - 0.246\, S^2$$

and accordingly we get, passing to the empirical values,

$$\mathfrak{E}^* = 32.3\, E^* - 24.6\, S^{*2}, \tag{18}$$

which should be expected to hold approximately for large values of n.

If the fifth moment is also available, the term $P_3(\Phi)$ may be added to the approximating expression for $F(x)$. The relation (18) is not changed, but (17) is transformed into

$$(19)\qquad \mathfrak{S}^* = 13.3\, S^* - 25.8\, S^{*3} + 66.5\, E^* S^* - 119.7\, T^*.$$

For the above materials A—E, and for some other statistical series taken from the publication of the Swedish State Institute for Race Biology quoted in Art. 4, the values of $\mathfrak{S}^*$ and $\mathfrak{E}^*$, together with the values of the right hand sides of (17) and (18), are given in the following table. The mean errors of $\mathfrak{S}^*$ and $\mathfrak{E}^*$ are calculated according to (16); for the mean error of the quantity $32.3\, E^* - 24.6\, S^{*2}$ we have used the expression

$$\varepsilon[32.3\, E^* - 24.6\, S^{*2}] = 32.3\,(1 + 5\, S^{*2})\, \varepsilon[E^*]$$

which is deduced in the same way as (13). — The order adopted in the table is that of increasing numerical values of $\mathfrak{S}^*$.

	n	$\mathfrak{S}^*$	$13.3\, S^*$	$\mathfrak{E}^*$	$32.3\, E^* - 24.6\, S^{*2}$
Head length	46 975	-0.26 ± 0.14	-0.21 ± 0.08	$+0.39 \pm 0.19$	$+0.40 \pm 0.09$
Face breadth	46 981	-0.26 ± 0.14	-0.31 ± 0.08	$+0.69 \pm 0.19$	$+1.70 \pm 0.09$
Trunk length	46 981	-0.29 ± 0.14	-0.31 ± 0.08	$+0.48 \pm 0.19$	$+1.93 \pm 0.09$
Arm length	46 976	-0.58 ± 0.14	-0.46 ± 0.08	$+0.17 \pm 0.19$	$+0.39 \pm 0.09$
Stature (A)	46 981	-0.68 ± 0.14	-0.80 ± 0.08	-0.43 ± 0.19	-0.12 ± 0.09
Head breadth	46 972	-1.00 ± 0.14	-0.80 ± 0.08	-0.50 ± 0.19	$+0.59 \pm 0.09$
Leg length (B)	46 982	-1.03 ± 0.14	-0.91 ± 0.08	-0.14 ± 0.19	$+0.04 \pm 0.09$
Face height (C)	46 982	-1.33 ± 0.14	-1.10 ± 0.08	$+0.85 \pm 0.19$	$+0.15 \pm 0.09$
Minimum frontal diameter	46 974	-1.46 ± 0.14	-0.95 ± 0.08	$+1.30 \pm 0.19$	$+1.37 \pm 0.09$
Beans (D)	12 000	-1.65 ± 0.28	-1.91 ± 0.15	$+0.43 \pm 0.38$	$+0.28 \pm 0.20$
Barometr. heights (E)	22 188	$+3.24 \pm 0.20$	$+2.77 \pm 0.11$	-2.22 ± 0.28	-1.52 ± 0.16

With regard to the mean errors, the agreement between $\mathfrak{S}^*$ and $13.3\, S^*$ is in most cases very good. For the excess, however, significant differences occur in several places.

For the material E, where the quantity T^* has a significant value, the skewness has been calculated also according to (19). The right hand side of this formula takes here the value $+ 3.08$, which agrees well with the value $\mathfrak{S}^* = + 3.24$.

Errata.

On the composition of elementary errors; first paper.

P. 24. The last sentence of Art. 6 should be deleted.

P. 45. In formula (41 a), for $f(t)$ read $f_n(t)$.

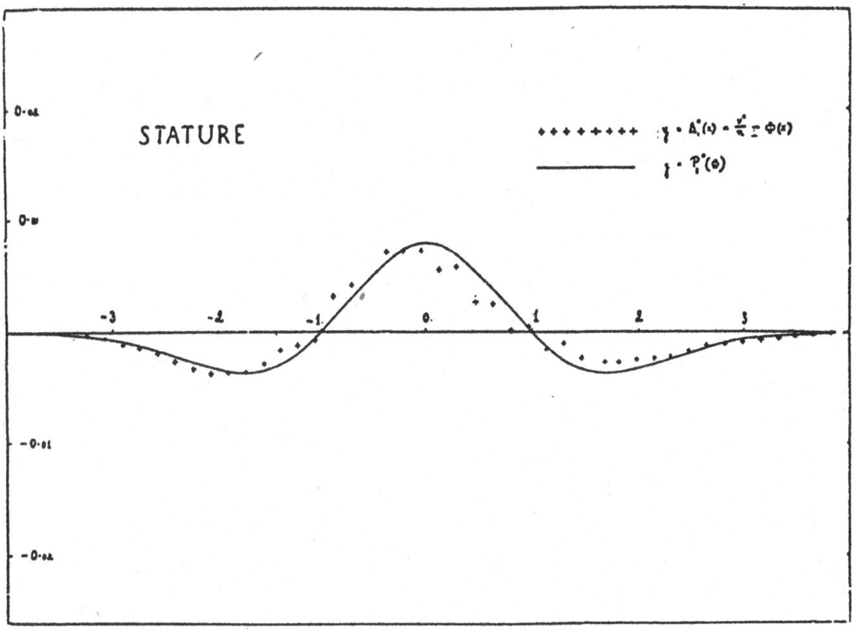

Fig. 1. Stature. Δ_1^* and P_1^*.

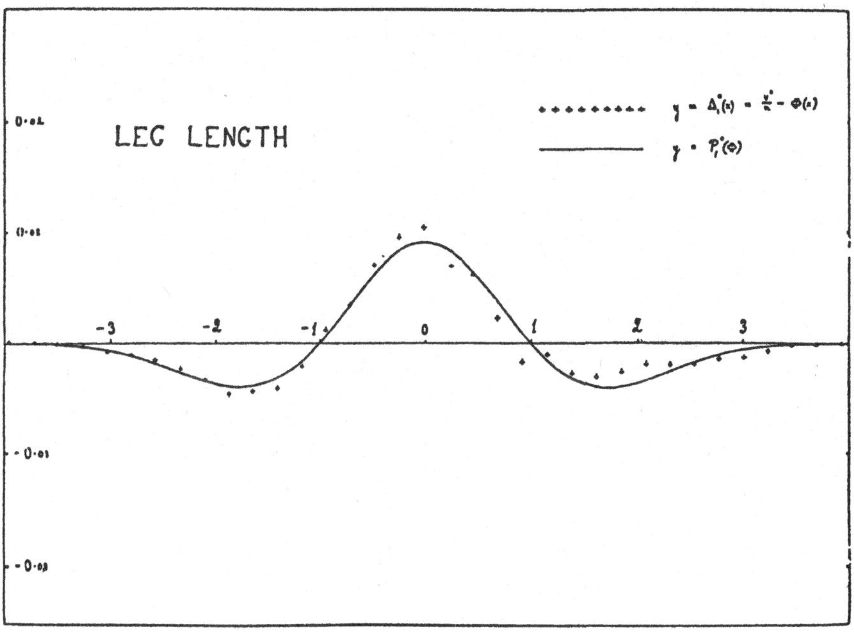

Fig. 2. Leg length. Δ_1^* and P_1^*.

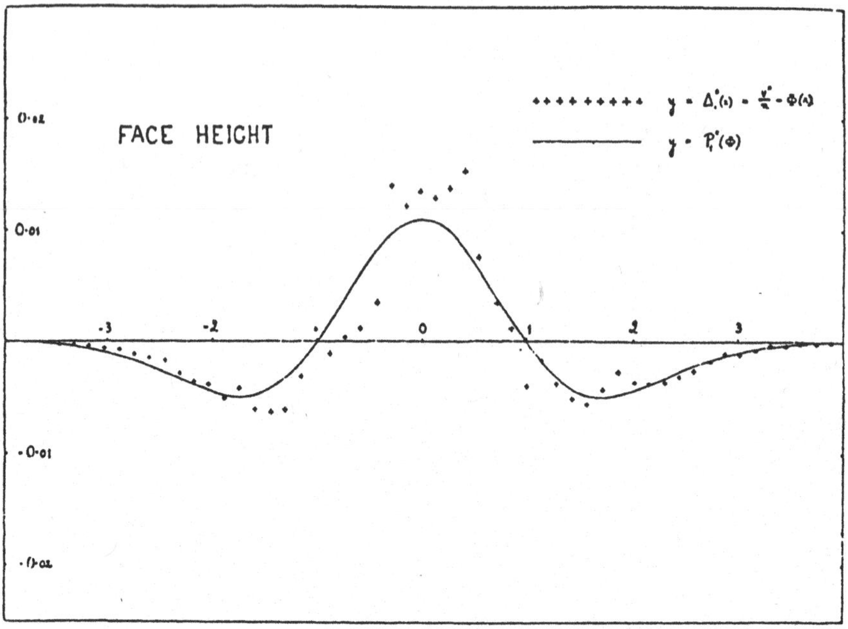

Fig. 3. Face Height. Δ_1^* and P_1^*.

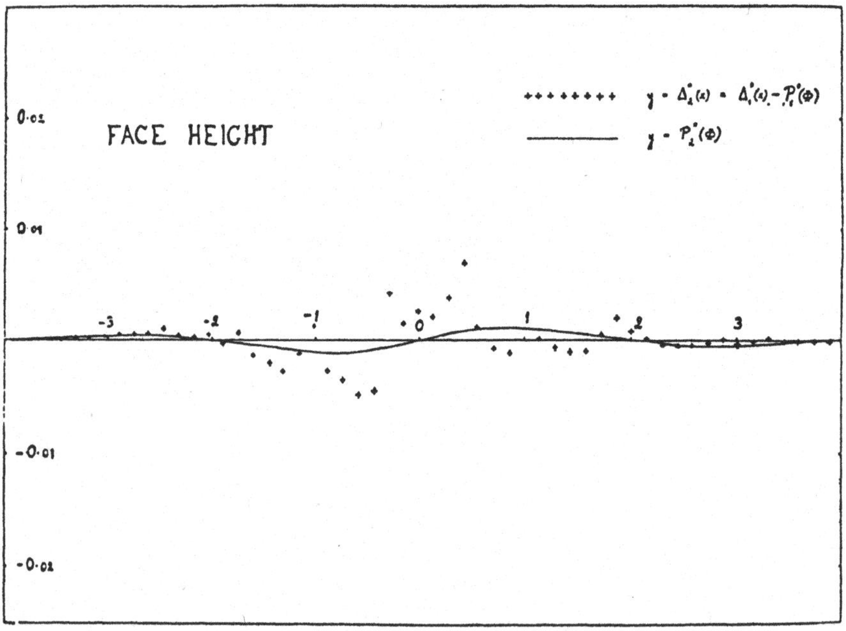

Fig. 4. Face Height. Δ_2^* and P_2^*.

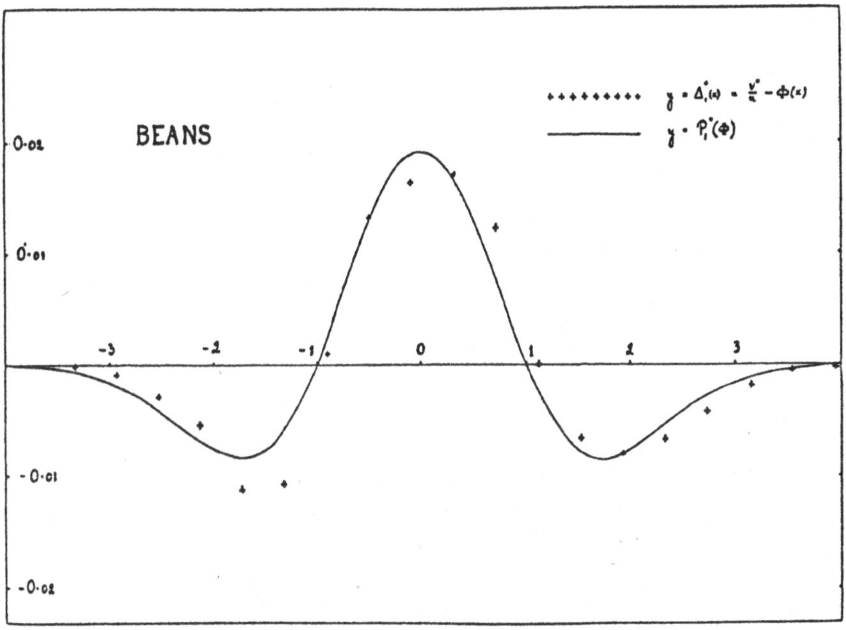

Fig. 5. Beans. Δ_1^* and P_1^*.

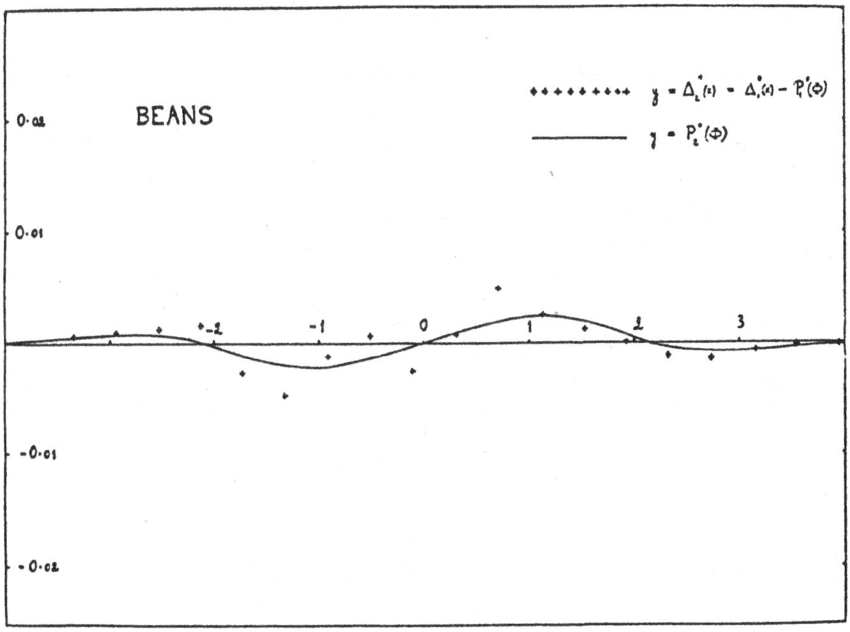

Fig. 6. Beans. Δ_2^* and P_2^*.

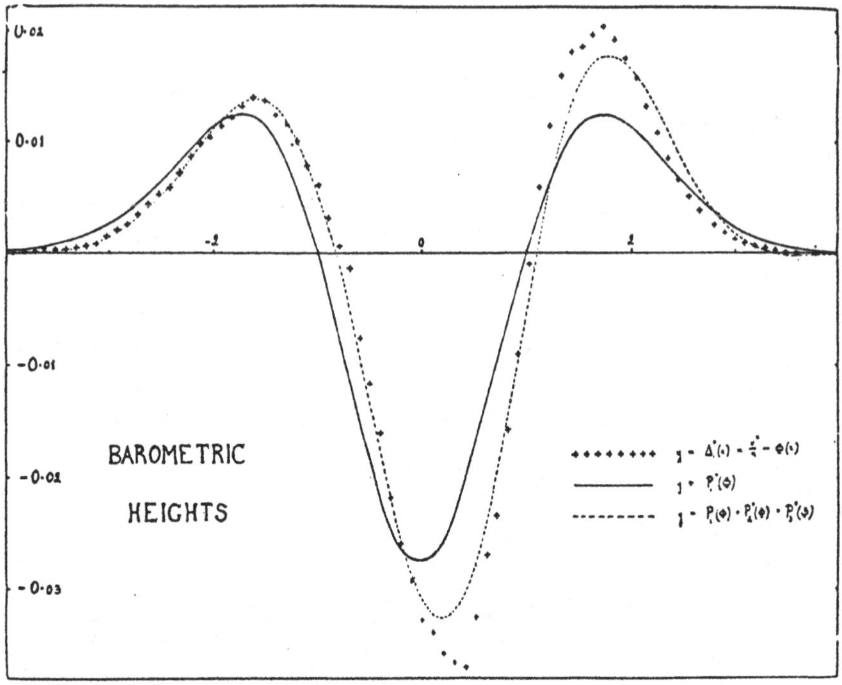

Fig. 7. Barometric Heights. Δ_1^* and P_1^*.

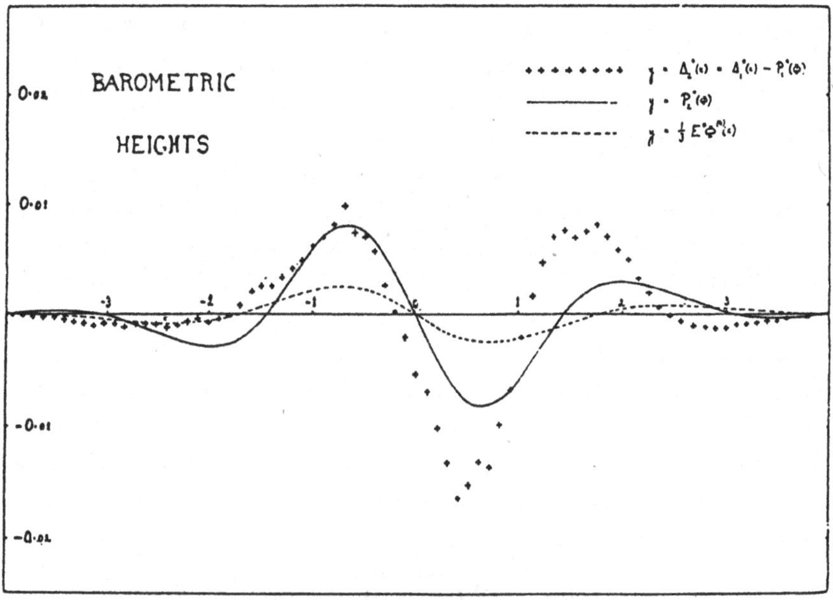

Fig. 8. Barometric Heights. Δ_2^* and P_2^*.

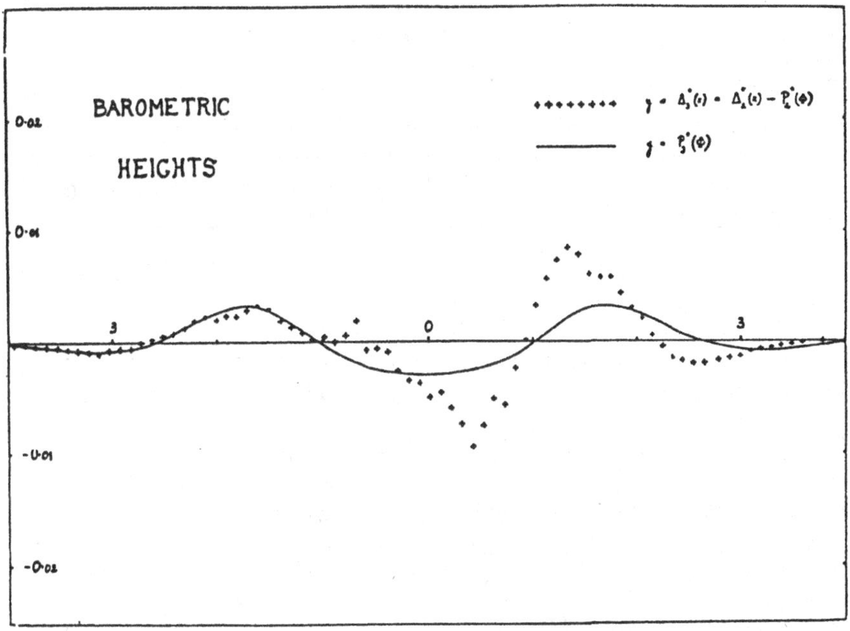

Fig. 9. Barometric Heights. Δ_3^* and P_3^*.

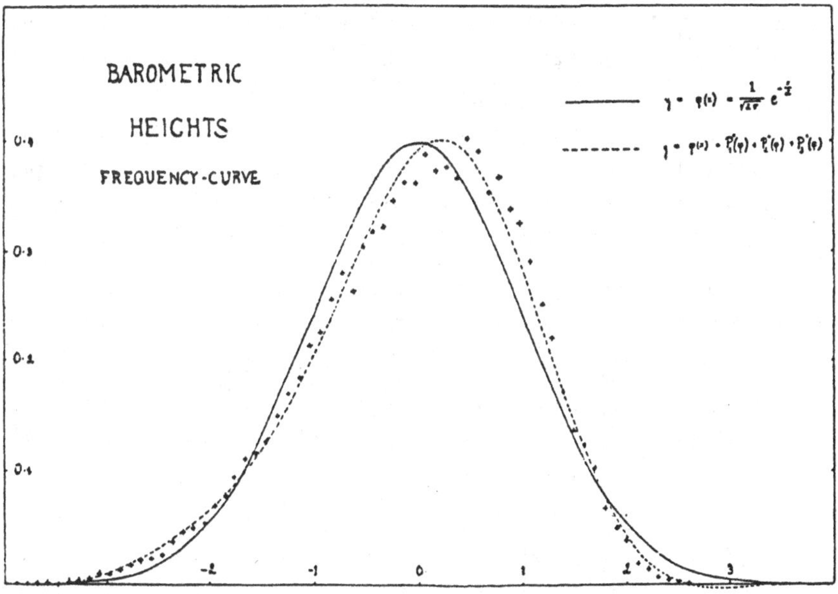

Fig. 10. Barometric Heights. Frequency-curve.

A. Stature.

x	$m^* = 172.228$ cm. $\sigma^* = 5.931$ cm.									
	ν^*	$10^4 \frac{\nu^*}{n}$	$10^4 \Phi$	$10^4 \Delta_1^*$	$10^4 P_1^*$	$10^4 \Delta_2^*$	$10^4 P_2^*$	$10^4 \Delta_3^*$	$10^4 P_3^*$	$10^4 \Delta_4^*$
−3.419	8	2	3	− 1	− 2	+ 1	0	+ 1	0	+ 1
3.250	15	3	6	− 3	− 4	+ 1	0	+ 1	−1	+ 2
3.082	21	4	10	− 6	− 6	0	0	0	−1	+ 1
2.913	34	7	18	−11	− 9	− 2	0	− 2	−1	− 1
2.744	77	16	30	−14	−12	− 2	−1	− 1	−1	0
2.576	146	31	50	−19	−16	− 3	−1	− 2	−1	− 1
2.407	256	54	80	−26	−21	− 5	−2	− 3	0	− 3
2.239	435	93	126	−33	−26	− 7	−2	− 5	0	− 5
2.070	727	155	192	−37	−31	− 6	−2	− 4	+1	− 5
1.901	1 177	251	287	−36	−34	− 2	−2	0	+2	− 2
1.733	1 784	380	415	−35	−35	0	−2	+ 2	+3	− 1
1.564	2 635	561	589	−28	−34	+ 6	−1	+ 7	+4	+ 3
1.396	3 747	798	814	−16	−28	+12	+1	+11	+5	+ 6
1.227	5 106	1 087	1 099	−12	−19	+ 7	+2	+ 5	+5	0
1.058	6 778	1 443	1 450	− 7	− 5	− 2	+4	− 6	+4	−10
0.890	8 924	1 899	1 867	+32	+11	+21	+6	+15	+2	+13
0.721	11 261	2 397	2 355	+42	+29	+13	+7	+ 6	0	+ 6
0.553	13 941	2 967	2 901	+66	+47	+19	+6	+13	−3	+16
0.384	16 806	3 577	3 505	+72	+63	+ 9	+5	+ 4	−5	+ 9
0.215	19 837	4 222	4 149	+73	+74	− 1	+3	− 4	−7	+ 3
−0.047	22 952	4 885	4 813	+72	+79	− 7	+1	− 8	−8	0
+0.122	26 036	5 542	5 486	+56	+78	−22	−2	−20	−8	−12
0.290	29 129	6 200	6 141	+59	+70	−11	−4	− 7	−6	− 1
0.459	31 917	6 794	6 769	+25	+56	−31	−6	−25	−4	−21
0.628	34 649	7 375	7 350	+25	+40	−15	−7	− 8	−2	− 6
0.796	36 981	7 871	7 870	+ 1	+21	−20	−6	−14	+1	−15
0.965	39 142	8 331	8 327	+ 4	+ 3	+ 1	−5	+ 6	+3	+ 3
1.133	40 868	8 699	8 714	−15	−12	− 3	−3	0	+4	− 4
1.302	42 400	9 025	9 035	−10	−24	+14	−1	+15	+5	+10
1.471	43 553	9 270	9 294	−24	−31	+ 7	0	+ 7	+5	+ 2
+1.639	44 479	9 467	9 494	−27	−35	+ 8	+1	+ 7	+4	+ 3

x	$m^* = 172.228$ cm. $\sigma^* = 5.931$ cm.									
	ν^*	$10^4 \frac{\nu^*}{n}$	$10^4 \Phi$	$10^4 \Delta_1^*$	$10^4 P_1^*$	$10^4 \Delta_2^*$	$10^4 P_2^*$	$10^4 \Delta_3^*$	$10^4 P_3^*$	$10^4 \Delta_4^*$
+1.808	45 198	9 620	9 647	−27	−35	+8	+2	+6	+3	+3
1.976	45 733	9 734	9 759	−25	−33	+8	+2	+6	+2	+4
2.145	46 118	9 816	9 840	−24	−29	+5	+2	+3	+1	+2
2.314	46 395	9 875	9 897	−22	−24	+2	+2	0	0	0
2.482	46 594	9 918	9 935	−17	−19	+2	+1	+1	0	+1
2.651	46 737	9 948	9 960	−12	−14	+2	+1	+1	−1	+2
2.819	46 817	9 965	9 976	−11	−10	−1	0	−1	−1	0
2.988	46 873	9 977	9 986	− 9	− 7	−2	0	−2	−1	−1
3.157	46 906	9 984	9 992	− 8	− 5	−3	0	−3	−1	−2
3.325	46 934	9 990	9 996	− 6	− 3	−3	0	−3	−1	−2
3.494	46 951	9 994	9 998	− 4	− 2	−2	0	−2	0	−2
3.662	46 964	9 996	9 999	− 3	− 1	−2	0	−2	0	−2
3.831	46 973	9 998	9 999	− 1	− 1	0	0	0	0	0
4.000	46 977	9 999	10 000	− 1	0	−1	0	−1	0	−1
4.168	46 980	10 000	10 000	0	0	0	0	0	0	0
+4.337	46 981	10 000	10 000	0	0	0	0	0	0	0

B. Leg Length.

x	$m^* = 92.020$ cm. $\sigma^* = 4.801$ cm.									
	ν^*	$10^4 \frac{\nu^*}{n}$	$10^4 \Phi$	$10^4 \Delta_1^*$	$10^4 P_1^*$	$10^4 \Delta_2^*$	$10^4 P_2^*$	$10^4 \Delta_3^*$	$10^4 P_3^*$	$10^4 \Delta_4^*$
−3.969	3	1	0	+ 1	0	+1	0	+1	0	+1
3.736	3	1	1	0	− 1	+1	+1	0	0	0
3.504	3	1	2	− 1	− 2	+1	+1	0	−1	+1
3.271	10	2	5	− 3	− 4	+1	+1	0	−1	+1
3.039	20	4	12	− 8	− 7	−1	+1	−2	−1	−1
2.806	65	14	25	−11	−12	+1	+2	−1	−1	0
2.574	162	34	50	−16	−19	+3	+2	+1	0	+1
2.341	337	72	96	−24	−26	+2	+1	+1	+1	0
−2.109	661	141	175	−34	−34	0	0	0	+2	−2

x	$m^* = 92.020$ cm.			$\sigma^* = 4.301$ cm.						
	ν^*	$10^4 \frac{\nu^*}{n}$	$10^4 \Phi$	$10^4 \Delta_1^*$	$10^4 P_1^*$	$10^4 \Delta_2^*$	$10^4 P_2^*$	$10^4 \Delta_3^*$	$10^4 P_3^*$	$10^4 \Delta_4^*$
−1.876	1 205	256	303	− 47	−39	− 8	−2	− 6	+3	− 9
1.644	2 147	457	501	− 44	−40	− 4	−3	− 1	+3	− 4
1.411	3 521	749	791	− 42	−33	− 9	−4	− 5	+3	− 8
1.179	5 503	1 171	1 192	− 21	−18	− 3	−4	+ 1	+2	− 1
0.946	8 144	1 733	1 721	+ 12	+ 6	+ 6	−4	+10	+1	+ 9
0.714	11 323	2 410	2 376	+ 34	+34	0	−3	+ 3	−1	+ 4
0.481	15 149	3 224	3 153	+ 71	+62	+ 9	−2	+11	−3	+14
0.249	19 321	4 112	4 017	+ 95	+82	+13	−1	+14	−5	+19
−0.016	23 685	5 041	4 936	+105	+91	+14	0	+14	−5	+19
+0.216	27 833	5 924	5 855	+ 69	+84	−15	+1	−16	−5	−11
0.449	31 925	6 795	6 733	+ 62	+65	− 3	+2	− 5	−4	− 1
0.681	35 444	7 544	7 521	+ 23	+39	−16	+3	−19	−1	−18
0.914	38 424	8 178	8 196	− 18	+10	−28	+4	−32	+1	−33
1.146	41 014	8 730	8 741	− 11	−15	+ 4	+4	0	+2	− 2
1.379	42 909	9 133	9 161	− 28	−32	+ 4	+4	0	+3	− 3
1.611	44 320	9 433	9 464	− 31	−39	+ 8	+3	+ 5	+3	+ 2
1.844	45 327	9 648	9 674	− 26	−40	+14	+2	+12	+3	+ 9
2.076	45 999	9 791	9 811	− 20	−35	+15	+1	+14	+2	+12
2.309	46 395	9 875	9 895	− 20	−27	+ 7	−1	+ 8	+1	+ 7
2.541	46 631	9 925	9 945	− 20	−20	0	−1	+ 1	0	+ 1
2.774	46 774	9 956	9 971	− 15	−13	− 2	−2	0	0	0
3.006	46 859	9 974	9 987	− 13	− 8	− 5	−2	− 3	−1	− 2
3.239	46 916	9 986	9 994	− 8	− 5	− 3	−1	− 2	−1	− 1
3.471	46 955	9 994	9 997	− 3	− 2	− 1	−1	0	−1	+ 1
3.704	46 965	9 996	9 999	− 3	− 1	− 2	−1	− 1	0	− 1
3.936	46 972	9 998	10 000	− 2	− 1	− 1	0	− 1	0	− 1
4.169	46 978	9 999	10 000	− 1	0	− 1	0	− 1	0	− 1
4.401	46 979	9 999	10 000	− 1	0	− 1	0	− 1	0	− 1
4.634	46 981	10 000	10 000	0	0	0	0	0	0	0
+4.866	46 982	10 000	10 000	0	0	0	0	0	0	0

C. Face Height.

x	$m^* = 126.574$ mm.		$\sigma^* = 6.937$ mm.							
	ν^*	$10^4 \frac{\nu^*}{n}$	$10^4 \Phi$	$10^4 \Delta_1^*$	$10^4 P_1^*$	$10^4 \Delta_2^*$	$10^4 P_2^*$	$10^4 \Delta_3^*$	$10^4 P_3^*$	$10^4 \Delta_4^*$
−3.470	2	0	3	− 3	− 3	0	+ 2	− 2	0	− 2
3.826	10	2	4	− 2	− 4	+ 2	+ 2	0	0	0
3.182	16	3	7	− 4	− 6	+ 2	+ 2	0	0	0
3.088	28	6	12	− 6	− 9	+ 3	+ 3	0	0	0
2.894	56	12	19	− 7	− 12	+ 5	+ 3	+ 2	+1	+ 1
2.750	87	19	30	− 11	− 16	+ 5	+ 4	+ 1	+1	0
2.606	146	31	46	− 15	− 21	+ 6	+ 4	+ 2	+1	+ 1
2.461	243	52	69	− 17	− 27	+10	+ 3	+ 7	+1	+ 6
2.817	351	75	104	− 29	− 33	+ 4	+ 3	+ 1	+1	0
2.173	532	113	149	− 36	− 39	+ 3	+ 2	+ 1	+1	0
2.029	812	173	212	− 39	− 44	+ 5	0	+ 5	+1	+ 4
1.885	1 157	246	297	− 51	− 48	− 3	− 1	− 2	0	− 2
1.741	1 720	366	408	− 42	− 49	+ 7	− 3	+10	−1	+11
1.596	2 305	491	552	− 61	− 48	−13	− 5	− 8	−2	− 6
1.452	3 146	670	733	− 63	− 43	−20	− 7	−13	−4	− 9
1.308	4 196	893	954	− 61	− 33	−28	− 9	−19	−4	−15
1.164	5 591	1 190	1 222	− 32	− 20	−12	−10	− 2	−4	+ 2
1.020	7 280	1 550	1 539	+ 11	− 3	+14	−11	+25	−4	+29
0.876	8 897	1 894	1 905	− 11	+ 17	−28	−11	−17	−3	−14
0.731	10 938	2 328	2 324	+ 4	+ 39	−35	−11	−24	−1	−23
0.587	13 147	2 798	2 786	+ 12	+ 61	−49	− 9	−40	0	−40
0.443	15 618	3 324	3 289	+ 35	+ 80	−45	− 8	−37	+3	−40
0.299	18 629	3 965	3 825	+140	+ 96	+44	− 5	+49	+5	+44
0.155	21 172	4 506	4 384	+122	+106	+16	− 3	+19	+6	+13
−0.011	23 930	5 093	4 956	+137	+110	+27	0	+27	+7	+20
+0.134	26 599	5 662	5 533	+129	+107	+22	+ 3	+19	+6	+13
0.278	29 283	6 233	6 095	+138	+ 98	+40	+ 5	+35	+5	+30
0.422	31 893	6 788	6 635	+153	+ 83	+70	+ 7	+63	+3	+60
0.566	33 915	7 219	7 143	+ 76	+ 64	+12	+ 9	+ 3	+1	+ 2
0.710	35 921	7 646	7 611	+ 35	+ 42	− 7	+10	−17	−1	−16
+0.854	37 792	8 044	8 034	+ 10	+ 21	−11	+11	−22	−3	−19

x	$m^* = 126.574$ mm. $\sigma^* = 6.937$ mm.									
	ν^*	$10^4 \frac{\nu^*}{n}$	$10^4 \Phi$	$10^4 \Delta_1^*$	$10^4 P_1^*$	$10^4 \Delta_2^*$	$10^4 P_2^*$	$10^4 \Delta_3^*$	$10^4 P_3^*$	$10^4 \Delta_4^*$
+0.999	39 328	8 371	8 411	−40	0	−40	+11	−51	−4	−47
1.143	40 961	8 718	8 735	−17	−18	+ 1	+11	−10	−4	− 6
1.287	42 151	8 972	9 010	−38	−32	− 6	+ 9	−15	−4	−11
1.431	43 161	9 187	9 238	−51	−41	−10	+ 8	−18	−3	−15
1.575	44 015	9 368	9 424	−56	−47	− 9	+ 6	−15	−3	−12
1.719	44 770	9 529	9 572	−43	−49	+ 6	+ 4	+ 2	−1	+ 3
1.863	45 387	9 661	9 688	−27	−48	+21	+ 2	+19	0	+19
2.008	45 765	9 741	9 777	−36	−45	+ 9	0	+ 9	+1	+ 8
2.152	46 065	9 805	9 843	−38	−40	+ 2	− 2	+ 4	+1	+ 3
2.296	46 303	9 855	9 892	−37	−34	− 3	− 3	0	+1	− 1
2.440	46 494	9 896	9 927	−31	−28	− 3	− 3	0	+1	− 1
2.584	46 631	9 925	9 951	−26	−22	− 4	− 4	0	+1	− 1
2.728	46 745	9 950	9 968	−18	−17	− 1	− 4	+ 3	+1	+ 2
2.873	46 830	9 968	9 980	−12	−13	+ 1	− 3	+ 4	+1	+ 3
3.017	46 870	9 976	9 987	−11	− 9	− 2	− 3	+ 1	0	+ 1
3.161	46 907	9 984	9 992	− 8	− 7	− 1	− 2	+ 1	0	+ 1
3.305	46 942	9 991	9 995	− 4	− 5	+ 1	− 2	+ 3	0	+ 3
3.449	46 948	9 993	9 997	− 4	− 3	− 1	− 2	+ 1	0	+ 1
3.593	46 957	9 995	9 998	− 3	− 2	− 1	− 1	0	0	0
3.738	46 964	9 996	9 999	− 3	− 1	− 2	− 1	− 1	0	− 1
3.882	46 969	9 997	9 999	− 2	− 1	− 1	− 1	0	0	0
4.026	46 971	9 998	10 000	− 2	− 1	− 1	0	− 1	0	− 1
4.170	46 974	9 998	10 000	− 2	0	− 2	0	− 2	0	− 2
4.314	46 977	9 999	10 000	− 1	0	− 1	0	− 1	0	− 1
4.458	46 979	9 999	10 000	− 1	0	− 1	0	− 1	0	− 1
4.603	46 981	10 000	10 000	0	0	0	0	0	0	0
+4.747	46 982	10 000	10 000	0	0	0	0	0	0	0

D. Beans.

x	$m^* = 8.5117$ mm. $\sigma^* = 0.6163$ mm.									
	ν^*	$10^4 \frac{\nu^*}{n}$	$10^4 \Phi$	$10^4 \Delta_1^*$	$10^4 P_1^*$	$10^4 \Delta_2^*$	$10^4 P_2^*$	$10^4 \Delta_3^*$	$10^4 P_3^*$	$10^4 \Delta_4^*$
−3.345	3	3	4	− 1	− 7	+ 6	+ 5	+ 1	+ 1	0
2.940	8	7	16	− 9	− 19	+10	+ 8	+ 2	+ 4	− 2
2.534	32	27	56	− 29	− 42	+13	+ 7	+ 6	+ 6	0
2.128	135	113	167	− 54	− 70	÷16	+ 1	+15	+ 3	+12
1.723	374	312	424	−112	− 85	−27	−11	−16	−11	− 5
1.317	998	832	939	−107	− 59	−48	−21	−27	−24	− 3
0.911	2 185	1 821	1 811	+ 10	+ 22	−12	−21	+ 9	−17	+26
0.506	3 835	3 196	3 064	+132	+125	+ 7	−13	+20	+12	+ 8
−0.100	5 718	4 765	4 602	+163	+189	−26	− 3	−23	+36	−59
+0.306	7 648	6 373	6 202	+171	+164	+ 7	+ 8	− 1	+27	−28
0.711	9 286	7 738	7 615	+123	+ 74	+49	+18	+31	− 4	+35
1.117	10 416	8 680	8 680	0	− 25	+25	+22	+ 3	−24	+27
1.523	11 153	9 294	9 361	− 67	− 79	+12	+17	− 5	−18	+13
1.928	11 580	9 650	9 731	− 81	− 81	0	+ 5	− 5	− 3	− 2
2.334	11 801	9 834	9 902	− 68	− 56	−12	− 5	− 7	+ 5	−12
2.740	11 911	9 926	9 969	− 43	− 29	−14	− 8	− 6	+ 5	−11
3.145	11 968	9 973	9 992	− 19	− 12	− 7	− 7	0	+ 3	− 7
3.551	11 992	9 993	9 998	− 5	− 4	− 1	− 4	+ 3	+ 1	− 1
3.956	11 998	9 998	10 000	− 2	− 1	− 1	− 1	0	0	− 1
+4.362	12 000	10 000	10 000	0	0	0	0	0	0	0

E. Barometric Heights.

x	$m^* = 760.318$ mm. $\sigma^* = 9.854$ mm.									
	ν^*	$10^4 \frac{\nu^*}{n}$	$10^4 \Phi$	$10^4 \Delta_1^*$	$10^4 P_1^*$	$10^4 \Delta_2^*$	$10^4 P_2^*$	$10^4 \Delta_3^*$	$10^4 P_3^*$	$10^4 \Delta_4^*$
−4.041	1	0	0	0	+1	−1	+1	−2	−2	0
3.939	3	1	0	+1	+2	−1	+1	−2	−2	0
3.838	5	2	1	+1	+2	−1	+1	−2	−3	+1
3.736	5	2	1	+1	+3	−2	+2	−4	−4	0
3.635	9	4	1	+3	+5	−2	+2	−4	−5	+1
3.533	12	5	2	+3	+6	−3	+2	−5	−6	+1
3.432	13	6	3	+3	+8	−5	+2	−7	−7	0
3.330	18	8	4	+4	+11	−7	+1	−8	−8	0
3.229	27	12	6	+6	+14	−8	+1	−9	−9	0
3.127	38	17	9	+8	+18	−10	0	−10	−9	−1
3.026	61	27	12	+15	+23	−8	−1	−7	−8	+1
2.925	81	37	17	+20	+29	−9	−3	−6	−7	+1
2.823	109	49	24	+25	+36	−11	−5	−6	−6	0
2.722	148	67	32	+35	+43	−8	−8	0	−4	+4
2.620	195	88	44	+44	+52	−8	−11	+3	−2	+5
2.519	247	111	59	+52	+61	−9	−15	+6	+2	+4
2.417	305	137	78	+59	+71	−12	−19	+7	+7	0
2.316	388	175	103	+72	+82	−10	−22	+12	+12	0
2.214	491	221	134	+87	+93	−6	−25	+19	+17	+2
2.113	601	271	173	+98	+102	−4	−27	+23	+22	+1
2.011	721	325	222	+103	+111	−8	−28	+20	+26	−6
1.910	876	395	281	+114	+118	−4	−28	+24	+28	−4
1.808	1 051	474	353	+121	+122	−1	−25	+24	+31	−7
1.707	1 265	570	439	+131	+123	+8	−21	+29	+31	−2
1.605	1 514	682	542	+140	+120	+20	−14	+34	+31	+3
1.504	1 776	800	663	+137	+112	+25	−5	+30	+28	+2
1.402	2 061	929	805	+124	+99	+25	+6	+19	+23	−4
1.301	2 398	1 081	966	+115	+82	+33	+19	+14	+17	−3
1.199	2 780	1 253	1 153	+100	+59	+41	+32	+9	+11	−2
1.098	3 195	1 440	1 361	+79	+31	+48	+46	+2	+4	−2
−0.996	3 676	1 657	1 596	+61	−1	+62	+58	+4	−2	+6

x	$m^* = 760.318$ mm. $\sigma^* = 9.854$ mm.									
	ν^*	$10^4 \frac{\nu^*}{n}$	$10^4 \Phi$	$10^4 \Delta_1^*$	$10^4 P_1^*$	$10^4 \Delta_2^*$	$10^4 P_2^*$	$10^4 \Delta_3^*$	$10^4 P_3^*$	$10^4 \Delta_4^*$
−0.895	4 184	1 886	1 854	+ 32	− 37	+ 69	+69	0	− 8	+ 8
0.793	4 762	2 146	2 139	+ 7	− 75	+ 82	+76	+ 6	−14	+20
0.692	5 394	2 431	2 445	− 14	−113	+ 99	+80	+19	−19	+38
0.590	5 988	2 699	2 776	− 77	−151	+ 74	+80	− 6	−22	+16
0.489	6 671	3 007	3 124	−117	−187	+ 70	+75	− 5	−25	+20
0.388	7 386	3 329	3 490	−161	−218	+ 57	+66	− 9	−27	+18
0.286	8 111	3 656	3 874	−218	−244	+ 26	+52	−26	−28	+ 2
0.185	8 889	4 006	4 266	−260	−262	+ 2	+36	−34	−28	− 6
−0.083	9 707	4 375	4 669	−294	−274	− 20	+16	−36	−28	− 8
+0.018	10 523	4 743	5 072	−329	−276	− 53	− 4	−49	−28	−21
0.120	11 400	5 138	5 478	−340	−271	− 69	−24	−45	−28	−17
0.221	12 242	5 517	5 875	−358	−257	−101	−42	−59	−28	−31
0.323	13 090	5 900	6 267	−367	−235	−132	−58	−74	−27	−47
0.424	13 915	6 271	6 642	−371	−207	−164	−70	−94	−27	−67
0.526	14 822	6 680	7 006	−326	−174	−152	−77	−75	−24	−51
0.627	15 702	7 077	7 347	−270	−138	−132	−81	−51	−21	−30
0.729	16 497	7 435	7 670	−235	− 99	−136	−79	−57	−18	−39
0.830	17 325	7 808	7 967	−159	− 61	− 98	−74	−24	−12	−12
0.932	18 087	8 152	8 243	− 91	− 24	− 67	−68	+ 1	− 6	+ 7
1.033	18 819	8 482	8 492	− 10	+ 11	− 21	−54	+33	0	+33
1.135	19 473	8 776	8 718	+ 58	+ 42	+ 16	−41	+57	+ 6	+51
1.236	20 040	9 032	8 918	+114	+ 68	+ 46	−27	+73	+13	+60
1.338	20 536	9 255	9 096	+159	+ 89	+ 70	−14	+84	+19	+65
1.439	20 923	9 430	9 249	+181	+105	+ 76	− 2	+78	+25	+53
1.541	21 230	9 568	9 383	+185	+116	+ 69	+ 8	+61	+29	+32
1.642	21 509	9 694	9 497	+197	+122	+ 75	+17	+58	+31	+27
1.744	21 739	9 798	9 594	+204	+123	+ 81	+23	+58	+31	+27
1.845	21 890	9 866	9 675	+191	+121	+ 70	+27	+43	+30	+13
1.947	22 001	9 916	9 742	+174	+116	+ 58	+28	+30	+27	+ 3
2.048	22 086	9 954	9 797	+157	+108	+ 49	+28	+21	+24	− 3
2.149	22 127	9 973	9 842	+131	+ 99	+ 32	+27	+ 5	+20	−15
2.251	22 158	9 986	9 878	+108	+ 89	+ 19	+24	− 5	+15	−20
+2.352	22 171	9 992	9 907	+ 85	+ 79	+ 6	+21	−15	+10	−25

x	$m^* = 760.318$ mm. $\sigma^* = 9.854$ mm.									
	ν^*	$10^4 \frac{\nu^*}{n}$	$10^4 \Phi$	$10^4 \Delta_1^*$	$10^4 P_1^*$	$10^4 \Delta_2^*$	$10^4 P_2^*$	$10^4 \Delta_3^*$	$10^4 P_3^*$	$10^4 \Delta^*$
+2.454	22 180	9 996	9 929	+67	+68	− 1	+17	−18	+5	−23
2.555	22 186	9 999	9 947	+52	+58	− 6	+14	−20	+1	−21
2.657	22 188	10 000	9 961	+39	+49	−10	+10	−20	−3	−17
2.800	22 188	10 000	9 974	+26	+37	−11	+ 6	−17	−6	−11
3.000	22 188	10 000	9 987	+13	+25	−12	+ 2	−14	−8	− 6
3.200	22 188	10 000	9 993	+ 7	+15	− 8	− 1	− 7	−9	+ 2
3.400	22 188	10 000	9 997	+ 3	+ 9	− 6	− 2	− 4	−7	+ 3
3.600	22 188	10 000	9 998	+ 2	+ 5	− 3	− 2	− 1	−5	+ 4
3.800	22 188	10 000	9 999	+ 1	+ 3	− 2	− 2	0	−3	+ 3
+4.000	22 188	10 000	10 000	0	+ 1	− 1	− 1	0	−2	+ 2

26.

On the mathematical theory of risk

Försäkringsaktiebolaget Skandia 1855–1930, Parts I and II, Stockholm 1930, 7–84

PART I. Object and Methods of the Theory.

Introduction.

1. The regular course of business of an Insurance Company may be disturbed by many different causes. There may be losses due to general economic fluctuations, unsuccessful investments, heavy expenses, and a great number of other causes which constitute a *commercial risk*, more or less common to all business enterprises.

But there are also other causes peculiar to insurance business, viz. those connected with the fluctuations of risk, measured by the deviations between the actual yearly amounts of claims and the average or expected ones. Among these fluctuations, we may distinguish two classes, according as they appear to us as the result of a definite external cause or not. The heavy mortality experienced by a Life Office during a war or an epidemic, as well as the increased amount of claims in Fire or Automobile Insurance during a period of general economic depression, obviously belongs to the former class. The fluctuations belonging to the latter class, i. e. those which cannot (at the actual state of our knowledge) be traced back to any definite cause, are called *random fluctuations*. A financial loss arising through random fluctuations may be due to an abnormally large number of claims, or to an unfavourable distribution of the amounts of claims, or to both incidents combined.

The object of the Theory of Risk is to give a mathematical analysis of the random fluctuations in an insurance business and to discuss the various means of protection against their inconvenient effects. The random fluctuations give rise to the element of *mathematical risk*, as opposed to commercial risk, and it is the task of the theory to find technical measures which reduce the mathematical risk to a minimum without reducing the bonus-earning power of the company more than absolutely necessary.

It is well known that the practical value of any such theory has been emphatically denied by certain authors. As a matter of fact, we find that practical insurance business has hitherto made little or no application of the results offered by the mathematical theory of risk. This may be partly due to the fact that the theory, in its present state, is really incapable of giving a satisfactory solution of most of the problems encountered in practice. If it were possible to supply an adequate theoretical treatment of these problems, it is not unlikely that the opinion of the practical value of the theory would change, and that some of its results would find application in practice.

In the present paper an attempt will be made to show that there are, in modern insurance practice, a number of important problems to be solved by a mathematical theory of risk, and to discuss the various methods that have been proposed for the solution of the problems.[1] We shall be chiefly concerned with the development of the mathematical theory, but we shall begin with a brief survey of the practical problems which the theory has to solve.

Practical Object of the Theory.

2. One of the principal arguments of those who deny the value of the theory of risk is that there is no need for special provisions intended to cover the risk arising from random fluctuations, since there are so many other dangers that may imperil the regular course of the business. Against these other dangers, we have no other means of protection than those imposed by practical common sense and general business principles, and thus, it is argued, we have no use for special precautions against one particular cause of disturbance.

If we look at the circumstances in practice we shall find, however, that *some* special precautions are generally taken. The premiums are calculated with a loading to provide a margin for fluctuations, insurance amounts exceeding a certain maximum are not retained on the Company's own risk but are reinsured, and in some cases even a special risk reserve is formed, with the object of smoothing the curve of yearly profits. — All these are special precautions against fluctuations of risk, and nobody will deny that, for a Company which takes no precautions whatever of this kind, the effect of purely random fluctuations may quite possibly prove disastrous. The *effectivity* of all these practical measures is, however, very incompletely known, from a theoretical as well as from an empirical point of view, and it is precisely the investigation of this question that constitutes one of the most important problems of the theory of risk. The value of any practical arrangement acting as a protection against fluctuations must be judged with regard to its effectivity and to the expenses or other inconveniences which it causes the Company. If we have at our disposal a theory that enables us to estimate the value of the different arrangements that might be proposed, it seems obvious that the results of such a theory ought not to be ignored. No doubt the element of commercial risk will require precautions of its own, but it must not be a priori expected that these will at the same time act as a satisfactory protection against random fluctuations.

[1] It must, of course, be understood that a considerable part of the paper consists in a recapitulation of the work of previous authors. Explicit references will, however, only exceptionally be given. The reader may be referred to the following works, where extensive references will be found to the literature on the subject: G. BOHLMANN, Die Theorie des mittleren Risikos, VI Intern. Kongr. f. Vers.-Wiss., Wien 1909, Bd 1; A. BERGER, Lebensversicherungstechnik, Bd 2, Berlin 1925. Cf also the forthcoming Transactions of the Ninth Intern. Congress of Actuaries, Stockholm 1930, Subject D.

3. Another argument intended to prove the needlessness of the theory of risk is the following one, which applies only to participating policies and in particular to Life Assurance.

The premiums charged in practice are always calculated so as to keep »on the safe side». If the extra loading is sufficiently high, it is to be expected that the fluctuations of risk will only cause a corresponding fluctuation in the bonus. Thus, as ALTENBURGER[1] puts it: »High initial premiums and rational participation of the insured in profits render theoretical examination of the mathematical risk unnecessary».

This statement does not seem quite clear. Does it imply that high premiums render reinsurance completely unnecessary? And what is to be understood by a »rational participation in profits»? How are we, in particular, to allow the risk fluctuations to be reflected in the bonus distribution? — When we try to analyze these questions, we are naturally led back to the fundamental points of view of the theory of risk.

The chief objection against ALTENBURGER's argument is, however, that the recent development, at least so far as Life Assurance Companies are concerned, seems to point in a direction essentially different from the one outlined in the quotation given above. Owing principally to the competition between the Companies and to reasons connected with taxation the premium level has in many cases been substantially lowered, which in the long run must cause a corresponding decrease of divisible surplus. At the same time, the bonus question has acquired a great importance for the writing of new business, and in many cases the Companies have issued Bonus Prospects which give, for every policy year, minimum values of the bonus that the Company expects to be able to pay to the policy-holder. Such prospects are largely used in underwriting business, and thus the habitual system of bonus rates adopted by the company is gradually becoming known to the public. It would cause considerable trouble to a Company if, for a certain year, the available surplus was not sufficient to allow the payment of the amounts stated in the Bonus Prospect used by the Company.

Under these circumstances, it must be considered as highly important to provide by all possible means for a smooth and regular course of business, and in particular to avoid considerable losses due to random fluctuations. If the regular bonus distribution of a Company is disturbed e. g. by a financial crisis, a war, or an epidemic, it is very likely that other Companies in the same country will be in a similar position, so that the event does not affect the relative position of the Companies. Variations in the annual surplus due to accidental fluctuations of risk, on the other hand, do not at all occur in the same uniform way, and a loss from such a cause will therefore be particularly unwelcome, since it will put the Company in a worse position when compared with other Companies. This state of things obviously tends to give an increased importance to the matter of random fluctuations and to the question of finding appropriate methods for neutralizing their effects.

---

[1] Verhandlungen d. VI. intern. Kongr. f. Vers.-Wiss., Wien 1909, Bd I, p. 962. Cf also BERGER's comments on ALTENBURGER's paper, l. c., p. 3.

4. We have seen that the fundamental question which practice puts before the theory of risk may be thus expressed: Which method ought a company to follow in order to eliminate as far as possible the inconvenient effects of random fluctuations?

No theory can, however, give a definite answer to a vague and unprecise question. Before we can set the theory to work on a practical problem, it is necessary to have a perfectly clear opinion as to the form in which the problem must be stated if it is to have any practical significance. What, exactly, do we mean by an »inconvenient effect«, and what is to be understood by the words »eliminate as far as possible«? In every particular case, the significance of these expressions must be clearly defined; otherwise we cannot expect any useful result from the theoretical treatment of the problems.

In investigations concerning the theory of risk, we generally find that the problem has been interpreted in the sense that we have to find measures which reduce as far as possible *the probability of the occurrence of certain losses due to random fluctuations.* Opinions divide, however, as to the question *which* losses should be kept in view in this connection. In order to make this clear we must ask ourselves at which point a loss due to random fluctuations begins to cause a real inconvenience to the Company.

It is obvious that we cannot hope to find a universal and definite answer to this question, and that every actual case must be considered on its own merits. But in general we may say that it is desirable for a Company to avoid every disturbance of the regular and habitual course of its business due to fluctuations such as those here considered. The Company may, e. g., be accustomed to pay every year certain dividends to its shareholders and certain amounts of bonus to policyholders with participating policies. If, for a certain year, the annual surplus is not sufficient to supply the means required for a distribution according to the ordinary rates of dividends and bonus adopted by the Company, this would signify a real inconvenience, and would impair the position of the Company in the competition, even if the deficit could be immediately covered by employing the available funds of the Company.

Accordingly we shall, in the sequel, take it as the object of the theory of risk to discuss the various means which a Company may, under given circumstances, employ in order to secure a smooth and stable course of profits, with the smallest possible probability of having its regular distribution of bonus (or dividends) disturbed by random fluctuations.

Mathematical Principles of the Theory.

5. From a mathematical point of view, the theory of risk is an application of Mathematical Probability. The fundamental problems of the theory of risk are intimately connected with the numerical evaluation of certain probabilities. Accordingly the analysis of our problems must be founded upon some hypothesis as to the probabilities which are at work in the groups of insurances to be considered.

The applicability of Mathematical Probability to these questions has been much discussed. Of course we must never forget that the assignation of a numerical probability to a certain event is nothing but a hypothesis, the consequences of which must be submitted to statistical verification. In several cases connected with insurance (especially as regards human mortality), statistical verifications have been made with tolerably good results, and it is to be hoped that such statistical work will be performed and published in various branches of insurance.

It is, of course, obvious that even in cases where the statistical verification of a hypothesis of probability has given a favourable result, the theory founded upon that hypothesis will not be capable of giving us more than an approximate and idealized picture of the real phenomena. Experience tells us, however, that in spite of this such a theory may, within the limits of common sense, be confidently used for statistical predictions. Suppose, taking an example from the theory of risk, that we have proved that the probability of a certain loss in an Insurance Company is equal to 0.01, but that after the introduction of some particular mode of reinsurance the same probability will become reduced to 0.0001. This result has been obtained by means of certain hypotheses on the basic probabilities connected with the various insurances of the Company. If, in similar cases, such hypotheses have been statistically verified with a tolerably good result, we may confidently recommend the adoption of the proposed mode of reinsurance.

The simplest hypothesis is to assume that the basic probabilities are constant. Thus in works concerning the applications of the theory of risk to Life Insurance, the theory is founded upon the assumption[1] of a fixed system of mortality rates or a fixed force of mortality μ_x, and similar hypotheses are made for the applications to other branches. Owing to its simplicity this is, of course, a hypothesis that presents itself most naturally, but it should by no means be considered as a theoretical necessity. It is quite possible, and even certain, that in some cases we can obtain a much better representation of known statistical facts by introducing some kind of variation in the basic probabilities. This may be done in different ways. In mortality statistics, e. g., it is often found that we can improve the agreement between theory and observations by introducing a secular trend in the mortality rates. In other cases, we might try the hypothesis of a probability which is itself subject to irregular variations due to some kind of random selection. Thus we see (and this is a point which should perhaps be somewhat emphasized) that the value of the theory of risk is not wholly dependent on the hypothesis of a constant basic probability. Even if, in a particular case, it will be found necessary to replace this hypothesis by one more complicated, it will still be possible to submit the random fluctuations to a theoretical treatment.

[1] We cannot agree with the view expressed by BOHLMANN (l. c.) that no more assumptions from the theory of probability are required for the theory of risk than those already required for the calculation of premiums. For the latter purpose, we only need a hypothesis about the *average* mortality among the insured persons; for the theory of risk, on the other hand, we want a hypothesis which can tell us something about the distribution of the *deviations from the average* experienced by a Company during the various years.

In this paper we shall, however, only deal with the simple hypothesis of constant basic probabilities. It is obviously advisable in the first place to work out a complete theory founded upon this simple hypothesis before proceeding to more general assumptions.

6. According to the discussion of the practical object of the theory that has been given above, we must require that the theory gives us the means of calculating the probability of the occurence of a loss of given magnitude in a Company working under given conditions. We must be able to show how this probability is affected by a change in the bonus system or the method of reinsurance, or by the formation of a special risk reserve.

The event »occurrence of a loss of given magnitude» may be interpreted in different ways. Fixing our attention at a period of a certain length t, we consider the continuous development of the business during this period. Thus we regard premiums and interest as being continuously paid to the Company, while insurance sums, bonuses etc. are paid out. The formation of the gain arising on the business during the period may then be continuously followed from the beginning to the end of the period. At every instant, the gain (a loss counting as a negative gain) formed during the past part of the period has a certain value, and the course of this gain can be graphically represented by a curve starting from the value 0 at the beginning of the period and ending at a value, which represents the result of the period taken as a whole. Suppose that we are interested in the occurrence of a loss exceeding some given value K. We may then ask for the probability that a loss of this magnitude occurs:

A. At the end of the period t.

B. At any moment during the period.

C. At the end of any business period (usually = calendar year) belonging to the period t.

It is obvious that it is the question C which is the relevant question in practice. But as a rule it will be found that this is a particularly difficult problem, and so we must confine ourselves to an investigation of A, or in some cases B. Even if we succeed only in solving the question A, this will still be of great value with regard to the practical problems, since by giving different values to t we may at least form an estimate of the answer to a question of the type C.

7. In previous works on the theory we may distinguish two fundamentally different points of view.

According to the ordinary method, we begin by considering the gain or loss arising during a certain period on one particular insurance policy. This is a quantity capable of assuming certain values with certain probabilities, which may be calculated. In the mathematical theory of Probability, such a quantity is generally denoted as a *variable*, or a *random variable*. By studying the *probability distribution* of such a variable gain we get an idea of the nature and extent of the *risk* of the Company associated with each single policy. We then remark that the total gain of the Company is equal to the sum of the gains (each taken with its proper sign) on the individual policies, and we try to estimate the probability distribution of this sum by the aid of the known distributions of the components. — In practically all works on the subject, this estimate has been performed by means of a remarkable theorem in Probability which asserts that, subject to certain general conditions, the sum of a large number of mutually independent variables is approximately distributed according to the so called *normal* or GAUSS-LAPLACE *probability function*. It will be argued in the present paper that, in many cases, the approximation obtained by using the normal function is not sufficiently good to justify the conclusions that have been drawn in this way. — We may remark that the additive composition of the total gain by means of individual gains only applies to question A of the preceding article. As regards the questions B and C, there is no similar simple relation connecting the total with the individual policies, and so it is obviously much more intricate to deal with these questions.

The above may be called the *individual risk theory*, since it is founded upon an investigation of the risk associated with an individual policy. In a remarkable series of papers,[1] Dr. F. LUNDBERG has developed a fundamentally different way of attacking the problems. His theory may be properly described as a *collective risk theory.*

The most striking feature of this theory is that LUNDBERG does not require any knowledge of the number or kind of the individual policies. Accordingly he does not apply the conception of the total gain of the Company as the sum of individual gains. He considers the risk business of the Company as a whole, and fixes his attention on the amount of *risk premium* which is being continuously paid to the Company, and on the distribution of this amount with respect to policies with different *sums at risk*. By these means, he is able to find a very remarkable method of approximating to the probability distribution of the Company's gain during a certain period. Mathematically speaking, this problem is not exactly equivalent to the one encountered when we regard the total gain as composed of individual gains. For the practical object of solving the problems of the theory of risk, however, both points of view seem to be equally entitled to our attention. It is particularly remarkable that in LUNDBERG's theory it is possible to treat also the question B of Art. 6 by relatively simple methods.

A detailed account of the fundamental ideas of the collective theory must be postponed to Part III. In Part II we shall discuss the mathematical theory according to the individual point of view. We shall begin by an introductory section devoted to some general theorems concerning random variables and their probability distributions.

[1] F. LUNDBERG, Über die Wahrscheinlichkeitsfunktion einer Risikenmasse, Skandinavisk Aktuarietidskrift 1930. In this paper, references are given to the earlier papers by LUNDBERG bearing on the subject. Cf also a Review by the present writer in Skandinavisk Aktuarietidskrift 1926, and the work by I. LAURIN: An Introduction into Lundberg's Theory of Risk, Skandinavisk Aktuarietidskrift 1930. By the kindness of Dr. LUNDBERG and Dr. LAURIN, I have been allowed to read their recent works here quoted in manuscript.

PART II. Individual Risk Theory.

Variables and Probability Distributions.

8. By a *variable* in the sense of the Theory of Probability we mean a quantity x, which may assume certain real values with certain probabilities.

Thus we may think e. g. of the number of points obtained in a throw with two dice, the number of fires occurring within the next year in a given district, the age at death of a man now aged 30, the gain or loss of an insurance company arising on a particular policy. All these are different instances of the general conception of a variable.

If x is a variable, the probability that x takes a value such that $x \leq t$ is generally a function of t, which we denote by $F(t)$ and call the *probability function* of x. Obviously $F(t)$ is a never decreasing function of the real variable t such that

$$F(-\infty) = 0, \qquad F(+\infty) = 1.$$

In the simplest case, there are only a finite or infinite number of discrete values $a_1, a_2, \ldots$, that are possible values for our variable x, and the probability that x takes the value a_i is equal to p_i, so that

$$p_1 + p_2 + \ldots = 1.$$

Such is the case in the two first examples given above. The graph of the function $F(t)$ then takes the form of a »staircase» with steps in the points a_i, the height of each step being equal to the corresponding p_i. The points a_i are points of discontinuity for $F(t)$; in one of these points, $F(t)$ always assumes the *upper* limiting value. — The same distribution may also be graphically represented by drawing simply through every point a_i an ordinate of the height p_i.

Another simple case of probability distribution is obtained when $F(t)$ has a derivative

$$f(t) = F'(t)$$

which is continuous everywhere except perhaps in a finite number of points. Then the probability that x takes a value such that $t_1 < x \leq t_2$ is equal to

$$F(t_2) - F(t_1) = \int_{t_1}^{t_2} f(t)\, dt.$$

Allowing the interval (t_1, t_2) to be infinitely small, this may be expressed by saying that the probability for our variable to fall in an infinitely small interval $(t, t + dt)$ is equal to

$$f(t)\, dt.$$

Thus $f(t)$ measures the frequency with which the variable tends to fall in the immediate vicinity of the point t, and accordingly $f(t)$ will be called the *frequency function* of the variable x. In the third example given above (the age at death of a man now aged 30), we are concerned with a variable possessing a finite frequency function. — The graph of the frequency function is a generally continuous curve with the equation $y = f(x)$ and the area enclosed between the curve, the axis of x and the ordinates through $x = a$ and $x = b$ is equal to the probability that the variable will fall between a and b. Thus when a tends to $-\infty$ and b to $+\infty$, it follows that the whole area situated between the curve and the axis of x is equal to 1.

The two simple types of probability distribution now considered will be referred to as distributions of the *first* and *second kind* respectively.

The most general type of probability distribution required in ordinary applications is obtained by a combination of the two simple types. Thus a general probability function $F(x)$ will be assumed to have a derivative $f(x) = F'(x)$ everywhere except at a finite or infinite number of points $x = a_i$, where $F(x)$ has a discontinuity with the saltus $p_i (i = 1, 2, \ldots)$. This means that our variable takes the values $x = a_i$ with the probabilities p_i and that, for a value x different from all the a_i, we have the probability $f(x)\,dx$ that the variable falls between x and $x + dx$. Obviously we must always have

$$\int_{-\infty}^{\infty} f(t)\,dt + \sum_i p_i = 1.$$

The graph of such a probability function $F(x)$ is shown in Fig. 1, while Fig. 2 gives another graphical representation of the same probability distribution by means of the frequency curve $y = f(x)$ and isolated ordinates of the height p_i through the points $x = a_i$. — In the sequel, a distribution of this general type will be called a *mixed distribution*.

9. The properties of a probability distribution may be illustrated by the following mechanical analogy. Let us consider a *distribution of mass* over the axis of x, such that the total quantity of mass on the axis is equal to one unit, and let $F(t)$ denote the quantity of mass situated to the left of the point $x = t$, or in this point itself. Obviously this function $F(t)$ will possess all the characteristic properties of a probability function, and there is a perfect analogy between the two types of distribution. A probability distribution of the *first kind* corresponds to a distribution of mass, where all the mass is concentrated in a finite or infinite number of isolated points, so that the point $x = a_i$ carries the mass p_i. A distribution of the *second kind*, on the other hand, corresponds to a continuous distribution of mass with a finite density $f(t)$ at each point t, so that $f(t)\,dt$ is the quantity of mass situated between t and $t + dt$. By a combination of these two kinds of distribution, we obtain a *mixed distribution*, one part of the mass being continuously distributed and the remaining part concentrated in isolated points.

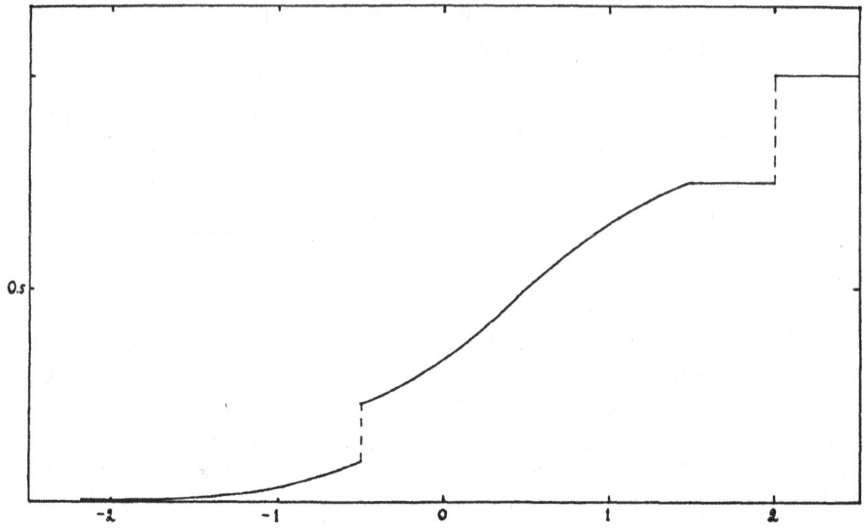

Fig. 1. Probability function of mixed type. Discontinuities in $x = -0.5$ (saltus 0.14) and $x = +2$ (saltus 0.25).

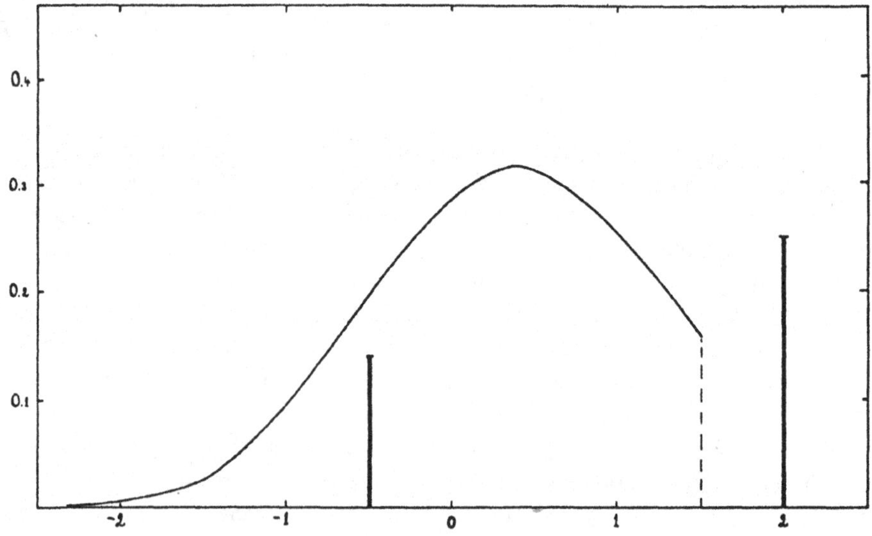

Fig. 2. Another representation of the same probability distribution as in Fig. 1.

16

The *mean value* $M[x]$ of the variable coincides by definition with the centre of gravity of the mass. Thus we have for a distribution of the first kind.

$$M[x] = \sum_i p_i a_i,$$

and for a distribution of the second kind

$$M[x] = \int_{-\infty}^{\infty} t f(t)\, dt.$$

For a general distribution, the mean value is given by an expression which contains two terms: one sum and one integral. Both these terms may, however, be united into one if we use the integral conception introduced by STIELTJES and write

$$M[x] = \int_{-\infty}^{\infty} t\, dF(t).$$

In a similar way, the higher moments of our variable are equal to the corresponding moments of the mass. In particular, the square of the *standard deviation* $\varepsilon[x]$ is equal to the moment of inertia of the mass with respect to a vertical axis through the centre of gravity. Thus if the mean $M[x]$ is equal to m, we define the moments ν_i about the mean by the relation

$$\nu_i = M[(x-m)^i] = \int_{-\infty}^{\infty} (t-m)^i\, dF(t),$$

and we have in particular

$$\nu_2 = M[(x-m)^2] = \varepsilon^2[x].$$

For a distribution of the first or second kind, the expression of ν_i evidently reduces to a sum or an ordinary integral.

According to a well-known theorem due to TCHEBYCHEFF, the probability that a variable x takes a value such that

$$\left| x - M[x] \right| > k\, \varepsilon[x]$$

is, for every positive k, less than $\frac{1}{k^2}$.

The variables x and y are *mutually independent*, if the probability of the »event» $x = t$ is independent of the value assumed by y and vice versâ. Whether the variables are independent or not, we always have

$$M[x+y] = M[x] + M[y].$$

If the variables are independent, we further have

$$M[xy] = M[x] \cdot M[y]$$

and

$$\varepsilon^2[x + y] = \varepsilon^2[x] + \varepsilon^2[y].$$

All these relations are immediately generalized to an arbitrary number of variables.

10. If x is a variable with the probability function $F(t)$, the mean value of any real or complex function $\psi(x)$ of x is defined by the integral

$$M[\psi(x)] = \int_{-\infty}^{\infty} \psi(u)\, dF(u).$$

The mean value of the particular function

$$\psi(x) = e^{-itx}$$

is called the *adjunct* of the probability function and is denoted by

$$g(t) = M[e^{-itx}] = \int_{-\infty}^{\infty} e^{-itu}\, dF(u).$$

The adjunct is finite and continuous for all values of the real variable t, and its modulus can never exceed 1.

If the mean and the standard deviation of x are denoted by m and σ, and if a new variable ξ is introduced by the relation

$$x = m + \sigma\xi, \tag{1}$$

then the probability function of ξ will be

$$F(m + \sigma t)$$

and the corresponding adjunct

$$e^{\frac{mit}{\sigma}} g\left(\frac{t}{\sigma}\right).$$

Let us now suppose that x is the sum of n mutually independent variables:

$$x = x_1 + x_2 + \ldots x_n,$$

and that x_ν has the mean m_ν, the standard deviation σ_ν and the probability funciom $U_\nu(t)$ with the adjunct $v_\nu(t)$. Then the mean m and the standard deviation σ of x will be given by

$$m = m_1 + m_2 + \ldots + m_n,$$

$$\sigma^2 = \sigma_1^2 + \sigma_2^2 + \ldots + \sigma_n^2,$$

18

and the corresponding adjunct will be

$$M\left[e^{-itx}\right] = M\left[e^{-itx_1}\right] \cdot M\left[e^{-itx_2}\right] \cdot \ldots M\left[e^{-itx_n}\right]$$
$$= v_1(t)\, v_2(t) \ldots v_n(t).$$

If, by the substitution (1), we introduce the new variable ξ, and denote its probability function by $F_n(t)$ and the corresponding adjunct by $g_n(t)$, we thus have

$$g_n(t) = e^{\frac{mit}{\sigma}} v_1\left(\frac{t}{\sigma}\right) v_2\left(\frac{t}{\sigma}\right) \ldots v_n\left(\frac{t}{\sigma}\right)$$

(2)

$$= \prod_{\nu=1}^{n} e^{\frac{m_\nu it}{\sigma}} v_\nu\left(\frac{t}{\sigma}\right).$$

11. The particular probability function

$$\Phi(t) = \frac{1}{\sqrt{2\pi}} \int_{-\infty}^{t} e^{-\frac{u^2}{2}}\, du$$

will be called the *normal probability function*, and its derivative

$$\varphi(t) = \Phi'(t) = \frac{1}{\sqrt{2\pi}} e^{-\frac{t^2}{2}},$$

the *normal frequency function.*

The adjunct of the normal probability function is, by a well known formula in the Integral Calculus,

$$\int_{-\infty}^{\infty} e^{-itu}\, d\,\Phi(u) = \frac{1}{\sqrt{2\pi}} \int_{-\infty}^{\infty} e^{-itu-\frac{u^2}{2}}\, du = e^{-\frac{t^2}{2}},$$

and from this relation we deduce by partial integration

(3) $$\int_{-\infty}^{\infty} e^{-itu}\, d\,\Phi^{(\nu)}(u) = (it)^\nu e^{-\frac{t^2}{2}}$$

for $\nu = 0, 1, 2, \ldots$

If x is a variable with the mean m and the standard deviation σ, and if the variable ξ introduced by the substitution (1) has the probability function $\Phi(t)$, then we shall say that the variable x is *normally distributed.* The probability function of x itself will then be

$$\Phi\left(\frac{t-m}{\sigma}\right)$$

and the corresponding adjunct

$$(4) \qquad e^{-mit-\frac{1}{2}\sigma^2 t^2}.$$

For a normally distributed variable, there is only a very small probability that the variable deviates from its mean value by more than 3 or 4 times the standard deviation. Thus e. g. the probalility that

$$x < m - 3\sigma$$

is equal to

$$\Phi(-3) = 0.00135.$$

If, as in the preceding article, x is the sum of n mutually independent components x_ν, and if these components are all normally distributed, then the adjunct which corresponds to the variable ξ is, according to (2) and (4)

$$g_n(t) = \prod_{\nu=1}^{n} e^{\frac{m_\nu it}{\sigma}} \cdot e^{-\frac{m_\nu it}{\sigma}-\frac{1}{2}\sigma_\nu^2\frac{t^2}{\sigma^2}}$$

$$= e^{-\frac{t^2}{2}}.$$

Since it can be shown that no adjunct can correspond to more than one probability distribution[1], this proves that *the sum of any number of normally distributed and mutually independent variables is itself normally distributed.*

It is a highly remarkable fact that it is possible to generalize this theorem by proving that, *even if the components* x_ν *are not normally distributed, their sum* x *will, under fairly general conditions with regard to the probability functions* $U_\nu(t)$ *of the components, be at least approximately normally distributed when* n *is a large number.*

By the expression »approximately normally distributed» we mean that the probability function $F_n(t)$ of the variable ξ will be, if not exactly, then at any rate approximately equal to $\Phi(t)$.

This theorem has played a fundamental part in the theory of risk. Since, later on in this paper, we shall be obliged to penetrate much deeper into these questions, we shall here only give an outline of the proof of the theorem in the special case when all the probability functions $U_\nu(t)$ of the components are identical.

In this case we have, using the same notations as in the preceding article,

$$m = n m_1, \qquad \sigma^2 = n\sigma_1^2,$$

$$g_n(t) = \left(e^{\frac{m_1 it}{\sigma}} \cdot v_1\left(\frac{t}{\sigma}\right)\right)^n.$$

[1] Cf LÉVY, Calcul des probabilités, Paris 1925, p. 166.

But

$$e^{m_1 it} v_1(t) = \int_{-\infty}^{\infty} e^{-it(x-m_1)} d\,U_1(x)$$

$$= \int_{-\infty}^{\infty} \left(1 - it(x - m_1) + \frac{(it)^2}{2!}(x - m_1)^2 - \vartheta \frac{(it)^3}{3!}(x - m_1)^3\right) d\,U_1(x)$$

where $|\vartheta| < 1$, since for every real a and every $k > 0$ we have

$$e^{ia} = \sum_{\nu=0}^{k-1} \frac{(ia)^\nu}{\nu!} + \vartheta \frac{(ia)^k}{k!}.$$

Assuming that the integral

$$\beta = \int_{-\infty}^{\infty} |x - m_1|^3 \, d\,U_1(x)$$

converges, we thus get

$$e^{m_1 it} v_1(t) = 1 - \frac{1}{2}\sigma_1^2 t^2 + \frac{1}{6}\vartheta\beta|t^3|$$

and

$$e^{\frac{m_1 it}{\sigma}} v_1\left(\frac{t}{\sigma}\right) = 1 - \frac{t^2}{2n} + \vartheta \frac{\beta}{6\sigma_1^3} \cdot \frac{|t|^3}{n\sqrt{n}},$$

$$g_n(t) = \left(1 - \frac{t^2}{2n}\right)^n \cdot \left(1 + \vartheta \frac{\beta|t|^3}{6\sigma_1^3 n\sqrt{n}\left(1 - \frac{t^2}{n}\right)}\right)^n.$$

The second factor on the right hand side obviously tends to 1 as n tends to infinity, while the first factor tends to $e^{-\frac{t^2}{2}}$, and thus we have

$$g_n(t) \to e^{-\frac{t^2}{2}}$$

as $n \to \infty$, uniformly in every finite interval with respect to t. Thus the adjunct of the probability function $F_n(t)$ tends uniformly towards the adjunct of the normal function $\Phi(t)$. Hence it can be deduced without difficulty that the corresponding relation

$$F_n(t) \to \Phi(t)$$

holds uniformly for all real t. We shall not enter here upon a discussion of this latter part of the proof, since we shall have to return to similar questions in the further developments that will be given below.

Risk of an Insurance Company. Mean Risk and Average Risk.

12. If we regard, during the time t, a group of n policies belonging to a certain Insurance Company, we can draw a revenue account for this group as a whole. On the income side, we shall have to put premiums and interest on funds, as well as payments from Reinsurance Companies. As expenditure, on the other hand, we count general expenses, reinsurance premiums, increase of technical reserves during the period, distribution of bonus to the policyholders and payment of the sums at risk for the claims arising during the period. As the result of this revenue account, a certain gain of the company will appear (a loss counting as a negative gain), and the value of this gain, referred to the beginning of the period, will be denoted by Γ.

From this revenue account, we can extract all the items connected with one single particular policy, and so form a separate revenue account for this particular policy. The same thing may be done for each policy in our group, and the resulting gains (always referred to the beginning of the period) may be denoted by $\gamma_1, \gamma_2, \ldots \gamma_n$. Obviously we then have

$$\Gamma = \gamma_1 + \gamma_2 + \cdots + \gamma_n,$$

which expresses that the total gain of the company during the period is equal to the sum of the gains on the individual policies.

We now suppose that, for every policy and at every instant during the period, we know the probability that the sum at risk will fall due. If the sum at risk may vary within certain limits (as e. g. in Fire Insurance), we shall suppose also that we know the probabilities with which the various possible values are taken. Then each partial gain γ_ν may be regarded as a variable with a known probability distribution, and the total gain Γ will be the sum of n such variables.

If our group of n policies includes all the policies in force with our company and if, in an actual case, the total gain Γ becomes negative, this means that the regular course of business of the company has been disturbed by random fluctuations. Now, as we have seen above, the fundamental problem to be considered by the theory of risk is precisely to find out what measures a company ought to take in order to reduce as far as possible the probability of such disturbances. For the treatment of this problem, it is obvious that we shall require a thorough numerical knowledge of the probability distribution of the variable Γ. This distribution must be investigated under various assumptions concerning the types of insurances represented in our group, the method of reinsurance and of bonus distribution etc. We must be able to show how the probability function of Γ is affected, if the company proposes to change its premiums or its bonus rates, to increase the maximum amount retained on the company's own risk or to adopt certain rules for the formation of a technical risk reserve. Further, as the choice of the period t has been wholly arbitrary, we must also investigate the influence which this choice may exert upon our conclusions.

Clearly it will be convenient in the first place to limit ourselves to the con-

22

sideration of general methods for the investigation of the probability distributions appearing in the risk problem. This task is naturally divided into two parts: first the probability functions of the partial gains γ_ν connected with the particular policies must be studied, and then the corresponding function attached to the total gain Γ must be formed.

When we have once succeeded in finding sufficiently accurate methods for the numerical analysis of the probability functions, we shall be well prepared to deal with the various risk problems occurring in practical insurance business.

We now proceed to an investigation of the probability functions. It is not necessary, or even desirable, that we should deal with this problem throughout quite so generally as it has been stated above, where we have taken into account all such elements as general expenses, reinsurance, bonus distribution etc. When making a special study of the influence of one particular element upon the course of business, it will be permitted, at least in a preliminary investigation, to make simplifying assumptions as to the other elements. Accordingly we shall begin in the following article by considering the probability function connected with one single policy under the assumption that the premiums are pure net premiums and that the only expenditure to be borne by the company is the payment of the insurance amount whenever a claim arises on the policy.

13. We begin with the following simple example: we have the probability q that the sum s falls due during the time t, and this time is so short that we may disregard the interest earned during the period. The net premium payable in advance for the whole period is equal to sq, and for the gain γ we have the two possible values

$$(5) \qquad \gamma = \begin{cases} +sq & \text{with the probability } p = 1 - q, \\ -sp & \text{» \quad » \qquad » \qquad } q. \end{cases}$$

This is a very simple case of a distribution of the *first kind*, according to the terminology used in Art. 8. A practical example will be given e. g. by an ordinary life policy, when t is equal to one year and s denotes the difference between the insurance amount and the reserve.

In several other branches of insurance, the sum to be paid in case of a claim is not a priori fixed, but may assume any value between 0 and s. If $\psi(x)\,dx$ denotes the probability that, during the period, a claim arises for an amount between sx and $s(x + dx)$, where $0 < x < 1$, then the total probability of a claim is (for the sake of simplicity we suppose that only one claim may arise)

$$p = \int_0^1 \psi(x)\,dx,$$

and the net premium of the policy is

$$sP = s\int_0^1 x\,\psi(x)\,dx$$

If there is a claim for the amount sx, the gain of the company is equal to

$$\gamma = s(P - x),$$

and thus we see that the probability that the gain of the company falls between γ and $\gamma + d\gamma$ is

$$\frac{1}{s}\psi\left(\frac{sP - \gamma}{s}\right) d\gamma,$$

where

$$s(P - 1) < \gamma < sP.$$

Further, we have the probability $1 - p$ that no claim at all arises during the period, and in that case the gain will be equal to the net premium sP. Thus in this case the gain γ gives rise to a *mixed* distribution according to the terminology of Art. 8.

If the period t extends over several years, the question becomes more complicated, as premiums are in most cases not paid for more than one year at a time, and it becomes necessary to take into account also the interest earned during the period. For the discussion of these questions, we shall choose as our examples some ordinary types of life policies, but there is nothing to prevent that e. g. the formulæ relating to a temporary life insurance may be taken to apply, mutatis mutandis, also to other branches of insurance.

If we observe during the next t years an insured person (x), who is now aged x, he may either die at some time during the period, or he may be alive at the end of the period. The probability that he will die after attaining the age $x + \tau$ $(0 < \tau < t)$ but before attaining the age $x + \tau + d\tau$ is

$$\frac{l_{x+\tau}}{l_x}\mu_{x+\tau}\,d\tau.$$

and the probability that he will still be alive after t years is

$$\frac{l_{x+t}}{t_x}.$$

Thus we see that *the length of that part of the period t, during which (x) is alive,* is a variable with a probability distribution of the mixed type. This variable may assume any value in the interval from 0 to t, and there is a finite probability for the particular value t, while all the other possible values give rise to the continuous part of the distribution.

Now, it is obvious that the gain or loss arising during the period t on some ordinary life policy upon the life of (x) is a function of the variable that we have just considered. Thus the probability distribution of any such gain, considered as a new variable, will also generally be of the mixed type. The possibility that

(x) may be alive at the end of the period gives rise to a finite probability for a certain particular value of the gain, while the possibility that he may die at some time τ, such that $0 < \tau < t$, corresponds to the continuous part of the distribution of the gain. (There are, of course, exceptional cases such as the case of a whole-life policy when we put $t = \infty$, which gives us a distribution of the second kind.)

We shall deduce here the formulæ which represent the distribution of the gain connected with some current types of life policies. It will be assumed throughout that the insurance amount is equal to 1 and that the period considered coincides with the first t years of the duration of the policy.[1] Similar formulæ can without difficulty be developped for other types of policies and for any period of t years. We shall denote generally by γ the gain of the Company (always referred to the beginning of the period) corresponding to the case that the insured dies after τ years, where $0 < \tau < t$. By $f(\gamma)\, d\gamma$ we shall denote the probability that the value of this gain is situated between γ and $\gamma + d\gamma$, and by γ_L the value of the gain in the case where the insured is alive at the end of the period, the probability of this last case being $\frac{l_{x+t}}{l_x}$.

For most ordinary types of life policies, the gain γ is negative for small values of τ and then steadily increases as τ increases from 0 to t. In all such cases we have

$$f(\gamma)\, d\gamma = \frac{l_{x+\tau}}{l_x} \mu_{x+\tau}\, d\tau$$

and hence

$$f(\gamma) = \frac{l_{x+\tau}}{l_x} \mu_{x+\tau} \cdot \frac{d\tau}{d\gamma}. \tag{6}$$

We consider first an ordinary *endowment policy*, duration n years. The premiums are supposed to be paid continuously, so that during the time $d\tau$ we obtain the premium $\bar{p}\, d\tau$, where

$$\bar{p} = \frac{1}{\bar{a}_{x\,\overline{n|}}} - \delta,$$

δ denoting the force of interest. Assuming that the rate of interest actually attained by the Company coincides with the one used for the calculation of premiums, the gain γ will be given by the relation (where as usual we put $v = e^{-\delta}$)

$$\gamma = \bar{p}\, \bar{a}_{\overline{\tau|}} - v^{\tau},$$

which expresses that γ is equal to the value of the premiums obtained until the

---

[1] If we consider an arbitrary insurance sum s. it is easily seen that in our formulæ γ and γ_L will be multiplied with s and $f(\gamma)$ divided by s; otherwise the formulæ will be unchanged.

moment of death, less the value of the insurance amount paid at death. By an easy transformation, we get

$$(7) \qquad \gamma = \frac{\bar{A}_{x\overline{n}|} - v^{\tau}}{\delta\, \bar{a}_{x\overline{n}|}},$$

and hence by (6)

$$(8) \qquad f(\gamma) = \bar{a}_{x\overline{n}|} v^{-\tau} \mu_{x+\tau} \frac{l_{x+\tau}}{l_x}.$$

If we consider τ as a parameter varying from 0 to t, the relations (7) and (8) give a representation of the »frequency curve» $y = f(\gamma)$ which forms the continuous part of the distribution of γ. — For the case where the insured is alive at the end of the period, we obtain

$$(9) \qquad \begin{aligned} \gamma_L &= \bar{p}\, \bar{a}_{\overline{t}|} - {}_tV_{x\overline{n}|} v^t \\ &= \frac{\bar{A}_{x\overline{n}|} - \bar{A}_{x+t\,\overline{n-t}|} v^t}{\delta\, \bar{a}_{x\overline{n}|}} \end{aligned}$$

For an *endowment policy with single premium*, we obtain in a similar way

$$\gamma = \bar{A}_{x\overline{n}|} - v^{\tau},$$

$$f(\gamma) = \frac{1}{\delta \bar{a}_{x\overline{n}|}} \cdot \bar{a}_{x\overline{n}|} v^{-\tau} \mu_{x+\tau} \frac{l_{x+\tau}}{l_x},$$

$$\gamma_L = \bar{A}_{x\overline{n}|} - \bar{A}_{x+t\,\overline{n-t}|} v^t.$$

For a *temporary life assurance* for n years, with continuous premiums, we get after some simple reductions

$$\gamma = \frac{D_x - D_{x+n}}{D_x} \cdot \frac{\bar{A}_{x\overline{n}|} - v^{\tau}}{\delta\, \bar{a}_{x\overline{n}|}} - \frac{D_{x+n}}{D_x},$$

$$f(\gamma) = \frac{D_x}{D_x - D_{x+n}} \cdot \bar{a}_{x\overline{n}|} v^{-\tau} \mu_{x+\tau} \frac{l_{x+\tau}}{l_x},$$

$$\gamma_L = \frac{D_x - D_{x+n}}{D_x} \cdot \frac{\bar{A}_{x\overline{n}|} - \bar{A}_{x+t\,\overline{n-t}|} v^t}{\delta\, \bar{a}_{x\overline{n}|}} - \frac{D_{x+n}}{D_x} + \frac{D_{x+n}}{D_{x+t}} v^t.$$

For an n-year *assurance »à terme fixe»*, the corresponding relations are

$$\gamma = v^n \frac{\bar{A}_{x\overline{n}|} - v^{\tau}}{\delta\, \bar{a}_{x\overline{n}|}},$$

$$f(\gamma) = v^{-n} \cdot \bar{a}_{x\overline{n}|} v^{-\tau} \mu_{x+\tau} \frac{l_{x+\tau}}{l_x},$$

$$\gamma_L = v^n \frac{\bar{A}_{x\overline{n}|} - \bar{A}_{x+t\,\overline{n-t}|} v^t}{\delta\, \bar{a}_{x\overline{n}|}}.$$

It is obvious that the expressions obtained in the four typical cases now considered are closely related to one another. Thus e. g., denoting by $\gamma^{(1)}$, $\gamma^{(2)}$, $\gamma^{(3)}$ and $\gamma^{(4)}$ respectively the expressions for γ corresponding to the four cases, we have

$$\gamma^{(2)} = \delta\, \bar{a}_{x\overline{n}|} \cdot \gamma^{(1)},$$

$$\gamma^{(3)} = \frac{D_x - D_{x+n}}{D_x} \cdot \gamma^{(1)} - \frac{D_{x+n}}{D_x},$$

$$\gamma^{(4)} = v^n \cdot \gamma^{(1)}.$$

Similar relations hold for $f(\gamma)$ and γ_L.

Fig. 3—6 illustrate the probability distributions in the four cases treated above The ultimate table of 1928 of the Swedish Life Assurance Companies and the rate of interest 4 % have been used. We have taken $x = 40$, $n = t = 20$, so that the period considered coincides with the total duration of the policy. The distribution is represented by the curve $y = f(\gamma)$ and the isolated ordinate $\frac{l_{x+t}}{l_x}$ through the point γ_L. It is highly instructive to compare such distributions for various values of t and various kinds of policies. In our figures, we observe in particular the great risk associated with the temporary assurance as compared with the other types. The probability of a considerable loss is much greater than in the other cases. The single premium and the »terme fixe» assurance, on the contrary, are much less dangerous than an ordinary endowment policy for the same amount.

As a final example, we shall choose an insurance with a varying amount, which has been especially constructed so as to give a mathematically simple form to the probability distribution of the gain. — Let us consider a whole-life policy on the life of (x), the sum to be paid if death occurs after τ years being

$$(10) \qquad s_\tau = \left(2\frac{l_{x+\tau}}{l_x} + 1\right) e^{\delta\tau} - 2.$$

Thus for $\tau = 0$, the insurance amount is unity, and then increases steadily with τ, if x is not too large. The continuous premium for this policy is independent of x:

$$\bar{p} = 2\,\delta,$$

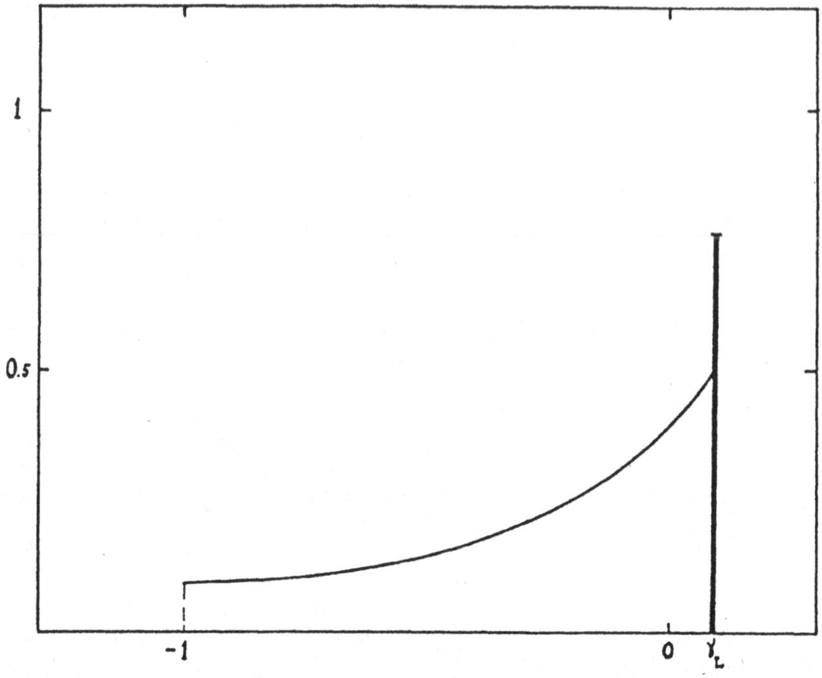

Fig. 3. Endowment Assurance. $x = 40$, $n = t = 20$.

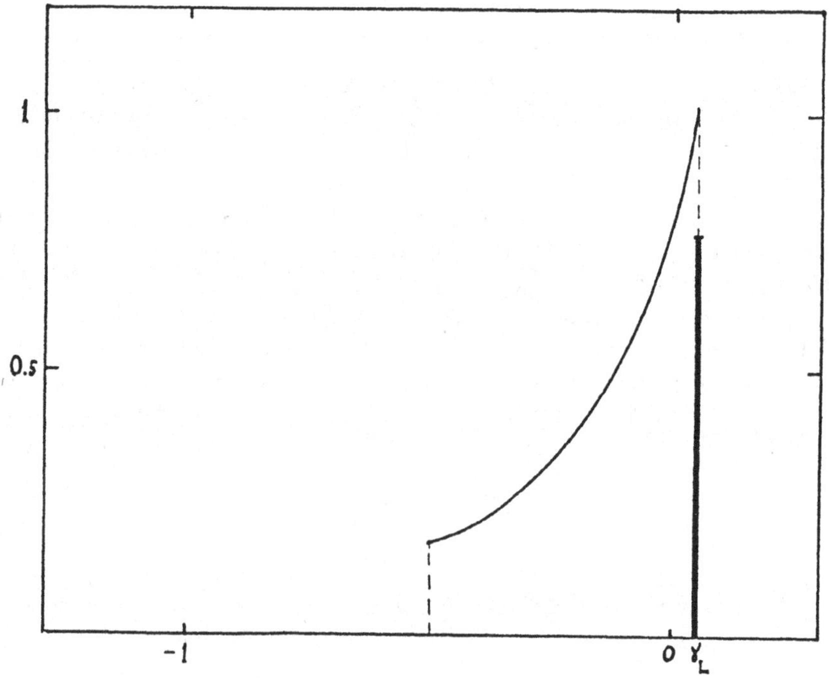

Fig. 4. Endowment Assurance with single premium. $x = 40$, $n = t = 20$.

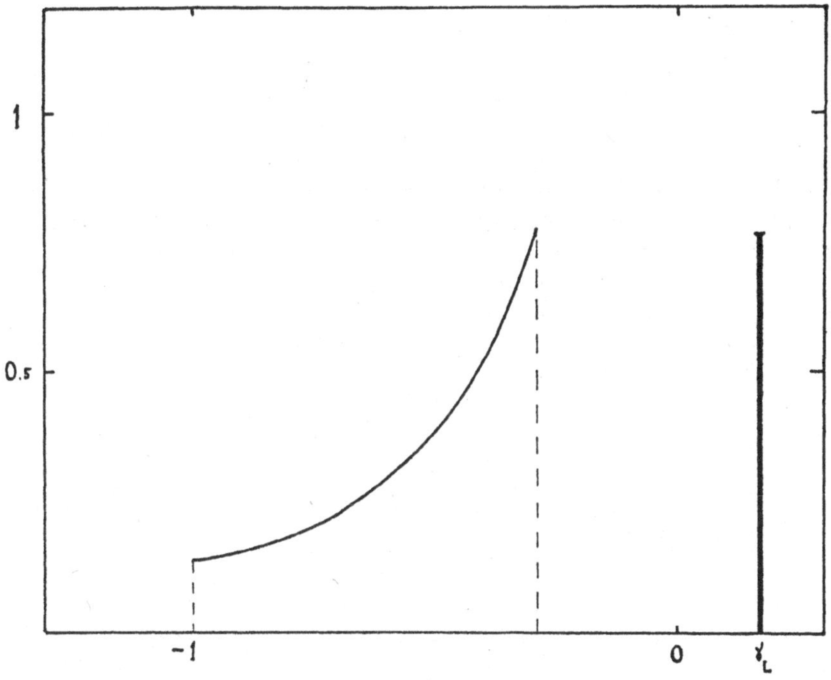

Fig. 5. Temporary Assurance. $x = 40$, $n = t = 20$.

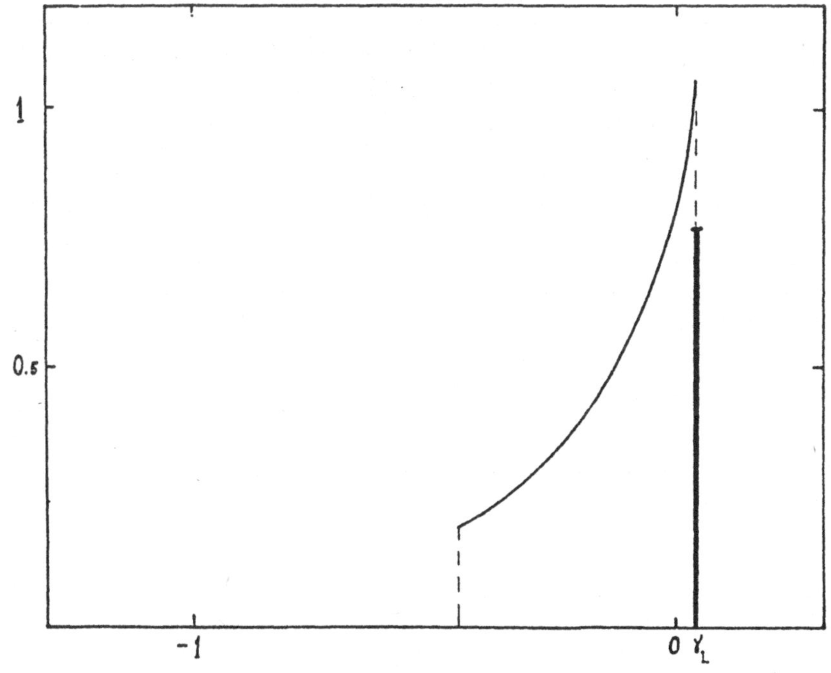

Fig. 6. Terme-Fixe Assurance. $x = 40$, $n = t = 20$.

and the reserve after τ years is

$$_{\tau}V_x = \left(\frac{l_{x+\tau}}{l_x} + 1\right) e^{\delta\tau} - 2.$$

Using the same assumptions as above with respect to mortality and interest, and putting the initial sum equal to 1 000 instead of 1, we get the following values for $x = 40$.

τ	s	V	τ	s	V
0	1 000	0	30	6 662	2 953
10	2 182	831	40	6 882	5 842
20	5 556	1 873	50	7 457	7 282

For the distribution of the gain we obtain the following expressions, using the same notations as above:

$$\gamma = 1 - 2\frac{l_{x+\tau}}{l_x},$$

$$f(\gamma) = \tfrac{1}{2},$$

$$\gamma_L = 1 - \frac{l_{x+t}}{l_x}.$$

Fig. 7 illustrates this distribution for a value of t such that $\frac{l_{x+t}}{l_x} = \frac{1}{2}$.

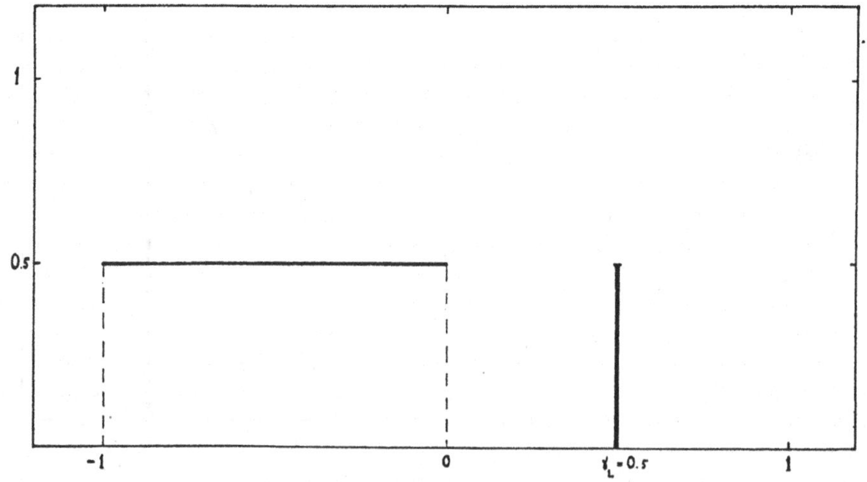

Fig. 7. Whole-Life Assurance, increasing amount. $x = 40$, $t = 30.9$.

14. When, as in the preceding article, we consider only net premiums, calculated according to the »true» probability of risk and the »true» rate of interest, it is clear that the mean value of the gain for any period must be equal to zero, so that we always have, according to the notation of Art. 8—9,

$$M[\gamma] = 0.$$

This is an immediate consequence of the general principles for the determination of net premiums and reserves, and may be verified without difficulty for all the examples given above.

Thus if the probability distribution of the gain is interpreted as a distribution of mass over the axis of γ (cf. Art. 9), the centre of gravity of the mass is situated in the point $\gamma = 0$. The moments of the distribution about the mean, and in particular the standard deviation, are easily calculated. Thus for the simple kind of policy which gives rise to the distribution defined by (5), the i:th moment ν_i about the mean is

(11) $$\nu_i = p\,(sq)^i + q\,(-sp)^i.$$

This formula applies to various branches of insurance other than life, and also to life policies, if s denotes the sum at risk and t is so small that we may neglect the interest earned during the period.

For larger values of t, we obtain for a life policy on the amount 1, belonging to one of the ordinary types,

$$\nu_i = \int \gamma^i f(\gamma)\,d\gamma + \gamma_L^i \frac{l_{x+t}}{l_x},$$

where the integral extends between the values of γ corresponding to the beginning and the end of the period t. Introducing the variable τ by means of (6), we have

(12) $$\nu_i = \int_0^t \gamma^i \frac{l_{x+\tau}}{l_x} \mu_{x+\tau}\,d\tau + \gamma_L^i \frac{l_{x+t}}{l_x}.$$

By means of this formula, the standard deviation and the higher moments may be calculated for any ordinary type of life policy and for any period of t years, not necessarily coinciding with the t first years of the duration of the policy. We have only to deduce expressions for γ and γ_L, similar to those occurring in the preceding article, and to substitute these expressions in (12). — If the insurance sum is equal to s instead of 1, the expression for ν_i is multiplied by s^i.

For an ordinary endowment policy, γ and γ_L are given by (7) and (9.) For the special case $t = n$, the period includes the total duration of the policy and (12) gives, after a partial integration,

$$(13) \qquad \nu_i = (-1)^i + \frac{i\delta}{(\delta \bar{a}_{x\overline{n}|})^i} \int_0^n (\bar{A}_{x\overline{n}|} - v^\tau)^{i-1} v^\tau \frac{l_{x+\tau}}{l_x} d\tau.$$

For a single-premium endowment, the right hand side of this expression should be multiplied by the factor $(\delta \bar{a}_{x\overline{n}|})^i$, and for an »à terme fixe» assurance by v^{in}. For a temporary assurance, the formula becomes a little more complicated, but there is no difficulty of principle involved in deducing a simular formula for any ordinary type of policy. — For the numerical calculation of ν_i according to (13), tables of temporary annuities with the forces of interest δ, 2δ, ... $i\delta$ are sufficient.

For the increasing assurance considered at the end of the preceding article we have, choosing t so that $\frac{l_{x+t}}{l_x} = \frac{1}{2}$,

$$(14) \qquad \begin{aligned} \nu_i &= \tfrac{1}{2} \int_{-1}^{0} \gamma^i \, d\gamma + (\tfrac{1}{2})^{i+1} \\ &= \tfrac{1}{2} \left[\frac{(-1)^i}{i+1} + (\tfrac{1}{2})^i \right]. \end{aligned}$$

15. The standard deviation

$$\varepsilon[\gamma] = \sqrt{\nu_2}$$

of the gain arising on a given insurance during a given period is usually called the *mean risk* of the insurance corresponding to the given period. This is a characteristic of the probability distribution of γ which, to a certain extent, may serve as a measure of the degree of dispersion about the mean.

The mean risk may be calculated by putting simply $i = 2$ in the general moment formulæ of the preceding article. Thus for a short-term insurance we obtain from (11)

$$\varepsilon[\gamma] = s\sqrt{pq},$$

s denoting the sum at risk. For a life policy, considered during a period of t years, we have to put $i = 2$ in the general formula (12). For an ordinary endowment policy, and for the special case $t = n$, we obtain from (13) by some easy calculation

$$(15) \qquad \varepsilon[\gamma] = \frac{1}{\delta \bar{a}_{x\overline{n}|}} \sqrt{\bar{A}'_{x\overline{n}|} - (\bar{A}_{x\overline{n}|})^2},$$

where $\bar{A}'_{x\overline{n}|}$ denotes the same insurance value as $\bar{A}_{x\overline{n}|}$, but calculated with the force of interest $2\,\delta$.

For a single-premium endowment, the right hand side of (15) should be multiplied by $\delta\,\bar{a}_{x\overline{n}|}$, and for an »à terme fixe» assurance by v^n. For the temporary assurance, we obtain the corresponding formula

$$\nu_2 = \varepsilon^2[\gamma] =$$

$$= \left(\frac{D_x - D_{x+n}}{D_x}\right)^2 \cdot \frac{\bar{A}'_{x\overline{n}|} - (\bar{A}_{x\overline{n}|})^2}{(\delta\,\bar{a}_{x\overline{n}|})^2} + \frac{D_{x+n}}{D_x}\left(2\frac{D_x - D_{x+n}}{D_x} \cdot \frac{\bar{a}_{\overline{n}|}}{\bar{a}_{x\overline{n}|}} + \frac{D_{x+n}}{D_x} + v^n - 2\right).$$

For the distributions represented in Fig. 3--6, we find the following numerical values of the mean risk per unit of assurance amount:

Endowment assurance (Fig. 3) .... $\varepsilon[\gamma] = 0.218$

» » , single premium (Fig. 4).................... $\varepsilon[\gamma] = 0.109$

Temporary assurance (Fig. 5)..................................... $\varepsilon[\gamma] = 0.313$

Assurance à terme fixe (Fig. 6).................................. $\varepsilon[\gamma] = 0.099$

For an arbitrary assurance sum s the value of the mean risk is, of course, multiplied by s.

The mean risk is, by definition, equal to the standard deviation or mean error of a certain random variable associated with one particular insurance. We now define the *mean risk of a group of n policies* in an analogous way, as the standard deviation of the total gain Γ arising in the group during a given period t. The total gain is, by Art. 12, equal to the sum of the partial gains:

$$\Gamma = \gamma_1 + \gamma_2 + \ldots + \gamma_n,$$

and so we have, by Art. 9,

$$M[\Gamma] = M[\gamma_1] + M[\gamma_2] + \ldots + M[\gamma_n] = 0,$$

since the mean value of each partial gain is equal to zero. Assuming that all the insurances are mutually independent, we have further by Art. 9

(16) $$\varepsilon^2[\Gamma] = \varepsilon^2[\gamma_1] + \varepsilon^2[\gamma_2] + \ldots + \varepsilon^2[\gamma_n],$$

so that the mean risk for the group can be immediately found from the mean risks of the individual policies. Similar, but more complicated relations also hold for the higher moments.

The mean of *the absolute value of the gain* corresponding to one single policy or a group of policies is called the *average risk* of that policy or group of policies.[1] Thus in our notation the average risks of the individual policies and the total group are $M[|\gamma_1|], \ldots M[|\gamma_n|]$ and $M[|\Gamma|]$ respectively. The average risk furnishes an alternative measure of the dispersion and may thus render services of the same kind as the mean risk. Each of the two measures has advantages of its own, and in particular it may be observed that, for one single insurance, the average risk is often more easily calculated than the mean risk. The latter has, however, the fundamental advantage that, according to (16), the mean risk of a group of mutually independent insurances is simply expressed by the individual mean risks. For the average risk, there is no simple relation of this kind, and this must be regarded as a serious drawback. Since in the present work we are especially interested in the probability distribution associated with a *group* of policies, we shall not find occasion to occupy ourselves with the average risk.

16. The calculation of the mean risk for a life policy may often be considerably simplified by means of a theorem known as HATTENDORFF's theorem. In a recent paper[2], STEFFENSEN pointed out that no rigorous proof of this theorem seems to have been given by previous authors, and gave three different rigorous proofs. We shall give below another proof, which is substantially equivalent to a proof published by CANTELLI[3]. The argument applies to any kind of life policy, but for the sake of simplicity we may fix our attention to an endowment assurance with continuous premiums.

Let us consider two periods during the duration of the policy: the period P_0 from duration $t = 0$ to $t = t_1$ and the period P_1 from $t = t_1$ to $t = t_2$, where $0 < t_1 < t_2 \leqq n$. The gain arising on the policy during P_0 and P_1 we denote by γ_0 and γ_1 respectively, both these gains being referred to the beginning of P_0, i. e. to the point $t = 0$. Then if Γ denotes the total gain arising during $P_0 + P_1$, we obviously have

$$\Gamma = \gamma_0 + \gamma_1. \tag{17}$$

The variables γ_0 and γ_1, are, of course, not mutually independent, and thus we cannot tell a priori if it is allowed to calculate the standard deviation of Γ according to the ordinary rule of addition of the squares. That this is in reality allowed is precisely the substance of HATTENDORFF's theorem, which we are now going to prove.

---

[1] Most authors define the average risk as *half* the mean value mentioned in the text. We agree, however, with the view expressed by STEFFENSEN (in a work quoted below) that the above definition should be preferred.

[2] J. F. STEFFENSEN, On Hattendorff's theorem in the theory of risk, Skandinavisk Aktuarietidskrift, 1929.

[3] F. P. CANTELLI, Un teoreme sulle variabili casuali dipendenti etc., Rivista Italiana di Statistica 1929.

Since the mean values of Γ and γ_0 are both equal to zero, it follows from (17) that the same thing must be true for γ_1. Thus we have

$$\begin{aligned}\varepsilon^2[\Gamma] &= M[(\gamma_0+\gamma_1)^2]\\ &= M[\gamma_0^2] + M[\gamma_1^2] + 2M[\gamma_0\gamma_1]\\ &= \varepsilon^2[\gamma_0] + \varepsilon^2[\gamma_1] + 2M[\gamma_0\gamma_1].\end{aligned}$$

With respect to γ_1 two cases are possible. Either the insured may die already during P_0; in that case the policy has ceased to exist before the beginning of P_1 and γ_1 is obviously equal to zero. Or the insured is alive at the beginning of P_1, the probability of this case being $\frac{l_{x+t_1}}{l_x}$. In the latter case, γ_0 has a certain constant value and γ_1 becomes equal to $v^{t_1}\gamma_1'$, where γ_1' denotes the gain arising during P_1 (and referred to the beginning of that period) on a policy which is in force at the beginning of P_1. The mean value of γ_1' being equal to zero, it follows that

$$M[\gamma_0\gamma_1] = 0$$

and

$$\varepsilon^2[\gamma_1] = \frac{l_{x+t_1}}{l_x} v^{2t_1} \varepsilon^2[\gamma_1'].$$

But $\varepsilon[\gamma_0]$, $\varepsilon[\gamma_1']$ and $\varepsilon[\Gamma]$ are the mean risks of the policy corresponding to the periods P_0, P_1 and $P_0 + P_1$ respectively, and so we have, in a notation which should be easily understood,

$$\varepsilon^2(P_0+P_1) = \varepsilon^2(P_0) + \frac{l_{x+t_1}}{l_x} v^{2t_1} \varepsilon^2(P_1).$$

It is obvious that the argument is perfectly general, so that we may write, without further explanations,

$$\varepsilon^2(P_0+P_1+\ldots+P_r) = \sum_{i=0}^{r} \frac{l_{x+t_i}}{l_x} v^{2t_i} \varepsilon^2(P_i)$$

where $t_0 = 0$ and the period P_i extends from $t = t_i$ to $t = t_{i+1}$. This is HATTENDORFF's theorem.

In order to use the theorem for practical calculation of the mean risk, we take the periods P_i rather short, e. g. equal to one year each. Then $t_i = i$ and according to Art. 15 we may put approximately[1] for the insurance sum 1

$$\varepsilon^2(P_i) = p_{x+i}\, q_{x+i}\, v\,(1 - {}_{i+\frac{1}{2}}V_{x\overline{n}|})^2.$$

It is obvious that it is in no way necessary to let the first period P_0 begin at the duration $t = 0$. Perfectly similar developments hold with respect to the mean risk corresponding to any period during the duration of the policy.

[1] If claims are paid at the end of the year, and premiums paid in advance, we may even get an exact formula in this way.

17. In the investigations contained in Arts. 13—16, we have been concerned with the net risk business only, and we have assumed that our net premiums have been calculated according to the »true» probability of risk and the rate of interest actually attained by the Company. For the practical applications it is, however, necessary to pay attention also to a number of other features of the business, such as loading of premiums, bonus distribution, reassurance etc., and so we must try to estimate the influence which this will exert on the probability distribution of the Company's gain.

It is obvious that the treatment of such questions must be adapted to the practical circumstances in every particular case, and we shall here only give some examples to illustrate the method. We shall neglect throughout the general expenses of the Company and the corresponding part of the loading, so that the loading of premiums considered in the sequel will always be the extra loading intended as a margin for fluctuations. This means simply that we consider the loading for expenses as belonging to a separate branch of the business, which has to carry all the general expenses of the Company, and that we take no account of the influence of random fluctuations upon this part of the business.

As our first example, we take the short-term policy in considered Arts. 13—15, where we have the probability q that the sum at risk falls due within the next year. Let us suppose that the total sum at risk under the policy is $s + \sigma$, and that the amount s is retained on the Company's own risk, while the amount σ has been reinsured with another Company. The total net premium for one year is $q(s + \sigma)$, and we suppose that we have further at our disposal a certain loading (loading for expenses not included, cf. above!) amounting to $\lambda q(s + \sigma)$. For the reinsurance, the net premium is $q\sigma$ and the loading $\lambda' q\sigma$ where, generally, we have $\lambda' > \lambda$ since at least a part of the loading for expenses will be included in the reinsurance premium. For the gain γ arising during the year, there are still two possible values

$$\gamma = \begin{cases} sq + \lambda sq - (\lambda' - \lambda)\sigma q & \text{with the probability } p = 1 - q, \\ -sp + \lambda sq + (\lambda' - \lambda)\sigma q & \text{» » » } q. \end{cases}$$

Comparing with (5), we see that the mean value of γ is altered so that we have now

$$M[\gamma] = \lambda sq - (\lambda' - \lambda)\sigma q,$$

but the standard deviation and the higher moments about the mean preserve their values as given by (11). Thus in particular the mean risk of the policy (by this expression, we always understand the standard deviation of the gain, even in the present more general case, when the mean value is not zero) is still equal to $s\sqrt{pq}$.

Next, we consider an endowment assurance with the sum $s + \sigma$, the amount s being kept on the Company's own risk and σ reinsured. The premium and the reserve are calculated according to a force of interest δ_1 lower than the interest actually attained by the Company. Thus the continuous premium per unit of insurance amount is given by the relation

$$\bar{p} = \frac{1}{\bar{a}_{x\,\overline{n}|}} - \delta_1,$$

the annuity $\bar{a}_{x\,\overline{n}|}$ being calculated with the force of interest δ_1. The Company further proposes to pay bonus according to a force of interest $\delta_2 > \delta_1$, while the actually attained force of interest is $\delta_3 > \delta_2$. During the infinitely small time dt, the bonus

$$(\delta_2 - \delta_1)\,{}_t\bar{V}_{x\,\overline{n}|}\,dt$$

is produced per unit of insurance amount, and we suppose that this is immediately paid out to the policyholder. For the reinsured part of the policy, the premium $\sigma\bar{p}\,dt$ is paid to the reinsurer for the time dt. We shall suppose, however, that the reinsurer undertakes to repay the amount $\sigma(\delta_2 - \delta_1)\,{}_t\bar{V}_{x\,\overline{n}|}\,dt$ out of his excess interest, so that he is only allowed to retain the amount $\sigma(\delta_3 - \delta_2)\,{}_t\bar{V}_{x\,\overline{n}|}\,dt$. Under these hypotheses, the reinsured part of the policy disappears completely from the expressions for the gain of the Company. If we consider the gain arising during the first t years of the duration of the policy we obtain, using the same notations as in Art. 13,

$$\gamma = s\left(\bar{p}\,\frac{1 - v_3^{\tau}}{\delta_3} - v_3^{\tau} - (\delta_2 - \delta_1)\int_0^{\tau} {}_z\bar{V}_{x\,\overline{n}|}\,v_3^{z}\,dz\right),$$

$$\gamma_L = s\left(\bar{p}\,\frac{1 - v_3^{t}}{\delta_3} - {}_t\bar{V}_{x\,\overline{n}|}\,v_3^{t} - (\delta_2 - \delta_1)\int_0^{t} {}_z\bar{V}_{x\,\overline{n}|}\,v_3^{z}\,dz\right),$$

where $v_3 = e^{-\delta_3}$. These relations express that the gain is equal to the value of the premiums less the value of the insurance sum (or the value of the reserve at the end of the period) and the value of the bonus paid to the policyholder.

Since the Company reserves a certain excess interest for its own account, the mean value of the gain must be equal to a positive quantity m, which is given by the expression

$$m = \int_0^t \gamma\,\frac{l_{x+\tau}}{l_x}\,\mu_{x+\tau}\,d\tau + \gamma_L\,\frac{l_{x+t}}{l_x}. \tag{18}$$

The general expression (12) for the moments about the mean must be modified, since m is no longer equal to zero, and we obtain

$$\nu_i = \int_0^t (\gamma - m)^i\,\frac{l_{x+\tau}}{l_x}\,\mu_{x+\tau}\,d\tau + (\gamma_L - m)^i\,\frac{l_{x+t}}{l_x}. \tag{19}$$

When we wish to make numerical applications of formulæ such as those here deduced, it is generally found that the standard deviation and higher characteristics of the distribution, which depend on the moments about the mean, are not considerably changed by introducing the complicated hypotheses about bonus, reinsurance etc. Thus it will, at least as a first approximation, be sufficient for the practical applications if we calculate the mean risk and higher characteristics as for the net risk business, while for the mean value m we use the exact formula. But of course it should always be kept in mind that this is only an approximation.

As soon as the details of a particular case are given, there is theoretically no difficulty involved in deducing the exact formulæ for γ and γ_L, following the lines given above, but their application to the numerical calculation of ν_2 and he higher moments usually requires a good deal of labour and patience.

Applications of the Elementary Theory.

18. We consider a large group of policies, e. g. all the policies of a certain Company in force at a given moment. We assume that the policies are all *mutually independent*, so that the probability that a claim occurs on one policy is in no way affected by whatever may happen to the others.

For the gain Γ arising in our group during a given period t we then have, using the same notations as above,

$$(20)\quad \left|\begin{aligned} \Gamma &= \gamma_1 + \gamma_2 + \cdots + \gamma_n, \\ M[\Gamma] &= M[\gamma_1] + M[\gamma_2] + \cdots + M[\gamma_n], \\ \varepsilon^2[\Gamma] &= \varepsilon^2[\gamma_1] + \varepsilon^2[\gamma_2] + \cdots + \varepsilon^2[\gamma_n], \end{aligned}\right.$$

the partial gains γ_i corresponding to the individual policies in the group. The gain on each policy may, as we have seen above, be defined in various ways according to the particular circumstances of the case under discussion. In a general way we shall let it be defined as the gain which is left for the Company's own account after the fulfilment of all regular payments such as those for claims, increase of reserve, bonus distribution according to the ordinary rates of the Company, etc. Then Γ is precisely the gain that, for the period t, is available to meet the effects of random fluctuations without causing any serious trouble to the Company, and our fundamental question is: »*what is the probability that Γ becomes negative?*» — It is this probability that must be made as small as possible.

According to the general theorem discussed in Art. 11 we may, subject to certain general conditions, assert that the variable Γ will be approximately normally distributed when n is large. This means that, for any real x, the probability that the inequality

$$\Gamma < M[\Gamma] + x\,\varepsilon[\Gamma]$$

will be satisfied, is approximately equal to

$$\Phi(x) = \frac{1}{\sqrt{2\pi}} \int_{-\infty}^{x} e^{-\frac{t^2}{2}} dt.$$

Putting

(21 $$x_0 = \frac{M[\Gamma]}{\varepsilon[\Gamma]},$$

we then see that the probability of $\Gamma < 0$ is approximately equal to $\Phi(-x_0)$.

Assuming that the error of approximation can be neglected, so that for practical purposes we may regard Γ as an exactly normally distributed variable, we thus see that it is sufficient to consider the quantity x_0. If, e. g., we have $x_0 > 3$, the probability of a negative value of Γ is less than $\Phi(-3) = 0.00135$, which means that there is only a very small chance for a disturbance of the regular bonus distribution, when the total result of the period is alone considered. The greater x_0, the smaller the probability of a loss due to random fluctuations. Thus it is important to investigate the conditions under which x_0 becomes greater than some conveniently fixed value, say 3. The practical measures concerning reinsurance etc. will have to be arranged so as to yield the greatest possible value of x_0, and the business should be so conducted that a value of x_0 once acquired should, generally speaking, not be allowed to decrease. If the value of x_0 with respect to a period of $t = 1$ year may, at a certain moment, be regarded as satisfactory, and if we know that x_0 is continually increasing, then according to our theoretical assumptions we should be allowed to conclude that satisfactory precautions have been taken against random fluctuations.

According to the above assumptions, it thus appears that the practical applications of the theory of risk can be entirely founded upon the principle that, in accordance with the normal probability function, deviations from the average course of business exceeding three or four times the mean risk are practically impossible. The applications of this principle are very numerous, and we now proceed to the discussion of a few simple examples intended to show the general character of the results obtained in this way.

19. Let us denote by q_i the probability that the ith policy in our group becomes a claim during the next year, and by $s_i + \sigma_i$ the total sum at risk under the policy, s_i being the part retained on the Company's own risk. (Thus we have $\sigma_i = 0$ in all cases where no part at all has been reinsured.) As in Art. 17 we suppose that a loading amounting to λ per unit of net premium may be disposed for the object of covering random fluctuations, and that the reinsurance premium contains a loading of λ' per unit of net premium. It must be kept in mind that, according to this definition, λ is generally only a small part of the loading contained in the office premium, while λ', on the other hand, denotes the total loading of the reinsurance premium. According to Art. 17 and the general relations (20) and (21), we then have

$$M[\Gamma] = \lambda \sum s_i q_i - (\lambda' - \lambda) \sigma_i q_i,$$

$$\varepsilon^2 [\Gamma] = \sum s_i^2 p_i q_i,$$

$$(22) \qquad x_0 = \frac{\lambda \sum s_i q_i - (\lambda' - \lambda) \sum \sigma_i q_i}{\sqrt{\sum s_i^2 p_i q_i}}$$

The expression for x_0 may be discussed from various points of view. We want a comparatively large value of x_0 in order to provide for a stable course of business. We also want a large value of $M[\Gamma]$, which means that we do not wish to employ reinsurance to a greater extent than necessary. On the other hand, it is obviously desirable to take λ as small as possible, so as to make the rest of the loading contained in the office premiums available for distribution of bonus or dividends. Accordingly, we may ask how the reinsured amounts σ_i should be fixed so as to combine as far as possible large values of x_0 and $M[\Gamma]$ with a small value of λ. If, on the other hand, the maximum of the Company and the premium rates for reinsurance are given, the question arises as to what part (λ) of the loading should be kept in hand in order to produce a sufficiently large value of x_0. Further, assuming that a satisfactory value of x_0 has been once attained, it is important to know how the business should be conducted so as to avoid a decrease of x_0.

It is hardly possible to obtain general results from a discussion of (22) without introducing some simplifying assumptions. For the further investigation, we shall suppose that all the probabilities q_i are equal, so that we obtain[1]

$$(22\text{ a}) \qquad x_0 = \sqrt{\frac{q}{p}} \cdot \frac{\lambda \sum s_i - (\lambda' - \lambda) \sum \sigma_i}{\sqrt{\sum s_i^2}}.$$

Thus it is obvious that the *distribution of the sums* will play an important part in the investigation. It has been suggested[2] that the remarkable law proposed by PARETO for the distribution of incomes in the general population should be applicable also to the problem here under discussion, and I have recently[3] given some figures which seem to show that, at least for the distribution of life assurance amounts, PARETO's law can be used as a first approximation.

---

[1] In the very special case when also all the s_i are equal and nothing is reinsured, (22 a) reduces to $x_0 = \sqrt{\frac{nq}{p}}$, so that x_0 can be made as large as we please by increasing sufficiently the number of policies.

[2] Cf. MEIDELL, VII Congrès Intern. des Actuaires, I, p. 85 and HAGSTRÖM, Skand. Aktuarietidskr. 1925, p. 65.

[3] Sjunde nordiska Lifförsäkringskongressen, Oslo 1926, p. 64.

According to this law, the number of sums $s_i + \sigma_i$ exceeding an arbitrary value t would be approximately given by the expression

$$F(t) = n\left(\frac{t}{m}\right)^{-a},$$

where m denotes the smallest sum occurring in the group, n is the total number of policies and a is a constant which, generally, seems to lie somewhere near 1.5.

Let us suppose that this law holds for the policies in our group, with a value of a between 1 and 2, and that the maximum amount retained by our Company on its own risk on one policy is equal to

$$R = \rho\, m.$$

Then we may say that the number of sums between the limits t and $t + dt$ is

$$-F'(t)\, dt = \frac{a\, n}{m}\left(\frac{t}{m}\right)^{-a-1} dt,$$

and for every policy with a total sum at risk exceeding R the amount retained by the Company is $s_i = R$. Thus we obtain

$$\Sigma s_i = \frac{a\, n}{m}\int_m^{\rho m} t\left(\frac{t}{m}\right)^{-a-1} dt + \frac{a\, n}{m}\int_{\rho m}^{\infty} \rho\, m\left(\frac{t}{m}\right)^{-a-1} dt$$

$$= \frac{a - \rho^{1-a}}{a - 1}\, n\, m,$$

$$\Sigma \sigma_i = \frac{a\, n}{m}\int_{\rho m}^{\infty} (t - \rho\, m)\left(\frac{t}{m}\right)^{-a-1} dt$$

$$= \frac{\rho^{1-a}}{a - 1}\, n\, m,$$

$$\Sigma s_i^2 = \frac{a\, n}{m}\int_m^{\rho m} t^2\left(\frac{t}{m}\right)^{-a-1} dt + \frac{a\, n}{m}\int_{\rho m}^{\infty} (\rho\, m)^2\left(\frac{t}{m}\right)^{-a-1} dt$$

$$= \frac{2\rho^{2-a} - a}{2 - a}\, n\, m^2.$$

Substituting these expressions in (22 a), we get

(23) $$x_0 = \frac{\sqrt{2-a}}{a-1}\cdot\sqrt{\frac{n\, q}{p}}\cdot\frac{a\,\lambda - \lambda'\,\rho^{1-a}}{\sqrt{2\rho^{2-a} - a}},$$

and for the »normal» value $\alpha = 1.5$ this reduces to

$$(23\text{ a}) \qquad x_0 = \sqrt{\frac{nq}{p}} \cdot \frac{3\lambda\rho^{\frac{1}{2}} - 2\lambda'}{\sqrt{\rho(4\rho^{\frac{1}{2}} - 3)}}.$$

Let us now suppose e. g. that λ' is given, which means that the premium rates for reinsurance are fixed. We then wish to take λ as small as possible and ρ as large as possible (since obviously a large maximum increases the mean value $M[\Gamma]$) without reducing x_0 below the value 3, or any other value that we may have chosen. In the following table some numerical values of x_0 are given, corresponding to the case $\lambda' = 0.5$, $\frac{nq}{p} = 1\,000$.

Table 1.

Values of x_0. $\lambda' = 0.5$, $\frac{nq}{p} = 1\,000$.

ρ \ λ	0.100	0.125	0.150	0.175	0.200	0.225	0.250
10	− 0.17	0.60	1.36	2.13	2.89	3.65	4.42
20	+ 0.63	1.24	1.86	2.48	3.08	3.70	4.32
30	0.85	1.39	1.94	2.49	3.03	3.58	4.12
40	0.97	1.46	1.96	2.46	2.96	3.46	3.97
50	1.00	1.47	1.94	2.41	2.89	3.36	3.83
60	1.02	1.47	1.92	2.37	2.82	3.26	3.71
70	1.03	1.46	1.89	2.33	2.76	3.18	3.61
80	1.04	1.45	1.87	2.29	2.70	3.11	3.52
90	1.04	1.44	1.85	2.25	2.65	3.05	3.45
100	1.04	1.43	1.82	2.22	2.61	2.99	3.38

According to this table, the smallest value of λ, such that it is possible by a suitable choice of ρ to make $x_0 > 3$, would be about $\lambda = 0.2$, and the largest corresponding value of the maximum R would be about $30\,m$. A comparatively small increase of λ would apparently allow a considerable increase of R (cf., however, Arts. 23 and 28).

For a given value of ρ (or R), the value of x_0 increases steadily with λ, and from a table like the above we can easily determine the smallest value of λ that makes $x_0 \geqq 3$. For a given value of λ, on the other hand, the values of x_0 generally show a maximum for a certain ρ, and the corresponding R is thus, from the point of view of stability, the most advantageous value that can be used in combination with the given value of λ. An inspection of the table shows that the position of this optimum value of ρ varies with λ. It may be deduced from (23 a) that the largest possible x_0 is obtained by putting

$$\rho = \left[\frac{\lambda'}{\lambda} + \sqrt{\frac{\lambda'}{\lambda}\left(\frac{\lambda'}{\lambda} - 1\right)}\right]^2.$$

It follows from (23) or (23 a) that x_0 is proportional to $\sqrt{n}$. Thus if, starting from the assumptions of Table 1, we wish to make $x_0 \geq 3$ for $\lambda = 0.1$, $\rho = 50$, by increasing n without altering λ' or q, the number of policies must be multiplied by nine.

It should be remarked that, when trying to apply PARETO's law to the distribution of insurance amounts, m should not be put equal to the smallest sum actually occurring. As a matter of fact, the law does not hold even approximately for sums falling below a certain limit. Thus the smallest sums would have to be left out, and the law applied to the sums exceeding the limit. According to experience relating to Swedish Life Assurance, it would seem that m could perhaps be put equal to 2 000 kr., so that the order of magnitude of R suggested above would not seem unreasonable.

20. If we have a group of policies of the kind considered in the preceding article, and if we know that the corresponding value of x_0 can be regarded as satisfactory (so that, e. g., $x_0 \geqq 3$), then we may ask how to manage reinsurance questions etc. for new policies so as to avoid a decrease of x_0. Is it possible, e. g., to increase the maximum retained by the Company for new policies?

For the discussion of this question, we shall suppose that the Company writes a number of new policies, such that the sums $\sum s_i$, $\sum \sigma_i$ and $\sum s_i^2$ are increased by the quantities δ_1, δ_2 and δ_3 respectively.

We suppose further that the quantities

$$\frac{\delta_1}{\sqrt{\sum s_i^2}}, \quad \frac{\delta_2}{\sqrt{\sum s_i^2}}, \quad \frac{\delta_3}{\sum s_i^2},$$

are small.

The value of x_0 for the total group then becomes, according to (22 a),

$$x_0 + \delta x_0 = \sqrt{\frac{q}{p}} \cdot \frac{\lambda(\sum s_i + \delta_1) - (\lambda' - \lambda)(\sum \sigma_i + \delta_2)}{\sqrt{\sum s_i^2 + \delta_3}},$$

and we obtain, neglecting quantities of the second order,

$$\delta x_0 = \sqrt{\frac{q}{p}} \cdot \frac{\lambda \delta_1 - (\lambda' - \lambda)\delta_2}{\sqrt{\sum s_i^2}} - \frac{\delta_3 x_0}{2 \sum s_i^2}. \tag{24}$$

Usually, the present question is treated so that only *one* new policy is considered, and it is asked: »Which is the highest sum that can be kept on own risk for a new policy without producing a decrease of x_0?». In this case, we have $\delta_1 = s$, $\delta_2 = 0$ and $\delta_3 = s^2$, where s denotes the sum at risk under the new policy. The largest possible value of s is determined by the condition $\delta x_0 \leqq 0$ which gives

$$s \leqq \frac{2\lambda}{x_0}\sqrt{\frac{q}{p}\sum s_i^2}. \tag{25}$$

In the case $\lambda' = \lambda$ this reduces, by (22 a), to the well-known inequality

$$s \leqq 2\frac{\sum s_i^2}{\sum s_i}.$$

It does, however, seem unnecessarily severe to require that the addition of *every* individual new policy to our group should bring an increase of x_0. It would seem to be quite sufficient to know, e. g., that the addition of *all* the new policies acquired during a period of, say, one year would cause such an increase. In reality, it can be easily shown by examples that the condition (25) leads to unreasonable consequences.

In order to show this, let us suppose that the sums are distributed according to PARETO's law as in the preceding article, with $\alpha = 1.5$. We further take $\frac{nq}{p} = 1\,000$, $\lambda' = 0.5$, $\lambda = 0.2$, and we suppose that the maximum kept on the Company's own risk is $R = 30\ m$. Then by Table 1 we have $x_0 = 3.03$, and so obtain from (25), substituting the expression for $\sum s_i^2$ under PARETO's law according to the preceding article,

$$s \leqq 18.2\ m.$$

In this case, our condition would thus lead to the result that the maximum for one single new policy should be considerably lower than the maximum previously employed by the Company, a result which is evidently absurd.

If, on the contrary, we consider all the n_1 new policies obtained during e. g. one year, and suppose that these sums are distributed according to the same law, we have by Art. 19, the maximum kept at own risk for the new policies being denoted by $R_1 = \rho_1 m$,

$$\delta_1 = \left(3 - 2\rho_1^{-\frac{1}{2}}\right) n_1 m,$$

$$\delta_2 = 2\rho_1^{-\frac{1}{2}} n_1 m,$$

$$\delta_8 = \left(4\rho_1^{\frac{1}{2}} - 3\right) n_1 m^2,$$

$$\Sigma s_i^2 = \left(4\rho^{\frac{1}{2}} - 3\right) n m^2.$$

Introducing these expressions in (24), the condition that the addition of *all* the new policies should make $\delta x_0 \geqq 0$ gives us after some reductions

$$\rho_1^{\frac{1}{2}}\left(\rho_1^{\frac{1}{2}} - \frac{3}{4}\right) + \frac{2\lambda' - 3\lambda\rho_1^{\frac{1}{2}}}{x_0}\sqrt{\frac{nq}{p}\left(\rho^{\frac{1}{2}} - \frac{3}{4}\right)} \leqq 0. \tag{26}$$

The left hand side of this inequality is a polynomial of the second order in $\rho_1^{\frac{1}{2}}$, and as soon as the quantities n, q, λ, λ' and ρ relating to the original group are given, x_0 may be calculated from (23 a) and then we can find the values of ρ_1 which satisfy (26). In the numerical example considered an instant ago, (26) becomes

$$\rho_1 - 14.4\,\rho_1^{\frac{1}{2}} + 22.7 \leqq 0,$$

and this inequality is satisfied within the approximate limits

$$3.3 < \rho_1 < 157.5$$

so that, according to this condition, a very considerable increase of the maximum would be allowed. It should, however, not be forgotten that this conclusion has been reached under the assumption that the new policies only form an insignificant part of the total group.

21. In the two preceding articles, we have treated some typical applications of the theory of the mean risk. Everything is based on the fundamental assumption that the gain of the Company can be regarded as a normally distributed variable, so that deviations exceeding three or four times the mean risk are practically impossible. So far, we have employed the mean risk for a period of one year only, and for a particularly simple type of policies. Quite similar considerations may, however, be made also for more complicated cases and for periods of any given length.

Taking, e. g., a group of life policies, the mean risk may be calculated for any period of length t by the methods of Arts. 13—17. Then we shall have to calculate x_0 according to the general expression (21), and to investigate the influence upon the value of x_0 exerted by various arrangements as to reassurance, loading of premiums etc., just as in the questions treated in the two preceding articles. It will also be important to study the effect of a variation of t. If, in an actual case, we know e. g. that $x_0 > 3$ for all values of t from one year up to ten years, we shall obviously be entitled to feel considerably more on the safe side than if we know the same thing for one single value of t only.

In the above discussion of the problems, it has been assumed throughout that the only means available to cover a loss due to random fluctuations during a certain period must be taken from the gain earned by the Company during the same period. As a matter of fact, it seems to be the case usually occurring in practice that a loss which could not be covered in this way would be very inconvenient to the Company. In some cases, however, there may be a special reserve constituted for this purpose, which can be disposed of without inconvenience. Such a reserve should, of course, be added to the mean value $M[\Gamma]$ of the gain when x_0 is calculated according to (21). Thus it is the sum of the special risk reserve and the mean value of the gain that should be made to exceed three or four times the mean risk.

Critical Remarks.

22. The mathematical treatment of the theory of risk as given above is in several respects open to criticism. Coming back to the preliminary discussion of the problems in Art. 6, we observe in the first place that, so far, we have only been able to approach the solution of the question A as stated in Art. 6. Thus we have considered a certain period t, and we have investigated the probability that the gain arising during this period, *taken as a whole*, will assume a negative value.

As far as short-term policies of the particular type considered in Arts. 19—20 are concerned, it is possible that we might be satisfied if we know that the probability just referred to is small for a period of $t =$ one year, and that the quantity denoted by x_0 is steadily increasing through the assimilation of new policies.

If, on the contrary, we consider a group of ordinary life policies, the probabilities of death and the sums at risk under the various policies are subject to a systematic variation from year to year, and so in this case it becomes important to investigate the risk for longer periods as well. Then we have to follow the continuous formation of the gain from the beginning to the end of the period, as indicated in Art. 6, and we are interested in the probability that a negative value of the gain will *never* occur during this time, or at least never at the end of any business period. Thus we are here led to the questions B and C as stated in Art. 6, and question A in itself has only a secondary interest. We may of course, as suggested in the preceding article, consider question A for a set of different values of t, and in this way obtain at least an upper limit for the probability required in question C, but generally we shall reach by this method only a rough estimate of the true value.

It seems, however, likely that we shall have to be satisfied with such a rough estimate, since a direct treatment of the questions B and C, under the assumptions of the individual risk theory, apart from the very simplest cases, appears to be a highly difficult problem. We shall get an idea of the degree of difficulty of the problem by considering the analogous problems in the theory of ordinary games of chance. In the first place, we may ask for the probability that, after a set of n games, the player has realized a certain gain. This corresponds to our question A, and is usually not a very complicated problem. Next, we suppose that our player starts with a given sum of money, and we ask for the probability that he will be ruined during the course of n games.

This is a form of the famous problem of the »Duration of Play», and obviously corresponds to our question B. It is well known that the problem of the Duration of Play is highly complicated, even if the conditions of each individual game are fairly simple. Now, the total risk business of an Insurance Company may be regarded as a continuous set of games of chance, every infinitely small time interval dt corresponding to a game, which may result either in the gain of an infinitesimal amount of risk premium corresponding to the time dt, or in a loss of the sum at risk under some of the policies of the Company. It will be easily understood that, under such complicated circumstances, the problem of Duration of Play must be exceedingly difficult. — We shall return to this question in Part III, in connection with the collective risk theory.

23. An even more serious objection against the theory is that the fundamental assumption concerning the normal distribution of the gain cannot be regarded as sufficiently well established. As we have seen above, all the practical applications are founded upon the assumption that the gain arising in a group of a large number of policies can be regarded as a normally distributed variable, and this assumption itself is derived from the general theorem on random variables discussed in Art. 11.

This theorem is concerned with a variable that, like our total gain

$$\Gamma = \gamma_1 + \gamma_2 + \cdots + \gamma_n$$

is equal to the sum of a large number of mutually independent components. What the theorem really asserts is simply that, subject to certain restrictions, the probability function of the normalized variable

$$\frac{\Gamma - M[\Gamma]}{\varepsilon[\Gamma]}$$

tends towards the normal function $\Phi(x)$ *as* n *tends to infinity.* It is perfectly obvious that this theorem does not give us any information at all as to the order of magnitude of the error committed when we assume that, already for a finite value of n, the variable Γ is normally distributed.

If we look at the various proofs of the theorem occurring in the literature, we shall easily find that in most cases the question of the error of approximation has been completely neglected, and that the results actually reached in this direction are not sufficient for our purpose. *Thus we are led to the conclusion that a considerable part of the applications of the theory of risk cannot be regarded as founded upon a sound mathematical basis.*

From the nature of the proofs of the general theorem, it may be suspected that the approximation will be particularly bad when the probability functions of the components are widely different, as e. g. when the sums at risk under the various policies show a great variation. Thus e. g. the conclusions drawn from a table such as our Table 1 (Art. 19), with respect to cases where the maximum amount at risk is very large, must be looked upon as particulary uncertain.

In order to construct a rigorous foundation of the theory of risk, it thus becomes necessary *to regard the numerical evaluation of the probability functions as an independent problem.* In the first place, the order of approximation furnished by the normal function should be investigated, and if satisfactory results are not reached in this way, we must try to find other expressions giving a closer approximation. In a recent paper[1], I have made some attempts in this direction. The results so far obtained have been of an essentially theoretical character; in the sequel I shall give some applications of the same method to the problem of numerical calculation of the probability functions occurring in the theory of risk. It is obvious that for such a purpose it will not be sufficient to consider only the mean risk or the average risk of the various policies. We shall be obliged to use also the higher moments or other parameters for a more intimate characterization of the probability functions.

On the Numerical Calculation of Probability Functions.

24. Using the same notations as in Arts. 10—11, we shall consider a sum

$$x = \sum_{1}^{n} x_\nu \tag{27}$$

of n mutually independent components x_ν. The probability function of x_ν will be denoted by $U_\nu(t)$ and the corresponding adjunct by $v_\nu(t)$, while m and σ are the mean value and the standard deviation of the sum x.

The probability function of the normalized variable

$$\xi = \frac{x - m}{\sigma}$$

is denoted by $F_n(t)$, and the corresponding adjunct by $g_n(t)$. It is the object of this Chapter to discuss a method for the approximate numerical calculation

[1] »On the Composition of Elementary Errors», I—II, Skandinavisk Aktuarietidskrift 1928, p. 13 and p. 141.

of $F_n(t)$. As this method will not give good results unless n is large, we shall suppose from the beginning that we have

$$n \geqq 1\,000.$$

It will be convenient to point out that no restriction is implied by assuming from the beginning that *the mean value of every* x_ν *is equal to zero (and consequently also that* $m = 0$*)*. As a matter of fact, if x_ν has the mean value m_ν, we need only replace (27) by the relation

$$x - m = \sum_1^n (x_\nu - m_\nu)$$

where, now, each component has the mean value zero. Every result obtained with respect to the probability distribution of $x - m$ gives immediately rise to a corresponding result about x.

Introducing the moments of the components x_ν:

$$\alpha_r^{(\nu)} = M[x_\nu^r] = \int_{-\infty}^{\infty} t^r \, d\, U_\nu(t),$$

we accordingly have

$$\alpha_1^{(\nu)} = 0$$

for $\nu = 1, 2, \ldots n$, and

$$\sigma^2 = \sum_1^n \alpha_2^{(\nu)}.$$

We further put

$$\beta_r^{(\nu)} = M[\,|x_\nu|^r] = \int_{-\infty}^{\infty} |t|^r \, d\, U_\nu(t),$$

(so that $\alpha_r^{(\nu)}$ and $\beta_r^{(\nu)}$ are identical for all even values of r), and

(28) $$B_r = \frac{1}{n} \sum_{\nu=1}^{n} \beta_r^{(\nu)},$$

(29) $$\lambda_r = \left(\frac{B_2^{\frac{1}{2}}}{B_r^{\frac{1}{r}}} \right)^6.$$

It follows from well-known properties of the moments (cf. e. g. my paper quoted above, p. 42 and p. 70) that we always have

(30) $$\begin{aligned} (\beta_r^{(\nu)})^{\frac{1}{r}} &\leqq (\beta_{r+1}^{(\nu)})^{\frac{1}{r+1}}, \\ (B_r)^{\frac{1}{r}} &\leqq (B_{r+1})^{\frac{1}{r+1}}, \end{aligned}$$

and consequently

$$1 \geqq \lambda_3 \geqq \lambda_4 \geqq \dots \tag{31}$$

In the particular case when all the probability functions $U_\nu(t)$ are equal, λ_r is obviously independent of n. *We shall suppose in the following that the moments* $\alpha_r^{(\nu)}$ *and* $\beta_r^{(\nu)}$ *of all the components are finite for* $r = 1, 2, \dots 5$.

For the sum x, we denote the r:th moment by

$$\mu_r = M[x^r].$$

so that μ_r can always be expressed in terms of the $\alpha_i^{(\nu)}$ for $i = 1, 2, \dots r$. In particular we have

$$\mu_2 = \sigma^2 = \sum_1^n \alpha_2^{(\nu)} = \sum_1^n \beta_2^{(\nu)} = n B_2 \tag{32}$$

Finally, we put

$$\begin{cases} \Phi_1(x) = \Phi(x) - \dfrac{1}{3!} \cdot \dfrac{\mu_3}{\sigma^3} \Phi^{(3)}(x), \\ \Phi_2(x) = \Phi(x) - \dfrac{1}{3!} \cdot \dfrac{\mu_3}{\sigma^3} \Phi^{(3)}(x) + \dfrac{1}{4!}\left(\dfrac{\mu_4}{\sigma^4} - 3\right)\Phi^{(4)}(x) + \dfrac{10}{6!}\left(\dfrac{\mu_3}{\sigma^3}\right)^2 \Phi^{(6)}(x) \end{cases} \tag{33}$$

where $\Phi(x)$ denotes, as usual, the normal probability function, and $\Phi^{(3)}, \dots$ are the successive derivatives of Φ.

It is our intention to investigate the approximation to $F_n(x)$ furnished by the expressions $\Phi_1(x)$ and $\Phi_2(x)$. For that purpose we put

$$\begin{cases} Q_n(x) = F_n(x) - \Phi_1(x), \\ R_n(x) = F_n(x) - \Phi_2(x), \end{cases} \tag{34}$$

and we introduce the notation

$$I^{(\omega)} f(x) = \frac{1}{\Gamma(\omega)} \int_{-\infty}^{x} (x-t)^{\omega - 1} f(t)\, dt$$

for the generalized (»RIEMANN-LIOUVILLE») integral of order $\omega > 0$ of a function $f(x)$. — The symbol $I^{(0)} f(x)$ will be defined as meaning the function $f(x)$ itself.

25. With the notations introduced in the preceding article, we have the following two fundamental inequalities:

$$|I^{(\omega)} Q_n(x)| < \frac{0.35}{\lambda_4 n} + \frac{0.44}{\sqrt[3]{\lambda_4 n}} e^{-\sqrt[3]{\lambda_4 n}} + \frac{1}{\pi} \int_{\sqrt{\frac{1}{8}\lambda_4 n}}^{\infty} |g_n(t)| \frac{dt}{t^{1+\omega}}, \tag{35}$$

$$(36) \qquad |I^{(\omega)} R_n(x)| < \frac{0.46}{\sqrt{(\lambda_5 n)^3}} + \frac{0.55}{\sqrt[3]{\lambda_5 n}} e^{-\sqrt[3]{\lambda_5 n}} + \frac{1}{\pi} \int\limits_{\sqrt{\frac{1}{6}\lambda_4 n}}^{\infty} |g_n(t)| \frac{dt}{t^{1+\omega}},$$

which hold for all real values of x under the assumptions $n \geqq 1\,000$, $\lambda_4 n \geqq 1$ (for (35)), $\lambda_5 n \geqq 1$ (for (36)) and $0 \leqq \omega \leqq 1$. — The value $\omega = 0$ gives, of course, finite values to the right hand sides of (35) and (36) if and only if the integral occurring in these relations is convergent for $\omega = 0$. For $\omega > 0$, it is readily seen that the integral is always convergent, since we have $|g_n(t)| \leqq 1$ for all values of t.

The proofs of (35) and (36) being absolutely similar, we shall restrict ourselves here to the proof of (35), which is slightly less complicated.

Putting

$$q_n(t) = \int\limits_{-\infty}^{\infty} e^{-ity} \, d\, Q_n(y),$$

we have by (3)

$$(37) \qquad q_n(t) = g_n(t) - e^{-\frac{t^2}{2}} + \frac{1}{3!} \cdot \frac{\mu_3}{\sigma^3} (it)^3 e^{-\frac{t^2}{2}}.$$

Denoting by $\Re z$ the real part of z, we shall first prove the identity

$$(38) \qquad I^{(\omega)} Q_n(x) = \frac{1}{\pi} \Re \int\limits_0^{\infty} \frac{e^{itx}}{(it)^{1+\omega}} q_n(t)\, dt.$$

The integral on the right hand side of this relation is absolutely convergent for $\omega > 0$. We shall assume for the proof that we have $0 < \omega < 1$, the result being extended to the value $\omega = 1$ by continuity, and to the value $\omega = 0$ in the same way if the integral is absolutely convergent for $\omega = 0$. For $0 < \omega < 1$ we have, the argument of $(it)^{1+\omega}$ being always taken between $\frac{1}{2}\pi$ and π,

$$(39) \qquad \frac{1}{\pi} \Re \int\limits_0^{\infty} \frac{e^{it(x-y)} - 1}{(it)^{1+\omega}} dt = \begin{cases} \dfrac{(x-y)^{\omega}}{\Gamma(1+\omega)} & \text{for } y < x \\ 0 & \text{» } y \geqq x. \end{cases}$$

Further, we obtain by a partial integration

$$I^{(\omega)} Q_n(x) = \frac{1}{\Gamma(\omega)} \int\limits_{-\infty}^{x} (x-y)^{\omega-1} Q_n(y)\, dy$$

$$= \frac{1}{\Gamma(1+\omega)} \int\limits_{-\infty}^{x} (x-y)^{\omega} \, d\, Q_n(y),$$

and by means of (39) this may be written

$$I^{(\omega)} Q_n(x) = \frac{1}{\pi} \Re \int_{-\infty}^{\infty} d\,Q_n(y) \int_0^{\infty} \frac{e^{it(x-y)} - 1}{(it)^{1+\omega}}\, dt.$$

The repeated integral occurring here is easily seen to be absolutely convergent, so that the order of integration may be reversed, and we obtain

$$I^{(\omega)} Q_n(x) = \frac{1}{\pi} \Re \int_0^{\infty} \frac{dt}{(it)^{1+\omega}} \int_{-\infty}^{\infty} (e^{it(x-y)} - 1)\, d\,Q_n(y)$$

$$= \frac{1}{\pi} \Re \int_0^{\infty} \frac{dt}{(it)^{1+\omega}} \int_{-\infty}^{\infty} e^{it(x-y)}\, d\,Q_n(y)$$

$$= \frac{1}{\pi} \Re \int_0^{\infty} \frac{e^{itx}}{(it)^{1+\omega}}\, q_n(t)\, dt,$$

so that (38) is proved.

From (37) and (38) we obtain

$$|I^{(\omega)} Q_n(x)| < A_1 + A_2 + A_3,$$

where

(40) $$A_1 = \frac{1}{\pi} \int_0^{\frac{3}{2}\sqrt[6]{\lambda_4 n}} |q_n(t)| \frac{dt}{t^{1+\omega}},$$

(41) $$A_2 = \frac{1}{\pi} \int_{\frac{3}{2}\sqrt[6]{\lambda_4 n}}^{\sqrt{\frac{1}{3}\lambda_4 n}} |g_n(t)| \frac{dt}{t^{1+\omega}} + \frac{1}{\pi} \int_{\frac{3}{2}\sqrt[6]{\lambda_4 n}}^{\infty} \left(1 + \frac{|\mu_3|}{6\sigma^3} t^3\right) e^{-\frac{t^2}{2}} \frac{dt}{t^{1+\omega}},$$

$$A_3 = \frac{1}{\pi} \int_{\sqrt{\frac{1}{3}\lambda_4 n}}^{\infty} |g_n(t)| \frac{dt}{t^{1+\omega}},$$

In order to prove (35), it will thus be sufficient to prove the inequalities

(42) $$A_1 < \frac{0.35}{\lambda_4 n}$$

and

(43) $$A_2 < \frac{0.44}{\sqrt[3]{\lambda_4 n}} e^{-\sqrt[3]{\lambda_4 n}}.$$

We begin with the proof of (42). — Since by hypothesis $m_\nu = 0$ for each component x_ν, we obtain from (2)

(44)
$$g_n(t) = \prod_1^n v_\nu\left(\frac{t}{\sigma}\right).$$

For every real a we have

$$e^{ia} = \sum_0^3 \frac{(ia)^r}{r!} + \vartheta\frac{a^4}{4!},$$

ϑ denoting here and in the following a real or complex number, such that $|\vartheta| < 1$. Hence we obtain

$$v_\nu\left(\frac{t}{\sigma}\right) = \int_{-\infty}^{\infty} e^{-\frac{itx}{\sigma}}\, d\, U_\nu(x)$$

(45)
$$= 1 + \frac{\alpha_2^{(\nu)}}{2!}\left(-\frac{it}{\sigma}\right)^2 + \frac{\alpha_3^{(\nu)}}{3!}\left(-\frac{it}{\sigma}\right)^3 + \vartheta\frac{\beta_4^{(\nu)}}{4!}\left(-\frac{it}{\sigma}\right)^4$$

$$= 1 + z_\nu,$$

(46)
$$\log v_\nu\left(\frac{t}{\sigma}\right) = z_\nu + \vartheta\left(\frac{|z_\nu|^2}{2} + \frac{|z_\nu|^3}{3} + \cdots\right)$$

$$= z_\nu + \frac{1}{2}\vartheta z_\nu^2\left(1 + \frac{2}{3}\cdot\frac{|z_\nu|}{1 - |z_\nu|}\right).$$

For $0 < t < \frac{3}{2}\sqrt[6]{\lambda_4 n}$ and for $r = 2, 3, 4$ we have by (28), (29), (30) and (32)

$$\left|\frac{\alpha_r^{(\nu)}}{r!}\left(-\frac{it}{\sigma}\right)^r\right| \leq \left(\frac{(\beta_4^{(\nu)})^{\frac{1}{4}}\, t}{\sigma}\right)^r$$

$$\frac{(\beta_4^{(\nu)})^{\frac{1}{4}}\, t}{\sigma} < \frac{(n B_4)^{\frac{1}{4}}\cdot\frac{3}{2}\sqrt[6]{\lambda_4 n}}{(n B_2)^{\frac{1}{2}}} = \frac{3}{2} n^{-\frac{1}{12}} < 0.844\ldots$$

Substituting these inequalities in (45), we obtain

$$z_\nu = \frac{1}{2}\vartheta\left(\frac{(\beta_4^{(\nu)})^{\frac{1}{4}}\, t}{\sigma}\right)^2\left(1 + \frac{0.844}{3} + \frac{(0.844)^2}{12}\right)$$

$$= 0.671\,\vartheta\left(\frac{(\beta_4^{(\nu)})^{\frac{1}{4}}\, t}{\sigma}\right)^2$$

$$= 0.478\,\vartheta,$$

$$z_\nu^2 = 0.451\,\vartheta\frac{\beta_4^{(\nu)}\, t^4}{\sigma^4}.$$

From (46) we get, since $|z_\nu| < 0.478$,

$$\log v_\nu\left(\frac{t}{\sigma}\right) = z_\nu + 0.806\,\vartheta\, z_\nu^2$$

and further, by means of the expressions for z_ν and z_ν^2,

$$\log v_\nu\left(\frac{t}{\sigma}\right) = -\frac{\alpha_2^{(\nu)}}{2!}\left(\frac{t}{\sigma}\right)^2 - \frac{\alpha_3^{(\nu)}}{3!}\left(\frac{it}{\sigma}\right)^3 + 0.406\,\vartheta\,\frac{\beta_4^{(\nu)}\,t^4}{\sigma^4}.$$

According to (44) we then have

$$\log g_n(t) = \sum_1^n \log v_\nu\left(\frac{t}{\sigma}\right)$$

$$= -\frac{1}{2}t^2 - \frac{1}{3!}\cdot\frac{\mu_3}{\sigma^3}(it)^3 + 0.406\,\vartheta\,\frac{n\,B_4\,t^4}{(n\,B_2)^2}$$

$$= -\frac{1}{2}t^2 - \frac{1}{3!}\,\frac{\mu_3}{\sigma^3}(it)^3 + 0.406\,\vartheta\,\frac{t^4}{\lambda_4^{\frac{2}{3}}\,n}$$

$$= -\frac{1}{2}t^2 + y,$$

(47) $$e^{\frac{t^2}{2}} g_n(t) = e^y = 1 + y + \frac{1}{12}\vartheta\, y^2 (5 + |y| + e^{|y|}),$$

the development of e^y being a simple consequence of the ordinary MAC LAURIN's series, valid for all complex values of y. Now we have by (29) and (31)

(48) $$\frac{\mu_3}{\sigma^3} = \vartheta\,\frac{n\,B_3}{(n\,B_2)^{\frac{3}{2}}} = \frac{\vartheta}{\sqrt{\lambda_3\,n}} = \frac{\vartheta}{\sqrt{\lambda_4\,n}},$$

and this gives us

$$y = -\frac{1}{3!}\cdot\frac{\mu_3}{\sigma^3}(it)^3 + 0.406\,\vartheta\,\frac{t^4}{\lambda_4^{\frac{2}{3}}\,n}$$

$$= \vartheta\,\frac{t^3}{\sqrt{\lambda_4\,n}}\left(\frac{1}{6} + 0.406\,\frac{t}{\sqrt[6]{\lambda_4\,n}\cdot\sqrt[3]{n}}\right)$$

$$= 0.228\,\vartheta\,\frac{t^3}{\sqrt{\lambda_4\,n}}$$

$$= 0.770\,\vartheta.$$

Introducing these inequalities in (47), we obtain in the first place

$$\frac{1}{12} y^2 (5 + |y| + e^{|y|}) = 0.035\, \vartheta \frac{t^6}{\lambda_4 n}$$

and hence, since $\lambda_4 \leqq 1$

$$e^{\frac{t^2}{2}} g_n(t) = 1 - \frac{1}{3!} \cdot \frac{\mu_3}{\sigma^3} (it)^3 + \frac{\vartheta}{\lambda_4 n} (0.406\, t^4 + 0.035\, t^6).$$

By (37), we thus have for $0 < t < \frac{3}{2} \sqrt[6]{\lambda_4 n}$

$$|q_n(t)| < \frac{0.406\, t^4 + 0.035\, t^6}{\lambda_4 n} e^{-\frac{t^2}{2}}$$

Substituting this in (40), we obtain for $0 \leqq \omega \leqq 1$

$$A_1 < \frac{1}{\lambda_4 n \pi} \left(0.406 \int_0^\infty t^{3-\omega} e^{-\frac{t^2}{2}} dt + 0.035 \int_0^\infty t^{5-\omega} e^{-\frac{t^2}{2}} dt \right)$$

$$= \frac{1}{\lambda_4 n \pi} \left[0.406 \cdot 2^{1-\frac{\omega}{2}} \Gamma\left(2 - \frac{\omega}{2}\right) + 0.035 \cdot 2^{2-\frac{\omega}{2}} \Gamma\left(3 - \frac{\omega}{2}\right) \right]$$

$$< \frac{1}{\lambda_4 n \pi} \left(0.406 \cdot 2 \cdot \Gamma(2) + 0.035 \cdot 2^2 \cdot \Gamma(3) \right)$$

$$< \frac{0.35}{\lambda_4 n},$$

so that (42) is proved.

It thus only remains to prove (43). According to (41). A_2 is the sum of two integrals, the upper limit of the first integral being $\sqrt{\frac{1}{6} \lambda_4 n}$. We shall begin by proving that, for $0 < t < \sqrt{\frac{1}{6} \lambda_4 n}$, we have

$$|g_n(t)| < e^{-\frac{4}{9} t^2} \tag{49}$$

We have

$$|v_\nu(t)|^2 = \int_{-\infty}^{\infty}\int_{-\infty}^{\infty} e^{it(x-y)}\, d\,U_\nu(x)\, d\,U_\nu(y)$$

$$= \int_{-\infty}^{\infty}\int_{-\infty}^{\infty} \cos t(x-y)\, d\,U_\nu(x)\, d\,U_\nu(y),$$

and

$$\cos\, t(x-y) \leqq 1 - \frac{1}{2}t^2(x-y)^2 + \frac{1}{24}t^4(x-y)^4$$

$$\leqq - 1\; \frac{1}{2}t^2(x^2 - 2\,xy + y^2) + \frac{1}{3}t^4(x^4 + y^4).$$

This gives us, since by hypothesis $m_\nu = 0$,

$$|v_\nu(t)|^2 \leqq 1 - \alpha_2^{(\nu)} t^2 + \frac{2}{3}\beta_4^{(\nu)} t^4$$

$$\leqq e^{-\alpha_2^{(\nu)} t^2 + \frac{2}{3}\beta_4^{(\nu)} t^4},$$

and thus we deduce from (44) and (31)

$$|g_n(t)| = \prod_1^n \left| v_\nu\left(\frac{t}{\sigma}\right)\right| \leqq e^{-\frac{1}{2}t^2 + \frac{1}{3}\cdot\frac{t^4}{\lambda_4 n}}$$

$$= e^{-\frac{1}{2}t^2\left(1 - \frac{2}{3}\cdot\frac{t^2}{\lambda_4 n}\right)},$$

and for $0 < t < \sqrt{\frac{1}{6}\lambda_4 n}$, we obtain (49).

By (48) and (49), we obtain from (41), since by hypothesis we have $\lambda_4 n \geqq 1$,

$$A_2 < \frac{2}{\pi}\int_{\frac{3}{2}\sqrt[6]{\lambda_4 n}}^{\infty} \frac{1}{t} e^{-\frac{4}{9}t^2}\, dt + \frac{1}{6\pi\sqrt{\lambda_4 n}}\int_{\frac{3}{2}\sqrt[6]{\lambda_4 n}}^{\infty} t^2 e^{-\frac{1}{2}t^2}\, dt.$$

By means of the easily obtained inequalities

$$\int_a^{\infty} \frac{1}{t} e^{-bt^2}\, dt < \frac{1}{2\,a^2 b} e^{-a^2 b},$$

$$\int_a^{\infty} t^2 e^{-\frac{1}{2}t^2}\, dt < \left(a + \frac{1}{a}\right) e^{-\frac{1}{2}a^2},$$

where $a > 0$, $b > 0$, we immediately deduce

$$A_2 < \frac{0.44}{\sqrt[8]{\lambda_4 n}} e^{-\sqrt[8]{\lambda_4 n}},$$

so that (43), and hereby also (35), is proved. — The second fundamental relation, (36), is proved in quite a similar way.

26. We now proceed to show how the inequalities (35) and (36) may be employed for the numerical calculation of the probability function $F_n(x)$.

If the probability distributions of the components x_ν are of the *second kind* (cf. Art. 8) and if the frequency functions are sufficiently regular, it is easily shown that the adjunct $v_\nu(t)$ corresponding to x_ν satisfies the inequality

$$|v_\nu(t)| < \frac{k_\nu}{t}$$

for all $t > 0$, k_ν being a constant. According to (44), it follows that the integral occurring in (35) and (36) is convergent for $\omega = 0$. Thus it is allowed to put $\omega = 0$ in (35) and (36), and so we directly obtain upper limits for the differences

$$Q_n(x) = F_n(x) - \Phi_1(x)$$

and

$$R_n(x) = F_n(x) - \Phi_2(x).$$

In all the examples of probability distributions treated above in connection with the theory of risk we have seen, however, that these have been either distributions of the *first kind* (short-term policies) or *mixed distributions* (various types of life policies, considered during an arbitrary period t). In such cases, it is not even true that the adjunct $v_\nu(t)$ tends to zero as t tends to infinity, and thus we shall obtain a divergent integral if we put $\omega = 0$ in (35) and (36). This means that we shall have to start from (35) or (36), with some positive value of ω, and then try to deduce an upper limit for the difference $Q_n(x)$ or $R_n(x)$ itself.

In the present article, it will be shown how this question can be solved if the distributions of the components x_ν are of the mixed type. In this case, the absolute value of the adjunct $v_\nu(t)$ attains its maximum 1 for the value $t = 0$ only[1], and in every finite or infinite interval which does not contain the point $t = 0$ we can assign to $|v_\nu(t)|$ an upper limit which is less than 1. The adjunct $g_n(t)$ being a product of n factors of this type, it is obvious that the upper limit obtained for $|g_n(t)|$ will generally be small when n is large.

Let us suppose that, in an actual case, we have proved the inequality

(50) $$|g_n(t)| < \frac{1}{2}\pi\varepsilon$$

[1] Cf. Skandinavisk Aktuarietidskr. 1928, p. 59—60.

for all $t > \sqrt{\frac{1}{6}\lambda_4 n}$, ε denoting a small positive number. For the sake of simplicity, we shall assume in the sequel $\lambda_4 n \geqq 6$. (If this condition is not satisfied, it is fairly obvious that neither (35) nor (36) will give any valuable results.) Then, putting

$$(51)\qquad \begin{cases} \frac{1}{2}\delta = \frac{0.35}{\lambda_4 n} + \frac{0.44}{\sqrt[3]{\lambda_4 n}} e^{-\sqrt[3]{\lambda_4 n}}, \\ \frac{1}{2}\delta' = \frac{0.46}{\sqrt{(\lambda_5 n)^3}} + \frac{0.55}{\sqrt[3]{\lambda_5 n}} e^{-\sqrt[3]{\lambda_5 n}}, \end{cases}$$

(35) and (36) may be written

$$(35\text{ a})\qquad |I^{(\omega)} Q_n(x)| < \frac{1}{2}\left(\delta + \frac{\varepsilon}{\omega}\right),$$

$$(36\text{ a})\qquad |I^{(\omega)} R_n(x)| < \frac{1}{2}\left(\delta' + \frac{\varepsilon}{\omega}\right).$$

Let us consider first (35 a). This may be written

$$\int_{-\infty}^{x} (x - t)^{\omega - 1} Q_n(t)\, dt = \frac{1}{2}\vartheta\, \Gamma(\omega)\left(\delta + \frac{\varepsilon}{\omega}\right),$$

and by a theorem due to M. Riesz[1] we may conclude that we have for any $h > 0$

$$\int_{-\infty}^{x-h} (x - t)^{\omega - 1} Q_n(t)\, dt = \frac{1}{2}\vartheta\, \Gamma(\omega)\left(\delta + \frac{\varepsilon}{\omega}\right),$$

ϑ always denoting a number such that $|\vartheta| < 1$. This gives us

$$\int_{x-h}^{x} (x - t)^{\omega - 1} Q_n(t)\, dt = \vartheta\, \Gamma(\omega)\left(\delta + \frac{\varepsilon}{\omega}\right),$$

and according to (34) this may be written

$$\int_{x-h}^{x} (x - t)^{\omega - 1} F_n(t)\, dt = \int_{x-h}^{x} (x - t)^{\omega - 1} \Phi_1(t)\, dt + \vartheta\, \Gamma(\omega)\left(\delta + \frac{\varepsilon}{\omega}\right).$$

[1] »Sur un théorème de la moyenne et ses applications», Acta Univ. Hungaricæ Franc.-Jos., Vol. I, 1923, p. 114. Cf. also Hardy and Riesz, »The general theory of Dirichlet's series» Cambridge Tracts in Mathematics, No. 18 (1915), Lemma 7, p. 28.

Now $F_n(t)$ is a never decreasing function of t, while according to (33) and (48) the modulus of the derivative $\Phi_1'(t)$ can never exceed unity. Thus we obtain

$$F_n(x)\int_{x-h}^{x}(x-t)^{\omega-1}\,dt > (\Phi_1(x)-h)\int_{x-h}^{x}(x-t)^{\omega-1}\,dt - \Gamma(\omega)\left(\delta+\frac{\varepsilon}{\omega}\right)$$

or

$$F_n(x) > \Phi_1(x) - h - h^{-\omega}\,\Gamma(\omega+1)\left(\delta+\frac{\varepsilon}{\omega}\right).$$

In the same way we get, replacing $x-h$ by x,

$$F_n(x) < \Phi_1(x) + h + h^{-\omega}\,\Gamma(\omega+1)\left(\delta+\frac{\varepsilon}{\omega}\right),$$

and finally, since $\Gamma(\omega+1)\leqq 1$ for $0<\omega\leqq 1$,

$$(52)\qquad |F_n(x)-\Phi_1(x)| < h + h^{-\omega}\left(\delta+\frac{\varepsilon}{\omega}\right).$$

The relation (35 b) may be treated in the same way, since it is easily shown that we have also $|\Phi_2'(t)|<1$, and we obtain

$$(53)\qquad |F_n(x)-\Phi_2(x)| < h + h^{-\omega}\left(\delta'+\frac{\varepsilon}{\omega}\right).$$

In (52) and (53), h denotes an absolutely arbitrary positive number, while ω is a positive number $\leqq 1$. By giving various numerical values to h and ω, we obtain from (52) and (53) numerical limits for the error committed when $F_n(x)$ is replaced by $\Phi_1(x)$ or $\Phi_2(x)$.

Putting, e. g.,

$$\omega = \frac{\varepsilon}{\delta},$$

$$h = \delta^{\frac{\delta}{\delta+\varepsilon}},$$

we obtain from (52)

$$(54)\qquad |F_n(x)-\Phi_1(x)| < 3\,\delta^{\frac{\delta}{\delta+\varepsilon}}.$$

Putting, on the other hand,

$$\omega = \frac{1}{\log\frac{1}{\varepsilon}},$$

$$h = \varepsilon,$$

we obtain

$$(55) \qquad |F_n(x) - \Phi_1(x)| < \varepsilon + e\left(\delta + \varepsilon \log \frac{1}{\varepsilon}\right).$$

In most cases, it will be found that (55) gives better results than (54).

By substituting δ' for δ, we can obviously deduce in the same way the inequalities

$$(56) \qquad |F_n(x) - \Phi_2(x)| < 3\,\delta'^{\frac{\delta'}{\delta'+\varepsilon}}$$

and

$$(57) \qquad |F_n(x) - \Phi_2(x)| < \varepsilon + e\left(\delta' + \varepsilon \log \frac{1}{\varepsilon}\right).$$

For the sake of clearness, we shall finally repeat the assumptions underlying the inequalities (52)—(57). These inequalities hold for all real values of x, if $n \geqq 1\,000$ and $\lambda_4 n \geqq 6$. In the relations concerned with Φ_2, we have also assumed $\lambda_5 n \geqq 1$.

27. In order to illustrate by a numerical example the application of the inequalities deduced in the preceding article, we shall choose the whole-life assurance with increasing amount considered in Art. 13 (cf. formula (10) and Fig. 7). If the initial assurance amount is s, the sum payable at death after t years under this policy is

$$s\left[\left(2\frac{l_{x+t}}{l_x} + 1\right)e^{\delta t} - 2\right].$$

We consider the gain arising during a period of t years, t being so chosen that

$$\frac{l_{x+t}}{l_x} = \tfrac{1}{2}.$$

For the case $s = 1$, the probability distribution is then represented by Fig. 7 and the moments are given by formula (14) of Art. 14. If we are given a group of n policies of this type, the initial assurance amounts being $s_1, s_2, \ldots s_n$, we thus have, according to the notations of Art. 24,

$$\alpha_r^{(\nu)} = \frac{1}{2} s_\nu^r \left[\frac{(-1)^r}{r+1} + \left(\frac{1}{2}\right)^r\right],$$

$$\beta_r^{(\nu)} = \frac{1}{2} s_\nu^r \left[\frac{1}{r+1} + \left(\frac{1}{2}\right)^r\right].$$

Putting

$$S_r = \frac{1}{n}\sum_{\nu=1}^{n} s_\nu^r,$$

we further have

$$B_r = \frac{1}{2}\, S_r \left[\frac{1}{r+1} + \left(\frac{1}{2}\right)^r\right],$$

and thus in particular

$$\sigma^2 = \frac{7}{24}\, n\, S_2,$$

$$\lambda_4 = 0.5218\, \frac{S_2^3}{S_4^{\frac{3}{2}}},$$

$$\lambda_5 = 0.3982\, \frac{S_2^3}{S_5^{\frac{6}{5}}}.$$

The adjunct $v_\nu(t)$ is given by the relation

$$v_\nu(t) = \frac{1}{2\, s_\nu} \int_{-s_\nu}^{0} e^{-itx}\, dx + \frac{1}{2}\, e^{-\frac{1}{2} i s_\nu t}$$

$$= \frac{1}{2}\left(e^{-\frac{1}{2} i s_\nu t} + e^{\frac{1}{2} i s_\nu t}\, \frac{\sin \frac{1}{2} s_\nu t}{\frac{1}{2} s_\nu t}\right),$$

and thus we have

$$|g_n(t)| = \prod_{1}^{n} \left| v_\nu\left(\frac{t}{\sigma}\right)\right|$$

$$< \prod_{1}^{n} \frac{1}{2}\left(1 + \left|\frac{\sin x_\nu t}{x_\nu t}\right|\right)$$

where

$$x_\nu = \frac{6\, s_\nu}{\sqrt{42\, n\, S_2}}.$$

Now it is easily proved that we have

$$\frac{1}{2}\left(1 + \left|\frac{\sin x}{x}\right|\right) < \begin{cases} e^{-\frac{x^2}{24}} & \text{for } 0 < x \leq 3 \\ e^{-\frac{3}{8}} & \text{» } x \geq 3. \end{cases}$$

It follows that, if for a certain value of ν the inequality

$$(58) \qquad x_\nu \sqrt{\frac{1}{6}\lambda_4 n} \leq 3$$

is satisfied, we have for all $t > \sqrt{\frac{1}{6}\lambda_4 n}$

$$\frac{1}{2}\left(1 + \left|\frac{\sin x_\nu t}{x_\nu t}\right|\right) < e^{-\frac{1}{144}\lambda_4 n x_\nu^2}$$

$$= e^{-\frac{\lambda_4 n}{168} \cdot \frac{s_\nu^2}{\Sigma s_\nu^2}}.$$

Hence we conclude, for $t > \sqrt{\frac{1}{6}\lambda_4 n}$.

$$(59) \qquad |g_n(t)| < e^{-M}$$

with

$$M = \frac{\lambda_4 n}{168} \cdot \frac{\Sigma' s_\nu^2}{\Sigma s_\nu^2},$$

the sum Σ' being extended over the values of ν which satisfy (58). Now (58) may be written

$$s_\nu^2 \leq \frac{63}{\lambda_4} S_2,$$

and by most distributions of assurance amounts likely to occur in practice this is satisfied by *all* the sums s_ν, so that we have $M = \frac{\lambda_4 n}{168}$. In particular, this will be the case in all the numerical examples given below.

Comparing (59) with (50), it follows that we can put

$$\varepsilon = \frac{2}{\pi} e^{-M}.$$

As soon as the sums s_ν are given, we can thus calculate the quantities δ, δ' and ε [δ and δ' being given by (51)], and then according to the inequalities deduced in the preceding article we can find an upper limit for the error committed when the probability function of the gain is replaced by $\Phi_1(x)$ or $\Phi_2(x)$.

In the numerical examples given in Table 2, we have first considered the case when the initial sums s_ν are all equal to s, so that we have $S_r = s$ for all r. The table gives, for the period considered, the probability of a loss exceeding 2σ, 3σ and 4σ, partly according to the normal function and partly according to the expression $\Phi_2(x)$. In the last column, the upper limit of the error $|F_n - \Phi_2|$, calculated by means of (56) and (57), is given with two significant digits. As soon as the number of policies n exceeds 5 000, it follows from the table that the probability function is obtained to four correct places of decimals, and for $n = 10\,000$ the error does not exceed one unit in the fifth place of decimals.

Further, the table gives the values of the corresponding probabilities and errors under the assumption that the initial amounts of the policies are distributed according to PARETO's law (cf. Art. 19) with $a = 1.5$, the smallest sum being m, and the largest initial amount kept on the Company's own risk being $R = \rho m$. If s_ν denotes the initial amount for that part of the ν:th policy which is kept on the Company's own risk, we then find by a simple generalization of the deductions of Art. 19

$$S_r = \frac{1}{n}\sum s_\nu^r = \frac{2r\rho^{r-\frac{3}{2}} - 3}{2r - 3} m^r .$$

In this part of the table, we have supposed $n = 100\,000$, and we have considered five different values of ρ. If $\rho = 1$, all the sums kept on own risk are equal, so that this case forms an extension of the first part of the table to the case $n = 100\,000$, and it is seen that the error can only amount to about three units in the seventh place of decimals. As soon as we pass to cases with $\rho > 1$ we see, however, that the limit of the error furnished by our theory rapidly increases. Already for $\rho = 20$ it is obvious that such a statement as e. g. the following: »The probability of a loss exceeding 3σ is approximately equal to $\Phi(-3) = 0.001\,35$», cannot be regarded as well-founded.

Table 2.

	Probability of loss exceeding						Maximum error of $\Phi_2(x)$
	2σ		3σ		4σ		
	Normal function	$\Phi_2(x)$	Normal function	$\Phi_2(x)$	Normal function	$\Phi_2(x)$	
1. *All s_ν equal.*							
$n = 2\,500$ ..	0.022 750	0.022 964	0.001 350	0.001 397	0.000 032	0.000 034	0.006 8
5 000 ..	»	0.022 901	»	0.001 383	»	0.000 034	0.000 035
7 500 ..	»	0.022 873	»	0.001 377	»	0.000 033	0.000 016
10 000 ..	»	0.022 857	»	0.001 373	»	0.000 033	0.000 010
2. *Sums distributed according to* PARETO's *law.* $n = 100\,000$							
$\rho = 1$	0.022 750	0.022 784	0.001 350	0.001 357	0.000 032	0.000 032	0.000 000 33
5	»	0.022 800	»	0.001 361	»	0.000 032	0.000 005 1
10	»	0.022 820	»	0.001 365	»	0.000 033	0.000 049
15	»	0.022 839	»	0.001 369	»	0.000 033	0.000 48
20	»	0.022 855	»	0.001 373	»	0.000 033	0.018

In none of the cases occurring in the table, the correction terms containing derivatives of $\Phi(x)$ occurring in $\Phi_2(x)$ show any remarkable influence on the numerical values of the probabilities. As far as our example goes, we can thus say that the normal function gives a remarkably good approximation in cases where we hawe some thousands of policies of nearly equal amounts. It cannot, however, be regarded as proved that the same thing is true in the very cases where we want to apply it, viz. those cases where the amounts of the various policies are distributed in such a way as is usually found in practice.

28. When the distributions of the components x_ν are all of the *first kind* the method of approximation developed in the preceding articles fails completely, since in this case it does not seem possible to obtain something corresponding to (50), with a sufficiently small value of ε. For this case, I know of no better general result than the inequality

$$|F_n(x) - \Phi(x)| < \frac{3 \log n}{\sqrt{\lambda_3 n}} \tag{60}$$

published by me some years ago.[1] This inequality holds for all $n > 1$ and for all values of x without any restrictions whatever. Generally, however, it does not give anything like so good results as are required for the applications occurring in the theory of risk.

Take, e. g., the case of a group of short-term policies considered in Arts. 19—20. In this case, we find

$$B_2 = pq\, S_2,$$

$$B_3 = pq(1 - 2pq)\, S_3,$$

$$\lambda_3 = \frac{pq}{(1 - 2pq)^2} \cdot \frac{S_2^3}{S_3^2},$$

S_r denoting as in the preceding article the arithmetic mean of the r:th powers of the sums s_ν. Consequently we obtain from (60)

$$|F_n(x) - \Phi(x)| < 3(1 - 2pq) \frac{\log n}{\sqrt{n pq}} \cdot \frac{S_3}{S_2^{\frac{3}{2}}}.$$

In the numerical example set out in Table 1, Art. 19, we have assumed $\frac{nq}{p} = 1\,000$.

[1] »Das Gesetz von GAUSS und die Theorie des RISIKOS», Skandinavisk Aktuarietidskrift 1923, p. 209. The notations of that paper differ from those used above, but the inequality here given is easily deduced from Satz 1 of the 1923 paper. Cf. also my paper in the same journal, 1928, p. 19.

Thus we can take, e. g, $n = 20\,000$ and $q = \frac{1}{21}$, and for this case the last inequality gives us

$$| F_n(x) - \Phi(x) | < 0.898 \frac{S_3}{S_2^{\frac{3}{2}}}.$$

Now, the factor $S_3 S_2^{-\frac{3}{2}}$ is always ≥ 1, and so the limit of the error furnished by the general theorem is of no value for the question of the order of approximation. Even by taking such a large number of policies as $n = 200\,000$ we shall only get the numerical factor 0.898 replaced by 0.350.

29. It appears from all the results obtained in this Chapter that we cannot, at the present state of the theory, prove that the probability functions occurring in the theory of risk are represented by the normal function with an order of approximation, which is sufficiently good to justify the applications usually made of the theory. It is possible, of course, that a further development of the theory may show that the limits of the errors obtained with our present methods are much too large and that, in reality, the approximation is satisfactory. It is, however, obvious that this question cannot be finally settled without further investigations into the problem of numerical calculation of probability functions. Such investigations, and also the detailed numerical treatment of particular examples, seem to be highly desirable for the development of the theory.

* *
*

PART III. Collective Risk Theory.

Fundamental Ideas.

30. In Part II of the present paper, the theory of risk has been treated according to the ordinary, individual, point of view. Beginning with an investigation of the probability distribution of the gain corresponding to one single policy, we have expressed the total gain of a Company as the sum of a large number of individual gains, and we have tried to reach in this way a satisfactory solution of the practical problems connected with the subject of random fluctuations. It has been shown, however, that the development of the theory along these lines meets with considerable difficulties which cannot, at the present state of our knowledge, be completely removed.

This being so, we may quite reasonably ask if it would not be possible to find some convenient way of modifying the *problems*, so as to make a rigorous solution more easily obtainable. This is, in reality, the way that has been followed by D:r F. LUNDBERG in his highly remarkable works on the theory of risk.[1] We shall give below some comments upon the fundamental ideas of LUNDBERG's theory, as compared with the ordinary point of view, and then proceed to a brief account of some of the main results of the theory. For further developments, the reader must be referred to the original papers.

As we have indicated in Art. 22, the total risk business of an Insurance Company can be regarded as a continuous set of games of chance. Every infinitely small time dt corresponds to one game, during which the totality of the policyholders pay to the Company a certain total amount of risk premium dP, while on the other hand the Company is obliged to pay the sum at risk under any policy that becomes a claim during the time dt.

Let us now suppose that $d\pi$ is the probability that a claim occurs during dt and that, *if such of claim occurs*, we have the probability $p(z)\,dz$ that the corresponding sum at risk falls between z and $z + dz$. Taking the average sum at risk as unit for z, we have

$$(61) \qquad \int_0^\infty z p(z) dz = \int_0^\infty p(z) dz = 1.$$

Since the mean value of the payments for claims during dt must be equal to the net risk premium dP, we further have

$$d\pi \int_0^\infty z p(z) dz = dP,$$

[1] Cf. Art. 7, footnote.

and thus

$$d\pi = dP.$$

The conditions of the »elementary game» played during dt are thus completely known: The Company receives the amount dP as a stake, and has the probability

$$p(z)dz\,dP$$

of being obliged to pay an amount between z and $z + dz$.

It is obvious that the total gain of the Company during a certain period is equal to the sum of the gains corresponding to all the elementary games played during the period. *According to the individual point of view, the successive elementary games cannot, however, be considered as strictly independent.* As a matter of fact, if we start with a given number of policies and consider the development of the business during a previously fixed period t, the amount of risk premium collected by the Company during a certain interval dt obviously depends on what has already happened to the group. Similarly, the form of the function $p(z)$ is dependent on the number and distributions of the claims that have already occurred in the group. — This makes it, of course, exceedingly difficult to investigate the probability distribution of the total gain by means of the distributions corresponding to the various elementary games, as long as we remain on the individual point of view.

By some reflection we shall find, however, that the consideration of a given number of policies and a fixed period t are not absolutely essential. On the contrary, when we wish to investigate the future development of the total risk business of a Company we should, strictly speaking, take into account also the changes in the stock of the Company which are incessantly caused by lapses and by the assimilation of new policies. From this point of view, it is obvious that the inter-dependence of the various elementary games becomes less important. Further, in an Insurance Company which has reached a certain state of equilibrium, the distribution of the sums at risk that have fallen due during various periods does not show any considerable variations, so that the form of the function $p(z)$ might in such cases, at least as a first approximation, be considered invariable.

In the collective theory of risk as given by LUNDBERG, *it is assumed that the elementary games (by* LUNDBERG *called »risk elements») are all mutually independent. The number and kind of the individual policies are left entirely out of consideration, and the accumulated risk premium P, instead of the time t, is chosen as our independent variable.*

According to the above, the total gain arising during a certain period will then be equal to a sum of mutually independent variables with known probability functions, so that the methods of Part II will still be applicable to the problem (after a passage to the limit, since we have to make dP infinitely small and thus get an infinite number of components in the sum).

It should be well observed that in LUNDBERG'S theory we are not allowed to ask such a question as the following: »What is the probability function of the gain arising during a given time t?» It is not t, but P, that is the relevant variable, and so the correct way of putting the question is: »What is the probability function of the gain arising during a certain period, when the total risk premium income for the same period is P?» Thus it is not at all necessary to consider explicitly the variable t, and we can talk of »the period P» or »the interval dP» without giving rise to any confusion.

As regards the function $p(z)$, we shall assume in the sequel that it is invariable. A certain number of the results of the theory hold true independently of this assumption, but we shall not enter here upon these questions.

31. If in a rectangular system of coordinates we represent the accumulated risk premium P on the horizontal axis and the accumulated gain x on the vertical axis, the course of the gain in an actual case is represented by a curve starting from the point $x = P = 0$. For the sake of brevity, this curve will be called below »*the line G*».

G is a broken line composed of straight lines parallel to the line $x = P$ or to the axis of x. As long as no claim arises, the gain is continuously increased by the amount of risk premium, and so the point representing the gain moves in a direction making an angle of 45° with the axis of P. At the occurrence of a claim for the amount z, the gain decreases discontinuously by the same amount, so that the moving point suddenly falls down along a vertical of the height z. (Cf. Fig. 8, Art. 34.)

The ordinate of the point of intersection between G and a vertical line through the point P on the horizontal axis represents the value of the accumulated gain *at the moment P*. Thus the probability that this gain falls between the limits x_1 and x_2 coincides with the probability that the ordinate of the point of intersection falls between the same limits.

Following LUNDBERG, we shall suppose that a certain part of the loading, amounting to λ per unit of risk premium[1], is available for the purpose of covering random fluctuations. The rest of the loading is employed for payment of expenses, distribution of bonus or dividends etc. We shall consider the pure risk business only, so that no other sources of income than the loaded risk premiums, and in particular no interest, are taken into account.

Thus for every period the net risk premium P and the loading λP are collected into a *risk reserve*, from which all claims are paid. If the initial value (for $P = 0$) of this reserve is denoted by u, and x is the gain on the net risk business according to the graphical representation discussed above, then the value of the risk reserve at the moment P clearly is

$$x + \lambda P + u.$$

Now, the following question arises: »What is the probability that the risk reserve will not become negative?» It is obvious that the arrangement cannot

[1] LUNDBERG usually denotes by $\lambda - 1$ the quantity which is here denoted by λ.

be regarded as satisfactory unless this probability is very nearly equal to unity. As we have seen in Art. 6, the question may be interpreted in various ways, according as we require the probability that the risk reserve will not become negative:

A. At a certain moment P.

B. At any moment during the interval $(0, P)$.

C. At the end of any business period during the interval $(0, P)$.

Drawing in our diagram the straight line L (cf. Fig. 8) with the equation

$$x + \lambda P = -u,$$

the question A is concerned with the probability that the point of intersection between the line G and a vertical through the point P falls above the line L. Question B, on the other hand, refers to the probability that the line G is situated above L *during the whole interval from 0 to P*, while according to C we ask for the probability that G falls above L in all the points P_1, P_2, . . . , which correspond to the ends of the business periods falling during the interval from 0 to P.

In questions B and C, we can allow P to tend to infinity, and so arrive e. g. at the question: »What is the probability that the risk reserve will never at any moment assume a negative value when P increases from 0 to infinity?» — It is particularly remarkable that, as we shall see below, it is really possible under the assumptions of LUNDBERG'S theory to find an approximate answer to this question.

32. It has been assumed above that the sums at risk in our collective risk business show a continuous distribution, characterized by the function $p(z)$. In cases where this condition is not satisfied, the corresponding formulæ should be applied »cum grano salis», which as a rule is not difficult.[1] As an illustration, we shall consider here the simple case when all the sums at risk are equal to 1.

Then the conditions of each elementary game are as follows: The Company receives the sum dP as a stake, and has the probability dP of being obliged to pay the sum 1. — Assuming first that dP is a finite quantity $= \frac{P}{n}$, the probability that exactly ν claims arise during the period P is

$$\binom{n}{\nu} (dP)^{\nu} (1 - dP)^{n-\nu}.$$

As n tends to infinity and dP becomes infinitely small, it is well known that this probability tends to the limit

$$\frac{P^{\nu}}{\nu!} e^{-P}, \tag{62}$$

[1] By the use of STIELTJES' integrals, it would of course be possible to obtain perfectly general formulæ.

which gives thus the exact expression of the probability that ν claims arise during the period P, under the hypothesis that the elementary games are strictly independent.

If q denotes the force of risk, and we have N units of »exposed to risk», counting all the individual policies that have been under observation, then we have $P = Nq$, and thus (62) becomes

$$(63) \qquad \frac{(Nq)^\nu}{\nu!} e^{-Nq}.$$

In the individual risk theory, the corresponding question would be as follows. We consider N policies, and for each policy we have the probability q that it becomes a claim during a certain period. What is the probability that exactly ν claims occur during the period? — The answer is given by the expression

$$(64) \qquad \binom{N}{\nu} q^\nu (1-q)^{N-\nu}.$$

When Nq is large and ν does not differ considerably from its mean value Nq, it is easily shown that both expressions (63) and (64) give approximately equal values of the probability, since both can be approximated by the normal frequency function. The fundamental difference is, however, that according to (64) we have necessarily $0 \leqq \nu \leqq N$, while (63) gives a positive probability for arbitrary large values of ν. This apparently paradoxical circumstance is due to the fact that, in the collective theory, we make no hypothesis at all as to the number of policies that have contributed to our »exposed to risk». There is a theoretical possibility that this number may have been extremely large, every policy having become a claim very soon after entering into observation, and it is this possibility which gives rise to the positive probabilities of large values of ν.

The Probability Function of the Gain.

33. Using the notations of Art. 30, our Company receives during each elementary game the amount dP as a stake, and has the probability $p(z)dz\,dP$ of being obliged to pay the amount z. The gain of the Company arising from such a game may assume

the value dP with the probability $1 - dP$,
» » $dP - z$ » » » $p(z)dz\,dP$.

The adjunct (cf. Art. 10) corresponding to this gain is

$$v(t) = (1 - dP)\, e^{-it\,dP} + dP \int_0^\infty e^{-it(dP - z)} p(z) dz$$
$$= e^{-it\,dP} \left(1 + dP \int_0^\infty (e^{itz} - 1) p(z)\, dz\right).$$

As in the preceding article, we shall first consider dP as a finite quantity, so that the interval from 0 to P is composed of a finite number of elementary games, each corresponding to a certain dP. The gain x arising during the period P, taken as a whole, is the sum of the gains in the various elementary games. Since by hypothesis these are all mutually independent, the adjunct corresponding to the total gain is (cf. Art. 10)

$$\Pi v(t) = e^{-itP} \Pi \left(1 + dP \int_0^\infty (e^{itz} - 1) p(z)\, dz\right).$$

Now, we allow every dP to tend to zero in such a way that the sum of all the dP is constantly equal to P, while obviously the number of the elementary games becomes infinite. Then it follows immediately that we have

$$\lim_{dP\to 0} \Pi v(t) = e^{-itP} \cdot e^{P\int_0^\infty (e^{itz}-1)p(z)\,dz},$$

$$= e^{w(t)},$$

where

$$w(t) = P\int_0^\infty (e^{itz} - 1 - itz)\, p(z)\, dz,$$

so that $e^{w(t)}$ is the adjunct which corresponds, under the hypotheses of the collective theory, to the total gain x arising during the period P. We put

(65)
$$p_\nu = \int_0^\infty z^\nu p(z)\, dz,$$

$$c_\nu = \frac{p_\nu}{p_2^{\frac{\nu}{2}}},$$

and shall, for the sake of simplicity, assume that p_ν is finite for all $\nu = 1, 2, \ldots$, and that the series $\sum p_\nu t^\nu$ is convergent for some positive value of t. Then we have for sufficiently small values of t

$$w(t) = P\left(-\frac{1}{2}p_2 t^2 - \frac{1}{6} i p_3 t^3 + \ldots\right),$$

$$e^{w(t)} = 1 - \frac{1}{2} P p_2 t^2 - \frac{1}{6} i P p_3 t^3 + \ldots$$

Hence it follows that the mean value and the standard deviation of the gain x are given by the expressions

$$M[x] = 0, \qquad \varepsilon[x] = \sqrt{P p_2}.$$

Denoting the probability function of the normalized gain $\frac{x}{\sqrt{P p_2}}$ by $F(t)$, and the corresponding adjunct by $g(t)$, we then have according to Art. 10

$$g(t) = \int_{-\infty}^{\infty} e^{-itx}\, dF(x)$$

$$= e^{w}\left(\frac{t}{\sqrt{P p_2}}\right).$$

We have, however,

$$w\left(\frac{t}{\sqrt{P p_2}}\right) = P \sum_{2}^{\infty} \frac{p_\nu}{\nu!}\left(\frac{it}{\sqrt{P p_2}}\right)^\nu$$

$$= -\frac{1}{2} t^2 + \sum_{3}^{\infty} \frac{c_\nu}{P^{\frac{\nu}{2}-1}} \cdot \frac{(it)^\nu}{\nu!},$$

and thus for large values of P there is an asymptotic expansion of $e^{\frac{t^2}{2}} g(t)$ in powers of $P^{-\frac{1}{2}}$, beginning in the following way:

$$e^{\frac{t^2}{2}} g(t) = 1 + \frac{c_3}{3!\, P^{\frac{1}{2}}}(it)^3 + \frac{c_4}{4!\, P}(it)^4 + \frac{10\, c_3^2}{6!\, P}(it)^6 + \ldots,$$

the neglected terms being at most of the order $P^{-\frac{3}{2}}$.

This shows in particular that we have

$$\lim_{P \to \infty} g(t) = e^{-\frac{t^2}{2}},$$

uniformly in every finite interval of t, and thus according to Art. 11 we may conclude

$$\lim_{P \to \infty} F(x) = \Phi(x),$$

so that $F(x)$, for large values of P, will be approximately represented by the normal function $\Phi(x)$.

Further, putting

$$\Phi_1(x) = \Phi(x) + \frac{c_3}{3!\, P^{\frac{1}{2}}} \Phi^{(3)}(x),$$

$$\Phi_2(x) = \Phi(x) + \frac{c_3}{3!\, P^{\frac{1}{2}}} \Phi^{(3)}(x) + \frac{c_4}{4!\, P} \Phi^{(4)}(x) + \frac{10\, c_3^2}{6!\, P} \Phi^{(6)}(x),$$

$$Q(x) = F(x) - \Phi_1(x),$$

$$R(x) = F(x) - \Phi_2(x),$$

we can deduce as in Art. 25 the identity

$$I^{(\omega)} Q(x) = \frac{1}{\pi} \Re \int_0^\infty \frac{e^{itx}}{(it)^{1+\omega}} \left(g(t) - e^{-\frac{t^2}{2}} - \frac{c_3}{3!\, P^{\frac{1}{2}}} (it)^3 e^{-\frac{t^2}{2}} \right) dt,$$

where $0 < \omega \leqq 1$, and a similar formula for $I^{(\omega)} R(x)$.

Thus the problem of approximating to $F(x)$ by means of the normal function and its derivatives can be treated by the same methods as those used in Arts. 24--26. The number n of the components in the sum $x = x_1 + \ldots + x_n$ is here replaced by the accumulated risk premium P. If we put

$$\lambda_4 = c_4^{-\frac{6}{4}}, \qquad \lambda_5 = c_5^{-\frac{6}{5}},$$

we can deduce for $0 \leqq \omega \leqq 1$ the inequalities

(66) $$|I^{(\omega)} Q(x)| < \frac{0.12}{\lambda_4 P} + \frac{0.44}{\sqrt[3]{\lambda_4 P}} e^{-\sqrt[3]{\lambda_4 P}} + \frac{1}{\pi} \int_{\sqrt{\lambda_4 P}}^{\infty} |g(t)| \frac{dt}{t^{1+\omega}},$$

(67) $$|I^{(\omega)} R(x)| < \frac{0.38}{\sqrt{(\lambda_5 P)^3}} + \frac{0.65}{\sqrt[3]{\lambda_5 P}} e^{-\sqrt[3]{\lambda_5 P}} + \frac{1}{\pi} \int_{\sqrt{\lambda_4 P}}^{\infty} |g(t)| \frac{dt}{t^{1+\omega}},$$

which are perfectly analogous to (35) and (36). These inequalities hold for all real values of x, if $\lambda_4 P \geqq 1$ (in (66)) or $\lambda_5 P \geqq 1$ (in (67)). Thus no assumption corresponding to $n \geqq 1\,000$ is required here.

The proofs of (66) and (67) are quite similar to the proofs of (35) and (36), and the further treatment in order to obtain upper limits for $Q(x)$ and $R(x)$ can be made according to the method shown in Art. 26 for distributions of the mixed type.

The Minimum Value of the Risk Reserve.

34. In Art. 31, we have considered a risk reserve into which the risk premium P and a certain part of the loading λP are collected, and from which all claims are paid. If we suppose that this reserve starts from a positive initial value u for $P = 0$, we may put the question: »What is the probability that the risk reserve will be negative for *some* positive value of P?»

We may, however, equally well consider a risk reserve which starts from the initial value zero, and then our question becomes: »What is the probability that the minimum value of the risk reserve, as P increases from 0 to ∞, is $< -u$?» Obviously this probability coincides with the probability required in the first form of the question.

The minimum value of the risk reserve, when the initial value is zero, obviously constitutes a random variable capable of assuming the value 0 or any negative

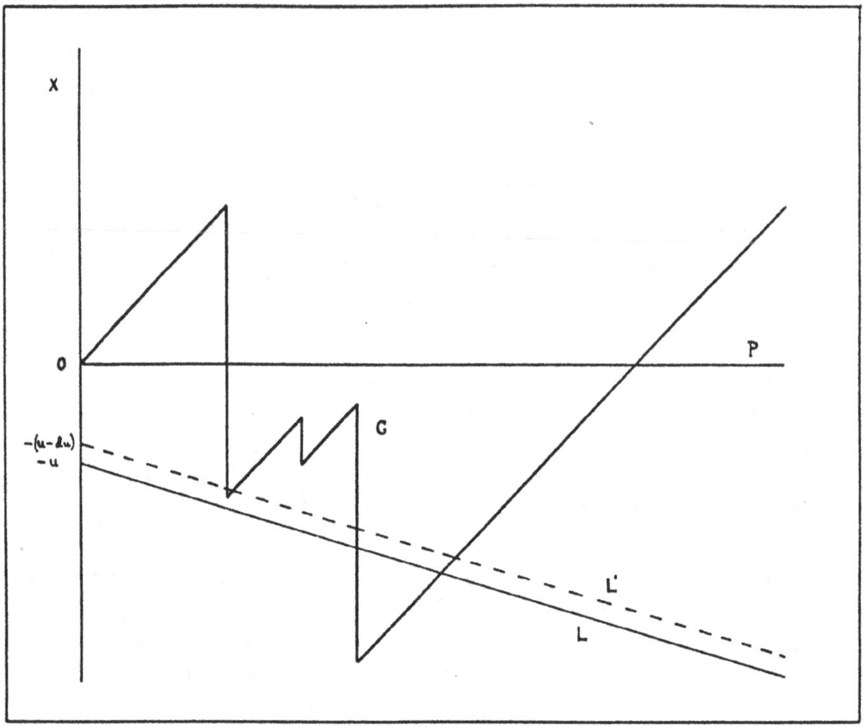

Fig. 8.

value. We shall denote by $\psi(u)$ *the probability that this minimum is* $< -u$, so that $\psi(u)$ is defined for all real u. Obviously $\psi(u)$ is a never negative and never increasing function of u, such that $\psi(u) = 1$ for $u < 0$. The function $\psi(-u)$ coincides, in all its points of continuity, with the ordinary probability function of the minimum value, as defined in Art. 8.

We shall begin by proving the inequality

$$\psi(u) < e^{-Ru} \tag{68}$$

for all $u > 0$, where R is a positive constant which, under certain general conditions, is uniquely determined by the factor λ and the function $p(z)$.

If the initial value of the risk reserve is zero, and x denotes as before the accumulated gain on the net risk business, then the value of the risk reserve at the moment P is clearly equal to $x + \lambda P$. Thus if we draw, in the system of coordinates considered in Art. 31, the line L with the equation

$$x + \lambda P = -u,$$

it follows that $\psi(u)$ is equal to the probability that there is at least one point of intersection between L and the line G which represents the course of the gain (cf. Fig. 8).

If there are several points of intersection, we consider the one corresponding to the smallest value of P. This point must belong to one of the vertical parts of G, and the lowest point of this vertical may lie at various distances from L.

Accordingly we introduce a new definition. By $\chi(u, v)\,dv$ we denote the probability of the following composed event: *the risk reserve will at some time be* $< -u$, *and the first time (i. e. for the smallest value of P) this happens, its value will fall between* $-(u+v)$ *and* $-(u+v+dv)$. It will be sufficient to consider, in this definition, only positive values of u and v.

Obviously we have for all $u > 0$

$$\psi(u) = \int_0^\infty \chi(u, v)\,dv. \tag{69}$$

The event which has the probability $\chi(u, v)\,dv$ may occur in two mutually exclusive ways. Drawing a line L' parallel to L with the equation

$$x + \lambda P = -(u - du),$$

the first points of intersection of G with L' and L may belong to the same vertical part of G, or there may already for a smaller value of P have been a point of intersection between G and L' (Fig. 8 corresponds to the latter case). du and dv being considered as differentials, this distinction gives rise to the relation

$$\chi(u, v)\,dv = \chi(u - du,\ v + du)\,dv + \chi(u - du, 0)\,du \cdot \chi(0, v)\,dv,$$

which leads to the partial differential equation

$$\frac{\partial \chi}{\partial u} - \frac{\partial \chi}{\partial v} = \chi(u, 0)\,\chi(0, v). \tag{70}$$

We may also consider another division of the same event into mutually exclusive events. Let us consider the first elementary game played while the risk premium increases from 0 to dP. The line G starts from the point $x = P = 0$, and according to the result of the first elementary game three different cases may occur, which may all lead to the realization of the desired event. These cases are as follows:

1. No claim occurs.

2. A claim occurs, of such a magnitude that the desired event is at once produced.

3. A claim occurs for an amount which is not sufficiently large to produce a point of intersection between G and L.

From this division we deduce

$$\chi(u, v)\,dv = (1 - dP)\,\chi\big(u + (1+\lambda)\,dP, v\big)\,dv +$$

$$+ dv\,dP\left[p\big(u + v + (1+\lambda)\,dP\big) + \int_0^{u+(1+\lambda)\,dP} \chi\big(u + (1+\lambda)\,dP - z, v\big)\,p(z)\,dz\right]$$

and hence, treating $d\,v$ and $d\,P$ like differentials,

$$(71)\qquad (1+\lambda)\frac{\partial \chi}{\partial u} - \chi(u, v) + p(u+v) + \int_0^u \chi(u-z, v)\, p(z)\, dz = 0.$$

The relations (70) and (71) will serve as starting point for the following investigation. Writing $\chi(0, v) = \nu(v)$, we shall first prove that we have

$$(72)\qquad \nu(v) = \chi(0, v) = \frac{1}{1+\lambda}\int_v^\infty p(z)\, dz.$$

Eliminating $\frac{\partial \chi}{\partial u}$ between (70) and (71) and putting $u = 0$, we obtain

$$(73)\qquad \nu'(v) + \left(\nu(0) - \frac{1}{1+\lambda}\right)\nu(v) + \frac{1}{1+\lambda}\, p(v) = 0.$$

Multiplying this equation with $d\,v$ and integrating from 0 to ∞, we get

$$\left(\nu(0) - \frac{1}{1+\lambda}\right)\left(\int_0^\infty \nu(v)\, d\,v - 1\right) = 0.$$

Let us first suppose that the second factor on the left hand side is equal to zero. By (69) and (72) we then have $\psi(0) = 1$, so that the line G must *certainly* cut every line L passing through the origin, however large the value of λ. This does not seem very plausible, and I have shown elsewhere[1] how it can be rigorously proved to be impossible. Thus the first factor must vanish, and then (73) gives

$$\nu(v) = \frac{1}{1+\lambda}\int_v^\infty p(z)\, d\,z,$$

so that (72) is proved. Hence we deduce, by means of (69) and (61), the important relation

$$\psi(0) = \int_0^\infty \nu(v)\, d\,v = \frac{1}{1+\lambda}.$$

Thus, considering a risk reserve starting from any given initial value, the probability that the risk reserve will at some time fall below this initial value, is independent of the function $p(z)$ and equal to $\frac{1}{1+\lambda}$.

[1] Skandinavisk Aktuarietidskrift 1926, p. 240—243.

From (72) we deduce

$$(74) \qquad \int_0^\infty e^{cv} \nu(v)\, dv = \frac{1}{(1+\lambda)c} \int_0^\infty (e^{cz} - 1) p(z)\, dz$$

for every $c > 0$ that renders the integral on the right hand side convergent. The right hand side of this identity is a function of c, which for $c = 0$ assumes the value $\frac{1}{1+\lambda} < 1$ and then steadily increases with c, as may be seen by considering the derivative. Generally the integral ceases to converge when c increases beyond a certain limit. We shall suppose, however, that there is at least one positive value of c such that the right hand side of (74) converges and takes a value $\geqq 1$. Then it follows that there is one uniquely determined value $c = R > 0$, such that we have

$$(75) \qquad \int_0^\infty e^{Rv} \nu(v)\, dv = \frac{1}{(1+\lambda)R} \int_0^\infty (e^{Rz} - 1) p(z)\, dz = 1.$$

Multiplying (70) with $e^{Rv}\, dv$ and integrating from 0 to ∞, we obtain after a partial integration

$$\frac{d}{du} \int_0^\infty e^{Rv} \chi(u,v)\, dv = -R \int_0^\infty e^{Rv} \chi(u,v)\, dv,$$

and thus

$$\int_0^\infty e^{Rv} \chi(u, v)\, dv = e^{-Ru},$$

$$\psi(u) = \int_0^\infty \chi(u, v)\, dv < e^{-Ru},$$

so that (68) is proved.

35. In this article, it will be shown that the function $\psi(u)$ satisfies an integral equation of VOLTERRA's type.

From (70) we obtain, multiplying with dv and integrating from 0 to ∞,

$$\psi'(u) + \chi(u, 0) = \chi(u, 0) \int_0^\infty \nu(v)\, dv$$

$$= \frac{1}{1+\lambda} \chi(u, 0),$$

and thus

$$Z(u, 0) = -\frac{1+\lambda}{\lambda} \psi'(u). \tag{76}$$

Further, substituting in (70)

$$Z(u, v) = \int_0^u Z(z, 0)\, Z(0, u+v-z)\, dz + f(u, v),$$

it follows that $f(u, v)$ must satisfy the equation

$$\frac{\partial f}{\partial u} - \frac{\partial f}{\partial v} = 0,$$

and thus $f(u, v)$ is a function of $u + v$ which may be denoted by $F(u + v)$. This gives us

$$Z(u, v) = \int_0^u Z(z, 0)\, Z(0, u+v-z)\, dz + F(u+v),$$

and, putting $u = 0$, we obtain

$$F(v) = Z(0, v) = \nu(v).$$

Thus we have

$$Z(u, v) = \int_0^u Z(z, 0)\, \nu(u+v-z)\, dz + \nu(u+v)$$

$$= \int_0^u Z(u-z, 0)\, \nu(v+z)\, dz + \nu(u+v),$$

and by (76)

$$Z(u, v) = -\frac{1+\lambda}{\lambda} \int_0^u \psi'(u-z)\, \nu(v+z)\, dz + \nu(u+v).$$

Multiplying here with dv and integrating from 0 to ∞, we obtain by (69)

$$\psi(u) = -\frac{1+\lambda}{\lambda} \int_0^u \psi'(u-z)\, dz \int_z^\infty \nu(v)\, dv + \int_u^\infty \nu(v)\, dv.$$

Transforming the first term on the right hand side by a partial integration, we obtain the result

$$\psi(u) = \int_0^u \psi(u-v)\, \nu(v)\, dv + \int_u^\infty \nu(v)\, dv. \tag{77}$$

This is an integral equation which determines uniquely the function $\psi(u)$ since according to (72) $\nu(v)$ is a known function expressible by means of λ and $p(z)$. The well-known methods for the solution of integral equations obviously lead to an explicit expression for $\psi(u)$.

36. Under certain conditions, it is possible to obtain from the integral equation (77) an asymptotic expression for $\psi(u)$, as we are now going to show. We put

$$\omega(s) = -\int_0^\infty e^{su}\,\psi'(u)\,du,$$

$$\xi(s) = \int_0^\infty e^{su}\,\nu(u)\,du,$$

s being a complex variable. Both integrals are convergent in a certain half-plane, situated to the left of a line parallel to the imaginary axis in the plane of s. We have

$$\int_0^\infty e^{su}\,\psi(u)\,du = \frac{1}{s}\left(\omega(s) - \frac{1}{1+\lambda}\right),$$

$$\int_0^\infty e^{su}\,du\int_u^\infty \nu(v)\,dv = \frac{1}{s}\left(\xi(s) - \frac{1}{1+\lambda}\right).$$

Multiplying (77) with $e^{su}\,du$ and integrating from 0 to ∞, we obtain

$$\int_0^\infty e^{su}\,\psi(u)\,du = \int_0^\infty e^{su}\,\psi(u)\,du \cdot \int_0^\infty e^{su}\,\nu(u)\,du + \int_0^\infty e^{su}\,du\int_u^\infty \nu(v)\,dv,$$

and thus

$$\omega(s) = \frac{1}{1+\lambda} - \frac{\xi(s) - \dfrac{1}{1+\lambda}}{\xi(s) - 1} = -\frac{\lambda}{1+\lambda}\cdot\frac{\xi(s)}{\xi(s)-1}. \tag{78}$$

In the domain of convergence of the integral $\xi(s)$, we thus see that $\omega(s)$ has no other singularities than the roots of the equation

$$\xi(s) = \int_0^\infty e^{su}\,\nu(u)\,du = 1. \tag{79}$$

Comparing this equation with (75), we see that $s = R$ is a root of this equation, and obviously any other root must have a real part $> R$. Let us suppose that there is a number $\alpha > 0$, such that in the half-plane $\Re(s) \leq R + \alpha$ the integral $\xi(s)$ is convergent and the equation (79) has no other root than $s = R$. Then

$\omega(s)$ is bounded in the same half-plane, exept for the point $s = R$, which is a pole of the first order with the residue

$$r = -\frac{\lambda}{1+\lambda}\cdot\frac{\xi(R)}{\xi'(R)} = \frac{\lambda R}{1+\lambda-\int\limits_0^\infty e^{Ru}\,up(u)\,du}.$$

According to well-known theorems in the theory of DIRICHLET's series and integrals we have, however (cf. the analogous relations deduced in Art. 25),

$$\frac{1}{2\pi i}\int\limits_{c-i\infty}^{c+i\infty}\frac{e^{-us}}{s^2}\,\omega(s)\,ds = \int\limits_u^\infty \psi(t)\,dt, \tag{80}$$

the integral on the left hand side being extended over the line $\Re(s) = c$ with $0 < c < R$. By an application of CAUCHY's theorem, the path of integration may be moved to the right to the line $\Re(s) = R + a$, allowing for the residue corresponding to the pole $s = R$, so that we obtain

$$\int\limits_u^\infty \psi(t)\,dt = -\frac{r}{R^2}e^{-Ru} + \frac{1}{2\pi i}\int\limits_{R+a-i\infty}^{R+a+i\infty}\frac{e^{-us}}{s^2}\,\omega(s)\,ds.$$

On the line $\Re(s) = R + a$, it follows from (78) that $|\omega(s)|$ is bounded, and we have $|e^{-us}| = e^{-(R+a)u}$, so that we obtain

$$\int\limits_u^\infty \psi(t)\,dt = -\frac{r}{R^2}e^{-Ru} + O\left(e^{-(R+a)u}\right),$$

the last term denoting a quantity, the modulus of which is less than $e^{-(R+a)u}$ multiplied with a constant. Hence we can deduce by a method similar to that used in Art. 26 (replacing u by $u + h$, and putting afterwards $h = e^{-\frac{1}{2}au}$) that we have

$$\psi(u) = -\frac{r}{R}e^{-Ru} + O\left(e^{-(R+\frac{1}{2}a)u}\right)$$

$$= \frac{\lambda}{\int\limits_0^\infty e^{Ru}\,up(u)\,du - 1 - \lambda}\,e^{-Ru} + O\left(e^{-(R+\frac{1}{2}a)u}\right). \tag{81}$$

As soon as λ and $p(u)$ are given, such that the above conditions are satisfied, we can thus find an asymptotic expression for the probability that the risk reserve will, for some value of P, fall below its initial value diminished by a given quantity u.

In order to illustrate this result by some simple examples, we take first

$$p(u) = e^{-u}.$$

Then (72) gives

$$\nu(u) = \frac{1}{1+\lambda} e^{-u},$$

so that equation (79) becomes

$$\xi(s) = \frac{1}{(1+\lambda)(1-s)} = 1$$

which has only one root, $s = \frac{\lambda}{1+\lambda}$, so that we have in this case

$$R = \frac{\lambda}{1+\lambda}.$$

Further, (78) becomes

$$\omega(s) = -\frac{\lambda}{(1+\lambda)^2} \cdot \frac{1}{s - \frac{\lambda}{1+\lambda}}.$$

Substituting this in (80), we obtain

$$\int_u^\infty \psi(t)\, dt = -\frac{\lambda}{2\pi i (1+\lambda)^2} \int_{c-i\infty}^{c+i\infty} \frac{e^{-us}}{s^2 \left(s - \frac{\lambda}{1+\lambda}\right)} ds.$$

The integral on the right hand side can be exactly evaluated, and we find

$$\int_u^\infty \psi(t)\, dt = \frac{1}{\lambda} e^{-\frac{\lambda u}{1+\lambda}},$$

$$\psi(u) = \frac{1}{1+\lambda} e^{-\frac{\lambda u}{1+\lambda}}.$$

As our second example, we take the case when all the sums at risk are equal to 1. The above formulæ are not directly applicable to this case, since they have been based upon the hypothesis of a continuous distribution of the sums according to the function $p(u)$. It is easily seen, however, that we may regard this as a limiting case, which may be uniformly approximated by means of continuous distributions. According to (72), we shall have to put

$$\nu(u) = \begin{cases} \frac{1}{1+\lambda} & \text{for } u < 1 \\ 0 & \text{» } u > 1 \end{cases}$$

Then equation (79) becomes

$$\xi(s) = \frac{e^s - 1}{(1 + \lambda) s} = 1.$$

This equation has one, and only one, positive root $s = R$, so that R is equal to the only positive root of the equation

$$e^R - 1 = (1 + \lambda) R.$$

Further, there is an infinite number of complex roots, and it is possible to find a number $\alpha > 0$ such that the real part of any complex root is $> R + \alpha$.

From (78) we obtain in this case

$$\omega(s) = -\frac{\lambda}{1 + \lambda} \cdot \frac{e^s - 1}{e^s - 1 - (1 + \lambda) s},$$

so that the residue in the pole $s = R$ is

$$r = -\frac{\lambda R}{(1 + \lambda) R - 1}.$$

According to (81) we then conclude

$$\psi(u) = \frac{\lambda}{(1 + \lambda) R - 1} e^{-Ru} + O(e^{-(R + \frac{1}{2}\alpha) u}).$$

Further Investigations.

37. In his recent papers, LUNDBERG has considerably developed the theory and has obtained a number of remarkable new results. We shall here briefly indicate some of the principal ones.

Let us denote by $f(x, P)$ the *frequency function* of the gain arising during the period P (i. e. while the accumulated risk premium increases from 0 to P). Then $f(x, P)\,dx$ is the probability that the line G (cf. Art. 31 and Fig. 8) cuts the vertical through the point P between the ordinates x and $x + dx$. It is sufficient to consider values of x such that $x < P$, since the line G can never rise above the line $x = P$.

Now, drawing a straight line K which passes through the origin and the point (P, x), it is seen that the probability $f(x, P)\,dx$ only differs by an infinitely small quantity of the second order from the probability that G cuts K for a value of the P-coordinate falling between the limits $P - \frac{P}{P - x} dx$ and P.

Denoting generally by $\mu(k, P)\,dP$ the probability that G cuts the line with the equation $x = kP$ for a value of the abscissa falling between the limits P and $P + dP$, we thus have

$$(82) \qquad f(x, P) = \frac{1}{1-k} \mu(k, P)$$

with

$$k = \frac{x}{P}.$$

By means of functional equations similar to (70) and (71) it can be proved that $\mu(k, P)$, regarded as a function of P, satisfies an integral equation of the same type as (77). In this way, we can deduce an explicit expression for the function

$$M(s) = \int_0^\infty e^{sP} \mu(k, P)\, dP.$$

From this expression, it follows that the integral which represents $M(s)$ is absolutely convergent as long as the real part of s is less than the quantity h defined by the relation

$$h = (1-k) R' - \int_0^\infty (e^{R'u} - 1)\, p(u)\, du,$$

where R' is the only real root of the equation

$$\int_0^\infty e^{R'u}\, u\, p(u)\, du = 1 - k.$$

In the point $s = h$, $M(s)$ has an algebraic singularity of the order $-\frac{1}{2}$ with the dominating term

$$\frac{1-k}{\sqrt{2\beta(h-s)}},$$

where

$$\beta = \int_0^\infty e^{R'u}\, u^2\, p(u)\, du.$$

All the above results can be rigorously proved. The asymptotic formula for $\mu(k, P)$ suggested by the properties of $M(s)$ would be

$$\mu(k, P) \sim \frac{1-k}{\sqrt{2\pi\beta P}} e^{-hP},$$

and according to (82) we should then have

$$f(x, P) \sim \frac{1}{\sqrt{2\pi\beta P}} e^{-hP}$$

where, in the relations which define h and β, we have to put $k = \frac{x}{P}$.

This asymptotic representation of $f(x, P)$ has not been rigorously proved, but LUNDBERG has shown by examples that it gives remarkably good results when applied to the numerical calculation of $f(x, P)$ and its integral, the probability function. In several cases, the approximation obtained in this way is very much better than the one offered by the normal function.

LUNDBERG has also given other methods for the numerical calculation of the frequency functions and probability functions. For these methods, that will not be discussed here, he uses the integral equations satisfied by $\psi(u)$, $\mu(k, P)$ and other similar functions, and shows that certain majorating and minorating expressions can be constructed for the solutions. The results obtained in this way seem to be even better than with the preceding method.

38. By the various methods discussed above, we can calculate the probability that the risk reserve, starting from any given initial value, will during the course of time become negative. If the initial risk reserve and the factor λ are sufficiently large, that probability will be very small. In such a case, we should thus expect a considerable increase of the risk reserve, and we may wish to make some part of this increase available for other purposes. LUNDBERG has discussed the influence of various rules for the reduction of an increasing risk reserve upon the probability that the reserve will become negative. He has also discussed the question C of Art. 31, and various other modifications of the problems. For all these questions the reader must, however, be referred to the original papers.

27.

The risk problem. Exordial review

Proc. Trans. 9th Internat. Congr. Actuaries, IV, 163–172 (1931)

The Question D, as it has been formulated by the Organizing Committee, is mainly concerned with the practical applications of the so called Mathematical Theory of Risk. Thus it is asked whether »any advantage for the practice of Life Assurance is to be expected from theoretical investigations concerning the mathematical risk and similar questions». Further, the following more particular question is added: »Are current methods of reassurance and of forming risk reserves insufficient to eliminate inconvenient effects of mortality fluctuations?»

It seems highly appropriate that such a question should be proposed for discussion at an Actuarial Congress. During the course of time, a large number of distinguished authors have worked out an extensive theory of the mathematical risk of a Life Assurance Company. Numerous applications of the theory have been made, concerning questions such as the extra loading of premiums, the method of reassurance and the fixation of the maximum amount on a Company's own risk, the formation of mortality reserves etc. Practical Life Assurance has, however, so far paid little or no attention to all these theoretical results. While it is universally agreed that the calculation of premiums, reserves, surrender values etc. should strictly conform to the mathematical theory, we find that e. g. all questions concerned with reassurance and the fixation of maximum amounts are as a rule dealt with from a pure business point of view, without the aid of any mathematical theory.

Trying to find the causes of this apparent discrepancy, we arrive at the general question proposed by the Organizing Committee. Is it possible that the mathematical theory of risk will ever find application in practice?

Among the many valuable papers on this subject which have been presented to the Congress, it seems to be the paper written by M. DUMAS that takes the most negative position towards the question. According to M. DUMAS, the correlation between the mortality rates of the various insured lifes which is manifested e. g. during a large epidemic renders it utterly impossible to apply the theory of mathematical probability to fluctuations of mortality, and thus in his opinion the practical applications of actuarial theory are essentially restricted to the calculation of premiums and reserves, while the investigations concerning the mathematical risk are deprived of practical value. Other authors, as e. g. M. HOCHART, hold the view that the mathematical theory may be safely applied to find measures of protection against mortality fluctuations, as long as only pure random fluctuations and slow »secular» variations are concerned, but that against the violent fluctuations of mortality caused by war, epidemics etc., we have no other defence than those inspired by common sense. M. HOCHART points out, however, that we must never forget that the results of a pure theory should always be tested by statistical experience before they can be considered to have any practical value. M. BAPTIST, in his report, has expressed similar views.

It will be generally admitted, I think, that the theory of risk is fundamentally incapable of dealing with the effects of violent mortality fluctuations due to definite external causes such as wars and epidemics. It is the *random fluctuations* of mortality that form the object of the theory.

For a variety of reasons, there has been in the recent development of Life Assurance in many countries a tendency of lowering the premium level. Such has been the case e. g. in this country, where the last ten years have brought a reduction of premiums hardly dreamt of before. At the same time, the bonus question has acquired a great importance for the writing of new business, and in many cases the Companies have issued Bonus Prospects which give, for every policy year, minimum values of the bonus that the Company expects to be able to pay to the policy-holder. Such prospects are largely used in underwriting business, and thus the habitual system of bonus rates adopted by the Company is gradually becoming known to the public. It would cause considerable trouble to a Company if, for a certain year, the available surplus was not sufficient to allow the distribution of bonus according to ordinary rates.

Under these circumstances, it must be considered as highly important to provide by all possible means for a smooth and regular course of business, and in particular to avoid considerable losses due to random fluctuations. If the regular bonus distribution of a Company is disturbed e. g. by a financial crisis, a war, or an epidemic, it is very likely that other Companies in the same country will be in a similar position, so that the event does not affect the relative position of the Companies. Losses due to random fluctuations of mortality, on the other hand, do not at all occur in the same uniform way, and will thus be particularly unwelcome, since they will put the Company in a worse position when compared with other Companies. This state of things obviously tends to give an increased importance to the matter of random fluctuations and to the question of finding appropriate methods for neutralizing their effects.

Even if we agree with M. DUMAS as to the impossibility of a theoretical treatment of fluctuations due to definite causes such as wars and epidemics, I think we may still defend the view that the random fluctuations can fairly well be dealt with on the lines of the mathematical theory of probability. This view has been expressed e. g. by M. CANTELLI, who gives in his report an interesting discussion of the relations betweeen the calculus of probability and general actuarial theory. Some statistical data bearing on the subject are presented by Dr. HINTIKKA, who has investigated the fluctuations of death claims in Finnish Life Assurance Companies. As Dr. HINTIKKA points out, the fluctuations in his material fall in practically all cases within the limits required by the ordinary theory; it appears, however, difficult to obtain sufficiently detailed information in such matters. Further statistical investigations along these lines seem to be highly desirable.

The fundamental practical problem to be solved by the mathematical theory of risk may be formulated in the following way. — A Company is working under given circumstances with respect to premiums, calculation of the mathematical reserves etc. The Company further undertakes to pay bonus according to a given system of bonus rates. What measures should the Company take in order to have the greatest possible probability of realizing successfully its bonus scheme, without any disturbance due to random fluctuations?

In close accordance with this point of view, Dr. VAJDA has calculated the maximum amount of bonus that can be paid out if certain conditions relating to the probability of a loss are to be satisfied.

The practical measures usually introduced as safeguards against mortality fluctuations are reassurance and extra loading of premiums.

The *effectivity* of these methods is, however, very incompletely known, from a theoretical as well as from an empirical point of view. What we want the theory to tell us in this respect is how the probability of a loss is affected by various methods of reassurance and of loading, or by the formation of a particular risk reserve. Considering a Model Office working under given conditions and with a fixed system of bonus rates, we should thus ask: What is the probability that, within a certain interval of time, random fluctuations will occur which upset the regular bonus distribution? And further: How is the numerical value of this probability affected by a change of the maximum amount kept on own risk, by the introduction of an extra loading of the premiums, or generally by any variation of the fundamental elements of the calculation?

Thus it appears that, for the practical applications of the theory of risk, methods for *the numerical calculation of certain probabilities* are above all required. In this respect, the methods of the *mean risk* or the *average risk* have been accepted in the majority of works concerning the theory.

In the first place, the gain or loss of the Company arising during a given period on one individual policy is considered. This is a variable quantity which may assume different values, and the probabilities of these various values may be calculated from the mortality table. The standard deviation of this variable gain has received the name of the *mean risk* of the policy corresponding to the given period, while the mean of the absolute value of the gain is called the *average risk* of the policy. The mean risk and the average risk are particular instances of the general conception of mean value in the sense of the calculus of probability, and the general rules for the calculation with mean values can thus be applied in the theory of risk.

A great part of the work laid down on the theory of risk has been concerned with the deduction of formulae for the calculation of mean risks and average risks for various types of policies, and with the discussion of the relative advantage of these two conceptions. In the reports presented to this Congress, we find that Dr. SCHÖNWIESE strongly objects to the use of the average risk, while Dr. GRUDER, Dr. JACOB, Dr. SMOLENSKY, Dr. GULDBERG and Dr FUHRICH seem to take the opposite position. Some of these authors, as e. g. Dr. JACOB, consider exclusively the average risk, while others, as Dr. GRUDER and Dr. SMOLENSKY, deal with both conceptions simultaneously. Dr. JACOB and Dr. SMOLENSKY point out that, in their opinion, the average risk has a particular importance for the selection of risks and the fixation of premiums for sub-standard lives, while Dr. SCHÖNWIESE makes some critical remarks upon this point.

M. CVETNIČ investigates in his report the mean risk connected with Fire Insurance, where it is necessary to take account of the variability of claims. He proposes a simple formula for the calculation of the mean risk in such cases.

The total gain arising in a group of policies during a certain time is equal to the sum of the gains corresponding to the individual policies. Accordingly, the mean risk for the total can be simply calculated from the mean risks of the individual policies, if these are considered as mutually independent. By a general theorem in the theory of probability, the sum of a large number of mutually independent variables is approximately distributed according to the *normal* or GAUSS-LAPLACE probability function. Applying this theorem to the total gain arising in a group of a large number of policies, it follows that fluctuations exceeding three or four times the mean risk are practically impossible. The majority of applications of the theory of risk to practical problems have been founded upon this theorem, and it is to this fact that the conception of the mean risk owes it particular importance. If the approximation obtained by using the normal law is really satisfactory, then we can solve all our problems by the calculation of the mean risk.

It appears, however, that the question of the order of the error of approximation has never been thoroughly examined. As Dr. RIETZ points out in his report, there are plenty of cases occurring in practice where the approximation is far from satisfactory. Dr. RIETZ considers particularly the case of a comparatively small Company, with a small expected number of claims, and he shows that in such cases the POISSON exponential function gives a much better representation of the probability than the normal function.

Personally, I am inclined to think that this question of the approximation furnished by the normal law is a point of fundamental importance. In a paper on the mathematical theory of risk, published a few weeks ago, I have endeavoured to reach some definite results in this direction. The results so far obtained are far from satisfactory, and I should like to draw your attention to the general problem of the numerical evaluation of the probability functions which occur in the theory of risk. Until this problem has received a satisfactory solution, the applications of the theory of risk cannot be considered to rest upon a sound mathematical basis In order to reach such a solution, it appears to be necessary to consider also other parameters than the mean risk and the average risk, such as e. g. the higher moments of the distributions.

A fundamentally different method of attacking the problems of the theory of risk has been developed by my fellow-countryman Dr. LUNDBERG. In his report to the Congress, LUNDBERG has explained the general principles of his theory, while extensive mathematical developments are given in his recent paper in the Skandinavisk Aktuarietidskrift. The main difference as against the ordinary way of proceeding lies in the fact that LUNDBERG does not begin by considering the gain arising on an individual policy. He tries to deal directly with the total risk business of a Company in all its complexity, and his theory may thus be adequately described as a *collective risk theory*, as opposed to the ordinary, individual, point of view. The results obtained by this method are highly remarkable, and in particular the problem of the numerical evaluation of the probability functions has been treated in a very interesting way in LUNDBERG's last paper. Surely this beautiful theory will awake a vivid interest among Actuaries.

In concluding this speech, I should like to emphazise three points which are, in my opinion, of a fundamental importance for the future development of the theory of risk. In the first place, the problems on which the theory is set to work should be more closely adapted to the requirements of the practice. Further, the mathematical treatment of the problems should always be rigorous, and in particular questions of numerical approximation should not be treated without due care. Finally, it is highly desirable that the results of the theory should be tested and completed by statistical investigations.

28.

The theory of risk in its application to life insurance problems

Proc. Trans. 9th Internat. Congr. Actuaries, II, 380–390 (1931)

1. Practical Life Insurance has hitherto paid but little attention to the considerations offered by the mathematical theory of risk. While it is universally agreed that the calculation of premiums, reserves, surrender values etc. has to be founded on a rigorous mathematical basis, the further claims of the theoreticians connected with the theory of risk have not been successful. In the vast majority of cases, the method of reinsurance, the rules for the management of risk reserves etc. are fixed from a pure business point of view, without the aid of any mathematical theory.

The different arguments that have been advanced in support of this position may be divided into two groups, according as they tend to show that the theory of risk is *unnecessary* for the practice, or that it is *inadequate* to its purpose.

Before proceeding to a detailed analysis of some of these arguments it may be stated that, in the opinion of the present writer, those belonging to the latter class are the most serious ones, whereas some arguments belonging to the former class appear to him as definitely preposterous. The object of this paper is to submit the thesis that, in the practice of modern Life Insurance, there are important problems to be solved by a mathematical theory of risk, and that, since the present state of this theory is a very primitive one, it is desirable that we should direct our efforts towards its perfection.

2. The theory of risk, in its application to Life Insurance, is concerned with the economic effects of random fluctuations of mortality. Starting from an insurance company working under given conditions,

the object of the theory is to investigate what measures — with respect to reinsurance, risk reserves, extra loading of premiums etc. — the company should take in order to eliminate as far as possible the inconvenient effects of accidental deviations between the actual and expected amounts of claims.

If an unusually large number or average amount of claims arises during a certain period, the company suffers a loss. In works concerning the theory, we usually find that the »mathematical risk» of the company connected with such losses is defined in some way or other, and it is endeavoured to find measures which reduce the mathematical risk to a minimum without reducing the bonus-earning power of the company more than absolutely necessary.

Now, the view is often expressed that such a theory is quite unnecessary for the practice, since the random mortality fluctuations only constitute one of the elements that may disturb the regular course of profits of a company. Besides the mathematical risk there is also a »commercial risk» connected with the investments and general management of the company. Further, there are losses from mortality due to war, epidemics etc., which cannot be placed under the heading »random fluctuations». Thus, it is concluded, there is no use for special provisions intended to cover the mathematical risk, and the only safeguards that are wanted are those imposed by practical common sense and general business principles.

This inference does, however, seem to be somewhat hasty. Let us suppose for a moment that we have at our disposal a well-established mathematical theory of risk, capable of telling us what measures we ought to take in order to counteract the influence of random mortality fluctuations in the best possible way. It is perfectly true that, besides these fluctuations, there are quite a number of other important factors at work, but nobody will deny that the effect of purely random fluctuations may quite possibly prove disastrous, if we consider a company that does not reinsure and has no extra mortality reserves. Everybody agrees that *some* measures should be taken, either in the way of reinsurance, or by forming a special risk reserve, or both. The value of any such arrangement depends partly on its efficiency as a precaution against mortality fluctuations, partly on the expenses or other inconveniences which it causes the company. If we possess a theory that enables us to estimate the value of the different arrangements that might be proposed, it seems fairly obvious that our practical common sense will be wrong in ignoring the results of such a theory.

No doubt the element of commercial risk will require precautions of its own, but it is only by our presumed theory of risk that we can find out if these may at the same time act as a satisfactory substitute for reinsurance or risk reserve.

3. Another argument intended to prove the needlessness of the theory of risk is the following one. The premiums charged in practice are always calculated so as to keep »on the safe side». If the extra loading is sufficiently high, it is to be expected that the mortality fluctuations will only cause a corresponding fluctuation in the bonus. Thus, as ALTENBURGER[1] puts it: »High initial premiums and rational participation of the insured in profits render theoretical examination of the mathematical risk unnecessary».

This phrase does not seem quite clear. Does it imply that high premiums render reinsurance completely unnecessary? And what is to be understood by a »rational participation in profits»? How are we, in particular, to allow mortality fluctuations to be reflected in the bonus distribution? — For the analysis of all these questions, the theory of risk, in its presumed perfect state, would be able to yield valuable contribution.[2]

The chief objection against ALTENBURGER's argument is, however, that the recent development of Life Insurance seems to point in a direction essentially different from the one outlined in the phrase quoted above. Owing principally to the competition between the companies and to reasons connected with taxation, the premium level has in many countries been substantially lowered, which in the long run must cause a corresponding decrease of divisible surplus.

Such has been the case, *e.g.*, in Sweden, where the last ten years have brought a reduction of premiums hardly dreamt of before. At the same time, the bonus question has acquired a great importance for the writing of new business, and the majority of the companies have adopted certain Bonus Schemes giving, for every policy year, *minimum values* of the bonus that the company expects to be able to pay to the policy-holder. These schemes have to be approved by the authorities in accordance with the Swedish Insurance Act, and are largely used in underwriting business. It would cause considerable trouble to a company if, for a certain year, the available surplus was not sufficient to

[1] ALTENBURGER, Verhandlungen d. VI. intern. Kongr. f. Versicherungs-Wissenschaft, Wien 1909, p. 962.

[2] Cf BERGER's comments on ALTENBURGER's paper: »Die Prinzipien der Lebensversicherungstechnik», Berlin 1925, Vol. 2, p. 3.

allow the payment of at least the amounts prescribed by the approved Bonus Scheme of the company.

More or less similar circumstances seem to be present also in several other countries. Comparatively low premiums and a habitual system of bonus rates known to the policy-holders make it highly important to provide by all possible means for a smooth and regular course of business. A fall in the rate of interest or a general mortality increase affects all the companies working in the same country in a more or less uniform way. Variations in the annual surplus due to accidental mortality fluctuations, on the contrary, do not at all occur in a uniform way, and a considerable loss from such a cause will therefore be particularly unwelcome, since it will put the company in a worse position when compared with other companies. This state of things obviously tends to give an increased importance to the matter of mortality fluctuations and to the question of finding appropriate methods for neutralizing their effects.

Even if this point is admitted, it might still be objected that the effects of mortality fluctuations are neutralized simply by reinsurance, and that the practical business man will easily find out the appropriate maximum to be retained on his company's own risk in order to ensure a sufficiently stable course of profits. We feel, however, that this is not quite so simple as it may seem. The efficiency of the customary methods of reinsurance is at present only very incompletely known, from an empirical as well as from a theoretical point of view. In particular, we know very little about the influence which a change in a company's maximum will exert upon the future course of the business. Thus it is doubtful if it is always possible to obtain by ordinary reinsurance a satisfactory protection against mortality fluctuations. In reality, the analysis of these and other related questions appears to be one of the most urgent problems for the theory of risk.

Thus we cannot but arrive at the conclusion that there are, in modern Life Insurance, important practical problems connected with the subject of random mortality fluctuations. In order to give a satisfactory treatment of these problems we shall have to construct a mathematical theory of the fluctuations in a Life business, and to test the results of our theory by statistical experience. — We now proceed to consider the present state of the theory of risk, and to see how far it has succeeded to solve its problems.

4. The theory of risk rests upon the fundamental assumption that

the mortality fluctuations may be treated along the lines of the mathematical theory of probability. Thus it is assumed that the »true» force of mortality is known and that the deviations of the actual amount of claims from the expected one can be regarded as errors of random sampling.

It is well known that this point of view has been sharply disputed, and of course it must be admitted that the theory of probability can never yield more than an idealized and approximate picture of the real phenomena. We shall never know the »true» force of mortality, simply because no such thing exists. In spite of this we know, however, that the hypothesis of a constant force of mortality allows us to construct a theory of mortality giving a fairly adequate representation of the principal facts known from statistical observations. We must regard this as a working hypothesis, and the theory of risk that is built upon this hypothesis must be taken as a first approximation, which may already be of some use. The theory will always have to be tested by statistical experience, and it is possible that we can obtain a still better accordance with the facts by some convenient modification of our hypothesis, *e.g.* by introducing a secular trend in the basic probabilities. In many cases it will, however, be found that the hypothesis of a constant force of mortality will be sufficient for the applications occurring in the theory of risk. The theory founded upon that simple hypothesis has as yet only reached a very imperfect state, and so it seems convenient for the present to limit our efforts to a perfection of this theory.

When making a special study of the element of mathematical risk, it will be permitted to make simplifying assumptions as to the other elements affecting the course of business. In the sequel, we shall thus always assume that the rate of interest attained by the company is constant, and we shall entirely overlook the expenses (with the exception of expenses for reinsurance, which must obviously be brought into consideration) and the corresponding part of the loading. This does not imply that our considerations will be deprived of practical value. By a more elaborate discussion, the simplifications could be at least partly avoided.

Bearing always these preliminary remarks in mind, the general problem to be considered by the theory of risk may now be thus enunciated: *We are given an insurance company working under known conditions as to the attained rate of interest and the average force of mortality. The premiums charged by the company and the rules for valuation are given. The company has adopted a certain bonus scheme which, on the average, is compatible with the given data. What measures should the company take in order to have the greatest possible chance of realizing successfully its bonus scheme, without any disturbances due to accidental fluctuations?*

This problem is, of course, much too general to allow of direct mathematical treatment. The general question will be approached by considering first the following more particular problem: *Under the same conditions as before, we further assume that our company has decided to employ a certain method of reinsurance or certain rules for the formation of a risk reserve. What is the probability that, within a given interval of time, fluctuations will occur which upset the regular bonus distribution?*

The discussion of general methods for the treatment of problems such as those here suggested, and the detailed numerical treatment of particular cases appear to constitute a highly recommendable line for future investigations into the mathematical theory of risk. In this report, it is not possible to enter upon a full discussion of these questions, and we shall have to confine ourselves to a few remarks of a general character.[1]

5. Let us suppose that, for the insurances now in force with our presumed company, we have at our disposal all the data concerning sums, ages, durations and types of insurance. If we consider one of these insurances individually during the next period of, say, t years, we can form a revenue account for this particular insurance, counting as income the premiums (loading for expenses not included, cf. above) and interest belonging to the period, and as expenditure the increase of the reserve during the period, the amounts of bonus provided by the bonus scheme and, if a claim arises during the period, the payment of the sum at risk. If the insurance has been partly reinsured, we shall have to introduce also the corresponding items connected with the reinsurance. According to this account, a certain gain or loss x arises for the company on this particular insurance. The quantity x may assume different values according to the different cases that are possible with regard to the life and death of the insured. Positive values of x correspond to a gain for the company, negative values to a loss.

Since by hypothesis we know the true force of mortality, we can

[1] A systematic account of the theory will be found in a paper by the present writer, published in the second volume of the work »Minnesskrift, utgiven av Försäkringsaktiebolaget Skandia», Stockholm 1930.

calculate the probability that x takes any given value, and thus we can regard x as a *random variable* with a known *probability function*. Hence we can also calculate the theoretical *mean value* of x, which we denote by $M[x]$, and the corresponding *standard deviation* $\varepsilon[x]$. It is well known that $\varepsilon[x]$ is the square root of the second moment of x about its mean. Obviously it will also be possible to calculate moments of higher orders if wanted.

Thus if we could observe during a period of t years a large number of insurances in every respect similar to the one considered above, we should expect that the mean value and the standard deviation of the empirically observed values of the gain x would lie pretty close to the theoretical values $M[x]$ and $\varepsilon[x]$.

In the particular case when the premiums are net premiums calculated from the true force of mortality and the actually attained rate of interest, and no bonus is allowed, $M[x]$ will obviously be equal to zero, since the expected values of the company's income and expenditure must be equal. This case is the one usually considered in the litterature of the subject, and the standard deviation $\varepsilon[x]$ has received the name of the *mean risk* of the insurance, with respect to the considered period of t years.

Under the circumstances occurring in practice, $M[x]$ must have a positive value, since a company will always choose its bonus scheme so as to be on the safe side. The standard deviation $\varepsilon[x]$ will serve as a measure of the order of magnitude to be expected of the deviations between the gain x and its mean value $M[x]$, and we may still call $\varepsilon[x]$ the mean risk of the insurance.

If $x_1, x_2, \ldots x_n$ signify the gains (or losses) on all the insurances of the company, the total gain (or loss) X of the company will be the sum

$$X = x_1 + x_2 + \ldots . + x_n . \tag{1}$$

The mean value $M[x]$ of the company's gain is then equal to

$$M[X] = M[x_1] + M[x_2] + \ldots . + M[x_n].$$

If it is assumed that the insurances are all mutually independent, which means that no person has more than one insurance, and that the chances of death of the insured persons are mutually independent, we have also

$$\varepsilon^2[X] = \varepsilon^2[x_1] + \varepsilon^2[x_2] + \ldots . + \varepsilon^2[x_n],$$

so that the mean risk for the company as a whole can be easily found from the mean risks for the individual insurances.

The total gain X is a random variable equal to the sum of the mutually independent variables $x_1, x_2, \ldots x_n$. Then we know from the work of many different authors that, under very general circumstances, X will be at least approximately distributed according to the so called *normal error function*, as soon as the number n of insurances is large. Thus if k is any given number, the probability that the gain X will assume a value such that

$$X < M[X] + k\,\varepsilon[X]$$

is approximately equal to

$$\Phi(k) = \frac{1}{\sqrt{2\pi}} \int_{-\infty}^{k} e^{-\frac{t^2}{2}} dt.$$

If we disregard the error of approximation, we thus find *e.g.* that the probability of the inequality

$$X < M[X] - 3\,\varepsilon[X]$$

is equal to

$$\Phi(-3) = 0.0013.$$

If, now, we have

$$\varepsilon[X] < \frac{1}{3} M[X], \tag{2}$$

it would follow that the probability of the inequality $X < 0$ is even smaller than 0.0013, which means that there is only a very small chance for the company of getting its regular bonus distribution disturbed by accidental mortality fluctuations during the time t. Thus if we can find some method of reinsurance or some other arrangement which allows the relation (2) to be satisfied, everything seems to be all right.

The above deduction may serve as an example of a type of reasoning very often employed in the theory of risk. In accordance herewith we find that the greatest part of the work laid down on this theory has been concerned with the mean risk or some other similar measure, as *e.g.* the *average risk* used by some authors. A considerable number of formulae have been deduced for the calculation of the mean risk and the average risk for various kinds of insurances. The use which has been made of these formulae generally rests upon the assumption that the total gain of the company is distributed according to the normal error function and that, consequently, deviations from the average course of business exceeding three or four times the mean risk are practically impossible.

This type of reasoning is, however, not very satisfactory. The investigations concerning the mean risk and the average risk are indeed very useful, but they constitute only a first step towards the solution of the problems that are practically important. The introduction of the normal error function is not legitimate until it has been shown that the error of approximation may actually be disregarded in cases occurring in practice. It is true that, under very general circumstances, we can prove that the probability function of the total gain X tends towards the normal function as n tends to infinity, but from this assertion we cannot infer anything about the size of the error of approximation for a finite value of n. As a matter of fact, the investigations connected with this error of approximation appear to be extremely difficult, and the work that has been made along this line seems to indicate that the order of approximation obtained when using the normal function is *not* sufficiently good to justify the conclusions alluded to above.

In order to construct a mathematically rigorous theory of risk, it will thus be necessary to find more satisfactory methods of approximating to the probability function of the total gain X. According to (1), this function is theoretically known, since the probability functions of the components $x_1, x_2, \ldots x_n$ are known. The general expression which relates the probability function of X to the corresponding functions of the components x_i is, however, a multiple integral unfit for numerical calculation, and for the purpose of the theory of risk it is precisely a numerical knowledge of the former function that is wanted. The problem of finding practicable methods for the attainment of such knowledge seems to be highly difficult, but as long as this problem is not solved, the applications made of the theory of risk will not rest upon a sound mathematical basis.

We cannot enter here upon a discussion of the various possible ways towards the solution of this problem. It is obvious, however, that it will not be sufficient to consider only the mean risk or the average risk of the insurances. We shall be obliged to use also the higher moments or other parameters for a more intimate characterization of the probability functions.

6. When we have once succeeded in finding a practicable solution of the numerical approximation problem indicated in the preceding

article, we shall be in a position to do something towards the solution of the practical insurance problems.

In the first place, we shall have to study the probability function of the total gain of the company under different assumptions concerning the distribution of insurance amounts and the maximum on the company's own risk. It would be especially interesting to see how the occurrence of exceptionally high insurance amounts affects the probability functions and how this influence is counteracted by reinsurance. Then it would be important to investigate the effect of a change in the bonus scheme or the introduction of a risk reserve acting as a sort of internal reinsurance provided by the company itself. The detailed numerical treatment of typical examples of these various kinds would be able to give us many a piece of useful information.

Many questions will, however, remain for further investigations. In the brief sketch of the mathematical treatment given above, we have considered the development of the course of business during a certain time t. The choice of t is, of course, wholly arbitrary, and it must be shown how this choice affects our conclusions. — Moreover, we have considered only the final gain X arising at the end of the period t. If X was positive, we have regarded this as an event indicating that the regular bonus distribution of the company was not disturbed. But obviously there might have been, during the course of the period, a disturbance which had afterwards disappeared. If we should want to find the probability that, *at every moment of the period* t, the company could balance its books with a positive gain, we should find ourselves confronted with a new mathematical problem presenting some analogy with the classical problem of the »duration of play», but obviously of a much more intricate character.

Finally, we must not forget that the whole theory has been founded upon the working hypothesis of a constant force of mortality, and the results of the theory must be tested by statistical experience.

7. A fundamentally different way of attacking the problems of the theory of risk has been followed by Dr. F. LUNDBERG. LUNDBERG considers the risk business of a company as a whole and does not require any knowledge of the individual policies. Consequently he does not regard the total gain of the company as composed of the gains of the individual insurances according to the relation (1). He fixes his attention on the amounts of *risk premium* continously earned by the company, and on the distribution of the *sums at risk*, that have fallen

due. By these means, he is able to find a very remarkable method of approximating to the probability function of the company's gain during a certain period. Mathematically speaking, this problem is not exactly equivalent to the one encountered when we regard the total gain as composed of individual gains. For the practical object of solving the problems of the theory of risk, however, both problems seem to be equally entitled to attract our attention. It is especially remarkable that in LUNDBERG's theory it is possible to treat the problem indicated in the preceding article: to find at least an estimate of the probability that, during a given period, the gain of the company never becomes negative.

For an account of this higly important and original theory, the reader is referred to Dr. LUNDBERG's report to the Congress and to the papers there quoted.

29.

Su un teorema relativo alla legge uniforme dei grandi numeri

Giorn. Ist. Ital. Attuari **5**, 3–15 (1934)

SUNTO. — L'A. considera la forma modificata della legge uniforme dei grandi numeri che si ottiene quando le somme

$$X_{(n)} = X_1 + X_2 + X_3 + \cdots + X_n$$

si formano per mezzo di differenti successioni di valori delle variabili componenti X_n e sono così mutuamente indipendenti. Egli dimostra, in questo caso che, sotto certe condizioni, la probabilità della coesistenza di tutte le ineguaglianze

$$\frac{|X_{(i)}|}{\sqrt{2A_i}} < \lambda_i \quad , \quad i = n, n+1, \cdots,$$

$\left(A_i\right.$ essendo il valore medio del quadrato delle $\left.X_{(i)}\right)$ tende all'unità per $n \to \infty$ se la serie $\sum \frac{1}{\lambda_i} e^{-\lambda_i^2}$ è convergente, ed è uguale a zero per qualsiasi valore di n se questa serie è divergente.

1. Indichiamo con $X_1, X_2, \cdots, X_n, \cdots$ una successione illimitata di variabili casuali *indipendenti tra loro.* Effettuando una prova su ciascuna variabile, otteniamo una successione di valori osservati $x_1, x_2, \cdots, x_n, \cdots$, essendo x_n il valore assunto dalla variabile X_n.

Supponendo

$$M(X_n) = 0$$

per ogni n, poniamo

$$[1] \qquad M(X_n^2) = a_n .$$

$$[2] \qquad A_n = a_1 + a_2 + \cdots + a_n \quad , \quad \bar{A}_n = \frac{1}{n} A_n ,$$

e, per $k > 2$,

$$[3] \qquad M(|X_n|^k) = b_{nk} ,$$

$$B_{nk} = b_{1k} + b_{2k} + \cdots + b_{nk} \quad , \quad \bar{B}_{nk} = \frac{1}{n} B_{nk} , \tag{4}$$

$$\rho_{nk} = \frac{\bar{B}_{nk}^{\frac{1}{k}}}{\bar{A}_n^{1/2}} . \tag{5}$$

Inoltre poniamo

$$X_{(n)} = X_1 + X_2 + \cdots + X_n . \tag{6}$$

In una recente Memoria [1] il Cantelli ha dimostrato il seguente teorema A, che generalizza dei teoremi precedentemente dati da Khintchine, Kolmogoroff e Lévy [2].

Teorema A. — *Supponiamo che per qualche valore di k, tale che sia $2 < k \leq 3$, si possano trovare due positivi α e β tali che, per ogni n, sia*

$$a_n > \alpha \quad , \quad b_{nk} < \beta .$$

Allora, $X_{(n)}$ essendo definita dalla [6], *la probabilità che siano soddisfatte simultaneamente le ineguaglianze*

$$\frac{|X_{(n)}|}{\sqrt{2 A_n}} < \sqrt{\log_2 n + c \log_3 n}$$

per ogni $n > n_0$ tende all'unità per $n_0 \to \infty$ se $c > 3/2$, è uguale a zero qualunque sia n_0 se $c \leq 1/2$.

È ovvio che le variabili $X_{(1)}, X_{(2)}, \cdots$, considerate in questo teorema, *non sono tra di loro indipendenti*. A pagg. 331–332 della sua Memoria, il Cantelli ha fatto un'osservazione interessante concernente il caso che le variabili $X_{(n)}$ siano dedotte per somma dalle variabili $X_1, X_2, \cdots$ in modo che riescano *fra di loro indipendenti*, come meglio sarà appresso specificato. Egli fa notare che in questo caso il suo primo teorema sulla legge uniforme dei grandi numeri [3] è pure valido, mentre ciò non può dirsi nè per i teoremi dati da Kintchine, Kolmogoroff e Lévy, nè per lo stesso teorema A.

---

1) F. P. Cantelli, *Considerazioni sulla legge uniforme dei grandi numeri e sulla generalizzazione di un fondamentale teorema del sig. Paul Lévy*, « Giornale dell'Istituto Italiano degli Attuari », anno IV, n. 3 (luglio 1933-XI).

2) Cfr. i lavori citati dal Cantelli nell' op. cit. [1].

3) F. P. Cantelli, *Sulla probabilità come limite della frequenza*, « Rendiconti della R. Accademia dei Lincei », vol. XXVI, 1917.

Nella presente Memoria mi propongo di considerare il caso indicato che le variabili $X_{(n)}$ siano tra di loro indipendenti e di dimostrare un teorema che corrisponde al teorema A del Cantelli.

2. Consideriamo ancora le variabili indipendenti $X_1, X_2, \cdots$ Una prima prova da farsi riguardi solo la variabile $X_1 = X_1^{(1)}$ e poniamo

$$X_{(1)} = X_1^{(1)} .$$

Una seconda prova sulle variabili della successione indicata consista in una nuova prova sulla variabile $X_1 = X_1^{(2)}$ ed in una prova sulla variabile $X_2 = X_2^{(2)}$; poniamo

$$X_{(2)} = X_1^{(2)} + X_2^{(2)} .$$

Continuando in questo modo otteniamo

[7] $$X_{(n)} = X_1^{(n)} + X_2^{(n)} + \cdots + X_n^{(n)} ,$$

cioè $X_{(n)}$ è sempre la somma delle variabili $X_1, X_2, \cdots, X_n$ indicate ma le variabili $X_{(n)}$ che vengono a considerarsi sono indipendenti tra loro e di valore medio nullo.

Tenendo conto che la variabile $X_n^{(i)}$ rappresenta sempre la stessa variabile X_n ne segue che i momenti a_n e b_{nk} relativi alle $X_n^{(i)}$ sono indipendenti dall'indice i. Così pure può ovviamente dirsi per le espressioni di A_n e B_{nk}, ecc. definite dalle relazioni [2] a [5].

Premesso quanto sopra, enuncierò il seguente teorema la dimostrazione del quale costituisce l'oggetto della presente Memoria.

TEOREMA B. — *Supponiamo di poter trovare una costante K tale che per tutti i valori di n sia*

[8] $$\rho_{n5} = \frac{\bar{B}_{n5}^{1/5}}{\bar{A}_n^{1/2}} < K .$$

Allora, $X_{(n)}$ essendo definita dalla [7], *la probabilità che siano soddisfatte simultaneamente le ineguaglianze*

$$\frac{|X_{(n)}|}{\sqrt{2A_n}} < \lambda_n \quad , \quad (\lambda_n > 0) ,$$

per ogni $n > n_0$, tende all'unità per $n_0 \to \infty$ se la serie

[9] $$\sum_{n=n_0}^{\infty} \frac{1}{\lambda_n} e^{-\lambda_n^2}$$

è convergente; è uguale a zero, qualunque sia n_0, se la serie [9] *è divergente.*

Quindi, ponendo per esempio

$$\lambda_n = \sqrt{\log n + c \log_2 n}\ , \tag{10}$$

si vede che la probabilità considerata nel teorema tende all'unità se $c > 1/2$, è uguale a zero se $c \leq 1/2$. È interessante paragonare questo risultato con quello del teorema A del Cantelli.

Osserviamo che la condizione [8] sarà sempre soddisfatta se sarà possibile trovare due positivi α e β tali che per ogni n sia

$$a_n > \alpha \quad , \quad b_{n5} < \beta .$$

Ciò perchè, in questo caso, essendo anche

$$\bar{A}_n > \alpha \quad , \quad \bar{B}_{n5} < \beta ,$$

la [8] risulta certamente soddisfatta. Se, in particolare, le leggi di ripartizione delle probabilità delle variabili X_n sono identiche, la [8] è soddisfatta qualora i momenti $b_{15} = b_{25} = \cdots$ siano limitati. A questo proposito è da osservare che in nessun caso il teorema B suppone che i momenti b_{nk} siano limitati quando sia $k > 5$.

3. Prima di procedere alla dimostrazione del teorema B, consideriamo i due teoremi A e B nel caso particolare dello schema bernoulliano.

In questo caso la variabile X_n assume il valore *uno* se si presenta pallina bianca in un'estrazione fatta da un'urna, altrimenti assume il valore *zero*; la probabilità di estrarre pallina bianca dall'urna si suppone costantemente eguale a p.

Il teorema A si riferisce al caso che, in n prove *successive*, appartenenti ad una stessa successione di prove, si presenti ν_n volte pallina bianca. Il teorema ci dice che la probabilità che siano soddisfatte simultaneamente le ineguaglianze

$$\left|\frac{\nu_n}{n} - p\right| < \sqrt{\frac{2pq}{n}(\log_2 n + c \log_3 n)}\,,$$

per ogni $n > n_0$, tende all'unità, per $n_0 \to \infty$, se $c > 3/2$; è uguale a *zero*, qualunque sia n_0, se $c \leq 1/2$.

D'altra parte consideriamo il caso che la frequenza ν_n sia ottenuta effettuando *n nuove prove indipendentemente da quelle che hanno fornito la frequenza* ν_{n-1}. In tal caso il teorema B ci dice (considerando il caso particolare in cui λ_n sia dato dalla [10]) che la probabilità che siano soddisfatte simultaneamente le ineguaglianze

$$\left|\frac{\nu_n}{n} - p\right| < \sqrt{\frac{2pq}{n}(\log n + c \log_2 n)},$$

per ogni $n > n_0$, tende all'unità, per $n_0 \to \infty$, se $c > 1/2$; è uguale a *zero*, qualunque sia n_0, se $c \leq 1/2$.

4. Per la dimostrazione del teorema B osserviamo dapprima che le variabili $X_{(1)}, X_{(2)}, \cdots, X_{(n)}, \cdots$ definite dalla [7] sono tra di loro indipendenti. Segue pertanto, dai lemmi I e II della Memoria del Cantelli, che il teorema B equivale all'affermazione che, se è soddisfatta la condizione [8], le due serie

$$\sum \Pr\left[\frac{|X_{(n)}|}{\sqrt{2A_n}} \geq \lambda_n\right] \quad , \quad \sum \frac{1}{\lambda_n} e^{-\lambda_n^2}$$

sono entrambe convergenti o entrambe divergenti. Il simbolo Pr significa qui «probabilità» in armonia colla notazione usata dal Cantelli.

D'altra parte si ha

$$\Pr\left[\frac{|X_{(n)}|}{\sqrt{2A_n}} \geq \lambda_n\right] = \Pr\left[\frac{X_{(n)}}{\sqrt{2A_n}} \geq \lambda_n\right] + \Pr\left[\frac{X_{(n)}}{\sqrt{2A_n}} \leq -\lambda_n\right].$$

Pertanto, applicando il seguente teorema C una volta alle variabili $X_{(n)}$ e un'altra volta alle variabili $-X_{(n)}$, si vede subito che il teorema B è contenuto nel teorema C, dimodochè basterà qui dimostrare il teorema C.

TEOREMA C. — *Se la condizione* [8] *è soddisfatta, le due serie*

[11] $$\sum \Pr\left[\frac{X_{(n)}}{\sqrt{2A_n}} \geq \lambda_n\right] \quad , \quad \sum \frac{1}{\lambda_n} e^{-\lambda_n^2} \, ,$$

($\lambda_n > 0$), *sono entrambe convergenti o entrambe divergenti.*

Ponendo, per tutti i valori reali di z,

$$F_n(z) = \Pr\left[\frac{X_{(n)}}{\sqrt{2A_n}} < z\right],$$

si ha

$$1 - F_n(\lambda_n) = \Pr\left[\frac{X_{(n)}}{\sqrt{2A_n}} \geqq \lambda_n\right];$$

pertanto le due serie [11] si possono scrivere sotto la forma

[12] $$\sum\left[1 - F_n(\lambda_n)\right], \quad \sum \frac{1}{\lambda_n} e^{-\lambda_n^2}.$$

5. È noto che, sotto condizioni molto generali, si ha

$$\lim_{n\to\infty} F_n(z) = \Phi(z) = \frac{1}{\sqrt{\pi}}\int_{-\infty}^{z} e^{-t^2}\,dt.$$

Supponiamo per un momento che si abbia *esattamente*

[13] $$F_n(z) = \Phi(z)$$

per valori di n sufficientemente grandi. È noto allora che, per tali valori di n, si ha

[14] $$1 - F_n(\lambda_n) = 1 - \Phi(\lambda_n) = \frac{1}{\sqrt{\pi}}\int_{\lambda_n}^{\infty} e^{-t^2}\,dt = \frac{1+\eta}{2\sqrt{\pi}} \cdot \frac{1}{\lambda_n} e^{-\lambda_n^2}$$

dove $\eta \to 0$ quando $\lambda_n \to \infty$, dimodochè *sotto l'ipotesi* [13], le due serie [12] sono entrambe convergenti o entrambe divergenti.

È pertanto ovvio che, per dimostrare il teorema C, è necessario considerare l'*ordine dell'errore* implicito nelle relazioni [13] e [14].

In base ad un ben noto teorema dovuto al Liapounoff (in una forma da me leggermente migliorata [4])) risulta che, per ogni $n > 1$ e per z reale qualunque si ha

[15] $$|F_n(z) - \Phi(z)| < 3\rho_{n3}^3 \frac{\log n}{\sqrt{n}}.$$

Poichè da note ineguaglianze tra i momenti risulta [5])

[16] $$\rho_{nk} \leqq \rho_{n,k+1},$$

---

4) *Das Gesetz von Gauss und die Theorie des Risikos*, «Skandinavisk Aktuarietidskrift», 1923, pag. 209–237. La relazione [15] può essere facilmente dedotta dal primo teorema di questa Memoria. Cfr. anche la mia Memoria citata nella nota in calce 7) a pag. 19.

5) Cfr. la mia Memoria cit. 7) a pag. 70.

segue che, sotto la condizione [8], il secondo membro della ineguaglianza [15] è minore di

$$3K^3 \frac{\log n}{\sqrt{n}} .$$

È però facile notare che questo limite superiore dell'errore non è sufficientemente basso per il nostro scopo e ciò perchè introduce la serie $\Sigma(\log n \cdot n^{-1/2})$ che è divergente [6]).

6. Dobbiamo pertanto cercare di ottenere una più precisa conoscenza del comportamento di $F_n(z)$ per grandi valori di n di quella che ci è consentita dal teorema di Liapounoff. Tale conoscenza può essere ottenuta per mezzo di un teorema [7]) da me pubblicato dapprima senza dimostrazione, nel 1925, e poi con una dimostrazione completa, nel 1928.

Detto teorema permette delle interessanti applicazioni a varie questioni di calcolo delle probabilità su cui mi propongo di ritornare in un altro lavoro. Qui mi limito a dimostrare il seguente lemma che è ovviamente sufficiente per stabilire immediatamente la validità del teorema C.

LEMMA. — *Se è soddisfatta la condizione* [8], *si ha*

$$1 - F_n(z) = \frac{1+\eta}{2\sqrt{\pi}} \cdot \frac{1}{z} e^{-z^2} + \varepsilon,$$

η *e* ε *essendo funzioni di* n *e* z, *tali che*

$$\eta = \eta(z, n) \to 0 \qquad \text{per} \quad z \to \infty,$$

uniformemente per ogni $n > 1$, *mentre*

$$\sum_{n=2}^{\infty} |\varepsilon| = \sum_{n=2}^{\infty} |\varepsilon(z, n)|$$

è uniformemente convergente per ogni $z \geqq 1$.

---

6) Cfr. le osservazioni del CANTELLI nella Memoria cit. 1), pag. 328–329.

7) *On some classes of series used in mathematical statistics*, Trans. of the 6th Scandinavian Congress of Math., Copenhagen 1925, pag. 399–425. – *On the composition of elementary errors*, I, «Skandinavisk Aktuarietidskrift», 1928, pag. 13–74. Il teorema citato nel testo è il teorema I della seconda Memoria. La dimostrazione di questo teorema è data a pag. 56–58 per il caso particolare che le leggi di distribuzione delle variabili X_i siano identiche, mentre il caso generale è trattato a pag. 69–71.

Rimane così solo da dimostrare questo lemma. *In seguito adopereremo la lettera* θ *per indicare un numero il cui valore assoluto sia minore di una costante.*

Secondo il mio teorema sopra citato, si ha per ogni $n > 1$ e per z reale qualunque, considerando il caso particolare $k = 5$ del teorema,

$$\int_{-\infty}^{z} (z - t) \cdot R_n(t) \cdot dt = \theta \cdot \rho_{n5}^{9} \cdot \frac{(\log n)^2}{n^{3/2}}.$$

Qui con $R_n(t)$ è indicata la funzione

[17] $$R_n(t) = F_n(t) - \Phi(t) - c_3 \Phi^{(3)}(t) - c_4 \Phi^{(4)}(t) - c_6 \Phi^{(6)}(t),$$

nella quale i coefficienti c_i, che sono i coefficienti che compaiono nella nota serie studiata da Charlier, Edgeworth ed altri autori, soddisfano alle relazioni

$$c_3 = \theta \frac{\rho_{n3}^{3}}{n^{1/2}} \quad , \quad c_4 = \theta \frac{\rho_{n4}^{4}}{n} \quad , \quad c_6 = \theta \frac{\rho_{n3}^{6}}{n}.$$

Se la condizione [8] è soddisfatta, si ha, avuto riguardo alla [16],

[18] $$\int_{-\infty}^{z} (z - t) \cdot R_n(t) \cdot dt = \theta \frac{(\log n)^2}{n^{3/2}},$$

[19] $$c_3 = \frac{\theta}{n^{1/2}} \quad , \quad c_4 = \frac{\theta}{n} \quad , \quad c_6 = \frac{\theta}{n}.$$

Formando la seconda differenza, si deduce dalla [18]

[20] $$\int_{z}^{z+h} (t - z) \cdot R_n(t)\, dt + \int_{z+h}^{z+2h} (z + 2h - t) \cdot R_n(t)\, dt = \theta \frac{(\log n)^2}{n^{3/2}}.$$

Se ora indichiamo con $g(t)$ la funzione definita, per $0 < t < 2h$, dalle relazioni

$$g(t) = t \quad , \quad 0 < t \leq h,$$

$$g(t) = 2h - t \quad , \quad h \leq t < 2h$$

avremo

$$\int_{0}^{2h} g(t)\, dt = h^2$$

e la [20] può essere scritta come segue

[21] $$\frac{1}{h^2}\int_0^{2h} g(t)\cdot R_n(z+t)\cdot dt = \theta\,\frac{(\log n)^2}{h^2\cdot n^{3/2}}\,.$$

Supposto

[22] $$z \geqq 1 \quad , \quad 0 < h \leqq \frac{1}{4z}\,,$$

si ha per $0 < t < 2h$ e per ogni ν assegnato

$$\Phi(z+t) = \Phi(z) + \theta h e^{-z^2},$$

$$\Phi^{(\nu)}(z+t) = \theta z^{\nu-1} e^{-z^2},$$

talchè le [17] e [19] forniscono

$$R_n(z+t) = F_n(z+t) - \Phi(z) + \theta\cdot\left(h e^{-z^2} + n^{-1/2} z^5 e^{-z^2}\right).$$

Segue così dalla [21]

$$\frac{1}{h^2}\int_0^{2h} g(t)\cdot F_n(z+t)\cdot dt = \Phi(z) + \theta\cdot\left(h e^{-z^2} + n^{-1/2} z^5 e^{-z^2} + \frac{(\log n)^2}{h^2\cdot n^{3/2}}\right).$$

Si conclude pertanto, essendo $F_n(z)$ una funzione non decrescente di z, che

[23] $$F_n(z) < \Phi(z) + \theta\cdot\left(h e^{-z^2} + n^{-\frac{1}{2}} z^5 e^{-z^2} + \frac{(\log n)^2}{h^2\cdot n^{3/2}}\right).$$

Cambiando ora nella [20] h in $-h$ e ragionando nello stesso modo si ottiene, sotto le condizioni [22],

[24] $$F_n(z) > \Phi(z) + \theta\cdot\left(h e^{-z^2} + n^{-1/2} z^5 e^{-z^2} + \frac{(\log n)^2}{h^2\, n^{3/2}}\right).$$

Pertanto in base alle [23] e [24] deve essere

[25] $$F_n(z) = \Phi(z) + \theta\cdot\left(h e^{-z^2} + n^{-\frac{1}{2}} z^5 e^{-z^2} + \frac{(\log n)^2}{h^2\, n^{3/2}}\right).$$

In questa relazione supponiamo z compreso nell'intervallo $(1\,,\,n^{1/14})$ e poniamo

$$h = \frac{1}{4z^2}\,;$$

allora le [22] sono ovviamente soddisfatte e risulta dalla [25]

[26] $$F_n(z) = \Phi(z) + \theta \cdot \left(\frac{1}{z^2} e^{-z^2} + \frac{(\log n)^2}{n^{17/14}}\right).$$

Resta così dimostrata l'esistenza di una funzione

$$\theta_1 = \theta_1(z, n),$$

soddisfacente alla [26], che è uniformemente limitata per tutti i valori di n e z tali che sia

[27] $$1 \leq z \leq n^{1/14} \quad , \quad (n = 2, 3, \cdots).$$

D'altra parte abbiamo

[28] $$1 - \Phi(z) = \frac{1 + \delta}{2\sqrt{\pi}} \cdot \frac{1}{z} e^{-z^2},$$

dove $\delta = \delta(z) \to 0$ per $z \to \infty$. Ponendo, per i valori di n e z soddisfacenti alle [27],

[29] $$\eta = \delta - \theta_1 \frac{2\sqrt{\pi}}{z},$$ $$\varepsilon = -\theta_1 \frac{(\log n)^2}{n^{17/14}},$$

si ricava, dalle [26] e [28],

[30] $$1 - F_n(z) = \frac{1 + \eta}{2\sqrt{\pi}} \cdot \frac{1}{z} e^{-z^2} + \varepsilon.$$

Per valori di n e z tali che sia

[31] $$z > n^{1/14} \quad , \quad (n = 2, 3, \cdots),$$

si ha

$$0 \leq 1 - F_n(z) \leq 1 - F_n(n^{1/14}),$$

e quindi dalla [26]

[32] $$1 - F_n(z) = \theta \frac{(\log n)^2}{n^{17/14}} = \frac{1}{2\sqrt{\pi}} \cdot \frac{1}{z} e^{-z^2} + \theta \frac{(\log n)^2}{n^{17/14}}.$$

Abbiamo così dimostrato l'esistenza di una funzione

$$\theta_2 = \theta_2(z, n),$$

soddisfacente alla [32], che è uniformemente limitata per tutti i valori di n e z che soddisfano alla [31]. Ponendo per questi valori di n e z

[33]
$$\eta = 0 ,$$
$$\varepsilon = \theta_2 \frac{(\log n)^2}{n^{17/14}} ,$$

la relazione [30] è ancora soddisfatta.

In base alle [29] e [33], η e ε sono definite per ogni $z \geqq 1$ e per $n = 2, 3, \cdots$ Dalle definizioni date di queste funzioni segue immediatamente

$$\eta = \eta(z, n) \to 0 \quad \text{per } z \to \infty ,$$

uniformemente per $n = 2, 3, \cdots$, mentre

$$\sum_{n=2}^{\infty} |\varepsilon| = \sum_{n=2}^{\infty} |\varepsilon(z, n)|$$

è uniformemente convergente per ogni valore di $z \geqq 1$. Pertanto è dimostrato il lemma e resta stabilita la validità dei teoremi B e C.

7. Come è stato mostrato dal Cantelli, [8] il suo teorema A rimane vero se noi sostituiamo la ineguaglianza

$$\frac{|X_{(n)}|}{\sqrt{2A_n}} < \sqrt{\log_2 n + c \log_3 n}$$

con l'altra

$$\frac{|X_{(n)}|}{\sqrt{2A_n}} < \sqrt{\log_2 A_n + c \log_3 A_n} \,.$$

Se nel teorema B, poniamo le condizioni

[34]
$$a_n > \alpha \quad , \quad b_{n5} < \beta ,$$

si dimostra facilmente (nello stesso modo che nel lavoro del Cantelli) che può sostituirsi A_n al posto di n, nel caso particolare indicato dalla [10].

Nel teorema B, già citato, noi abbiamo, tuttavia introdotto, invece della [34], la condizione più generale

[35]
$$\rho_{n5} < K ,$$

8) Loc. cit. [1]).

e ora dimostreremo, con un esempio, che vi sono casi che soddisfano la [35] ma tali che la sostituzione di A_n al posto di n, nella [10], non è consentita.

Consideriamo, per esempio, le variabili $X_1, X_2, \cdots$, tali che X_n possa assumere i due valori $\pm \log n$, ciascuno con probabilità $1/2$.

Allora abbiamo

$$[36] \qquad \begin{aligned} A_n &= \sum_1^n (\log n)^2 \sim n (\log n)^2, \\ B_{n5} &= \sum_1^n (\log n)^5 \sim n (\log n)^5 \end{aligned}$$

così che, in base alla [5], risulta

$$\rho_{n5} \to 1$$

per $n \to \infty$. Così la [35] è soddisfatta ed il teorema B risulta applicabile a queste variabili $X_1, X_2, \cdots$

D'altra parte otteniamo dalla [36]

$$\log A_n + c \log_2 A_n = \log n + (c+2) \log_2 n + \varepsilon_n,$$

dove $\varepsilon_n \to 0$ per $n \to \infty$ e ponendo, nel teorema B,

$$\lambda_n = \sqrt{\log A_n + c \log_2 A_n}$$

segue che la probabilità considerata nel teorema tenderà all'unità per $c > -3/2$ e sarà uguale a zero per $c \leq -3/2$. È ovvio che questo risultato differisce da quello precedentemente ottenuto per il caso

$$\lambda_n = \sqrt{\log n + c \log_2 n}\,.$$

8. Come il prof. Cantelli mi ha gentilmente comunicato noi possiamo, invece della [10], considerare il caso in cui sia

$$\lambda_n = \sqrt{\log n + \frac{1}{2} \log_2 n + \log_3 n + \cdots + c \log_k n}$$

per qualsiasi $k \geq 3$ e verificare facilmente che, secondo il teorema B, la probabilità considerata in questo teorema tende all'unità per $n_0 \to \infty$ se $c > 1$, mentre la stessa probabilità è uguale a zero per qualunque n_0 se $c \leq 1$. È interessante paragonare questo risultato completo con alcune attinenti osservazioni fatte dal Lévy [9] a proposito dello schema bernoulliano corrispondente al caso particolare del teorema A.

---

9) *Sur un théoréme de M. Khintchine*, « Bull. des Sc. Mathématiques », 2e Série, T. LV, Paris, 1931.

30.

with H. Wold

On the development of the mortality of the adult Swedish population since 1800

Nordic Statistical Journal 5, 3–22 (1934)

In 1932, a Committee of experts appointed by the Swedish Government published a Report[1] concerning the question of adopting a new mortality table for life annuity insurance. This Report contains i. a. a detailed investigation into the development of the mortality of the various age groups of the Swedish population ever since 1850, the proposed new mortality table "R 32" being constructed by means of an extrapolation of the sinking mortality.

One of the main lines of investigation followed in the Report is intimately connected with the well-known *Makeham formula* for the force of mortality. For ages above 30, the force of mortality μ_x may usually be well represented by the Makeham formula

$$\mu_x = \alpha + \beta c^x.$$

The method of investigation referred to consists in a determination of the *Makeham constants* α, β and c for the various periods considered. The tables and diagrams representing the values of these constants for, say, quinquennial periods give a very good idea of the general trend of mortality and seem to form a convenient basis for a careful extrapolation.

This part of the investigation was (cf. the Report, p. 8) founded upon some research work made at the Department of Actuarial Mathematics and Mathematical Statistics, University of Stockholm, under the direction of one of the present authors (H. C.), who was a member of the Committee. Since the publication of the Report, we have continued the investigation on an extended scale and with considerably improved methods. The unique material furnished by Swedish Population Statistics seems capable of giving still many interesting results. We now

[1] Dödlighetsantaganden för livränteförsäkring, Statens Offentliga utredningar 1932:4. An account of this Report has been given by the Chairman of the Committee, Dr. O. A. Åkesson, in this Journal, Vol. 3, p. 231. Cf. also B. E. Meurk, Skandinavisk Aktuarietidskrift, 1932, p. 251.

propose to give in this note a preliminary account of the results so far obtained, reserving a detailed account of the methods and further analysis of the results for a forthcoming publication.

The investigation has been extended to the whole period 1801—1930. For every quinquennial period, the numbers of exposed to risk and the numbers of deaths have been compiled from official statistics for quinquennial age groups between ages 30 and 90, separately for males and females. A schematic representation of the quinquennial death rates which are thus included in our material is given in diagram 1.

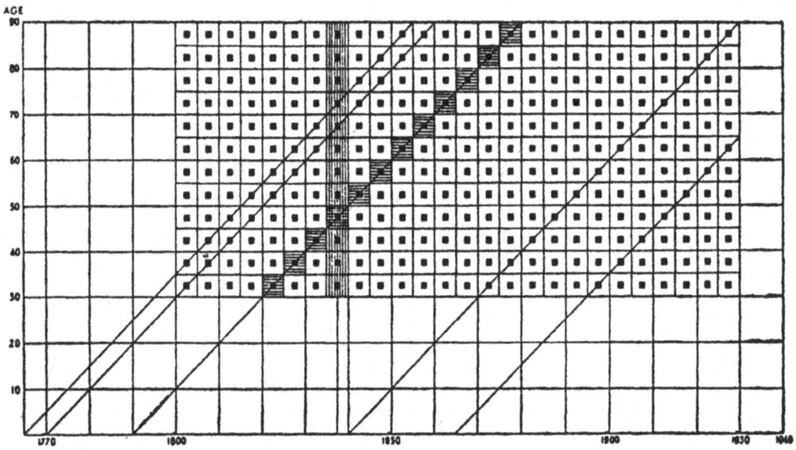

DIAGRAM 1. Schematic representation of the material. Mortality by periods and by generations.

In order to graduate a series of these death rates by means of the Makeham formula we may proceed in two fundamentally different ways, the death rates being arranged either by *calendar periods* or by *generations*. Thus as indicated in diagram 1 the observed death rates along a *vertical* give the mortality during a certain calendar period, e. g. the period 1836—1840, while the death rates along a *diagonal* represent the mortality of a certain generation, e. g. a generation, whose central birth year is 1790.

Our investigation includes both these lines. Thus in the first place, studying the *period mortality*, we denote by $\mu(x, t)$ the force of mortality for a person aged x during the calendar year t, and we put

$$\mu(x, t) = \alpha_t + \beta_t \cdot c_t^x,$$

the "Makeham constants" α_t, β_t and c_t being regarded as parameters which are functions of t only. These parameters are determined from the statistics for quinquennial intervals and are afterwards graduated.

On the other hand, when we are concerned with the *generation mortality*, we denote by $\mu(x, \tau)$ the force of mortality for a person aged x and born in the calendar year τ, and we put

$$\mu(x, \tau) = \alpha_\tau + \beta_\tau \cdot c_\tau^x,$$

the parameters α_τ, β_τ and c_τ being now regarded as functions of τ only. As in the previous case, these parameters are determined for quinquennial intervals and graduated. It is easily seen from diagram 1 that for generations from $\tau = 1770$ up to $\tau = 1840$ our material includes death rates for all ages between $x = 30$ and $x = 90$. The generations between $\tau = 1715$ and $\tau = 1770$, as well as between $\tau = 1840$ and $\tau = 1895$ are, however, also in-

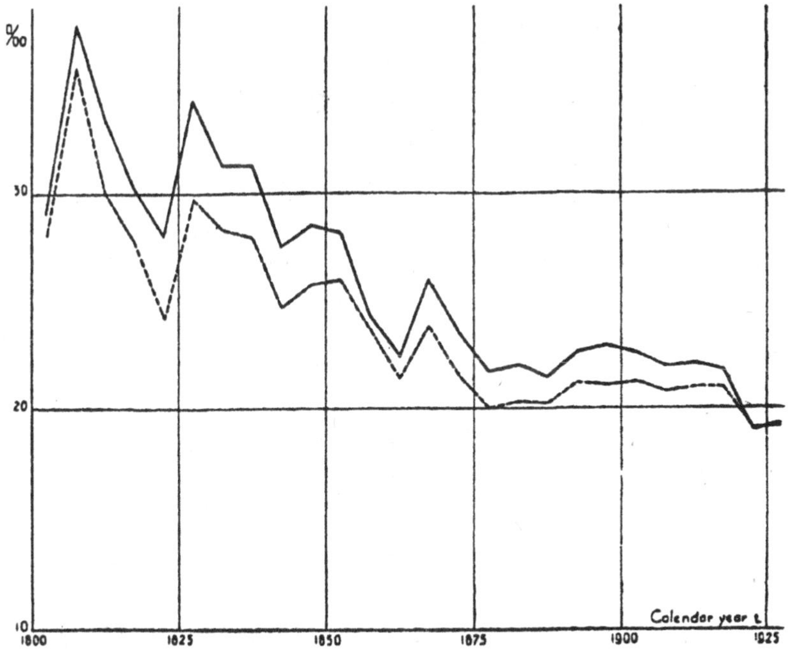

DIAGRAM 2 a. Periods. General death rates for ages 30—90.
Males: ———————— Females: — — — — — —

cluded in the material, but only for a part of the age interval. The constants corresponding to these "incomplete" generations have not been used for our graduations.

It will thus be seen that by our method we get in two fundamentally different ways a *two-dimensional graduation* of the material. Both these graduations may be used for extrapolation, and it will obviously be interesting to compare the "mortality forecasts" obtained in this way.

A preliminary idea of the development of the mortality in our material is given by diagrams 2a and 2b, where the general death rate for the whole age interval 30—90 is represented, diagram 2a giving the rates for calendar periods and diagram 2b for generations. Both these diagrams show a general decrease of mortality; there is, however, a marked contrast between the diagrams with respect to the smoothness of the curves. As might reasonably be expected, the curves corres-

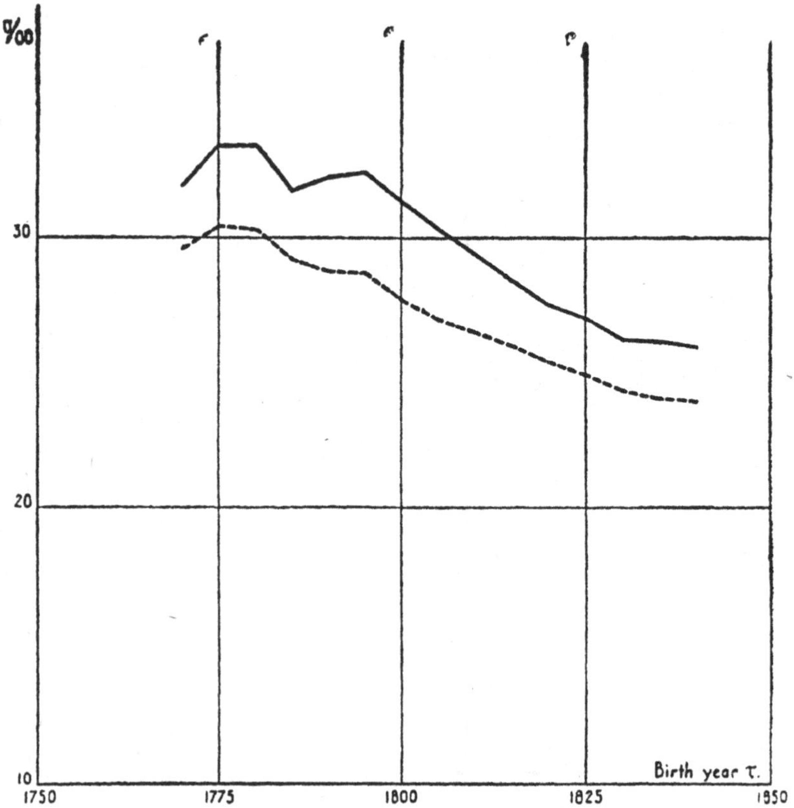

DIAGRAM 2 b. Generations. General death rates for ages 30—90.
Males: ———————— Females: — — — — — —

ponding to the death rates for generations run much smoother than the curves for calendar periods.

In order to graduate the experience by means of the Makeham formula, we may first try to give to log c a constant value and use this value for all periods and generations. Putting log $c = 0.045$ and using the method of moments for the determination of α and β, we obtain the values of these constants shown in diagrams 3a and 3b. It follows from these diagrams that both α and β show a decreasing trend, which might fairly well be represented by a straight line. In the case of β, a closer inspection of the curves shows, however, an obvious tendency to stagnation of the decrease in the latter part of the material.

The goodness of fit of our graduations has been tested by the χ^2-method. Thus, denoting by n_x and d_x the numbers of exposed to risk and of deaths in an age group with the central age x, we have calculated

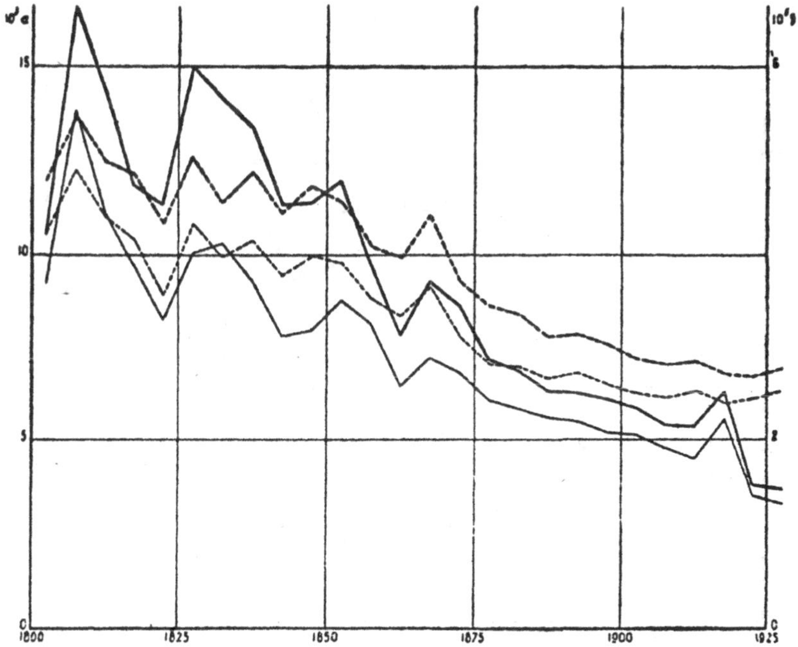

DIAGRAM 3 a. Periods. α and β for log c = 0,045.
α: Males ————, Females ————
β: Males — — — —, Females — — — — —

$$\chi^2 = \sum \frac{(d_x - n_x \mu_x)^2}{n_x \mu_x},$$

the summation being extended to all twelve quinquennial age groups between 30 and 90. Thus, the graduated value μ_x corresponding to the central age of the group has been identified with the central death rate of the group.

The values of χ^2 obtained at the preliminary graduation with $\log c = 0.045$ are given in tables 1 and 2, cols. 2—3. In most cases, these values are rather large, thus indicating according to current principles a very bad fit. Taking as an example the period 1876—1880, we have for males $\chi^2 = 24.2$, and thus obtain from the well-known Elderton tables a value of P which is approximately 0.002. For generations, the χ^2-values are on the whole larger than for periods, a fact which must be

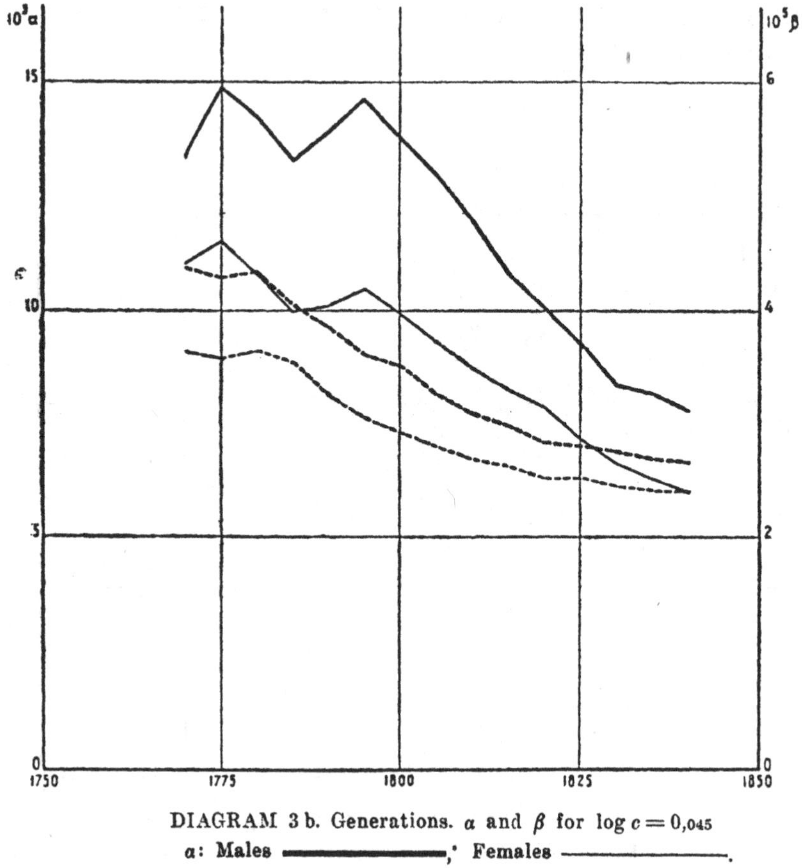

DIAGRAM 3 b. Generations. α and β for $\log c = 0{,}045$

α: Males ———, Females ———.

β: Males — — — —, Females - - - - -

ascribed to the circumstance that accidental causes, such as wars, epidemics etc. influence the mortality of a certain period in the same sense for all ages, while this is not true for a generation.

It is, however, possible to improve the fit in a very considerable way by allowing log c to vary. If, for a given period or a given generation, α and β are calculated by the method of moments for a set of different values of log c, and χ^2 is computed according to the formula given above, it will be found that χ^2, considered as a function of log c, is very well represented by a second-degree parabola. Contrary to an opinion which has sometimes been expressed in the litterature, we have found that in the cases investigated by us this parabola has a rather high curvature, thus giving for a certain value of log c a well distinguished minimum value of χ^2 (and a corresponding maximum of the goodness of fit). An example of this is given in diagram 4, which shows that even a small deviation of log c

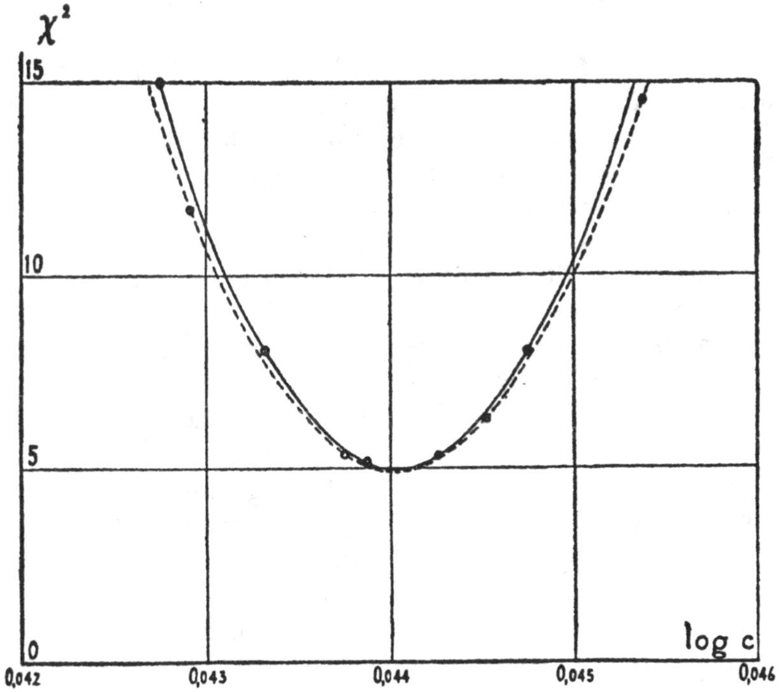

DIAGRAM 4. Period 1921—25, males. Values of χ^2 for different values of log c.
Simplified χ^2-minimum method as used in this investigation: ————
Rigorous χ^2-minimum method: — — — ....

from the "best" value gives rise to a very considerable increase in χ^2.

In the way now described, the parabola representing χ^2 has been determined for all periods and for all generations, and the value of log c corresponding to the minimum value of χ^2 has been determined for each parabola. In the sequel, we shall refer to this method for the determination of log c and the corresponding values of α and β as the *χ^2-minimum method.*

With respect to this method, it may be objected that a rigorous application of the ordinary χ^2-minimum principle obviously requires that, for any given value of log c, we should not determine α and β by the method of moments, but by the condition that χ^2 be a minimum. We have, however, found that the values of χ^2 obtained in this way only differ by quite insignificant amounts from the values given by the simpler method described above. Further, the values of log c corresponding to the minimum values of χ^2 are practically identical. These facts are illustrated by diagram 4, where the dotted curve shows the χ^2-values obtained by the more stringent method.

The values of χ^2 determined by our method are given in tables 1 and 2, cols. 4—5. It is readily seen that in practically all cases the fit has been very considerably improved as against the fit obtained by giving a constant value to log c. Thus in the case of the period 1876—1880 referred to above, χ^2 has been reduced from 24.2 to 7.3, which for $n' = 9$ gives $P = 0.506$. In many cases the fit is excellent; we find several values of χ^2 less than 6, corresponding to $P > 0.65$. It is interesting to find that for males the fit obtained is in most cases much better than for females. Further, the fit for periods is markedly better than for generations which, according to the above, might well have been expected.

The values of log c given by the χ^2-minimum method are shown in tables 1 and 2, cols. 6—7, and in diagrams 5a and 5b. At an early stage of our investigations, our material was confined to the period 1850—1930, and we endeavoured to graduate the log c values by various polynomial and exponential functions. When, later on, we considered the whole period 1800—1930, our diagrams suggested at once that *logistic curves* should be used for the graduation. The equation of the ordinary logistic curve may be put in the form

$$y = \frac{A + Be^{k(x-x_0)}}{1 + e^{k(x-x_0)}},$$

where A and B are the ordinates of the asymptotes, x_0 is the abscissa of the inflexion point and k a constant which, for given A, B and x_0 determines the slope of the curve. We have worked out a method for fitting this curve, according to the method of least squares, to statistical data, which as far as we know seems to be new, and which requires a comparatively small amount of work. We do not, however, propose to give an account of the method in this preliminary note.

The graduated values of log c are found in tables 1 and 2, cols. 8—9, and in diagrams 5a and 5b. The fit is very good, especially for periods. As a matter of fact, the logistic curves in diagram 5a, which covers the whole period 1800—1930, give for the period 1850—1930 a better fit than any of the curves used at the stage when we were dealing with the latter period alone.

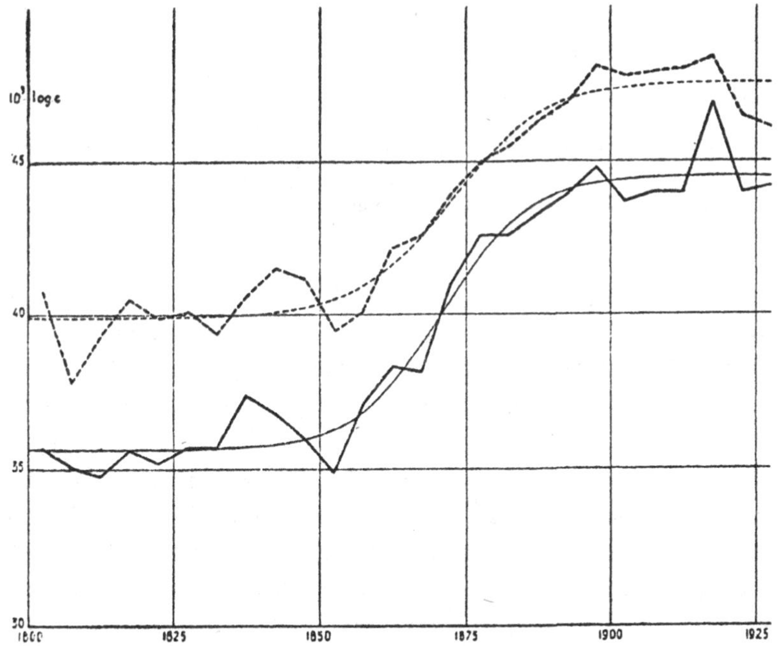

DIAGRAM 5 a. **Periods. Values of log c, calculated by χ^2-minimum method and graduated by logistic curve.**

Males: ———————— Females: — — — — — —

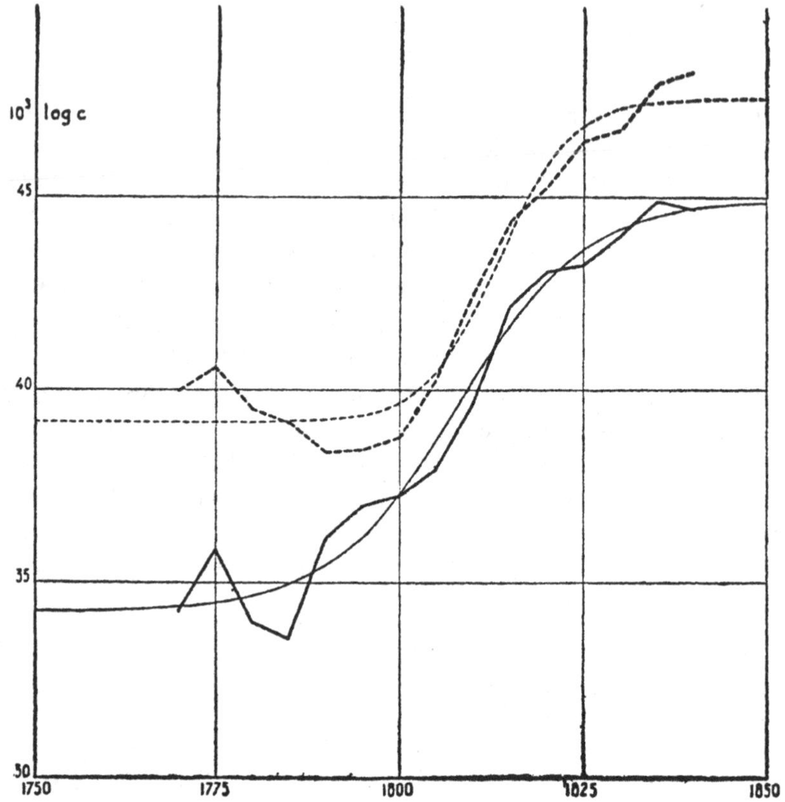

DIAGRAM 5 b. Generations. Values of log *c*, calculated by χ^2-minimum method and graduated by logistic curve.

Males: ——————— Females: — — — — — —

The values of a and β given by the χ^2-minimum method are shown in tables 3 and 4, cols. 2—5 and in diagrams 6 and 7. These values were not as in the case of log c directly graduated, as it was found more convenient to proceed in the following way.

From the graduated log *c*-values, β was calculated by the method of moments. The corresponding values of log β were then graduated by logistic curves, and the graduated values of β are given in tables 3 and 4, cols. 8—9, and diagrams 6a and 6b. In the diagrams, the β-values directly calculated by the χ^2-minimum method are shown together with those calculated from the graduated values of log *c*. The last-mentioned values show a particularly good agreement with the logistic curves.

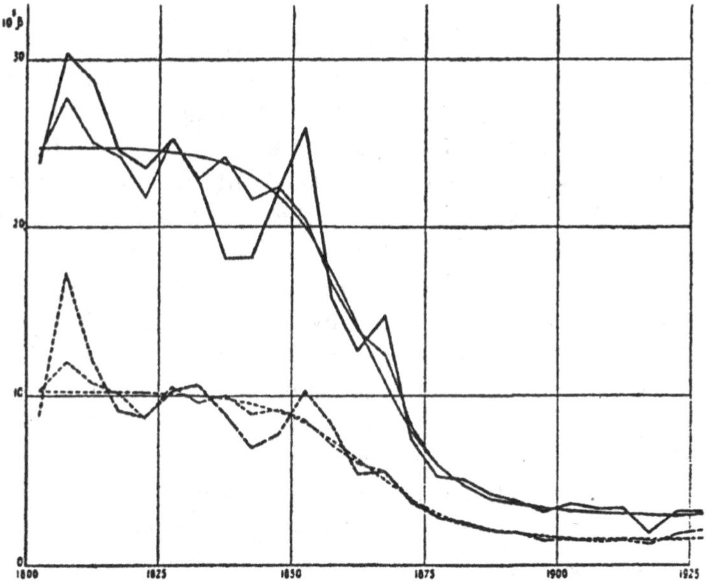

DIAGRAM 6 a. Periods. Values of β, directly calculated by χ^2-minimum method (thick lines), calculated from graduated values of log c (thin lines), and graduated by logistic curve.

Males: ———————— Females: — — — — — —

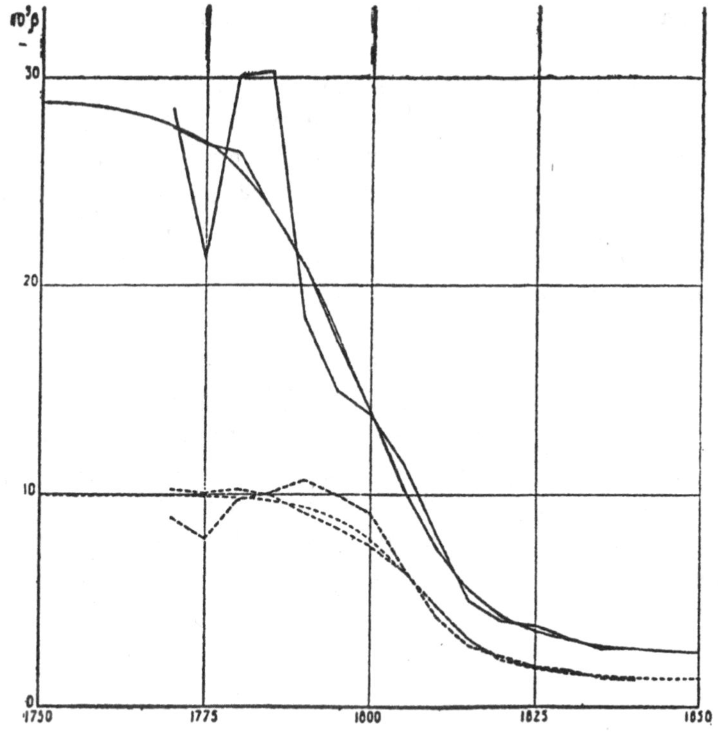

DIAGRAM 6 b. Generations. Values of β, calculated and graduated as in diagram 6 a.

Finally, from the graduated values of log c and β, α was calculated so as to make χ^2 a minimum. For periods, these α-values could be well graduated by a straight line, whereas for generations such graduation was only possible for females. For males a straight line could be used only for the generations born after 1800, while for the earlier generations a parabola was used. The graduated values of α are given in tables 3 and 4, cols. 6—7, and in diagrams 7a—7d. The diagrams show the α-values directly calculated by the χ^2-minimum method as well as those calculated from the graduated values of log c and those obtained in the final calculation by means of graduated values of log c *and* β. All three series give a fairly good fit to the straight lines, especially the last one.

We have now obtained analytic graduations of all the three parameters α, β and log c in the Makeham formula, separately for periods and for generations. The two-dimensional graduation formula for the force of mortality thus obtained contains nine constants, viz. four in the logistic expression for log c, three in the corresponding expression for β (where we use the same value of x_0 as for log c) and two in the linear expression for α.

The values of χ^2 obtained for the two-dimensional graduation are given below for three different cases. Under *Case I* we give the values obtained when, in the expression for χ^2, the summation is extended to all the 312 quinquennial groups contained in the material (cf. diagram 1). As the great epidemics during the years 1918 and 1919 cause particularly large deviations for the period 1916—1920, we give under *Case II* the values of χ^2 obtained when the 36 groups corresponding to the period 1916—1930 are omitted. For the "generation mortality" we have, however, in both these cases used extrapolated figures, as the constants in the graduation formula are determined solely by means of the experience for the generations between 1770 and 1840. For this reason, we give under *Case III* the values of χ^2 obtained when the summation is extended to the 180 quinquennial groups belonging to these generations.

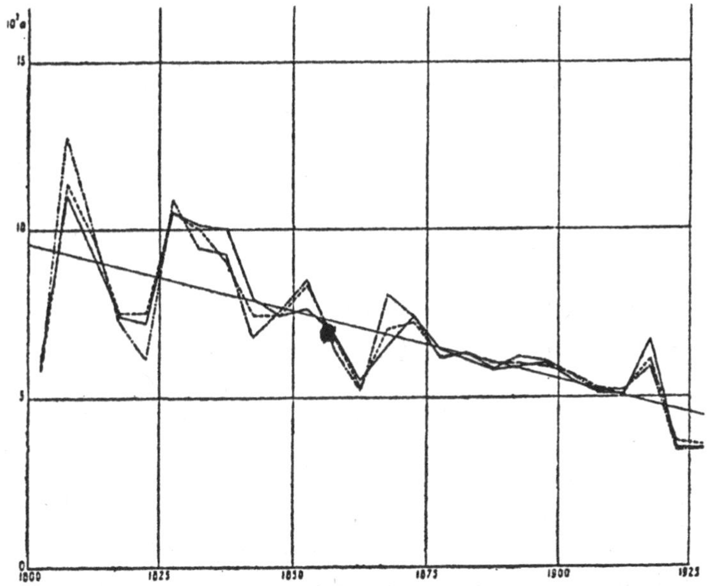

DIAGRAM 7 a. Periods, males. Values of a, directly calculated by χ^2-minimum method (————————), calculated from graduated values of log c (— — — — — —), from graduated values of log c *and* β (— · — · — · —) and graduated by straight line.

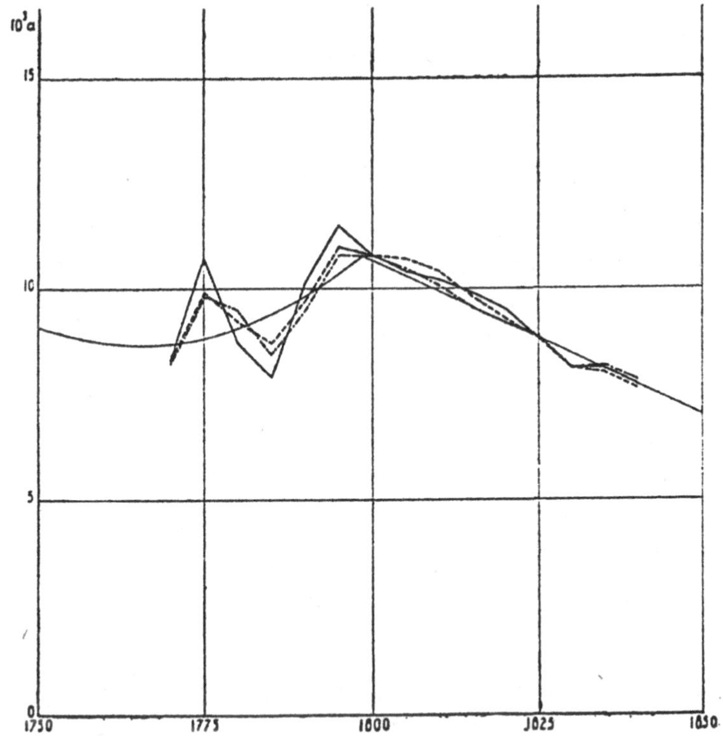

DIAGRAM 7 b. Generations, males, Values of a, calculated and graduated as in diagram 7 a.

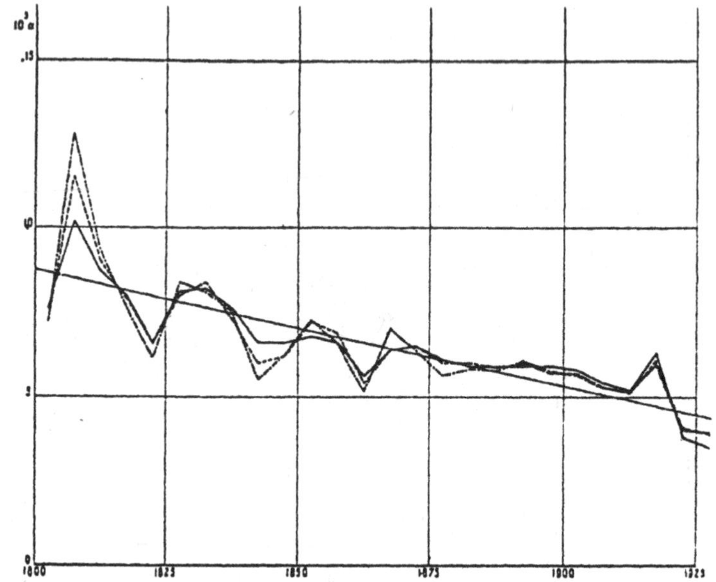

DIAGRAM 7 c. Periods, females. Values of a, calculated and graduated as in diagram 7 a.

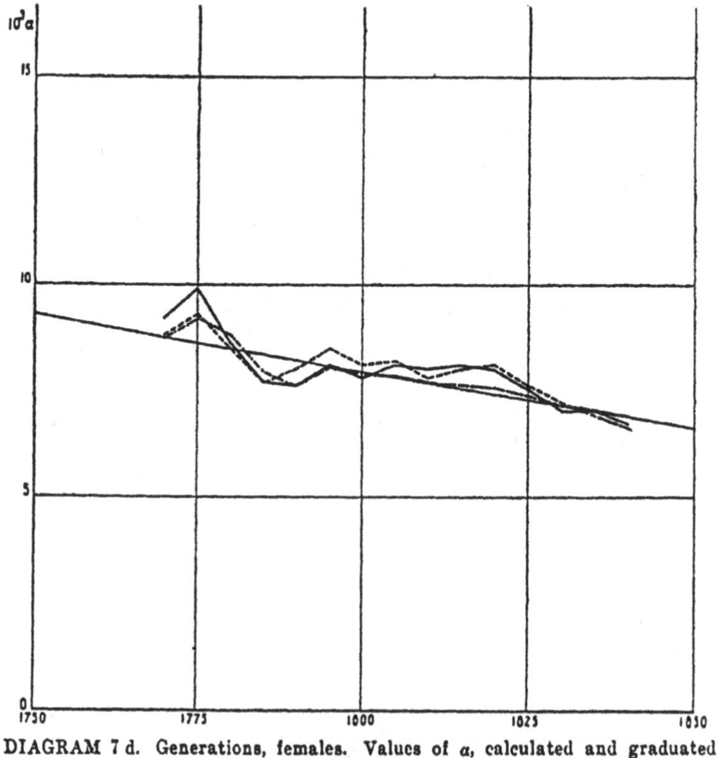

DIAGRAM 7 d. Generations, females. Values of a, calculated and graduated as in diagram 7 a.

	χ^2 according to graduation by			
	Periods 1800—1930		Generations 1770—1840	
	Males	Females	Males	Females
Case I	3099	3209	3127	5412
Case II	2411	2695	2446	4274
Case III	1732	1403	1751	1499

It is interesting to find that, according to these figures, the two different methods of graduation give in most cases nearly equal values of χ^2. For males this holds true in all three cases, and for females in Case III. For females in Cases I and II, on the other hand, the graduation by generations gives considerably less good results than the graduation by periods. It must, however, be borne in mind that in these cases extrapolated figures are used in the calculation of χ^2 for the graduation by generations. Still, the graduation by generations for females seems to be less satisfactory than the others, and we hope to be able to improve the results for this case.

In table 5 and in diagrams 8a—8d, we give for some age groups the observed and graduated values of μ_x. The graduated values have been extrapolated by means of the analytic formulae used for the graduation, and the extrapolated values are also given. For the graduation by periods, the extrapolation begins simply where the curve representing the observed figures ends. For the graduation by generations, on the other hand, the observed values were according to the above not all used for the graduation, so that in this case the extrapolation may to a certain extent be compared with the experience. We have marked on the generation curves in the diagrams the points between which the curves are to be considered as graduations. Outside these points, the curves represent the extrapolated values.

In table 5, we have given for comparison also the μ_x-values according to the tables "A. F. K.", "P. F. K." and "R 32". The values given by the last-mentioned table are generally somewhat lower than our extrapolated values for the year 1952.5.

H. Cramér and H. Wold

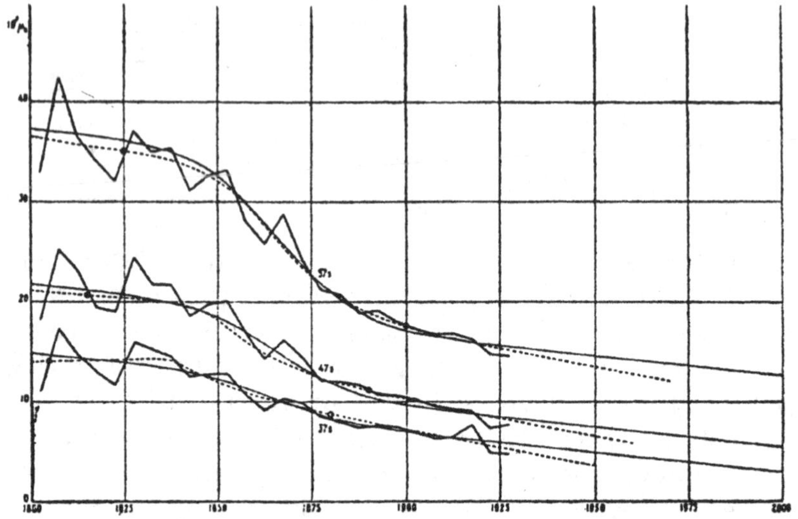

DIAGRAM 8 a. Males. Observed and graduated values of μ_x for ages 37,5, 47,5 and 57,5.
Graduation by periods ——————, by generations — — — — —

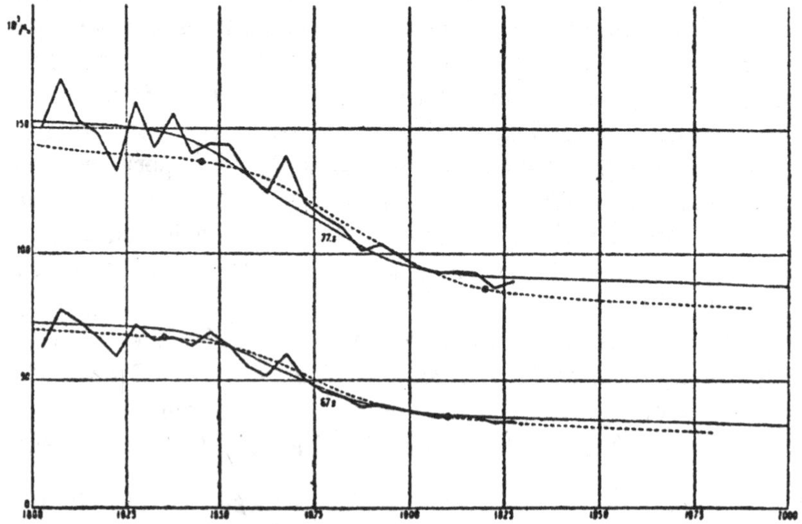

DIAGRAM 8 b. Males. Observed and graduated values of μ_x for ages 67,5 and 77,5.
Graduation by periods ——————, by generations — — — — —

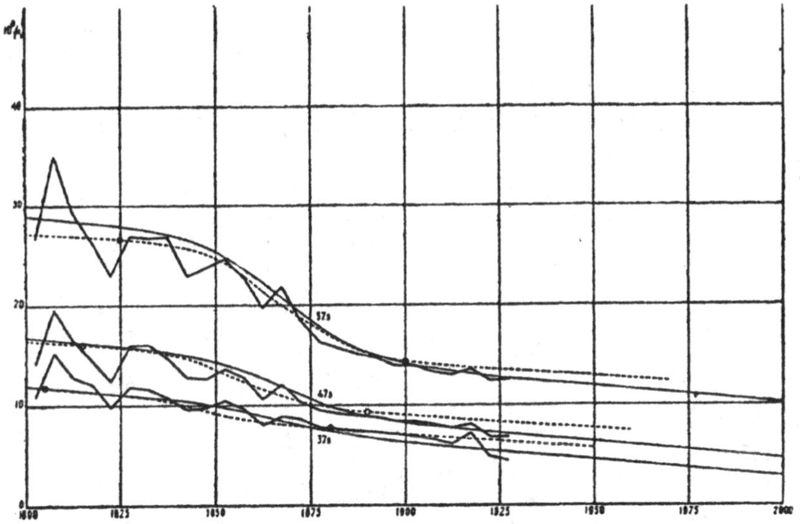

DIAGRAM 8 c. Females. Observed and graduated values of μ_x for ages 37,5, 47,5 and 57,5.

Graduation by periods ——————, by generations – – – – – –

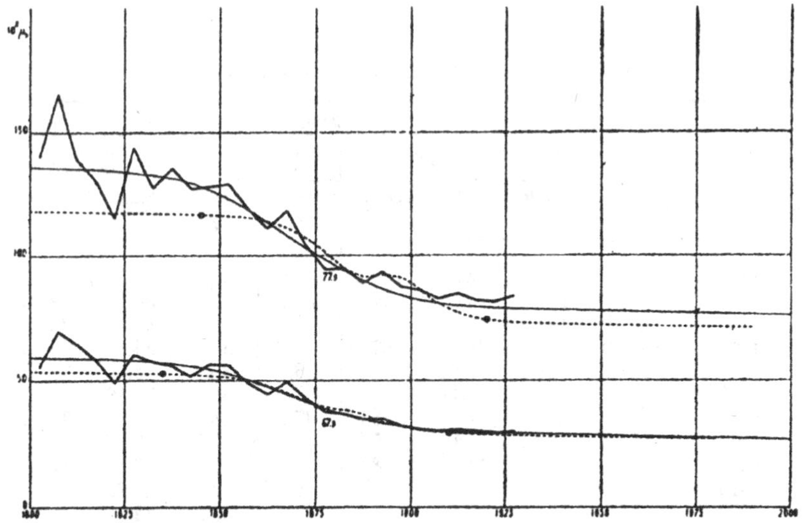

DIAGRAM 8 d. Females. Observed and graduated values of μ_x for ages 67,5 and 77,5.

Graduation by periods ——————, by generations – – – – – –

TABLE 1. PERIODS, χ^2 AND log c.

Period	χ^2 for log c = 0.045		χ^2-minimum-method					
			χ^2		10^3 log c, directly calculated		10^3 log c, graduated	
	Males	Females	Males	Females	Males	Females	Males	Females
1	2	3	4	5	6	7	8	9
1801—05	196.9	73.3	25.4	31.7	35.7	40.8	35.63	39.92
1806—10	219.4	166.1	12.4	25.0	35.1	37.8	35.63	39.92
1811—15	202.5	103.6	10.0	20.1	34.8	39.3	35.63	39.92
1816—20	185.2	80.1	16.6	31.8	35.6	40.5	35.64	39.92
1821—25	184.6	84.5	20.7	27.0	35.2	39.9	35.64	39.93
1826—30	206.4	86.0	36.7	19.9	35.7	40.1	35.65	39.94
1831—35	183.0	91.4	24.0	12.6	35.7	39.4	35.68	39.97
1836—40	146.2	78.6	29.1	22.9	37.4	40.6	35.72	40.01
1841—45	158.8	66.3	25.2	32.3	36.8	41.5	35.81	40.10
1846—50	232.1	81.2	48.3	35.1	36.0	41.2	35.98	40.25
1851—55	267.8	147.7	31.6	49.7	34.9	39.5	36.29	40.51
1856—60	154.4	105.7	12.0	30.1	37.1	40.1	36.85	40.95
1861—65	128.0	40.7	12.7	15.9	38.3	42.2	37.76	41.64
1866—70	155.4	36.3	8.5	15.1	38.1	42.6	39.04	42.60
1871—75	47.4	33.4	5.0	29.8	41.0	44.0	40.53	43.75
1876—80	24.2	20.4	7.3	20.4	42.6	45.0	41.93	44.90
1881—85	33.9	26.6	16.4	25.6	42.6	45.5	42.99	45.86
1886—90	33.4	28.3	24.2	22.0	43.3	46.3	43.68	46.55
1891—95	8.0	48.9	3.4	34.0	43.9	46.9	44.08	46.99
1896—00	12.8	61.0	12.5	21.8	44.8	48.1	44.29	47.25
1901—05	19.5	57.7	12.6	23.4	43.7	47.8	44.41	47.40
1906—10	15.7	61.3	11.5	20.9	44.0	47.9	44.47	47.48
1911—15	10.2	70.9	5.2	23.6	44.0	48.0	44.50	47.53
1916—20	77.7	125.5	58.7	67.1	46.9	48.4	44.51	47.55
1921—25	10.4	27.6	4.9	12.8	44.0	46.5	44.52	47.57
1926—30	8.9	26.3	4.6	18.4	44.2	46.1	44.53	47.57

TABLE 2. GENERATIONS, χ^2 AND log c.

Central Birth Year	χ^2 for log c = 0.045		χ^2-minimum-method					
			χ^2		10^3 log c, directly calculated		10^3 log c, graduated	
	Males	Females	Males	Females	Males	Females	Males	Females
1	2	3	4	5	6	7	8	9
1770	558.3	252.6	354.5	192.3	34.3	40.0	34.37	39.16
1775	240.9	190.4	77.3	137.7	35.8	40.6	34.47	39.16
1780	362.9	176.3	115.9	88.8	34.0	39.5	34.65	39.17
1785	395.9	181.2	162.8	87.4	33.5	39.1	34.94	39.18
1790	315.8	241.9	163.2	114.9	36.1	38.4	35.43	39.23
1795	235.3	206.3	93.4	73.0	37.0	38.4	36.19	39.35
1800	199.8	177.2	65,0	55.9	37.2	38.7	37.28	39.66
1805	185.3	100.4	78.1	35.2	38.0	40.2	38.68	40.42
1810	181.4	96.6	117.6	78·3	39.7	42.5	40.22	41.93
1815	137.5	65.9	116.6	64.4	42.1	44.3	41.66	44.01
1820	88.0	39.1	76.0	38.9	43.1	45.2	42.82	45.82
1825	83.9	62.7	73.5	53.8	43.3	46.4	43.64	46.85
1830	71.0	77.5	67.4	65.3	44.0	46.7	44.18	47.31
1835	17.5	88.3	17.5	48.6	44.9	48.0	44.50	47.48
1840	20.7	72.8	20.2	24.8	44.7	48.3	44.69	47.55

TABLE 3. PERIODS, α AND β.

Period	$10^3\,\alpha$				$10^5\,\beta$			
	By χ^2-minimum-method		Final Graduation		By χ^2-minimum-method		Final Graduation	
	Males	Females	Males	Females	Males	Females	Males	Females
1	2	3	4	5	6	7	8	9
1801—05	6.1	7.6	9.486	8.672	23.86	8.85	24.780	10.228
1806—10	11.0	10.2	9.283	8.501	30.36	17.29	24.766	10.221
1811—15	9.2	8.7	9.081	8.329	28.73	12.04	24.740	10.209
1816—20	7.4	8.0	8.879	8.158	24.64	9.12	24.694	10.187
1821—25	7.2	6.6	8.676	7.987	23.55	8.71	24.613	10.151
1826—30	10.5	8.1	8.474	7.816	25.30	10.23	24.469	10.089
1831—35	10.1	8.2	8.271	7.645	22.67	10.62	24.215	9.985
1836—40	10.0	7.6	8.069	7.473	18.06	8.97	23.775	9.811
1841—45	7.9	6.6	7.867	7.302	18.16	6.95	23.027	9.528
1846—50	7.4	6.6	7.664	7.131	22.24	7.72	21.802	9.079
1851—55	7.6	6.8	7.462	6.960	25.96	10.23	19.913	8.404
1856—60	6.9	6.6	7.259	6.789	15.83	8.39	17.265	7.469
1861—65	5.5	5.6	7.057	6.618	12.63	5.44	14.032	6.310
1866—70	6.5	6.4	6.854	6.446	14.71	5.60	10.727	5.070
1871—75	7.4	6.5	6.652	6.275	7.49	3.74	7.940	3.942
1876—80	6.4	6.1	6.450	6.104	5.26	2.83	5.963	3.065
1881—85	6.1	5.9	6.247	5.933	5.09	2.54	4.721	2.462
1886—90	5.8	5.9	6.045	5.762	4.21	2.12	3.992	2.080
1891—95	5.9	5.9	5.842	5.590	3.83	1.95	3.578	1.849
1896—00	6.0	5.9	5.640	5.419	3.17	1.50	3.345	1.712
1901—05	5.5	5.8	5.438	5.248	3.66	1.51	3.215	1.631
1906—10	5.1	5.4	5.235	5.077	3.36	1.45	3.142	1.584
1911—15	5.1	5.2	5.033	4.906	3.42	1.47	3.101	1.556
1916—20	6.7	6.3	4.830	4.734	1.92	1.30	3.079	1.540
1921—25	3.5	3.8	4.628	4.563	3.20	1.85	3.066	1.531
1926—30	3.5	3.5	4.426	4.392	3.22	2.09	3.059	1.525

TABLE 4. GENERATIONS, α AND β.

Central Birth Year	$10^3\,\alpha$				$10^5\,\beta$			
	By χ^2-minimum-method		Final Graduation		By χ^2-minimum-method		Final Graduation	
	Males	Females	Males	Females	Males	Females	Males	Females
1	2	3	4	5	6	7	8	9
1770	8.3	9.2	8.683	8.751	28.51	8.88	27.705	9.934
1775	10.7	9.9	8.819	8.618	21.25	7.89	26.873	9.900
1780	8.7	8.6	9.053	8.485	30.06	9.75	25.573	9.828
1785	7.9	7.7	9.383	8.352	30.32	10.08	23.634	9.682
1790	10.1	7.6	9.811	8.219	18.41	10.66	20.935	9.388
1795	11.5	8.1	10.336	8.086	14.91	9.96	17.543	8.828
1800	10.8	7.8	10.662	7.953	13.96	9.07	13.812	7.852
1805	10.4	8.1	10.294	7.820	11.48	6.63	10.311	6.407
1810	10.2	8.0	9.927	7.687	8.06	4.28	7.516	4.730
1815	9.9	8.1	9.559	7.554	5.00	2.96	5.568	3.289
1820	9.5	8.0	9.191	7.421	4.02	2.42	4.331	2.351
1825	8.8	7.5	8.823	7.288	3.84	1.93	3.585	1.842
1830	8.1	7.0	8.456	7.154	3.30	1.79	3.144	1.588
1835	8.1	7.0	8.088	7.021	2.74	1.39	2.884	1.466
1840	7.7	6.7	7.720	6.888	2.82	1.32	2.731	1.407

TABLE 5. VALUES OF $10^3 \mu_x$.

Males.

Age	1802.5			1852.5			1902.5			1952.5 (Extrapolated)		Å.F.K.	P.F.K.	R 32
	Observed.	Graduated by		Observed.	Graduated by		Observed.	Graduated by		Graduated by				
		Periods	Gener.		Periods	Gener.		Periods	Gener.	Periods	Gener.			
32.5	9.4	13.1	12.3	10.5	10.5	10.3	6.2	6.3	6.2	4.3	2.6	5.6	3.8	3.4
37.5	11.0	14.9	14.1	12.8	12.0	11.6	7.0	6.9	7.1	4.8	3.4	6.0	4.1	3.9
42.5	14.3	17.6	16.9	16.5	14.4	13.8	8.2	7.9	8.3	5.8	4.6	7.0	4.9	4.6
47.5	18.2	21.7	21.0	20.1	18.0	17.4	10.3	9.6	10.1	7.4	6.4	8.6	6.2	5.9
52.5	25.7	27.9	27.2	25.9	23.5	23.2	13.3	12.3	12.9	10.1	9.1	11.1	9.0	8.0
57.5	32.9	37.2	36.4	33.2	31.8	31.5	17.3	16.9	17.3	14.5	13.3	14.7	13.4	11.6
62.5	45.1	51.3	49.9	44.1	44.4	44.1	24.8	24.6	24.7	21.9	20.2	20.7	20.4	17.8
67.5	63.0	72.5	69.9	63.9	63.5	63.3	37.0	37.4	37.2	34.3	31.6	31.7	31.5	28.3
72.5	103.5	104.4	99.3	99.1	92.6	92.2	57.9	58.8	58.6	55.0	50.4	51.5	50.5	46.1
77.5	150.1	152.5	142.7	143.7	136.8	134.9	94.8	94.3	95.2	89.6	81.6	83.4	84.5	76.3
82.5	226.5	225.0	206.8	200.0	203.8	198.4	154.5	153.7	156.8	147.3	134.0	150.0	136.8	127.6
87.5	297.6	334.4	301.6	296.1	305.7	292.8	246.6	252.6	255.6	243.7	221.5	246.2	229.1	214.8

Females.

Age	1802.5			1852.5			1902.5			1952.5 (Extrapolated)		Å.F.K.	P.F.K.	R 32
	Observed.	Graduated by		Observed.	Graduated by		Observed.	Graduated by		Graduated by				
		Periods	Gener.		Periods	Gener.		Periods	Gener.	Periods	Gener.			
32.5	9.2	10.7	10.6	8.6	8.7	8.1	6.4	5.8	6.6	4.1	5.2	6.0	3.8	3.4
37.5	10.8	11.9	11.8	10.5	9.7	9.0	6.8	6.2	7.0	4.5	5.7	6.1	4.2	3.8
42.5	13.4	13.8	13.6	12.3	11.4	10.6	7.5	6.9	7.8	5.1	6.5	6.8	5.0	4.4
47.5	14.2	16.7	16.4	13.6	14.0	13.1	8.4	8.2	9.0	6.3	7.6	7.4	6.3	5.6
52.5	20.3	21.4	20.6	18.1	18.2	17.4	10.4	10.3	10.9	8.3	9.6	8.8	8.7	7.5
57.5	26.6	28.9	27.2	24.6	24.9	24.2	13.9	13.9	14.3	11.8	12.8	12.1	12.1	10.7
62.5	39.2	40.6	37.5	36.2	35.6	34.8	20.0	20.2	20.1	17.8	18.3	17.9	17.9	16.3
67.5	55.5	59.3	53.5	56.3	52.6	51.1	30.8	31.1	30.5	28.2	27.8	27.3	28.3	25.7
72.5	91.1	88.8	78.6	87.2	79.6	76.5	50.7	49.8	49.9	46.3	44.0	44.1	46.4	41.6
77.5	140.0	135.6	117.9	128.8	122.8	115.9	86.1	82.2	86.0	77.4	72.0	79.9	79.9	68.9
82.5	219.8	209.7	179.6	184.1	191.7	177.7	139.2	138.0	149.1	131.4	120.3	134.5	135.4	115.1
87.5	292.8	326.9	276.3	268.2	301.5	274.6	220.2	234.3	240.9	224.6	203.9	218.6	212.0	193.5